VOLUME ONE HUNDRED AND THIRTY ONE

PROGRESS IN MOLECULAR BIOLOGY AND TRANSLATIONAL SCIENCE

Molecular and Cell Biology of Pain

VOLUME ONE HUNDRED AND THIRTY ONE

Progress in MOLECULAR BIOLOGY AND TRANSLATIONAL SCIENCE

Molecular and Cell Biology of Pain

Edited by

THEODORE J. PRICE
Behavioral and Brain Sciences,
The University of Texas at Dallas, Richardson, Texas, USA

GREGORY DUSSOR
Behavioral and Brain Sciences,
The University of Texas at Dallas, Richardson, Texas, USA

ELSEVIER

AMSTERDAM • BOSTON • HEIDELBERG • LONDON
NEW YORK • OXFORD • PARIS • SAN DIEGO
SAN FRANCISCO • SINGAPORE • SYDNEY • TOKYO
Academic Press is an imprint of Elsevier

Academic Press is an imprint of Elsevier
225 Wyman Street, Waltham, MA 02451, USA
525 B Street, Suite 1800, San Diego, CA 92101-4495, USA
125 London Wall, London, EC2Y 5AS, UK
The Boulevard, Langford Lane, Kidlington, Oxford OX5 1GB, UK

First edition 2015

ISBN: 978-0-12-801389-2
ISSN: 1877-1173

For information on all Academic Press publications
visit our website at store.elsevier.com

CONTENTS

Contributors *xiii*
Preface *xvii*

1. An Introduction to Pain Pathways and Pain "Targets" **1**
Vaskar Das

1. An Introduction to Pain and Pain Pathways 2
2. Ion Channels, Receptors, and Other "Targets" for Persistent Inflammatory or Neuropathic Pain 9
3. Summary 18

Acknowledgments 18
References 18

2. Peripheral Scaffolding and Signaling Pathways in Inflammatory Pain **31**
Nathaniel A. Jeske

1. Introduction 32
2. Inflammatory Mediators 32
3. Signaling Mechanisms 40
4. Scaffolding Structures 42
5. Concluding Remarks 46

References 47

3. Contribution of Mechanosensitive Ion Channels to Somatosensation **53**
Reza Sharif-Naeini

1. The Evolution of Mechanosensing 54
2. MSCs in Nociceptors 55
3. Difficulties in Identifying Genes Encoding MSCs 57
4. Gating Mechanisms of MSCs 59
5. Role of MSCs in the Transmission of Noxious Mechanical Inputs 60
6. Sensitization of MSCs Is Necessary for the Induction of Mechanical Allodynia 62
7. Potassium-Selective MSCs in Nociceptors 65

References 65

4. Sensory TRP Channels: The Key Transducers of Nociception and Pain **73**

Aaron D. Mickle, Andrew J. Shepherd, and Durga P. Mohapatra

1. Introduction 74
2. Ion Channels in the TRPV Subfamily 78
3. Ion Channels in the TRPM Subfamily 93
4. Ion Channels in the TRPA Subfamily 98

Acknowledgments 105

References 105

5. The Contribution of Mitochondria to Sensory Processing and Pain **119**

Sarah J.L. Flatters

1. Introduction 120
2. Role of Mitochondria in Normal Sensory Processing and Acute Pain 121
3. Clinical Evidence of Mitochondrial Dysfunction in Painful Conditions 126
4. Preclinical Evidence of Mitochondrial Dysfunction in Chronic Pain States 129
5. Future Perspectives 140

Acknowledgments 141

References 141

6. Regulation of Gene Expression and Pain States by Epigenetic Mechanisms **147**

Sandrine M. Géranton and Keri K. Tochiki

1. Introduction 149
2. Epigenetic Mechanisms and Gene Expression 149
3. Epigenetic Regulation of Synaptic Plasticity in Learning and Memory 156
4. Epigenetic Mechanisms and Pain States 163
5. Future Directions 173

Acknowledgments 175

References 175

7. Translational Control of Chronic Pain **185**

Ohannes K. Melemedjian and Arkady Khoutorsky

1. Introduction 186
2. Repair of Injured Tissue 186
3. Case for Axonal Translation 188
4. Regulation of Protein Synthesis 189
5. Transport and Storage of mRNAs 199

6. Conclusions and Perspectives 200
7. Translational Control of Central Sensitization 201
8. Conclusions 206
References 207

8. MicroRNA Biology and Pain 215
Marguerite K. McDonald and Seena K. Ajit
1. Introduction 216
2. History and Biogenesis 216
3. Genomic Location of miRNAs 218
4. miRNA Nomenclature 219
5. miRNA Target Determination 220
6. miRNA Detection, Quantification, and Functional Studies 221
7. Circulating miRNAs 224
8. miRNAs and Pain 227
9. Challenges Associated with miRNA Research 240
10. Conclusions 242
Acknowledgments 242
References 242

9. Role of Extracellular Damage-Associated Molecular Pattern Molecules (DAMPs) as Mediators of Persistent Pain 251
Jungo Kato and Camilla I. Svensson
1. Introduction 252
2. High-Mobility Group Box 1 254
3. S100 Proteins 262
4. Heat-Shock Proteins 264
5. microRNA 266
6. Purine Metabolites 267
7. Summary and Perspective 268
References 269

10. mGluRs Head to Toe in Pain 281
Benedict J. Kolber
1. Introduction 282
2. Metabotropic Receptors 282
3. Analysis of mGluRs at Each Level of the Pain Neuraxis 288
4. Conclusion 316
References 316

11. Nonneuronal Central Mechanisms of Pain: Glia and Immune Response **325**

E. Alfonso Romero-Sandoval and Sarah Sweitzer

1. Introduction 326
2. The Central Multipartite Synapse in the Mechanisms of Chronic Pain 326
3. Glial Activation or Reactivity: A Change in Cellular Phenotype in Pain States 330
4. Triggers of the Change of Glial Phenotype in Pain States 333
5. Drugs with Glial Modulation Capabilities and Efficacy in Preclinical Models of Pain 336
6. What It Is Known of Glial Cells in Humans with Pain Conditions? 338
7. Glial Cells: Human Versus Rodents and Other Species 342
8. Clinical Data Using Glial Modulating Agents 343
9. Conclusions 345

Acknowledgments 349
References 349

12. Synaptic Inhibition and Disinhibition in the Spinal Dorsal Horn **359**

Steven A. Prescott

1. Introduction 360
2. Synthesis and Handling of Inhibitory Neurotransmitters 360
3. GABA and Glycine Receptors 362
4. Driving Force and Ion Flux 364
5. Regulation of Intracellular Anion Concentrations 366
6. Mechanisms of Inhibition 369
7. Inhibitory Interneurons and Dorsal Horn Circuitry 370
8. Pathological Changes in Inhibition 372
9. Implications for Therapeutic Interventions 374
10. Summary 375

References 376

13. Dendritic Spine Dysgenesis in Neuropathic Pain **385**

Andrew Michael Tan

1. A Brief History 386
2. Dendritic Spine Morphology 387
3. Dendritic Spines in Synaptic Function 389
4. Dendritic Spines in Pathology 390
5. Dendritic Spines in Neuropathic Pain 391
6. Spinal Cord Injury 392

7. Peripheral Nerve Injury 394
8. Diabetes Mellitus 394
9. Thermal Burn Injury 395
10. Molecular Control of Dendritic Spine Dysgenesis 396
11. Spinal Memory Mechanism in Neuropathic Pain 400
12. Final Perspectives 401
References 402

14. Commonalities Between Pain and Memory Mechanisms and Their Meaning for Understanding Chronic Pain 409

Theodore J. Price and Kufreobong E. Inyang

1. Introduction: Pain Plasticity and "Pain Memory" 410
2. The Adaptive Nature of Pain Plasticity 412
3. The Peripheral Nociceptor as the Locus of "Pain Memory," Focus on Translation Control 413
4. "Pain Memory" in the Spinal Dorsal Horn 420
5. Clinical Implications of Pain Memory 427
6. Conclusions 428
Acknowledgments 428
References 428

15. Dietary Influence on Pain via the Immune System 435

Stacie K. Totsch, Megan E. Waite, and Robert E. Sorge

1. Cytokines and Immunity 436
2. Cytokines and Chronic Pain 439
3. Obesity and Pain 440
4. Diet and Inflammation 441
5. Omega-3 and Omega-6 PUFAs 441
6. Saturated Fats 444
7. Grape Seed Extract 444
8. Green Tea Extract 445
9. Soy Products 446
10. Broccoli 447
11. Carotenoids 448
12. Ginger 449
13. Ginseng 451
14. Caffeine 452
15. Grains and Gluten 453
16. Ketogenic Diets 453

17. Dietary Interventions 454
18. Conclusions 456
References 457

16. **Chemotherapy-Induced Peripheral Neuropathy** **471**
Jill C. Fehrenbacher
1. Chemotherapy-Induced Neuropathy: An Introduction 472
2. Classes of Chemotherapeutics 473
3. Clinical Assessment of CIPN 476
4. Experimental Studies: Animal Models of CIPN 478
5. Experimental Studies: *In Vitro* Models of CIPN 481
6. Proposed Mechanisms Underlying CIPN 483
7. Challenges to CIPN Research 492
References 493

17. **Stress and Chronic Pelvic Pain** **509**
Angela N. Pierce and Julie A. Christianson
1. Introduction 509
2. Central and Peripheral Regulation of the Stress Pathway 510
3. Consequences of Early Life Stress 513
4. Irritable Bowel Syndrome 514
5. Interstitial Cystitis/Painful Bladder Syndrome 517
6. Vulvodynia 520
7. Chronic Prostatitis/Chronic Pelvic Pain Syndrome 522
8. Conclusions 524
References 524

18. **Meningeal Afferent Signaling and the Pathophysiology of Migraine** **537**
Carolina Burgos-Vega, Jamie Moy, and Gregory Dussor
1. Introduction 538
2. Features of Migraine 538
3. Current Migraine Treatments 541
4. Anatomy of the Meningeal Afferent System 542
5. Migraine Pathophysiology 544
6. Potential Mechanisms of Dural Afferent Activation 545
7. Ion Channels and Dural Afferent Activation 546
8. Dural Afferent Input May Lead to Neuroplasticity 549
9. Neuroplasticity and Migraine 550

10. CGRP and Migraine Plasticity 553
11. BDNF and Migraine 554
12. Conclusion 555
Acknowledgments 556
References 557

19. Chronic Pain Syndromes, Mechanisms, and Current Treatments 565

Justin Sirianni, Mohab Ibrahim, and Amol Patwardhan

1. Cancer Pain 566
2. Central Poststroke Pain 572
3. Chronic Pelvic Pain 573
4. Complex Regional Pain Syndrome 575
5. Failed Back Syndrome 576
6. Fibromyalgia 578
7. Irritable Bowel Syndrome 579
8. Migraine 581
9. Multiple Sclerosis 582
10. Peripheral Neuropathic Pain 583
11. Persistent Postsurgical Pain 586
12. Phantom Limb Pain 587
13. Postherpetic Neuralgia 588
14. Rheumatoid Arthritis 589
15. Trigeminal Neuralgia 589
16. Future Directions 590
References 591

Index *613*

CONTRIBUTORS

Seena K. Ajit
Department of Pharmacology & Physiology, Drexel University College of Medicine, Philadelphia, Pennsylvania, USA

E. Alfonso Romero-Sandoval
Department of Pharmaceutical and Administrative Sciences, Presbyterian College School of Pharmacy, Clinton, South Carolina, USA

Carolina Burgos-Vega
Behavioral and Brain Sciences, The University of Texas at Dallas, Richardson, Texas, USA

Julie A. Christianson
Department of Anatomy and Cell Biology, School of Medicine, University of Kansas Medical Center, Kansas City, Kansas, USA

Vaskar Das
Behavioral and Brain Sciences, The University of Texas at Dallas, Richardson, Texas, USA

Gregory Dussor
Behavioral and Brain Sciences, The University of Texas at Dallas, Richardson, Texas, USA

Jill C. Fehrenbacher
Department of Pharmacology and Toxicology; Stark Neuroscience Research Institute, and Department of Anesthesiology, Indiana University School of Medicine, Indianapolis, Indiana, USA

Sarah J.L. Flatters
Wolfson Centre for Age-Related Diseases, King's College London, London, United Kingdom

Sandrine M. Géranton
Department of Cell and Developmental Biology, University College London, London, United Kingdom

Mohab Ibrahim
Department of Anesthesiology, University of Arizona, Tucson, Arizona, USA

Kufreobong E. Inyang
Behavioral and Brain Sciences, The University of Texas at Dallas, Richardson, Texas, USA

Nathaniel A. Jeske
Department of Oral and Maxillofacial Surgery, UT Health Science Center, San Antonio, Texas, USA

Jungo Kato
Department of Physiology and Pharmacology, Karolinska Institutet, Stockholm, Sweden

Arkady Khoutorsky
Department of Biochemistry, Rosalind and Morris Goodman Cancer Research Centre, McGill University, Montréal, Quebec, Canada

Benedict J. Kolber
Department of Biological Sciences, Duquesne University, Pittsburgh, Pennsylvania, USA

Marguerite K. McDonald
Department of Pharmacology & Physiology, Drexel University College of Medicine, Philadelphia, Pennsylvania, USA

Ohannes K. Melemedjian
Department of Neural and Pain Sciences, School of Dentistry, University of Maryland, Baltimore, Maryland, USA

Aaron D. Mickle
Department of Pharmacology, The University of Iowa Roy J. and Lucile A. Carver College of Medicine, Iowa City, Iowa, and Department of Anesthesiology, Washington University School of Medicine, St. Louis, Missouri, USA

Durga P. Mohapatra
Department of Pharmacology; Department of Anesthesia, The University of Iowa Roy J. and Lucile A. Carver College of Medicine, Iowa City, Iowa, and Department of Anesthesiology, Washington University School of Medicine, St. Louis, Missouri, USA

Jamie Moy
Behavioral and Brain Sciences, The University of Texas at Dallas, Richardson, Texas, USA

Amol Patwardhan
Department of Anesthesiology, University of Arizona, Tucson, Arizona, USA

Angela N. Pierce
Department of Anatomy and Cell Biology, School of Medicine, University of Kansas Medical Center, Kansas City, Kansas, USA

Steven A. Prescott
Neurosciences and Mental Health, The Hospital for Sick Children, and Department of Physiology, University of Toronto, Toronto, Ontario, Canada

Theodore J. Price
Behavioral and Brain Sciences, The University of Texas at Dallas, Richardson, Texas, USA

Reza Sharif-Naeini
Department of Physiology and Cell Information Systems Group, McGill University, Montreal, Quebec, Canada

Andrew J. Shepherd
Department of Pharmacology, The University of Iowa Roy J. and Lucile A. Carver College of Medicine, Iowa City, Iowa, and Department of Anesthesiology, Washington University School of Medicine, St. Louis, Missouri, USA

Justin Sirianni
Department of Anesthesiology, University of Arizona, Tucson, Arizona, USA

Robert E. Sorge
Department of Psychology, University of Alabama at Birmingham, Birmingham, Alabama, USA

Camilla I. Svensson
Department of Physiology and Pharmacology, Karolinska Institutet, Stockholm, Sweden

Sarah Sweitzer
Department of Pharmaceutical and Administrative Sciences, Presbyterian College School of Pharmacy, Clinton, South Carolina, USA

Andrew Michael Tan
Department of Neurology and Center for Neuroscience and Regeneration Research, Yale University School of Medicine, New Haven; Rehabilitation Research Center, Veterans Affairs Connecticut Healthcare System, West Haven, and Hopkins School, New Haven, Connecticut, USA

Keri K. Tochiki
Department of Cell and Developmental Biology, University College London, London, United Kingdom

Stacie K. Totsch
Department of Psychology, University of Alabama at Birmingham, Birmingham, Alabama, USA

Megan E. Waite
Department of Psychology, University of Alabama at Birmingham, Birmingham, Alabama, USA

PREFACE

When we were approached to assemble and edit this volume, we were immediately faced with a dilemma: there are many excellent pain textbooks that already exist, why should we set out to create a new one? After some time, and some perusing of the venerable *Textbook of Pain*, which has just received a fresh update, it occurred to us that there was indeed room for a new textbook on pain. While the existing texts are unquestionably excellent, they largely fail to touch on exciting new areas of research emerging from young investigators in the field. There are many young investigators who were on the frontlines of pain research in the laboratories of well-known figures in the field and now have their own independent laboratories continuing their exciting lines of investigation. Therefore, we decided to set out to create a textbook with a slightly different agenda. Rather than focusing on specific topics, we decided to assemble a group of the leading young investigators in the field of pain research, give them some guidance on our overall goals, and set them loose to create the chapters they would like to see based on their most exciting new areas of research.

The title of this book is: "Molecular and Cell Biology of Pain." A book with such a title could have 100 chapters and take up an entire bookshelf. This volume is not meant to be comprehensive, not by any stretch of the imagination, but it is meant to be exciting and new. We hope that these are topics that are largely not covered in existing texts in the field but we also hope that these topics will have staying power in the field. To that end, the group of investigators assembled for this volume have already agreed, in principle, to update this volume periodically as our research areas continue to progress. We hope that the evolution of this title over the coming years will give a sort of history of the research endeavors engaged by the authors of these chapters.

We are indebted to our mentors, who are many and whom we assume know who they are by this point. We are also indebted to the institutions, University of Arizona and University of Texas at Dallas that have given us the opportunity to do this work. We would like to thank the production team at Elsevier and especially Helene Kabes for her work and collaboration on this project. Most importantly, we want to thank our colleagues who agreed to take on this project and without whom this volume most certainly would not exist. We value your academic collaboration, your scientific accomplishments at such an early career stage, and your continued friendship.

Finally, it is our sincere hope that this book will find its way into lecture halls throughout the world. We have aimed the material at graduate and medical students and we think it can give these students an interesting snapshot of the forefront of the field. If you are a student reading this foreword, we hope you find something in this book that inspires discovery. All of us started in your shoes aspiring to learn and make a contribution to mankind through scientific discovery. We wish you the best, and we look forward to reading the volume that will be written by the coming generations.

THEODORE J. PRICE
GREGORY DUSSOR

CHAPTER ONE

An Introduction to Pain Pathways and Pain "Targets"

Vaskar Das[1]
Behavioral and Brain Sciences, The University of Texas at Dallas, Richardson, Texas, USA
[1]Corresponding author: e-mail address: vxd150530@utdallas.edu

Contents

1. An Introduction to Pain and Pain Pathways 2
   1.1 Neuropathic pain 5
   1.2 Inflammatory Pain 7
2. Ion Channels, Receptors, and Other "Targets" for Persistent Inflammatory or Neuropathic Pain 9
   2.1 Ion channels 9
   2.2 Sodium channels 10
   2.3 Calcium channels 12
   2.4 K^+ channels 12
   2.5 Receptors 13
   2.6 Purinergic receptors 13
   2.7 Toll-like receptors 13
   2.8 PAR receptors 14
   2.9 Glutamate receptors 14
   2.10 AMPA receptors 15
   2.11 NMDA receptors 15
   2.12 Metabotropic glutamate receptors 15
   2.13 Opioid receptors 16
   2.14 TRPV receptors 16
   2.15 Prostaglandin (prostanoid) E_2 17
   2.16 Pronociceptive neurotransmitters 17
3. Summary 18
Acknowledgments 18
References 18

Abstract

The purpose of this chapter is to provide a brief introduction to the anatomy and physiology of pain pathways from peripheral nociceptors to central nervous system areas involved in the perception and modulation of pain. This chapter also provides a short introduction to major types of persistent pain: neuropathic and inflammatory persistent pain, and gives an overview of some important molecular targets that are thought to mediate these types of pain. These targets, which include ion channels, receptors, and

Progress in Molecular Biology and Translational Science, Volume 131
ISSN 1877-1173
http://dx.doi.org/10.1016/bs.pmbts.2015.01.003

some neurotransmitters, are further discussed in the context of their relevance as potential drug targets for the better treatment of pain in patients with persistent pain. Finally, this chapter introduces several important concepts in pain research that will be primary topics for chapters that come later in the book.

1. AN INTRODUCTION TO PAIN AND PAIN PATHWAYS

The International Association for the Study of Pain (IASP) has defined pain as "an unpleasant sensory and emotional experience associated with actual or potential tissue damage, or described in terms of such damage."[1] When asked to describe their pain, individuals variously described it in terms of severity (mild, moderate, severe), duration (acute or chronic), and type (nociceptive, inflammatory, neuropathic).[2]

Nociceptive pain is the normal acute pain sensation produced by activation of nociceptors in skin, viscera, and other internal organs in the absence of sensitization.[3–7] It may occur as a result of mechanical, thermal, or chemical noxious stimulation and is variously described as an aching or throbbing kind of pain.[5,6,8,9] Nociceptive pain comprises four main stages: transduction (i.e., action at receptors in the periphery), transmission (i.e., action potentials along axons), perception (i.e., cortical processing of nociceptive input), and modulation (i.e., engagement of descending circuits).[4,10–12] Noxious stimuli are first detected by mechanical, thermal, and chemical nociceptors found on specialized nerve endings present in skin (cutaneous), viscera, and other internal or external organs.[8,9,13,14] Nociceptive impulses are transmitted from the periphery to the spinal cord via primary afferent nerve fibers which may be unmyelinated or myelinated.[3,15–20] The central nervous system (CNS) components of this pathway constitute particular anatomical connections in the spinal cord, brain stem, thalamus, and cortex (the "pain pathway"), linking the sensory inflow generated in high threshold primary afferents with those parts of the CNS responsible for conscious awareness of painful sensations[21] (Fig. 1). Unmyelinated nerve fibers are small diameter C-fibers with diameters in the range 0.4–1.2 μm.[22,23] Myelinated primary afferent nerve fibers are the Aδ-fibers (2–6 μm diameter), whereas the thinly myelinated nerve fibers are the Aβ-fibers (>10 μm diameter).[23,24] Primary afferent C-fibers and Aδ-fibers are responsible for transmission of noxious stimuli whereas Aβ-fibers transmit innocuous, mechanical stimuli such as touch.[21–24] Put simply, nociceptors collect information from noxious stimuli which are transmitted by C-fibers and

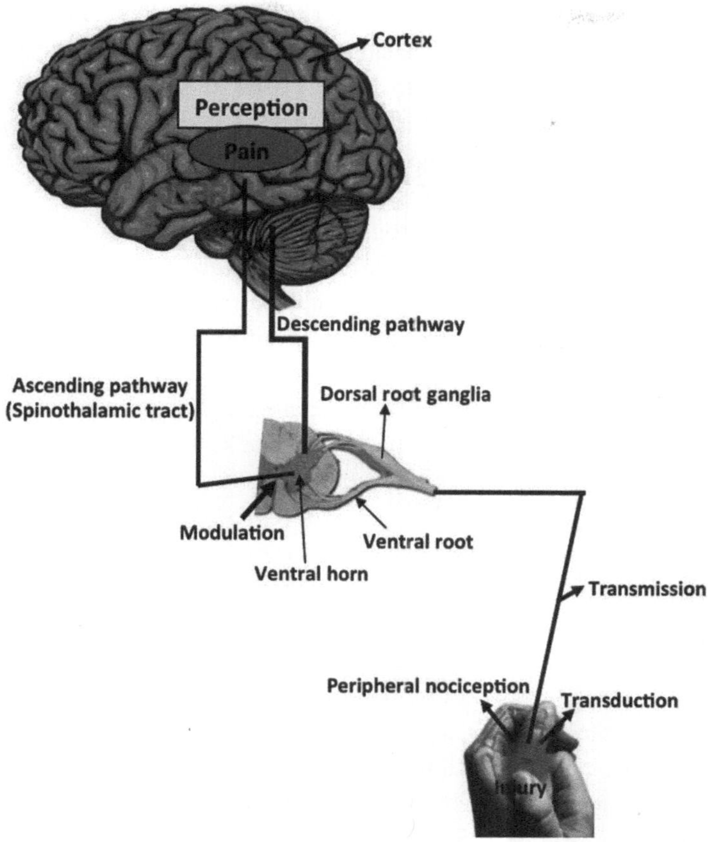

Figure 1 Simplified schematic diagram of the pain pathway. Pain begins with detection of damage or potentially damaging stimuli by nociceptive neurons in the periphery that can transduce this signal into transmission toward the CNS. The first synapse in this pathway is in the dorsal horn, where these projection neurons can send pain-related information onto multiple brain areas. Pain perception occurs in the brain and can be modulated by different centers in the brain. The brain also sends modulatory inputs back down to the spinal cord to induce pain modulation.

Aδ-fibers through the dorsal root ganglia to the superficial laminae I/II of the dorsal horn of the spinal cord.[20,23] Aδ-fibers transmit impulses from the dorsal horn to deeper laminae (III–IV) of the spinal cord and onto higher centers in the brain via the spinothalamic tracts.[20] Dorsal horn neurons comprise (i) projection neurons, (ii) local interneurons, and (iii) propriospinal neurons.[20,25] Although projection neurons are the primary means for transferring sensory information from the spinal cord to the brain, they are only a small fraction of the total number of cells in the dorsal horn.[23,26] Many projection neurons have axons that cross the midline and ascend to multiple areas of the brain including the thalamus, periaqueductal gray matter, lateral parabrachial area of the pons, and various parts of the medullary reticular formation.[27] These neurons are also involved in activation of endogenous descending inhibitory pathways that modulate dorsal horn neurons.[26] Activity-dependent synaptic plasticity in the spinal cord that generates postinjury pain hypersensitivity together with the cellular and molecular mechanisms responsible for this form of neuronal plasticity are termed "central sensitization."[21] Neuroplastic changes relating to the function, chemical profile, or structure of the peripheral nervous system are encompassed by the term "peripheral sensitization" and encompass changes in receptor, ion-channel, and neurotransmitter expression levels.[28,29] Central sensitization in the spinal cord includes sensitization and disinhibition mechanisms, and supraspinally there are functional changes such as enlargement of receptive fields.[30,31] In the CNS, there are also changes in the dynamic interplay between neuronal structures and activated glial cells,[30,32,33] a topic covered in depth in Chapter "Nonneuronal Central Mechanisms of Pain: Glia and Immune Response" by E. Alfonso Romero-Sandoval and Sarah Sweitzer.

Following tissue injury and inflammation, vasoactive mediators such as histamine, substance P (SP), serotonin (5-HT), nitric oxide (NO), prostaglandins (PGs), and bradykinin are released which activate nociceptors resulting in nociception.[13] This in turn can induce release of pronociceptive neurotransmitters such as SP, calcitonin gene-related peptide (CGRP), dynorphin (Dyn), neurokinin A (NKA), glutamate, adenosine triphosphate (ATP), NO, PGs, and neurotrophins such as brain-derived neurotropic factor (BDNF), from primary afferents either in the periphery or at the first synapse in the dorsal horn of the spinal cord.[13,20,22,34,35] More recently, the important role of proinflammatory cytokines (e.g., tumor necrosis factor-alpha (TNF-α), interleukin-1β, interleukin-18, etc.) in peripheral and central sensitization mechanisms associated with persistent pain states has begun to be appreciated.[36]

Many C-fibers express transient receptor potential vanilloid 1 (TRPV1) receptors and hence are sensitive to the vanilloid, capsaicin, which is a high-affinity ligand for TRPV1 receptors.[37] TRPV1-expressing C-fibers may be further subdivided into two major classes:

(i) those that contain the neuropeptides, SP, and CGRP, express the high-affinity nerve growth factor (NGF) receptor, TrkA, and are developmentally dependent on NGF,[34,38,39] and

(ii) those that express isolectin B4, (IB-4), the $P2X_3$ purinergic receptor,[40] fluoride-resistant acid phosphatase, do not contain SP or CGRP,[41] and are dependent on glial cell line-derived neurotrophic factor (GDNF).[34]

1.1 Neuropathic pain

The IASP has defined neuropathic pain as "Pain initiated or caused by a primary lesion or dysfunction in the nervous system."[1,42] Neuropathic pain is variously described by patients as having one or more of the following qualities: burning, tingling, electric shock like, and stabbing or pins and needles.[5,42,43] The appearance of abnormal sensory signs such as allodynia in response to innocuous (nonnoxious) stimulation and/or hyperalgesia in response to noxious stimulation is common.[43] When neuropathic pain is evoked, it may be classified as having dysesthetic, hyperalgesic, or allodynic properties depending upon the dynamic or static characteristics of the stimulus.[44]

In recent years, it has begun to be appreciated that the pathobiology of various neuropathic pain subtypes may differ.[45,46] Hence, multiple research groups have focussed on developing and validating rodent models of each of these neuropathic pain conditions.[47,48] These more relevant rodent models of neuropathic pain have considerable potential not only in terms of unraveling the neurobiology of each of these neuropathic pain subtypes but also in terms of identifying novel targets for discovery of new efficacious and well-tolerated analgesics to improve relief of these persistent pain conditions.[43,49]

Inflammatory and neuroimmune mechanisms contribute to both peripheral and central sensitization that underpin the pathobiology of neuropathic pain.[50] Following peripheral nerve injury, inflammatory cells including mast cells, neutrophils, macrophages, and T-lymphocytes contribute to peripheral sensitization and hyperexcitability of injured and adjacent noninjured primary afferent nerve fibers.[50] In the CNS, activation of glial cells including microglia and astrocytes leads to the production and secretion of various proinflammatory mediators that promote neuroimmune activation and can sensitize the central terminals of primary afferent and second-order neurons to increase the intensity and duration of pain.[50–53]

However, rodent models of neuropathic pain allow behavioral pain responses such as mechanical allodynia in response to application of a nonnoxious stimulus (light pressure) to the hindpaws and/or hyperalgesia in response to application of noxious stimuli (pressure, heat, cold) to the hindpaws, to be quantified.[43,49] The three most commonly used rodent models of neuropathic pain involve induction of a unilateral chronic constriction injury of the sciatic nerve,[54] partial sciatic nerve ligation,[55] and L5 spinal nerve ligation.[56] Over the past two decades, these models have been used to identify numerous neuropathic pain "targets" encompassing various receptors, ion channels, and enzymes, as well as to assess novel pain therapeutics in development.[43,49]

More recently, rodent models of varicella zoster virus-induced neuropathic pain, antiretroviral drug-induced neuropathic pain, cancer chemotherapy-induced neuropathic pain, bone cancer pain, and multiple sclerosis-induced neuropathic pain have been developed.[43,57,58] It is hoped that through use of these more sophisticated rodent pain models, it will be possible to gain an enhanced understanding of the pathobiology of these conditions. Additionally, it may be possible to identify novel "druggable" targets for drug discovery aimed at producing novel analgesics with improved efficacy and reduced adverse-event profiles for improved relief of these chronic pain conditions in the clinical setting.[49,59]

Persistent ongoing pain secondary to nerve injury is underpinned by considerable complexity and plasticity at multiple levels of the neuraxis.[49,58,60–62] Following peripheral nerve injury, ectopic firing of injured and uninjured afferents induces neuroplastic changes and "central sensitization" in the spinal cord and the brain, underpinned by both neuronal and nonneuronal mechanisms[49] (Fig. 2). Intensive research over the past two decades has revealed a large number of receptors, enzymes, and ion channels as potential novel targets for drug discovery programs aimed at

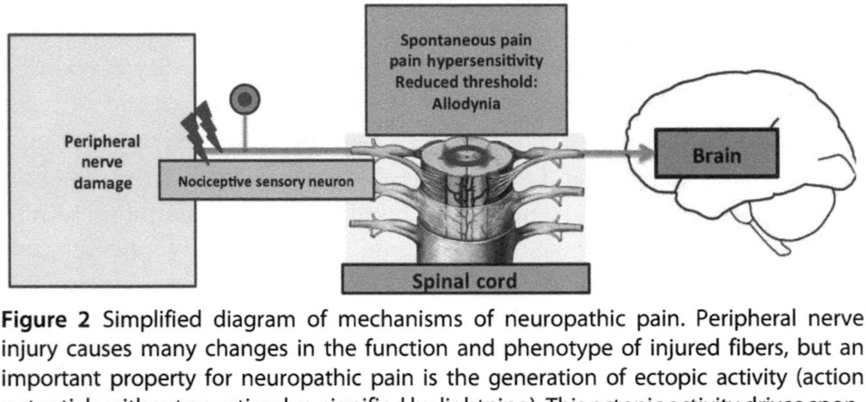

Figure 2 Simplified diagram of mechanisms of neuropathic pain. Peripheral nerve injury causes many changes in the function and phenotype of injured fibers, but an important property for neuropathic pain is the generation of ectopic activity (action potentials without any stimulus, signified by lightning). This ectopic activity drives spontaneous pain and plasticity in the dorsal horn and brain that may underlie clinical features like allodynia.

producing new drugs for the relief of neuropathic pain.[63–65] Although several molecules have entered preclinical and clinical development, very few have been approved by regulatory agencies for clinical use.[66] Hence, this large unmet medical need is driving research in this field.

1.2 Inflammatory Pain

Inflammatory pain is precipitated by an insult to the integrity of tissues at the cellular level. One of the characteristic features of inflammatory states is that normally innocuous stimuli can produce pain.[67] Inflammation is classically associated with pain (dolor), heat (calor), redness (rubor), swelling (tumor), and loss-of-function (function laesa).[67,68] Examples of inflammatory pain include pain secondary to tissue injury and infection as well as rheumatoid arthritis.[69,70] Following tissue injury, nociceptors in the affected tissue become sensitized due to the release of proinflammatory mediators from damaged cells and blood vessels at the site of injury as well as from immune cells that invade the injured site.[71] This topic is covered in detail in Chapter "Peripheral Scaffolding and Signaling Pathways in Inflammatory Pain by Jeske Nathan."

Inflammatory mediators including protons, 5HT, histamine, adenosine, bradykinin, prostaglandin E_2 (PGE_2), NO, IL-1, TNF-α, interleukin-6 (IL-6), leukemia inhibitory factor, and NGF[5,72,73] contribute to nociceptor sensitization so that innocuous stimuli are detected as painful (allodynia) or there is an exaggerated response to noxious stimuli (hyperalgesia).[34,74] The central terminals of primary afferent nerve fibers (first-order neurons) are located in the superficial layers (laminae I/II) of the dorsal horn of the spinal cord.[13] Synaptic input from these terminals to second-order neurons in the spinal cord transfers information created by action potentials in primary afferents secondary to peripheral noxious stimuli (depending on intensity and duration), to the thalamus, and then onto the cerebral cortex in the brain.[13] Synaptic function at the central terminals of first-order neurons is regulated by neurotransmitter release, primarily involving glutamate, and neuroactive peptides like substance P and CGRP.[60–62,75–78]

In inflammatory pain, peripheral inflammation induces a phenotypic switch in primary sensory neurons to induce a change in their neurochemical character and properties.[62] This topic is covered in some depth in Chapters "Translation Control of Chronic Pain by Ohannes K. Melemedjian and Arkady Khoutorsky," "Regulation of Gene Expression and Pain States by Epigenetic Mechanisms by S.M. Géranton and K.K. Tochiki," and "Commonalities between Pain and Memory Mechanisms and their Meaning for Understanding Chronic Pain by Theodore J Price and Kufreobong E Inyang." In brief, this is underpinned by alterations in transcription and translation of various receptors and ion channels to induce central sensitization by virtue of a change in the level of synaptic input produced by the sensitized afferent nerve fibers[62] (Fig. 3). Put simply, continuous inputs from sensitized nociceptive afferents can activate or trigger central sensitization that is characterized by a reduced threshold of dorsal horn neurons to noxious stimulation[16,79–84] (Fig. 3). There is expansion of the receptive fields of dorsal horn neurons,[85,86] and temporal summation of slow postsynaptic potentials resulting in a cumulative depolarization and a prolonged after discharge or "wind up" of dorsal horn neurons.[15] There is also increased excitability of the flexion reflex in response to peripheral stimulation.[82,87] The neural mechanisms underlying central sensitization involve excitatory amino acids (EAAs, e.g., mainly L-glutamate) acting at AMPA and NMDA receptors with the net result being persistent activation of the NMDA receptor[84] to allow Ca^{2+} entry into neurons and activation of numerous intracellular signaling pathways.[62]

SP-expressing C-fibers and BDNF-expressing dorsal root ganglion (DRG) neurons both have a significant role in inflammation-induced central sensitization after exposure of their peripheral terminals to inflammatory mediators and NGF released from immune cells.[62,88,89] Ectopic firing of sensitized terminals increase Aβ-mediated synaptic input to superficial dorsal horn neurons[62,90] and induction of cyclooxygenase-2 (COX-2) expression levels to drive production of PGE_2.[62,91,92] Retrograde transport of NGF from the peripheral terminals of C-fibers to the dorsal root ganglia induces upregulation of TRPV1 expression and activation (phosphorylation) of p38 MAPK.[39,93] This in turn leads to upregulated synthesis and release of

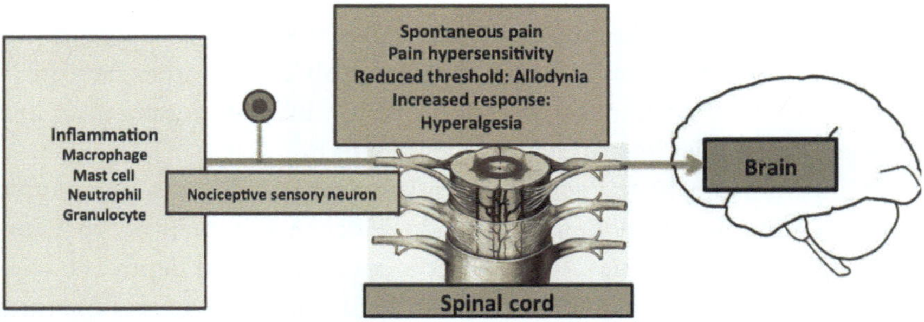

Figure 3 Simplified diagram of mechanisms of inflammatory pain. Activation of immune cells during inflammation leads to the release of inflammatory mediators that act on nociceptors. Many of these inflammatory mediators from immune cells directly activate or modulate the activity of nociceptors. This can drive spontaneous pain and plasticity in the dorsal horn and brain that underlies clinical features including allodynia and hyperalgesia.

proinflammatory cytokines among which IL-1β and TNF-α contribute to the development of central sensitization by enhancing excitatory and reducing inhibitory currents, and by activating induction of COX-2.[62,91,94] As a result, GluR2-containing AMPA receptor activation allows entry of Ca^{2+} into neurons,[62,95] and this mechanism generates as much Ca^{2+} influx as occurs with NMDA receptor activation during inflammatory pain.[62,95]

2. ION CHANNELS, RECEPTORS, AND OTHER "TARGETS" FOR PERSISTENT INFLAMMATORY OR NEUROPATHIC PAIN

Intensive research over the past two decades has revealed a vast array of ion channels, receptors, transporters, and enzymes that are potential "druggable" targets for use in discovery programs aimed at developing the next generation of analgesic drugs.[66] Examples of "pain targets" for potential modulation by novel analgesic agents include voltage-gated sodium channels ($Na_v1.3$, $Na_v1.7$, and $Na_v1.8$),[12] voltage-gated potassium channels (Kv1.4) is the sole Kv1 subunit expressed in smaller diameter sensory neurons[96–98] suggesting that homomeric Kv1.4 channels predominate in Aδ and C-fibers arising from these cells.[97] By contrast, larger diameter neurons associated with mechanoreception and proprioception express high levels of Kv1.1 and Kv1.2 without Kv1.4 or other Kv1 subunits, suggesting that heteromers of these subunits predominate on large, myelinated afferent axons that extend from these cells.[97,99–102] Additional "pain targets" include voltage-gated calcium channels (VGCC) (α2δ subunits; $Ca_v2.2$, $Ca_v3.1$, $Ca_v3.2$, $Ca_v3.3$),[103] acid-sensing ion channels (ASICs)[104] covered to some extent in Chapter "Meningeal Afferent Signaling and the Pathophysiology of Migraine by Carolina Burgos-Vega, Jamie Moy and Greg Dussor," NMDA receptors, TRPV1 receptors, covered in Chapter "Sensory TRP Channels: The Key Transducers of Nociception and Pain by Aaron D. Mickle, Andrew J. Shepherd and Durga P. Mohapatra," NKA, purinergic receptors, toll-like receptors (TLRs), protease-activated receptors (PAR) receptors, opioid receptors, the norepinephrine transporter, and cyclooxygenases (COX-1/2).[105] Many of these "pain targets" are described briefly in the following sections and in more detail throughout this volume, as noted and summarized in Fig. 4.

2.1 Ion channels

Ion channels play a key role in nociception and are involved in sensory transduction (TRPV1), regulation of neuronal excitability (potassium

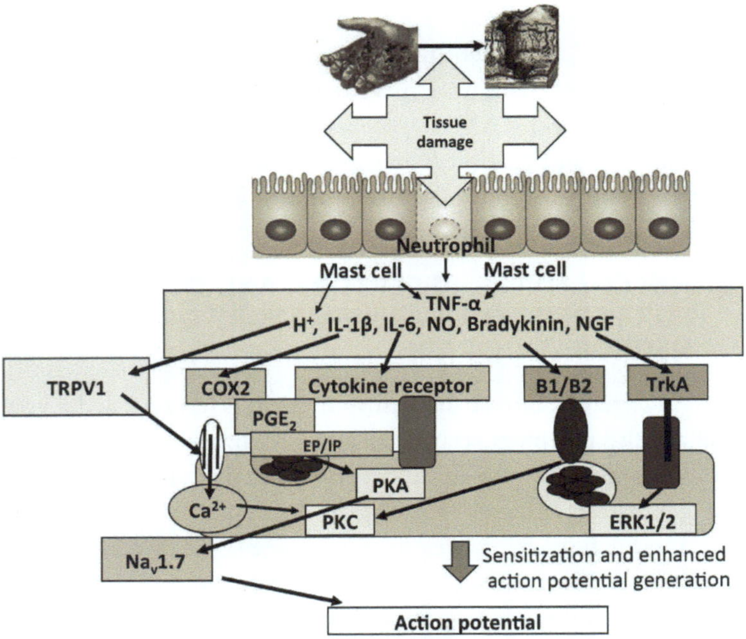

Figure 4 Mechanisms of inflammation-induced pain in the periphery. Tissue damage causes an immune response and the release of inflammatory mediators that act on nociceptors. As described in the text, peripheral nociceptors are finely tuned to detect mediators released by immune cells via the expression of receptors that bind these ligands and the presence of ion channels that are altered by the signaling pathways downstream of activation of these receptors.

channels), action potential propagation (sodium channels, ATP-gated channels, ASICs), and presynaptic release of various neurotransmitters (calcium channels).[106]

2.2 Sodium channels

Voltage-gated sodium channels are considered a major target for the development of novel therapies for improving pain management, as the ectopic firing of primary afferents is associated with abnormal sodium channel regulation.[107–109]

Sodium channels comprise an α-subunit containing a voltage-gated sodium-selective aqueous pore and one or two smaller ancillary β-subunits.[110] Sodium channel subtypes differ in their sensitivity to block by tetrodotoxin (TTX) with six isoforms being sensitive (TTXs: $Na_v1.1$, $Na_v1.2$, $Na_v1.3$, $Na_v1.4$, $Na_v1.6$, $Na_v1.7$) to block by nanomolar concentrations of TTX and three isoforms being resistant to micromolar concentrations of TTX ($Na_v1.5$, $Na_v1.8$, $Na_v1.9$).[111]

Acute inflammatory and neuropathic pain can be attenuated or abolished by local treatment with sodium channel blockers,[112] showing that peripheral nociceptive input is dependent on the presence of functional voltage-gated sodium channels. Four voltage-gated sodium channel subtypes ($Na_v1.3$, $Na_v1.7$, $Na_v1.8$, and $Na_v1.9$) are of greatest interest in pain due to their selective expression in peripheral nerves.[107,113–128] For first-order sensory neurons, $Na_v1.8$ is expressed by the cell body, peripheral terminals, and central terminals within the dorsal horn of the spinal cord.[110] Anatomical and electrophysiological evidence indicates that expression of $Na_v1.9$ is largely restricted to nociceptive Aδ- and C-fibers.[110]

Highlighting the importance of $Na_v1.7$ in inflammatory pain (Fig. 4), levels of expression of $Na_v1.7$ are increased in sensory nerve terminals by inflammation[12,128] and following ablation of $Na_v1.7$ in nociceptive neurons, inflammatory pain responses are greatly reduced.[129] Additionally, dominant gain-of-function mutations in *SCN9A*, the gene encoding $Na_v1.7$ to increase DRG neuron excitability are thought to be causal in two inherited chronic pain disorders in humans, viz, erythromelalgia, characterized by burning pain and skin redness in the extremities, and paroxysmal extreme pain disorder, characterized by skin flushing, rectal, periocular, and perimandibular pain evoked principally by mechanical stimuli.[109,130] Additionally, congenital indifference to pain due to rare recessive loss-of-function mutations in *SCN9A* mean that although individuals so-affected are of normal intelligence, they often fail to recognize and report pain in response to injury or infection which can lead to early mortality.[131]

Conditional knockout of *SCN9A* in mice abolished mechanical pain, inflammatory pain, and reflex withdrawal responses to noxious heat recapitulating the pain-free phenotype in humans with *SCN9A* loss-of-function mutations.[128] Interestingly, as conditional knockout of *SCN9A* in both sensory and sympathetic neurons in mice with spinal nerve transection, markedly reduced neuropathic pain behavior in these animals, neuropathic pain therefore appears to involve interaction between sensory and sympathetic neurons.[128]

In neuropathic pain, levels of expression of $Na_v1.3$ are increased in damaged peripheral nerves and this is highly correlated with the appearance of a rapidly repriming sodium current in small DRG neurons consistent with the notion that $Na_v1.3$ channels make a key contribution to neuronal hyperexcitability in neuropathic pain.[12,105,110,132–135]

2.3 Calcium channels

VGCCs play an important role in synaptic transmission and nociceptive signaling.[103,136–138] N-type ($Ca_v2.2$) VGCCs are expressed in high density in the cell bodies of primary afferents in the dorsal root ganglia and on presynaptic terminals that form synapses with second-order neurons in the dorsal horn.[139–144] $Ca_v2.2$ channels regulate calcium entry, modulate neurotransmitter release, and lead to changes in sensory nerve excitability for the modulation of pain.[67,145–147] Mice lacking $Ca_v2.2$ channels have reduced pain responses secondary to inflammation and peripheral nerve injury.[95,103,148,149]

T-type VGCCs ($Ca_v3.1$, $Ca_v3.2$, and $Ca_v3.3$) are expressed on the cell bodies and peripheral terminals of primary afferents, where they contribute to initiation of the action potential to regulate neuronal excitability.[103,150–152]

Multiple studies in rodent pain models have implicated the $\alpha_2\delta_1$ subunit of presynaptic calcium channels as having an important role in persistent pain states.[153] This is emphasized by clinical studies showing that the anti-neuropathic drugs, gabapentin, and pregabalin that are ligands at the $\alpha_2\delta_1$ subunit, have efficacy for the relief of neuropathic pain.[154–157] The mechanisms of action of these drugs are still controversial, but their widespread use for neuropathic pain disorders is highlighted in Chapter "Chronic Pain Syndromes, Mechanisms, and Current Treatments by Justin Sirianni, Mohab Ibrahim and Amol Patwardhan."

2.4 K^+ channels

Voltage-gated K^+ (Kv) channel subunits are expressed in DRG neurons and have an important physiological role in the regulation of membrane potentials in excitable tissues including nociceptive neurons.[101,102,158–160] The Kv channel subunit Kv1.4 is the sole Kv1 α subunit expressed in smaller diameter neurons, suggesting that homomeric Kv1.4 channels predominate in Aδ and C-fibers arising from these cells.[97,99,100] Additionally, these neurons are presumably nociceptors, because they also express the TRPV1 capsaicin receptor, CGRP, and/or Na^+ channel SNS/PN3/$Na_v1.8$.[97,99,161–164] However, larger diameter neurons associated with mechanoception and proprioception express high levels of Kv1.1 and Kv1.2 without Kv1.4 or other Kv1 α subunits, suggesting that heteromers of these subunits predominate on large, myelinated afferent axons that extend from these cells.[97,99,100,164] As the opening of K^+ channels leads to hyperpolarization of the cell membrane and so decreased nerve cell excitability, several Kv channels are implicated as possible targets for novel pain therapeutics. For example, A-type potassium currents contribute significantly to neuronal excitability and central sensitization in the dorsal horn of the spinal cord in inflammatory pain.[100,165–169]

However, abnormal hyperexcitability of primary sensory neurons plays an important role in neuropathic pain.[164,166] Kv channels regulate neuronal excitability by affecting the resting membrane potential and influencing the repolarization and frequency of the action potential and may therefore play a key role in ectopic activity that develops in peripheral nerves driving neuropathic pain.[163,164] Additionally, diabetes primarily reduces Kv channel activity in medium and large DRG neurons.[164] Increased BDNF activity in these neurons likely contributes to the reduction in Kv channel function through TrkB receptor stimulation in painful diabetic neuropathy.[164]

2.5 Receptors

In inflammatory pain, multiple receptor classes located on nociceptors are modulated by vasoactive mediators released from damaged tissues and immune cells that invade the inflamed tissues.[170]

2.6 Purinergic receptors

ATP activates P2X purinergic receptors, especially $P2X_1$, $P2X_3$, or $P2X_7$ receptors, to produce pain.[171–173] Currently, a range of preclinical studies are investigating a role for P2X receptors in pain, inflammation, osteoporosis, multiple sclerosis, spinal cord injury, and bladder dysfunction.[173] Some of these have been progressed into clinical trials for rheumatoid arthritis, pain, and cough.[173]

$P2X_3$ receptors, located exclusively on small diameter nociceptive-fibers, are implicated in inflammatory pain and $P2X_4$ receptors on microglia in the dorsal horn of the spinal cord are implicated in the pathogenesis of neuropathic pain.[174] Antagonists at the $P2X_7$ receptor also reduce pain behaviors in rodent models of inflammatory and neuropathic pain, again highlighting that purinergic glial–neuronal interactions are important modulators of noxious nociceptive neurotransmission.[175] Hence, $P2X_3$, $P2X_4$, and $P2X_7$ receptors are potential targets for novel therapeutics for the treatment of inflammatory and neuropathic pain conditions.[176,177]

2.7 Toll-like receptors

Proinflammatory central immune signaling contributes significantly to the initiation and maintenance of heightened pain states because recent discoveries have implicated the innate immune system, in particular, pattern recognition TLRs in triggering these proinflammatory central immune signaling events.[178] There is considerable interest in the targeting of TLRs on immune cells for the prevention and treatment of cancer, infection, inflammation, and autoimmune diseases.[179] In neuropathic pain, "TLR4 receptors on peripheral immune cells (e.g., monocytes/macrophages, dendritic cells, and immune-related cells such as keratinocytes)"[170] as well as "activated microglia in the CNS" appear to have a key role in the establishment of this condition.[180–183] Acute TLR4 antagonism attenuates neuropathic pain behavior and potentiates opioid antinociception.[184] Hence, TLR4 appears to be a possible target for therapeutic intervention for relief of neuropathic pain and for augmenting opioid analgesia. As already noted in an earlier section of this literature review, dysregulation of chemokines (Section 1) and their receptors (Section 2.1; Fig 4), particularly fractalkine and its CX3CR1 receptor, appear to play a key role in neuroimmune signaling that contributes significantly to the pathobiology of neuropathic pain.[185] This target is discussed in more detail in Chapter "Role of Extracellular Damage-Associated Molecular Pattern Molecules (DAMPS) as Mediators of Persistent Pain by Jungo Kato and Camilla I Svensson."

2.8 PAR receptors

PARs are G-protein coupled receptors (GPCRs) that have a unique activation mechanism involving specific proteolytic cleavage of the amino-terminal sequence by serine proteases.[186–188] PAR_1 is expressed by primary afferent neurons and can modulate nociception.[189] PAR_2 is expressed by SP- and CGRP-containing primary afferents.[189] Activation of PAR_2 induces the release of the pronociceptive neurotransmitters, SP, and CGRP from both peripheral and central terminals of primary DRG neurons.[189] PAR_4 modulates nociceptive responses in normal and inflammatory conditions such that a PAR_4 agonist alleviated inflammatory pain in rats.[159,188,190–196]

2.9 Glutamate receptors

Glutamate is the major EAA neurotransmitter in the central nervous system and is found in at least 70% of sensory neurons in the DRGs.[197] It is released from the central terminals of primary afferents and has an important role in nociceptive neurotransmission.[197] Glutamate acts via two main receptor classes, iGluRs (ionotropic), and mGluRs (metabotropic) with iGluRs further subdivided into AMPA, NMDA, and kainate receptors.[72,198] AMPA and NMDA receptors are directly coupled to cation-permeable ion channels whereas metabotropic glutamate receptors (mGluRs) are coupled via G-proteins to soluble second messengers.[53,72,198–200] Brief nociceptive stimuli primarily activate AMPA receptors whereas stimuli of more prolonged duration activate NMDA receptors.[35,201] Astroglial cells remove excess glutamate from the extracellular space and express the glutamate reuptake transporters, GLAST/EEAT-1 (excitatory amino acid transporter) and glutamate transporters EEAT-2/GLT-1.[202–207]

2.10 AMPA receptors

AMPA receptors have an important role in acute spinal processing of nociceptive and nonnociceptive inputs.[208] Activation of AMPA receptors by glutamate results in potent depolarization of dorsal horn neurons to remove the Mg^{2+} block, and hence activate NMDA receptors resulting in calcium influx and initiation of a cascade of downstream signaling events.[209] Because AMPA receptors also have roles in many other CNS functions,[208] they are generally regarded as being unsuitable targets for development of novel pain therapeutics.[208]

2.11 NMDA receptors

NMDA receptors (Fig. 1) are located in the superficial and deeper laminae of the spinal dorsal horn on the central terminals of primary afferents as well as on membranes that are postsynaptic to the primary afferent.[210] All NMDA receptors display a certain degree of voltage-dependent Mg^{2+} block and marked permeability to Ca^{2+} after removal of the Mg^{2+} block (Fig. 1).[198,211–214] Spinal NMDA receptors have an important role in "central sensitization" in the spinal cord in persistent pain states.[215] Although molecules targeting NMDA receptors have potential for the relief of persistent pains such as neuropathic pain, NMDA receptors are involved in normal physiological functions and so the first generation of these agents were hampered by CNS side effects in the analgesic dose range.[216]

2.12 Metabotropic glutamate receptors

Group I mGluRs in laminae I/II of the spinal dorsal horn play an important role in the transduction of nociceptive input from C-fibers.[217] There are three mGluR classes containing eight cloned mGluRs and *in vivo* studies show that these are not involved in acute nociceptive signaling.[218–220] Group I mGluRs (mGlu1 and 5) are implicated in central sensitization and persistent nociception.[221] By contrast, activation of group II mGluRs (mGlu2/3) alleviates neuropathic and inflammatory pain.[221] This topic is discussed in more detail in Chapter "mGluRs Head to Toe in Pain by Benedict J. Kolber."

2.13 Opioid receptors

Opioid receptors are members of the superfamily of 7-transmembrane spanning region GPCRs that are coupled to intracellular effectors via G-proteins, mainly of the inhibitory type, i.e., $G_{i,o.}$[25,222,223] There are high densities of opioid receptors in various brain regions, the dorsal horn of the spinal cord, on peripheral nerve terminals, in peripheral tissues including the gastrointestinal tract and on immune cells (Fig. 2).[198,224,225]

In the 1990s, three opioid receptor types, viz, μ (MOP), δ (DOP), and κ (KOP) were cloned, and more recently multiple splice variants of these receptors, particularly the MOP receptor, have been identified.[222,226–228] The endogenous ligands for opioid receptors include the endomorphins (highly MOP selective), β-endorphin (equal MOP and DOP selectivity), met- and leu-enkephalin (more selective at DOP than MOP), and Dyn (KOP selective).[229–231] Opioid agonists activate opioid receptors to produce potent analgesia by activation of the descending inhibitory system to inhibit ascending excitatory nociceptive transmission (Fig. 2).[224,232,233] Studies using MOP receptor knockout mice show that the antinociceptive and other effects of morphine and most other clinically available opioid analgesics, are produced secondary to activation of the MOP receptor.[222,223,234–239] Studies using rodents show that peripheral opioid actions are increased in inflammation, suggesting that peripherally selective opioid analgesics may have benefit as future analgesic agents that are devoid of CNS side effects.[25,222]

2.14 TRPV receptors

TRPV1 receptor (Fig. 4) is a ligand-gated nonselective cation channel expressed on primary afferent sensory neurones that can be activated by exogenous agents (e.g., capsaicin), endogenous substances (e.g., bradykinin, ethanol, nicotine, anandamide, and insulin) as well as by heat (>43 °C) and low pH.[240–243] Following activation of TRPV1, there is a rapid increase in intracellular Ca^{2+} concentrations resulting in nociceptive signal transduction via C-fibers to produce pain in humans and pain behaviors in animals.[243,244] TRPV1 knockout mice exhibit reduced thermal nociception and a loss of inflammatory thermal hyperalgesia.[242,245] However, both NGF and GDNF elicit thermal hyperalgesia during peripheral inflammation via an increase in TRPV1 expression.[246] Expression of these two growth factors follows different time courses and they act on distinctive subpopulations of DRG neurons.[246] Intradermal injection of capsaicin and NGF produce heat hyperalgesia via activation of their respective receptors, viz, TRPV1 and TrkA on sensory nerve terminals. Moreover, PI3K induces heat hyperalgesia, possibly by regulating TRPV1 activity, in an ERK-dependent manner, and the PI3K pathway also appears to play a role that is distinct from ERK (Fig. 4) by regulating the early onset of inflammatory pain.[39,93,247,248] TRP channels are discussed in more detail in Chapter "Sensory TRP Channels: The Key Transducers of Nociception and Pain by Aaron D. Mickle, Andrew J. Shepherd and Durga P. Mohapatra."

2.15 Prostaglandin (prostanoid) E_2

Inflammatory pain hypersensitivity is regulated by prostaglandin receptors (EP1, EP2, EP3, EP4 receptors; Fig. 4).[249] At the site of inflammation, PGE_2 sensitizes peripheral nociceptors via activation of EP2 receptors that are present on the peripheral terminals of high threshold sensory nerve fibers by reducing the nerve firing threshold and increasing responsiveness, which is the key phenomenon of peripheral sensitization.[249,250]

Following tissue injury, the synthesis of PGE_2 in the spinal cord[91] contributes to central sensitization[251] and increased excitability of spinal dorsal horn neurons.[249] NSAIDs inhibit prostaglandin synthesis through nonselective inhibition of constitutively expressed cyclooxygenase COX-1 as well as the inducible isoform COX-2.[252–254]

2.16 Pronociceptive neurotransmitters

2.16.1 Nitric oxide

In the CNS, NO is synthesized primarily from the precursor, L-arginine, by the enzyme nitric oxide synthase (NOS).[255] There are three NOS isoforms, viz, neuronal (nNOS), endothelial (eNOS), and inducible nitric oxide synthase (iNOS), with nNOS having a role in the modulation of nociceptive transmission in the spinal cord.[255,256] Following glial cell activation in the CNS, NO is produced by iNOS, to further sensitize nociceptive neurones and contribute to the maintenance of central sensitization in persistent pain conditions.[256] NO produced in excess by iNOS and nNOS is implicated in inflammatory and neuropathic pain, and so iNOS and nNOS inhibitors have been investigated as potential novel agents for alleviation of these chronic pain conditions.[257]

2.16.2 Nerve growth factor

NGF levels increase during inflammation[258–260] resulting in sensitization of primary afferents to noxious thermal, mechanical, and chemical stimuli via upregulated synthesis of TRPV1 receptors as well as SP, CGRP, and bradykinin receptors[261–265] (Fig. 4). Anti-NGF therapeutics have advanced to clinical trials in humans and have shown broad efficacy but also severe, albeit rare, side effects. NGF signaling is discussed in more detail in Chapters "Translation Control of Chronic Pain by Ohannes K. Melemedjian and Arkady Khoutorsky" and "Commonalities between Pain and Memory Mechanisms and their Meaning for Understanding Chronic Pain by Theodore J Price and Kufreobong E Inyang."

3. SUMMARY

Chronic pain is notoriously difficult to treat and so there is a large unmet clinical need for new treatments to alleviate pain. Rodent models of neuropathic and inflammatory pain have had a huge impact on our basic understanding of pain and pain plasticity, a feature that will be discussed throughout this volume. These basic science investigations have led to the development of a broad variety of pain targets, some of which were discussed briefly here and some of which will be discussed in great detail throughout the rest of this book. The goal of this line of research is to develop a new generation of pain therapies with broad efficacy and limited side effects.

ACKNOWLEDGMENTS

Dr. Das is grateful to Professors Maree Smith at University of Queensland and Theodore Price at University of Texas at Dallas for editing of the chapter.

REFERENCES

1. Merskey B. *Classification of Chronic Pain. Descriptions of Chronic Pain Syndromes and Definitions of Pain Terms*. 2nd ed. Seattle: IASP; 1994:1–222.
2. Rainville P. Brain mechanisms of pain affect and pain modulation. *Curr Opin Neurobiol.* 2002;12:195–204.
3. Besson JM, Rivot JP. Spinal interneurones involved in presynaptic controls of supraspinal origin. *J Physiol.* 1973;230:235–254.
4. Carr DB, Goudas LC. Acute pain. *Lancet.* 1999;353:2051–2058.
5. Woolf CJ, Mannion RJ. Neuropathic pain: aetiology, symptoms, mechanisms, and management. *Lancet.* 1999;353:1959–1964.
6. Besson J. The neurobiology of pain. *Lancet.* 1999;353:1610–1615.
7. Woolf CJ. Pain: moving from symptom control toward mechanism-specific pharmacologic management. *Ann Intern Med.* 2004;140:441–451.

8. Bessou P, Perl E. Response of cutaneous sensory units with unmyelinated fibers to noxious stimuli. *J Neurophysiol.* 1969;32:1025–1043.
9. Beitel R, Dubner R. Response of unmyelinated (C) polymodal nociceptors to thermal stimuli applied to monkey's face. *J Neurophysiol.* 1976;39:1160–1175.
10. Mense S, Simons DG, Jon Russell MI. *Muscle Pain: Understanding Its Nature, Diagnosis, and Treatment.* Pennsylvania, USA: Wolters Kluwer Health; 2001.
11. Berry PH, CHPN C, Paice JA. Patient care: pain management. In: *Core Curriculum for the Generalist Hospice and Palliative Nurse.* Dubuque, IA, USA: Kendall Hunt; 2002:63.
12. Wood JN, Boorman JP, Okuse K, Baker MD. Voltage-gated sodium channels and pain pathways. *J Neurobiol.* 2004;61:55–71.
13. Millan MJ. The induction of pain: an integrative review. *Prog Neurobiol.* 1999;57:1–164.
14. Jesse CR, Savegnago L, Nogueira CW. Effect of a metabotropic glutamate receptor 5 antagonist, MPEP, on the nociceptive response induced by intrathecal injection of excitatory amino acids, substance P, bradykinin or cytokines in mice. *Pharmacol Biochem Behav.* 2008;90:608–613.
15. Mendell LM. Physiological properties of unmyelinated fiber projection to the spinal cord. *Exp Neurol.* 1966;16:316–332.
16. Price DD, Hull CD, Buchwald NA. Intracellular responses of dorsal horn cells to cutaneous and sural nerve A and C fiber stimuli. *Exp Neurol.* 1971;33:291–309.
17. Gregor M, Zimmermann M. Characteristics of spinal neurones responding to cutaneous myelinated and unmyelinated fibres. *J Physiol.* 1972;221:555–576.
18. Handwerker H, Iggo A, Zimmermann M. Segmental and supraspinal actions on dorsal horn neurons responding to noxious and non-noxious skin stimuli. *Pain.* 1975;1:147–165.
19. Calvillo O, Madrid J, Rudomin P. Presynaptic depolarization of unmyelinated primary afferent fibers in the spinal cord of the cat. *Neuroscience.* 1982;7:1389–1409.
20. Willis Jr WD, Coggeshall RE. *Sensory Mechanisms of the Spinal Cord. Primary Afferent Neurons and the Spinal Dorsal Horn;* Vol. 1. New York, USA: Springer-Verlag; 2004.
21. Woolf CJ. Central sensitization: implications for the diagnosis and treatment of pain. *Pain.* 2011;152:S2–S15.
22. Millan MJ. Descending control of pain. *Prog Neurobiol.* 2002;66:355–474.
23. Urch C. Normal pain transmission. *Rev Pain.* 2007;1:2–6.
24. Millan MJ. Evidence that an alpha 2A-adrenoceptor subtype mediates antinociception in mice. *Eur J Pharmacol.* 1992;215:355–356.
25. Smith MT, South SM. *Pain Pharmacology and Analgesia.* In: Majewski H, ed. Oxford, UK: EOLSS Publishers; 2008:1–80.
26. Woolf C, Salter M. Plasticity and pain: role of the dorsal horn. In: *Wall and Melzack's Textbook of Pain.* 5th ed. Philadelphia, USA: Elsevier; 2006:91–105.
27. Todd AJ, Koerber HR. Neuroanatomical substrates of spinal nociception. In: *Wall and Melzack's Textbook of Pain.* Philadelphia, USA: Elsevier; 2006:73–90.
28. Bartlett S, Bonci A, Haass-Koffler C, Naeemuddin M. Ion channel compositions and uses thereof. 2010. WO Patent App. PCT/US2010/049,149.
29. Hicks Jr EC, Dutton CT. Methods for chronic pain management and treatment using HCG. 2012. US Patent 20,120,265,712.
30. May A. Chronic pain may change the structure of the brain. *Pain.* 2008;137:7–15.
31. Carlton SM, Du J, Tan HY, et al. Peripheral and central sensitization in remote spinal cord regions contribute to central neuropathic pain after spinal cord injury. *Pain.* 2009;147:265–276.
32. Joksimovic M, Yun BA, Kittappa R, et al. Wnt antagonism of Shh facilitates midbrain floor plate neurogenesis. *Nat Neurosci.* 2009;12:125–131.
33. Yong VW, Marks S. The interplay between the immune and central nervous systems in neuronal injury. *Neurology.* 2010;74:S9–S16.

34. Dubner R. The neurobiology of persistent pain and its clinical implications. *Suppl Clin Neurophysiol.* 2004;57:3–7.
35. Yaksh TL. Calcium channels as therapeutic targets in neuropathic pain. *J Pain.* 2006;7: S13–S30.
36. Griffis CA, Compton P, Doering L. The effect of pain on leukocyte cellular adhesion molecules. *Biol Res Nurs.* 2006;7:297–312.
37. Rami HK, Gunthorpe MJ. The therapeutic potential of TRPV1 (VR1) antagonists: clinical answers await. *Drug Discov Today.* 2004;1:97–104.
38. Basbaum AI. Spinal mechanisms of acute and persistent pain. *Reg Anesth Pain Med.* 1999;24:59–67.
39. Smith MT, Woodruff TM, Wyse BD, Muralidharan A, Walther T. A small molecule angiotensin II type 2 receptor (AT2R) antagonist produces analgesia in a Rat model of neuropathic pain by inhibition of p38 mitogen-activated protein kinase (MAPK) and p44/p42 MAPK activation in the dorsal root ganglia. *Pain Med.* 2013;14(10):1557–1568.
40. Vulchanova L, Riedl MS, Shuster SJ, et al. P2X3 is expressed by DRG neurons that terminate in inner lamina II. *Eur J Neurosci.* 1998;10:3470–3478.
41. Silverman JD, Kruger L. Acid phosphatase as a selective marker for a class of small sensory ganglion cells in several mammals: spinal cord distribution, histochemical properties, and relation to fluoride-resistant acid phosphatase (FRAP) of rodents. *Somatosens Mot Res.* 1988;5:219–246.
42. Dworkin RH. An overview of neuropathic pain: syndromes, symptoms, signs, and several mechanisms. *Clin J Pain.* 2002;18:343–349.
43. Bridges D, Thompson S, Rice A. Mechanisms of neuropathic pain. *Br J Anaesth.* 2001;87:12–26.
44. Rasmussen PV, Sindrup SH, Jensen TS, Bach FW. Symptoms and signs in patients with suspected neuropathic pain. *Pain.* 2004;110:461–469.
45. Maier C, Baron R, Tölle T, et al. Quantitative sensory testing in the German Research Network on Neuropathic Pain (DFNS): somatosensory abnormalities in 1236 patients with different neuropathic pain syndromes. *Pain.* 2010;150:439–450.
46. Jensen TS, Baron R, Haanpää M, et al. A new definition of neuropathic pain. *Pain.* 2011;152:2204–2205.
47. Walker K, Fox AJ, Urban LA. Animal models for pain research. *Mol Med Today.* 1999;5:319–321.
48. Mogil JS. Animal models of pain: progress and challenges. *Nat Rev Neurosci.* 2009;10:283–294.
49. Campbell JN, Meyer RA. Mechanisms of neuropathic pain. *Neuron.* 2006;52:77–92.
50. Moalem G, Tracey DJ. Immune and inflammatory mechanisms in neuropathic pain. *Brain Res Rev.* 2006;51:240–264.
51. Basbaum AI, Jessell TM. The perception of pain. In: Kandel ER, Schwartz JH, Jessell TM, eds. *Principles of Neural Science.* New York: McGraw-Hill; 2000:472–491.
52. Saadé NE, Massaad CA, Kanaan SA, Safieh-Garabedian B, Jabbur SJ, Atweh SF. Pain and neurogenic inflammation: a neural substrate for neuroendocrine-immune interactions. In: Saadé NE, Apkarian AV, Jabbur SJ, eds. *Pain and Neuroimmune Interactions.* Springer; 2000:111–123.
53. Block BM, Hurley RW, Raja SN. Mechanism-based therapies for pain. *Drug News Perspect.* 2004;17:172–186.
54. Bennett GJ, Xie Y-K. A peripheral mononeuropathy in rat that produces disorders of pain sensation like those seen in man. *Pain.* 1988;33:87–107.
55. Seltzer Z, Dubner R, Shir Y. A novel behavioral model of neuropathic pain disorders produced in rats by partial sciatic nerve injury. *Pain.* 1990;43:205–218.

56. Kim SH, Chung JM. An experimental model for peripheral neuropathy produced by segmental spinal nerve ligation in the rat. *Pain*. 1992;50:355–363.
57. Hasnie F, Breuer J, Parker S, et al. Further characterization of a rat model of varicella zoster virus–associated pain: relationship between mechanical hypersensitivity and anxiety-related behavior, and the influence of analgesic drugs. *Neuroscience*. 2007;144:1495–1508.
58. Jaggi AS, Jain V, Singh N. Animal models of neuropathic pain. *Fundam Clin Pharmacol*. 2011;25:1–28.
59. Burgess G, Williams D. The discovery and development of analgesics: new mechanisms, new modalities. *J Clin Invest*. 2010;120:3753.
60. Woolf CJ, Salter MW. Neuronal plasticity: increasing the gain in pain. *Science*. 2000;288:1765–1769.
61. Ji R-R, Kohno T, Moore KA, Woolf CJ. Central sensitization and LTP: do pain and memory share similar mechanisms? *Trends Neurosci*. 2003;26:696–705.
62. Latremoliere A, Woolf CJ. Central sensitization: a generator of pain hypersensitivity by central neural plasticity. *J Pain*. 2009;10:895–926.
63. Searle R, Simpson K. Management of chronic pain: current and future options. *Fut Prescriber*. 2010;11:11–14.
64. Inoue K, Tsuda M. P2X4 receptors of microglia in neuropathic pain. *CNS Neurol Disord Drug Targets*. 2012;11:699–704.
65. Kulkantrakorn K. Emerging concepts and treatment in neuropathic pain. *Neurol Asia*. 2012;17:265–271.
66. Rice AS, Cimino-Brown D, Eisenach JC, Kontinen VK, Lacroix-Fralish ML, Machin I. Animal models and the prediction of efficacy in clinical trials of analgesic drugs: a critical appraisal and call for uniform reporting standards. *Pain*. 2008;139:243–247.
67. Kidd BL, Urban LA. Mechanisms of inflammatory pain. *Br J Anaesth*. 2001;87:3–11.
68. Marchand F, Perretti M, McMahon SB. Role of the immune system in chronic pain. *Nat Rev Neurosci*. 2005;6:521–532.
69. Tracey KJ. The inflammatory reflex. *Nature*. 2002;420:853–859.
70. Staud R. Fibromyalgia pain: do we know the source? *Curr Opin Rheumatol*. 2004;16:157–163.
71. Woolf CJ, Ma Q. Nociceptors—noxious stimulus detectors. *Neuron*. 2007;55:353–364.
72. Smith M. Neurobiology of pain. In: *Molecular Pathomechanisms and New Trends in Drug Research*. London: Taylor & Francis; 2003:302–318.
73. Sessle BJ. Peripheral and central mechanisms of orofacial inflammatory pain. *Int Rev Neurobiol*. 2011;97:179–206.
74. Smith MT, Wyse BD, Edwards SR. Small molecule angiotensin II type 2 receptor (AT2R) antagonists as novel analgesics for neuropathic pain: comparative pharmacokinetics, radioligand binding, and efficacy in rats. *Pain Med*. 2013;14(5):692–705.
75. Woolf CJ, Thompson SW. The induction and maintenance of central sensitization is dependent on N-methyl-D-aspartic acid receptor activation; implications for the treatment of post-injury pain hypersensitivity states. *Pain*. 1991;44:293–299.
76. Cervero F. Visceral pain: mechanisms of peripheral and central sensitization. *Ann Med*. 1995;27:235–239.
77. Cervero F, Laird J. Mechanisms of touch-evoked pain (allodynia): a new model. *Pain*. 1996;68:13–23.
78. Cervero F. *Understanding Pain*. Cambridge, MA, USA: The MIT Press; 2012.
79. Perl ER, Kumazawa T, Lynn B, Kenins P. Sensitization of high threshold receptors with unmyelinated (C) afferent fibers. *Prog Brain Res*. 1976;43:263–277.
80. Price DD, Hayes RL, Ruda M, Dubner R. Spatial and temporal transformations of input to spinothalamic tract neurons and their relation to somatic sensations. *J Neurophysiol*. 1978;41(4):933–947.

81. Kenshalo Jr DR, Leonard RB, Chung JM, Willis WD. Responses of primate spinothalamic neurons to graded and to repeated noxious heat stimuli. *J Neurophysiol.* 1979;42:1370–1389.
82. Woolf CJ. Evidence for a central component of post-injury pain hypersensitivity. *Nature.* 1983;306:686–688.
83. Coderre TJ, Melzack R. Increased pain sensitivity following heat injury involves a central mechanism. *Behav Brain Res.* 1985;15:259–262.
84. Coderre TJ, Melzack R. The contribution of excitatory amino acids to central sensitization and persistent nociception after formalin-induced tissue injury. *J Neurosci.* 1992;12:3665–3670.
85. Cook AJ, Woolf CJ, Wall PD, McMahon SB. Dynamic receptive field plasticity in rat spinal cord dorsal horn following C-primary afferent input. *Nature.* 1987;325:151–153.
86. Hylden JL, Nahin RL, Traub RJ, Dubner R. Expansion of receptive fields of spinal lamina I projection neurons in rats with unilateral adjuvant-induced inflammation: the contribution of dorsal horn mechanisms. *Pain.* 1989;37:229–243.
87. Woolf CJ, Wall PD. Relative effectiveness of C primary afferent fibers of different origins in evoking a prolonged facilitation of the flexor reflex in the rat. *J Neurosci.* 1986;6:1433–1442.
88. Mannion RJ, Doubell TP, Coggeshall RE, Woolf CJ. Collateral sprouting of uninjured primary afferent A-fibers into the superficial dorsal horn of the adult rat spinal cord after topical capsaicin treatment to the sciatic nerve. *J Neurosci.* 1996;16:5189–5195.
89. Neumann S, Doubell TP, Leslie T, Woolf CJ. Inflammatory pain hypersensitivity mediated by phenotypic switch in myelinated primary sensory neurons. *Nature.* 1996;384:360–364.
90. Baba H, Doubell TP, Woolf CJ. Peripheral inflammation facilitates Abeta fiber-mediated synaptic input to the substantia gelatinosa of the adult rat spinal cord. *J Neurosci.* 1999;19:859–867.
91. Samad TA, Moore KA, Sapirstein A, Billet S, Allchorne A, Poole S. Interleukin-1beta-mediated induction of Cox-2 in the CNS contributes to inflammatory pain hypersensitivity. *Nature.* 2001;410:471–475.
92. Vardeh D, Wang D, Costigan M, Lazarus M, Saper CB, Woolf CJ. COX2 in CNS neural cells mediates mechanical inflammatory pain hypersensitivity in mice. *J Clin Invest.* 2009;119:287–294.
93. Ji R-R, Samad TA, Jin S-X, Schmoll R, Woolf CJ. p38 MAPK activation by NGF in primary sensory neurons after inflammation increases TRPV1 levels and maintains heat hyperalgesia. *Neuron.* 2002;36:57–68.
94. Kawasaki Y, Zhang L, Cheng J-K, Ji R-R. Cytokine mechanisms of central sensitization: distinct and overlapping role of interleukin-1beta, interleukin-6, and tumor necrosis factor-alpha in regulating synaptic and neuronal activity in the superficial spinal cord. *J Neurosci.* 2008;28:5189–5194.
95. Luo C, Seeburg PH, Sprengel R, Kuner R. Activity-dependent potentiation of calcium signals in spinal sensory networks in inflammatory pain states. *Pain.* 2008;140:358–367.
96. Cooper EC, Milroy A, Jan YN, Jan LY, Lowenstein DH. Presynaptic localization of Kv1. 4-containing A-type potassium channels near excitatory synapses in the hippocampus. *J Neurosci.* 1998;18:965–974.
97. Rasband MN, Park EW, Vanderah TW, Lai J, Porreca F, Trimmer JS. Distinct potassium channels on pain-sensing neurons. *Proc Natl Acad Sci USA.* 2001;98:13373–13378.
98. Chi XX, Nicol GD. Manipulation of the potassium channel Kv1. 1 and its effect on neuronal excitability in rat sensory neurons. *J Neurophysiol.* 2007;98:2683–2692.
99. Takeda M, Kitagawa J, Takahashi M, Matsumoto S. Activation of interleukin-1β receptor suppresses the voltage-gated potassium currents in the small-diameter trigeminal ganglion neurons following peripheral inflammation. *Pain.* 2008;139:594–602.

100. Takeda M, Tanimoto T, Nasu M, Matsumoto S. Temporomandibular joint inflammation decreases the voltage-gated K+ channel subtype 1.4-immunoreactivity of trigeminal ganglion neurons in rats. *Eur J Pain.* 2008;12:189–195.
101. Takeda M, Matsumoto S, Sessle BJ, Shinoda M, Iwata K. Peripheral and central mechanisms of trigeminal neuropathic and inflammatory pain. *J Oral Biosci.* 2011;53:318–329.
102. Takeda M, Tsuboi Y, Kitagawa J, Nakagawa K, Iwata K, Matsumoto S. Potassium channels as a potential therapeutic target for trigeminal neuropathic and inflammatory pain. *Mol Pain.* 2011;7:1–8.
103. Doan L. Voltage-gated calcium channels and pain. *Techniques in Regional Anesthesia and Pain Management.* 2010;14:42–47.
104. Gu Q, Lee L-Y. Acid-sensing ion channels and pain. *Pharmaceuticals.* 2010;3: 1411–1425.
105. McMahon S, Koltzenburg M, Tracey I, Turk DC. *Wall and Melzack's Textbook of Pain: Expert Consult-Online.* Philadelphia, USA: Elsevier Health Science; 2013.
106. Cregg R, Momin A, Rugiero F, Wood JN, Zhao J. Pain channelopathies. *J Physiol.* 2010;588:1897–1904.
107. Clare JJ, Tate SN, Nobbs M, Romanos MA. Voltage-gated sodium channels as therapeutic targets. *Drug Discov Today.* 2000;5:506–520.
108. Devor M. Centralization, central sensitization and neuropathic pain. Focus on "sciatic chronic constriction injury produces cell-type-specific changes in the electrophysiological properties of rat substantia gelatinosa neurons" *J Neurophysiol.* 2006;96:522–523.
109. Fischer TZ, Waxman SG. Familial pain syndromes from mutations of the NaV1. 7 sodium channel. *Ann N Y Acad Sci.* 2010;1184:196–207.
110. Amir R, Argoff CE, Bennett GJ, et al. The role of sodium channels in chronic inflammatory and neuropathic pain. *J Pain.* 2006;7:1–29.
111. Catterall WA, Perez-Reyes E, Snutch TP, Striessnig J. International Union of Pharmacology. XLVIII. Nomenclature and structure-function relationships of voltage-gated calcium channels. *Pharmacol Rev.* 2005;57:411–425.
112. Bhattacharya A, Wickenden AD, Chaplan SR. Sodium channel blockers for the treatment of neuropathic pain. *Neurotherapeutics.* 2009;6:663–678.
113. Lifton RP, Gharavi AG, Geller DS. Molecular mechanisms of human hypertension. *Cell.* 2001;104:545–556.
114. Hille B. *Ion Channels of Excitable Membranes.* Sunderland, MA: Sinauer; 2001.
115. Wang S-Y, Wang GK. Voltage-gated sodium channels as primary targets of diverse lipid-soluble neurotoxins. *Cell Signal.* 2003;15:151–159.
116. Felix JP, Williams BS, Priest BT, et al. Functional assay of voltage-gated sodium channels using membrane potential-sensitive dyes. *Assay Drug Dev Technol.* 2004;2: 260–268.
117. Nassar MA, Levato A, Stirling LC, Wood JN. Neuropathic pain develops normally in mice lacking both Na(v)1.7 and Na(v)1.8. *Mol Pain.* 2005;1:24.
118. Garcia ML, Kaczorowski GJ. Potassium channels as targets for therapeutic intervention. *Sci STKE.* 2005;2005(302):pe46.
119. Brochu RM, Dick IE, Tarpley JW, et al. Block of peripheral nerve sodium channels selectively inhibits features of neuropathic pain in rats. *Mol Pharmacol.* 2006; 69:823–832.
120. Cummins TR, Sheets PL, Waxman SG. The roles of sodium channels in nociception: implications for mechanisms of pain. *Pain.* 2007;131:243–257.
121. Dib-Hajj SD, Cummins TR, Black JA, Waxman SG. From genes to pain: Na_v1.7 and human pain disorders. *Trends Neurosci.* 2007;30:555–563.

122. Dick IE, Brochu RM, Purohit Y, Kaczorowski GJ, Martin WJ, Priest BT. Sodium channel blockade may contribute to the analgesic efficacy of antidepressants. *J Pain.* 2007;8:315–324.
123. Kaczorowski GJ, McManus OB, Priest BT, Garcia ML. Ion channels as drug targets: the next GPCRs. *J Gen Physiol.* 2008;131:399–405.
124. Schmalhofer WA, Calhoun J, Burrows R, et al. ProTx-II, a selective inhibitor of NaV1. 7 sodium channels, blocks action potential propagation in nociceptors. *Mol Pharmacol.* 2008;74:1476–1484.
125. England S, de Groot MJ. Subtype-selective targeting of voltage-gated sodium channels. *Br J Pharmacol.* 2009;158:1413–1425.
126. Dib-Hajj SD, Binshtok AM, Cummins TR, Jarvis MF, Samad T, Zimmermann K. Voltage-gated sodium channels in pain states: role in pathophysiology and targets for treatment. *Brain Res Rev.* 2009;60:65–83.
127. Dib-Hajj SD, Cummins TR, Black JA, Waxman SG. Sodium channels in normal and pathological pain. *Annu Rev Neurosci.* 2010;33:325–347.
128. Minett MS, Nassar MA, Clark AK, et al. Distinct Nav1. 7-dependent pain sensations require different sets of sensory and sympathetic neurons. *Nat Commun.* 2012;3:791.
129. Nassar MA, Stirling LC, Forlani G, et al. Nociceptor-specific gene deletion reveals a major role for Nav1.7 (PN1) in acute and inflammatory pain. *Proc Natl Acad Sci USA.* 2004;101:12706–12711.
130. Estacion M, Dib-Hajj S, Benke P, et al. NaV1. 7 gain-of-function mutations as a continuum: A1632E displays physiological changes associated with erythromelalgia and paroxysmal extreme pain disorder mutations and produces symptoms of both disorders. *J Neurosci.* 2008;28:11079–11088.
131. Goldberg Y, MacFarlane J, MacDonald M, et al. Loss-of-function mutations in the Nav1. 7 gene underlie congenital indifference to pain in multiple human populations. *Clin Genet.* 2007;71:311–319.
132. Rogers M, Tang L, Madge DJ, Stevens EB. The role of sodium channels in neuropathic pain. *Semin Cell Dev Biol.* 2006;17(5):571–581 [Elsevier].
133. Cheng J-K, Ji R-R. Intracellular signaling in primary sensory neurons and persistent pain. *Neurochem Res.* 2008;33:1970–1978.
134. Costigan M, Scholz J, Woolf CJ. Neuropathic pain: a maladaptive response of the nervous system to damage. *Annu Rev Neurosci.* 2009;32:1.
135. Baron R, Binder A, Wasner G. Neuropathic pain: diagnosis, pathophysiological mechanisms, and treatment. *Lancet Neurol.* 2010;9:807–819.
136. Snutch TP. Targeting chronic and neuropathic pain: the N-type calcium channel comes of age. *NeuroRx.* 2005;2:662–670.
137. Altier C, Dale CS, Kisilevsky AE, et al. Differential role of N-type calcium channel splice isoforms in pain. *J Neurosci.* 2007;27:6363–6373.
138. Zamponi GW, Lewis RJ, Todorovic SM, Arneric SP, Snutch TP. Role of voltage-gated calcium channels in ascending pain pathways. *Brain Res Rev.* 2009;60:84–89.
139. Nowycky MC, Fox AP, Tsien RW. Three types of neuronal calcium channel with different calcium agonist sensitivity. *Nature.* 1985;316:440–443.
140. Cao YQ, Mantyh PW, Carlson EJ, Gillespie AM, Epstein CJ, Basbaum AI. Primary afferent tachykinins are required to experience moderate to intense pain. *Nature.* 1998;392:390–394.
141. Tsien RW, Wheeler DB. Voltage-gated calcium channels. In: Carafoli E, Klee CB, eds. *Calcium as a Cellular Regulator.* New York, USA: Oxford University Press; 1999:171–199.
142. Miljanich G. Ziconotide: neuronal calcium channel blocker for treating severe chronic pain. *Curr Med Chem.* 2004;11:3029–3040.

143. Wermeling DP. Ziconotide, an intrathecally administered n-type calcium channel antagonist for the treatment of chronic pain. *Pharmacotherapy*. 2005;25:1084–1094.
144. Cao Y-Q. Voltage-gated calcium channels and pain. *Pain*. 2006;126:5–9.
145. Maggi CA, Tramontana M, Cecconi R, Santicioli P. Neurochemical evidence for the involvement of N-type calcium channels in transmitter secretion from peripheral endings of sensory nerves in guinea pigs. *Neurosci Lett*. 1990;114:203–206.
146. Westenbroek RE, Hell JW, Warner C, Dubel SJ, Snutch TP, Catterall WA. Biochemical properties and subcellular distribution of an N-type calcium hannel α1 subunit. *Neuron*. 1992;9:1099–1115.
147. Berridge MJ, Lipp P, Bootman MD. The versatility and universality of calcium signalling. *Nat Rev Mol Cell Biol*. 2000;1:11–21.
148. Saegusa H, Kurihara T, Zong S, et al. Suppression of inflammatory and neuropathic pain symptoms in mice lacking the N-type Ca^{2+} channel. *The EMBO J*. 2001; 20:2349–2356.
149. Luo Z, Calcutt N, Higuera E, et al. Injury type-specific calcium channel α2δ-1 subunit up-regulation in rat neuropathic pain models correlates with antiallodynic effects of gabapentin. *J Pharmacol Exp Ther*. 2002;303:1199–1205.
150. Jevtovic-Todorovic V, Todorovic SM. The role of peripheral T-type calcium channels in pain transmission. *Cell Calcium*. 2006;40:197–203.
151. Jagodic MM, Pathirathna S, Nelson MT, et al. Cell-specific alterations of T-type calcium current in painful diabetic neuropathy enhance excitability of sensory neurons. *J Neurosci*. 2007;27:3305–3316.
152. Jagodic MM, Pathirathna S, Joksovic PM, Lee W, Nelson MT, Naik AK. Upregulation of the T-type calcium current in small rat sensory neurons after chronic constrictive injury of the sciatic nerve. *J Neurophysiol*. 2008;99:3151–3156.
153. Maneuf Y, Luo Z, Lee K. α2δ and the mechanism of action of gabapentin in the treatment of pain. *Semin Cell Dev Biol*. 2006;17(5):565–570 [Elsevier].
154. Finnerup NB, Sindrup SH, Bach FW, Johannesen IL, Jensen TS. Lamotrigine in spinal cord injury pain: a randomized controlled trial. *Pain*. 2002;96:375–383.
155. Camerino DC, Desaphy J-F. Grand challenge for ion channels: an underexploited resource for therapeutics. *Front Pharmacol*. 2010,1:113.
156. Yang C-M, Chen N-C, Shen H-C, et al. Guideline of neuropathic pain treatment and dilemma from neurological point of view. *Acta Neurol Taiwan*. 2012;21:136–144.
157. Mindruta I, Cobzaru A-M, Bajenaru OA. Overview of neuropathic pain diagnosis and assessment—an approach based on mechanisms. In: *Cyprian Chukwunonye Udeagha (ED), InTech, 1–23, Neuropathic Pain*. Romania: University of Medicine and Pharmacy of Bucharest; 2012:1.
158. Kim DS, Choi JO, Rim HD, Cho HJ. Downregulation of voltage-gated potassium channel α gene expression in dorsal root ganglia following chronic constriction injury of the rat sciatic nerve. *Mol Brain Res*. 2002;105:146–152.
159. Kim MS, Jo H, Um JY, et al. Agonists of proteinase-activated receptor 2 induce TNF-α secretion from astrocytoma cells. *Cell Biochem Funct*. 2002;20:339–345.
160. Binshtok AM. Mechanisms of nociceptive transduction and transmission: a machinery for pain sensation and tools for selective analgesia. *Int Rev Neurobiol*. 2011;97:143–177.
161. Varga AW, Yuan L-L, Anderson AE, et al. Calcium–calmodulin-dependent kinase II modulates Kv4. 2 channel expression and upregulates neuronal A-type potassium currents. *J Neurosci*. 2004;24:3643–3654.
162. Tanimoto T, Takeda M, Nasu M, Kadoi J, Matsumoto S. Immunohistochemical co-expression of carbonic anhydrase II with Kv1. 4 and TRPV1 in rat small-diameter trigeminal ganglion neurons. *Brain Res*. 2005;1044:262–265.

163. Tan Z, Donnelly D, LaMotte R. Effects of a chronic compression of the dorsal root ganglion on voltage-gated Na+and K+ currents in cutaneous afferent neurons. *J Neurophysiol.* 2006;95:1115–1123.
164. Cao XH, Byun HS, Chen SR, Cai YQ, Pan HL. Reduction in voltage-gated K+ channel activity in primary sensory neurons in painful diabetic neuropathy: role of brain-derived neurotrophic factor. *J Neurochem.* 2010;114:1460–1475.
165. Kajander KC, Wakisaka S, Bennett GJ. Spontaneous discharge originates in the dorsal root ganglion at the onset of a painful peripheral neuropathy in the rat. *Neurosci Lett.* 1992;138:225–228.
166. Rudy B, Chow A, Lau D, et al. Contributions of Kv3 channels to neuronal excitability. *Ann N Y Acad Sci.* 1999;868:304–343.
167. Watanabe S, Hoffman DA, Migliore M, Johnston D. Dendritic K+ channels contribute to spike-timing dependent long-term potentiation in hippocampal pyramidal neurons. *Proc Natl Acad Sci USA.* 2002;99:8366–8371.
168. Hu H-J, Carrasquillo Y, Karim F, et al. The kv4. 2 potassium channel subunit is required for pain plasticity. *Neuron.* 2006;50:89–100.
169. Hu H-J, Gereau RW. Metabotropic glutamate receptor 5 regulates excitability and Kv4. 2-containing K+ channels primarily in excitatory neurons of the spinal dorsal horn. *J Neurophysiol.* 2011;105:3010–3021.
170. Ren K, Dubner R. Interactions between the immune and nervous systems in pain. *Nature Med.* 2010;16:1267–1276.
171. Burnstock G, Wood JN. Purinergic receptors: their role in nociception and primary afferent neurotransmission. *Curr Opin Neurobiol.* 1996;6:526–532.
172. Tsuda M, Koizumi S, Kita A, Shigemoto Y, Ueno S, Inoue K. Mechanical allodynia caused by intraplantar injection of P2X receptor agonist in rats: involvement of heteromeric P2X2/3 receptor signaling in capsaicin-insensitive primary afferent neurons. *J Neurosci.* 2000;20:RC90.
173. North RA, Jarvis MF. P2X receptors as drug targets. *Mol Pharmacol.* 2013;83:759–769.
174. Burnstock G. Purinergic receptors and pain. *Curr Pharm Des.* 2009;15:1717–1735.
175. Donnelly-Roberts D, McGaraughty S, Shieh C-C, Honore P, Jarvis MF. Painful purinergic receptors. *J Pharmacol Exp Ther.* 2008;324:409–415.
176. Dray A. Neuropathic pain: emerging treatments. *Br J Anaesth.* 2008;101:48–58.
177. Gunosewoyo H, Kassiou M. P2X purinergic receptor ligands: recently patented compounds. *Expert Opin Ther Pat.* 2010;20:625–646.
178. Nicotra L, Loram LC, Watkins LR, Hutchinson MR. Toll-like receptors in chronic pain. *Exp Neurol.* 2012;234:316–329.
179. Abadie J. How can the clinical picture guide appropriate laboratory drug testing in the treatment of pain clinic patients with opioid analgesics? *Pain Med.* 2012;13:857–859.
180. Lehnardt S, Massillon L, Follett P, et al. Activation of innate immunity in the CNS triggers neurodegeneration through a Toll-like receptor 4-dependent pathway. *Proc Natl Acad Sci USA.* 2003;100:8514–8519.
181. Jou I, Lee JH, Park SY, Yoon HJ, Joe E-H, Park EJ. Gangliosides trigger inflammatory responses via TLR4 in brain glia. *Am J Pathol.* 2006;168:1619–1630.
182. Tang S-C, Arumugam TV, Xu X, et al. Pivotal role for neuronal Toll-like receptors in ischemic brain injury and functional deficits. *Proc Natl Acad Sci USA.* 2007;104:13798–13803.
183. Buchanan MM, Hutchinson M, Watkins LR, Yin H. Toll-like receptor 4 in CNS pathologies. *J Neurochem.* 2010;114:13–27.
184. Hutchinson MR, Bland ST, Johnson KW, Rice KC, Maier SF, Watkins LR. Opioid-induced glial activation: mechanisms of activation and implications for opioid analgesia, dependence, and reward. *Sci World J.* 2007;7:98–111.
185. Clark AK, Malcangio M. Fractalkine/CX3CR1 signaling during neuropathic pain. *Front Cell Neurosci.* 2014;8:121.

186. Nguyen C, Coelho A-M, Grady E, et al. Colitis induced by proteinase-activated receptor-2 agonists is mediated by a neurogenic mechanism. *Can J Physiol Pharmacol.* 2003;81:920–927.
187. Grant AD, Cottrell GS, Amadesi S, et al. Protease-activated receptor 2 sensitizes the transient receptor potential vanilloid 4 ion channel to cause mechanical hyperalgesia in mice. *J Physiol.* 2007;578:715–733.
188. Vergnolle N. Protease-activated receptors as drug targets in inflammation and pain. *Pharmacol Ther.* 2009;123:292–309.
189. Dale C, Vergnolle N. Protease signaling to G protein-coupled receptors: implications for inflammation and pain. *J Recept Signal Transduct Res.* 2008;28:29–37.
190. Garavilla LD, Vergnolle N, Young SH, et al. Agonists of proteinase-activated receptor 1 induce plasma extravasation by a neurogenic mechanism. *Br J Pharmacol.* 2001;133:975–987.
191. Macfarlane SR, Seatter MJ, Kanke T, Hunter GD, Plevin R. Proteinase-activated receptors. *Pharmacol Rev.* 2001;53:245–282.
192. Hollenberg MD, Compton SJ. International union of pharmacology. XXVIII. Proteinase-activated receptors. *Pharmacol Rev.* 2002;54:203–217.
193. Asfaha S, Brussee V, Chapman K, Zochodne DW, Vergnolle N. Proteinase-activated receptor-1 agonists attenuate nociception in response to noxious stimuli. *Br J Pharmacol.* 2002;135:1101–1106.
194. Coelho A-M, Vergnolle N, Guiard B, Fioramonti J, Bruno L. Proteinases and proteinase-activated receptor 2: a possible role to promote visceral hyperalgesia in rats. *Gastroenterology.* 2002;122:1035–1047.
195. Steinhoff M, Stander S, Seeliger S, Ansel JC, Schmelz M, Luger T. Modern aspects of cutaneous neurogenic inflammation. *Arch Dermatol.* 2003;139:1479.
196. Asfaha S, Cenac N, Houle S, et al. Protease-activated receptor-4: a novel mechanism of inflammatory pain modulation. *Br J Pharmacol.* 2007;150:176–185.
197. Hokfelt T. Neuropeptides in perspective: the last ten years. *Neuron.* 1991;7:867–879.
198. Dickenson A. NMDA receptor antagonists: interactions with opioids. *Acta Anaesthesiol Scand.* 1997;41:112–115.
199. Yaksh TL, Malmberg A. Central pharmacology of nociceptive transmission. In: Wall PD, Melzack R, eds. *Textbook of Pain.* 3rd ed. Churchill Livingstone: Edinburgh; 1994:165–200.
200. Merighi A, Carmignoto G, Gobbo S, Lossi L, Salio C, Maria VA. Neurotrophins in spinal cord nociceptive pathways. *Prog Brain Res.* 2004;146:291–321.
201. South SM, Kohno T, Kaspar BK, et al. A conditional deletion of the NR1 subunit of the NMDA receptor in adult spinal cord dorsal horn reduces NMDA currents and injury-induced pain. *J Neurosci.* 2003;23:5031–5040.
202. Riedel G. Function of metabotropic glutamate receptors in learning and memory. *Trends Neurosci.* 1996;19:219–224.
203. Hansson E, Ronnback L. Glial-neuronal signaling and astroglial swelling in physiology and pathology. *Adv Exp Med Biol.* 2004;559:313–323.
204. Hansson E, Rönnbäck L. Altered neuronal–glial signaling in glutamatergic transmission as a unifying mechanism in chronic pain and mental fatigue. *Neurochem Res.* 2004;29:989–996.
205. Adolph O, Köster S, Räth M, et al. Rapid increase of glial glutamate uptake via blockade of the protein kinase A pathway. *Glia.* 2007;55:1699–1707.
206. Jablonski MM, Freeman NE, Orr WE, et al. Genetic pathways regulating glutamate levels in retinal Müller cells. *Neurochem Res.* 2011;36:594–603.
207. Lane DJ, Lawen A. The glutamate aspartate transporter (GLAST) mediates L-glutamate-stimulated ascorbate-release via swelling-activated anion channels in cultured neonatal rodent astrocytes. *Cell Biochem Biophys.* 2013;65:107–119.

208. Garry EM, Fleetwood-Walker SM. A new view on how AMPA receptors and their interacting proteins mediate neuropathic pain. *Pain*. 2004;109:210–213.
209. Raouf RK. *Functional Regulation of N-methyl-D-Aspartate Receptors by Serine Threonine Protein Kinases*. Canada: University of Toronto; 1997.
210. Randic M, Jiang M, Cerne R. Long-term potentiation and long-term depression of primary afferent neurotransmission in the rat spinal cord. *J Neurosci*. 1993; 13:5228–5241.
211. Dickenson AH, Sullivan AF. Evidence for a role of the NMDA receptor in the frequency dependent potentiation of deep rat dorsal horn nociceptive neurones following C fibre stimulation. *Neuropharmacology*. 1987;26:1235–1238.
212. Hollmann M, Boulter J, Maron C, et al. Zinc potentiates agonist-induced currents at certain splice variants of the NMDA receptor. *Neuron*. 1993;10:943–954.
213. Hollmann M, Heinemann S. Cloned glutamate receptors. *Annu Rev Neurosci*. 1994; 17:31–108.
214. Hollmann M, Maron C, Heinemann S. N-glycosylation site tagging suggests a three transmembrane domain topology for the glutamate receptor GluR1. *Neuron*. 1994;13:1331–1343.
215. Petrenko AB, Yamakura T, Baba H, Shimoji K. The role of N-methyl-D-aspartate (NMDA) receptors in pain: a review. *Anesth Analg*. 2003;97:1108–1116.
216. Childers WE, Baudy RB. N-methyl-D-aspartate antagonists and neuropathic pain: the search for relief. *J Med Chem*. 2007;50:2557–2562.
217. Fundytus ME, Fisher K, Dray A, Henry JL, Coderre TJ. In vivo antinociceptive activity of anti-rat mGluR1 and mGluR5 antibodies in rats. *Neuroreport*. 1998;9:731–735.
218. Schoepp DD, Jane DE, Monn JA. Pharmacological agents acting at subtypes of metabotropic glutamate receptors. *Neuropharmacology*. 1999;38:1431–1476.
219. Schoepp DD. Unveiling the functions of presynaptic metabotropic glutamate receptors in the central nervous system. *J Pharmacol Exp Ther*. 2001;299:12–20.
220. Chiechio S, Nicoletti F. Metabotropic glutamate receptors and the control of chronic pain. *Curr Opin Pharmacol*. 2012;12:28–34.
221. Chiechio S, Copani A, Zammataro M, Battaglia G, Nicoletti F. Transcriptional regulation of type-2 metabotropic glutamate receptors: an epigenetic path to novel treatments for chronic pain. *Trends Pharmacol Sci*. 2010;31:153–160.
222. Smith MT. Differences between and combinations of opioids re-visited. *Curr Opin Anesthiol*. 2008;21:596–601.
223. Yekkirala AS, Lunzer MM, McCurdy CR, et al. N-naphthoyl-beta-naltrexamine (NNTA), a highly selective and potent activator of μ/kappa-opioid heteromers. *Proc Natl Acad Sci USA*. 2011;108:5098–5103.
224. Fields HL. Pain modulation: expectation, opioid analgesia and virtual pain. *Prog Brain Res*. 1999;122:245–253.
225. Fields H. State-dependent opioid control of pain. *Nat Rev Neurosci*. 2004;5:565–575.
226. Snyder SH, Pasternak GW. Historical review: opioid receptors. *Trends Pharmacol Sci*. 2003;24:198–205.
227. Pasternak GW. Molecular biology of opioid analgesia. *J Pain Symptom Manag*. 2005;29:2–9.
228. Pasternak GW. Opioids and their receptors: are we there yet? *Neuropharmacology*. 2014;76(Part B):198–203.
229. Raynor K, Kong H, Chen Y, et al. Pharmacological characterization of the cloned kappa-, delta-, and mu-opioid receptors. *Mol Pharmacol*. 1994;45:330–334.
230. Smith KM. *Investigation of Natural Product Scaffolds for the Development of Opioid Receptor Ligands*. USA: Medicinal Chemistry and the Graduate Faculty of the University of Kansas; 2012.

231. Li Y, Lefever MR, Muthu D, Bidlack JM, Bilsky EJ, Polt R. Opioid glycopeptide analgesics derived from endogenous enkephalins and endorphins. *Future Med Chem.* 2012;4:205–226.
232. Yaksh T. Central pharmacology of nociceptive transmission. In: Wall PD, Melzack R, eds. *Textbook of Pain*. Edinburg, UK: Churchill Livingstone; 1999:253–308.
233. Whiteside GT, Boulet JM, Walker K. The role of central and peripheral μ opioid receptors in inflammatory pain and edema: a study using morphine and DiPOA ([8-(3, 3-diphenyl-propyl)-4-oxo-1-phenyl-1, 3, 8-triaza-spiro [4.5] dec-3-yl]-acetic acid). *J Pharmacol Exp Ther.* 2005;314:1234–1240.
234. Hargreaves KM, Pardridge WM. Neutral amino acid transport at the human blood-brain barrier. *J Biol Chem.* 1988;263:19392–19397.
235. Joris J, Costello A, Dubner R, Hargreaves KM. Opiates suppress carrageenan-induced edema and hyperthermia at doses that inhibit hyperalgesia. *Pain.* 1990;43:95–9103.
236. Zhu Y, King MA, Schuller AG, et al. Retention of supraspinal delta-like analgesia and loss of morphine tolerance in <i> δ opioid receptor </i> knockout mice. *Neuron.* 1999;24:243–252.
237. Kieffer BL, Gavériaux-Ruff C. Exploring the opioid system by gene knockout. *Prog Neurobiol.* 2002;66:285–306.
238. Hack SP, Bagley EE, Chieng BC, Christie MJ. Induction of δ-opioid receptor function in the midbrain after chronic morphine treatment. *J Neurosci.* 2005;25:3192–3198.
239. Gaveriaux-Ruff C, Karchewski LA, Hever X, Matifas A, Kieffer BL. Inflammatory pain is enhanced in delta opioid receptor-knockout mice. *Eur J Neurosci.* 2008;27:2558–2567.
240. Tominaga M, Caterina MJ, Malmberg AB, et al. The cloned capsaicin receptor integrates multiple pain-producing stimuli. *Neuron.* 1998;21:531–543.
241. Szallasi A, Szabo T, Biro T, Modarres S, Blumberg PM, Krause JE. Resiniferatoxin-type phorboid vanilloids display capsaicin-like selectivity at native vanilloid receptors on rat DRG neurons and at the cloned vanilloid receptor VR1. *Br J Pharmacol.* 1999;128:428–434.
242. Caterina MJ, Leffler A, Malmberg AB, et al. Impaired nociception and pain sensation in mice lacking the capsaicin receptor. *Science.* 2000;288:306–313.
243. Cortright DN, Szallasi A. Biochemical pharmacology of the vanilloid receptor TRPV1. An update. *Eur J Biochem.* 2004;271:1814–1819.
244. Cui M, Honore P, Zhong C, et al. TRPV1 receptors in the CNS play a key role in broad-spectrum analgesia of TRPV1 antagonists. *J Neurosci.* 2006;26:9385–9393.
245. Davis JB, Gray J, Gunthorpe MJ, et al. Vanilloid receptor-1 is essential for inflammatory thermal hyperalgesia. *Nature.* 2000;405:183–187.
246. Amaya F, Shimosato G, Nagano M, et al. NGF and GDNF differentially regulate TRPV1 expression that contributes to development of inflammatory thermal hyperalgesia. *Eur J Neurosci.* 2004;20:2303–2310.
247. Zhuang Z-Y, Xu H, Clapham DE, Ji R-R. Phosphatidylinositol 3-kinase activates ERK in primary sensory neurons and mediates inflammatory heat hyperalgesia through TRPV1 sensitization. *J Neurosci.* 2004;24:8300–8309.
248. Costa B, Comelli F, Bettoni I, Colleoni M, Giagnoni G. The endogenous fatty acid amide, palmitoylethanolamide, has anti-allodynic and anti-hyperalgesic effects in a murine model of neuropathic pain: involvement of CB₁TRPV1 and PPARγ receptors and neurotrophic factors. *Pain.* 2008;139:541–550.
249. Lin C-R, Amaya F, Barrett L, et al. Prostaglandin E2 receptor EP4 contributes to inflammatory pain hypersensitivity. *J Pharmacol Exp Ther.* 2006;319:1096–1103.
250. Omote K, Kawamata T, Nakayama Y, Yamamoto H, Kawamata M, Namiki A. Effects of a novel selective agonist for prostaglandin receptor subtype EP4 on hyperalgesia and inflammation in monoarthritic model. *Anesthesiology.* 2002;97:170–176.

251. Minami T, Nakano H, Kobayashi T, et al. Characterization of EP receptor subtypes responsible for prostaglandin E2-induced pain responses by use of EP1 and EP3 receptor knockout mice. *Br J Pharmacol.* 2001;133:438–444.
252. Zhang Y, Shaffer A, Portanova J, Seibert K, Isakson PC. Inhibition of cyclooxygenase-2 rapidly reverses inflammatory hyperalgesia and prostaglandin E2 production. *J Pharmacol Exp Ther.* 1997;283:1069–1075.
253. Turini ME, DuBois RN. Cyclooxygenase-2: a therapeutic target. *Annu Rev Med.* 2002;53:35–57.
254. Rao P, Knaus EE. Evolution of nonsteroidal anti-inflammatory drugs (NSAIDs): cyclooxygenase (COX) inhibition and beyond. *J Pharm Pharm Sci.* 2008;11:81s–110s.
255. Freire MAM, Guimarães JS, Leal WG, Pereira A. Pain modulation by nitric oxide in the spinal cord. *Front Neurosci.* 2009;3:175.
256. Covey WC, Ignatowski TA, Renauld AE, Knight PR, Nader ND, Spengler RN. Expression of neuron-associated tumor necrosis factor alpha in the brain is increased during persistent pain. *Reg Anesth Pain Med.* 2002;27:357–366.
257. Miclescu A, Gordh T. Nitric oxide and pain: 'something old, something new'. *Acta Anaesthesiol Scand.* 2009;53:1107–1120.
258. Shu X, Mendell LM. Nerve growth factor acutely sensitizes the response of adult rat sensory neurons to capsaicin. *Neurosci Lett.* 1999;274:159–162.
259. Zhu W, Galoyan SM, Petruska JC, Oxford GS, Mendell LM. A developmental switch in acute sensitization of small dorsal root ganglion (DRG) neurons to capsaicin or noxious heating by NGF. *J Neurophysiol.* 2004;92:3148–3152.
260. Malin SA, Molliver DC, Koerber HR, et al. Glial cell line-derived neurotrophic factor family members sensitize nociceptors in vitro and produce thermal hyperalgesia in vivo. *J Neurosci.* 2006;26:8588–8599.
261. Okuse K. Pain signalling pathways: from cytokines to ion channels. *Int J Biochem Cell Biol.* 2007;39:490–496.
262. Basbaum AI, Bautista DM, Scherrer G, Julius D. Cellular and molecular mechanisms of pain. *Cell.* 2009;139:267–284.
263. Premkumar LS. Targeting TRPV1 as an alternative approach to narcotic analgesics to treat chronic pain conditions. *AAPS J.* 2010;12:361–370.
264. Schumacher MA. Transient receptor potential channels in pain and inflammation: therapeutic opportunities. *Pain Practice.* 2010;10:185–200.
265. Pethő G, Reeh PW. Sensory and signaling mechanisms of bradykinin, eicosanoids, platelet-activating factor, and nitric oxide in peripheral nociceptors. *Physiol Rev.* 2012;92:1699–1775.

CHAPTER TWO

Peripheral Scaffolding and Signaling Pathways in Inflammatory Pain

Nathaniel A. Jeske[1]
Department of Oral and Maxillofacial Surgery, UT Health Science Center, San Antonio, Texas, USA
[1]Corresponding author: e-mail address: jeske@uthscsa.edu

Contents

1. Introduction 32
2. Inflammatory Mediators 32
   2.1 Bradykinin receptor 36
   2.2 Prostaglandin receptor 36
   2.3 Serotonin receptor 37
   2.4 Purinergic receptors 37
   2.5 Receptor tyrosine kinases 38
   2.6 Neurokinin receptor 38
   2.7 Glutamate receptors 38
   2.8 Protease-activated receptors 39
   2.9 Calcitonin receptor-like receptor and receptor activity-modifying protein 1 39
   2.10 Endothelin receptors 40
3. Signaling Mechanisms 40
4. Scaffolding Structures 42
   4.1 A-kinase anchoring protein 79/150 42
   4.2 β-Arrestin 45
5. Concluding Remarks 46
References 47

Abstract

Peripheral injury precipitates the release and accumulation of extracellular molecules at the site of injury. Although these molecules exist in various forms, they activate specific receptor classes expressed on primary afferent neurons to mediate cellular and behavioral responses to both nonpainful and painful stimuli. These inflammatory mediators and subsequent receptor-mediated effects exist to warn an organism of future injury, thereby resulting in protection and rehabilitation of the wounded tissue. In this chapter, inflammatory mediators, their target receptor classes, and downstream signaling pathways are identified and discussed within the context of inflammatory hyperalgesia. Furthermore, scaffolding mechanisms that exist to support inflammatory signaling in

Progress in Molecular Biology and Translational Science, Volume 131
ISSN 1877-1173
http://dx.doi.org/10.1016/bs.pmbts.2014.11.016

peripheral afferent neuronal tissues specifically are identified and discussed. Together, the mediators, pathways, and scaffolding mechanisms involved in inflammatory hyperalgesia provide a unique knowledge point from which new therapeutic targets can be understood.

1. INTRODUCTION

Human physiology integrates multiple redundant homeopathic signaling mechanisms that maintain tissue structure and function. One such mechanism, inflammation, occurs in response to tissue damage and is stimulated by the generation of an immune response. Following injury, neutrophils, leukocytes, and other immune cells release and stimulate the release of signaling factors from adjacent tissues. These inflammatory signaling factors activate receptors located on terminal endings of primary afferent fibers, which in turn activate intracellular mechanisms that sensitize additional receptors and ion channels. In the end, a state of inflammatory hyperalgesia, or increased sensitivity to pain following an inflammatory response, promotes homeostatic behavior through stimulus avoidance of the inflamed and/or the injured tissue.

The activation of proinflammatory receptor systems creates hyperalgesic conditions to reduce the likelihood of further injury to the damaged peripheral tissue. Most mediators released into the extracellular space surrounding injured tissue activate receptor systems expressed by primary afferent neurons that sensitize those neurons to physical and chemical stimuli. In most cases, this occurs through posttranslational modifications to ion channels, reducing either the ligand dependency or the voltage dependency required for activation and eventual depolarization of the neuron. The phenotype of inflammatory hyperalgesia is marked by increased sensitivity to a normally painful stimulus and is typically supported by reversible changes to intraneuronal biochemistry. In this chapter, these molecules, their receptors, signaling pathways, and scaffolding complexes will be identified and characterized for their roles in inflammatory hyperalgesia.

2. INFLAMMATORY MEDIATORS

Injury to peripheral tissue elicits changes to the extracellular biochemistry surrounding the tissue and associated primary afferents that innervate the tissue. Damaged epithelial cells, in concert with responding immune cells, release small peptides and molecules that target receptors expressed throughout the periphery. These receptors control peripheral vasodilation, microglial response, epithelial cell repair, and nociceptor sensitivity. As inflammatory mediators such as bradykinin and endothelin (ET) increase vasodilation, plasma extravasation brings additional immune cells and inflammatory mediators into the area of the tissue injury, creating a feed-forward inflammatory response. The activation of microglia by inflammatory mediators can result in the release of molecules such as prostaglandins (PGs) and adenosine triphosphate (ATP), both of which serve to activate specific receptors expressed on primary afferent neurons. The activation of these receptor systems, along with others that are activated by numerous other inflammatory mediators, stimulates intracellular signaling pathways that increase nociceptor sensitivity, resulting in hypersensitization. In such an environment, the nociceptive terminal ending is highly sensitive to reduced thresholds for somatosensory and chemical activation, causing increased pain. In Table 1, a list of inflammatory mediators and their target receptor systems is outlined.

The process of peripheral inflammation is a feed-forward mechanism in its execution and end point. The local invasion of immune cells in response to tissue injury results in the release of molecules such as serotonin and serine proteases. These initial immune factors act upon their respective receptors expressed on terminal endings of nociceptors to sensitize neuronal depolarization and the subsequent release of substance P, CGRP, PGs, and glutamate. These nociceptor-originated mediators then act to perpetuate the sensitization of terminal nociceptors through receptor activation and also by stimulating vasodilation, thereby increasing the delivery of circulating immune cells and inflammatory peptides to the area of injury. As depicted in Table 1, most inflammatory mediators target specific isoforms of receptors expressed by nociceptors and other cell types. Receptor activation is what drives inflammatory sensitization of peripheral tissue, as ligands make contact with specific receptors that dictate downstream signaling systems.

Inflammatory sensitization of primary afferent nociceptors can only occur if the receptor system that the inflammatory mediator is directed toward is mated to a stimulatory signaling cascade. Indeed, several specific isoforms of the serotonin receptor have been identified as regulating the sensitizing behavior of serotonin on nociceptor functions. Depending on the isoform activated, either Gαq- or Gαs-driven downstream signaling would occur, resulting in the sensitization of differentially unique targets within the nociceptor itself. Other inflammatory mediator receptor isoforms do not

Table 1 Inflammatory mediators

Mediator	Source	Tissue target	Receptor	Signaling mechanism	Somatic target
Bradykinin	Immune cells, plasma	Blood vessels, nociceptors	Bradykinin type 1 and 2 receptors (B1R, B2R)	$G\alpha q \rightarrow PLC \rightarrow IP_3$ receptor $\rightarrow$ PKC COX/LOX	Thermal, mechanical
Prostaglandin	Microglia, nociceptors	Nociceptors	EP_1, EP_2, EP_{3B}, EP_{3C}, EP_4, DP_1, and DP2, receptors	$G\alpha s \rightarrow AC \rightarrow PKA$	Thermal, mechanical, chemical
Serotonin	Immune cells	Nociceptors	5-HT1A, 2A, 2B, 3, 4, 7	$G\alpha q \rightarrow PLC \rightarrow IP_3$ receptor $\rightarrow$ PKC $G\alpha s \rightarrow AC \rightarrow PKA$	Thermal, mechanical, chemical
ATP	Microglia, nociceptors	Nociceptors, immune cells, microglia	P2X2, 3, 4, 7	NA^+, K^+, Ca^{2+} influx	Thermal, mechanical
NGF	Immune cells, nociceptors	Immune cells, nociceptors	TrkA, GPR30	Tyr-kinase	Thermal, mechanical
Substance P	Nociceptors	Immune cells, blood vessels, nociceptors	NK1R	$G\alpha q \rightarrow PLC \rightarrow IP_3$ receptor $\rightarrow$ PKC	Thermal, mechanical
Glutamate	Microglia, nociceptors	Nociceptors	mGluR1, mGluR2/3, mGluR5	$G\alpha q \rightarrow PLC \rightarrow IP_3$ receptor $\rightarrow$ PKC	Thermal, mechanical
Serine proteases	Immune cells, nociceptors	Nociceptors	PAR1, PAR2, PAR4	$G\alpha q \rightarrow PLC \rightarrow IP_3$ receptor $\rightarrow$ PKC	Thermal, mechanical

CGRP	Nociceptors	Blood vessels, nociceptors, endothelial cells	Calcitonin receptor-like receptor (CRLR), receptor activity-modifying protein 1 (RAMP1)	Gαs ≫ AC ≫ PKA	Thermal, mechanical
Endothelin	Endothelial cells	Blood vessels, nociceptors	ET_AR, ET_BR	Gαq → PLC → IP_3 receptor → PKC	Thermal, mechanical, chemical

A cumulative list of known inflammatory mediators that modulate peripheral nociceptor sensitivities to somatosensory stimulation. As listed, the inflammatory mediator is released from its source(s) and targets a specific cell type to activate the noted receptor isoform(s), stimulating the identified signaling pathway to sensitize the associated somatosensation.

share this unique characteristic, as the whole receptor family signals through the same pathway, including those for glutamate (mGluR1,[1] mGluR2/3,[2] mGluR5[3]). Still, other mediators are more likely to activate certain isoforms. For instance, bradykinin differentially activates one of two isoforms of the bradykinin receptor, based upon expression profiles.[4] Similarly, several isoforms for the purinergic receptor activated by ATP are also differentially expressed on primary afferent terminals, including P2X2, P2X3,[5] P2X4,[6] and P2X7.[7] However, each of the isoforms in these receptor classes signals through similar downstream cascades. Further, the specificity of these receptors to their cascades and eventual targets remain the important links that connect inflammatory mediators to nociceptor sensitization. Here, we will outline the major classes of receptors activated by inflammatory mediators and directly tied to inflammatory hyperalgesia.

2.1. Bradykinin receptor

Two isoforms of the bradykinin receptor have been characterized: bradykinin type 1 and 2 receptors (B1R and B2R, respectively). B1R is expressed minimally in naïve states and is upregulated following inflammation. In contrast, B2R is constitutively expressed and undergoes no change in expression following inflammation. Therefore, initial inflammatory insult to peripheral tissues innervated by nociceptive neurons will most likely be sensitized via B2R signaling mechanisms prior to B1R having significant effect, due to differential expression profiles. Furthermore, it should be noted that the affinity of bradykinin, created by proteolytic breakdown of endogenous kinins within the blood stream and extracellular space, for B2R (0.54 Ki[8]) is significantly greater than for B1R (>10,000 Ki[9]). Both isoforms of bradykinin receptor propagate their signals through a Gαq downstream signaling mechanism.

2.2. Prostaglandin receptor

PGs are formed from the metabolism of arachidonic acid molecules released from the plasma membrane in multiple cell types. Prostaglandin E2 (PGE2) is one the most abundant PGs produced and has multiple physiological roles depending on receptor activation and site of action. After formation, PGE2 diffuses across the plasma membrane to act locally on four receptors (EP1–EP4) expressed by sensory neurons.[10] EP3 and EP4 receptors have the highest affinity for PGE2 and generally couple to Gαs signaling pathways. On the other hand, EP1 receptors couple to Gαq signaling pathway, stimulating PLC activity intraneuronally. Previous studies identify EP4 upregulation in a subset of DRG neurons following peripheral inflammation, revealing a potential target for treating inflammatory hyperalgesia.[11]

Prostaglandin D2 (PGD2) also accumulates in peripheral tissues following inflammatory injury[12] and binds to DP1 and DP2 receptors. Both isoforms are expressed in DRG neurons[13]: DP1 couples to Gαs and DP2 couples to Gαi signaling pathways. Although DP1 activation in DRG neurons enhances voltage-gated sodium currents and DP2 activation has an opposite effect, PGD2 treatment still yields a net enhancement.[13] Interestingly, the PGD2 metabolite, 15*d*PGJ2, directly activates nociceptors via TRPA1 in a non-neuromodulatory role.[14]

2.3. Serotonin receptor

Multiple serotonin receptor isoforms participate in inflammatory hyperalgesia. To date, isoforms 5-HT_{1A} and 5-HT_{2A},[15] 5-HT_{2B},[16] 5-HT_3,[17] 5-HT_4,[18] and 5-HT_7[19] have been implicated. However, conflicting findings over the years have failed to identify significant isoforms responsible for various modes of inflammatory hyperalgesia. For instance, 5-HT3 has been shown to contribute to chemical pain,[20] while other studies indicate that only 5-HT1A agonists mimic serotonin's ability to induce inflammatory hyperalgesia.[21] Alternative studies identify that only 5-HT2A agonists behave similarly to serotonin-induced thermal hyperalgesia following inflammation,[22] while the 5-HT4 receptor subtype enhances inflammatory pain.[18] It should also be noted that both Gαs and Gαq signaling pathways have been implicated in serotonin's effects as an inflammatory mediator, providing for a difficult assessment on the exact downstream mechanisms responsible.

2.4. Purinergic receptors

ATP released from sensory neurons and adjacent microglia can activate any one of four purinergic receptor isoforms that are expressed in nociceptive DRG neurons, including P2X2, P2X3, P2X4, and P2X7.[23] P2X receptors function as nonselective cation channels, such that their activation stimulates multiple cation-dependent signaling pathways in sensory neurons, including activation of voltage-gated calcium channels and sensitization of the ligand-gated ion channel TRPV1.[24] Interestingly, the extracellular degradation of ATP by membrane-bound nucleotidases limits the inflammatory role of ATP molecules *in vivo.*[25] Despite this, multiple P2X mechanisms likely contribute to either the activation or the sensitization of primary afferent nociceptors in inflammatory hyperalgesia.

2.5. Receptor tyrosine kinases

Nerve growth factor (NGF) is the most understood growth factor in pain sensitivity and inflammatory hyperalgesia. NGF activates two receptors: the p75 neurotrophin receptor ($p75^{NTR}$) and the tropomyosin-related kinase A (TrkA) receptor, a receptor tyrosine kinase. Studies on TrkA activation of NGF have defined both its homeostatic and inflammatory roles in pain sensation. Congenital loss-of function mutation of TrkA results in a fatal disorder known as congenital insensitivity to pain, detailed by an individual's inability to sense either superficial or deep painful stimuli.[26] As a mediator, NGF contributes to the development of inflammatory thermal and mechanical hyperalgesia.[27,28] Indeed, NGF is upregulated in peripheral tissues following inflammatory injury[29,30] and contributes to the sensitization of both myelinated and nonmyelinated sensory neurons through TrkA activation. TrkA receptor activation then stimulates Src-tyrosine kinases, PI-3-K, and PLC, leading to both posttranslational and transcriptional upregulation of channel activities, such as TRPV1.[31,32] In addition, other growth factors including glial cell line-derived neurotrophic factor and artemin have shown similar sensitizing behaviors on peripheral sensory systems following inflammation.[33,34]

2.6. Neurokinin receptor

The primary ligand of the neurokinin 1 receptor (NK1R) is substance P, a peptide generated by and released from sensory neurons. Peptide release does not occur by simple diffusion across the plasma membrane, but by vesicular release following neuronal depolarization. Therefore, activation of sensory neurons stimulates the release of substance P, which can then stimulate NK1R in an autocrine fashion, activating Gαq-coupled signaling pathways. NK1R is expressed in sensory afferent endings,[35] immune cells,[36] and both epithelial and larger blood vessels,[37] leading some to categorize the peptide as a broadly efficacious neuropathic inflammatory mediator. Furthermore, due to increased expression of substance P and NK1R within and adjacent to inflamed skeletal joints,[38,39] many studies suggest that substance P may be an important modulator in rheumatoid arthritis.

2.7. Glutamate receptors

Many types of glutamate receptors, both ionotropic and metabotropic, are expressed throughout the body. However, in inflammatory hyperalgesia, the periphery is predominantly populated by metabotropic glutamate receptors (mGluRs). Based on sequence homology, mGluRs can be categorized into three main groups. Group I mGluRs, including isoforms mGluR1 and mGluR5, are expressed on afferent sensory neurons[40] and couple to Gαq signaling pathways. The group II subclass, mGluR2/3, preferentially couples to the Gαi signaling pathway and is also functionally expressed in sensory neurons.[40] Little-to-no evidence supports the functional expression of group III mGluR isoforms, mGluR45, mGluR6, mGluR7, or mGluR8, in peripheral sensory neurons. Despite the dichotomous functional coexpression of group I and group II mGluRs in the periphery, studies demonstrate that glutamate maintains inflammatory pain,[41] as it is released from microglia and nociceptors and activates sensory neurons in an autocrine fashion.

2.8. Protease-activated receptors

Following injury and/or inflammation, serine proteases and trypsins are released into the affected area by immune cells and nociceptors (respectively) to help activate three possible isoforms of protease-activated receptors (PARs) expressed in sensory neurons, including PAR1, PAR2, and PAR4.[42] Coupled to Gαq signaling pathways, PAR is activated by proteolytic degradation of its extracellular amino acid sequence, which produces a peptide ligand capable of activating the very receptor from which it came. Importantly, DRG expression of the PAR2 isoform colocalizes with substance P and CGRP, such that PAR2 activation drives neurogenic inflammation.[43] Similarly, coexpression with TRPV1 in sensory neurons allows for PAR2 activation to induce sensitization of CAP responses in a protein kinase C (PKC)-dependent manner.[44,45]

2.9. Calcitonin receptor-like receptor and receptor activity-modifying protein 1

Similar to Substance P, calcitonin gene-related peptide (CGRP) is released by depolarized sensory neurons and acts in a paracrine fashion to induce neurogenic inflammation following neuronal depolarization. However, unlike the substance P receptor, the receptor for CGRP is composed of a heterodimer between a small, one transmembrane receptor activity-modifying protein 1 (RAMP1), and the larger, more conventional GPCR calcitonin receptor-like receptor (CRLR).[46] When heterodimerized and bound by CGRP, Gαs signaling pathways are activated to stimulate multiple physiological effects in adjacent blood vessels, endothelial cells, and nociceptors. Indeed, CRLR and RAMP1 are coexpressed in 36% of sensory neurons of the trigeminal ganglia, yet CGRP is only expressed in small-to-medium trigeminal neurons negative for expression of either receptor.[47] These findings indicate that autoreceptor regulation (autocrine CGRP activation) is low in trigeminal sensory neurons, suggesting that paracrine CGRP inflammatory sensitization affects sensory modalities not transduced by c-type nociceptors.

2.10. Endothelin receptors

Inflammatory hyperalgesia marked by ET accumulation in the periphery is governed by the activation of two ET receptor isoforms, ET type A receptor (ET_AR) and ET type B receptor (ET_BR). Both GPCR isoforms couple to Gαq signaling pathways, which can sensitize all PKC-sensitive receptors and channels that are coexpressed. In sensory neurons, ET_ARs coexpress strongly with TRPV1 in sensory neurons, sensitizing nociceptor responses to chemical, mechanical, and thermal stimuli.[48–50] Of clinical importance, ETAR sensitization of TRPV1 in trigeminal sensory neurons has also been reported.[51] Trigeminal sensory neurons innervate the orofacial cavity, a tissue that is sometimes populated by oral squamal cell carcinomas (OSCCs), one of leading oral cancers with the highest mortality rate. OSCCs produce significant amounts of ET,[52] consequentially supporting inflammatory hyperalgesia[53] in one of the most sensitive tissues of the body.

3. SIGNALING MECHANISMS

The intracellular biochemical environment in primary afferent terminals of nociceptors is similar to that of most other cells on a general level. G-protein-coupled receptors (GPCRs) coupled to Gαq and Gαs signaling pathways are responsible for the majority of inflammatory mediator signaling actions within nociceptive neurons. As shown in Fig. 1, the activation of Gαs by associated GPCRs stimulates adenylyl cyclase to increase cAMP at local, plasma membrane sites stimulating protein kinase A (PKA) activity and subsequent phosphorylation of PKA consensus motifs on known protein targets, including ligand-gated ion channels such as TRPV1.[54] Similar activation of Gαq by its ligand-bound GPCR stimulates phospholipase C generation of inositol triphosphate (IP_3) and diacylglycerol (DAG) from plasma membrane-associated phosphatidylinositol-4,5-bisphosphate (PIP_2). IP_3 activates inositol triphosphate receptors on the sarcoplasmic reticulum to release internal calcium stores into the cytoplasm, while newly created DAG can activate DAG-sensitive PKC isoforms, including both conventional and

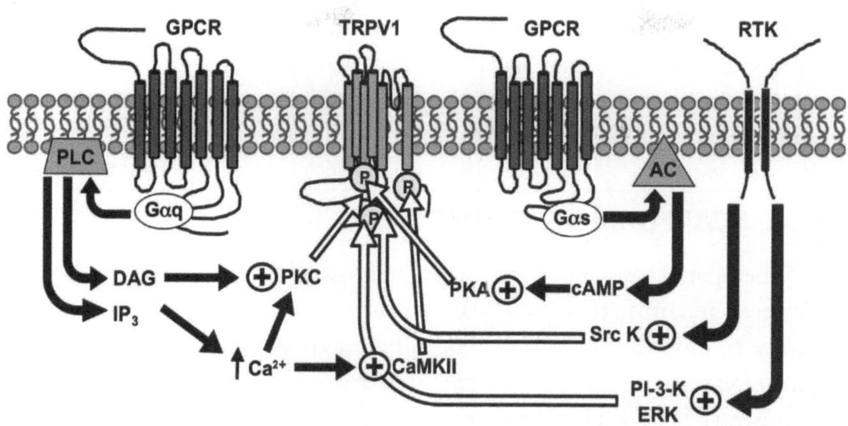

Figure 1 Signaling pathways that phosphorylate and sensitize TRPV1. Inflammatory mediator activation of its respective GPCR or RTK results in the activation of identified and mapped signaling pathways that posttranslationally modify ligand-gated ion channels, such as TRPV1.

novel isoforms. In combination with conventional PKC isoforms that require Ca^{2+} for activation, the generation of DAG can superactivate many PKC isoforms in nociceptive neurons. These PKC isoforms,[55–57] along with calcium-dependent calmodulin kinase type 2[58], target plasma membrane proteins in nociceptive neurons, including ligand-gated ion channels such as TRPV1. Together, Gαs- and Gαq-coupled GPCRs are responsible for much of the posttranslational inflammatory modulation of nociceptive ion channels expressed by primary afferent neurons.

In addition to GPCRs, other inflammatory mediators activate another type of receptor. However, this receptor class is activated by growth factors such as NGF and belongs to the receptor tyrosine kinase class of receptors due to its specialized structure and downstream signaling pathways. NGF association with one of the receptor "arms" as shown in Fig. 1 results in dimerization and subsequent phosphorylation of the intracellular termini, catalyzing downstream signaling through Src-tyrosine kinase, phosphatidylinositol-3-kinase (PI-3-K), and extracellular-regulated kinase (ERK). Importantly, each of these signaling pathways has been implicated in inflammatory sensitization of the ligand-gated ion channel TRPV1 in nociceptive neurons: Src kinase,[32,59] PI-3-K, and ERK.[60,61] Recent studies have identified other growth factors as inflammatory mediators capable of sensitizing nociceptive neurons, including brain-derived neurotrophic factor and glial-derived neurotrophic factor.[62] Indeed, inflammatory ramifications are not only dependent upon mediator availability but also dependent upon receptor coexpression, meaning that not all ligand-gated ion channels can be modulated by all inflammatory receptors. However, coexpression profile studies have revealed that specific groups of neurons coexpress certain receptors, providing for sensitive modulation of nociceptive information to inflammatory injury.

4. SCAFFOLDING STRUCTURES

The spatial and temporal organizations of signaling mechanisms in response to inflammatory injury are dynamic and congested in nature. If one were to envision the number of inflammatory mediators that accumulate from various sources following tissue injury, and then consider the mechanism(s) by which signals are propagated from receptor activation, to pathway stimulation, to end-point signal manifestation, all within a short time frame and under tight physiologic regulation, one must assume that an organizational structure exists intracellularly to maintain signal "flow." Indeed, signal organization and maintenance do exist and dynamically control the "flow," as well as strength, of signaling pathways by spatially and temporally controlling posttranslational modifications to end-point targets, via protein/protein interactions between scaffolding proteins and the target itself. In this section, we will discuss several scaffolding proteins that operate to dynamically control signaling pathways and dictate primary afferent neuron sensitivity to painful stimuli.

4.1. A-kinase anchoring protein 79/150

Identified in 1995 as a scaffolding protein for PKA,[63,64] A-kinase anchoring protein 79/150 (AKAP 79/150) functions to localize certain enzymes to target substrates on a subcellular level. In terms of its isoform designations, AKAP79 refers to the human isoform of the scaffolding proteins (79 kDa), while AKAP150 is the rodent isoform (150 kDa). Considerable research has identified multiple substrates and enzymes that AKAP scaffolds to establish organized and efficient signal transduction. AKAP150 is predominantly anchored at the plasma membrane in multiple cell types, including neurons, and scaffolds PKA, PKC, and calcineurin (CaN, PP2B)[65] along identified amino acid motifs within its tertiary structure.[66] These enzymes are then capable to effectively act upon protein targets that AKAP150 associates with, including AMPA receptor,[67] NMDA receptor,[67] L-type voltage-gated channels,[68] and the KCNQ2 channel.[69] Additional studies have also confirmed that AKAP150 targets and associates with multiple TRP channels, including TRPV1.[70–72] AKAP150/TRPV1 scaffolding association allows primary afferent sensory neurons the opportunity to dynamically respond to various inflammatory mediators and stimuli that precipitate inflammatory hyperalgesia.

AKAP150 association with TRPV1 was first established in experiments investigating PKA catalytic subunit proximity to neuronal plasma membranes following inflammatory mediator administration.[73] TRPV1 acts as the primary, PKA-sensitive heat transducer responsible for inflammatory hypersensitivity to a thermal stimulus, although the hypersensitivity was sensitive to St-Ht31, a peptide that blocks PKA association with AKAP150. Additionally, adenylyl cyclase sensitization of TRPV1 currents in DRG neurons was negatively affected by St-Ht31, indicating that AKAP150 participates in the TRPV1-signaling complex that mediates inflammatory sensitization of the ligand-gated ion channel. Despite no information on any physical interaction between Gαs subunit, AKAP150, and TRPV1, scaffolding proteins were found to exert physiologic significance by dynamically affecting receptor sensitivities to proalgesic effectors of inflammatory signaling pathways.

AKAP scaffolds multiple enzymes and substrate proteins in specific subcellular compartments in order to organize signal propagation. Certain AKAP isoforms scaffold enzymes including phosphodiesterase 4 D3 (PDE4D3),[74,75] protein kinase N,[76] and protein phosphatase 2A,[76] while AKAP150 primarily orients PKA, PKC, and CaN with plasma membrane substrates, including TRPV1 (Fig. 2). Indeed, genetic knockdown of AKAP150 blocks inflammatory PKA phosphorylation and PKC phosphorylation and sensitization of TRPV1.[70,77] Therefore, AKAP150 functionally supports certain inflammatory receptor-activated signaling pathways that utilize these kinases, including PG[71] and bradykinin[72] (see Table 1). Conversely, TRPV1 dephosphorylation and desensitization serves as a feedback mechanism for controlling inflammatory sensitization of the ligand-gated ion channel and the primary afferent neurons itself. Yet, the primary mechanism for dephosphorylation is calcineurin (CaN) dependent,[78,79] and is not dependent on AKAP150,[80] suggesting that CaN may be attracted to TRPV1 by other means, such as by a potential calmodulin scaffolding mechanism.[81] Kinase scaffolding by AKAP150 regulates TRPV1 sensitization by multiple inflammatory mediators. However, recent work demonstrates that inflammatory mediators also control the association of AKAP150 with TRPV1 in sensory neurons.

Multiple reports provide indirect evidence of physical protein–protein interaction(s) between AKAP150 and TRPV1, including

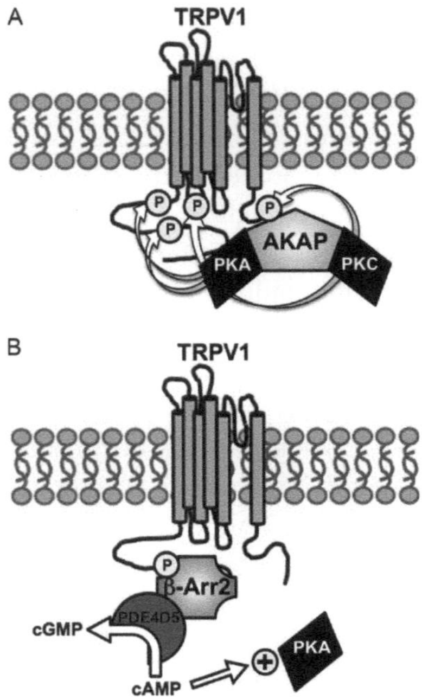

Figure 2 Scaffolding mechanisms. (A) AKAP scaffolding protein associates with TRPV1 and brings PKA and PKC into close spatial proximity with TRPV1 via associative binding. (B) β-Arrestin2 (β-Arr2) brings phosphodiesterase 4D5 (PDE4D5) into close spatial proximity to TRPV1, allowing for hydrolysis of cAMP, thereby controlling PKA activity and subsequent phosphorylation of targets, including TRPV1.

coimmunoprecipitation from transfected homologous cell culture models as well as primary sensory neuron cultures.[70–72] Furthermore, confocal analysis of cotransfected cells using total internal reflective fluorescence–Forster resonance energy transfer (TIRF–FRET) findings indicate strong association at the plasma membrane.[82] Indeed, recent work demonstrates that AKAP150 association with TRPV1 is controlled by anchoring to PIP_2 in the plasma membrane. Following seminal studies that define targeting of AKAP to the plasma membrane via its association with certain phosphoinositides,[83] recent findings indicate that PLC activation releases AKAP150 from its plasma membrane anchorage to associate with adjacent, membrane-associated proteins, such as TRPV1.[84] Consequently, these findings provide an explanation to reported diverging roles for PIP_2 in TRPV1 regulation.[85–87] Furthermore, activation of Gαq-coupled GPCRs by inflammatory mediators such as bradykinin and ET are now understood as strong inflammatory mediators in the sensitization of thermal and chemical pain through their abilities to stimulate AKAP150 association with TRPV1 (via PLC activation) and PKC phosphorylation of TRPV1 (via PKC activation). Culminated research can now explain the intense reactions caused by multiple inflammatory mediators known to couple to Gαq-signaling pathways.

Fortunately, AKAP150 only mediates phosphorylation and sensitization of TRPV1, preventing redundancy within the system. TIRF–FRET studies have revealed that AKAP150 association with TRPV1 is calcium sensitive, demonstrating that calcium-activated calmodulin significantly reduces ion channel association with the scaffolding protein.[82] Importantly, the competitive interaction between calmodulin:TRPV1 and AKAP150:TRPV1 may exist as a desensitization mechanism allowing the calmodulin-scaffolded phosphatase CaN to dephosphorylate TRPV1, while blocking AKAP150-bound PKA and PKC from phosphorylating and resensitizing TRPV1. There also exists another exciting mechanism that mediates TRPV1 desensitization, using a scaffolding mechanism shared with GPCRs.

4.2. β-Arrestin

Historically, β-arrestin molecules are charged with regulating GPCR desensitization in most cell types. Following agonist activation and subsequent G-protein phosphorylation and activation, certain kinases, such as G-protein receptor kinase 2, phosphorylate the GPCR, marking it for association with β-arrestin. Once the arrestin molecule binds to the GPCR, a concert of molecular events that includes clathrin association and plasma membrane invagination results in the internalization of the GPCR, bound for recycling to the membrane or degradation by proteolysis. In either case, the GPCR is desensitized from additional agonist activation, making β-arrestin a significant player in neuronal physiology. However, recent studies have identified new, scaffolding roles for β-arrestin, including several roles as an associative scaffolding protein with ligand-gated ion channels in nociceptive neuron populations.

AKAP150 association with TRPV1 does not effectively regulate the desensitization of the ion channel.[88] Recent studies that utilize peptide blockers of AKAP150 and TRPV1 support this posit, showing no change in pharmacological desensitization,[89] a process demonstrated to be dependent upon CaN[90]. However, work on the molecule β-arrestin revealed interesting findings. Classically, β-arrestin proteins are the primary mediators of GPCR desensitization and subsequent receptor internalization. Newer studies have even revealed that β-arrestins serve as end-point effectors of by signaling pathways in several cell models. However, some have demonstrated that β-arrestins are capable of mediating the activities of ligand-gated ion channels through scaffolding capabilities.

It is has been well established that β-arrestin molecules serve as scaffolds in a variety of cellular circumstances. c-Jun N-terminal Kinase 3 activation is regulated by MAPK scaffolding to β-arrestin,[91] while ubiquitin scaffolding to β-arrestin regulates β2 adrenergic receptor signaling.[92] Additionally, β-arrestin scaffolding of ubiquitin modulates TRPV4 activation,[93] providing the first example of arrestin-scaffolded regulation of an ion channel. Although TRPV4 has little to do with sensory perception by primary afferent neurons, β-arrestin molecules were also found to target TRPV1 ion channels with scaffolding phosphodiesterase 4D5 (PDE4D5,[94]). Similar scaffolding of PDE4 by β-arrestin effectively regulates cAMP recruitment following β-adrenergic receptor activation.[95] Therefore, a scaffolding mechanism was found to exist that regulates the desensitization of a ligand-gated ion channel sensitive to inflammatory mediators.

Subsequent work on TRPV1 and β-arrestin found that scaffolded PDE4D5 reduced PKA phosphorylation of TRPV1, through the hydrolysis of available cAMP in the plasma membrane compartment. Given that multiple inflammatory mediators signal through Gas-coupled pathways to increase cellular cAMP via adenylyl cyclase activation, β-arrestin scaffolding of TRPV1 is an important mechanism to control channel activity. As depicted in Fig. 2B, PDE4D5 scaffolding by β-arrestin to TRPV1 results in reduced PKA phosphorylation of the channel, maintaining TRPV1 in a desensitized state. Furthermore, additional results demonstrate that β-arrestin recruitment to TRPV1 is dependent upon TRPV1 phosphorylation at amino acid residues posttranslationally modified by kinases activated by inflammatory mediators.[96] Therefore, phosphorylation of TRPV1 by PKA and/or PKC following inflammatory insult results in the recruitment of a β-arrestin scaffolding complex capable of reducing PKA activity in the plasma membrane compartment, reducing TRPV1 phosphorylation, through a negative feedback mechanism.

5. CONCLUDING REMARKS

Peripheral injury and/or insult create an environment that supports sensitization of primary afferent neurons responsible for communicating somatosensory input to the CNS. This dynamic sensitization drives neuroplastic changes in afferent terminal endings to increase pain sensitivity by reducing stimulus threshold, resulting in homeostatic hyperalgesia. The concert of biochemical signaling events that dictate these changes are well organized and efficient in nature. However, isoform-specific subtleties still exist, perhaps contributing to individual cases in the clinic in which abnormal responses are witnessed and do not respond to normal pain treatments. In this vein, studies on the peripheral inflammasome will continue to characterize the multitude of signaling molecules that contribute to inflammatory pain.

REFERENCES

1. Miyata M, Kashiwadani H, Fukaya M, et al. Role of thalamic phospholipase C[beta]4 mediated by metabotropic glutamate receptor type 1 in inflammatory pain. *J Neurosci.* 2003;23:8098–8108.
2. Carlton SM, Zhou S, Govea R, Du J. Group II/III metabotropic glutamate receptors exert endogenous activity-dependent modulation of TRPV1 receptors on peripheral nociceptors. *J Neurosci.* 2011;31:12727–12737.
3. Hu HJ, Bhave G, Gereau RWt. Prostaglandin and protein kinase A-dependent modulation of vanilloid receptor function by metabotropic glutamate receptor 5: potential mechanism for thermal hyperalgesia. *J Neurosci.* 2002;22:7444–7452.
4. Leeb-Lundberg LM, Marceau F, Muller-Esterl W, Pettibone DJ, Zuraw BL. International union of pharmacology. XLV. Classification of the kinin receptor family: from molecular mechanisms to pathophysiological consequences. *Pharmacol Rev.* 2005;57:27–77.
5. Xu GY, Huang LY. Peripheral inflammation sensitizes P2X receptor-mediated responses in rat dorsal root ganglion neurons. *J Neurosci.* 2002;22:93–102.
6. Guo LH, Trautmann K, Schluesener HJ. Expression of P2X4 receptor by lesional activated microglia during formalin-induced inflammatory pain. *J Neuroimmunol.* 2005;163:120–127.
7. Hughes JP, Hatcher JP, Chessell IP. The role of P2X(7) in pain and inflammation. *Purinergic Signal.* 2007;3:163–169.
8. Hess JF, Borkowski JA, Macneil T, et al. Differential pharmacology of cloned human and mouse B2 bradykinin receptors. *Mol Pharmacol.* 1994;45:1–8.
9. Bastian S, Loillier B, Paquet JL, Pruneau D. Stable expression of human kinin B1 receptor in 293 cells: pharmacological and functional characterization. *Br J Pharmacol.* 1997;122:393–399.
10. Southall MD, Vasko MR. Prostaglandin receptor subtypes, EP3C and EP4, mediate the prostaglandin E2-induced cAMP production and sensitization of sensory neurons. *J Biol Chem.* 2001;276:16083–16091.
11. Lin CR, Amaya F, Barrett L, et al. Prostaglandin E2 receptor EP4 contributes to inflammatory pain hypersensitivity. *J Pharmacol Exp Ther.* 2006;319:1096–1103.
12. Schuligoi R, Schmidt R, Geisslinger G, Kollroser M, Peskar BA, Heinemann A. PGD2 metabolism in plasma: kinetics and relationship with bioactivity on DP1 and CRTH2 receptors. *Biochem Pharmacol.* 2007;74:107–117.
13. Ebersberger A, Natura G, Eitner A, Halbhuber KJ, Rost R, Schaible HG. Effects of prostaglandin D2 on tetrodotoxin-resistant Na + currents in DRG neurons of adult rat. *Pain.* 2011;152:1114–1126.

14. Taylor-Clark TE, Undem BJ, Macglashan Jr DW, Ghatta S, Carr MJ, McAlexander MA. Prostaglandin-induced activation of nociceptive neurons via direct interaction with transient receptor potential A1 (TRPA1). *Mol Pharmacol.* 2008;73:274–281.
15. Sufka KJ, Schomburg FM, Giordano J. Receptor mediation of 5-HT-induced inflammation and nociception in rats. *Pharmacol Biochem Behav.* 1992;41:53–56.
16. Lin SY, Chang WJ, Lin CS, Huang CY, Wang HF, Sun WH. Serotonin receptor 5-HT2B mediates serotonin-induced mechanical hyperalgesia. *J Neurosci.* 2011;31:1410–1418.
17. Zeitz KP, Guy N, Malmberg AB, et al. The 5-HT3 subtype of serotonin receptor contributes to nociceptive processing via a novel subset of myelinated and unmyelinated nociceptors. *J Neurosci.* 2002;22:1010–1019.
18. Doak GJ, Sawynok J. Formalin-induced nociceptive behavior and edema: involvement of multiple peripheral 5-hydroxytryptamine receptor subtypes. *Neuroscience.* 1997;80:939–949.
19. Brenchat A, Romero L, Garcia M, et al. 5-HT7 receptor activation inhibits mechanical hypersensitivity secondary to capsaicin sensitization in mice. *Pain.* 2009;141:239–247.
20. Eschalier A, Kayser V, Guilbaud G. Influence of a specific 5-HT3 antagonist on carrageenan-induced hyperalgesia in rats. *Pain.* 1989;36:249–255.
21. Taiwo YO, Levine JD. Serotonin is a directly-acting hyperalgesic agent in the rat. *Neuroscience.* 1992;48:485–490.
22. Tokunaga A, Saika M, Senba E. 5-HT2A receptor subtype is involved in the thermal hyperalgesic mechanism of serotonin in the periphery. *Pain.* 1998;76:349–355.
23. Donnelly-Roberts D, McGaraughty S, Shieh CC, Honore P, Jarvis MF. Painful purinergic receptors. *J Pharmacol Exp Ther.* 2008;324:409–415.
24. Moriyama T, Iida T, Kobayashi K, et al. Possible involvement of P2Y2 metabotropic receptors in ATP-induced transient receptor potential vanilloid receptor 1-mediated thermal hypersensitivity. *J Neurosci.* 2003;23:6058–6062.
25. Burnstock G. Physiology and pathophysiology of purinergic neurotransmission. *Physiol Rev.* 2007;87:659–797.
26. Indo Y, Tsuruta M, Hayashida Y, et al. Mutations in the TRKA/NGF receptor gene in patients with congenital insensitivity to pain with anhidrosis. *Nat Genet.* 1996;13:485–488.
27. Andreev N, Dimitrieva N, Koltzenburg M, McMahon SB. Peripheral administration of nerve growth factor in the adult rat produces a thermal hyperalgesia that requires the presence of sympathetic post-ganglionic neurones. *Pain.* 1995;63:109–115.
28. Lewin GR, Ritter AM, Mendell LM. Nerve growth factor-induced hyperalgesia in the neonatal and adult rat. *J Neurosci.* 1993;13:2136–2148.
29. Nicol GD, Vasko MR. Unraveling the story of NGF-mediated sensitization of nociceptive sensory neurons: ON or OFF the Trks? *Mol Interv.* 2007;7:26–41.
30. Woolf CJ, Safieh-Garabedian B, Ma QP, Crilly P, Winter J. Nerve growth factor contributes to the generation of inflammatory sensory hypersensitivity. *Neuroscience.* 1994;62:327–331.
31. Xue Q, Jong B, Chen T, Schumacher MA. Transcription of rat TRPV1 utilizes a dual promoter system that is positively regulated by nerve growth factor. *J Neurochem.* 2007;101:212–222.
32. Zhang X, Huang J, McNaughton PA. NGF rapidly increases membrane expression of TRPV1 heat-gated ion channels. *Embo J.* 2005;24:4211–4223.
33. Albers KM, Woodbury CJ, Ritter AM, Davis BM, Koerber HR. Glial cell-line-derived neurotrophic factor expression in skin alters the mechanical sensitivity of cutaneous nociceptors. *J Neurosci.* 2006;26:2981–2990.

34. Elitt CM, McIlwrath SL, Lawson JJ, et al. Artemin overexpression in skin enhances expression of TRPV1 and TRPA1 in cutaneous sensory neurons and leads to behavioral sensitivity to heat and cold. *J Neurosci*. 2006;26:8578–8587.
35. Sluka KA, Milton MA, Willis WD, Westlund KN. Differential roles of neurokinin 1 and neurokinin 2 receptors in the development and maintenance of heat hyperalgesia induced by acute inflammation. *Br J Pharmacol*. 1997;120:1263–1273.
36. Suzuki H, Miura S, Liu YY, Tsuchiya M, Ishii H. Substance P induces degranulation of mast cells and leukocyte adhesion to venular endothelium. *Peptides*. 1995;16:1447–1452.
37. Brain SD. Sensory neuropeptides: their role in inflammation and wound healing. *Immunopharmacology*. 1997;37:133–152.
38. Garrett NE, Kidd BL, Cruwys SC, Tomlinson DR. Changes in preprotachykinin mRNA expression and substance P levels in dorsal root ganglia of monoarthritic rats: comparison with changes in synovial substance P levels. *Brain Res*. 1995;675:203–207.
39. Iwasaki A, Inoue K, Hukuda S. Distribution of neuropeptide-containing nerve fibers in the synovium and adjacent bone of the rat knee joint. *Clin Exp Rheumatol*. 1995;13:173–178.
40. Crawford JH, Wainwright A, Heavens R, et al. Mobilisation of intracellular Ca2+ by mGluR5 metabotropic glutamate receptor activation in neonatal rat cultured dorsal root ganglia neurones. *Neuropharmacology*. 2000;39:621–630.
41. Picker MJ, Daugherty D, Henry FE, Miller LL, Dykstra LA. Metabotropic glutamate antagonists alone and in combination with morphine: comparison across two models of acute pain and a model of persistent, inflammatory pain. *Behav Pharmacol*. 2011;22:785–793.
42. Zhu WJ, Yamanaka H, Obata K, et al. Expression of mRNA for four subtypes of the proteinase-activated receptor in rat dorsal root ganglia. *Brain Res*. 2005;1041:205–211.
43. Steinhoff M, Vergnolle N, Young SH, et al. Agonists of proteinase-activated receptor 2 induce inflammation by a neurogenic mechanism. *Nat Med*. 2000;6:151–158.
44. Amadesi S, Cottrell GS, Divino L, et al. Protease-activated receptor 2 sensitizes TRPV1 by protein kinase Cepsilon- and A-dependent mechanisms in rats and mice. *J Physiol*. 2006;575:555–571.
45. Amadesi S, Nie J, Vergnolle N, et al. Protease-activated receptor 2 sensitizes the capsaicin receptor transient receptor potential vanilloid receptor 1 to induce hyperalgesia. *J Neurosci*. 2004;24:4300–4312.
46. McLatchie LM, Fraser NJ, Main MJ, et al. RAMPs regulate the transport and ligand specificity of the calcitonin-receptor-like receptor. *Nature*. 1998;393:333–339.
47. Eftekhari S, Salvatore CA, Calamari A, Kane SA, Tajti J, Edvinsson L. Differential distribution of calcitonin gene-related peptide and its receptor components in the human trigeminal ganglion. *Neuroscience*. 2010;169:683–696.
48. Balonov K, Khodorova A, Strichartz GR. Tactile allodynia initiated by local subcutaneous endothelin-1 is prolonged by activation of TRPV-1 receptors. *Exp Biol Med*. 2006;231:1165–1170.
49. Plant TD, Zollner C, Mousa SA, Oksche A. Endothelin-1 potentiates capsaicin-induced TRPV1 currents via the endothelin A receptor. *Exp Biol Med*. 2006;231:1161–1164.
50. Plant TD, Zollner C, Kepura F, et al. Endothelin potentiates TRPV1 via ETA receptor-mediated activation of protein kinase C. *Mol Pain*. 2007;3:35.
51. Yamamoto T, Ono K, Hitomi S, et al. Endothelin receptor-mediated responses in trigeminal ganglion neurons. *J Dent Res*. 2013;92:335–339.
52. Pickering V, Jordan RC, Schmidt BL. Elevated salivary endothelin levels in oral cancer patients—a pilot study. *Oral Oncol*. 2007;43:37–41.
53. Schmidt BL, Pickering V, Liu S, et al. Peripheral endothelin A receptor antagonism attenuates carcinoma-induced pain. *Eur J Pain*. 2007;11:406–414.

54. Bhave G, Zhu W, Wang H, Brasier DJ, Oxford GS, Gereau RWt. cAMP-dependent protein kinase regulates desensitization of the capsaicin receptor (VR1) by direct phosphorylation. *Neuron*. 2002;35:721–731.
55. Bhave G, Hu HJ, Glauner KS, et al. Protein kinase C phosphorylation sensitizes but does not activate the capsaicin receptor transient receptor potential vanilloid 1 (TRPV1). *Proc Natl Acad Sci USA*. 2003;100:12480–12485.
56. Cesare P, Dekker LV, Sardini A, Parker PJ, McNaughton PA. Specific involvement of PKC-epsilon in sensitization of the neuronal response to painful heat. *Neuron*. 1999;23:617–624.
57. Premkumar LS, Ahern GP. Induction of vanilloid receptor channel activity by protein kinase C. *Nature*. 2000;408:985–990.
58. Jung J, Shin JS, Lee SY, et al. Phosphorylation of vanilloid receptor 1 by Ca2+/calmodulin-dependent kinase II regulates its vanilloid binding. *J Biol Chem*. 2004;279:7048–7054.
59. Jeske NA, Patwardhan AM, Henry MA, Milam SB. Fibronectin stimulates TRPV1 translocation in primary sensory neurons. *J Neurochem*. 2009;108:591–600.
60. Zhu W, Oxford GS. Phosphoinositide-3-kinase and mitogen activated protein kinase signaling pathways mediate acute NGF sensitization of TRPV1. *Mol Cell Neurosci*. 2007;34:689–700.
61. Zhuang ZY, Xu H, Clapham DE, Ji RR. Phosphatidylinositol 3-kinase activates ERK in primary sensory neurons and mediates inflammatory heat hyperalgesia through TRPV1 sensitization. *J Neurosci*. 2004;24:8300–8309.
62. Price TJ, Louria MD, Candelario-Soto D, et al. Treatment of trigeminal ganglion neurons in vitro with NGF, GDNF or BDNF: effects on neuronal survival, neurochemical properties and TRPV1-mediated neuropeptide secretion. *BMC Neurosci*. 2005;6:4.
63. Bregman DB, Hirsch AH, Rubin CS. Molecular characterization of bovine brain P75, a high affinity binding protein for the regulatory subunit of cAMP-dependent protein kinase II beta. *J Biol Chem*. 1991;266:7207–7213.
64. Carr DW, Stofko-Hahn RE, Fraser ID, Cone RD, Scott JD. Localization of the cAMP-dependent protein kinase to the postsynaptic densities by A-kinase anchoring proteins. Characterization of AKAP 79. *J Biol Chem*. 1992;267:16816–16823.
65. Coghlan VM, Perrino BA, Howard M, et al. Association of protein kinase A and protein phosphatase 2B with a common anchoring protein. *Science*. 1995;267:108–111.
66. Hoshi N, Langeberg LK, Scott JD. Distinct enzyme combinations in AKAP signalling complexes permit functional diversity. *Nat Cell Biol*. 2005;7:1066–1073.
67. Colledge M, Dean RA, Scott GK, Langeberg LK, Huganir RL, Scott JD. Targeting of PKA to glutamate receptors through a MAGUK-AKAP complex. *Neuron*. 2000;27:107–119.
68. Gao T, Yatani A, Dell'Acqua ML, et al. cAMP-dependent regulation of cardiac L-type Ca2+ channels requires membrane targeting of PKA and phosphorylation of channel subunits. *Neuron*. 1997;19:185–196.
69. Hoshi N, Zhang JS, Omaki M, et al. AKAP150 signaling complex promotes suppression of the M-current by muscarinic agonists. *Nat Neurosci*. 2003;6:564–571.
70. Jeske NA, Diogenes A, Ruparel NB, et al. A-kinase anchoring protein mediates TRPV1 thermal hyperalgesia through PKA phosphorylation of TRPV1. *Pain*. 2008;138:604–616.
71. Schnizler K, Shutov LP, Van Kanegan MJ, et al. Protein kinase A anchoring via AKAP150 is essential for TRPV1 modulation by forskolin and prostaglandin E2 in mouse sensory neurons. *J Neurosci*. 2008;28:4904–4917.
72. Zhang X, Li L, McNaughton PA. Proinflammatory mediators modulate the heat-activated ion channel TRPV1 via the scaffolding protein AKAP79/150. *Neuron*. 2008;59:450–461.

73. Rathee PK, Distler C, Obreja O, et al. PKA/AKAP/VR-1 module: a common link of Gs-mediated signaling to thermal hyperalgesia. *J Neurosci*. 2002;22:4740–4745.
74. Dodge KL, Khouangsathiene S, Kapiloff MS, et al. mAKAP assembles a protein kinase A/PDE4 phosphodiesterase cAMP signaling module. *EMBO J*. 2001;20:1921–1930.
75. Tasken KA, Collas P, Kemmner WA, Witczak O, Conti M, Tasken K. Phosphodiesterase 4D and protein kinase a type II constitute a signaling unit in the centrosomal area. *J Biol Chem*. 2001;276:21999–22002.
76. Takahashi M, Shibata H, Shimakawa M, Miyamoto M, Mukai H, Ono Y. Characterization of a novel giant scaffolding protein, CG-NAP, that anchors multiple signaling enzymes to centrosome and the golgi apparatus. *J Biol Chem*. 1999;274:17267–17274.
77. Jeske NA, Patwardhan AM, Ruparel NB, Akopian AN, Shapiro MS, Henry MA. A-kinase anchoring protein 150 controls protein kinase C-mediated phosphorylation and sensitization of TRPV1. *Pain*. 2009;146:301–307.
78. Docherty RJ, Yeats JC, Bevan S, Boddeke HW. Inhibition of calcineurin inhibits the desensitization of capsaicin-evoked currents in cultured dorsal root ganglion neurones from adult rats. *Pflugers Arch*. 1996;431:828–837.
79. Jeske NA, Patwardhan AM, Gamper N, Price TJ, Akopian AN, Hargreaves KM. Cannabinoid WIN 55,212-2 regulates TRPV1 phosphorylation in sensory neurons. *J Biol Chem*. 2006;281:32879–32890.
80. Por ED, Samelson BK, Belugin S, Akopian AN, Scott JD, Jeske NA. PP2B/calcineurin-mediated desensitization of TRPV1 does not require AKAP150. *Biochem J*. 2010;432:549–556.
81. Numazaki M, Tominaga T, Takeuchi K, Murayama N, Toyooka H, Tominaga M. Structural determinant of TRPV1 desensitization interacts with calmodulin. *Proc Natl Acad Sci USA*. 2003;100:8002–8006.
82. Chaudhury S, Bal M, Belugin S, Shapiro MS, Jeske NA. AKAP150-mediated TRPV1 sensitization is disrupted by calcium/calmodulin. *Mol Pain*. 2011;7:34.
83. Dell'Acqua ML, Faux MC, Thorburn J, Thorburn A, Scott JD. Membrane-targeting sequences on AKAP79 bind phosphatidylinositol-4, 5-bisphosphate. *Embo J*. 1998;17:2246–2260.
84. Jeske NA, Por ED, Belugin S, et al. A-kinase anchoring protein 150 mediates transient receptor potential family v type 1 sensitivity to phosphatidylinositol-4,5-bisphosphate. *J Neurosci*. 2011;31:8681–8688.
85. Lukacs V, Thyagarajan B, Varnai P, Balla A, Balla T, Rohacs T. Dual regulation of TRPV1 by phosphoinositides. *J Neurosci*. 2007;27:7070–7080.
86. Prescott ED, Julius D. A modular PIP2 binding site as a determinant of capsaicin receptor sensitivity. *Science*. 2003;300:1284–1288.
87. Rohacs T. Regulation of TRP channels by PIP(2). *Pflugers Arch*. 2007;453:753–762.
88. Por ED, Samelson BK, Belugin S, Akopian AN, Scott JD, Jeske NA. PP2B/calcineurin-mediated desensitization of TRPV1 does not require AKAP150. *The Biochemical journal*. 2010;432:549–556.
89. Fischer MJ, Btesh J, McNaughton PA. Disrupting sensitization of transient receptor potential vanilloid subtype 1 inhibits inflammatory hyperalgesia. *J Neurosci*. 2013;33:7407–7414.
90. Patwardhan AM, Jeske NA, Price TJ, Gamper N, Akopian AN, Hargreaves KM. The cannabinoid WIN 55,212–2 inhibits transient receptor potential vanilloid 1 (TRPV1) and evokes peripheral antihyperalgesia via calcineurin. *Proc Natl Acad Sci U S A*. 2006;103:11393–11398.
91. McDonald PH, Chow CW, Miller WE, et al. Beta-arrestin 2: a receptor-regulated MAPK scaffold for the activation of JNK3. *Science*. 2000;290:1574–1577.

92. Shenoy SK, McDonald PH, Kohout TA, Lefkowitz RJ. Regulation of receptor fate by ubiquitination of activated beta 2-adrenergic receptor and beta-arrestin. *Science*. 2001;294:1307–1313.
93. Shukla AK, Kim J, Ahn S, et al. Arresting a transient receptor potential (TRP) channel: beta-arrestin 1 mediates ubiquitination and functional down-regulation of TRPV4. *J Biol Chem*. 2010;285:30115–30125.
94. Por ED, Bierbower SM, Berg KA, et al. Beta-Arrestin-2 desensitizes the transient receptor potential vanilloid 1 (TRPV1) channel. *J Biol Chem*. 2012;287:37552–37563.
95. Baillie GS, Sood A, McPhee I, et al. Beta-Arrestin-mediated PDE4 cAMP phosphodiesterase recruitment regulates beta-adrenoceptor switching from Gs to Gi. *Proc Natl Acad Sci U S A*. 2003;100:940–945.
96. Por ED, Gomez R, Akopian AN, Jeske NA. Phosphorylation regulates TRPV1 association with beta-arrestin-2. *Biochem J*. 2013;451:101–109.

CHAPTER THREE

Contribution of Mechanosensitive Ion Channels to Somatosensation

Reza Sharif-Naeini[1]
Department of Physiology and Cell Information Systems Group, McGill University, Montreal, Quebec, Canada
[1]Corresponding author: e-mail address: Reza.Sharif@mcgill.ca

Contents

1. The Evolution of Mechanosensing 54
2. MSCs in Nociceptors 55
3. Difficulties in Identifying Genes Encoding MSCs 57
   3.1 Piezo2 as a mechanosensitive channel in sensory neurons 58
4. Gating Mechanisms of MSCs 59
5. Role of MSCs in the Transmission of Noxious Mechanical Inputs 60
6. Sensitization of MSCs Is Necessary for the Induction of Mechanical Allodynia 62
7. Potassium-Selective MSCs in Nociceptors 65
References 65

Abstract

Mechanotransduction, the conversion of a mechanical stimulus into an electrical signal, is a central mechanism to several physiological functions in mammals. It relies on the function of mechanosensitive ion channels (MSCs). Although the first single-channel recording from MSCs dates back to 30 years ago, the identity of the genes encoding MSCs has remained largely elusive. Because these channels have an important role in the development of mechanical hypersensitivity, a better understanding of their function may lead to the identification of selective inhibitors and generate novel therapeutic pathways in the treatment of chronic pain. Here, I will describe our current understanding of the role MSCs may play in somatosensation and the potential candidate genes proposed to encode them.

We can sense mechanical changes in our outer environment, such as touch and sounds, and in our inner environment, such as fluid flow and osmotic shifts, because these stimuli are detected by sensory cells and converted into electrical signals. This conversion occurs in less than a millisecond and is therefore thought to involve mechanosensitive ion channels (MSCs) rather than engage second messenger intermediates.[1–3] Interestingly, these MSCs

Progress in Molecular Biology and Translational Science, Volume 131
ISSN 1877-1173
http://dx.doi.org/10.1016/bs.pmbts.2014.11.011

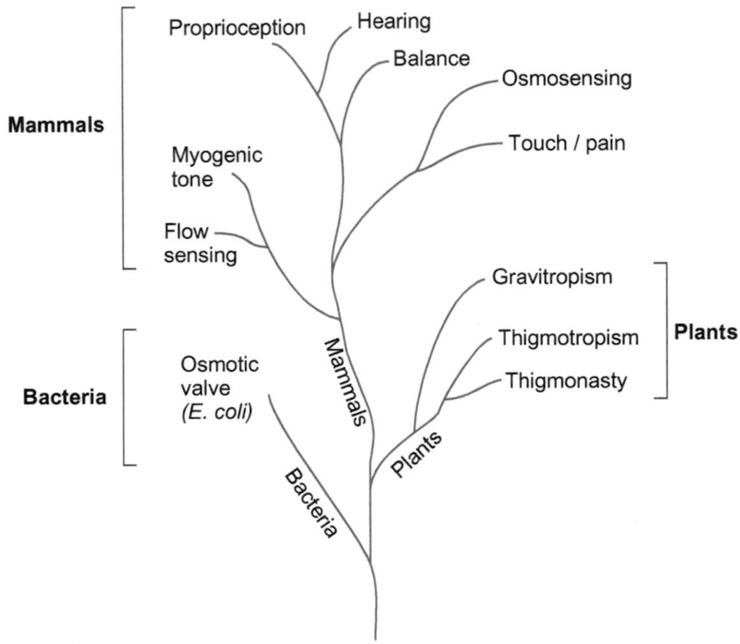

Figure 1 Evolution of MSCs across species. Schematic diagram of the expression of MSCs in different kingdoms, along with the proposed function of the MSCs in each of these kingdoms.

exist in a wide variety of organisms and must have therefore been present in the early stages of life on the planet. Indeed, mechanotransduction is one of the earliest functions developed by living organisms, appearing billions of years ago. While the function of MSCs remained simple in the early life forms, it became more and more complex through evolution. In mammals, these channels are expressed in several tissues and are involved in a multitude of essential physiological functions (Fig. 1). In the next sections, I will review several aspects of MSCs, ranging from their evolution, gating mechanisms, role in cell excitability, and tissue distribution. This review will emphasize the role these channels play in the transmission of painful mechanical stimuli in sensory neurons during physiological and pathological conditions.

1. THE EVOLUTION OF MECHANOSENSING

Mechanosensing dates back about 3.8 billion years, when microbes one of the first forms of life appeared on the planet.[4] In these cells, the primary function of MSCs is to act as emergency valves releasing excess osmolytes when their extracellular environment became hypoosmotic, such as during rainfall.[5] As evolution progressed, these channels remained expressed in higher organisms (Fig. 1), such as plants, where their roles have been associated with root gravity sensing,[6] as well as the thigmotropic[7] and thigmonastic[8] behaviors of the vine and the flytrap, respectively. In mammals, these channels attained their greatest functional diversity and underlie several essential physiological functions. In kidney epithelial cells, MSCs protect against mechanical stresses such as compression caused by cyst growth.[9] In vascular smooth muscle cells, MSCs act as sensors of membrane tension and trigger the myogenic tone of resistance arteries.[10–12] In endothelial cells, they can sense blood flow and can stimulate arterial vasodilation.[13,14] These channels are expressed in the nerve terminals of nodose ganglia neurons, located in the wall of the aortic branch, where they sense blood pressure and can trigger the baroreflex.[15–17] In the supraoptic nucleus of the hypothalamus, these channels are responsible for the intrinsic osmosensitivity of vasopressin neurons[18,19] and trigger the release of the antidiuretic hormone to induce water reabsorption. In hair cells of the inner ear, these channels are responsible for the detection of movement in the cilia, responsible for our capacity to detect sounds.[20–22] Finally in the nerve terminals of our sensory neurons, located in the skin and other organs, these channels underlie our capacity to detect mechanical stimuli such as touch and pain.[23–28]

2. MSCs IN NOCICEPTORS

The ability of neurons to detect and transduce mechanical stimuli impinging on them is a fundamental process that underlies hearing, balance, touch, and pain.[20,25,29] It is generally accepted that this sensitivity is conferred by the presence of MSCs on the plasma membrane of these neurons.[20,24,25] An increase in the sensitivity of MSCs, or in their expression at the plasma membrane, may therefore result in enhanced mechanosensation. Remarkably, although mechanical hypersensitivity in nociceptors is a cause of chronic inflammatory pain, including in rheumatoid arthritis and osteoarthritis,[30–32] the role of MSCs in these pathologies has not been examined. This is due in part to our lack of knowledge of the properties of MSCs in nociceptors, and how the function of these channels can be modulated by accessory proteins.

Advances in our understanding of MSCs come from electrophysiological studies performed *in vitro* on isolated sensory neurons. All these experiments have been done by recording mechanically activated (MA) whole-cell or single-channel currents at the cell soma of isolated neurons. Given the inaccessibility of nociceptor peripheral terminals, somatic recording from isolated nociceptors is the only approach currently available to study MSCs.[24,26–28,33–57] Somatic recording is a standard approach in molecular pain research because the membrane of the cell body expresses all the molecules that determine the function of these neurons.[58–62] The clearest demonstration that receptors in the nerve terminals are also expressed in the cell body is that capsaicin, which selectively activates the transient receptor potential, vanilloid-1 (TRPV1)-containing nociceptors when injected in the skin,[63] can selectively activate nociceptors when applied to their soma in *in vitro* experiments.[63–74] Furthermore, inflammatory conditions associated with increased TRPV1 responses in peripheral terminals are also associated with an increased responsiveness at the cell body.[75,76] If the same distribution occurs for MSCs in nociceptors, we would expect that the somatically recorded currents elicited by mechanical stimulation of the nociceptor neurites or soma would be similar. Indeed, it has been reported that mechanical stimulation of nociceptors' soma or distant neurites over 70 μm away produces comparable inward currents.[23] This indicates that mechanisms that activate MSCs in distant neurites are the same as those in the soma.

These experiments have helped identify important characteristics of MSCs in nociceptors. Whole-cell voltage-clamp recordings from nociceptors indicate that these neurons display a MA slowly adapting (SA) current.[38,40,41,55,77] In current-clamp experiments, these neurons fire tonically in response to a mechanical stimulation, irrespective of the stimulus velocity or duration.[41] This slow adaptation kinetics might be a critical aspect of nociceptive signal generation[41] that ensures that the pain signal gets transmitted to the spinal cord and brain. Consequently, if the adaptation kinetics of the SA current was slower, or removed, one might expect that these nociceptors might fire more action potentials and generate a longer lasting pain signal. The adaptation in these neurons is likely an intrinsic feature of the MSCs because electrophysiological experiments have demonstrated that the adaptation is almost completely removed when the recording is performed at positive membrane potentials.[34] Moreover, whether the adaptation kinetics is modified in the setting of chronic pain remains to be determined. In single-channel recordings from nociceptors, it was demonstrated that these neurons express a 14-picoSiemens, high-threshold mechanosensitive channel[24,26,27,53] that is voltage-independent and has a reversal potential near zero mV, consistent with a nonselective cation channel.[26,27] This channel has been speculated to be responsible for the mechanically induced SA current in nociceptors,[40,41] but experimental evidence is lacking. Finally, pharmacological characterization indicates that the SA current is blocked by the trivalent ion gadolinium, the inorganic dye ruthenium red, and by a novel conopeptide named NMB-1.[24,26,27,53]

Despite these observations, advances in our understanding of MSC function in nociceptors have been slow primarily because the genes encoding these channels or their modulators are unknown. Indeed, genetic studies have implicated genes necessary for the mechanosensory function of invertebrate nociceptors, notably *osm9* in the nematode and *Painless* and Dm*piezo* in *Drosophila* larvae,[36,78,79] yet no valid candidate genes encoding a nonselective cationic MSCs or MSC modulators in mammalian nociceptors have emerged.

3. DIFFICULTIES IN IDENTIFYING GENES ENCODING MSCs

The identification of candidate proteins forming an MSC is difficult for several reasons. First and foremost, most, if not all, cells known to date express an MSC. This renders a functional cloning approach such as the one used to identify the capsaicin receptor almost impossible. Second, purification from mechanosensitive cells is difficult because MSCs are not expressed in large amounts. For instance, each hair cell is thought to have about 50–100 transduction channels.[80–82] A similar number is also believed to be present on osmosensitive vasopressin neurons. Third, establishing that a candidate protein is intrinsically mechanosensitive is a difficult task because it has to be expressed in heterologous systems, which already have their own set of MSCs. One would therefore have to look at an increase in mechanically evoked current above the endogenous level. This can be bypassed by a labor-intensive protein purification and reconstitution in artificial lipids, devoid of any other protein that may interfere with the recording. Nonetheless, the chances of success remain low because this approach still makes the assumption that eukaryotic MSCs are made of homomeric subunits, whereas in reality, the MSC may be a mechanotransduction complex requiring several accessory subunits, including specialized elements of the cytoskeleton and the extracellular matrix, as is the case for the MSC of the nematode *Caenorhabditis elegans*.[3,83,84] Because of these difficulties, very few candidate proteins have been identified as MSCs. In *Escherichia coli*, the genes encoding the MSCs were identified some 20 years ago and have since been extensively studied.[85] But they have no eukaryotic homologs. In mammals, until 5 years ago, the only known *bona fide* MSC was the potassium-selective (hyperpolarizing) MSC named TREK-1 (also known as KCNK2),[86,87] a member of the two-pore potassium (K2P) channel family. More recently, two novel genes, named *Fam38A* and *Fam38B*, encoding depolarizing cationic MSCs Piezo1 and Piezo2 were identified.[34–37]

3.1. Piezo2 as a mechanosensitive channel in sensory neurons

A significant advancement in the field of mechanotransduction came with the identification of Piezo1 and Piezo2 by the Patapoutian lab in 2010. In the mouse, Piezo1 is a 30-transmembrane domain protein expressed primarily in the lung, bladder, skin, colon, and kidney. Piezo2, on the other hand, is a 34-transmembrane domain protein that is highly expressed in dorsal root ganglia, lung, and bladder. Such high expression level of Piezo2 in sensory neurons made it an exciting candidate as the mechanosensor in nociceptors. Using *in situ* hybridization combined with immunohistochemistry, Coste and colleagues demonstrated that Piezo2 mRNA is expressed in a subset of peripherin (60%)- and neurofilament 200 (28%)-immunopositive neurons, typically associated with large diameter, touch-sensitive Aβ neurons. Additionally, Piezo2 was also expressed in TRPV1-immunoreactive neurons, which are nociceptors. These observations lead the authors to speculate that Piezo2 was likely involved in the sensation of noxious mechanical stimuli.

Although the *Drosophila melanogaster* homolog of Piezo, which shares only 24% homology with the mouse *piezo* genes, was shown to be involved in mechanosensory nociception, the involvement of Piezo2 in mammalian somatosensation seems to be restricted to the detection of touch stimuli. Indeed, knocking down Piezo2 in cultured mouse sensory neurons did not affect the slow-adapting MA current, found in nociceptors,[25,38,77] but only reduced the rapid adapting MA current, found in touch-sensitive fibers.[55,88] This is also supported by experiments demonstrating that the zebrafish homolog of piezo2, which shares 63% sequence homology, is specifically expressed in a subset of neurons responsible for detecting light touch, and its knockdown leads to deficits in the response to light touch without affecting the responses to noxious chemical or mechanical stimuli.[89]

The role of Piezo2 in the mammalian sense of touch is further supported by its expression in Merkel cells, the gentle touch receptor in the skin that enables us to distinguish the fine details of objects.[90–94] Indeed, a recent report by Woo and colleagues demonstrated that the intrinsic mechanosensitivity of Merkel cells is due to the expression of Piezo2.[92,95] Deletion of this gene in Merkel cells abolishes their mechanosensitivity and the mice behavioral response to gentle touch becomes impaired.[95] The identity of the mammalian mechanosensitive channel of nociceptor therefore remains elusive.

4. GATING MECHANISMS OF MSCs

MSCs are embedded in a lipid bilayer where they are sensitive to local, as well as global, stress in the bilayer. When a mechanical stimulus impinges on a cell's membrane, the stress is distributed to all components including the bilayer, the cytoskeleton (CSK), and the extracellular matrix (ECM), which converge on the MSC and induce a transition from the closed to open state. How the convergence of this stress induces the transition has been debated over the years and has led to two possible mechanisms: the bilayer model and the tether model[96] (Fig. 2). Most of what we know about the bilayer model comes from studies of the bacterial MSCs.[29] In this model, the stresses

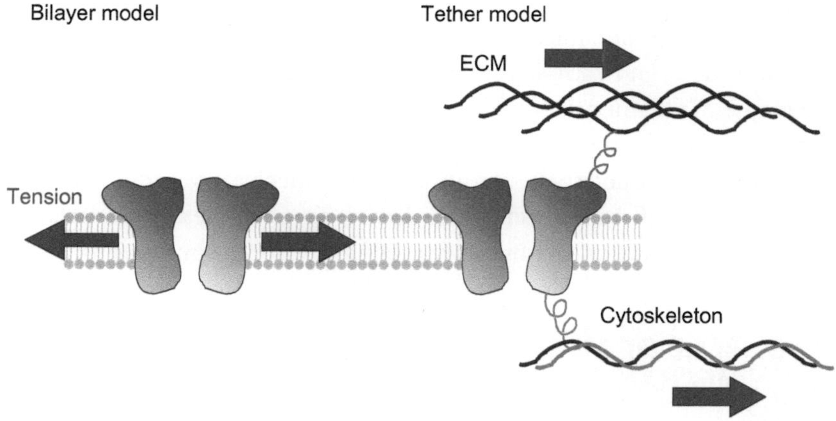

Figure 2 Models of the gating of MSCs. Two of the most popular models of MSC gating by mechanical stimuli. In the bilayer model (left panel), an increase in membrane tension is believed to cause a thinning of the plasma membrane, thus exposing hydrophobic residues of the channels that were previously embedded in the bilayer. This forces the channel subunits to tilt and induces an opening. In the tether model (model to the right), the membrane stress is conveyed to the channel by means of accessory proteins, such as the cytoskeleton or the extracellular matrix. These are thought to directly interact with the channel subunit.

absorbed by membrane (or cytoskeleton or ECM) cause an increase in the tension in the bilayer, which modifies the interaction between the hydrophobic portion of the protein and the lipid bilayer and causes the protein to change its configuration and tilt its membrane helices, inducing its opening.[97,98] This gating mechanism is supported by experimental data obtained in artificial bilayer experiments.

In the tether model, the stresses absorbed by the membrane are transmitted to the channel via the interaction with ECM or CSK elements.[29] These additional structures are therefore essential components of the transduction machinery. This model was first proposed for hair cell transduction[96] but has since been observed in other systems including osmosensory neurons[99–101] and pain-sensing neurons.[27] In hair cells, the transduction channel is tethered to an external tip link that connects the stereocilia to the actin cytoskeleton. Bending of this stereocilia stretches the channel and causes its opening.[96] Loss of tip links can effectively abolish transduction.[102] In vasopressin neurons of the hypothalamus, osmosensing relies on a stretch-inactivated ion channel that is linked to the actin and tubulin cytoskeleton.[99,101] Disrupting either filaments with depolymerization agents can blunt osmosensory transduction, whereas increasing filament density with stabilizing agents increases the coupling between mechanical stresses and channel opening, i.e., increases the sensitivity of the system. Finally, depolymerization of either microtubules or F-actin reduced the sensitivity of MSCs in touch- and pain-sensing neurons.[26,27]

Because of the importance of the integrity of the ECM or the CSK in the tethered model, any signaling pathway that converges on these elements has the potential to modulate the sensitivity of the mechanotransduction apparatus. Examples of these mechanisms have been proposed to underlie the effect of angiotensin on the osmosensitivity of vasopressin neurons.[100]

5. ROLE OF MSCs IN THE TRANSMISSION OF NOXIOUS MECHANICAL INPUTS

The neuronal processes leading to the sensation of pain can be divided into four distinct steps: transduction, conduction, processing, and interpretation (Fig. 3). The first step is the transduction of the mechanical stimulus into an electric signal. The activation of the MSCs in the terminals of the nociceptive afferents by a noxious stimulus generates a membrane depolarization termed receptor potential. Mechanosensitive channels in touch-sensitive neurons have a low threshold for activation by mechanical stimuli

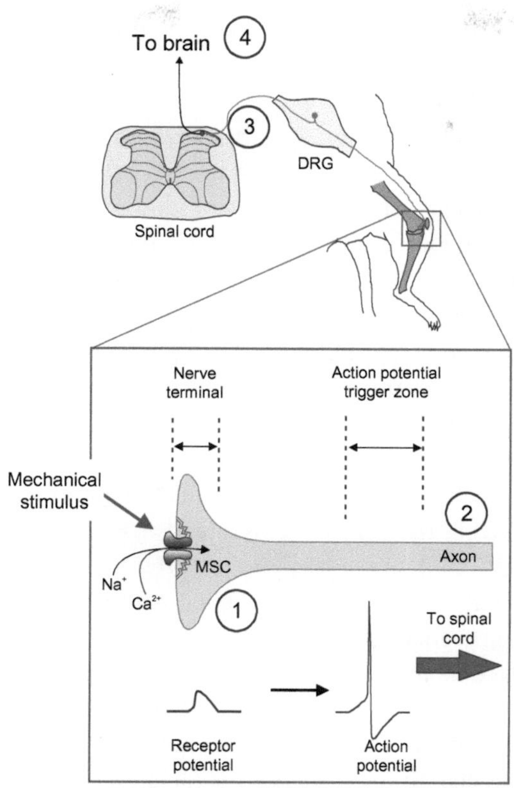

Figure 3 Schematic diagram of the transmission of painful mechanical information from the periphery to the brain. The mechanical stimulus is first detected by the MSCs (step 1), which transforms the mechanical energy into an electrical signal. The conduction (step 2) of the signal is enabled by the presence voltage-gated sodium channels, which carry the electrical input to the spinal cord for its processing (step 3). There, the signal is processed before it is transmitted to projection neurons, which carry the information to supraspinal centers (step 4).

whereas those expressed in nociceptors have a high threshold.[27] If this wave of depolarization is large enough when it reaches the spike trigger zone, a region enriched in voltage-gated sodium channels, then an action potential is generated and conducted (step 2) along the sensory nerve toward the spinal cord. The third step occurs in the spinal cord, where the signal is processed by interneurons or directly transmitted to projection neurons. The latter will transmit the signal to the brain (step 4), where it is interpreted as painful.

During chronic pain conditions associated with a mechanical sensitization of nociceptors, such as in rheumatoid arthritis or osteoarthritis, the pathology underlying the hypersensitivity can be in all four steps.[30,32]

Indeed, the release of inflammatory cytokines in the joint is likely the primary cause of nociceptor sensitization.[103,104] Although the downstream target of these cytokines is unknown, it has been shown that blocking the NaV1.8 subset of sodium channel in the setting of OA can attenuate both the hyperexcitability of nociceptors and the behavioral pain symptoms in a preclinical OA model.[105] The processing of sensory information in the dorsal horn of the spinal cord can also be affected, as shown in rodent models of OA, where microglia and astrocytes show significant hyperactivity in the later stages of the disease.[106] Finally, changes at supraspinal sites can take place (step 4). In a surgical model of OA, an increased functional connectivity of supraspinal neuronal networks was observed in the nucleus accumbens and ventral posterior lateral thalamus.[107] Furthermore, when compared to controls, patients with chronic pain due to hip osteoarthritis had a characteristic decrease in gray matter in the regions of the anterior cingulate cortex (ACC), right insular cortex and operculum, dorsolateral prefrontal cortex (DLPFC), amygdala, and brainstem.[108] Of those that became pain-free following hip surgery, the gray matter size increased in the DLPFC, ACC, amygdala, and brainstem.[108] This suggests that supraspinal changes take place in the setting of OA pain, though the mechanisms are currently elusive. The remainder of this review will focus on the peripheral mechanisms underlying the sensitization of nociceptors that produces mechanical allodynia.

Recent therapeutic approaches have targeted the transmission of the mechanical stimulus to the spinal cord (step 2) through the development of selective blockers of voltage-gated sodium channels.[109–111] An alternative approach could be to identify the MSCs responsible for the transduction of the noxious mechanical stimulus, thus blocking the activation of the nociceptor. In healthy individuals, the role of MSCs in nociceptors is to warn the organism about any potential harmful mechanical stimuli. However, during chronic inflammatory pain, nociceptors become sensitized to mechanical stimuli;[30,31,112] and become activated by innocuous, touch-like, stimuli, a phenomenon that underlies part of the mechanical allodynia seen in these conditions.

6. SENSITIZATION OF MSCs IS NECESSARY FOR THE INDUCTION OF MECHANICAL ALLODYNIA

During inflammation, proinflammatory mediators sensitize the nerve terminals of nociceptors to mechanical stimuli,[30] causing these afferents to become activated by innocuous stimuli. This increased excitability can be due to changes in the function of voltage-gated ion channels or of the mechanotransduction apparatus in the nerve terminals.[52] A decrease in expression of background potassium channels, responsible for maintaining the resting membrane potential at a hyperpolarized level, could cause a slight membrane depolarization,[113–115] thus helping a mechanically induced depolarizing wave to reach the action potential threshold (APT; Fig. 4). By itself, this mechanism would enhance the excitation of the nociceptor only if a mechanically applied stimulus is of high enough (noxious) intensity,

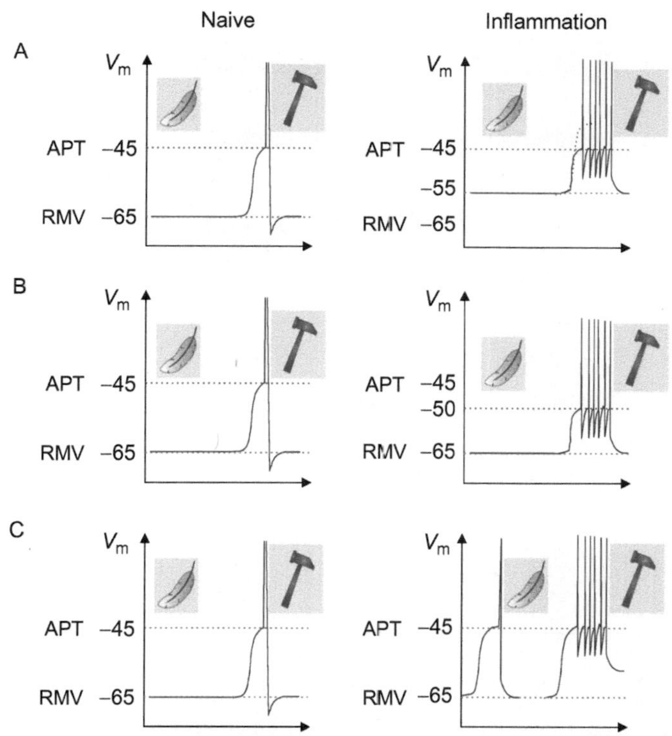

Figure 4 Proposed contribution of MSCs to the development of mechanical allodynia. The activation of MSCs in nociceptors can only be done by high-intensity stimuli. During inflammation, both hyperalgesia and allodynia are visible. However, without the sensitization of MSCs, it will be impossible to generate mechanical allodynia. (A) A decrease in background potassium conductances will depolarize the plasma membrane and facilitate the generation of an action potential by mechanically evoked depolarization. (B) A decrease in the action potential threshold would also facilitate the generation of an action potential. However, in both situations (top two rows), the precipitation of mechanical allodynia would not be possible without the sensitization of MSCs (lower panel). It is expected that only if the activation threshold of the MSC is reduced (C), there will be a receptor potential generated that can elicit an action potential. This sensitization would explain mechanical allodynia.

producing a hyperalgesia-like state. Alternatively, increased expression of $Na_V1.9$ channels,[113,116–118] which have a hyperpolarized activation threshold,[119,120] would lower the APT and facilitate firing in nociceptors (Fig. 4B). Once again, this enhancement of the firing can occur only if the mechanical stimulus can open the MSC, otherwise the depolarization will not occur. A sensitization of the MSCs, on the other hand, will lead

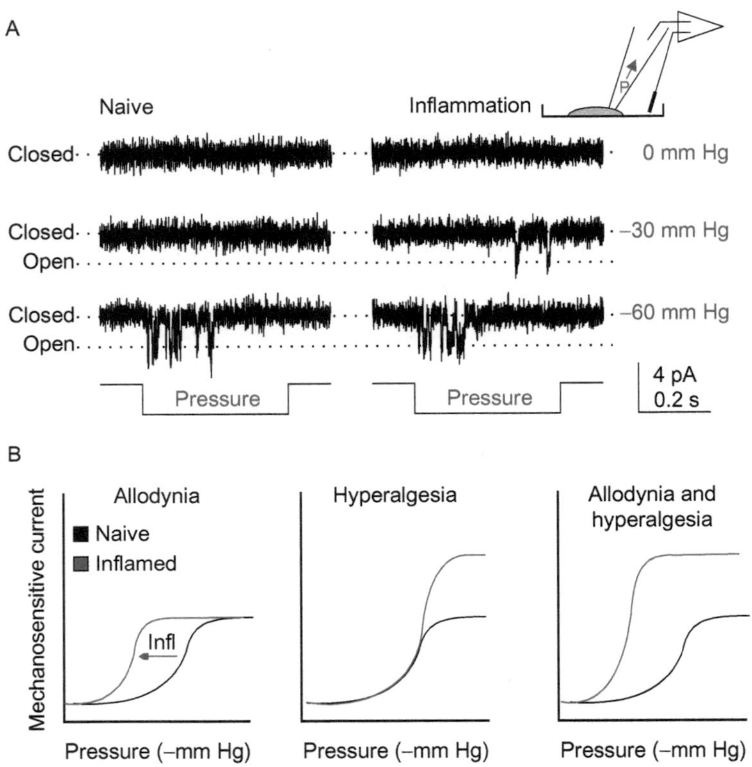

Figure 5 Example of sensitization of MSCs during inflammation. Illustration of the behavior of MSCs in nociceptors from naïve or inflamed animals. (A) The activation threshold of MSCs is typically high under normal conditions (−60 mm Hg in the example). The downward deflections represent openings of single MSCs in response to increased tension produced by steps of negative pressure pulses through the recording electrode. During inflammatory pain, MSCs can open at lower thresholds through mechanisms that have yet to be determined. (B) Three different scenarios can take place during the sensitization of MSCs. Their maximum activity can remain the same, indicating that there is no upregulation in the number of channels, but those present on the membrane now open at lower mechanical stimuli, predisposing the system for mechanical allodynia (left panel). There may be an increased expression of the MSCs, without altering their activation threshold, thus enabling the development of mechanical hyperalgesia (middle panel). Finally, a combination of both mechanisms can occur, which would generate both allodynia and hyperalgesia (right panel).

to the generation of a receptor potential with low-intensity stimuli and will be able to elicit mechanical allodynia (Fig. 4C). Although these three mechanisms are described individually here, they are likely to occur at the same time to produce both mechanical allodynia and hyperalgesia during inflammation (Fig. 5).

7. POTASSIUM-SELECTIVE MSCs IN NOCICEPTORS

Another level of complexity is added to the role of MSCs in nociceptors by the presence of potassium-selective MSCs on the same terminals.[121,122] Activation of the latter leads to membrane hyperpolarization and prevents the cell from firing action potentials. This means that when a mechanical stimulus impinges on the nerve terminal of a nociceptor, it will activate the depolarizing, nonselective cationic MSC, as well as activate the hyperpolarizing, potassium-selective MSC. The perception of pain at supraspinal sites would therefore indicate that the level of depolarizing exceeded the level of hyperpolarizing. However, conditions with decreased expression of Kv1.1 or K2Ps would therefore lead to a mechanical hypersensitivity. Evidence for this has been previously reported, whereby mice with a gene deletion for TREK-1 have significantly lower mechanosensitivity.[122]

In summary, several candidates have been proposed to form the MSC in nociceptors, but definite proof is lacking. Physiological studies in animal models of inflammatory pain have provided hints that the activity of MSCs in nociceptors might be impaired. However, no report has clearly examined, at the single-channel level, how the properties of these channels would be affected in conditions such as rheumatoid arthritis or osteoarthritis. Further studies will be required to answer these questions.

REFERENCES

1. Corey DP, Hudspeth AJ. Response latency of vertebrate hair cells. *Biophys J.* 1979;26(3):499–506.
2. Walker RG, Willingham AT, Zuker CS. A Drosophila mechanosensory transduction channel. *Science.* 2000;287(5461):2229–2234.
3. O'Hagan R, Chalfie M, Goodman MB. The MEC-4 DEG/ENaC channel of Caenorhabditis elegans touch receptor neurons transduces mechanical signals. *Nat Neurosci.* 2005;8(1):43–50.
4. Martinac B, Kloda A. Evolutionary origins of mechanosensitive ion channels. *Prog Biophys Mol Biol.* 2003;82(1–3):11–24.
5. Booth IR, Blount P. The MscS and MscL families of mechanosensitive channels act as microbial emergency release valves. *J Bacteriol.* 2012;194(18):4802–4809.

6. Tatsumi H, Furuichi T, Nakano M, et al. Mechanosensitive channels are activated by stress in the actin stress fibres, and could be involved in gravity sensing in plants. *Plant Biol.* 2014;16(suppl 1):18–22.
7. Strohm AK, Baldwin KL, Masson PH. Molecular mechanisms of root gravity sensing and signal transduction. *Wiley Interdiscip Rev Dev Biol.* 2012;1(2):276–285.
8. Escalante-Perez M, Krol E, Stange A, et al. A special pair of phytohormones controls excitability, slow closure, and external stomach formation in the Venus flytrap. *Proc Natl Acad Sci U S A.* 2011;108(37):15492–15497.
9. Peyronnet R, Sharif-Naeini R, Folgering JH, et al. Mechanoprotection by polycystins against apoptosis is mediated through the opening of stretch-activated K(2P) channels. *Cell Rep.* 2012;1(3):241–250.
10. Sharif-Naeini R, Folgering JH, Bichet D, et al. Polycystin-1 and -2 dosage regulates pressure sensing. *Cell.* 2009;139(3):587–596.
11. Sharif-Naeini R, Dedman A, Folgering JH, et al. TRP channels and mechanosensory transduction: insights into the arterial myogenic response. *Pflugers Arch.* 2008;456(3): 529–540.
12. Folgering JH, Sharif-Naeini R, Dedman A, Patel A, Delmas P, Honore E. Molecular basis of the mammalian pressure-sensitive ion channels: focus on vascular mechanotransduction. *Prog Biophys Mol Biol.* 2008;97(2–3):180–195.
13. Nauli SM, Kawanabe Y, Kaminski JJ, Pearce WJ, Ingber DE, Zhou J. Endothelial cilia are fluid shear sensors that regulate calcium signaling and nitric oxide production through polycystin-1. *Circulation.* 2008;117(9):1161–1171.
14. AbouAlaiwi WA, Takahashi M, Mell BR, et al. Ciliary polycystin-2 is a mechanosensitive calcium channel involved in nitric oxide signaling cascades. *Circ Res.* 2009;104(7):860–869.
15. Sullivan MJ, Sharma RV, Wachtel RE, et al. Non-voltage-gated Ca2+ influx through mechanosensitive ion channels in aortic baroreceptor neurons. *Circ Res.* 1997;80(6): 861–867.
16. Snitsarev V, Whiteis CA, Abboud FM, Chapleau MW. Mechanosensory transduction of vagal and baroreceptor afferents revealed by study of isolated nodose neurons in culture. *Auton Neurosci Basic Clin.* 2002;98(1–2):59–63.
17. Kraske S, Cunningham JT, Hajduczok G, Chapleau MW, Abboud FM, Wachtel RE. Mechanosensitive ion channels in putative aortic baroreceptor neurons. *Am J Physiol.* 1998;275(4 pt 2):H1497–H1501.
18. Oliet SH, Bourque CW. Mechanosensitive channels transduce osmosensitivity in supraoptic neurons. *Nature.* 1993;364(6435):341–343.
19. Sharif Naeini R, Witty MF, Seguela P, Bourque CW. An N-terminal variant of Trpv1 channel is required for osmosensory transduction. *Nat Neurosci.* 2006;9(1): 93–98.
20. Vollrath MA, Kwan KY, Corey DP. The micromachinery of mechanotransduction in hair cells. *Annu Rev Neurosci.* 2007;30:339–365.
21. Holt JR, Corey DP. Two mechanisms for transducer adaptation in vertebrate hair cells. *Proc Natl Acad Sci U S A.* 2000;97(22):11730–11735.
22. Eatock RA, Corey DP, Hudspeth AJ. Adaptation of mechanoelectrical transduction in hair cells of the bullfrog's sacculus. *J Neurosci.* 1987;7(9):2821–2836.
23. Hu J, Lewin GR. Mechanosensitive currents in the neurites of cultured mouse sensory neurones. *J Physiol.* 2006;577(pt 3):815–828.
24. Hu J, Milenkovic N, Lewin GR. The high threshold mechanotransducer: a status report. *Pain.* 2006;120(1–2):3–7.
25. Lewin GR, Moshourab R. Mechanosensation and pain. *J Neurobiol.* 2004;61(1): 30–44.
26. Cho H, Koo JY, Kim S, Park SP, Yang Y, Oh U. A novel mechanosensitive channel identified in sensory neurons. *Eur J Neurosci.* 2006;23(10):2543–2550.

27. Cho H, Shin J, Shin CY, Lee SY, Oh U. Mechanosensitive ion channels in cultured sensory neurons of neonatal rats. *J Neurosci.* 2002;22(4):1238–1247.
28. Delmas P, Hao J, Rodat-Despoix L. Molecular mechanisms of mechanotransduction in mammalian sensory neurons. *Nat Rev Neurosci.* 2011;12(3):139–153.
29. Kung C. A possible unifying principle for mechanosensation. *Nature.* 2005;436(7051): 647–654.
30. McDougall JJ. Arthritis and pain. Neurogenic origin of joint pain. *Arthritis Res Ther.* 2006;8(6):220.
31. Schaible HG, Richter F, Ebersberger A, et al. Joint pain. *Exp Brain Res.* 2009; 196(1):153–162.
32. Schaible HG, Grubb BD. Afferent and spinal mechanisms of joint pain. *Pain.* 1993; 55(1):5–54.
33. Raouf R, Rugiero F, Kiesewetter H, et al. Sodium channels and mammalian sensory mechanotransduction. *Mol Pain.* 2012;8:21.
34. Coste B, Mathur J, Schmidt M, et al. Piezo1 and Piezo2 are essential components of distinct mechanically activated cation channels. *Science.* 2010;330(6000):55–60.
35. Coste B, Xiao B, Santos JS, et al. Piezo proteins are pore-forming subunits of mechanically activated channels. *Nature.* 2012;483(7388):176–181.
36. Kim SE, Coste B, Chadha A, Cook B, Patapoutian A. The role of Drosophila Piezo in mechanical nociception. *Nature.* 2012;483(7388):209–212.
37. Dubin AE, Schmidt M, Mathur J, et al. Inflammatory signals enhance Piezo2-mediated mechanosensitive currents. *Cell Rep.* 2012;2(3):511–517.
38. McCarter GC, Reichling DB, Levine JD. Mechanical transduction by rat dorsal root ganglion neurons in vitro. *Neurosci Lett.* 1999;273(3):179–182.
39. McCarter GC, Levine JD. Ionic basis of a mechanotransduction current in adult rat dorsal root ganglion neurons. *Mol Pain.* 2006;2:28.
40. Hao J, Delmas P. Recording of mechanosensitive currents using piezoelectrically driven mechanostimulator. *Nat Protoc.* 2011;6(7):979–990.
41. Hao J, Delmas P. Multiple desensitization mechanisms of mechanotransducer channels shape firing of mechanosensory neurons. *J Neurosci.* 2010;30(40):13384–13395.
42. Roudaut Y, Lonigro A, Coste B, Hao J, Delmas P, Crest M. Touch sense: functional organization and molecular determinants of mechanosensitive receptors. *Channels (Austin).* 2012;6(4):234–245.
43. Dubreuil AS, Boukhaddaoui H, Desmadryl G, et al. Role of T-type calcium current in identified D-hair mechanoreceptor neurons studied in vitro. *J Neurosci.* 2004;24(39): 8480–8484.
44. Heidenreich M, Lechner SG, Vardanyan V, et al. KCNQ4 K(+) channels tune mechanoreceptors for normal touch sensation in mouse and man. *Nat Neurosci.* 2012;15(1): 138–145.
45. Heppenstall PA, Lewin GR. A role for T-type Ca2+ channels in mechanosensation. *Cell Calcium.* 2006;40(2):165–174.
46. Hu J, Chiang LY, Koch M, Lewin GR. Evidence for a protein tether involved in somatic touch. *EMBO J.* 2010;29(4):855–867.
47. Lechner SG, Frenzel H, Wang R, Lewin GR. Developmental waves of mechanosensitivity acquisition in sensory neuron subtypes during embryonic development. *EMBO J.* 2009;28(10):1479–1491.
48. Martinez-Salgado C, Benckendorff AG, Chiang LY, et al. Stomatin and sensory neuron mechanotransduction. *J Neurophysiol.* 2007;98(6):3802–3808.
49. Stucky CL, Lewin GR. Isolectin B(4)-positive and -negative nociceptors are functionally distinct. *J Neurosci.* 1999;19(15):6497–6505.
50. Wang R, Lewin GR. The Cav3.2T-type calcium channel regulates temporal coding in mouse mechanoreceptors. *J Physiol.* 2011;589(pt 9):2229–2243.

51. Wetzel C, Hu J, Riethmacher D, et al. A stomatin-domain protein essential for touch sensation in the mouse. *Nature*. 2007;445(7124):206–209.
52. Di Castro A, Drew LJ, Wood JN, Cesare P. Modulation of sensory neuron mechanotransduction by PKC- and nerve growth factor-dependent pathways. *Proc Natl Acad Sci U S A*. 2006;103(12):4699–4704.
53. Drew LJ, Rugiero F, Cesare P, et al. High-threshold mechanosensitive ion channels blocked by a novel conopeptide mediate pressure-evoked pain. *PLoS One*. 2007;2(6):e515.
54. Drew LJ, Wood JN. FM1-43 is a permeant blocker of mechanosensitive ion channels in sensory neurons and inhibits behavioural responses to mechanical stimuli. *Mol Pain*. 2007;3:1.
55. Drew LJ, Wood JN, Cesare P. Distinct mechanosensitive properties of capsaicin-sensitive and -insensitive sensory neurons. *J Neurosci*. 2002;22(12):RC228.
56. Quick K, Zhao J, Eijkelkamp N, et al. TRPC3 and TRPC6 are essential for normal mechanotransduction in subsets of sensory neurons and cochlear hair cells. *Open Biol* 2012;2(5):120068.
57. Rugiero F, Wood JN. The mechanosensitive cell line ND-C does not express functional thermoTRP channels. *Neuropharmacology*. 2009;56(8):1138–1146.
58. Basbaum AI, Bautista DM, Scherrer G, Julius D. Cellular and molecular mechanisms of pain. *Cell*. 2009;139(2):267–284.
59. Hucho T, Levine JD. Signaling pathways in sensitization: toward a nociceptor cell biology. *Neuron*. 2007;55(3):365–376.
60. Rush AM, Cummins TR, Waxman SG. Multiple sodium channels and their roles in electrogenesis within dorsal root ganglion neurons. *J Physiol*. 2007;579(pt 1):1–14.
61. Bautista DM, Siemens J, Glazer JM, et al. The menthol receptor TRPM8 is the principal detector of environmental cold. *Nature*. 2007;448(7150):204–208.
62. Peier AM, Moqrich A, Hergarden AC, et al. A TRP channel that senses cold stimuli and menthol. *Cell*. 2002;108(5):705–715.
63. Caterina MJ, Leffler A, Malmberg AB, et al. Impaired nociception and pain sensation in mice lacking the capsaicin receptor. *Science*. 2000;288(5464):306–313.
64. Baccaglini PI, Hogan PG. Some rat sensory neurons in culture express characteristics of differentiated pain sensory cells. *Proc Natl Acad Sci U S A*. 1983;80(2):594–598.
65. Brenneis C, Kistner K, Puopolo M, et al. Phenotyping the function of TRPV1-expressing sensory neurons by targeted axonal silencing. *J Neurosci*. 2013;33(1): 315–326.
66. Spahn V, Fischer O, Endres-Becker J, Schafer M, Stein C, Zollner C. Opioid withdrawal increases transient receptor potential vanilloid 1 activity in a protein kinase A-dependent manner. *Pain*. 2013;154(4):598–608.
67. Puopolo M, Binshtok AM, Yao GL, Oh SB, Woolf CJ, Bean BP. Permeation and block of TRPV1 channels by the cationic lidocaine derivative QX-314. *J Neurophysiol*. 2013;109(7):1704–1712.
68. Siemens J, Zhou S, Piskorowski R, et al. Spider toxins activate the capsaicin receptor to produce inflammatory pain. *Nature*. 2006;444(7116):208–212.
69. Bohlen CJ, Priel A, Zhou S, King D, Siemens J, Julius D. A bivalent tarantula toxin activates the capsaicin receptor, TRPV1, by targeting the outer pore domain. *Cell*. 2010;141(5):834–845.
70. Bohlen CJ, Chesler AT, Sharif-Naeini R, et al. A heteromeric Texas coral snake toxin targets acid-sensing ion channels to produce pain. *Nature*. 2011;479(7373): 410–414.
71. Caterina MJ, Schumacher MA, Tominaga M, Rosen TA, Levine JD, Julius D. The capsaicin receptor: a heat-activated ion channel in the pain pathway. *Nature*. 1997;389(6653):816–824.

72. Tominaga M, Caterina MJ, Malmberg AB, et al. The cloned capsaicin receptor integrates multiple pain-producing stimuli. *Neuron*. 1998;21(3):531–543.
73. Bhave G, Zhu W, Wang H, Brasier DJ, Oxford GS. Gereau RWt. cAMP-dependent protein kinase regulates desensitization of the capsaicin receptor (VR1) by direct phosphorylation. *Neuron*. 2002;35(4):721–731.
74. Ahern GP, Brooks IM, Miyares RL, Wang XB. Extracellular cations sensitize and gate capsaicin receptor TRPV1 modulating pain signaling. *J Neurosci*. 2005;25(21): 5109–5116.
75. Li D, Ren Y, Xu X, Zou X, Fang L, Lin Q. Sensitization of primary afferent nociceptors induced by intradermal capsaicin involves the peripheral release of calcitonin gene-related Peptide driven by dorsal root reflexes. *J Pain*. 2008;9(12): 1155–1168.
76. Ji RR, Samad TA, Jin SX, Schmoll R, Woolf CJ. p38 MAPK activation by NGF in primary sensory neurons after inflammation increases TRPV1 levels and maintains heat hyperalgesia. *Neuron*. 2002;36(1):57–68.
77. Drew LJ, Rohrer DK, Price MP, et al. Acid-sensing ion channels ASIC2 and ASIC3 do not contribute to mechanically activated currents in mammalian sensory neurones. *J Physiol*. 2004;556(pt 3):691–710.
78. Tobin DM, Bargmann CI. Invertebrate nociception: behaviors, neurons and molecules. *J Neurobiol*. 2004;61(1):161–174.
79. Tracey Jr WD, Wilson RI, Laurent G, Benzer S. painless, a Drosophila gene essential for nociception. *Cell*. 2003;113(2):261–273.
80. Howard J, Hudspeth AJ. Compliance of the hair bundle associated with gating of mechanoelectrical transduction channels in the bullfrog's saccular hair cell. *Neuron*. 1988;1(3):189–199.
81. Crawford AC, Evans MG, Fettiplace R. Activation and adaptation of transducer currents in turtle hair cells. *J Physiol*. 1989;419:405–434.
82. Holton T, Hudspeth AJ. The transduction channel of hair cells from the bull-frog characterized by noise analysis. *J Physiol*. 1986;375:195–227.
83. Chelur DS, Ernstrom GG, Goodman MB, et al. The mechanosensory protein MEC-6 is a subunit of the C. elegans touch-cell degenerin channel. *Nature*. 2002;420(6916): 669–673.
84. Goodman MB, Ernstrom GG, Chelur DS, O'Hagan R, Yao CA, Chalfie M. MEC-2 regulates C. elegans DEG/ENaC channels needed for mechanosensation. *Nature*. 2002;415(6875):1039–1042.
85. Sukharev SI, Blount P, Martinac B, Blattner FR, Kung C. A large-conductance mechanosensitive channel in *E. coli* encoded by mscL alone. *Nature*. 1994;368(6468): 265–268.
86. Maingret F, Patel AJ, Lesage F, Lazdunski M, Honore E. Mechano- or acid stimulation, two interactive modes of activation of the TREK-1 potassium channel. *J Biol Chem*. 1999;274(38):26691–26696.
87. Patel AJ, Honore E, Maingret F, et al. A mammalian two pore domain mechano-gated S-like K+ channel. *EMBO J*. 1998;17(15):4283–4290.
88. Werner G, Mountcastle VB. Neural activity in mechanoreceptive cutaneous afferents: stimulus-response relations, weber functions, and information transmission. *J Neurophysiol*. 1965;28:359–397.
89. Faucherre A, Nargeot J, Mangoni ME, Jopling C. piezo2b regulates vertebrate light touch response. *J Neurosci*. 2013;33(43):17089–17094.
90. Lumpkin EA, Caterina MJ. Mechanisms of sensory transduction in the skin. *Nature*. 2007;445(7130):858–865.

91. Maksimovic S, Baba Y, Lumpkin EA. Neurotransmitters and synaptic components in the Merkel cell-neurite complex, a gentle-touch receptor. *Ann N Y Acad Sci*. 2013; 1279:13–21.
92. Maksimovic S, Nakatani M, Baba Y, et al. Epidermal Merkel cells are mechanosensory cells that tune mammalian touch receptors. *Nature*. 2014;509(7502):617–621.
93. Maricich SM, Wellnitz SA, Nelson AM, et al. Merkel cells are essential for light-touch responses. *Science*. 2009;324(5934):1580–1582.
94. Nakatani M, Maksimovic S, Baba Y, Lumpkin EA. Mechanotransduction in epidermal Merkel cells. *Pflugers Arch*. 2015;467(1):101–108.
95. Woo SH, Ranade S, Weyer AD, et al. Piezo2 is required for Merkel-cell mechanotransduction. *Nature*. 2014;509(7502):622–626.
96. Chalfie M. Neurosensory mechanotransduction. *Nat Rev Mol Cell Biol*. 2009;10(1): 44–52.
97. Vasquez V, Sotomayor M, Cordero-Morales J, Schulten K, Perozo E. A structural mechanism for MscS gating in lipid bilayers. *Science*. 2008;321(5893):1210–1214.
98. Wang W, Black SS, Edwards MD, et al. The structure of an open form of an E. coli mechanosensitive channel at 3.45 A resolution. *Science*. 2008;321(5893): 1179–1183.
99. Zhang Z, Kindrat AN, Sharif-Naeini R, Bourque CW. Actin filaments mediate mechanical gating during osmosensory transduction in rat supraoptic nucleus neurons. *J Neurosci*. 2007;27(15):4008–4013.
100. Zhang Z, Bourque CW. Amplification of transducer gain by angiotensin II-mediated enhancement of cortical actin density in osmosensory neurons. *J Neurosci*. 2008; 28(38):9536–9544.
101. Prager-Khoutorsky M, Khoutorsky A, Bourque CW. Unique interweaved microtubule scaffold mediates osmosensory transduction via physical interaction with TRPV1. *Neuron*. 2014;83(4):866–878.
102. Assad JA, Shepherd GM, Corey DP. Tip-link integrity and mechanical transduction in vertebrate hair cells. *Neuron*. 1991;7(6):985–994.
103. Sofat N, Ejindu V, Kiely P. What makes osteoarthritis painful? The evidence for local and central pain processing. *Rheumatology*. 2011;50(12):2157–2165.
104. Boettger MK, Weber K, Grossmann D, et al. Spinal tumor necrosis factor alpha neutralization reduces peripheral inflammation and hyperalgesia and suppresses autonomic responses in experimental arthritis: a role for spinal tumor necrosis factor alpha during induction and maintenance of peripheral inflammation. *Arthritis Rheum*. 2010;62(5): 1308–1318.
105. Schuelert N, McDougall JJ. Involvement of Nav 1.8 sodium ion channels in the transduction of mechanical pain in a rodent model of osteoarthritis. *Arthritis Res Ther*. 2012;14(1):R5.
106. Sagar DR, Burston JJ, Hathway GJ, et al. The contribution of spinal glial cells to chronic pain behaviour in the monosodium iodoacetate model of osteoarthritic pain. *Mol Pain*. 2011;7:88.
107. Upadhyay J, Baker SJ, Rajagovindan R, et al. Pharmacological modulation of brain activity in a preclinical model of osteoarthritis. *Neuroimage*. 2013;64:341–355.
108. Rodriguez-Raecke R, Niemeier A, Ihle K, Ruether W, May A. Brain gray matter decrease in chronic pain is the consequence and not the cause of pain. *J Neurosci*. 2009;29(44):13746–13750.
109. Dib-Hajj SD, Cummins TR, Black JA, Waxman SG. Sodium channels in normal and pathological pain. *Annu Rev Neurosci*. 2010;33:325–347.
110. Priest BT. Future potential and status of selective sodium channel blockers for the treatment of pain. *Curr Opin Drug Discov Devel*. 2009;12(5):682–692.

111. Waxman SG. Painful Na-channelopathies: an expanding universe. *Trends Mol Med.* 2013;19(7):406–409.
112. Brenn D, Richter F, Schaible HG. Sensitization of unmyelinated sensory fibers of the joint nerve to mechanical stimuli by interleukin-6 in the rat: an inflammatory mechanism of joint pain. *Arthritis Rheum.* 2007;56(1):351–359.
113. Feng X, Zhou YL, Meng X, et al. Hydrogen sulfide increases excitability through suppression of sustained potassium channel currents of rat trigeminal ganglion neurons. *Mol Pain.* 2013;9:4.
114. Acosta C, Djouhri L, Watkins R, Berry C, Bromage K, Lawson SN. TREK2 expressed selectively in IB4-binding C-fiber nociceptors hyperpolarizes their membrane potentials and limits spontaneous pain. *J Neurosci.* 2014;34(4):1494–1509.
115. Pollema-Mays SL, Centeno MV, Ashford CJ, Apkarian AV, Martina M. Expression of background potassium channels in rat DRG is cell-specific and down-regulated in a neuropathic pain model. *Mol Cell Neurosci.* 2013;57:1–9.
116. Wang JG, Strong JA, Xie W, Zhang JM. Local inflammation in rat dorsal root ganglion alters excitability and ion currents in small-diameter sensory neurons. *Anesthesiology.* 2007;107(2):322–332.
117. Lolignier S, Amsalem M, Maingret F, et al. Nav1.9 channel contributes to mechanical and heat pain hypersensitivity induced by subacute and chronic inflammation. *PLoS One.* 2011;6(8):e23083.
118. Qiu F, Jiang Y, Zhang H, Liu Y, Mi W. Increased expression of tetrodotoxin-resistant sodium channels Nav1.8 and Nav1.9 within dorsal root ganglia in a rat model of bone cancer pain. *Neurosci Lett.* 2012;512(2):61–66.
119. Cummins TR, Dib-Hajj SD, Black JA, Akopian AN, Wood JN, Waxman SG. A novel persistent tetrodotoxin-resistant sodium current in SNS-null and wild-type small primary sensory neurons. *J Neurosci.* 1999;19(24):RC43.
120. Waxman SG. The molecular pathophysiology of pain: abnormal expression of sodium channel genes and its contributions to hyperexcitability of primary sensory neurons. *Pain.* 1999;suppl 6:S133–S140.
121. Hao J, Padilla F, Dandonneau M, et al. Kv1.1 channels act as mechanical brake in the senses of touch and pain. *Neuron.* 2013;77(5):899–914.
122. Alloui A, Zimmermann K, Mamet J, et al. TREK-1, a K+ channel involved in polymodal pain perception. *EMBO J.* 2006;25(11):2368–2376.

CHAPTER FOUR

Sensory TRP Channels: The Key Transducers of Nociception and Pain

Aaron D. Mickle*[,‡,1], Andrew J. Shepherd*[,‡,1], Durga P. Mohapatra*[,†,‡,2]

*Department of Pharmacology, The University of Iowa Roy J. and Lucile A. Carver College of Medicine, Iowa City, Iowa, USA

†Department of Anesthesia, The University of Iowa Roy J. and Lucile A. Carver College of Medicine, Iowa City, Iowa, USA

‡Department of Anesthesiology, Washington University School of Medicine, St. Louis, Missouri, USA

[2]Corresponding author: e-mail address: mohapatrad@anest.wustl.edu

Contents

1. Introduction 74
2. Ion Channels in the TRPV Subfamily 78
   2.1 Transient receptor potential vanilloid 1 78
   2.2 Transient receptor potential vanilloid 2, 3, and 4 91
3. Ion Channels in the TRPM Subfamily 93
   3.1 Transient receptor potential melastatin 3 93
   3.2 Transient receptor potential melastatin 8 94
4. Ion Channels in the TRPA Subfamily 98
   4.1 Transient receptor potential ankyrin 1 98
Acknowledgments 105
References 105

Abstract

Peripheral detection of nociceptive and painful stimuli by sensory neurons involves a complex repertoire of molecular detectors and/or transducers on distinct subsets of nerve fibers. The majority of such molecular detectors/transducers belong to the transient receptor potential (TRP) family of cation channels, which comprise both specific receptors for distinct nociceptive stimuli, as well as for multiple stimuli. This chapter discusses the classification, distribution, and functional properties of individual TRP channel types that have been implicated in various nociceptive and/or painful conditions.

[1] These authors contributed equally to this work.

Progress in Molecular Biology and Translational Science, Volume 131
ISSN 1877-1173
http://dx.doi.org/10.1016/bs.pmbts.2015.01.002

1. INTRODUCTION

The neural mechanisms and pathways for the encoding and processing of noxious stimuli, widely termed "nociception,"[1] constitute the physiological and/or pathophysiological bases of the somatic and trigeminal sensory system. The nociceptive neurons in different sensory ganglia send peripheral afferents to the somatic, visceral, and trigeminal regions and also connect to the spinal cord (SC) and brain stem, thereby serving as the mediator of sensory signal transmission between the peripheral and central nervous systems (PNS and CNS). These neurons express a variety of receptors and ion channels in the plasma membrane throughout the soma and fibers, in order to detect various noxious stimuli, convert those to electrical signals, and subsequent transmission to the CNS. The transient receptor potential (TRP) family of receptor ion channels constitutes the major class of detectors and transducers in nociceptive neurons. Members of the TRP channel superfamily are Ca^{2+}-permeable, nonselective cation channels found in metazoans and fungi (reviewed in Refs. 2,3). Broadly speaking, they act as primary transducers of multiple noxious stimuli, although this belies the diversity of functions they play in eliciting physiological responses to changes in the environment, encompassing the five traditional somatosensory modalities (vision, hearing, olfaction, taste, and touch), as well as sensing changes in temperature and osmolarity (reviewed in Refs. 3–5). The TRP terminology is derived from the *Drosophila* mutant of the same name, which displays an abnormally transient response to bright light, a defect in the prototypical channel, and its role in phototransduction.[6,7] In the decades since this initial discovery, the superfamily has grown to include 28 members across 6 subfamilies in mammals, categorized as canonical (TRPC), vanilloid (TRPV), ankyrin (TRPA), melastatin (TRPM), polycystin (TRPP), and mucolipin (TRPML; reviewed in Refs. 2,3). To date, members of three of these subfamilies have been implicated in the sensory detection transduction of nociception and pain: TRPV, TRPA, and TRPM (Fig. 1).

TRP channels exhibit enormous structural and functional diversity, albeit there are several common features that unite them under one family. Functional TRP channels consist of tetramers of 6-transmembrane (6-TM) segment polypeptide subunits, similar to that of the α-subunits of voltage-gated K^+ (K_v) channels. A hydrophilic loop between TM segments S5 and S6 of each subunit constitutes the ion conduction pore domain in tetrameric

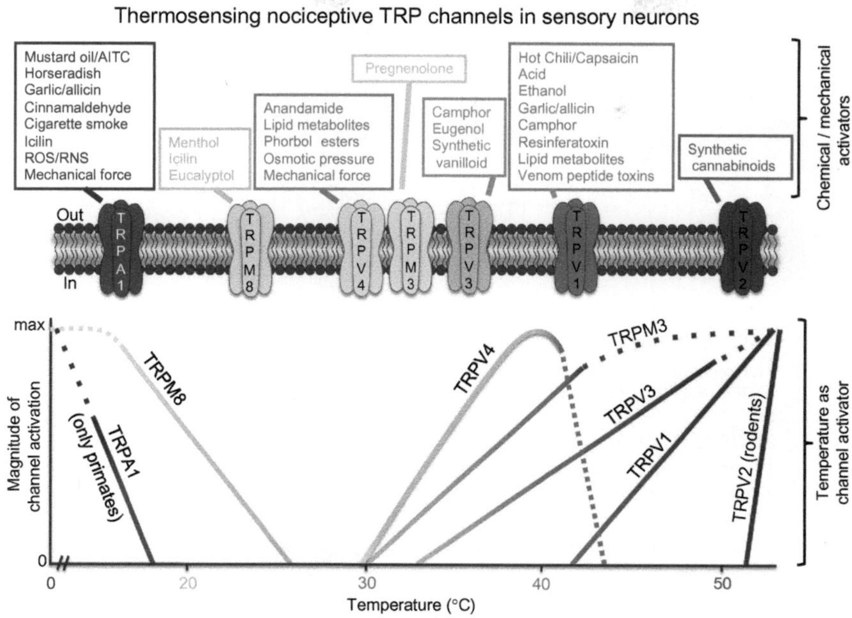

Figure 1 Classification of thermosensing nociceptive TRP channels in mammalian sensory neurons. The upper row of individual boxes denotes chemical/mechanical activators of marked TRP channels. The lower panel depicts the magnitude of channel activity upon activation by temperature of the independent TRP channels shown. (See the color plate.)

channels (called the pore loop; Fig. 2). The precise positioning of amino acid residues within the pore-loop region dictates the ion selectivity of individual channels for various cations. Although the majority of TRP channels conduct cations in an outwardly rectifying manner, a number of individual channels exhibit linear (e.g., TRPC2 and TRPM2) or inwardly rectifying cation conduction (e.g., TRPV6). TRP channels are generally nonselective cation channels; however, there is considerable variability in divalent cation (Ca^{2+} and Mg^{2+}) selectivity between channels, with TRPV5, TRPV6, and the majority of TRPC family channels exhibiting very high selectivity for Ca^{2+} (reviewed in Refs. 3,10,11).

By far the greatest level of sequence diversity among TRP channels has been observed in the cytoplasmic N- and C-termini. The N-terminus often contains ankyrin repeat domains (ARDs; particularly in TRPC, TRPV, and TRPA channels; Fig. 3), the precise role of which in channel assembly and function is not well understood. The so-called TRP box, a 6-residue sequence in the proximal C-terminus, is also a conserved motif among the members of TRPV, TRPC, TRPM, and TRPA families (Fig. 3).

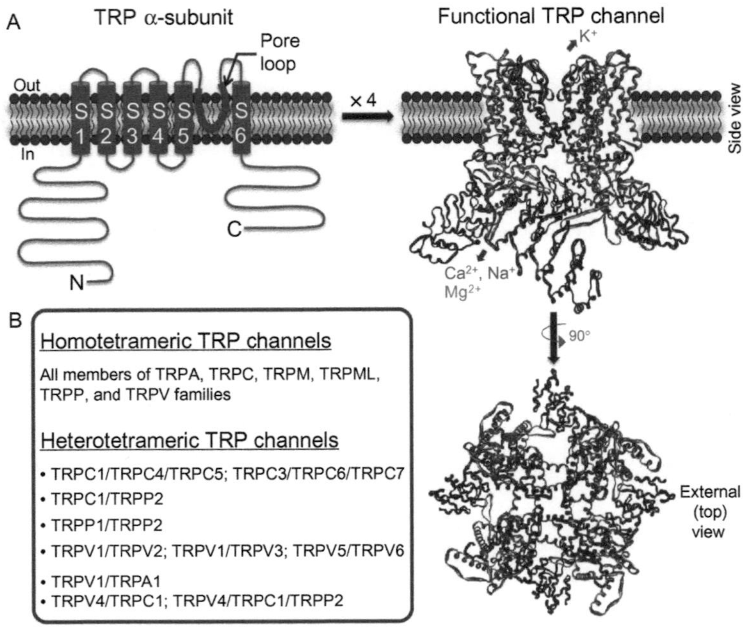

Figure 2 Structural features of functional TRP channels. (A) Model of predicted secondary structure topology of a monomeric TRP channel subunit (called α-subunits) on a plasma membrane lipid bilayer. S1–S6 denote transmembrane segments/domains 1 through 6, with a hydrophilic pore loop between S5 and S6 transmembrane segments, and N and C denote cytoplasmic amino- and carboxy-terminals, respectively. Four monomeric α-subunits assemble in a three-dimensional arrangement to make a functional channel, which is shown on the right. The structure was constructed from the PDB structure file obtained from the protein data bank, originally submitted by Refs. 8,9 from their cryoelectron microscopic structure of TRPV1. (B) List of homo- and heterotetrameric TRP channels formed within individual TRP subfamily member channels, as well as in combination of inter-subfamily TRP channel members. (See the color plate.)

Several studies have presented evidence that this region, at least in TRPM and TRPV channels, is crucial for the coupling of ligand sensing to channel gating.[12] In addition, there are a number of domains/motifs, such as coiled-coil domain, kinase domain, NUDIX domain, EF hand Ca^{2+}-binding motif, CIRB motif, and PDZ motif in the N- and C-termini of various TRP channels (Fig. 3). Despite the structural resemblance to classical voltage-gated Na^{+}, Ca^{2+} (Na_v, Ca_v) and K_v channels, TRP channels are only weakly voltage sensitive, presumably due to a relative paucity of positively charged residues in the TM segment S4. However, it has been proposed that modulation of voltage sensitivity of TRPV1 and TRPM8 channels occurs in response to changes in temperature and agonist exposure.[13] Such data

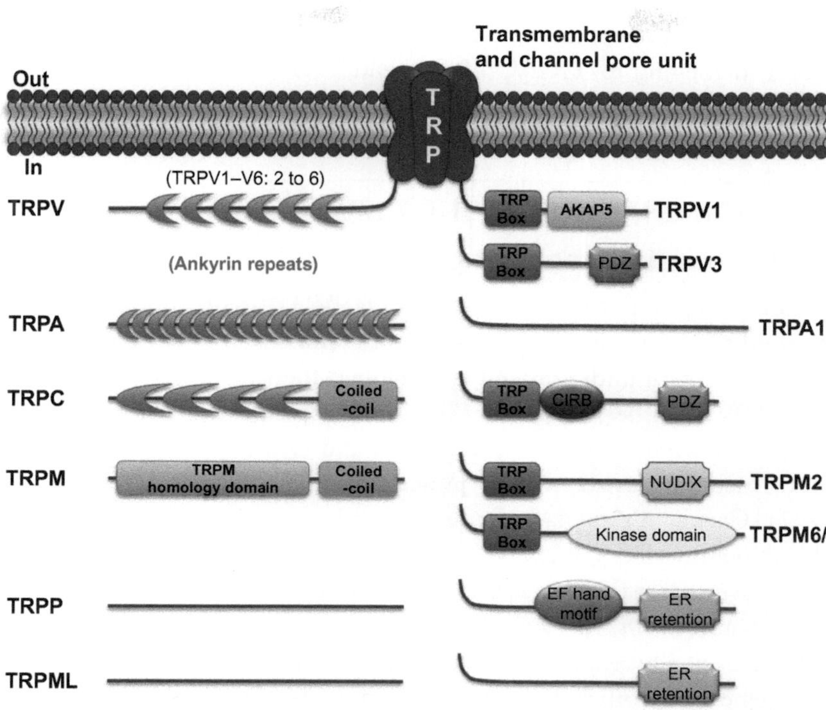

Figure 3 Structural–functional features of cytoplasmic regions in TRP channels from individual subfamilies.

serve to highlight the broad range of environmental stimuli that are sensed by these channels, and the enormous technical challenge that has been encountered in attempting to understand the function and modulation of these channels.

Most of the functional TRP channels are homotetramers of an individual channel member, although a number of reports have also suggested the formation of heterotetrameric TRP channels. Several members of TRPC, TRPV, TRPM, and TRPP families form heterotetramers (Fig. 2) with distinct functional properties, when expressed in mammalian cell lines and *Xenopus* oocytes.[11,14–18] However, the existence of such heterotetrameric channels in sensory neurons *in vivo* and their physiological relevance is elusive. Although activation of nociceptive TRP channels by specific noxious and/or pain-producing stimuli constitutes the principal detection for nociception and pain under physiological conditions, modulation of channel activation/inactivation, intracellular trafficking/targeting, and protein expression is highly critical for nociception and pain under pathophysiological conditions. This chapter presents a summary of the expression, distribution, and structure–function information of key nociceptive TRP channels belonging to TRPV, TRPA, and TRPM subfamilies, as well as their role in various nociception and pain conditions.

2. ION CHANNELS IN THE TRPV SUBFAMILY

2.1 Transient receptor potential vanilloid 1

Transient receptor potential vanilloid 1 (TRPV1), also previously known as "vanilloid receptor subtype-1" (VR1), is the most well-characterized TRP channel and the founding member of cloned thermosensitive TRP channels. TRPV1 can be activated by noxious temperatures and acidic pH, as well as by a wide variety of endogenous and exogenous algogenic compounds including capsaicin, the pungent ingredient in hot chili peppers (Fig. 1). Capsaicin's excitatory effects on neurotransmission, nociception, and pain were well documented before the molecular identity of the receptor was revealed, as far back as the 1960s with most of the work reported in the 1980s and 1990s. In 1997, the molecular signature of the receptor was finally identified by expression cloning of the TRPV1 gene using a cDNA library generated from rodent sensory neurons, and functional identification was performed with dye-based Ca^{2+} imaging upon capsaicin and heat activation.[19] Subsequent characterization revealed that TRPV1 in sensory neurons (and also under heterologous expression) could be activated by multiple noxious stimuli (see Fig. 1 for agonists of TRPV1; reviewed in Ref. 20). Follow-up investigations utilizing *in vivo* pharmacological approaches and TRPV1 gene knockdown in mice suggested the critical role of this channel in inflammatory pain conditions (reviewed in Ref. 20). This key information, along with the availability and development of specific physicochemical agonists (and antagonists) of TRPV1, led to a massive expansion in studies aimed at characterizing the channel and its role in nociception and various pain states.

2.1.1 Expression and distribution in nervous system

TRPV1 is predominantly expressed in the neurons of the PNS and CNS (reviewed in Ref. 20). In the PNS, TRPV1 is primarily expressed in small- and medium-diameter nociceptive neurons in the dorsal root, nodose, sympathetic, and trigeminal ganglia (DRG, NG, SG, and TG, respectively), such as peptidergic and nonpeptidergic C-fibers, as well as in some Aδ fibers.[21–26] TRPV1 expression has also been detected in sensory nerve fibers innervating bladder, lung/airways, and cochlea, where it is involved in bladder function, sensing airway irritants, cough reflex, and hearing.[27]

In addition to the PNS, TRPV1 is also expressed in different CNS regions, specifically in the laminae I and II (and in the laminae V and X to some extent) of spinal cord (SC) dorsal horn, where it plays an integral role in synaptic transmission/modulation of peripheral nociceptive signal input.[28,29] The majority of TRPV1 expressed in SC dorsal horn is presynaptic on the central axon terminals; however, there is also evidence of postsynaptic expression of TRPV1 in some neurons of SC dorsal horn, although their functionality and role in nociception are not understood.[30] Presynaptic TRPV1 can act to modulate synaptic transmission in different inflammatory models and conditions.[31] Genetically engineered mice with expression of fluorescent reporters, such as eukaryotic yellow fluorescent protein and tdTomato, under the control of TRPV1 promoter led to more convincing information on TRPV1 expression in different regions of the nervous system. Fluorescence labeling was prominent in small- and medium-diameter sensory nerve fibers in the skin, cornea, and bladder; small- and medium-diameter neurons in the DRG and TG; SC dorsal horn and trigeminal tract; and in some restricted regions in the brain, such as the brain stem, nucleus of the solitary tract, nucleus caudalis, nucleus ambiguous, olfactory bulb, and parabrachial nucleus.[32,33] Also, some scattered neurons in the cortex exhibited fluorescence labeling, and other major regions in the forebrain, such as hypothalamus and hippocampus, did not show any detectable signal in these mice, suggesting no TRPV1 expression in these brain regions.[32,33] In addition to PNS and CNS, TRPV1 mRNA (in some instances functional channels) has also been detected in a number of nonneuronal tissues such as bladder, lungs, keratinocytes, various cell types in dental tissue, fibroblasts, mast cells, hair follicles, blood vessels, testis, and ovary; however, some degree of controversy still remains regarding the extent of expression and function of TRPV1 in these tissues (reviewed in Refs. 20,34).

2.1.2 Structure

Secondary structure prediction of cloned TRPV1 suggested a 6-TM polypeptide, with a hydrophilic pore loop between S5 and S6, as well as tetrameric assembly (Fig. 2) resembling to the α-subunit of K_v channels.[19] Subsequently, the first evidence of tetrameric assembly of TRPV1 came from biochemical experiments on heterologously expressed recombinant TRPV1.[35] Accurately resolving the structures of complex membrane proteins is technically challenging, and TRP channels have proven no exception. The first structural information on TRPV1 came from the X-ray crystallographic structure of the ARD at the cytoplasmic N-terminus of the channel protein, suggesting six ARDs, and their role in adenosine triphosphate (ATP) and calmodulin (CaM) binding.[36] Subsequently, the three-dimensional (3D) structure of purified recombinant TRPV1 channel protein expressed in yeast was determined by using single-particle electron cryo-microscopy (cryo-EM) at a resolution of 19 Å. This structure revealed a tetrameric channel conformation with bulky open basket-like domain formed by the cytoplasmic N- and C-termini, and a compact membrane domain that has several resemblances to the transmembrane domain architectures of K_v channels that were determined by X-ray crystallography.[37] More recently, high-resolution cryo-EM allowed the determination of tetrameric TRPV1 structure at a near-atomic resolution, 3.4 Å, exhibiting a four-fold symmetrical TM segments formed by S1–S4 of each subunit that surrounds the central ion conduction pathway constituted by S5, the pore loop, and S6 (Fig. 2).[8,9] Structural and functional predictions based on prior evidence were largely confirmed by this new structural model, overall suggesting that TRPV1 does indeed bear a structural resemblance to K_v channels.[38] However, unlike K_v channels, the S4 segment does not demonstrate strongly voltage-dependent movement. Rather there appear to be two distinct "gating" regions: the outer one near the pore helix at the inner selectivity filter region of the channel and the inner gate closer to the cytoplasmic side, which upon vanilloid binding leads to shifting of the lower S6 and S4–S5 linker regions to open the channel.[9,39]

2.1.3 Functional properties of the channel

TRPV1 is the best example of a polymodal receptor channel that can be physicochemically activated by a large number of intracellular and extracellular ligands from both endogenous and exogenous sources (Fig. 1). Endogenous chemical ligands that can directly activate TRPV1 channel include extracellular acidic environment (~pH 6.0 or less),[19,24,40] intracellular basic environment (~pH 7.8 or more)[41], and several endovanilloids and endocannabinoids that are generated by various lipid metabolism pathways under inflammatory conditions.[20,42–44] Most of these are fatty acid derivatives, including amines (anandamide, N-arachidonoyldopamine, N-oleoylethanolamine, N-arachidonolylserine), oxygenated eicosatetraenoic acids (lipoxygenase products and their hydroxylic analogs, prostaglandins and leukotrienes), and lysophosphatidic acid.[20,45,46] Reactive oxygen and nitrogen species (ROS and RNS) have also been suggested to activate

TRPV1 channel directly, although most of these findings are restricted to recombinant channels expressed in heterologous system, and require further confirmation in native sensory neurons (reviewed in Ref. 47). Noxious temperature (>42 °C) constitutes the endogenous physical/thermal ligand for TRPV1 (reviewed in Ref. 20).

Capsaicin, the pungent ingredient of hot chili peppers, is the first known exogenous ligand for TRPV1, and in fact, this knowledge was highly critical for the molecular identification and expression cloning of TRPV1 from mammalian sensory neurons.[19] In addition to capsaicin, there are a number of exogenous compounds that activate TRPV1. Other pungent plant products that can activate TRPV1 include piperine, found in black pepper; allicin, found in garlic; and eugenol, found in clove oil (reviewed in Ref. 20). In addition to these compounds, there are also a number of plant and animal toxins that target TRPV1. Resiniferatoxin (RTX), derived from the cactus *Euphorbia resinifera*, is a highly potent agonist of TRPV1 (3–4 times more than capsaicin).[48] There are a number of potent spider toxins that bind and activate TRPV1 including the vanillotoxins (VaTx)-1, 2, 3, and "double-knot toxin" (DkTx).[49,50] In addition to these algogenic compounds of plant and animal toxins, TRPV1 can also be activated by ethanol.[51]

Extensive structure–function studies on recombinant TRPV1 channels have led to the identification of molecular determinants of channel activation by specific ligands. These studies utilized site-directed mutagenesis and chimeric TRPV1/2/M8 channel generation/expression approaches with altered TRPV1 regions and/or single amino acid substitution approaches. Extracellular acidic pH activates TRPV1 by protonation of extracellular residues E600 and E648, whereas intracellular alkalization activates the channel that is dependent on the residue H378.[40,41] Capsaicin, RTX and most of the endovanilloids and endocannabinoids, which easily penetrate through the plasma membrane, due to their lipophilic nature, activate the channel from the intracellular side. With the cloning of TRPV1 gene from multiple mammalian species, it was found that the channel in birds is insensitive to capsaicin, whereas the channel in rabbit is mildly sensitive to capsaicin, without any alteration in channel activation by acidic pH and noxious temperature. Based on this information, and subsequent experimentation on mutational/chimeric channels, the capsaicin (and RTX) activation of TRPV1 was localized to regions between the TM S3–S4 segments of the protein, specifically involving residues Y511, S512, M547, and T550.[52,53] On the other hand, the toxins VaTx and DkTx activate TRPV1 by directly binding to the external region of the pore loop and lead to prolonged and irreversible channel open durations.[49] With the recent unraveling of high-resolution cryo-EM structure of TRPV1, without and with RTX-/DkTx-bound conditions, it became clear that the RTX-binding pocket in the S3–S4 region involves the residues Y511, S512, M547, and T550, and this pocket sits just above the residue E570 in the S4–S5 linker, with proximity to residue L669 in the S6 of neighboring subunit.[8] The S4–S5 linker and the nearby proline-valine-proline (PVP) region in the S6 segment of $K_v1.2$ channel have been established to play the critical role of a channel "gate" in response to voltage-dependent movement of S4 segment.[54,55] Therefore, it is suggested that RTX (or capsaicin) binding pulls the S4–S5 linker of one subunit, which then twists the S6 segment of the neighboring subunit to open the lower gate, and subsequently the upper gate of the channel.[8] In contrast, the DkTx interacts with TRPV1 at the extracellular pore region by intercalating between the top of the pore helix of one subunit and the outer pore-loop proximal to the S6 segment of the neighboring subunit, thereby locking the channel in the open conformation following activation.[8] However, the precise mechanism by which DkTx binding leads to the displacement/opening of both the upper and lower gate of the channel is not well understood from these studies. Regarding heat activation of TRPV1 channels, several studies have convincingly showed that this channel is intrinsically heat sensitive, and increased temperatures shift the voltage dependence of channel activation to physiological membrane potentials, which otherwise do not undergo voltage-dependent activation at these potentials.[39,56] Subsequent studies revealed that the pore region is critical to channel activation at noxious temperatures, and prolonged heat activation leads to pore dilation.[57] However, the precise structural mechanisms that lead to channel gate opening upon exposure to noxious temperatures still remain elusive. Other intracellular ligands, such as ATP and allicin, are shown to activate the channel by interacting with the ARD in cytoplasmic N-terminus of the channel protein; specifically, ATP intercalates between fingers 1–2 of ARD, whereas allicin interacts with the C157 residue in the same region.[36,41]

Upon ligand binding (or exposure to noxious temperatures), the channel gate opens, and under physiological conditions, it allows the influx of Na^+ and Ca^{2+} through the central ion conduction pore, with relatively high permeability for Ca^{2+} (three- to nine-fold).[10,19,57,58] The residues D636 in the pore region, and Y671 in the S6 segment have been suggested to be the critical determinants of high relative Ca^{2+} permeability of

TRPV1 channel. Rapid influx of Na^+ and Ca^{2+} through TRPV1 results in depolarization of neuronal plasma membrane potential and opening of Na_v/Ca_v channels, which leads to action potential (AP) firing. Immediately following activation, the TRPV1 channel undergoes desensitization in the presence of the ligand (analogous to voltage-dependent inactivation of $Na_v/Ca_v/K_v$ channels), which results in diminished firing upon prolonged or subsequent TRPV1 activation.[19,24,58–60] Desensitization of TRPV1 under physiological conditions is dependent on Ca^{2+} influx through the channel,[58,59] which presumably operates via activation of the protein phosphatase 2B (PP2B), leading to channel dephosphorylation and desensitization.[61,62] Additionally, Ca^{2+}/CaM binding and interaction of phosphatidylinositol-bis-phosphate (PIP2) with the N-/C-termini of TRPV1 protein have also shown to be critical structural determinants for channel desensitization.[36,63] Although the precise structural mechanisms underlying TRPV1 channel desensitization following activation are not well understood, the residue Y671 in the S6 segment has been proposed to be involved in the structural arrangements in the channel pore that governs channel desensitization.[58] In fact, the recently solved high-resolution cryo-EM structure of TRPV1 shows the Y671 residues in the inner pore cavity[9]; however, the precise structural changes surrounding this residue upon prolonged activation/desensitization awaits further exploration.

2.1.4 Modulation of channel expression and function

Activation of TRPV1 in peripheral sensory nerve fibers not only leads to AP firing but also releases of neuropeptides, such as calcitonin gene-related peptide (CGRP) and neurokinins or substance P (SP). Sustained TRPV1 activation in these fibers leads to increased CGRP and SP release, which leads to vasodilatation and activation of a variety of immune and other cell types in the skin, leading to pro-inflammatory mediator release that result in a positive signaling feedback loop causing potentiation of TRPV1 channel activation and nociceptive signaling (reviewed in Ref. 20). At the same time, massive Ca^{2+} influx resulting from prolonged/episodic TRPV1 activation leads to increased expression of several nociceptive and other genes. Tissue injury and inflammatory conditions have been shown to increase TRPV1 expression in sensory neurons (both at mRNA and protein level), as well as increase/recruitment of functional TRPV1-expressing sensory neurons (reviewed in Ref. 20). More recent studies have also shown that proinflammatory mediators like interleukin-6 induce rapid translation of TRPV1 mRNA at peripheral sensory fibers, thereby increasing the magnitude of nociceptor excitation.[64,65] Such modulation of TRPV1 gene transcription and translation has been suggested to underlie enhanced nociceptive/pain sensitivity (mainly thermal) *in vivo*.

The majority of TRPV1 modulation relates to functional modulation of the channel protein. Cross-sensitization or agonist-induced potentiation of channel activation properties by multiple agonists is one of the hallmarks of functional modulation in TRPV1. Under physiological conditions (~pH 7.3), TRPV1 in sensory neurons is activated at noxious temperatures (>42 °C). The temperature activation threshold of TRPV1 channel drops to ~35–37 °C upon extracellular acidification (~pH 7.0 to 6.5), a pH range that does not induce substantial channel activation at room temperature.[24] This phenomenon is critical under tissue injury and inflammation, where there is significant tissue acidosis in the inflamed microenvironment (~pH range 7.2 to 6.4), thereby leading to robust activation of TRPV1 at body temperature.[24] Similar cross-sensitization of TRPV1 has also been seen with capsaicin and temperature activation of the channel.[24,58] The amino acid residues E648 and Y671 have been identified as the molecular determinants of TRPV1 underlying such cross-sensitization and modulation of channel activation properties.[24,58] On the other hand, the channel undergoes diverse and extensive posttranslational modifications that leads to multiple levels of functional modulation TRPV1 that are widely observed under tissue injury, inflammation, and a wide variety of pathological conditions (reviewed in Ref. 20).

A wide variety of pro-inflammatory mediators, cytokines, chemokines, growth factors, pruritogens, ATP, lipid metabolism products, ROS, RNS, neurotransmitters, and neuropeptides are released at elevated levels during injury/inflammation/pathological conditions. These mediators activate their respective receptors expressed on sensory neuron plasma membrane, which mainly belong to the group of G protein-coupled receptors and growth factor tyrosine kinase receptors (TKs). Activation of these receptors leads to a wide variety of intracellular signaling pathways that result in the activation of (1) protein kinases such as protein kinase A (PKA), protein kinase C (PKC), mitogen-activated protein kinases (MAPKs), extracellular signal-regulated kinases, Ca^{2+}–CaM-activated protein kinases (CaMKs), cyclin-dependent kinase-5 (Cdk5), *Src* kinase, and phosphoinositide kinases (PI3/4/5Ks); (2) phospholipases such as phospholipase C-β (PLC-β) and PLC-γ; and (3) protein phosphatases such as protein phosphatase 1/2A (PP1/2A), PP2B, and protein tyrosine phosphatases. TRPV1 channel activity can be modulated by most of these kinases/phosphatases. TRPV1 protein has been shown to be directly phosphorylated at residues S6, S116, T144, T370, S502, S774, and S820 by PKA; residues S502, T704, and S800 by PKC; residue Y200 by Src; residues S502 and T704 by CaMKII; and residue T407 by Cdk5 (based on rat TRPV1 amino acid sequence).[66–73] Phosphorylation of TRPV1 protein by PKA and PKC has been shown to shift the thermal activation of the channel from 43 to 38–33 °C,[70,72] thereby opening the channel at body temperature, a mechanism that has been suggested to underlyie inflammatory thermal pain. In addition, TRPV1 protein phosphorylation by PKA and PKC shifts the proton activation of TRPV1 from a pH range of 5.8–6.2 to a pH range of 6.4–7.0, which also serves as a mechanism for nociceptor sensitization during tissue injury/inflammations.[60,66,74] On the other hand, phosphorylation of TRPV1 by CaMKII regulates capsaicin and other vanilloid binding to the channel protein.[69] In addition to direct modulation of channel activity by PKA/PKC phosphorylation of channel protein, activation of these kinases and p38 MAPK has been shown to upregulate the expression of TRPV1 protein,[75] which provides the neurons with a long-term sensitization mechanism. Furthermore, phosphorylation of TRPV1 by the *Src*-type TK increases the trafficking of TRPV1-containing vesicles to the plasma membrane, thereby providing the neurons with increased numbers of surface channel protein.[68,73,76] Collectively, activation of various protein kinases in sensory neurons upon activation of surface receptors for pro-inflammatory mediators, growth factors, and receptors sensitize nociceptors via (1) direct modulation of TRPV1 channel activation at body temperatures and the mildly acidic environment of inflamed/injured tissue, (2) increased number of TRPV1 channels on the plasma membrane, and (3) increased expression of TRPV1 proteins.

In addition to phosphorylation-dependent modulation, PLC regulation of TRPV1 channel activity has also been suggested to provide a molecular mechanism underlying nociceptor sensitization, although conflicting observations have been reported in this regard.[77–79] PLC has been shown to both activate and desensitize TRPV1. On one hand, PLC activation results in hydrolysis of PIP2 to inositol triphosphate and diacylglycerol, and then later activates PKC to sensitize TRPV1 (reviewed in Ref. 20). On the other hand, it has been shown that PIP2 constitutively binds to TRPV1 protein in the plasma membrane to inhibit channel opening, and hydrolysis of PIP2 by PLC releases the channel from constitutive inhibition, thereby enhancing channel opening.[77] In addition, results from other studies have shown that plasma membrane PIP2 is necessary for TRPV1 channel activation.[78] One major reason for these discrepancies could be the use of different cell types, conditions, and test parameters used therein, as well as the possible involvement of specific PLC isoforms in individual cell types,[79] but as a result a definitive understanding of PIP2 regulation of TRPV1 is currently lacking.

2.1.5 Channel desensitization

Under physiological conditions, activation of TRPV1 causes a Ca^{2+}-dependent desensitization of channel activation. This is an important physiological property of TRPV1 and is critical to the efficacy of TRPV1 agonists to induce pain. Removal of extracellular Ca^{2+} reduces desensitization elicited by capsaicin, but does not completely abolish desensitization[58,59]; a small part of TRPV1 desensitization is therefore Ca^{2+} independent. It has been suggested that Ca^{2+} entering into neurons via TRPV1 channels presumably binds to CaM, and this complex activates the protein phosphatase calcineurin (or PP2B), which dephosphorylates the channel protein, resulting in channel desensitization.[62] CaM has also been reported to physically interact with the cytoplasmic C-terminus of the channel protein, thereby regulating Ca^{2+}-dependent desensitization of TRPV1; however, it primarily affects desensitization to short/repeated agonist applications, but not to prolonged agonist applications.[63,80] Phosphorylation by PKA has been shown to reduce the Ca^{2+}-dependent desensitization of TRPV1.[66,74] In addition to Ca^{2+}/CaM, PIP2 has also been implicated in TRPV1 channel desensitization, although it remains a highly debated matter. Finally, long-term desensitization due to repeated activation of TRPV1 can activate endocytosis and liposomal degradation of the channel.[81]

2.1.6 Involvement in pain conditions

TRPV1 is involved or proposed to be involved in a wide variety of pain conditions including migraine, dental pain, cancer pain, inflammatory pain, neuropathic pain, visceral pain, and osteoarthritis. This section briefly describes the suspected role of TRPV1 in these disease states. The first and foremost evidence of the role of TRPV1 in peripheral pain processing came from observations on the attenuation of inflammatory thermal hyperalgesia behaviors in rodents injected with capsazepine, a competitive vanilloid antagonist of TRPV1 channel. Furthermore, mice lacking functional TRPV1 (*Trpv1*$^{-/-}$ mice) exhibited no change in response to noxious temperatures, but they did show dramatic attenuation of thermal hyperalgesia elicited by a number of inflammatory mediators.[82,83] These observations not only prompted the development of small-molecule antagonists of TRPV1 as new-generation efficacious analgesics but also made it possible to thorough investigate the critical role of TRPV1 in the detection/transduction of pain in several pathologies.

One of the best-studied pain modalities using animal models is inflammatory pain. With the strong evidence from *Trpv1*$^{-/-}$ mice exhibiting a lack of inflammatory thermal hyperalgesia, various rodent models of inflammatory pain, including the widely used complete Freund's adjuvant (CFA)-induced inflammation, were subsequently utilized to determine the role of TRPV1 therein.[82] Similarly, studies on *Trpv1*$^{-/-}$ mice revealed its significant role in neuropathic pain conditions, including diabetic neuropathy (reviewed in Refs. 20,84).

In animal models, both genetic ablation and pharmacological blockade of TRPV1 can reduce arthritis-like symptoms.[85,86] Additionally, a variant of TRPV1 has been correlated with increased risk of knee osteoarthritis, as evident from observations on increased expression of TRPV1 in knee synovium in patients with osteoarthritis.[86,87] Furthermore, pharmacological evidence also suggests the critical role of TRPV1 in arthritis and/or joint pain. Topical capsaicin creams have long been used to alleviate arthritis-like joint pain, and this treatment works by desensitizing and/or excitotoxic degeneration of TRPV1-positive nociceptive terminals (reviewed in Refs. 20,84). Additionally, *Trpv1*$^{-/-}$ mice exhibit no edema or hypersensitivity following joint inflammation, another piece of evidence suggesting the involvement of TRPV1 in the pathogenesis of arthritis-like inflammatory conditions.[85] Altogether, this supports a potentially critical role for TRPV1 in arthritis and joint pain conditions, and that both TRPV1 agonism and antagonism presumably constitute analgesic targets for these conditions.

TRPV1 has long been suspected to be involved in bone cancer pain, and in fact, pharmacological inhibition of TRPV1 significantly decreased pain associated with an animal model of primary bone cancer.[88,89] There are a number of different mediators found in the bone tumor microenvironment including prostaglandins, bradykinin, and nerve growth factor (NGF), which could presumably sensitize TRPV1 through posttranslational modifications, thereby leading to nociceptor sensitization.[89] In addition to these mediators, the acidic tumor microenvironment has also been suggested to enhance the proton activation of TRPV1, thereby providing yet another TRPV1-dependent mechanism for constitutive nociceptor sensitization and pain in bone cancers.[89]

The pathophysiology of migraine is not well understood, but a number of studies suggest the involvement of TRPV1. Anandamide, an endogenous endocannabinoid, can activate TRPV1 and increase CGRP release, leading to neurogenic inflammation and vasodilation, which are key elements to migraine.[90] Additionally, alcohol is known to increase migraine symptoms, and ethanol has been shown to both directly activate and potentiate the activation of TRPV1 channel.[91–93] Such activation/potentiation of TRPV1 could also lead to nociception as well as further release of CGRP to potentiate migraine symptoms.

TRPV1 activation at noxious temperatures and the ability of different inflammatory mediators to reduce the thermal activation of the channel at temperatures below body temperature also suggest its involvement in dental pain conditions. TRPV1 is highly expressed in a large percentage (45–85%) of primary sensory afferents that innervate the tooth pulp.[94,95] Recent studies have demonstrated that TRPV1 expression is upregulated in a rat model of pulpitis induced by lipopolysaccharide (LPS).[96] While the mechanistic evidence is weak, TRPV1 still remains an attractive target for pharmaceutical intervention of dental pain.[97]

Both pharmacological and genetic inhibition of TRPV1 has been shown to exhibit decreased responses to colorectal distention in mice before and after inflammation.[98] Additional studies have shown that TRPV1 inhibition can reduce such disease severity in animal models, and evidence suggests that it may play a critical role in both the initiation and the maintenance of visceral hypersensitivity after injury.[99] In patients with colorectal disease, TRPV1 expression correlates positively with clinical score of severity of colorectal function.[100] In support of these observations, *Trpv1*$^{-/-}$ mice exhibit reduced visceral pain, including reduced mechanical hypersensitivity in models of cystitis[101] and inflammatory bowel disease.[102,103] Additionally, in *Trpv1*$^{-/-}$ mice increased number and levels of inflammatory markers have been detected after bowel inflammation as compared to wild-type controls, which further complicates understanding of the precise role of TRPV1 in inflammatory bowel conditions.[104,105] TRPV1-positive sensory nerve fibers also innervate the bladder and are important for the enhancement of bladder contractility after inflammation.[106–108] In this context, ablation of TRPV1-positive neurons in mice with RTX can decrease bladder contractions in animals with bladder inflammation.[106,109] Altogether, this indicates that both TRPV1 agonists and/or antagonists may be useful in treating specific types of inflammatory bowel diseases, interstitial cystitis, and painful bladder syndromes.

2.1.7 Involvement in other physiological and pathological conditions

In addition to its critical role in nociception, TRPV1 has been suspected to play a role in the regulation of body temperature. Initial observations from

Trpv1$^{-/-}$ mice showing no alteration in body temperature as compared to their wild-type littermates suggested that TRPV1 might not be involved in body temperature regulation.[82,83] However, subsequent reports showed that systemic administration of small-molecule antagonists of TRPV1 leads to induction of transient hyperthermia,[110] which suggests a critical role for this channel in body temperature regulation. Lack of such observations in *Trpv1*$^{-/-}$ mice could be attributed to some as yet unidentified compensatory mechanisms that are introduced under genetic knockout of TRPV1. Peripheral TRPV1 expressed in sensory afferents was thought to be the critical component of peripheral detection of noxious temperatures. However, ablation of TRPV1-expressing cells in the SC dorsal horn with high dose of intrathecal capsaicin leads to a complete loss of responsiveness to noxious heat, suggesting the critical role of spinal TRPV1 in the detection of noxious temperatures.[21] Although TRPV1 is predominantly expressed in peripheral sensory afferents and SC dorsal horn, expression of TRPV1 mRNA and protein has also been reported in the brain and brain stem. In fact, the critical role of TRPV1 in thermoregulation has been suggested to be mediated in part by TRPV1 in the brain stem. Whether TRPV1 in brain plays any role in nociceptive recognition is yet to be elucidated. TRPV1 mRNA and/or protein is expressed in periaqueductal gray, insular cortex, somatosensory cortex, hippocampus, thalamus, cortex, and amygdala region of brain (reviewed in Ref. 111). The function of TRPV1 in brain neurons is somewhat varied depending on the brain regions studied, with several reports showing its involvement in long-term potentiation, long-term depression, and synaptic plasticity (reviewed in Ref. 111).

TRPV1 has also been reported to be expressed in several nonneuronal tissues and has therefore been implicated in a number of nonpain-related pathologies, including diseases of the respiratory tract, cancer, and diabetes. TRPV1 is highly expressed in sensory neurons innervating the airways, and evidence suggests that it is involved in the cough reflex. Furthermore, antagonism of TRPV1 has been shown to reduce cough in some animal models.[112–115] TRPV1 protein expression is also upregulated in asthma and gastroesophageal reflux disease.[113,114] Part of the mechanism behind the activation of TRPV1 in the cough reflex may be due to the activation of protease-activated receptor-2 (PAR-2), which can be activated by extracellular proteases released during inflammation. Activation of PAR-2 has been shown to sensitize TRPV1 channel activity,[116,117] leading to nociceptor sensitization. TRPV1 also appears to be important for certain types of cancers, specifically one report showing that increased consumption of hot chilies (capsaicin) could lead to higher rates of stomach cancer.[114]

TRPV1 has also been suggested to be involved in diabetes; specifically, *Trpv1*$^{-/-}$ mice exhibit increased sensitivity to insulin.[118] Current evidence suggests that TRPV1 activation plays a protective role in type I diabetes, although another study suggested that channel activation could be detrimental in type 2 diabetes conditions.[114] Therefore, further in-depth research is needed to better understand the role of TRPV1 in multiple disease conditions, which could warrant the utilization of TRPV1 antagonists not only for treating painful conditions but also for the treatment of other pathological conditions such as asthma, pancreatitis, cancers, and diabetes.

2.1.8 Drug development targeting TRPV1

Since its discovery, TRPV1 has been at the forefront of pharmaceutical developments as an attractive analgesic drug target. Predominant expression of TRPV1 in sensory neurons is advantageous because an antagonist could theoretically inhibit pain without producing significant central side effects. Unfortunately, most of the first-generation small-molecule antagonists developed against TRPV1 failed at different stages of preclinical trials, mainly due to their hyperthermia-inducing effects.[20,110,119] These results in fact led to the further understanding of the role of TRPV1 in body temperature regulation, which is thought to be mediated by the channels expressed in the SC and brain stem.[110] The second-generation TRPV1 antagonists have attempted to target different modalities of channel activation, mainly avoiding the blockade of heat activation and focusing on inhibiting the proton activation of the channel.[120] At least one such drug is currently being investigated for its efficacy in advanced clinical trial phases as a new-generation analgesic.

On the other hand, TRPV1 agonists have also shown great promise in the treatment of moderate to severe pain. By rapidly and strongly activating TRPV1, an agonist can cause fast and prolonged channel desensitization, and/or eventual degeneration/death of nociceptive fibers due to excitotoxicity. This is the same rationale behind the use of topical capsaicin creams for the treatment of arthritis pain but has also been extended to more severe cancer pain. Studies using intrathecal injection of RTX showed that it can decrease pain behaviors in dogs with osteosarcoma, and in this line preliminary studies in human patients with bone cancer pain have been promising. However, the results from experiments on humans also showed that RTX does not work in all pain conditions; for example, intravesicular injection of RTX was not effective at relieving symptoms of pain associated with interstitial cystitis.[121–123] Overall, the debate continues over the use of agonists or antagonists of TRPV1 as effective pain-relieving mechanism; however, this channel still remains a hot spot for drug discovery in pain.

2.2 Transient receptor potential vanilloid 2, 3, and 4

The TRPV2 channel was identified/cloned soon after the cloning of TRPV1 and was primarily characterized as a molecular transducer of higher noxious temperatures (>52 °C) in rodent sensory neurons.[46,124] TRPV2 is expressed in mechano- and thermoresponsive neurons in the DRG and TG, and SC, as well as in a number of nonneuronal cells, including immune cells.[46,124] In addition to higher noxious temperatures, certain synthetic cannabinoids, such as cannabidiol, can activate heterologously expressed recombinant TRPV2 channels (Fig. 1), although specific physiological and natural chemical agonist(s) of this channel still remains to be identified.[46] TRPV2 shares high sequence similarity to TRPV1 (50% identical) and forms functional homotetrameric channels, as well as heterotetrameric channels with TRPV1, conducting both Ca^{2+} and Na^{+} (at a ratio of 3:1).[124] Although some modulation of TRPV2 channel expression and/or function by PKA has been shown in immune cells,[125] and TRPV2's role in macrophage function, phagocytosis,[126] so far no reports have shown its direct involvement in any painful conditions. Furthermore, mice lacking functional TRPV2 display normal thermal and mechanical nociceptive behaviors,[127] which is suggestive of no significant contribution of this channel to pathological pain/nociception conditions.

The TRPV3 channel was cloned, based on sequence homology cloning strategy utilizing the sequence data for other TRPV channels (TRPV1, TRPV2, TRPV5, TRPV6), simultaneously from human sensory neurons and from rodent keratinocytes.[128,129] Accordingly, its expression has been detected in human DRG and TG neurons, SC, brain, and keratinocytes, although some controversy exists regarding its expression in both rodent PNS and CNS neurons.[46,128–130] As opposed to the previous two TRPV channels, TRPV3 can be activated at warm temperatures, ranging from 32 to 40 °C, and by a number of chemical ligands such as camphor and the pungent ingredient of clove, eugenol (Fig. 1), as well as by a number of synthetic vanilloid compounds. TRPV3 shares 40% sequence similarity with TRPV1 and forms functional homotetrameric channels, as well as heterotetrameric channels with TRPV1, conducting both Ca^{2+} and Na^{+} with high relative Ca^{2+} permeability.[130] TRPV3 channel activity can also be enhanced by inflammatory signaling, lipid metabolites, and PKC,[46,130] which could contribute to the development of inflammatory pain. Due to species differences in the expression, tissue distribution, and mode of activation of TRPV3, precise identification of its role in pain and pathological nociception remains controversial and unclear. However, small-molecule antagonists of TRPV3 have been shown to provide some attenuation of inflammatory pain, skin hypersensitivity, and itch conditions.[46,130] Therefore, further in-depth studies, taking into account the species differences in the expression, distribution, and functional properties of TRPV3, are required to define its role in pain and nociception.

The TRPV4 channel was one of the first channels identified as the molecular detector of osmotic changes, pressure, and shear stress, both in neurons and in muscles.[84,131,132] Following cloning and characterization, TRPV4 was found to be expressed ubiquitously and contributes to intracellular Ca^{2+} signaling, transduction of osmotic and mechanical pressure, temperature sensing to some extent, cell volume regulation, and maintenance of energy homeostasis.[84,131,133] In line with the expression and function of TRPV4 in the sensory neurons of both DRG and TG, the heterologously expressed recombinant channel can be activated by increasing temperatures ($\geq$33 °C), polyunsaturated fatty acids (PUFAs) such as arachidonic acid metabolites, exogenous chemical ligands including synthetic phorbol esters such as 4α-phorbol 12,13-didecanoate, and plant extract bisandrographolide A.[84,134] Activation of TRPV4 leads to the influx of Ca^{2+} and Na^{+} (at a ratio of 2:1) in sensory neurons, leading to membrane depolarization and subsequent AP firing.[3,84] Like other TRPV channels, functional TRPV4 channels are formed by homotetramers. Although heterotetrameric TRPV channels involving TRPV4 monomeric subunits have not been reported yet, heterologous expression of functional TRPV4–TRPC1 and TRPV4–TRPC1–TRPP2 heterotetrameric channels has been shown.[135] With regard to modulation of functional properties of TRPV4, both PGE_2 and PAR-2 signaling have been shown to sensitize channel activity, presumably via PKA and *Src* phosphorylation of the channel protein.[136–139] Although several speculations have been made suggesting upregulation of TRPV4 channel activity by PKC, convincing experimental evidence in support of these have not yet been presented.[84] Functional expression of TRPV4 in DRG and TG neurons led to the suggestion that TRPV4 could be involved in the transduction of thermal and mechanical stimuli at both somatic and visceral tissue levels that lead to pain. TRPV4 has been suggested to be involved in nociceptive responses to osmotic stimuli, both hypotonic and hypertonic, leading to mechanical hyperalgesia, as evidenced from reduced mechanical hyperalgesia upon *in vivo* silencing of this channel in mouse DRG neurons.[137,140] Mice lacking functional TRPV4 (*Trpv4*$^{-/-}$) exhibit impaired sensitivity to high-threshold mechanical stimuli, and some degree of alterations in selective nociceptive responses to temperature and acid; however, no change has been observed in their responses to noxious temperatures and high-threshold mechanical stimuli.[141–143] Interestingly, following tissue injury and inflammation *Trpv4*$^{-/-}$ mice exhibit increased latency to escape from hot plate, suggesting some contribution of TRPV4 in the development of thermal hyperalgesia.[144] Furthermore, attenuated nociceptive responses in experimental rodent models of pancreatitis and irritable bowel syndrome (IBS) in *Trpv4*$^{-/-}$ mice have also been suggested.[145,146] Taken together, TRPV4 presumably serves as a molecular transducer of nociceptive/pain hypersensitivity associated with a number of inflammatory and neuropathic conditions, although more in-depth studies are required to determine the critical and specific role(s) of this channel in specific pain conditions.

3. ION CHANNELS IN THE TRPM SUBFAMILY

3.1 Transient receptor potential melastatin 3

TRPM3 was initially identified and cloned as a nonselective cation channel activated by pregnenolone sulfate; however, more recent reports on activation of TRPM3 at warm/noxious temperatures have brought this channel into the nociceptive TRP channel category.[46,147,148] In the nervous system, TRPM3 expression has been reported in both CNS and PNS, including DRG neurons, as well as in a variety of nonneuronal tissues.[147,148] Structurally, TRPM3 is a homotetrameric channel with a distinct ~700 amino acid long TRPM-specific domain in the cytoplasmic N-terminus of the protein. Like TRPV1, TRPM3 is a nonselective cation channel with relatively high Ca^{2+} permeability and has a strong outwardly rectifying current–voltage relationship.[147,148] The nociceptive role of TRPM3 came to light more recently with the finding that mice lacking functional TRPM3 (*Trpm3*$^{-/-}$) showed significant attenuation of inflammatory thermal hyperalgesia.[147] Studies at the cellular level further identified that TRPM3 could be activated by noxious temperatures (>30 °C). Furthermore, in experimental models of CFA-induced inflammatory pain, *Trpm3*$^{-/-}$ mice exhibited significant attenuation of thermal nociceptive behaviors, as compared to their wild-type littermates,[147] suggesting that TRPM3 constitutes yet another noxious heat sensor.

3.2 Transient receptor potential melastatin 8

It had been previously established that neurons in the DRG and TG exhibited cold- and menthol-activated currents, and it was also postulated that both cold and menthol presumably activate the same receptor.[149] After the discovery of TRPV1 and TRPV2 as heat-activated ion channels, the continued search for cold-sensing ion channels on sensory neurons by expression cloning led to the identification of TRPM8.[150,151] TRPM8 was first identified in prostate epithelial cells as a prostate-specific transcriptional marker; however, its role in sensory neurons was not realized until 2002, when it was identified/cloned from rodent DRG neurons. TRPM8 can be activated by both innocuous or cooling (26–15 °C) and noxious cold (15–8 °C) temperatures, as well as by a number of "cooling agents," such as menthol and the more potent icilin (reviewed in Refs. 20,152).

3.2.1 *Expression and distribution in nervous system*

TRPM8 is predominantly expressed in PNS neurons. Antibody-based immunocytochemical analysis in DRGs showed that TRPM8 is expressed in sensory neurons giving rise to mostly C and Aδ fibers.[20,84,150,151] Further studies in transgenic mice with EGFP expressed in TRPM8-positive cells have shown that TRPM8-positive neurons project to skin, oral cavity epithelium, regions in the tooth pulp, dentine, and the tongue as well as visceral organs innervated by the pelvic nerve (including the colon) and vagal nerve, such as bronchopulmonary tissue.[153–156] TRPM8 is expressed in approximately 5–10% of DRGs and 10–15% of TGs.[150,151,157] At the subcellular level, TRPM8 has been shown localized to flotillin-1-containing lipid rafts on the plasma membrane.[158] Based on immunocytochemical analysis of brain sections from TRPM8/EGFP reporter mice, no expression of TRPM8 has been detected in brain neurons. Other than sensory neurons, TRPM8 is also expressed in prostate, bladder, lungs, and the urogenital tract.[152] However, the functional role of TRPM8 channels in tissues/cells other than sensory neurons is not as well understood as its function in neurons.

3.2.2 *Structure*

Like other TRP channels, functional TRPM8 channels are homotetramers, with a characteristic long cytoplasmic N-terminus, without any ankyrin repeat, rather comprising of a ~700 amino acid residue TRPM domain (Fig. 3), the function of which still remains unclear. While it has been well established that cold temperature sensitivity of TRPM8 is intrinsic to the channel protein, the exact structural determinants in the channel critical for cooling/cold temperature activation are not well established. Heterologous expression studies on recombinant chimeric TRPV1, TRPV2, and TRPM8 channels have suggested that the cytoplasmic C-terminus is important for temperature activation of TRPM8 (reviewed in Ref. 20). On the other hand, it has also been postulated that the TRPM8 interacts with membrane lipids and those interactions may change under cold temperatures, thereby activating the channel. More is known about the interaction and activation of TRPM8 with its chemical agonists, menthol, and icilin. Mutational analysis and functional Ca^{2+} imaging in heterologous expression systems have shown that S2 and the TRP domain are critical to activation by icilin and menthol but not cold.[159]

3.2.3 Functional properties of the channel

As mentioned previously, TRPM8 can be activated by innocuous cooling to noxious cold temperatures (8–26 °C), as well as by cooling agents such as methanol and icilin. Purified TRPM8 proteins incorporated into artificial membranes as well as in excised patches of mammalian cells expressing recombinant TRPM8 retain its activation by cooling/cold temperatures, suggesting that TRPM8 is inherently temperature sensitive.[160–162] Menthol and icilin activate TRPM8 by shifting the temperature activation threshold of the channel to higher temperatures, increase the open probability, and cause a hyperpolarizing shift in the voltage dependence of channel activation.[56] These combined actions cause channel activation under physiological conditions, which like other TRP channels is a nonselective cation channel that allows Ca^{2+} and Na^{+} (at ~3:1 ratio) into the cell, leading to membrane depolarization and AP firing.[150,151] One caveat to several studies using menthol is that it not only activates TRPM8 but also activates $GABA_A$,[163] TRPA1,[164,165] and TRPV3[166] and also has inhibitory actions on Na_v and certain Ca_v channels.[167,168] All these could contribute to the known effects of menthol-induced reduction in experimental thermal and mechanical pain hypersensitivities. Icilin is also not a specific agonist for TRPM8; it activates TRPA1 with a similar affinity. A number of specific agonists for TRPM8 have been developed for research use (WS-12 and WS-3), which aid in distinguishing the specific role of TRPM8 in sensory neurons, thereby eliminating the off-target effects of using menthol or icilin. Repeated activation of TRPM8 by cooling and/or menthol does not lead to channel desensitization; however, icilin activation leads to channel desensitization, suggesting distinct structural mechanisms of channel activation and desensitization. Overall, the precise mechanisms underlying channel activation still remain unclear.

3.2.4 Modulation of channel expression and function

TRPM8 channel exhibits different threshold of temperature activation in native neurons versus heterologous expression in cells, as well as in purified channels incorporated into liposomes, suggesting that there may be different modulators of the channel under different conditions and in different cell types.[152] Like most TRP channels, TRPM8 can be regulated by its interaction with PIP2. In general, an increase in PLC activity by increased Ca^{2+} influx through the channel or by activation of other cellular signaling cascade (e.g., $G\alpha_q$ signaling) enhances hydrolysis of PIP2, leading to a decrease in channel activity. In other words, PLC-modulated channels exhibit greater desensitization upon activation. This mode of TRPM8 regulation is in contrast to PIP2 action on TRPV1 activity where decreasing TRPV1/PIP2 association leads to increased current and channel sensitization.[79,152,160,161] In addition to PLC activity, some reports suggest that increased PKC activity could downregulate TRPM8 activity.[169–171] However, a recent study suggests that PKC inhibitors cannot decrease TRPM8 desensitization.[172] Studies on various inflammatory mediators such as bradykinin, histamine, serotonin, and ATP, which signal through $G\alpha_{q/11}$-coupled receptors, show that these mediators inhibit TRPM8 channel activity via direct interaction of $G\alpha_q$ subunit to the channel protein. Such a mechanism is thought to constitute a mechanism underlying reduced cold sensation under injury/inflammatory conditions. Unlike TRPV1, cAMP and PKA have no effect on TRPM8.[173–176] However, regarding channel desensitization, similar to observations on TRPV1 CaM has been suggested to be involved in acute desensitization of TRPM8.[177] Endovanniloids and endocannabinoids have also been shown to positively regulate TRPM8.[178] Phospholipase A2 (PLA_2) activity can increase TRPM8 channel activity.[179,180] Arachidonic acid and lysophospholipids are products of PLA_2 activity and the latter has been shown to increase the threshold of temperature activation of TRPM8 closer to body temperature.[179] When PLA_2 activity is inhibited, it decreases icilin but not menthol hypersensitivity. In contrast, arachidonic acid can decrease TRPM8 activity but it is thought that the effects of lysophospholipids are stronger and thus the end result is increased channel activity.[152] Overall, posttranslational modifications in TRPM8 channel protein and their impact on channel activity are not well understood.

3.2.5 Involvement in pain conditions

The exact role of TRPM8 in cold hypersensitivity associated with chronic inflammatory pain conditions like neuropathic pain has not been well established. Most of the work has been done in animal models and extrapolated to human conditions. There is some evidence that chemotherapy drug oxaliplatin-induced neuropathic cold allodynia may be dependent on TRPM8; however, recent evidence now suggests that the noxious cold-sensing channel TRPA1 might be a critical factor in such painful conditions.[181] TRPM8 has been suggested to also play a role in visceral pain as peppermint oil (containing menthol) can decrease visceral pain in IBS patients, in some cases but as mentioned before this could be due to effects on other channels mediating excitability ($GABA_A$) that could decrease pain.[182] Additionally, TRPM8 has been proposed to play a role in dry eye, airway irritation, and cold-induced urticaria (hives), but the evidence is very preliminary and primarily in rodent models.[181]

It seems logical that orofacial/dental pain resulting in cold hypersensitivity could involve TRPM8 since TRPM8 is functionally expressed in TG innervating the orofacial region. However, the only evidence supporting this claim is that cold-activated currents in neurons innervating important orofacial structures appear to be dependent on both TRPM8 and TRPA1 activity.[183]

Interestingly, there is a TRPM8 gene variant associated with increases in migraine susceptibility only in women.[184] The exact role of this variant is unknown, as is the association of TRPM8 and migraine but yet this provides intriguing evidence of another pathophysiological role for TRPM8.

TRPM8 knockout (*Trpm8*$^{-/-}$) mice exhibit no apparent deficiencies besides a decrease in avoidance behavior to moderately cold temperatures.[185–187] *Trpm8*$^{-/-}$ mice do not completely lack sensation to noxious cold, they still avoid temperatures lower than 0 °C. This residual cold sensation could be due to other cold sensitive channels. Two of the channels implicated are K^+ leak channels (TRAAK and TREK1), which close at very low temperatures to increase excitability, and TRPA1, which some groups have shown to be cold sensitive.[188,189] As a side note, while these animals have no change in core body temperature compared to wild-type littermate controls, antagonists of the channel can cause transient changes in body temperature, suggesting that TRPM8 is in fact involved in regulation of body temperature and *Trpm8*$^{-/-}$ undergoes adaption to control body temperature without functional TRPM8.[190] It is also important to point out that all menthol sensitivity is not lost in multiple *Trpm8*$^{-/-}$ mice. This could be due to high concentrations and ability of menthol to activate TRPA1 and TRPV3 as mentioned before (reviewed in Ref. 181).

In addition to the normal physiologic changes in temperature sensation, these *Trpm8*$^{-/-}$ mice also have attenuated cold hypersensitivity after injury. In two different pain models, chronic constriction injury[186] and the second phase of CFA injection[186] cold hypersensitivity is attenuated in *Trpm8*$^{-/-}$ mice compared to wild-type controls. Additionally, there is some evidence that TRPM8 antagonists can reduce visceral pain.[191] It is clear that TRPM8 does not only mediate noxious cold but also is important for cold hypersensitivity under injury conditions.

In contrast to TRPM8's role in cold hypersensitivity, it is also critical to cold and menthol-induced analgesia. In wild-type mice, mild cooling and menthol can reduce acute and inflammatory pain in the formalin injection pain model. In mice that lack functional TRPM8, the analgesic effects of mild cooling are absent for the inflammatory pain phase after formalin injection but still present for the acute phase.[187]

3.2.6 Drug development targeting TRPM8

Menthol and other cooling agents, as well as cold temperature, are commonly used for their analgesic properties. TRPM8 is expressed on sensory afferents that appear to be nociceptive (some TRPM8 coexpresses with TRPV1[153]). However, it may also be expressed in different populations of neurons that trigger inhibition of pain signals or increase central inhibition models.[192] No specific antagonists for TRPM8 are currently on the market or in the process of clinical trials for the treatment of pain conditions. TRPM8 agonists and antagonists still may be useful in treating an array of diseases from dry eye to cold hyperalgesia but more basic research using more specific pharmacological tools is needed, as well as consideration of TRPM8's role in thermoregulation.

4. ION CHANNELS IN THE TRPA SUBFAMILY

4.1 Transient receptor potential ankyrin 1

The transient receptor potential subfamily A member 1 (TRPA1) was originally cloned from human lung fibroblasts in 1999.[193] While certain key features of the channel were immediately apparent (an abundance of ankyrin repeats in the N-terminus; the 6-TM region and the resemblance to other TRP-related proteins), it was not until 2003 that "ANKTM1" (as TRPA1 was then known) was cloned and became the sixth thermosensitive TRP channel to be identified (after TRPV1–4 and TRPM8).[189]

4.1.1 Expression and distribution in the nervous system

TRPA1 expression in the nervous system is detectable in DRG, TG, and NG.[189,194,198] In the periphery, TRPA1 (much like TRPV1) is predominantly expressed in small-diameter, unmyelinated and partially myelinated C- and Aδ-fibers. Approximately 25% of neurons found to express TRPA1 are also peptidergic, and almost half are isolectin-B4 (IB4)-positive (i.e., nonpeptidergic) in the PNS.[194] In terms of coexpression with TRPV1, it appears that sensory afferents do coexpress with TRPV1, but the amount of "overlap" varies considerably between target tissues; colonic afferents have been shown to be predominantly responsive to both mustard oil (a TRPA1 agonist) and capsaicin (a TRPV1 agonist). However, in the skin, only 10% of neurons expressed both channels, a similar proportion expressed TRPA1 alone, and approximately 20% expressed TRPV1 alone.[189,194,198] It is important to bear in mind, however, that these figures do not take into account the density of innervation a particular organ receives. In addition to colonic afferents, enterochromaffin cells in the gastrointestinal tract express TRPA1, where it may be involved in manifestation of inflammatory disorders such as Crohn's disease and colitis.[195,196] TRPA1 is also expressed in skin, where keratinocytes showed changes in inflammatory mediator secretion in response to TRPA1 activation.[197] Sensory afferents expressing TRPA1 also innervate the bladder and prostate, raising the possibility that TRPA1 is involved in cystitis-associated pain.[196]

The central projections of TRPA1-expressing sensory neurons extend into the superficial laminae of the SC, where expression is detectable in presynaptic dendrites.[198] Consistent with this observation, TRPA1 activation in lamina II of the SC was found to increase glutamate release and evokes excitatory postsynaptic currents.[199] In the CNS, TRPA1 expression has been detected in the cerebellum, hippocampus, and forebrain.[200] Although the function of TRPA1 in the brain is poorly understood, one potential explanation was put forward when it was reported that TRPA1 regulates resting calcium levels in astrocytes, thereby enhancing extracellular GABA levels and influencing the efficacy of inhibitory synapses.[201] TRPA1 expression is found in the autonomic nervous system, where the vasodilation and consequent drop in blood pressure it causes could play a role in autonomic reflexes and disorders associated with neurovascular responses.[202] Other internal organs where there is evidence of TRPA1 function include the pancreas,[201] the inner ear (where it is known to be involved in mechanotransduction[202]), dental pulp (where a role in mechanotransduction and cold allodynia has been proposed[203]), and vascular endothelia and airway epithelial cells[202,206], where it modulates chemokine secretion and thereby inflammation.

4.1.2 Structure

The lengthy N-terminal portion of TRPA1 (some 700 of the 1100 total residues) contains as many as 18 ARDs, a number far in excess of the 4–6 repeats typically seen in TRPC and TRPV channels.[207,208] Our understanding of the contribution of these repeats to TRPA1 function is far from complete. However, there are suggestions that TRPA1 trafficking to the plasma membrane or insertion therein is compromised by the deletion of ARDs,[209] or that (in nonmammalian systems at least) they are vital mediators of mechanotransduction. Until recently, it was assumed that functional TRPA1 channels exist in a homotetrameric configuration. However, TRPV1–TRPA1 concatemers appear to produce functional heterotetrameric channels in heterologous expression systems.[16] Whether this occurs endogenously is yet to be determined.

Four potential disulfide bonds have been detected between five cysteine residues in the N-terminus of TRPA1.[207] Covalent modification at these sites by electrophilic compounds causes conformational changes, offering a glimpse into the mechanisms by which such compounds modify channel function.[13,208,209] Also contained within the N-terminus are three Ca^{2+}-binding EF hand domains, one of which has been shown to be responsible for activating the channel in response to elevated intracellular Ca^{2+}.[210] While still categorized as a non-selective cation channel, TRPA1 becomes more Ca^{2+}-permeable upon agonist stimulation.[211]

4.1.3 Functional properties of the channel

The list of TRPA1 gating modulators, regulators of function, and agonists is extensive. These compounds can be broadly separated into electrophilic (positively charged) and nonelectrophilic modulators. As previously mentioned, the N-terminus of TRPA1 is rich in cysteine residues. It is the negatively charged thiol group in the cysteine side chain with which electrophilic modulators react, resulting in TRPA1 activation.[212] The precise structural mechanism by which this activation is elicited remains unclear. One major group of electrophilic modulators is typified by mustard oil (the active component of which is allyl isothiocyanate) and other isothiocyanates, such as those found in wasabi and horseradish.[212,215]

ROS such as hydrogen peroxide are elevated during injury and inflammation and also cause cysteine oxidation or formation of disulfide bonds between cysteine residues. For example, it has been reported that UVA irradiation prompts production of ROS which results in TRPA1 activation.[215] RNS modify the channel by a process known as S-nitrosylation. One such species, nitric oxide, appears to be particularly important for pain sensitization, not only because of its ability to directly modify TRPA1 but also through its actions on cyclic GMP production, as well as generation of the fatty acid product nitro-oleic acid, another known modifier of TRPA1.[212,215] In addition to nitric oxide, another "gasotransmitter," hydrogen sulfide, increases CGRP release in a TRPA1-dependent fashion.[216] Certain prostaglandins, such as 15d-PGJ2 are metabolized into electrophilic products, which contain a reactive carbonyl group and are therefore capable of directly activating TRPA1.[212] Lipid peroxidation products, such as 4-hydroxynonenal, are produced downstream of ROS and elicit neuropeptide release and neurogenic inflammation. They act through the same covalent cysteine modification mechanism as other electrophilic modulators.[217]

Human TRPA1 has been shown to be sensitive to acidic pH (interestingly, the same is not true of rodent TRPA1[218]). Gating of the channel was found to increase in a dose-dependent fashion with increasing proton concentration associated with CO_2 or other weak organic acids.[219,220] Other compounds broadly characterized as "irritants" are electrophilic in nature, particularly formalin[221] and acrolein an unsaturated aldehyde found in vehicle exhaust fumes and tobacco smoke.[221,222] Similarly, TRPA1 activation by noxious cold temperatures is also dependent on species. While primate TRPA1 (macaque and human) could be activated at noxious cold temperatures, rodent TRPA1 channels are insensitive to cold temperatures, which is conferred by amino acid residues in the S5–S6 TM domains and the pore loop.[223]

Anesthetics often have paradoxical pronociceptive effects by acting through TRPA1.[224] Propofol and isoflurane excite sensory neurons through acting on TRPA1 and TRPV1.[225] It has been suggested that similar effects observed with other general anesthetics could be a substantial contributor to post-operative pain.[226] Local anesthetics, such as lidocaine, not only block Na_v channels but also activate TRPA1, leading to enhanced spinal release of glutamate.[227] Various alcohols can, at millimolar concentrations, activate TRPA1 in a carbon chain length-dependent manner.[228] Menthol, classically defined as an agonist of TRPM8, has also been shown to cause activation of TRPA1 at micromolar concentrations, while blocking the channel at higher concentrations.[164] The irritant effect of topical nicotine has been attributed to activation of TRPA1.[229]

An intriguing class of TRPA1 modulators are the PUFAs[230]; however, the physiological relevance of these noncovalent interactions is unknown. Interestingly, given their structural resemblance to prostaglandins, the fenamate class of nonsteroidal anti-inflammatory drugs were discovered to be TRPA1 agonists and potentiated the effects of other TRPA1 activators, such as AITC.[231]

TRPA1 was originally described as a cold-sensing ion channel,[232] but cinnamaldehyde (which provides cold sensation upon ingestion), a TRPA1 agonist, did not sensitize the response to noxious cold. Furthermore, blockade of TRPA1 in rats had no effect on cold responses.[233] However, other groups have reported cold-mediated TRPA1 activation consistent with the original finding.[234] As mentioned above, cold temperature activation of TRPA1 was subsequently found to the species-dependent, and is localized to the S5-S6 TM domains and the pore-loop of the channel protein.[223] Furthermore, the ankyrin repeat 6 has been suggested to be critical for the thermal sensitivity of TRPA1; specifically three point mutations in this region can individually confer heat sensitivity upon the channel, while leaving chemical sensitivity unaffected.[235]

TRPA1 is known to be required for responses to noxious chemical stimuli, but there is also a requirement for TRPA1 activity in mechanotransduction.[236] The precise ion channel repertoire contributing to mechanosensation and the modulation involved in pain perception are not fully understood. It is clear that although mechanical hypersensitivity is not completely blocked by inhibiting TRPA1 activity, it plays a significant role, since afferents shown to become hypersensitive to mechanical stimuli are TRPA1 positive.[236] This is true in skin as well as in the GI tract, where TRPA1 is implicated in normal nociception as well as inflammatory pain.[237]

4.1.4 Modulation of channel expression and function

There is Ca^{2+} involvement in both potentiation and desensitization of TRPA1 activity. Initial potentiation relies upon entry of Ca^{2+} through the pore, whereupon elevation of intracellular Ca^{2+} concentration produces channel inactivation.[238] Also, depletion of the phosphoinositide PIP2 renders TRP channels inactive, and applying PIP2 in the presence of agonists prolongs desensitization. Conversely, scavenging PIP2 accelerates desensitization. This led to the proposal that PIP2 reduces TRPA1's sensitivity to agonists.[239] It has been reported that the gating properties and sensitivities of TRPA1 are altered with coexpression of TRPV1,[240] although whether this relates to Ca^{2+}-induced potentiation and desensitization remains to be established.

Arachidonic acid metabolites are an important class of TRP channel modulators. Epoxyeicosatrienoic acid, for example, is known to sensitize TRPA1 in inflammation. It is released by activated sensory neurons and activates second-order neurons in the superficial laminae of the dorsal horn of the SC.[241] Lipoxygenases are another proinflammatory lipid species that have been shown to activate TRPA1 as well as TRPV1.[242] TRPA1 is known to be sensitized by bradykinin,[232] which presumably causes activation of PKC, but the exact mechanism and TRPA1 residues involved are unknown.

4.1.5 Involvement in pain conditions

It is abundantly clear that TRPA1 is modulated by known regulators of sensory neuron sensitivity, such as NGF, glial cell-derived neurotrophic factor, and artemin. These mediators can enhance the expression of TRPV1 and TRPA1 and thereby noxious cold sensitivity.[243] The TRPA1 knockout ($Trpa1^{-/-}$) mice display behavioral deficits in response to mechanical stimulation, AITC, and cold temperatures.[244] While it seems reasonable to assume these observations also apply to human TRPA1 (for example, AITC also causes mechanical hyperalgesia in humans[245]), caution must be exercised. Primate and rodent TRPA1 share approximately 80% amino acid identity and are reported to differ in their responses to menthol and caffeine.[246] They also differ in which cysteine residues are modified by electrophiles such as AITC.[212]

TRPA1 activity is involved in a number of inflammatory pain states, typified by allergic contact dermatitis,[247] where neurogenic inflammation is a key pathophysiological driver. Neurogenic inflammation is also a symptom of pathological TRPA1 activation in bladder, whereupon inflammatory pain ensues.[248] The prototypical pruritogen chloroquine is known to exert its effects at least in part by acting on TRPA1, an observation consistent with TRPA1's proposed role in chronic itch.[249] In chronic itch models, TRPA1 is proposed to be involved not only in the transduction of the itching sensation but also in the pathophysiological changes in skin associated with chronic itch.[250]

Familial episodic pain syndrome (FEPS) is associated with a mutation in TRPA1 at the amino acid residue asparagine at 855. This residue in the TM domain S4 is mutated to a serine, resulting in a fivefold increase in inward current.[251] While somewhat less clear than the FEPS mutation, there does also appear to be an epigenetic phenomenon related to TRPA1. People with elevated pain sensitivity show differential DNA methylation in the vicinity of the *TRPA1* gene. It is possible that such differences are a contributor to individual differences in pain sensitivity.[252]

Intriguingly, changes in expression of TRPA1 splice variants have been observed in mouse neuropathic pain models, but it has yet to be determined if similar process occurs in humans.[253] Many conditions that result in oxidative stress and enhanced production of ROS lead to pain through activation of TRPA1. Diabetes is one such example.[254,255] The cancer chemotherapeutics paclitaxel and oxaliplatin also lead to ROS production and TRPA1 activation, culminating in mechanical allodynia.[255,256] Furthermore, it has been suggested that ROS may originate from macrophages infiltrating nerves and activating TRPA1 in vincristine-induced allodynia.[253,257]

TRPA1 is involved in both irritant-induced and migraine headache.[258] This appears to be based on activation of TRPA1 in dural afferents.[259] TRPA1 expression in trigeminal ganglion neurons is consistent with its importance in orofacial and dental pain.[205]

In osteoarthritis models, activation of TRPA1, along with TRPV1, leads to elevated levels of TNFα, a proinflammatory cytokine crucial to the development of osteoarthritis.[260] TRPA1-expressing visceral afferents are involved in visceral hypersensitivity, i.e., IBS and colitis, where channel activation leads to neuropeptide-mediated neurogenic inflammation.[261] TRPA1, in combination with TRPV1, contributes to pain downstream of PAR-2-stimulated pancreatitis.[262]

Curiously, the antinociceptive effects of acetaminophen and tetrahydrocannabinol are lost in the TRPA1 knockout mouse, offering an insight into the mechanisms through which these compounds exert analgesia.[263]

A remarkable finding was reported recently where LPS, the bacterial cell wall component, was characterized as a novel "irritant" molecule acting directly on TRPA1, independent of any canonical Toll-like receptor signaling.[264]

Although not necessarily painful, TRPA1 is a key part of chemosensory airway reflexes in response to irritants. Trigeminal afferents expressing TRPA1 are important in mediating sneezing and coughing.[265]

4.1.6 Drug development targeting TRPA1

Given the diverse pain states in which TRPA1 has been implicated, and the relative absence of serious adverse effects of TRPA1 blockade/deletion, particularly an absence of thermoregulatory side effects, it stands to reason that it represents an attractive target for analgesic drug development. There are nonelectrophilic antagonists, such as HC-030031,[266] and electrophilic antagonists, such as AP-18 and A967079,[267,268] all of which have demonstrated efficacy in rodent models of inflammatory and neuropathic pain. Indeed, spinally delivered blockade of TRPA1 reduced low-intensity mechanical pain and inflammation[269]; the same approach also proved effective in a rat osteoarthritis model.[119] GRC17536, a TRPA1 antagonist developed by Glenmark pharmaceuticals, has recently passed phase I clinical trials as an antitussive therapeutic.[270]

ACKNOWLEDGMENTS

Research work on TRP channels in nociception and in our laboratory is supported by research grants from the US National Institutes of Health (NIH/NINDS-NS069869 to D.P.M.; NIH/NCI-CA171927 and NIH/NINDS-NS045549 to A.D.M.), US Department of Defense (DoD/PCRP-PC101096 to D.P.M.), American Pain Society (APS-FLP1483 to D.P.M.), and the International Association for the Study of Pain (IASP-1881650 to D.P.M.). We would like to apologize to the authors of several research and review papers, as well as book chapters in the TRP channel sensory biology area, whose work could not be cited in this chapter due to space limitations.

REFERENCES

1. Loeser JD, Treede RD. The Kyoto protocol of IASP basic pain terminology. *Pain.* 2008;137:473–477.
2. Venkatachalam K, Montell C. TRP channels. *Annu Rev Biochem.* 2007;76:387–417. http://dx.doi.org/10.1146/annurev.biochem.75.103004.142819.
3. Wu LJ, Sweet TB, Clapham DE. International Union of Basic and Clinical Pharmacology. LXXVI. Current progress in the mammalian TRP ion channel family. *Pharmacol Rev.* 2010;62:381–404.
4. Clapham DE. TRP channels as cellular sensors. *Nature.* 2003;426:517–524.
5. Montell C. Physiology, phylogeny, and functions of the TRP superfamily of cation channels. *Sci STKE.* 2001;2001:re1.
6. Cosens DJ, Manning A. Abnormal electroretinogram from a Drosophila mutant. *Nature.* 1969;224:285–287.
7. Minke B. The history of the Drosophila TRP channel: the birth of a new channel superfamily. *J Neurogenet.* 2010;24:216–233.
8. Cao E, Liao M, Cheng Y, Julius D. TRPV1 structures in distinct conformations reveal activation mechanisms. *Nature.* 2013;504:113–118.
9. Liao M, Cao E, Julius D, Cheng Y. Structure of the TRPV1 ion channel determined by electron cryo-microscopy. *Nature.* 2013;504:107–112.
10. Owsianik G, Talavera K, Voets T, Nilius B. Permeation and selectivity of TRP channels. *Annu Rev Physiol.* 2006;68:685–717.
11. Schaefer M. Homo- and heteromeric assembly of TRP channel subunits. *Pflugers Arch.* 2005;451:35–42.
12. Valente P, García-Sanz N, Gomis A, et al. Identification of molecular determinants of channel gating in the transient receptor potential box of vanilloid receptor I. *FASEB J.* 2008;22:3298–3309.
13. Matta JA, Ahern GP. Voltage is a partial activator of rat thermosensitive TRP channels. *J Physiol.* 2007;585:469–482.
14. Akopian AN. Regulation of nociceptive transmission at the periphery via TRPA1–TRPV1 interactions. *Curr Pharm Biotechnol.* 2011;12:89–94.

15. Cheng W, Yang F, Takanishi CL, Zheng J. Thermosensitive TRPV channel subunits coassemble into heteromeric channels with intermediate conductance and gating properties. *J Gen Physiol.* 2007;129:191–207.
16. Fischer MJ, Balasuriya D, Jeggle P, et al. Direct evidence for functional TRPV1/TRPA1 heteromers. *Pflugers Arch.* 2014;466:2229–2241.
17. Hellwig N, Albrecht N, Harteneck C, Schultz G, Schaefer M. Homo- and heteromeric assembly of TRPV channel subunits. *J Cell Sci.* 2005;118:917–928.
18. Kobori T, Smith GD, Sandford R, Edwardson JM. The transient receptor potential (TRP) channels TRPP2 and TRPC1 form a heterotetramer with a 2:2 stoichiometry and an alternating subunit arrangement. *J Biol Chem.* 2009;284: 35507–35513.
19. Caterina MJ, Schumacher MA, Tominaga M, Rosen TA, Levine JD, Julius D. The capsaicin receptor: a heat-activated ion channel in the pain pathway. *Nature.* 1997;389:816–824.
20. Julius D. TRP channels and pain. *Annu Rev Cell Dev Biol.* 2013;29:355–384.
21. Cavanaugh DJ, Lee H, Lo L, et al. Distinct subsets of unmyelinated primary sensory fibers mediate behavioral responses to noxious thermal and mechanical stimuli. *Proc Natl Acad Sci USA.* 2009;106:9075–9080.
22. Hwang SJ, Oh JM, Valtschanoff JG. Expression of the vanilloid receptor TRPV1 in rat dorsal root ganglion neurons supports different roles of the receptor in visceral and cutaneous afferents. *Brain Res.* 2005;1047:261–266.
23. Kobayashi K, Fukuoka T, Obata K, et al. Distinct expression of TRPM8, TRPA1, and TRPV1 mRNAs in rat primary afferent neurons with adelta/c-fibers and colocalization with trk receptors. *J Comp Neurol.* 2005;493:596–606.
24. Tominaga M, Caterina MJ, Malmberg AB, et al. The cloned capsaicin receptor integrates multiple pain-producing stimuli. *Neuron.* 1998;21:531–543.
25. Yu L, Yang F, Luo H, et al. The role of TRPV1 in different subtypes of dorsal root ganglion neurons in rat chronic inflammatory nociception induced by complete Freund's adjuvant. *Mol Pain.* 2008;4:61.
26. Zacharova G, Palecek J. Parvalbumin and TRPV1 receptor expression in dorsal root ganglion neurons after acute peripheral inflammation. *Physiol Res.* 2009;58:305–309.
27. Brito R, Sheth S, Mukherjea D, Rybak LP, Ramkumar V. TRPV1: a potential drug target for treating various diseases. *Cells.* 2014;3:517–545.
28. Jeffry JA, Yu SQ, Sikand P, Parihar A, Evans MS, Premkumar LS. Selective targeting of TRPV1 expressing sensory nerve terminals in the spinal cord for long lasting analgesia. *PLoS One.* 2009;4:e7021.
29. Spicarova D, Palecek J. The role of the TRPV1 endogenous agonist N-Oleoyldopamine in modulation of nociceptive signaling at the spinal cord level. *J Neurophysiol.* 2009;102:234–243.
30. Spicarova D, Palecek J. The role of spinal cord vanilloid (TRPV1) receptors in pain modulation. *Physiol Res.* 2008;57(Suppl 3):S69–S77.
31. Spicarova D, Nerandzic V, Palecek J. Update on the role of spinal cord TRPV1 receptors in pain modulation. *Physiol Res.* 2014;63(Suppl 1):S225–S236.
32. Cavanaugh DJ, Chesler AT, Braz JM, Shah NM, Julius D, Basbaum AI. Restriction of transient receptor potential vanilloid-1 to the peptidergic subset of primary afferent neurons follows its developmental downregulation in nonpeptidergic neurons. *J Neurosci.* 2011;31:10119–10127.
33. Mishra SK, Tisel SM, Orestes P, Bhangoo SK, Hoon MA. TRPV1-lineage neurons are required for thermal sensation. *EMBO J.* 2011;30:582–593.
34. Vodo S, Arcelli D, Fiorenzani P, et al. Gonadal ERalpha/beta, AR and TRPV1 gene expression: modulation by pain and morphine treatment in male and female rats. *Physiol Behav.* 2013;110–111:80–86.

35. Kedei N, Szabo T, Lile JD, et al. Analysis of the native quaternary structure of vanilloid receptor 1. *J Biol Chem*. 2001;276:28613–28619.
36. Lishko PV, Procko E, Jin X, Phelps CB, Gaudet R. The ankyrin repeats of TRPV1 bind multiple ligands and modulate channel sensitivity. *Neuron*. 2007;54:905–918.
37. Moiseenkova-Bell VY, Stanciu LA, Serysheva II, Tobe BJ, Wensel TG. Structure of TRPV1 channel revealed by electron cryomicroscopy. *Proc Natl Acad Sci USA*. 2008;105:7451–7455.
38. Henderson R. Structural biology: ion channel seen by electron microscopy. *Nature*. 2013;504:93–94.
39. Cao E, Cordero-Morales Julio F, Liu B, Qin F, Julius D. TRPV1 channels are intrinsically heat sensitive and negatively regulated by phosphoinositide lipids. *Neuron*. 2013;77:667–679.
40. Jordt SE, Tominaga M, Julius D. Acid potentiation of the capsaicin receptor determined by a key extracellular site. *Proc Natl Acad Sci USA*. 2000;97:8134–8139.
41. Dhaka A, Uzzell V, Dubin AE, et al. TRPV1 is activated by both acidic and basic pH. *J Neurosci*. 2009;29:153–158.
42. Huang SM, Bisogno T, Trevisani M, et al. An endogenous capsaicin-like substance with high potency at recombinant and native vanilloid VR1 receptors. *Proc Natl Acad Sci USA*. 2002;99:8400–8405.
43. Hwang SW, Cho H, Kwak J, et al. Direct activation of capsaicin receptors by products of lipoxygenases: endogenous capsaicin-like substances. *Proc Natl Acad Sci USA*. 2000;97:6155–6160.
44. Zygmunt PM, Petersson J, Andersson DA, et al. Vanilloid receptors on sensory nerves mediate the vasodilator action of anandamide. *Nature*. 1999;400:452–457.
45. Nieto-Posadas A, Picazo-Juarez G, Llorente I, et al. Lysophosphatidic acid directly activates TRPV1 through a C-terminal binding site. *Nat Chem Biol*. 2012;8:78–85.
46. Vriens J, Appendino G, Nilius B. Pharmacology of vanilloid transient receptor potential cation channels. *Mol Pharmacol*. 2009;75:1262–1279.
47. Kozai D, Ogawa N, Mori Y. Redox regulation of transient receptor potential channels. *Antioxid Redox Signal*. 2014;21:971–986.
48. Szallasi A, Szabo T, Biro T, et al. Resiniferatoxin-type phorboid vanilloids display capsaicin-like selectivity at native vanilloid receptors on rat DRG neurons and at the cloned vanilloid receptor VR1. *Br J Pharmacol*. 1999;128:428–434.
49. Bohlen CJ, Priel A, Zhou S, King D, Siemens J, Julius D. A bivalent tarantula toxin activates the capsaicin receptor, TRPV1, by targeting the outer pore domain. *Cell*. 2010;141:834–845.
50. Siemens J, Zhou S, Piskorowski R, et al. Spider toxins activate the capsaicin receptor to produce inflammatory pain. *Nature*. 2006;444:208–212.
51. Trevisani M, Patacchini R, Nicoletti P, et al. Hydrogen sulfide causes vanilloid receptor 1-mediated neurogenic inflammation in the airways. *Br J Pharmacol*. 2005;145:1123–1131.
52. Gavva NR, Klionsky L, Qu Y, et al. Molecular determinants of vanilloid sensitivity in TRPV1. *J Biol Chem*. 2004;279:20283–20295.
53. Jordt SE, Julius D. Molecular basis for species-specific sensitivity to "hot" chili peppers. *Cell*. 2002;108:421–430.
54. Long SB, Campbell EB, Mackinnon R. Voltage sensor of Kv1.2: structural basis of electromechanical coupling. *Science*. 2005;309:903–908.
55. Pathak MM, Yarov-Yarovoy V, Agarwal G, et al. Closing in on the resting state of the Shaker K(+) channel. *Neuron*. 2007;56:124–140.
56. Voets T, Droogmans G, Wissenbach U, Janssens A, Flockerzi V, Nilius B. The principle of temperature-dependent gating in cold- and heat-sensitive TRP channels. *Nature*. 2004;430:748–754.

57. Chung M-K, Guler AD, Caterina MJ. TRPV1 shows dynamic ionic selectivity during agonist stimulation. *Nat Neurosci.* 2008;11:555–564.
58. Mohapatra DP, Wang SY, Wang GK, Nau C. A tyrosine residue in TM6 of the vanilloid receptor TRPV1 involved in desensitization and calcium permeability of capsaicin-activated currents. *Mol Cell Neurosci.* 2003;23:314–324.
59. Koplas PA, Rosenberg RL, Oxford GS. The role of calcium in the desensitization of capsaicin responses in rat dorsal root ganglion neurons. *J Neurosci.* 1997;17:3525–3537.
60. Loo L, Shepherd AJ, Mickle AD, et al. The C-type natriuretic peptide induces thermal hyperalgesia through a noncanonical Gbetagamma-dependent modulation of TRPV1 channel. *J Neurosci.* 2012;32:11942–11955.
61. Docherty RJ, Yeats JC, Bevan S, Boddeke HW. Inhibition of calcineurin inhibits the desensitization of capsaicin-evoked currents in cultured dorsal root ganglion neurones from adult rats. *Pflugers Arch.* 1996;431:828–837.
62. Mohapatra DP, Nau C. Regulation of Ca^{2+}-dependent desensitization in the vanilloid receptor TRPV1 by calcineurin and cAMP-dependent protein kinase. *J Biol Chem.* 2005;280:13424–13432.
63. Numazaki M, Tominaga T, Takeuchi K, Murayama N, Toyooka H, Tominaga M. Structural determinant of TRPV1 desensitization interacts with calmodulin. *Proc Natl Acad Sci USA.* 2003;100:8002–8006.
64. Melemedjian OK, Asiedu MN, Tillu DV, et al. IL-6- and NGF-induced rapid control of protein synthesis and nociceptive plasticity via convergent signaling to the eIF4F complex. *J Neurosci.* 2010;30:15113–15123.
65. Melemedjian OK, Tillu DV, Moy JK, et al. Local translation and retrograde axonal transport of CREB regulates IL-6-induced nociceptive plasticity. *Mol Pain.* 2014;10:45.
66. Bhave G, Hu H-J, Glauner KS, et al. Protein kinase C phosphorylation sensitizes but does not activate the capsaicin receptor transient receptor potential vanilloid 1 (TRPV1). *Proc Natl Acad Sci USA.* 2003;100:12480–12485.
67. Bhave G, Zhu W, Wang H, Brasier DJ, Oxford GS, Gereau Iv RW. cAMP-dependent protein kinase regulates desensitization of the capsaicin receptor (VR1) by direct phosphorylation. *Neuron.* 2002;35:721–731.
68. Jin X, Morsy N, Winston J, Pasricha PJ, Garrett K, Akbarali HI. Modulation of TRPV1 by nonreceptor tyrosine kinase, c-Src kinase. *Am J Physiol Cell Physiol.* 2004;287: C558–C563.
69. Jung J, Shin JS, Lee SY, et al. Phosphorylation of vanilloid receptor 1 by Ca^{2+}/calmodulin-dependent kinase II regulates its vanilloid binding. *J Biol Chem.* 2004;279:7048–7054.
70. Numazaki M, Tominaga T, Toyooka H, Tominaga M. Direct phosphorylation of capsaicin receptor VR1 by protein kinase Cepsilon and identification of two target serine residues. *J Biol Chem.* 2002;277:13375–13378.
71. Pareek TK, Keller J, Kesavapany S, et al. Cyclin-dependent kinase 5 modulates nociceptive signaling through direct phosphorylation of transient receptor potential vanilloid 1. *Proc Natl Acad Sci USA.* 2007;104:660–665.
72. Vellani V, Mapplebeck S, Moriondo A, Davis JB, McNaughton PA. Protein kinase C activation potentiates gating of the vanilloid receptor VR1 by capsaicin, protons, heat and anandamide. *J Physiol.* 2001;534:813–825.
73. Zhang X, Huang J, McNaughton PA. NGF rapidly increases membrane expression of TRPV1 heat-gated ion channels. *EMBO J.* 2005;24:4211–4223.
74. Mohapatra DP, Nau C. Desensitization of capsaicin-activated currents in the vanilloid receptor TRPV1 is decreased by the cyclic AMP-dependent protein kinase pathway. *J Biol Chem.* 2003;278:50080–50090.

75. Ji RR, Samad TA, Jin SX, Schmoll R, Woolf CJ. p38 MAPK activation by NGF in primary sensory neurons after inflammation increases TRPV1 levels and maintains heat hyperalgesia. *Neuron*. 2002;36:57–68.
76. Zhang X, McNaughton PA. Why pain gets worse: the mechanism of heat hyperalgesia. *J Gen Physiol*. 2006;128:491–493.
77. Chuang HH, Prescott ED, Kong H, et al. Bradykinin and nerve growth factor release the capsaicin receptor from PtdIns(4,5)P2-mediated inhibition. *Nature*. 2001;411:957–962.
78. Lukacs V, Yudin Y, Hammond GR, Sharma E, Fukami K, Rohacs T. Distinctive changes in plasma membrane phosphoinositides underlie differential regulation of TRPV1 in nociceptive neurons. *J Neurosci*. 2013;33:11451–11463.
79. Rohacs T. Regulation of transient receptor potential channels by the phospholipase C pathway. *Adv Biol Regul*. 2013;53:341–355.
80. Tominaga M, Tominaga T. Structure and function of TRPV1. *Pflugers Arch*. 2005;451:143–150.
81. Sanz-Salvador L, Andres-Borderia A, Ferrer-Montiel A, Planells-Cases R. Agonist- and Ca^{2+}-dependent desensitization of TRPV1 channel targets the receptor to lysosomes for degradation. *J Biol Chem*. 2012;287:19462–19471.
82. Caterina MJ, Leffler A, Malmberg AB, et al. Impaired nociception and pain sensation in mice lacking the capsaicin receptor. *Science*. 2000;288:306–313.
83. Davis JB, Gray J, Gunthorpe MJ, et al. Vanilloid receptor-1 is essential for inflammatory thermal hyperalgesia. *Nature*. 2000;405:183–187.
84. Bourinet E, Altier C, Hildebrand ME, Trang T, Salter MW, Zamponi GW. Calcium-permeable ion channels in pain signaling. *Physiol Rev*. 2014;94:81–140.
85. Barton NJ, McQueen DS, Thomson D, et al. Attenuation of experimental arthritis in TRPV1R knockout mice. *Exp Mol Pathol*. 2006;81:166–170.
86. Kelly S, Chapman RJ, Woodhams S, et al. Increased function of pronociceptive TRPV1 at the level of the joint in a rat model of osteoarthritis pain. *Ann Rheum Dis*. 2015;74:252–259.
87. Valdes AM, De Wilde G, Doherty SA, et al. The Ile585Val TRPV1 variant is involved in risk of painful knee osteoarthritis. *Ann Rheum Dis*. 2011;70:1556–1561.
88. Ghilardi JR, Rohrich H, Lindsay TH, et al. Selective blockade of the capsaicin receptor TRPV1 attenuates bone cancer pain. *J Neurosci*. 2005;25:3126–3131.
89. Mantyh P. Bone cancer pain: causes, consequences, and therapeutic opportunities. *Pain*. 2013;154(Suppl 1):S54–S62.
90. Akerman S, Kaube H, Goadsby PJ. Anandamide acts as a vasodilator of dural blood vessels in vivo by activating TRPV1 receptors. *Br J Pharmacol*. 2004;142:1354–1360.
91. Geppetti P, Capone JG, Trevisani M, Nicoletti P, Zagli G, Tola MR. CGRP and migraine: neurogenic inflammation revisited. *J Headache Pain*. 2005;6:61–70.
92. Trevisani M, Gazzieri D, Benvenuti F, et al. Ethanol causes inflammation in the airways by a neurogenic and TRPV1-dependent mechanism. *J Pharmacol Exp Ther*. 2004;309:1167–1173.
93. Trevisani M, Smart D, Gunthorpe MJ, et al. Ethanol elicits and potentiates nociceptor responses via the vanilloid receptor-1. *Nat Neurosci*. 2002;5:546–551.
94. Kim HY, Chung G, Jo HJ, et al. Characterization of dental nociceptive neurons. *J Dent Res*. 2011;90:771–776.
95. Park CK, Kim MS, Fang Z, et al. Functional expression of thermo-transient receptor potential channels in dental primary afferent neurons: implication for tooth pain. *J Biol Chem*. 2006;281:17304–17311.
96. Chung MK, Lee J, Duraes G, Ro JY. Lipopolysaccharide-induced pulpitis up-regulates TRPV1 in trigeminal ganglia. *J Dent Res*. 2011;90:1103–1107.

97. Chung G, Oh SB. TRP channels in dental pain. *Open Pain J.* 2013;6:31–36.
98. Rong W, Hillsley K, Davis JB, Hicks G, Winchester WJ, Grundy D. Jejunal afferent nerve sensitivity in wild-type and TRPV1 knockout mice. *J Physiol.* 2004;560:867–881.
99. Kimball ES, Wallace NH, Schneider CR, D'Andrea MR, Hornby PJ. Vanilloid receptor 1 antagonists attenuate disease severity in dextran sulphate sodium-induced colitis in mice. *Neurogastroenterol Motil.* 2004;16:811–818.
100. Chan CL, Facer P, Davis JB, et al. Sensory fibres expressing capsaicin receptor TRPV1 in patients with rectal hypersensitivity and faecal urgency. *Lancet.* 2003;361:385–391.
101. Wang ZY, Wang P, Merriam FV, Bjorling DE. Lack of TRPV1 inhibits cystitis-induced increased mechanical sensitivity in mice. *Pain.* 2008;139:158–167.
102. Jones 3rd RC, Otsuka E, Wagstrom E, Jensen CS, Price MP, Gebhart GF. Short-term sensitization of colon mechanoreceptors is associated with long-term hypersensitivity to colon distention in the mouse. *Gastroenterology.* 2007;133:184–194.
103. Jones 3rd RC, Xu L, Gebhart GF. The mechanosensitivity of mouse colon afferent fibers and their sensitization by inflammatory mediators require transient receptor potential vanilloid 1 and acid-sensing ion channel 3. *J Neurosci.* 2005;25:10981–10989.
104. Massa F, Sibaev A, Marsicano G, Blaudzun H, Storr M, Lutz B. Vanilloid receptor (TRPV1)-deficient mice show increased susceptibility to dinitrobenzene sulfonic acid induced colitis. *J Mol Med.* 2006;84:142–146.
105. Storr M. TRPV1 in colitis: is it a good or a bad receptor? A viewpoint. *Neurogastroenterol Motil.* 2007;19:625–629.
106. Cruz F, Guimaraes M, Silva C, Reis M. Suppression of bladder hyperreflexia by intravesical resiniferatoxin. *Lancet.* 1997;350:640–641.
107. Dang K, Bielefeldt K, Gebhart GF. Differential responses of bladder lumbosacral and thoracolumbar dorsal root ganglion neurons to purinergic agonists, protons, and capsaicin. *J Neurosci.* 2005;25:3973–3984.
108. Premkumar LS, Sikand P. TRPV1: a target for next generation analgesics. *Curr Neuropharmacol.* 2008;6:151–163.
109. Dinis P, Charrua A, Avelino A, Cruz F. Intravesical resiniferatoxin decreases spinal c-fos expression and increases bladder volume to reflex micturition in rats with chronic inflamed urinary bladders. *BJU Int.* 2004;94:153–157.
110. Gavva NR, Treanor JJ, Garami A, et al. Pharmacological blockade of the vanilloid receptor TRPV1 elicits marked hyperthermia in humans. *Pain.* 2008;136:202–210.
111. Liu MG, Zhuo M. No requirement of TRPV1 in long-term potentiation or long-term depression in the anterior cingulate cortex. *Mol Pain.* 2014;7:27.
112. Bhattacharya A, Scott BP, Nasser N, et al. Pharmacology and antitussive efficacy of 4-(3-trifluoromethyl-pyridin-2-yl)-piperazine-1-carboxylic acid (5-trifluoromethyl-pyridin-2-yl)-amide (JNJ17203212), a transient receptor potential vanilloid 1 antagonist in guinea pigs. *J Pharmacol Exp Ther.* 2007;323:665–674.
113. Geppetti P, Materazzi S, Nicoletti P. The transient receptor potential vanilloid 1: role in airway inflammation and disease. *Eur J Pharmacol.* 2006;533:207–214.
114. Gunthorpe MJ, Szallasi A. Peripheral TRPV1 receptors as targets for drug development: new molecules and mechanisms. *Curr Pharm Des.* 2008;14:32–41.
115. Jia Y, McLeod RL, Hey JA. TRPV1 receptor: a target for the treatment of pain, cough, airway disease and urinary incontinence. *Drug News Perspect.* 2005;18:165–171.
116. Amadesi S, Cottrell GS, Divino L, et al. Protease-activated receptor 2 sensitizes TRPV1 by protein kinase Cepsilon- and A-dependent mechanisms in rats and mice. *J Physiol.* 2006;575:555–571.
117. Gatti R, Andre E, Amadesi S, et al. Protease-activated receptor-2 activation exaggerates TRPV1-mediated cough in guinea pigs. *J Appl Physiol.* 2006;101:506–511.

118. Razavi R, Chan Y, Afifiyan FN, et al. TRPV1 + sensory neurons control beta cell stress and islet inflammation in autoimmune diabetes. *Cell*. 2006;127:1123–1135.
119. McGaraughty S, Chu K, Perner R, DiDomenico S, Kort M, Kym P. TRPA1 modulation of spontaneous and mechanically evoked firing of spinal neurons in uninjured, osteoarthritic, and inflamed rats. *Mol Pain*. 2010;6:14.
120. Reilly RM, McDonald HA, Puttfarcken PS, et al. Pharmacology of modality-specific transient receptor potential vanilloid-1 antagonists that do not alter body temperature. *J Pharmacol Exp Ther*. 2012;342:416–428.
121. Brown DC, Iadarola MJ, Perkowski SZ, et al. Physiologic and antinociceptive effects of intrathecal resiniferatoxin in a canine bone cancer model. *Anesthesiology*. 2005;103:1052–1059.
122. Iadarola MJ, Gonnella GL. Resiniferatoxin for pain treatment: an interventional approach to personalized pain medicine. *Open Pain J*. 2013;6:95–107.
123. Payne CK, Mosbaugh PG, Forrest JB, et al. Intravesical resiniferatoxin for the treatment of interstitial cystitis: a randomized, double-blind, placebo controlled trial. *J Urol*. 2005;173:1590–1594.
124. Caterina MJ, Rosen TA, Tominaga M, Brake AJ, Julius D. A capsaicin-receptor homologue with a high threshold for noxious heat. *Nature*. 1999;398:436–441.
125. Stokes AJ, Shimoda LM, Koblan-Huberson M, Adra CN, Turner H. A TRPV2-PKA signaling module for transduction of physical stimuli in mast cells. *J Exp Med*. 2004;200:137–147.
126. Link TM, Park U, Vonakis BM, Raben DM, Soloski MJ, Caterina MJ. TRPV2 has a pivotal role in macrophage particle binding and phagocytosis. *Nat Immunol*. 2010;11:232–239.
127. Park U, Vastani N, Guan Y, Raja SN, Koltzenburg M, Caterina MJ. TRP vanilloid 2 knock-out mice are susceptible to perinatal lethality but display normal thermal and mechanical nociception. *J Neurosci*. 2011;31:11425–11436.
128. Peier AM, Reeve AJ, Andersson DA, et al. A heat-sensitive TRP channel expressed in keratinocytes. *Science*. 2002;296:2046–2049.
129. Xu H, Ramsey IS, Kotecha SA, et al. TRPV3 is a calcium-permeable temperature-sensitive cation channel. *Nature*. 2002;418:181–186.
130. Yang P, Zhu MX. Trpv3. *Handb Exp Pharmacol*. 2014;222:273–291.
131. Ho TC, Horn NA, Huynh T, Kelava L, Lansman JB. Evidence TRPV4 contributes to mechanosensitive ion channels in mouse skeletal muscle fibers. *Channels (Austin)*. 2012;6:246–254.
132. Nilius B, Voets T. The puzzle of TRPV4 channelopathies. *EMBO Rep*. 2013;14:152–163.
133. Ye L, Kleiner S, Wu J, et al. TRPV4 is a regulator of adipose oxidative metabolism, inflammation, and energy homeostasis. *Cell*. 2012;151:96–110.
134. Watanabe H, Vriens J, Prenen J, Droogmans G, Voets T, Nilius B. Anandamide and arachidonic acid use epoxyeicosatrienoic acids to activate TRPV4 channels. *Nature*. 2003;424:434–438.
135. Du J, Ma X, Shen B, Huang Y, Birnbaumer L, Yao X. TRPV4, TRPC1, and TRPP2 assemble to form a flow-sensitive heteromeric channel. *FASEB J*. 2014;28:4677–4685.
136. Alessandri-Haber N, Dina OA, Joseph EK, Reichling DB, Levine JD. Interaction of transient receptor potential vanilloid 4, integrin, and SRC tyrosine kinase in mechanical hyperalgesia. *J Neurosci*. 2008;28:1046–1057.
137. Alessandri-Haber N, Yeh JJ, Boyd AE, et al. Hypotonicity induces TRPV4-mediated nociception in rat. *Neuron*. 2003;39:497–511.
138. Denadai-Souza A, Martin L, de Paula MA, et al. Role of transient receptor potential vanilloid 4 in rat joint inflammation. *Arthritis Rheum*. 2012;64:1848–1858.

139. Wegierski T, Lewandrowski U, Muller B, Sickmann A, Walz G. Tyrosine phosphorylation modulates the activity of TRPV4 in response to defined stimuli. *J Biol Chem*. 2009;284:2923–2933.
140. Alessandri-Haber N, Joseph E, Dina OA, Liedtke W, Levine JD. TRPV4 mediates pain-related behavior induced by mild hypertonic stimuli in the presence of inflammatory mediator. *Pain*. 2005;118:70–79.
141. Liedtke W, Friedman JM. Abnormal osmotic regulation in trpv4$^{-/-}$ mice. *Proc Natl Acad Sci USA*. 2003;100:13698–13703.
142. Liedtke W, Tobin DM, Bargmann CI, Friedman JM. Mammalian TRPV4 (VR-OAC) directs behavioral responses to osmotic and mechanical stimuli in Caenorhabditis elegans. *Proc Natl Acad Sci USA*. 2003;100(Suppl 2):14531–14536.
143. Suzuki M, Mizuno A, Kodaira K, Imai M. Impaired pressure sensation in mice lacking TRPV4. *J Biol Chem*. 2003;278:22664–22668.
144. Todaka H, Taniguchi J, Satoh J, Mizuno A, Suzuki M. Warm temperature-sensitive transient receptor potential vanilloid 4 (TRPV4) plays an essential role in thermal hyperalgesia. *J Biol Chem*. 2004;279:35133–35138.
145. Ceppa E, Cattaruzza F, Lyo V, et al. Transient receptor potential ion channels V4 and A1 contribute to pancreatitis pain in mice. *Am J Physiol Gastrointest Liver Physiol*. 2010;299:G556–G571.
146. D'Aldebert E, Cenac N, Rousset P, et al. Transient receptor potential vanilloid 4 activated inflammatory signals by intestinal epithelial cells and colitis in mice. *Gastroenterology*. 2011;140:275–285.
147. Vriens J, Owsianik G, Hofmann T, et al. TRPM3 is a nociceptor channel involved in the detection of noxious heat. *Neuron*. 2011;70:482–494.
148. Wagner TF, Loch S, Lambert S, et al. Transient receptor potential M3 channels are ionotropic steroid receptors in pancreatic beta cells. *Nat Cell Biol*. 2008;10:1421–1430.
149. Reid G, Flonta ML. Physiology. Cold current in thermoreceptive neurons. *Nature*. 2001;413:480.
150. McKemy DD, Neuhausser WM, Julius D. Identification of a cold receptor reveals a general role for TRP channels in thermosensation. *Nature*. 2002;416:52–58.
151. Peier AM, Moqrich A, Hergarden AC, et al. A TRP channel that senses cold stimuli and menthol. *Cell*. 2002;108:705–715.
152. Yudin Y, Rohacs T. Regulation of TRPM8 channel activity. *Mol Cell Endocrinol*. 2012;353:68–74.
153. Dhaka A, Earley TJ, Watson J, Patapoutian A. Visualizing cold spots: TRPM8-expressing sensory neurons and their projections. *J Neurosci*. 2008;28:566–575.
154. Harrington AM, Hughes PA, Martin CM, et al. A novel role for TRPM8 in visceral afferent function. *Pain*. 2011;152:1459–1468.
155. Takashima Y, Daniels RL, Knowlton W, Teng J, Liman ER, McKemy DD. Diversity in the neural circuitry of cold sensing revealed by genetic axonal labeling of transient receptor potential melastatin 8 neurons. *J Neurosci*. 2007;27:14147–14157.
156. Xing H, Ling JX, Chen M, et al. TRPM8 mechanism of autonomic nerve response to cold in respiratory airway. *Mol Pain*. 2008;4:22.
157. Abe J, Hosokawa H, Okazawa M, et al. TRPM8 protein localization in trigeminal ganglion and taste papillae. *Brain Res Mol Brain Res*. 2005;136:91–98.
158. Morenilla-Palao C, Pertusa M, Meseguer V, Cabedo H, Viana F. Lipid raft segregation modulates TRPM8 channel activity. *J Biol Chem*. 2009;284:9215–9224.
159. Bandell M, Dubin AE, Petrus MJ, et al. High-throughput random mutagenesis screen reveals TRPM8 residues specifically required for activation by menthol. *Nat Neurosci*. 2006;9:493–500.
160. Liu B, Qin F. Functional control of cold- and menthol-sensitive TRPM8 ion channels by phosphatidylinositol 4,5-bisphosphate. *J Neurosci*. 2005;25:1674–1681.

161. Rohacs T, Lopes CM, Michailidis I, Logothetis DE. PI(4,5)P2 regulates the activation and desensitization of TRPM8 channels through the TRP domain. *Nat Neurosci.* 2005;8:626–634.
162. Zakharian E, Cao C, Rohacs T. Gating of transient receptor potential melastatin 8 - (TRPM8) channels activated by cold and chemical agonists in planar lipid bilayers. *J Neurosci.* 2010;30:12526–12534.
163. Tani M, Onimaru H, Ikeda K, Kawakami K, Homma I. Menthol inhibits the respiratory rhythm in brainstem preparations of the newborn rats. *Neuroreport.* 2010;21:1095–1099.
164. Karashima Y, Damann N, Prenen J, et al. Bimodal action of menthol on the transient receptor potential channel TRPA1. *J Neurosci.* 2007;27:9874–9884.
165. Xiao B, Dubin AE, Bursulaya B, Viswanath V, Jegla TJ, Patapoutian A. Identification of transmembrane domain 5 as a critical molecular determinant of menthol sensitivity in mammalian TRPA1 channels. *J Neurosci.* 2008;28:9640–9651.
166. Macpherson LJ, Hwang SW, Miyamoto T, Dubin AE, Patapoutian A, Story GM. More than cool: promiscuous relationships of menthol and other sensory compounds. *Mol Cell Neurosci.* 2006;32:335–343.
167. Gaudioso C, Hao J, Martin-Eauclaire MF, Gabriac M, Delmas P. Menthol pain relief through cumulative inactivation of voltage-gated sodium channels. *Pain.* 2012;153:473–484.
168. Swandulla D, Carbone E, Schafer K, Lux HD. Effect of menthol on two types of Ca currents in cultured sensory neurons of vertebrates. *Pflugers Arch.* 1987;409:52–59.
169. Abe J, Hosokawa H, Sawada Y, Matsumura K, Kobayashi S. Ca^{2+}-dependent PKC activation mediates menthol-induced desensitization of transient receptor potential M8. *Neurosci Lett.* 2006;397:140–144.
170. Daniels RL, Takashima Y, McKemy DD. Activity of the neuronal cold sensor TRPM8 is regulated by phospholipase C via the phospholipid phosphoinositol 4,5-bisphosphate. *J Biol Chem.* 2009;284:1570–1582.
171. Premkumar LS, Raisinghani M, Pingle SC, Long C, Pimentel F. Downregulation of transient receptor potential melastatin 8 by protein kinase C-mediated dephosphorylation. *J Neurosci.* 2005;25:11322–11329.
172. Yudin Y, Lukacs V, Cao C, Rohacs T. Decrease in phosphatidylinositol 4,5-bisphosphate levels mediates desensitization of the cold sensor TRPM8 channels. *J Physiol.* 2011;589:6007–6027.
173. Bavencoffe A, Gkika D, Kondratskyi A, et al. The transient receptor potential channel TRPM8 is inhibited via the alpha 2A adrenoreceptor signaling pathway. *J Biol Chem.* 2010;285:9410–9419.
174. De Petrocellis L, Starowicz K, Moriello AS, Vivese M, Orlando P, Di Marzo V. Regulation of transient receptor potential channels of melastatin type 8 (TRPM8): effect of cAMP, cannabinoid CB(1) receptors and endovanilloids. *Exp Cell Res.* 2007;313:1911–1920.
175. Linte RM, Ciobanu C, Reid G, Babes A. Desensitization of cold- and menthol-sensitive rat dorsal root ganglion neurones by inflammatory mediators. *Exp Brain Res.* 2007;178:89–98.
176. Sarria I, Gu J. Menthol response and adaptation in nociceptive-like and nonnociceptive-like neurons: role of protein kinases. *Mol Pain.* 2010;6:47.
177. Sarria I, Ling J, Zhu MX, Gu JG. TRPM8 acute desensitization is mediated by calmodulin and requires PIP(2): distinction from tachyphylaxis. *J Neurophysiol.* 2011;106:3056–3066.
178. De Petrocellis L, Vellani V, Schiano-Moriello A, et al. Plant-derived cannabinoids modulate the activity of transient receptor potential channels of ankyrin type-1 and melastatin type-8. *J Pharmacol Exp Ther.* 2008;325:1007–1015.

179. Andersson DA, Nash M, Bevan S. Modulation of the cold-activated channel TRPM8 by lysophospholipids and polyunsaturated fatty acids. *J Neurosci.* 2007;27:3347–3355.
180. Vanden Abeele F, Zholos A, Bidaux G, et al. Ca^{2+}-independent phospholipase A2-dependent gating of TRPM8 by lysophospholipids. *J Biol Chem.* 2006;281:40174–40182.
181. Fernández-Peña C, Viana F. Targeting TRPM8 for pain relief. *Open Pain J.* 2013;6:154–164.
182. van Zanten SV. Review: fibre, antispasmodics, and peppermint oil are all effective for irritable bowel syndrome. *Evid Based Med.* 2009;14:84.
183. El Karim IA, Linden GJ, Curtis TM, et al. Human odontoblasts express functional thermo-sensitive TRP channels: implications for dentin sensitivity. *Pain.* 2011;152:2211–2223.
184. Chasman DI, Schurks M, Anttila V, et al. Genome-wide association study reveals three susceptibility loci for common migraine in the general population. *Nat Genet.* 2011;43:695–698.
185. Bautista DM, Siemens J, Glazer JM, et al. The menthol receptor TRPM8 is the principal detector of environmental cold. *Nature.* 2007;448:204–208.
186. Colburn RW, Lubin ML, Stone Jr DJ, et al. Attenuated cold sensitivity in TRPM8 null mice. *Neuron.* 2007;54:379–386.
187. Dhaka A, Murray AN, Mathur J, Earley TJ, Petrus MJ, Patapoutian A. TRPM8 is required for cold sensation in mice. *Neuron.* 2007;54:371–378.
188. Noel J, Zimmermann K, Busserolles J, et al. The mechano-activated K^+ channels TRAAK and TREK-1 control both warm and cold perception. *EMBO J.* 2009;28:1308–1318.
189. Story GM, Peier AM, Reeve AJ, et al. ANKTM1, a TRP-like channel expressed in nociceptive neurons, is activated by cold temperatures. *Cell.* 2003;112:819–829.
190. Gavva NR, Davis C, Lehto SG, Rao S, Wang W, Zhu DX. Transient receptor potential melastatin 8 (TRPM8) channels are involved in body temperature regulation. *Mol Pain.* 2012;8:36.
191. Lashinger ES, Steiginga MS, Hieble JP, et al. AMTB, a TRPM8 channel blocker: evidence in rats for activity in overactive bladder and painful bladder syndrome. *Am J Physiol Renal Physiol.* 2008;295:F803–F810.
192. McKemy DD. How cold is it? TRPM8 and TRPA1 in the molecular logic of cold sensation. *Mol Pain.* 2005;1:16.
193. Jaquemar D, Schenker T, Trueb B. An ankyrin-like protein with transmembrane domains is specifically lost after oncogenic transformation of human fibroblasts. *J Biol Chem.* 1999;274:7325–7333.
194. Barabas ME, Kossyreva EA, Stucky CL. TRPA1 is functionally expressed primarily by IB4-binding, non-peptidergic mouse and rat sensory neurons. *PLoS One.* 2012;7: e47988.
195. Engel MA, Leffler A, Niedermirtl F, et al. TRPA1 and substance P mediate colitis in mice. *Gastroenterology.* 2011;141:1346–1358.
196. DeBerry JJ, Schwartz ES, Davis BM. TRPA1 mediates bladder hyperalgesia in a mouse model of cystitis. *Pain.* 2014;155:1280–1287.
197. Atoyan R, Shander D, Botchkareva NV. Non-neuronal expression of transient receptor potential type A1 (TRPA1) in human skin. *J Invest Dermatol.* 2009;129:2312–2315.
198. Kim YS, Son JY, Kim TH, et al. Expression of transient receptor potential ankyrin 1 (TRPA1) in the rat trigeminal sensory afferents and spinal dorsal horn. *J Comp Neurol.* 2010;518:687–698.
199. Uta D, Furue H, Pickering AE, et al. TRPA1-expressing primary afferents synapse with a morphologically identified subclass of substantia gelatinosa neurons in the adult rat spinal cord. *Eur J Neurosci.* 2010;31:1960–1973.

200. Fernandes ES, Fernandes MA, Keeble JE. The functions of TRPA1 and TRPV1: moving away from sensory nerves. *Br J Pharmacol.* 2012;166:510–521.
201. Shigetomi E, Tong X, Kwan KY, et al. TRPA1 channels regulate astrocyte resting calcium and inhibitory synapse efficacy through GAT-3. *Nat Neurosci.* 2011;15:70–80.
202. Pozsgai G, Bodkin JV, Graepel R, et al. Evidence for the pathophysiological relevance of TRPA1 receptors in the cardiovascular system in vivo. *Cardiovasc Res.* 2010;87:760–768.
203. Cao D-S, Zhong L, T-h Hsieh, et al. Expression of transient receptor potential ankyrin 1 (TRPA1) and its role in insulin release from rat pancreatic beta cells. *PLoS One.* 2012;7:e38005.
204. Corey DP, Garcia-Anoveros J, Holt JR, et al. TRPA1 is a candidate for the mechanosensitive transduction channel of vertebrate hair cells. *Nature.* 2004;432:723–730.
205. Haas ET, Rowland K, Gautam M. Tooth injury increases expression of the cold sensitive TRP channel TRPA1 in trigeminal neurons. *Arch Oral Biol.* 2011;56:1604–1609.
206. Nassini R, Pedretti P, Moretto N, et al. Transient receptor potential ankyrin 1 channel localized to non-neuronal airway cells promotes non-neurogenic inflammation. *PLoS One.* 2012;7:e42454.
207. Gaudet R. A primer on ankyrin repeat function in TRP channels and beyond. *Mol BioSyst.* 2008;4:372–379.
208. Li J, Mahajan A, Tsai M-D. Ankyrin repeat: a unique motif mediating protein–protein interactions. *Biochemistry.* 2006;45:15168–15178.
209. Nilius B, Prenen J, Owsianik G. Irritating channels: the case of TRPA1. *J Physiol.* 2011;589:1543–1549.
210. Wang L, Cvetkov TL, Chance MR, Moiseenkova-Bell VY. Identification of in vivo disulfide conformation of TRPA1 ion channel. *J Biol Chem.* 2012;287:6169–6176.
211. Cvetkov TL, Huynh KW, Cohen MR, Moiseenkova-Bell VY. Molecular architecture and subunit organization of TRPA1 ion channel revealed by electron microscopy. *J Biol Chem.* 2011;286:38168–38176.
212. Macpherson LJ, Dubin AE, Evans MJ, et al. Noxious compounds activate TRPA1 ion channels through covalent modification of cysteines. *Nature.* 2007;445:541–545.
213. Zurborg S, Yurgionas B, Jira JA, Caspani O, Heppenstall PA. Direct activation of the ion channel TRPA1 by Ca^{2+}. *Nat Neurosci.* 2007;10:277–279.
214. Karashima Y, Prenen J, Talavera K, Janssens A, Voets T, Nilius B. Agonist-induced changes in Ca^{2+} permeation through the nociceptor cation channel TRPA1. *Biophys J.* 2010;98:773–783.
215. Takahashi N, Mizuno Y, Kozai D, et al. Molecular characterization of TRPA1 channel activation by cysteine-reactive inflammatory mediators. *Channels.* 2008;2:287–298.
216. Pozsgai G, Hajna Z, Bagoly T, et al. The role of transient receptor potential ankyrin 1 (TRPA1) receptor activation in hydrogen-sulphide-induced CGRP-release and vasodilation. *Eur J Pharmacol.* 2012;689:56–64.
217. Trevisani M, Siemens J, Materazzi S, et al. 4-Hydroxynonenal, an endogenous aldehyde, causes pain and neurogenic inflammation through activation of the irritant receptor TRPA1. *Proc Natl Acad Sci USA.* 2007;104:13519–13524.
218. de la Roche J, Eberhardt MJ, Klinger AB, et al. The molecular basis for species-specific activation of human TRPA1 protein by protons involves poorly conserved residues within transmembrane domains 5 and 6. *J Biol Chem.* 2013;288:20280–20292.
219. Wang YY, Chang RB, Allgood SD, Silver WL, Liman ER. A TRPA1-dependent mechanism for the pungent sensation of weak acids. *J Gen Physiol.* 2011;137:493–505.
220. Wang YY, Chang RB, Liman ER. TRPA1 is a component of the nociceptive response to CO2. *J Neurosci.* 2010;30:12958–12963.
221. McNamara CR, Mandel-Brehm J, Bautista DM, et al. TRPA1 mediates formalin-induced pain. *Proc Natl Acad Sci USA.* 2007;104:13525–13530.

222. Bautista DM, Jordt S-E, Nikai T, et al. TRPA1 mediates the inflammatory actions of environmental irritants and proalgesic agents. *Cell*. 2006;124:1269–1282.
223. Chen J, Kang D, Xu J, et al. Species differences and molecular determinant of TRPA1 cold sensitivity. *Nat Commun*. 2013;4:2501.
224. Leffler A, Lattrell A, Kronewald S, Niedermirtl F, Nau C. Activation of TRPA1 by membrane permeable local anesthetics. *Mol Pain*. 2011;7:62.
225. Fischer MJM, Leffler A, Niedermirtl F, et al. The general anesthetic propofol excites nociceptors by activating TRPV1 and TRPA1 rather than GABAA receptors. *J Biol Chem*. 2010;285:34781–34792.
226. Matta JA, Cornett PM, Miyares RL, Abe K, Sahibzada N, Ahern GP. General anesthetics activate a nociceptive ion channel to enhance pain and inflammation. *Proc Natl Acad Sci USA*. 2008;105:8784–8789.
227. Piao L-H, Fujita T, Jiang C-Y, et al. TRPA1 activation by lidocaine in nerve terminals results in glutamate release increase. *Biochem Biophys Res Commun*. 2009;379:980–984.
228. Komatsu T, Uchida K, Fujita F, Zhou Y, Tominaga M. Primary alcohols activate human TRPA1 channel in a carbon chain length-dependent manner. *Pflugers Arch*. 2012;463:549–559.
229. Talavera K, Gees M, Karashima Y, et al. Nicotine activates the chemosensory cation channel TRPA1. *Nat Neurosci*. 2009;12:1293–1299.
230. Motter AL, Ahern GP. TRPA1 is a polyunsaturated fatty acid sensor in mammals. *PLoS One*. 2012;7:e38439.
231. Hu H, Tian J, Zhu Y, et al. Activation of TRPA1 channels by fenamate nonsteroidal anti-inflammatory drugs. *Pflugers Arch*. 2010;459:579–592.
232. Bandell M, Story GM, Hwang SW, et al. Noxious cold ion channel TRPA1 is activated by pungent compounds and bradykinin. *Neuron*. 2004;41:849–857.
233. Dunham J, Leith J, Lumb B, LF D. Transient receptor potential channel A1 and noxious cold responses in rat cutaneous nociceptors. *Neuroscience*. 2010;165:1412–1419.
234. Karashima Y, Talavera K, Everaerts W, et al. TRPA1 acts as a cold sensor in vitro and in vivo. *Proc Natl Acad Sci USA*. 2009;106:1273–1278.
235. Jabba S, Goyal R, Sosa-Pagán Jason O, et al. Directionality of temperature activation in mouse TRPA1 ion channel can be inverted by single-point mutations in ankyrin repeat six. *Neuron*. 2014;82:1017–1031.
236. Brierley SM, Castro J, Harrington AM, et al. TRPA1 contributes to specific mechanically activated currents and sensory neuron mechanical hypersensitivity. *J Physiol*. 2011;589:3575–3593.
237. Brierley SM, Hughes PA, Page AJ, et al. The ion channel TRPA1 is required for normal mechanosensation and is modulated by algesic stimuli. *Gastroenterology*. 2009;137:2084–2095, e2083.
238. Wang YY, Chang RB, Waters HN, McKemy DD, Liman ER. The nociceptor ion channel TRPA1 is potentiated and inactivated by permeating calcium ions. *J Biol Chem*. 2008;283:32691–32703.
239. Kim D, Cavanaugh EJ, Simkin D. Inhibition of transient receptor potential A1 channel by phosphatidylinositol-4,5-bisphosphate. *Am J Physiol Cell Physiol*. 2008;295: C92–C99.
240. Salas MM, Hargreaves KM, Akopian AN. TRPA1-mediated responses in trigeminal sensory neurons: interaction between TRPA1 and TRPV1. *Eur J Neurosci*. 2009;29:1568–1578.
241. Sisignano M, Park C-K, Angioni C, et al. 5,6-EET is released upon neuronal activity and induces mechanical pain hypersensitivity via TRPA1 on central afferent terminals. *J Neurosci*. 2012;32:6364–6372.

242. Gregus AM, Doolen S, Dumlao DS, et al. Spinal 12-lipoxygenase-derived hepoxilin A3 contributes to inflammatory hyperalgesia via activation of TRPV1 and TRPA1 receptors. *Proc Natl Acad Sci USA*. 2012;109:6721–6726.
243. Elitt CM, McIlwrath SL, Lawson JJ, et al. Artemin overexpression in skin enhances expression of TRPV1 and TRPA1 in cutaneous sensory neurons and leads to behavioral sensitivity to heat and cold. *J Neurosci*. 2006;26:8578–8587.
244. Kwan KY, Allchorne AJ, Vollrath MA, et al. TRPA1 contributes to cold, mechanical, and chemical nociception but is not essential for hair-cell transduction. *Neuron*. 2006;50:277–289.
245. Koltzenburg M, Lundberg LER, Torebjörk HE. Dynamic and static components of mechanical hyperalgesia in human hairy skin. *Pain*. 1992;51:207–219.
246. Bianchi BR, Zhang X-F, Reilly RM, Kym PR, Yao BB, Chen J. Species comparison and pharmacological characterization of human, monkey, rat, and mouse TRPA1 channels. *J Pharmacol Exp Ther*. 2012;341:360–368.
247. Liu B, Escalera J, Balakrishna S, et al. TRPA1 controls inflammation and pruritogen responses in allergic contact dermatitis. *FASEB J*. 2013;27:3549–3563.
248. Nakamura Y, Une Y, Miyano K, et al. Activation of transient receptor potential ankyrin 1 evokes nociception through substance P release from primary sensory neurons. *J Neurochem*. 2012;120:1036–1047.
249. Than JY-XL, Li L, Hasan R, Zhang X. Excitation and modulation of TRPA1, TRPV1, and TRPM8 channel-expressing sensory neurons by the pruritogen chloroquine. *J Biol Chem*. 2013;288:12818–12827.
250. Wilson SR, Nelson AM, Batia L, et al. The ion channel TRPA1 is required for chronic itch. *J Neurosci*. 2013;33:9283–9294.
251. Kremeyer B, Lopera F, Cox JJ, et al. A gain-of-function mutation in TRPA1 causes familial episodic pain syndrome. *Neuron*. 2010;66:671–680.
252. Bell JT, Loomis AK, Butcher LM, et al. Differential methylation of the TRPA1 promoter in pain sensitivity. *Nat Commun*. 2014;5:2978.
253. Zhou Y, Suzuki Y, Uchida K, Tominaga M. Identification of a splice variant of mouse TRPA1 that regulates TRPA1 activity. *Nat Commun*. 2013;4:2399.
254. Kim S, Hwang S. Emerging roles of TRPA1 in sensation of oxidative stress and its implications in defense and danger. *Arch Pharm Res*. 2013;36:783–791.
255. Barrière DA, Rieusset J, Chanteranne D, et al. Paclitaxel therapy potentiates cold hyperalgesia in streptozotocin-induced diabetic rats through enhanced mitochondrial reactive oxygen species production and TRPA1 sensitization. *Pain*. 2012;153:553–561.
256. Nassini R, Gees M, Harrison S, et al. Oxaliplatin elicits mechanical and cold allodynia in rodents via TRPA1 receptor stimulation. *Pain*. 2011;152:1621–1631.
257. Old EA, Nadkarni S, Grist J, et al. Monocytes expressing CX3CR1 orchestrate the development of vincristine-induced pain. *J Clin Invest*. 2014;124:2023–2036.
258. Benemei S, De Cesaris F, Fusi C, Rossi E, Lupi C, Geppetti P. TRPA1 and other TRP channels in migraine. *J Headache Pain*. 2013;14:71.
259. Edelmayer RM, Le LN, Yan J, et al. Activation of TRPA1 on dural afferents: a potential mechanism of headache pain. *Pain*. 2012;153:1949–1958.
260. Fernandes ES, Russell FA, Spina D, et al. A distinct role for transient receptor potential ankyrin 1, in addition to transient receptor potential vanilloid 1, in tumor necrosis factor α-induced inflammatory hyperalgesia and Freund's complete adjuvant-induced monarthritis. *Arthritis Rheum*. 2011;63:819–829.
261. Yang J, Li Y, Zuo X, Zhen Y, Yu Y, Gao L. Transient receptor potential ankyrin-1 participates in visceral hyperalgesia following experimental colitis. *Neurosci Lett*. 2008;440:237–241.

262. Terada Y, Fujimura M, Nishimura S, et al. Contribution of TRPA1 as a downstream signal of proteinase-activated receptor-2 to pancreatic pain. *J Pharmacol Sci.* 2013;123:284–287.
263. Andersson DA, Gentry C, Alenmyr L, et al. TRPA1 mediates spinal antinociception induced by acetaminophen and the cannabinoid Δ9-tetrahydrocannabiorcol. *Nat Commun.* 2011;2:551.
264. Meseguer V, Alpizar YA, Luis E, et al. TRPA1 channels mediate acute neurogenic inflammation and pain produced by bacterial endotoxins. *Nat Commun.* 2014;5:3125.
265. Bessac BF, Sivula M, von Hehn CA, Escalera J, Cohn L, Jordt S-E. TRPA1 is a major oxidant sensor in murine airway sensory neurons. *J Clin Invest.* 2008;118:1899–1910.
266. Eid S, Crown E, Moore E, et al. HC-030031, a TRPA1 selective antagonist, attenuates inflammatory- and neuropathy-induced mechanical hypersensitivity. *Mol Pain.* 2008;4:48.
267. Chen J, Joshi SK, DiDomenico S, et al. Selective blockade of TRPA1 channel attenuates pathological pain without altering noxious cold sensation or body temperature regulation. *Pain.* 2011;152:1165–1172.
268. Petrus M, Peier A, Bandell M, et al. A role of TRPA1 in mechanical hyperalgesia is revealed by pharmacological inhibition. *Mol Pain.* 2007;3:40.
269. Pertovaara A, Koivisto A. TRPA1 ion channel in the spinal dorsal horn as a therapeutic target in central pain hypersensitivity and cutaneous neurogenic inflammation. *Eur J Pharmacol.* 2011;666:1–4.
270. Mukhopadhyay I, Kulkarni A, Aranake S, et al. Transient receptor potential ankyrin 1 receptor activation *in vitro* and *in vivo* by pro-tussive agents: GRC 17536 as a promising anti-tussive therapeutic. *PLoS One.* 2014;9:e97005.

CHAPTER FIVE

The Contribution of Mitochondria to Sensory Processing and Pain

Sarah J.L. Flatters[1]
Wolfson Centre for Age-Related Diseases, King's College London, London, United Kingdom
[1]Corresponding author: e-mail address: sarah.flatters@kcl.ac.uk

Contents

1. Introduction 120
2. Role of Mitochondria in Normal Sensory Processing and Acute Pain 121
3. Clinical Evidence of Mitochondrial Dysfunction in Painful Conditions 126
4. Preclinical Evidence of Mitochondrial Dysfunction in Chronic Pain States 129
   4.1 Chemotherapy-induced painful neuropathy 129
   4.2 Diabetic painful neuropathy 134
   4.3 HIV-associated painful neuropathy 138
   4.4 Other chronic pain states 139
5. Future Perspectives 140
Acknowledgments 141
References 141

Abstract

Mitochondria have a variety of essential functions within neurons including oxygen consumption, ATP generation, calcium buffering, and reactive oxygen species (ROS) generation. Despite extensive research into the contribution of mitochondrial function in other neurological disorders such as Parkinson's disease, the role of mitochondrial function in sensory processing and pain has been relatively unexplored until recent years. As this area of pain research is in its infancy, this review will be a descriptive summary—rather than a critical review—of data that suggests mitochondrial function/dysfunction as a causal or contributory mechanism of normal sensory processing and chronic pain. Evidence for mitochondrial dysfunction from both chronic pain patients and animal models of chronic pain will be described. Such evidence involves different aspects of mitochondria and their function including mitochondrial ultrastructure, distribution, oxygen consumption, oxidative phosphorylation, calcium buffering, ROS, and ATP levels. Most recently, substantial amounts of data have demonstrated mitochondrial involvement in painful peripheral neuropathies evoked by chemotherapy, diabetes, and HIV and these topics will be particularly highlighted in this review.

Progress in Molecular Biology and Translational Science, Volume 131
ISSN 1877-1173
http://dx.doi.org/10.1016/bs.pmbts.2014.12.004

1. INTRODUCTION

Mitochondria are known as "the powerhouse of the cell" due to their vital ability to generate adenosine triphosphate (ATP), which is essential to virtually all cellular processes. Oxidative phosphorylation (OXPHOS) involves the transfer of electrons through the mitochondrial electron transport chain (mETC) and uptake of molecular oxygen resulting in the generation of ATP. Electrons can leak from the mETC react with molecular oxygen generating superoxide radicals, one of several subspecies of reactive oxygen species (ROS). Therefore, mitochondria are also a major source of ROS. Mitochondria have many other roles aside from ATP generation; for example, they play an important role in calcium buffering within cells. Mitochondria also have their own separate DNA—mitochondrial DNA (mtDNA) of which only 5% is not also encoded in the nuclear genome.

Despite extensive research into the contribution of mitochondrial function in other neurological disorders such as Parkinson's disease and Alzheimer's disease, the role of mitochondrial function in sensory processing and pain has been relatively unexplored until the last decade. This chapter will detail research articles that have examined the role of mitochondria in sensory transduction with particular focus on those reports investigating the contribution of mitochondrial function/dysfunction to chronic pain states. As this area of pain research is in its infancy, this chapter will be a descriptive summary, rather than a critical review, of the diverse data published so far that links mitochondrial function/dysfunction as a causal pain mechanism. The wealth of research articles investigating the role of mitochondria in neurodegeneration and apoptosis, which may have some implications in degenerative chronic pain states, will not be described here. This chapter will describe data examining (a) the role of mitochondria in normal sensory processing and acute pain, (b) evidence for mitochondrial dysfunction from patients with painful conditions, and (c) evidence for mitochondrial dysfunction in animal models of chronic pain. In each of these sections, different aspects of mitochondria and their function will be described according to the literature published to date. Such aspects will include mitochondrial ultrastructure, distribution, oxygen consumption, OXPHOS, calcium buffering, ROS, and ATP levels. The experimental techniques utilized to assess these different aspects will also be briefly described.

2. ROLE OF MITOCHONDRIA IN NORMAL SENSORY PROCESSING AND ACUTE PAIN

ATP demand varies substantially between different cell types, due to their different functions. Mitochondria are therefore distributed accordingly to meet such variation in ATP requirements. In general, neurons have a reasonably high energetic demand overall. However, when the total size of the neuron—cell body and length of processes—is considered along with the constantly changing activity of different parts of the neuron, energy demand will be variable between structures within any one neuron. Therefore, a common assumption is that mitochondria are not randomly distributed within a neuron and their distribution will change in response to ATP requirements. Most of the primary evidence to support this comes from work examining mitochondrial function in central nervous system (CNS) neurons particularly in the brain. Quantitative analysis of mitochondria in the neuropil of the primate visual cortex demonstrated that over 60% of the total number of mitochondria and mitochondrial area were found in the dendrites.[1] Estimates of the signaling-related energy usage in the gray matter of the rat brain also showed that the predicted energy usage of axons and terminals (~42%) was less than the predicted energy usage of dendrites and soma (~52%).[2] Therefore, despite the greater volume of axons compared to the somatodendritic compartment, they contain less mitochondria and require less energy.

Less work has been reported on mitochondrial distribution and function in peripheral neurons, but there are reports relevant to the role of mitochondria in normal sensory processing. Using electron microscopy and tomographic reconstruction, the 3D structure of mitochondria in axons and Schwann cells in spinal roots of normal mice were described.[3] Condensed mitochondria with dilated cristae comprise roughly two-thirds of axonal spinal root mitochondria, with orthodox mitochondria constituting the remainder. Condensed mitochondria were present in higher concentrations in the nodal region. In comparison, the mitochondria in Schwann cells have a classical orthodox structure only.[3] Studies have examined the distribution of mitochondria and mitochondrial activity in human peripheral sensory nerves finding marked differences between myelinated and unmyelinated axons. Lehmann *et al.* determined the distribution of mitochondria within sensory nerve fibers of sural nerve specimens from humans and primates. In human sural nerves, approximately 55% of mitochondria are found in myelinated axons, 30% in unmyelinated axons, and 15% in Schwann cells. Similarly, in primates, the majority of mitochondria were found in myelinated axons; approximately 70%, 25%, and 5% of mitochondria were found in myelinated axons, unmyelinated axons, and Schwann cells, respectively.[4] However, when the number of mitochondria was calculated per axonal volume, these authors report that unmyelinated axons have "almost twice" the number of mitochondria compared to myelinated axons. Similar differences have also been observed in the human optic nerve, low cytochrome *c* activity was found in myelinated axons, and high cytochrome *c* activity was found in unmyelinated axons.[5] These findings suggest that unmyelinated axons have a higher energy requirement compared to myelinated axons. Animal studies confirm such a suggestion. Using brain white matter from a variety of different animal species, ranging from shrew to macaque, Wang and colleagues have estimated that unmyelinated axons consume 2.5- to 10-fold more energy per action potential evoked compared to myelinated axons.[6]

Studies of local glucose utilization have shown that energy metabolism alters relative to neuronal activity in a unidirectional manner.[7] Specifically, electrical stimulation of the sciatic nerve in anesthetized rats resulted in a frequency-dependent linear increase in glucose utilization in the lumbar spinal cord.[8] However, these experiments also showed that glucose utilization was unaltered in the dorsal root ganglia (DRG) by sciatic nerve stimulation.[8] This suggests that functional activity is dictated by glucose utilization in nerve terminals rather than cell bodies. More recently, using *in vivo* confocal imaging of saphenous nerves in anesthetized mice, mitochondrial transport has been examined in sensory axons in relation to neuronal activity.[9] Electrical stimulation of myelinated axons increased mitochondrial transport and the velocity of moving mitochondria. At low-frequency stimulation (1 Hz), mitochondrial movement was increased in both anterograde and retrograde directions. In comparison, at high-frequency stimulation (50 Hz), only anterograde mitochondrial movement and velocity was markedly increased. Sajic *et al.* also showed that with increasing electrical stimulation mitochondria accumulate in the skin, but not the DRG, demonstrating that mitochondria accumulate in distal portions of sensory axons with increasing neuronal activity (see Fig. 1). In myelinated axons of the CNS, there is greater accumulation of mitochondria at nodes of Ranvier and decreased mitochondrial transport during increased axonal electrical activity.[10] However, in myelinated axons of the saphenous nerve, no such nodal accumulation of mitochondria during increased electrical stimulation was observed.[9]

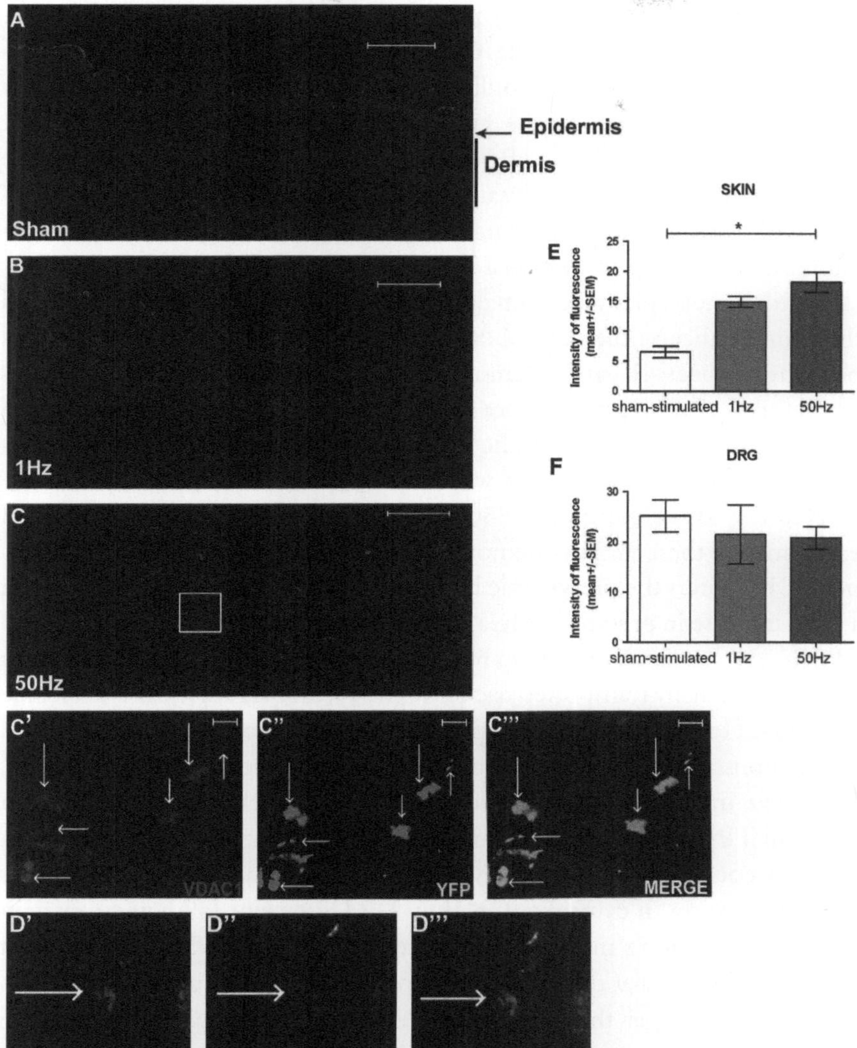

Figure 1 In normal saphenous nerve axons stimulated at high frequency (50 Hz), mitochondria accumulate at the peripheral sensory terminals. Skin innervated by saphenous nerve was immunohistochemically labeled with VDAC1 (red) and neuron-specific β-III-tubulin (green). In comparison with sham-stimulated animals (A), axons were strongly labeled for VDAC1 in nerves stimulated at 1 Hz (B) and 50 Hz (C). (C′–C‴) Saphenous nerve from a YFP^+ mouse (shown in low power in C), which expresses YFP (green) in a proportion of fibers, was stimulated with 50 Hz and labeled with VDAC1 (red). (D′–D‴) Saphenous nerve from a YFP^- mouse was stimulated with 50 Hz and double labeled with VDAC1 (red) and β-III-tubulin. The two markers were

(Continued)

These differences could be due to different methodologies or reflect differences in mitochondrial transport in CNS versus peripheral axons. Using capsaicin application to the skin to directly activate unmyelinated axons, there was a threefold increase in the number of moving mitochondria, but not their velocity, in isolectin B4-positive C-fibers.[9] These effects were observed 15 min following capsaicin application, and this temporally correlates with capsaicin-evoked firing of C-nociceptors,[11] again linking mitochondrial trafficking to neuronal activity in sensory afferents.

Mitochondria play an essential role in calcium buffering within cells, and their importance in the control of intracellular calcium, $[Ca^{2+}]_i$, in sensory neurons has been elegantly demonstrated by the work of Yuriy Usachev's laboratory.[12,13] Using a coculture of DRG and spinal cord neurons to model the first sensory synapse, cytosolic and mitochondrial Ca^{2+} in DRG neurons following capsaicin application were measured in conjunction with excitatory postsynaptic currents (EPSCs) in spinal cord neurons. In a series of experiments, these authors demonstrated that following capsaicin application, Ca^{2+} enters the presynaptic bouton through TRPV1 channels causing a large increase in presynaptic $[Ca^{2+}]_i$ which triggered glutamate release and EPSCs. Ca^{2+} is rapidly taken up by mitochondria in the presynaptic bouton and then gradually transported back into the cytosol via Na^+/Ca^{2+} exchange as opposed to the mitochondrial permeability transition pore (mPTP). This was demonstrated pharmacologically by comparing the effect of CGP37157, a selective inhibitor of mitochondrial Na^+/Ca^{2+} exchange, to cyclosporin A, an mPTP inhibitor. This calcium-buffering mechanism by mitochondria which is coupled to Ca^{2+} efflux via Na^+/Ca^{2+} exchange explains the prolonged elevation in cytosolic $[Ca^{2+}]_i$ in DRG neurons following capsaicin application resulting in further glutamate release and EPSCs. Medvedeva *et al.* also showed that the greater the amount of Ca^{2+} sequestered by mitochondria, the longer the duration of elevated presynaptic Ca^{2+}, glutamate release, and postsynaptic activity.[12] Prolonged increases in $[Ca^{2+}]_i$ are also routinely observed in DRG neurons following KCl application. Svichar *et al.* showed that the lipophilic cation tetraphenylphosphonium (TPP^+) which blocks both Na^+-dependent and Na^+-independent mitochondrial

Figure 1—Cont'd often found to colocalize. (E) Intensity of VDAC1 labeling was significantly higher within cutaneous fibers of saphenous nerve stimulated at 50 Hz ($n=5$, $*p<0.05$) than in sham-stimulated ($n=3$) or fibers stimulated at 1 Hz ($n=3$). (F) There was no difference in VDAC1-labeling intensity in DRGs of saphenous nerves between the groups. Scale bars in (A–C) = 100 μm, in (C′–C‴) = 20 μm, and in (D′–D‴) = 10 μm. *Taken from Ref. 9 figure 3—doi: http://dx.doi.org/10.1371/journal.pbio.1001754.g003.*

Ca^{2+} efflux markedly inhibited KCl-evoked elevations in $[Ca^{2+}]_i$ in both small and large DRG neurons.[14] This further demonstrates the importance of mitochondrial Na^+/Ca^{2+} exchange in presynaptic calcium control in sensory neurons. Further work using cocultures of DRG and spinal cord neurons showed that presynaptic mitochondria also sequestered Ca^{2+} following electrical stimulation.[13] Examining the potential mechanisms of presynaptic Ca^{2+} clearance in capsaicin-sensitive DRG neurons following electrical stimulation showed that presynaptic Ca^{2+} clearance was dependent upon plasma membrane Ca^{2+}-ATPase, mitochondria, plasmalemmal Na^+/Ca^{2+} exchange, and sarco–endoplasmic reticulum Ca^{2+}-ATPase by approximately 47%, 40%, 12%, and 0%, respectively. Mitochondrial calcium cycling in DRG neurons was also shown to promote the activation of nuclear factor of activated T-cells,[15] which has been implicated in neuronal survival, synaptic plasticity, and axonal growth (discussed in Ref. 15).

The specific effects of mitochondrial calcium uptake on NMDA and capsaicin-evoked pain behaviors and spinal long-term potentiation were assessed using Ru360, a specific inhibitor of the mitochondrial calcium uniporter (MCU).[16] Pretreatment with intrathecal Ru360 significantly inhibited the development of NMDA-evoked mechanical hypersensitivity in the hind paw. However, Ru360 had no effect on established NMDA-evoked mechanical hypersensitivity. NMDA evoked a 2.5-fold increase in mitochondrial Ca^{2+} levels (visualized with Rhod2/AM) in the superficial and deep dorsal horn which was markedly inhibited by pretreatment, but not posttreatment, with Ru360.[16] Ru360 also inhibited NMDA-evoked increases in mitochondrial superoxide levels in the superficial and deep dorsal horn. Using spinal cord slice preparations, Ru360 was shown to block spinal LTP despite large increases in cytosolic Ca^{2+} levels (which were unaffected by Ru360).[16] This indicates that high levels of Ca^{2+} cannot directly trigger LTP unless Ca^{2+} is first sequestered by mitochondria. Blockade of mitochondrial Ca^{2+} uptake also prevented the phosphorylation of several protein kinases in the spinal cord that have been implicated in synaptic plasticity and pain processing, namely protein kinase C, protein kinase A (PKA), and extracellular signal-related kinase (ERK).[16] In recent years, different components of mitochondrial Ca^{2+} transport have been identified in major breakthroughs for the field of mitochondrial research.[17–19] Using RT-PCR of RNA extracted from neuron-enriched DRG cultures, transcripts of the MCU, mitochondrial calcium uptake 1, $Na^+/Ca^{2+}/Li^+$ exchanger, and leucine zipper EF hand-containing transmembrane protein (Letm1) were all observed, providing further evidence for mitochondrial calcium buffering in sensory neurons.[13]

The measurement and manipulation of ROS levels have been performed under both normal conditions and acute pain conditions. In normal mice, intrathecal administration of rotenone (complex I inhibitor) produced a significant mechanical hypersensitivity from 5 h postinjection lasting until 48 h postinjection.[20] In comparison, systemic administration of rotenone had no effect on withdrawal responses to mechanical stimulation in normal rats.[21] Pretreatment with intrathecal rotenone markedly inhibited both NMDA- and capsaicin-evoked mechanical hypersensitivity in mice.[16] Rotenone also prevented NMDA-evoked increases in mitochondrial superoxide levels in the superficial and deep dorsal horn.[16] Intrathecal administration of antimycin A in normal mice also elicited a significant mechanical hypersensitivity from 3 h postinjection lasting until 24 h postinjection.[20] Antimycin A-evoked mechanical hypersensitivity was also shown to be reversible by systemic administration of ROS scavengers.[20] Significant increases in superoxide levels (visualized with MitoSOX) in both the superficial and deep dorsal horn were evident 9 h following intrathecal antimycin A.[20] Intraplantar capsaicin evokes an acute pain syndrome, evidenced with behavioral primary and secondary hyperalgesias that emerge and resolve within 24 h.[22,23] The development and maintenance of capsaicin-evoked secondary hyperalgesia were shown to be ROS-mediated, as ROS scavengers could alleviate capsaicin-evoked secondary, but not primary, hyperalgesia.[22] Capsaicin also increased mitochondrial superoxide levels in the superficial and deep dorsal horn.[22,23] In further work by the same group, superoxide dismutase (SOD-2) activity in the spinal cord was decreased following intraplantar capsaicin.[23] Mice overexpressing SOD-2 did not develop capsaicin-evoked secondary hyperalgesia nor show increased levels of mitochondrial superoxide in the dorsal horn.[23] In contrast, SOD-2 knockout mice developed enhanced capsaicin-evoked primary and secondary hyperalgesias and showed larger increases in mitochondrial superoxide levels, particularly in the deep dorsal horn (laminae III–V).[23] Mitochondrial superoxide has also been implicated in the modulation of neuronal excitability of amygdala neurons and amygdala-mediated behavior via an mGluR5-IP_3-ROS-ERK/PKA signaling cascade.[24]

3. CLINICAL EVIDENCE OF MITOCHONDRIAL DYSFUNCTION IN PAINFUL CONDITIONS

There is a plethora of inherited mitochondrial diseases, some of which are associated with neuropathy as a diagnostic criterion. The prevalence of pain caused by mitochondrial disease is not well reported compared to motor consequences of the neuropathies caused by these devastating disorders. Discussion of mitochondrial abnormalities in the context of peripheral neuropathies in mitochondriopathies, per se, is outside the scope of this chapter (for recent review on this topic see Ref. 25). There is debate as to whether polyneuropathy seen in mitochondriopathy is primarily due to mitochondriopathy or a consequence of simultaneously occurring diabetes for example. In a study of 108 patients with mitochondriopathy and neuropathy, the neuropathy in 38 of these patients was found to be only attributable to the mitochondriopathy itself.[26] Of those 38 patients, sensory fibers, motor fibers, and both sensory and motor fibers were predominantly affected in 3, 14, and 21 patients, respectively.[26] Paresthesia, dysesthesia, and hypesthesia were reported in 4, 4, and 10 of these 38 patients, respectively. Muscle ache was reported in 11 of these 38 patients, but no other details of pain symptoms were reported.[26] Detailed descriptive analysis of the changes to the ultrastructure of neuronal mitochondria in sural nerve biopsies from patients with mitochondrial myopathy, hereditary, and nonhereditary neuropathies has also been reported and compared.[27] Charcot–Marie–Tooth neuropathy type 2A, an hereditary axonal peripheral sensorimotor neuropathy that predominantly affects the ankles and feet, is caused by missense mutations in mitochondrial fusion protein mitofusin 2 (MFN2)—a key protein that regulates the mitochondrial network architecture by fusion of mitochondria.[28]

Two case reports describe morphological examination of sural nerve biopsies from patients with chemotherapy-induced neuropathy evoked by paclitaxel[29] and docetaxel.[30] Although not specifically described in these reports, examination of the electron micrographs presented demonstrates the occurrence of swollen, vacuolated mitochondria in sensory axons. Sural nerve biopsies from seven patients with adult-onset diabetes and progressive worsening of their associated neuropathy contained swollen, vacuolated mitochondria in myelinating Schwann cells.[31] In comparison, the mitochondria in nonmyelinating Schwann cells appeared normal; however, glycogen accumulations in the outer compartment of axonal mitochondria (often known as glycogenosomes) were occasionally observed.[31] HIV and one of its treatments have also been found to elicit effects on mitochondria in sensory axons.[32] Sural nerve biopsies were taken from three different groups of patients: HIV patients treated with 2′3′-dideoxycytidine (ddC), HIV patients with neuropathy but without ddC treatment (HIV+), and HIV patients with non-HIV-related axonal polyneuropathy (HIV−).

The percentage of abnormal (swollen and vacuolated) mitochondria found in both myelinated and unmyelinated axons and Schwann cells was then quantified. Approximately 55%, 28%, and 9% of mitochondria in the axoplasm were abnormal in ddC, HIV+, and HIV− patients, respectively. In addition, approximately 47%, 19%, and 14% of mitochondria in Schwann cells were abnormal in ddC, HIV+, and HIV− patients, respectively.[32] These authors also reported a significant decrease in mtDNA in sural nerves from ddC patients compared to HIV− controls. This demonstrates that ddC-induced painful neuropathy (and HIV-induced neuropathy to a lesser extent) can be correlated with a significant increase in abnormal neuronal mitochondria in sensory axons. Lehmann *et al.* compared human DRG and sural nerves postmortem specimens from HIV patients with painful neuropathy (HIV-PN), HIV patients without painful neuropathy (HIV+), and people without HIV (control), for the levels of mtDNA deletion mutations and mitochondrial proteins.[4] Levels of the common $mtDNA^{4977}$ deletion mutation (4977 is indicative of 4977 base pairs)—a marker of mitochondrial damage—were significantly increased in sural nerves from the HIV-PN group compared to HIV+ and controls. However, levels of $mtDNA^{4977}$ deletion in the DRG were similar between the three groups.[4] Further analysis of the sural nerve samples revealed that there was a marked decrease in a mitochondrial glycoprotein in both HIV-PN and HIV+ groups.[4] Moreover, the ratio of mitochondrial glycoprotein to the catalytic subunit I of complex IV (COX-I) was only decreased in the HIV-PN group.[4] This suggests that during HIV painful neuropathy, there is an increase in mtDNA damage and a loss of mitochondria in distal sensory axons.

Skin biopsies from 32 patients with painful sensory distal polyneuropathy caused by various etiologies were double-stained with PGP9.5 and complex IV antibodies, to quantify the number of intraepidermal nerve fibers (IENFs) and mitochondria (OXPHOS-immunoreactivity, OXPHOS-ir), respectively.[33] Based on their IENF density, the biopsies were split into two categories—definite small fiber neuropathy (SFN definite) and SFN borderline—and compared to biopsies from control subjects without neuropathy (control). In biopsies from control subjects without neuropathy, OXPHOS-ir colocalized with IENFs and there was a positive correlation between OXPHOS-ir mitochondria and PGP-positive axons.[33] However, this colocalization and correlation were lost in both the SFN definite and SFN borderline biopsies suggesting that the loss of mitochondria in cutaneous afferents is an early sign of painful peripheral neuropathy.[33]

There is also some clinical evidence suggesting that mitochondrial dysfunction could be contributory to intervertebral disk degeneration and back pain.[34] Cultured annulus cells from human disks of varying degenerative stages (II–V on Thompson Grade) were assessed for mitochondrial morphology, oxygen consumption, and citrate synthase content. There was some descriptive observation of morphological changes, for example, the presence of "abnormally shaped branching mitochondria with small cristae." Both basal respiration and mean mitochondrial respiration were significantly higher in pooled cells from more degenerate Grade III–V disks compared to cells from Grade II disks. Citrate synthase activity (which can considered as a measure of mitochondrial mass) significantly decreased with increasing degree of disk degeneration.[34]

4. PRECLINICAL EVIDENCE OF MITOCHONDRIAL DYSFUNCTION IN CHRONIC PAIN STATES

In recent years, data exploring the mechanistic contribution of mitochondrial function to the genesis and maintenance of chronic pain have started to emerge. Such data have mainly arrived from investigations using animal models of chemotherapy-induced, HIV-associated, and diabetic neuropathies. Studies have examined not only mitochondrial function/dysfunction, per se, but also the consequences of mitochondrial dysfunction in terms of the generation of ROS leading to oxidative stress/damage. The following sections will divide this literature in terms of reports detailing the evidence for mitochondrial dysfunction and oxidative stress in different preclinical chronic pain states.

4.1 Chemotherapy-induced painful neuropathy

Several widely used first-line chemotherapeutic drugs are associated with painful peripheral neuropathy. This side effect is dose dependent and often dose limiting as there are no treatments to prevent the development of chemotherapy-induced painful neuropathy (CIPN). In addition, many analgesics have failed to show efficacy in patients with CIPN. Chemotherapeutics that are associated with painful neuropathy include taxanes, e.g., paclitaxel; vinca alkaloids, e.g., vincristine; and platinum drugs, e.g., oxaliplatin, epothilones, bortezomib, and thalidomide. For reviews on chemotherapy-induced neuropathies from clinical viewpoints, see Refs. 35,36.

The first evidence that mitochondrial dysfunction is associated with CIPN came from an extensive electron microscopy study comparing saphenous nerves from paclitaxel- and vehicle-treated rats.[37] Saphenous nerves were compared at three key time points in the paclitaxel-induced pain syndrome: Day 7, 24 h after the last of four systemic injections of paclitaxel—prior to pain onset, day 27—at the peak of paclitaxel-induced pain behavior, and day 160—at the resolution of paclitaxel-induced pain behavior. Atypical neuronal mitochondria were identified as vacuolated (>50%) and/or markedly swollen and were present in C-fibers and myelinated axons of both paclitaxel- and vehicle-treated rats (Fig. 2). The striking findings of this study were a 64% and 128% increase in the prevalence of atypical mitochondria in C-fibers of paclitaxel-treated rats at day 7 and day 27, respectively. Furthermore, similar findings were observed in myelinated axons, with a 137% and 159% increase in the prevalence of atypical mitochondria day 7 and day 27, respectively. In contrast, at day 160, there were no significant changes in neuronal mitochondria in C-fibers or myelinated axons. Therefore, these paclitaxel-induced changes in neuronal mitochondria could be correlated to the development and maintenance of paclitaxel-induced pain syndrome, i.e., present prior to and during paclitaxel-induced pain, but absent when the pain syndrome had resolved. These low-dose paclitaxel-induced mitochondrial changes in C-fibers and myelinated axons have been further confirmed in other reports.[38–40] Further investigation using electron microscopy identified increased incidence of swollen/vacuolated mitochondria in C-fibers and A-fibers of the dorsal root at the peak of paclitaxel-induced pain behavior, but no change in the A-fibers of the ventral root.[39] No changes in the incidence of atypical mitochondria were observed in the Schwann cells (myelinating and nonmyelinating) of the saphenous nerve, dorsal, or ventral roots.[39]

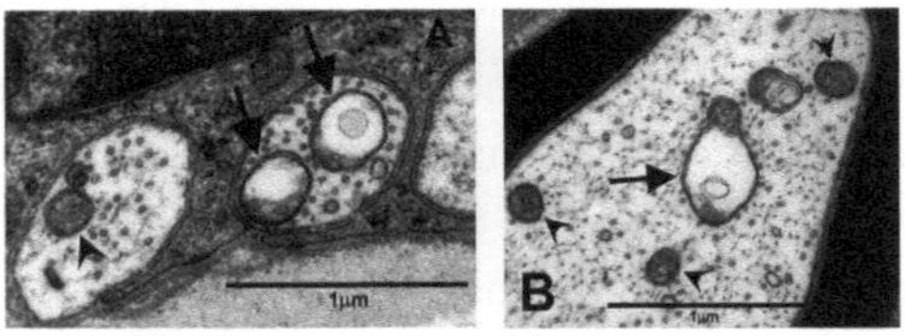

Figure 2 Examples of atypical neuronal mitochondria found in the saphenous nerve of paclitaxel-treated rats at the peak of paclitaxel-induced pain behavior. (A) C-fibers and (B) small, thinly myelinated Aδ-fiber from a paclitaxel-treated rat at day 27 posttreatment initiation. NB: Atypical mitochondria were observed in both small and large diameter myelinated axons. Magnification: 44,400×. *Taken from Ref. 37, figures 9 and 10. The figures have been reproduced with permission of the International Association for the Study of Pain® (IASP). The figures may NOT be reproduced for any other purpose without permission.*

Swollen/vacuolated mitochondria were also observed in DRG neuronal cell bodies following low-dose paclitaxel.[41] Collectively, these data show that only mitochondria in sensory axons are susceptible to paclitaxel perhaps providing some explanation as to why paclitaxel produces a predominantly sensory neuropathy. Swollen/vacuolated mitochondria have also been observed in saphenous nerves from rat models of oxaliplatin-induced painful neuropathy[42] and bortezomib-induced painful neuropathy.[43] In both studies, nerves were only examined at the peak of the oxaliplatin-/bortezomib-induced pain behaviors (~day 35 after first injection of oxaliplatin/bortezomib) and found significant increases in atypical axonal mitochondria in C-fibers and A-fibers. Similar to paclitaxel, there was also no significant change in the mitochondria in myelinating and nonmyelinating Schwann cells of saphenous nerves from oxaliplatin- and bortezomib-treated rats.[42,43]

Although the identification of swollen/vacuolated neuronal mitochondria provides evidence for mitochondrial dysfunction, it does not indicate the nature of the mitochondrial dysfunction that has occurred. Assays of mitochondrial function are required to decipher what aspects of mitochondrial function have been compromised and their consequences. Using a Clark-type oxygen electrode with high-resolution respirometry (Oroboros Oxygraph-2k), oxygen consumption was measured from sciatic nerve preparations from paclitaxel-, oxaliplatin-, and bortezomib-treated rats.[43,44] There were significant decreases in complex I-stimulated and complex II-stimulated respiration in nerves from paclitaxel-, oxaliplatin-, and bortezomib-treated rats compared to nerves from vehicle-treated rats. These effects were observed prior to chemotherapy-induced pain onset and at the peak of chemotherapy-induced pain behavior.[43,44] Using oxygen and pH-sensing fluorophores in a multiwell plate format (XF24 Extracellular Flux Analyzer), we have simultaneously measured both oxygen consumption and glycolysis in DRG cells from paclitaxel- and vehicle-treated rats.[45] By pharmacologically modulating the mETC, we measured basal respiration and glycolysis, ATP-linked respiration, maximal respiration, spare reserve capacity, and glycolytic capacity. There was a significant decrease in maximal respiration and spare reserve capacity (the ability of mitochondria to respond to stress) in DRG cells from paclitaxel-treated rats compared to DRG cells from vehicle-treated rats at day 7 prior to the emergence of pain behavior. At the peak of paclitaxel-induced pain behavior, there was a significant increase in basal glycolysis and glycolytic capacity in DRG cells from paclitaxel-treated rats. At the resolution of paclitaxel-induced pain behavior, there was no difference in oxygen consumption or glycolysis measures between paclitaxel- and vehicle-treated rats.[45]

Mitochondria play a fundamental role in the generation of ATP through OXPHOS. Basal ATP levels in sciatic nerves from paclitaxel-, oxaliplatin-, and bortezomib-treated rats were unaltered compared to those from vehicle-treated rats.[43,44] However, there was a reduction in the amount of ATP generated following stimulation (addition of complexes I and II substrates plus ADP) in sciatic nerves from paclitaxel-, oxaliplatin-, and bortezomib-treated rats relative to those from vehicle-treated rats. These effects were also observed prior to and at the peak of chemotherapy-induced pain behavior.[43,44] The ATP content (or lack of) within a cell, per se, is not necessarily indicative of mitochondrial (dys)function, as extensively discussed elsewhere.[46] The adenine nucleotide pool and its equilibrium should also be considered.[46] ATP synthase (complex V) function can be reversed, so less observed ATP could be accompanied with increased ADP, for example. We have observed decreases in ATP, but not ADP, levels in DRG from paclitaxel-treated rats compared to vehicle-treated rats at the peak of paclitaxel-induced pain behavior.[45]

There are a variety of pharmacological tools that can selectively inhibit/modulate different aspects of the ETC and mitochondrial function. In a series of experiments, rats with vincristine-induced mechanical hyperalgesia received intradermal administration of mETC inhibitors for complexes I–V.[47] All inhibitors produced modest reduction in mechanical thresholds over a 3-h period. The magnitude of reduction observed was between approximately 15–38% and was most pronounced at 30 min postadministration of all inhibitors. The smallest effect was observed with rotenone (complex I inhibitor) and the largest effect with sodium cyanide (complex IV inhibitor).[47] In other studies by a different group, 30 min following systemic rotenone, there was approximately 20–25% enhancement of the mechanical hypersensitivity induced by paclitaxel and oxaliplatin.[21] Paclitaxel- and oxaliplatin-induced mechanical hypersensitivity was also increased by 20–35%, 30 min following systemic oligomycin (an inhibitor of ATP synthase). Oligomycin also increased discharge frequency in A-fibers and C-fibers (to a larger extent) in both paclitaxel- and oxaliplatin-treated rats. These facilitatory effects were seen from 5 min post-injection, peaked at 15–20 min and were close to preinjection levels again by 30 min.[21] We have shown the causal role of ROS in chemotherapy-induced pain behaviors (discussed below). Complexes I and III are sources of electron leak from the mETC resulting in ROS generation. Therefore, we have administered rotenone (complex I inhibitor) and antimycin A (complex III inhibitor) in acute and prophylactic dosing paradigms in paclitaxel-treated rats to assess the effects of complex I and III inhibition on the maintenance and development of paclitaxel-induced pain.[48] We found that rotenone and antimycin A inhibited paclitaxel-induced mechanical hypersensitivity 1 h following administration. In addition, we found that prophylactic antimycin A attenuated the development of paclitaxel-induced mechanical hypersensitivity, whereas rotenone had no effect. In summary, these behavioral studies have shown that direct modulation of the mETC *in vivo* can have various consequences on chemotherapy-induced pain behaviors. The differences in observed effects may be explained by route of drug administration, inhibitor in question or time point examined post administration.

Other pharmacological studies in rodents displaying chemotherapy-induced pain behaviors have examined the role of ROS *in vivo*. Phenyl *N-tert*-butylnitrone (PBN), a nonspecific ROS scavenger, reversed established paclitaxel-induced mechanical hypersensitivity[49,50] and paclitaxel-induced cold allodynia.[50] PBN also markedly inhibited the development of paclitaxel-induced mechanical hypersensitivity, when administered in a prophylactic paradigm which that encompasses the week immediately following paclitaxel administration.[49,50] 4-Hydroxy-2,2,6,6-tetramethylpiperidine-1-oxyl (TEMPOL), an SOD-2 mimetic, inhibited established paclitaxel-induced mechanical hypersensitivity at higher doses than PBN but was ineffective on cold allodynia and the development of paclitaxel-induced mechanical hypersensitivity.[50] Peroxynitrite decomposition catalysts have been shown to reverse established paclitaxel-induced mechanical hypersensitivity[51] and to also prevent the development of mechanical hypersensitivity induced by paclitaxel, oxaliplatin, and bortezomib.[51,52] Recently, an increase in ROS and reactive nitrogen species (RNS) levels was seen in lumbar DRG following chronic oxaliplatin treatment in mice.[53] SS-31, a novel mitochondria-targeted antioxidant, reversed this oxaliplatin-evoked ROS/RNS increase and also attenuated oxaliplatin-induced cold and mechanical hypersensitivities.[53] Collectively, these data suggest that mitochondrial ROS is causal to the development and maintenance of (CIPN). Acetyl-L-carnitine (ALC) is involved in free fatty acid oxidation and also acts as an antioxidant. Initial studies demonstrated that prophylactic ALC administration could prevent the development of paclitaxel-induced mechanical hypersensitivity.[54] Closer analysis of the saphenous nerves from these animals showed that ALC had prevented the paclitaxel-evoked increase in swollen/vacuolated mitochondria in

C-fibers.[38] Prophylactic ALC administration was also reported to prevent paclitaxel-, oxaliplatin-, and bortezomib-evoked compromises in complex I-stimulated and complex II-stimulated respiration in sciatic nerves.[43,44] Despite promising reports of ALC efficacy from preclinical studies[54–57] and an open-label phase II trial of CIPN patients,[58] a recent placebo-controlled randomized clinical trial did not support the use of ALC as a treatment for CIPN.[59] Breast cancer patients received daily doses of ALC or placebo for 24 weeks from the beginning of their paclitaxel chemotherapy. At 12 weeks, there was no difference between ALC and placebo; however, at 24 weeks, CIPN and functional status of patients who had received ALC had significantly worsened.[59]

As discussed previously, there are a variety of pharmacological agents that specifically modulate different complexes of the mETC. There are other pharmacological reagents that directly interact with mitochondria and their function in different ways. Two such reagents have been evaluated for their effects on chemotherapy-induced pain behaviors.[60–62] Cholest-4-en-3-one, Oxime (TRO19622/Olesoxime), is a cholesterol-like compound that directly binds to two components of the mPTP: the voltage-dependent anion channel and the translocator protein 18 kDa (or peripheral benzodiazepine receptor).[63] Olesoxime had no effect on mechanical thresholds in normal rats.[60] In contrast, olesoxime produced ~50% reversal of established vincristine- and paclitaxel-induced mechanical hypersensitivity.[60,61] Olesoxime also attenuated both the development of paclitaxel-induced mechanical hypersensitivity and loss of IENFs. However, olesoxime had no effect on paclitaxel-induced spontaneous discharge in C- and A-fibers.[61] Another pharmacological reagent that interacts with a different aspect of mitochondrial function is mdivi-1, which is a selective Drp1 inhibitor. Dynamin-related protein 1 (Drp1) is a GTPase which catalyzes mitochondrial fission (the process of mitochondria dividing to increase mitochondria numbers). Intradermal injections of mdivi-1 significantly attenuated oxaliplatin-induced mechanical hyperalgesia.[62]

4.2 Diabetic painful neuropathy

Diabetic polyneuropathy is the most common complication associated with Type I and Type II diabetes that has been estimated to affect around two-thirds of patients in their lifetime. Diabetic polyneuropathy is a "dying-back," length-dependent neuropathy characterized by a loss of neurons which begins with the most distal axons in the feet and progresses up the legs. Diabetic polyneuropathy causes ongoing pain, sensory loss, foot ulceration, poor wound healing, and infection potentially leading to gangrene and amputation. Poor glucose control was thought to be the key factor in the genesis of diabetic polyneuropathy; however, it has been recently highlighted that glucose control has a large effect on prevention of neuropathy caused by Type I diabetes, but only a small effect on prevention of neuropathy caused by Type II diabetes.[64]

The majority of preclinical research into diabetic painful neuropathy has used chemically induced diabetes evoked by a single injection of streptozotocin (STZ), although there are many other models of diabetes (for review, see Ref. 65). STZ causes rapid ablation of pancreatic beta cells resulting in a loss of insulin production and hyperglycemia. STZ-treated rodents have been reported to display both mechanical and thermal hyperalgesia and hypoalgesia, with hyperalgesia behaviors often preceding hypoalgesia behaviors.[66,67] Different pharmacological tools which directly interact with mitochondria and modulate their function have been assessed for their impact on STZ-evoked mechanical hypersensitivity. Intradermal administration of mETC inhibitors for complexes I–V reduced STZ-evoked mechanical hyperalgesia by approximately 12–35%, 30 min following administration.[47] The smallest reduction was obtained by rotenone and the largest reduction by sodium cyanide.[47] Oral administration of olesoxime (TRO19622) inhibited STZ-evoked mechanical hypersensitivity following acute and repeated administration and also inhibited STZ-evoked heat hyperalgesia.[60] Olesoxime also significantly reduced STZ-evoked slowing of compound muscle action potentials, but had no effect on the sensory nerve conduction velocities.[60] The effects of ROS scavenging compounds have also been evaluated in STZ-evoked behaviors. *N*-Acetylcysteine inhibited thermal hyperalgesia and improved motor coordination in STZ rats.[68] Peroxynitrite decomposition catalysts (Fe(III) tetramesitylporphyrin octasulfonate—FeTMPS and Fe(III) tetrakis-2-(*N*-triethylene glycol monomethyl ether)pyridyl porphyrin—FP15) attenuated mechanical hypoalgesia, heat hypoalgesia, and mechanical hyperalgesia, but had no significant effect on mechanical allodynia in STZ mice.[69,70] Alleviation of both sciatic motor nerve conduction velocities and hind-limb digital sensory nerve conduction velocities was also observed following treatment with FeTMPS.[70] An array of biochemical changes associated with diabetic neuropathy occurs in STZ-treated rodents. Increased lipid peroxidation in sciatic nerves and increases in nitrotyrosine levels in the sciatic nerve, DRG, and spinal cord are reported.[68–70] Decreases in phospholipids,

Na^+–K^+-ATPase activity, thiols (nonprotein—GSH and protein), and antioxidant enzyme (SOD-2, catalase, glutathione peroxidase, glutathione reductase, glutathione-*S*-transferase) activity levels were all observed in sciatic nerves of STZ rats.[68] *N*-Acetylcysteine significantly corrected all of these biochemical changes except nitrotyrosine levels (which was not examined).[68] FeTMPS and FP15 treatment significantly reduced STZ-evoked increases in nitrotyrosine levels in the sciatic nerve, DRG, and spinal cord.[69,70]

Studies assessing neuronal mitochondria and mitochondrial dysfunction in preclinical models of diabetic neuropathy have examined several different aspects including ultrastructure, mitochondrial membrane potential, oxygen consumption, mitochondrial proteome, mitochondrial ROS, and mitochondrial calcium levels. Mitochondrial ultrastructure has been examined in several models of Type I diabetes. In diabetic BioBreeding/Worcester rats, there was no morphological evidence of mitochondrial abnormality in the DRG.[71] In nonobese diabetic and STZ-induced diabetic mice, "swollen neurites containing nearly pure collections of mitochondria" were observed in superior mesenteric–celiac ganglia.[72] In the spontaneously genetic diabetic Akita mouse (another model of Type I diabetes), tiny hyperchromatic dense mitochondria were found in the prevertebral sympathetic ganglia.[73] In comparison, in a model of Type II diabetes (db/db mice) when diabetic neuropathy had developed, there was an increase in the number of mitochondria in both myelinated and unmyelinated axons of the dorsal root.[74] Interestingly, the number of mitochondria in the ventral root did not change in db/db mice at the same time point or in younger diabetic mice prior to the emergence of diabetic neuropathy.[74] Enhanced mitochondrial biogenesis and fission have also been indicated in sensory neurons in db/db mice by increases in mtDNA and mitochondrial density and decreased mitochondrial diameter.[75]

The mitochondrial membrane potential ($\Delta\psi_m$) can be estimated by loading cells with a fluorescent probe (e.g., rhodamine 123, TMRM) and then applying an ionophore uncoupler, such as FCCP, to collapse $\Delta\psi_m$. The ratio between the fluorescent value prior to FCCP and following FCCP application is directly proportional to the status of $\Delta\psi_m$. DRG neurons isolated from STZ rats and loaded with rhodamine 123 showed that there was a loss in $\Delta\psi_m$ indicating that the mitochondrial membrane potential was depolarized during diabetic neuropathy.[76,77] Using TMRM, there was also evidence that neuronal mitochondria were depolarized in the axons of cultured DRG neurons from STZ rats.[78] Using the Oroboros Oxygraph-2k, oxygen consumption was measured from lumbar DRG tissue homogenates and mitochondria isolated from lumbar DRG of STZ rats.[79] Although basal respiration was unaffected, significant deficits in coupled OXPHOS and complex IV-mediated respiration were found in lumbar DRG tissue homogenates of STZ rats. Similarly, isolated mitochondria from 22-week-old STZ rats also showed deficits in coupled OXPHOS, uncoupled respiration, and complex IV-mediated respiration, whereas basal respiration was normal.[79] In further work, using the XF24 Extracellular Flux Analyzer, this group also demonstrated marked deficits in maximal respiration and spare reserve capacity in intact DRG neurons isolated from 22-week-old STZ rats compared to age-matched controls.[80] Recently, deficits in maximal respiration and spare reserve capacity of sensory neurons have also been associated with sensory hypoalgesia evoked by Type II diabetes.[81] Interestingly, these authors also report that STZ-evoked sensory hypoalgesia in mice precedes the deficit in spare reserve capacity of sensory neurons suggesting that mitochondrial dysfunction in more closely correlated with diabetic neuropathy evoked by Type II diabetes than Type I diabetes.[81] The deficits in mitochondrial membrane potential and oxygen consumption in DRG neurons from STZ rats could be a consequence of decreased expression and activity of mETC components. Protein levels and enzymatic activity levels of complexes I and IV, but not complex V, were decreased in the DRG of STZ rats compared to controls.[79,80] Citrate synthase activity and PGC-1α (considered a major regulator of mitochondrial biogenesis and cellular energy metabolism) expression were also significantly decreased in DRG from STZ rats.[79,80] An extensive proteomic analysis of DRG from 22-week-old STZ rats revealed that proteins associated with OXPHOS and mitochondrial dysfunction, e.g., subunits of complexes I, IV, and V, and SOD-2 were significantly downregulated.[78]

ROS levels in sensory neurons from diabetic rodents have also been examined with different fluorescent probes.[74,78,79] ROS levels—assessed by both DHR 123 and DCF fluorescence—were similar in acutely isolated DRG neurons from STZ and control rats.[79] Hydrogen peroxide levels (detected with Amplex Red) were also similar in isolated mitochondria from the DRG of STZ and control rats.[79] Axons of DRG cultures from STZ rats showed increased levels of DCF fluorescence and 4-HNE adducts, but significant decreases in mitochondrial superoxide levels—imaged with MitoSOX Red.[78] The increased ROS found in axons from STZ cultures was attributed to activity of the polyol pathway (also known as the sorbitol–aldose reductase pathway—an alternative pathway by which unused glucose is metabolized), as opposed to deficient mETC function.[78] DRG cultures from db/db mice showed increased mitochondrial superoxide levels in neurites and also in neuronal cell bodies following an hour's exposure to high glucose.[74] Finally, there is some evidence[82] showing altered mitochondrial calcium buffering in small DRG neurons of STZ and db/db mice, which could contribute to deregulated calcium homeostasis in diabetic neuropathy (for reviews on this topic, see Refs. 83,84).

4.3 HIV-associated painful neuropathy

HIV-associated painful neuropathy originates from two aspects—the HIV infection itself and the antiretroviral drugs used to control HIV. Regardless of the specific origin patients present with the same neuropathic pain symptoms including ongoing pain and mechanical allodynia/hyperalgesia in both feet.[85] HIV-associated painful neuropathy is a typical "dying-back," length-dependent neuropathy, with symptoms starting in the toes and then progressing up the legs. Once at mid-leg level, symptoms may then also start in the fingertips.[85]

Similar to observations from HIV patient samples, sural nerves from macaques infected with simian immunodeficiency virus (SIV) showed reductions in mitochondrial glycoprotein and the mitochondrial glycoprotein:COX-I ratio compared to controls.[4] Additional experiments using isolated mitochondria from sural and sciatic nerves of SIV-infected, highly active antiretroviral therapy (HAART)-treated, and control macaques provided further evidence of mitochondrial dysfunction in sensory neurons.[4] Increased susceptibility to mitochondrial permeability, increased hydrogen peroxide production, and a loss of mitochondrial membrane potential were observed in the mitochondria from sural, but not sciatic, nerves of SIV-infected and HAART-treated macaques.[4] In addition, sural nerves from SIV-infected macaques showed increased oxidative stress evidenced by increased carbonyl groups and nitrotyrosine levels.[4]

Nucleoside analog reverse-transcriptase inhibitors (NRTIs) were some of the first drugs used to control HIV and provide the backbone of combination HAART regimens. 2′-3′-Dideoxycytidine (zalcitabine), 2′,3′-dideoxyinosine (ddI—didanosine), and 2′,3′-didehydro-2′,3′-dideoxythymidine (d4T—stavudine) are all NRTIs and are particularly associated with painful peripheral neuropathy. There are only a few reports using animal models of HIV-treatment-associated painful neuropathy that explore the potential connection of mitochondrial (dys)function to this disorder. Rats with ddC-induced mechanical hyperalgesia received intradermal administration of mETC inhibitors for complexes I–V[47] in separate experiments. Similar to the effects seen in vincristine-treated rats, mechanical hyperalgesia in ddC-treated rats was reduced by approximately 15–38% and was most pronounced at 30 min postadministration of all inhibitors. The smallest reduction was elicited by rotenone and the largest reduction by sodium cyanide.[47] Intrathecal administration of antisense against Drp1 reversed ddC-evoked mechanical hyperalgesia.[62] In addition, intradermal administration of mdivi-1, a selective Drp1 inhibitor, also reversed ddC-evoked mechanical hyperalgesia.[62] The effects of feline immunodeficiency virus (FIV) and ddI, when given alone and in combination, on mitochondrial COX-I levels in DRG neurons were compared under *in vitro* and *in vivo* conditions.[86] *In vitro*, only combined FIV and ddI exposure significantly reduced COX-I mRNA, but not DNA, levels.[86] *In vivo*, FIV and ddI significantly decreased expression of COX-I DNA and protein levels.[86] No further decrease in COX-I DNA and protein levels was seen in DRGs from FIV-infected cats treated with ddI, although such additive effects were seen in COX-I mRNA levels.[86] Two intravenous injections of d4T (on days 0 and 4, in the absence of immunodeficiency virus) produced evoked and spontaneous pain behaviors in rats.[87] Proteomics on sural nerves harvested at day 7 found 10 proteins that were downregulated with at least 10-fold changes in d4T animals.[87] Three of these proteins were related to mitochondrial function, namely mitochondrial long-chain specific acyl-CoA dehydrogenase, NADH dehydrogenase (ubiquinone) 1 alpha subcomplex subunit 4, and mitochondrial ATP synthase subunit D.[87] Further analysis of sural nerves with Western blotting showed a 50% reduction of mitochondrial long-chain-specific acyl-CoA dehydrogenase following d4T.[87]

4.4 Other chronic pain states

Neuropathic pain and persistent inflammatory pain are typically modeled in rodents using peripheral nerve trauma via surgical techniques and Complete Freund's Adjuvant (CFA) injection into the hind paw, respectively. Using the spared nerve injury (SNI) and CFA mice models, the numbers of mitochondria in the dorsal horn were visualized following intrathecal administration of a fluorescent mitochondrial marker.[88] Seven days following CFA and 14 days following SNI, there were significant increases in the number of mitochondria in laminae I–V of the ipsilateral side of the spinal cord compared to the contralateral side.[88] Up to day 7 following partial sciatic nerve ligation (PSNL), an increase in cytochrome *c* levels in the spinal cord was reported[89] indicating mitochondrial dysfunction as cytochrome *c* is normally contained within the inner and outer mitochondrial membranes. Furthermore, systemic administration of cyclosporin A, an mPTP blocker, pre- and post-PSNL inhibited PSNL-evoked mechanical allodynia.[89]

The majority of the studies that suggest a potential role of mitochondrial dysfunction in neuropathic and inflammatory pain have focussed on ROS and oxidative stress. CCI-evoked heat hyperalgesia could be inhibited by antioxidants TEMPOL,[90] *N*-acetylcysteine,[91] and tirilazad.[92] In addition, increased SOD-2 levels were found in the axotomized sciatic nerve.[93] PBN, a nonspecific ROS scavenger, inhibited mechanical hypersensitivity evoked by spinal nerve ligation,[94–96] capsaicin-induced inflammation,[22,97] and visceral inflammation.[98] Other nonspecific ROS scavengers, 5,5-dimethylpyrroline-*N*-oxide and nitrosobenzene also relieved neuropathic pain behaviors.[94] Reagents that mimic SOD-2 activity inhibited hypersensitivity to mechanical/heat stimuli evoked by either peripheral nerve injury[90,96] or inflammation.[22,97,99,100] Finally, mitochondrial ROS-producing profiles, visualized through *in vivo* delivery of a fluorescent mitochondrial marker, were increased in the spinal cord following peripheral nerve injury[101] or an inflammatory stimulus.[22,23]

5. FUTURE PERSPECTIVES

In this chapter, I have summarized the evidence that suggests a contributory role of mitochondrial function to normal sensory processing and chronic pain. This particular area of research is novel to the pain research field with most studies only dating back over the last decade or so. The majority of these investigations have highlighted evidence for a specific contributory/causal role of mitochondria in painful peripheral neuropathies evoked by chemotherapy, diabetes, and HIV. The functions of mitochondria are diverse, and at present, there are many questions remaining as to how these different mitochondrial functions are contributory to pain states and also how different mitochondrial functions are modulated and/or impaired during chronic pain conditions. There are also several possibilities as to how dysfunctional mitochondria may mechanistically cause pain via, for example, ATP deficiency, and/or excessive ROS, and/or impaired calcium buffering. Further research using a range of assays of mitochondrial function will elucidate these mechanisms and aid analgesic drug development. Although modulating mitochondrial function could potentially have knock-on detrimental effects on cellular respiration, selective and careful targeting of mitochondrial function may well yield novel analgesics. Further research will also identify if mitochondrial modulation could be beneficial to chronic pain disorders of various etiologies or specifically to drug-induced neuropathies such as CIPN.

ACKNOWLEDGMENTS

S. J. L. F., PhD, is a Principal Investigator and Lecturer in Pharmacology at the Wolfson Centre for Age-Related Diseases, Institute of Psychiatry, Psychology & Neuroscience, King's College London, United Kingdom. S. J. L. F. is funded by the Higher Education Funding Council for England (HEFCE). Her research laboratory is currently funded by The Wellcome Trust, the Medical Research Council (MRC), Eli Lilly, and the British Pharmacological Society (BPS).

REFERENCES

1. Wong-Riley MT. Cytochrome oxidase: an endogenous metabolic marker for neuronal activity. *Trends Neurosci.* 1989;12(3):94–101.
2. Attwell D, Laughlin SB. An energy budget for signaling in the grey matter of the brain. *J Cereb Blood Flow Metab.* 2001;21(10):1133–1145.
3. Perkins GA, Ellisman MH. Mitochondrial configurations in peripheral nerve suggest differential ATP production. *J Struct Biol.* 2011;173(1):117–127.
4. Lehmann HC, Chen W, Borzan J, Mankowski JL, Hoke A. Mitochondrial dysfunction in distal axons contributes to human immunodeficiency virus sensory neuropathy. *Ann Neurol.* 2011;69(1):100–110.
5. Bristow EA, Griffiths PG, Andrews RM, Johnson MA, Turnbull DM. The distribution of mitochondrial activity in relation to optic nerve structure. *Arch Ophthalmol.* 2002;120(6):791–796.
6. Wang SS, Shultz JR, Burish MJ, et al. Functional trade-offs in white matter axonal scaling. *J Neurosci.* 2008;28(15):4047–4056.
7. Sokoloff L. Sites and mechanisms of function-related changes in energy metabolism in the nervous system. *Dev Neurosci.* 1993;15(3–5):194–206.
8. Kadekaro M, Crane AM, Sokoloff L. Differential effects of electrical stimulation of sciatic nerve on metabolic activity in spinal cord and dorsal root ganglion in the rat. *Proc Natl Acad Sci USA.* 1985;82(17):6010–6013.
9. Sajic M, Mastrolia V, Lee CY, et al. Impulse conduction increases mitochondrial transport in adult mammalian peripheral nerves in vivo. *PLoS Biol.* 2013;11(12):e1001754. http://dx.doi.org/10.1371/journal.pbio.1001754.
10. Ohno N, Kidd GJ, Mahad D, et al. Myelination and axonal electrical activity modulate the distribution and motility of mitochondria at CNS nodes of Ranvier. *J Neurosci.* 2011;31(20):7249–7258.
11. Lynn B, Ye W, Cotsell B. The actions of capsaicin applied topically to the skin of the rat on C-fibre afferents, antidromic vasodilatation and substance P levels. *Br J Pharmacol.* 1992;107(2):400–406.
12. Medvedeva YV, Kim MS, Usachev YM. Mechanisms of prolonged presynaptic Ca2+ signaling and glutamate release induced by TRPV1 activation in rat sensory neurons. *J Neurosci.* 2008;28(20):5295–5311.

13. Shutov LP, Kim MS, Houlihan PR, Medvedeva YV, Usachev YM. Mitochondria and plasma membrane Ca2+-ATPase control presynaptic Ca2+ clearance in capsaicin-sensitive rat sensory neurons. *J Physiol.* 2013;591(Pt 10):2443–2462.
14. Svichar N, Shishkin V, Kostyuk P. Mitochondrial participation in modulation of calcium transients in DRG neurons. *Neuroreport.* 1999;10(6):1257–1261.
15. Kim MS, Usachev YM. Mitochondrial Ca2+ cycling facilitates activation of the transcription factor NFAT in sensory neurons. *J Neurosci.* 2009;29(39):12101–12114.
16. Kim HY, Lee KY, Lu Y, et al. Mitochondrial Ca(2+) uptake is essential for synaptic plasticity in pain. *J Neurosci.* 2011;31(36):12982–12991.
17. Baughman JM, Perocchi F, Girgis HS, et al. Integrative genomics identifies MCU as an essential component of the mitochondrial calcium uniporter. *Nature.* 2011;476(7360):341–345.
18. De Stefani D, Raffaello A, Teardo E, Szabo I, Rizzuto R. A forty-kilodalton protein of the inner membrane is the mitochondrial calcium uniporter. *Nature.* 2011;476(7360):336–340.
19. Palty R, Silverman WF, Hershfinkel M, et al. NCLX is an essential component of mitochondrial Na+/Ca2+ exchange. *Proc Natl Acad Sci USA.* 2010;107(1):436–441.
20. Kim HY, Chung JM, Chung K. Increased production of mitochondrial superoxide in the spinal cord induces pain behaviors in mice: the effect of mitochondrial electron transport complex inhibitors. *Neurosci Lett.* 2008;447(1):87–91.
21. Xiao WH, Bennett GJ. Effects of mitochondrial poisons on the neuropathic pain produced by the chemotherapeutic agents, paclitaxel and oxaliplatin. *Pain.* 2012;153(3):704–709.
22. Schwartz ES, Lee I, Chung K, Chung JM. Oxidative stress in the spinal cord is an important contributor in capsaicin-induced mechanical secondary hyperalgesia in mice. *Pain.* 2008;138(3):514–524.
23. Schwartz ES, Kim HY, Wang J, et al. Persistent pain is dependent on spinal mitochondrial antioxidant levels. *J Neurosci.* 2009;29(1):159–168.
24. Li Z, Ji G, Neugebauer V. Mitochondrial reactive oxygen species are activated by mGluR5 through IP3 and activate ERK and PKA to increase excitability of amygdala neurons and pain behavior. *J Neurosci.* 2011;31(3):1114–1127.
25. Vital A, Vital C. Mitochondria and peripheral neuropathies. *J Neuropathol Exp Neurol.* 2012;71(12):1036–1046.
26. Finsterer J. Mitochondrial neuropathy. *Clin Neurol Neurosurg.* 2005;107(3):181–186.
27. Schroder JM, Sommer C. Mitochondrial abnormalities in human sural nerves: fine structural evaluation of cases with mitochondrial myopathy, hereditary and non-hereditary neuropathies, and review of the literature. *Acta Neuropathol.* 1991;82(6):471–482.
28. Zuchner S, Mersiyanova IV, Muglia M, et al. Mutations in the mitochondrial GTPase mitofusin 2 cause Charcot-Marie-Tooth neuropathy type 2A. *Nat Genet.* 2004;36(5):449–451.
29. Sahenk Z, Barohn R, New P, Mendell JR. Taxol neuropathy. Electrodiagnostic and sural nerve biopsy findings. *Arch Neurol.* 1994;51(7):726–729.
30. Fazio R, Quattrini A, Bolognesi A, et al. Docetaxel neuropathy: a distal axonopathy. *Acta Neuropathol.* 1999;98(6):651–653.
31. Kalichman MW, Powell HC, Mizisin AP. Reactive, degenerative, and proliferative Schwann cell responses in experimental galactose and human diabetic neuropathy. *Acta Neuropathol.* 1998;95(1):47–56.
32. Dalakas MC, Semino-Mora C, Leon-Monzon M. Mitochondrial alterations with mitochondrial DNA depletion in the nerves of AIDS patients with peripheral neuropathy induced by 2′3′-dideoxycytidine (ddC). *Lab Invest.* 2001;81(11):1537–1544.

33. Casanova-Molla J, Morales M, Garrabou G, et al. Mitochondrial loss indicates early axonal damage in small fiber neuropathies. *J Peripher Nerv Syst.* 2012;17(2):147–157.
34. Gruber HE, Watts JA, Riley FE, Fulkerson MB, Norton HJ, Hanley Jr EN. Mitochondrial bioenergetics, mass, and morphology are altered in cells of the degenerating human annulus. *J Orthop Res.* 2013;31(8):1270–1275.
35. Argyriou AA, Bruna J, Marmiroli P, Cavaletti G. Chemotherapy-induced peripheral neurotoxicity (CIPN): an update. *Crit Rev Oncol Hematol.* 2012;82(1):51–77.
36. Farquhar-Smith P. Chemotherapy-induced neuropathic pain. *Curr Opin Support Palliat Care.* 2011;5(1):1–7.
37. Flatters SJL, Bennett GJ. Studies of peripheral sensory nerves in paclitaxel-induced painful peripheral neuropathy: evidence for mitochondrial dysfunction. *Pain.* 2006;122(3):245–257. http://dx.doi.org/10.1016/j.pain.2006.01.037.
38. Jin HW, Flatters SJL, Xiao WH, Mulhem HL, Bennett GJ. Prevention of paclitaxel-evoked painful peripheral neuropathy by acetyl-L-carnitine: effects on axonal mitochondria, sensory nerve fiber terminal arbors, and cutaneous Langerhans cells. *Exp Neurol.* 2008;210:229–237.
39. Xiao WH, Zheng H, Zheng FY, Nuydens R, Meert TF, Bennett GJ. Mitochondrial abnormality in sensory, but not motor, axons in paclitaxel-evoked painful peripheral neuropathy in the rat. *Neuroscience.* 2011;199:461–469.
40. Wu Y, Li J, Zhou J, Feng Y. Dynamic long-term microstructural and ultrastructural alterations in sensory nerves of rats of paclitaxel-induced neuropathic pain. *Chin Med J (Engl).* 2014;127(16):2945–2952.
41. Barriere DA, Rieusset J, Chanteranne D, et al. Paclitaxel therapy potentiates cold hyperalgesia in streptozotocin-induced diabetic rats through enhanced mitochondrial reactive oxygen species production and TRPA1 sensitization. *Pain.* 2012;153(3):553–561.
42. Xiao WH, Zheng H, Bennett GJ. Characterization of oxaliplatin-induced chronic painful peripheral neuropathy in the rat and comparison with the neuropathy induced by paclitaxel. *Neuroscience.* 2012;203:194–206.
43. Zheng H, Xiao WH, Bennett GJ. Mitotoxicity and bortezomib-induced chronic painful peripheral neuropathy. *Exp Neurol.* 2012;238(2):225–234.
44. Zheng H, Xiao WH, Bennett GJ. Functional deficits in peripheral nerve mitochondria in rats with paclitaxel- and oxaliplatin-evoked painful peripheral neuropathy. *Exp Neurol.* 2011;232(2):154–161.
45. Duggett NA, Flatters SJL. Paclitaxel alters mitochondrial function, glycolytic capacity and adenine nucleotides in sensory neurones. In: *15th World Congress on Pain*; 2014, IASP Abstracts: PH339.
46. Brand MD, Nicholls DG. Assessing mitochondrial dysfunction in cells. *Biochem J.* 2011;435(2):297–312.
47. Joseph EK, Levine JD. Mitochondrial electron transport in models of neuropathic and inflammatory pain. *Pain.* 2006;121(1–2):105–114.
48. Griffiths LA, Flatters SJL. Pharmacological modulation of mitochondrial-derived reactive oxygen species in paclitaxel-induced painful peripheral neuropathy. In: *4th International Conference on Neuropathic Pain, NeupSIG*; 2013, NeupSIG Abstracts: P17.
49. Kim HK, Zhang YP, Gwak YS, Abdi S. Phenyl N-tert-butylnitrone, a free radical scavenger, reduces mechanical allodynia in chemotherapy-induced neuropathic pain in rats. *Anesthesiology.* 2010;112(2):432–439.
50. Fidanboylu M, Griffiths LA, Flatters SJL. Global inhibition of reactive oxygen species (ROS) inhibits paclitaxel-induced painful peripheral neuropathy. *PLoS One.* 2011;6(9): e25212.

51. Doyle T, Chen Z, Muscoli C, et al. Targeting the overproduction of peroxynitrite for the prevention and reversal of paclitaxel-induced neuropathic pain. *J Neurosci.* 2012;32(18):6149–6160.
52. Janes K, Doyle T, Bryant L, et al. Bioenergetic deficits in peripheral nerve sensory axons during chemotherapy-induced neuropathic pain resulting from peroxynitrite-mediated post-translational nitration of mitochondrial superoxide dismutase. *Pain.* 2013;154(11):2432–2440.
53. Toyama S, Shimoyama N, Ishida Y, Koyasu T, Szeto HH, Shimoyama M. Characterization of acute and chronic neuropathies induced by oxaliplatin in mice and differential effects of a novel mitochondria-targeted antioxidant on the neuropathies. *Anesthesiology.* 2014;120(2):459–473.
54. Flatters SJ, Xiao WH, Bennett GJ. Acetyl-L-carnitine prevents and reduces paclitaxel-induced painful peripheral neuropathy. *Neurosci Lett.* 2006;397(3):219–223.
55. Pisano C, Pratesi G, Laccabue D, et al. Paclitaxel and cisplatin-induced neurotoxicity: a protective role of acetyl-L-carnitine. *Clin Cancer Res.* 2003;9(15):5756–5767.
56. Ghirardi O, Lo Giudice P, Pisano C, et al. Acetyl-L-carnitine prevents and reverts experimental chronic neurotoxicity induced by oxaliplatin, without altering its antitumor properties. *Anticancer Res.* 2005;25(4):2681–2687.
57. Ghirardi O, Vertechy M, Vesci L, et al. Chemotherapy-induced allodynia: neuroprotective effect of acetyl-L-carnitine. *In Vivo.* 2005;19(3):631–637.
58. Bianchi G, Vitali G, Caraceni A, et al. Symptomatic and neurophysiological responses of paclitaxel- or cisplatin-induced neuropathy to oral acetyl-L-carnitine. *Eur J Cancer.* 2005;41(12):1746–1750.
59. Hershman DL, Unger JM, Crew KD, et al. Randomized double-blind placebo-controlled trial of acetyl-L-carnitine for the prevention of taxane-induced neuropathy in women undergoing adjuvant breast cancer therapy. *J Clin Oncol.* 2013;31(20):2627–2633.
60. Bordet T, Buisson B, Michaud M, et al. Specific antinociceptive activity of cholest-4-en-3-one, oxime (TRO19622) in experimental models of painful diabetic and chemotherapy-induced neuropathy. *J Pharmacol Exp Ther.* 2008;326(2):623–632.
61. Xiao WH, Zheng FY, Bennett GJ, Bordet T, Pruss RM. Olesoxime (cholest-4-en-3-one, oxime): analgesic and neuroprotective effects in a rat model of painful peripheral neuropathy produced by the chemotherapeutic agent, paclitaxel. *Pain.* 2009;147(1–3):202–209.
62. Ferrari LF, Chum A, Bogen O, Reichling DB, Levine JD. Role of Drp1, a key mitochondrial fission protein, in neuropathic pain. *J Neurosci.* 2011;31(31):11404–11410.
63. Bordet T, Buisson B, Michaud M, et al. Identification and characterization of cholest-4-en-3-one, oxime (TRO19622), a novel drug candidate for amyotrophic lateral sclerosis. *J Pharmacol Exp Ther.* 2007;322(2):709–720.
64. Callaghan BC, Hur J, Feldman EL. Diabetic neuropathy: one disease or two? *Curr Opin Neurol.* 2012;25(5):536–541.
65. King AJ. The use of animal models in diabetes research. *Br J Pharmacol.* 2012;166(3):877–894.
66. Chu Q, Moreland R, Yew NS, Foley J, Ziegler R, Scheule RK. Systemic insulin-like growth factor-1 reverses hypoalgesia and improves mobility in a mouse model of diabetic peripheral neuropathy. *Mol Ther.* 2008;16(8):1400–1408.
67. Courteix C, Eschalier A, Lavarenne J. Streptozocin-induced diabetic rats: behavioural evidence for a model of chronic pain. *Pain.* 1993;53(1):81–88.
68. Kamboj SS, Vasishta RK, Sandhir R. N-acetylcysteine inhibits hyperglycemia-induced oxidative stress and apoptosis markers in diabetic neuropathy. *J Neurochem.* 2010;112(1):77–91.

69. Drel VR, Pacher P, Vareniuk I, et al. A peroxynitrite decomposition catalyst counteracts sensory neuropathy in streptozotocin-diabetic mice. *Eur J Pharmacol.* 2007;569(1–2):48–58.
70. Drel VR, Pacher P, Vareniuk I, et al. Evaluation of the peroxynitrite decomposition catalyst Fe(III) tetra-mesitylporphyrin octasulfonate on peripheral neuropathy in a mouse model of type 1 diabetes. *Int J Mol Med.* 2007;20(6):783–792.
71. Kamiya H, Zhang W, Sima AA. Degeneration of the Golgi and neuronal loss in dorsal root ganglia in diabetic BioBreeding/Worcester rats. *Diabetologia.* 2006;49(11):2763–2774.
72. Schmidt RE, Dorsey DA, Beaudet LN, et al. Non-obese diabetic mice rapidly develop dramatic sympathetic neuritic dystrophy: a new experimental model of diabetic autonomic neuropathy. *Am J Pathol.* 2003;163(5):2077–2091.
73. Schmidt RE, Green KG, Snipes LL, Feng D. Neuritic dystrophy and neuronopathy in Akita (Ins2(Akita)) diabetic mouse sympathetic ganglia. *Exp Neurol.* 2009;216(1):207–218.
74. Vincent AM, Edwards JL, McLean LL, et al. Mitochondrial biogenesis and fission in axons in cell culture and animal models of diabetic neuropathy. *Acta Neuropathol.* 2010;120(4):477–489.
75. Edwards JL, Quattrini A, Lentz SI, et al. Diabetes regulates mitochondrial biogenesis and fission in mouse neurons. *Diabetologia.* 2010;53(1):160–169.
76. Huang TJ, Price SA, Chilton L, et al. Insulin prevents depolarization of the mitochondrial inner membrane in sensory neurons of type 1 diabetic rats in the presence of sustained hyperglycemia. *Diabetes.* 2003;52(8):2129–2136.
77. Huang TJ, Sayers NM, Verkhratsky A, Fernyhough P. Neurotrophin-3 prevents mitochondrial dysfunction in sensory neurons of streptozotocin-diabetic rats. *Exp Neurol.* 2005;194(1):279–283.
78. Akude E, Zherebitskaya E, Chowdhury SK, Smith DR, Dobrowsky RT, Fernyhough P. Diminished superoxide generation is associated with respiratory chain dysfunction and changes in the mitochondrial proteome of sensory neurons from diabetic rats. *Diabetes.* 2011;60(1):288–297.
79. Chowdhury SK, Zherebitskaya E, Smith DR, et al. Mitochondrial respiratory chain dysfunction in dorsal root ganglia of streptozotocin-induced diabetic rats and its correction by insulin treatment. *Diabetes.* 2010;59(4):1082–1091.
80. Roy Chowdhury SK, Smith DR, Saleh A, et al. Impaired adenosine monophosphate-activated protein kinase signalling in dorsal root ganglia neurons is linked to mitochondrial dysfunction and peripheral neuropathy in diabetes. *Brain.* 2012;135(Pt 6):1751–1766.
81. Ma J, Farmer KL, Pan P, et al. Heat shock protein 70 is necessary to improve mitochondrial bioenergetics and reverse diabetic sensory neuropathy following KU-32 therapy. *J Pharmacol Exp Ther.* 2014;348(2):281–292.
82. Kostyuk E, Svichar N, Shishkin V, Kostyuk P. Role of mitochondrial dysfunction in calcium signalling alterations in dorsal root ganglion neurons of mice with experimentally-induced diabetes. *Neuroscience.* 1999;90(2):535–541.
83. Fernyhough P, Calcutt NA. Abnormal calcium homeostasis in peripheral neuropathies. *Cell Calcium.* 2010;47(2):130–139.
84. Verkhratsky A, Fernyhough P. Calcium signalling in sensory neurones and peripheral glia in the context of diabetic neuropathies. *Cell Calcium.* 2014;56(5):362–371.
85. Keswani SC, Pardo CA, Cherry CL, Hoke A, McArthur JC. HIV-associated sensory neuropathies. *AIDS.* 2002;16(16):2105–2117.
86. Zhu Y, Antony JM, Martinez JA, et al. Didanosine causes sensory neuropathy in an HIV/AIDS animal model: impaired mitochondrial and neurotrophic factor gene expression. *Brain.* 2007;130(Pt 8):2011–2023.

87. Huang W, Calvo M, Karu K, et al. A clinically relevant rodent model of the HIV antiretroviral drug stavudine induced painful peripheral neuropathy. *Pain.* 2013;154(4):560–575.
88. Guo BL, Sui BD, Wang XY, et al. Significant changes in mitochondrial distribution in different pain models of mice. *Mitochondrion.* 2013;13(4):292–297.
89. Kim WS, Park JY, Kim TK, Baik SW. The changes of mitochondrial cytochrome c and GABAergic neuron in neuropathic pain model. *Korean J Anesthesiol.* 2012;62(4): 365–370.
90. Tal M. A novel antioxidant alleviates heat hyperalgesia in rats with an experimental painful peripheral neuropathy. *Neuroreport.* 1996;7(8):1382–1384.
91. Wagner R, Heckman HM, Myers RR. Wallerian degeneration and hyperalgesia after peripheral nerve injury are glutathione-dependent. *Pain.* 1998;77(2):173–179.
92. Khalil Z, Liu T, Helme RD. Free radicals contribute to the reduction in peripheral vascular responses and the maintenance of thermal hyperalgesia in rats with chronic constriction injury. *Pain.* 1999;79(1):31–37.
93. Rosenfeld J, Cook S, James R. Expression of superoxide dismutase following axotomy. *Exp Neurol.* 1997;147(1):37–47.
94. Kim HK, Park SK, Zhou JL, et al. Reactive oxygen species (ROS) play an important role in a rat model of neuropathic pain. *Pain.* 2004;111(1–2):116–124.
95. Gao X, Kim HK, Chung JM, Chung K. Reactive oxygen species (ROS) are involved in enhancement of NMDA-receptor phosphorylation in animal models of pain. *Pain.* 2007;131(3):262–271.
96. Tanabe M, Nagatani Y, Saitoh K, Takasu K, Ono H. Pharmacological assessments of nitric oxide synthase isoforms and downstream diversity of NO signaling in the maintenance of thermal and mechanical hypersensitivity after peripheral nerve injury in mice. *Neuropharmacology.* 2009;56(3):702–708.
97. Lee I, Kim HK, Kim JH, Chung K, Chung JM. The role of reactive oxygen species in capsaicin-induced mechanical hyperalgesia and in the activities of dorsal horn neurons. *Pain.* 2007;133(1–3):9–17.
98. Wang J, Cochran V, Abdi S, Chung JM, Chung K, Kim HK. Phenyl N-t-butylnitrone, a reactive oxygen species scavenger, reduces zymosan-induced visceral pain in rats. *Neurosci Lett.* 2008;439(2):216–219.
99. Wang ZQ, Porreca F, Cuzzocrea S, et al. A newly identified role for superoxide in inflammatory pain. *J Pharmacol Exp Ther.* 2004;309(3):869–878.
100. Khattab MM. TEMPOL, a membrane-permeable radical scavenger, attenuates peroxynitrite- and superoxide anion-enhanced carrageenan-induced paw edema and hyperalgesia: a key role for superoxide anion. *Eur J Pharmacol.* 2006;548(1–3):167–173.
101. Park ES, Gao X, Chung JM, Chung K. Levels of mitochondrial reactive oxygen species increase in rat neuropathic spinal dorsal horn neurons. *Neurosci Lett.* 2006;391(3):108–111.

CHAPTER SIX

Regulation of Gene Expression and Pain States by Epigenetic Mechanisms

Sandrine M. Géranton[1], Keri K. Tochiki
Department of Cell and Developmental Biology, University College London, London, United Kingdom
[1]Corresponding author: e-mail address: ucgasmg@ucl.ac.uk

Contents

1. Introduction 149
2. Epigenetic Mechanisms and Gene Expression 149
   2.1 Introduction 149
   2.2 Histone modifications 152
   2.3 DNA methylation 153
   2.4 Technical advances in the investigation of the epigenome 155
3. Epigenetic Regulation of Synaptic Plasticity in Learning and Memory 156
   3.1 Introduction 156
   3.2 Histones modifications and memory formation 157
   3.3 DNA methylation and memory formation 160
   3.4 The complexity of the interpretation of the epigenome 163
4. Epigenetic Mechanisms and Pain States 163
   4.1 Introduction 163
   4.2 Injury induces changes to the epigenetic machinery and the epigenome 164
   4.3 Targeting histone modifications and DNA methylation to improve injury-induced hypersensitivity 166
   4.4 Epigenetic mechanisms: Pharmacological targets for the treatment of persistent pain states? 171
5. Future Directions 173
   5.1 Long-lasting epigenetic modifications: The key to the transition from acute to chronic pain states? 173
   5.2 Could an early life injury leave long-lasting epigenetic marks responsible for long-term changes in pain sensitivity? 174
Acknowledgments 175
References 175

Abstract

The induction of inflammatory or neuropathic pain states is known to involve molecular activity in the spinal superficial dorsal horn and dorsal root ganglia, including intracellular signaling events which lead to changes in gene expression. These changes

ISSN 1877-1173
http://dx.doi.org/10.1016/bs.pmbts.2014.11.012

ultimately cause alterations in macromolecular synthesis, synaptic transmission, and structural architecture which support central sensitization, a process required for the establishment of long-term pain states. Epigenetic mechanisms are essential for long-term synaptic plasticity and modulation of gene expression. This is because epigenetic modifications are known to regulate gene transcription by aiding the physical relaxation or condensation of chromatin. These processes are therefore potential regulators of the molecular changes underlying permanent pain states. A handful of studies have emerged in the field of pain epigenetics; however, the field is still very much in its infancy. This chapter draws upon other specialities which have extensively investigated epigenetic mechanisms, such as learning and memory and oncology. After defining epigenetics as well as the recent field of "neuroepigenetics" and the main molecular mechanisms involved, this chapter describes the role of these mechanisms in the synaptic plasticity seen in learning and memory, and address those epigenetic mechanisms that have been linked with the development of acute and prolonged pain states. Finally, the idea that long-lasting epigenetic modifications could contribute to the transition from acute to chronic pain states by supporting maladaptive molecular changes is discussed.

ABBREVIATIONS

5-aza 5-azacytidine
AA anacardic acid
BDNF brain-derived neurotrophic factor
CCI chronic constriction injury
CFA Complete Freund's adjuvant
CREB cAMP response element-binding protein
DNMT DNA methyltransferase
DNMTis DNMT inhibitors
ERK extracellular signal-regulated kinases
HAT histone acetyltransferase
HDAC histone deacetylase
HDACis HDAC inhibitors
i.p. intraperitoneal
i.t. intrathecal
IEGs immediate early genes
LTP long-term potentiation
MAPK mitogen-activated protein kinase
NaB sodium butyrate
NMDA *N*-methyl-D-aspartate receptor
PNL partial sciatic nerve ligation
s.c. subcutaneous
SAHA suberoyl anilide hydroxamic acid
SNL spinal nerve ligation
SNT spinal nerve transection
TSA trichostatin A
VPA valproate

1. INTRODUCTION

Pain caused by injury, disease, or surgery is usually temporary but involves molecular alterations, such as changes in gene expression, within the nociceptive circuitry, both in the peripheral and the central nervous system. These molecular changes lead to a sensitized state where incoming signals are amplified to prevent further damage while the body is repairing. There is a strong rationale to support the idea that molecular changes governed by epigenetic mechanisms play an important role in the development of pain states. First, epigenetic mechanisms control high-order chromatin structure crucial to the regulation of gene expression. Second, an abundance of evidence exists within the field of learning and memory to implicate the involvement of epigenetic mechanisms in memory formation, a process requiring similar cellular activity as that seen in sensitized states following injury. Finally, neurotrophins such as nerve growth factor (NGF) and brain-derived neurotrophic factor (BDNF), and signaling pathways such as mitogen-activated protein/extracellular signal-regulated kinases (MAPK/ERK), which are involved in pain signaling, induce epigenetic modifications. While the field of pain and epigenetics is still in its infancy, this chapter will draw upon other specialities which have extensively investigated epigenetic mechanisms, such as learning and memory and oncology (Fig. 1). First, we will define epigenetics as well as the recent field of "neuroepigenetics" and the main molecular mechanisms involved. We will then describe the role of these mechanisms in the synaptic plasticity seen in learning and memory, and will address those epigenetic mechanisms that have been linked with the development of acute and prolonged pain states. Finally, we will discuss the idea that long-lasting epigenetic modifications could contribute to the transition from acute to chronic pain states by supporting maladaptive molecular changes.

2. EPIGENETIC MECHANISMS AND GENE EXPRESSION

2.1. Introduction

The term epigenetics can be traced back to the early 1940s and refers to molecular events occurring above or upon ("epi" in greek) the DNA sequence. Conrad Waddington introduced the "epigenetic landscape" to illustrate genome–environment interactions leading to variations in

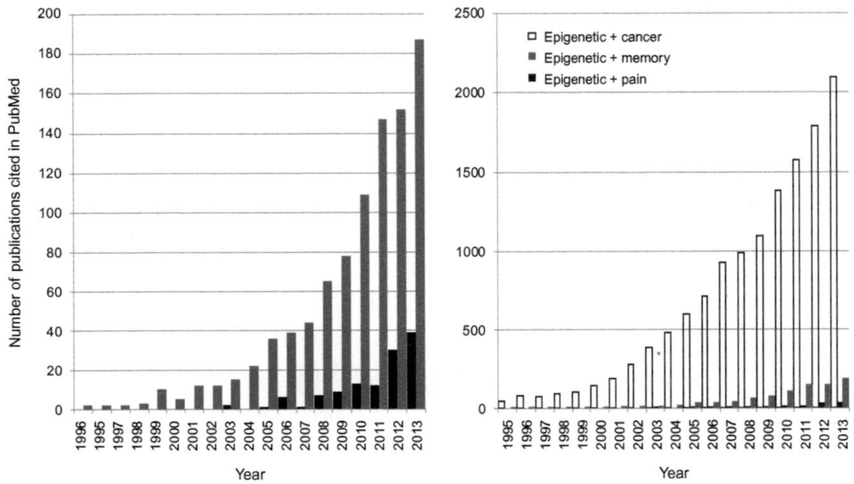

Figure 1 The field of pain and epigenetics is still in its infancy. However, it can draw upon works from other specialities such as oncology and learning and memory, where pharmacological tools that target epigenetic mechanisms have been used for many years.

phenotype, namely, the factors involved in the determination of cell fate into a diversity of cell types (e.g., muscle cell vs. neuron) during development.[1] Today, the word epigenetics is commonly used to describe phenotypic differences between individuals that are independent from their genetic code; for example, when identical twins show different personality or vulnerability to diseases.

At the molecular level, epigenetic mechanisms are responsible for a set of modifications that regulate the structure of chromatin—the combination of DNA, histones, and other proteins that make up chromosomes. These modifications lead to the dynamic remodeling of chromatin and the differential expression of genes without altering the DNA sequence itself. Epigenetic mechanisms include DNA methylation and posttranslational modification of histones (Fig. 2), and the sum of these modifications manifested as chemical marks are referred to as the epigenome. Many of the epigenetic mechanisms that operate during development remain active in terminally differentiated neurones. It is now widely accepted that these mechanisms can be engaged by environmental experience, such as early life stress, learning, or addiction,[2–4] indicating that activity-dependent experiences can be imprinted onto the fixed DNA sequence.[5–7] Consequently, the emerging field of "neuroepigenetics," or "behavioral epigenetics," refers to epigenetic mechanisms and processes allowing dynamic experience-dependent

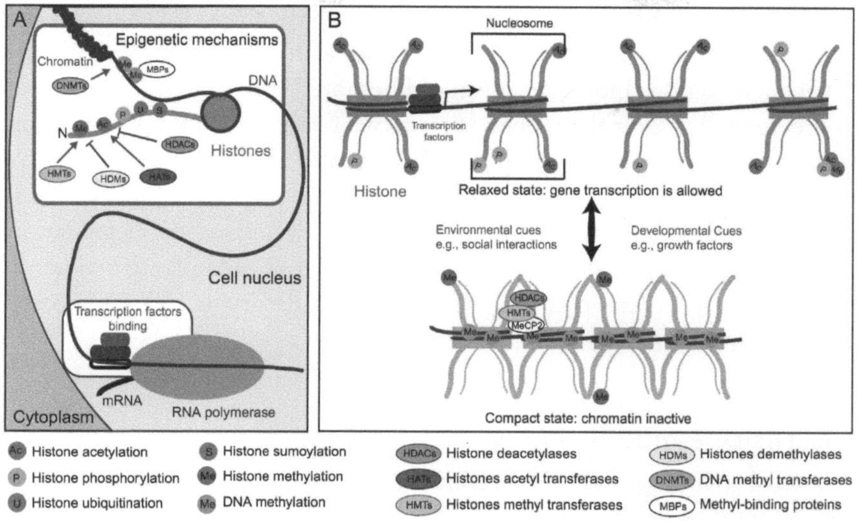

Figure 2 Regulation of gene expression via changes in chromatin compaction. (A) Epigenetic mechanisms control the compaction of chromatin by regulating DNA methylation and modifying histone tails by addition or removal of chemical groups. As a direct result, the binding of transcription factors is permitted or prevented. (B) While histone acetylation and phosphorylation have been shown to lead to an open state of the chromatin and therefore facilitate gene expression, histone methylation can have positive and negative impact on gene expression. DNA methylation can inhibit transcription by recruiting methylated DNA-binding proteins such as MeCP2, a member of the methyl-CpG DNA-binding protein family, and other proteins from the chromatin. (See the color plate.)

regulation of the epigenome in nondividing cells[7,8] and has important implications for postmitotic neural function in the central nervous system. Epigenetic mechanisms can therefore be considered as a bridge between the genes and the environment that refines neural networks according to experience and allows the environment to elicit lifelong biological changes in both gene expression and behavior.

Importantly, deregulation of epigenetic processes has been associated with a broad spectrum of neurological disorders,[9–11] the most well studied being Rett syndrome, an X-linked autism spectrum disorder. MeCP2, a member of the methyl-CpG DNA-binding protein family, was identified as the causative gene of the condition. Remarkably, in a mouse model of the disease, phenotypes associated with Rett syndrome could be rectified by restoration of the MeCP2 gene.[12] This suggests that unlike the hardwired information encoded in the DNA sequence, epigenetic programs are dynamic and reversible even in adulthood and thus ideal targets for therapeutic intervention.

We will now briefly describe the two main epigenetic mechanisms that regulate chromatin dynamics: posttranslational histone modifications and DNA methylation. However, it should be noted that it is widely accepted that epigenetic control involves a variety of facets including noncoding RNA activity and retrotransposable elements.[7,13,14] Noncoding RNAs including microRNAs (miRNAs) can have a profound effect in controlling gene expression by targeting mRNA for inhibition or degradation. Recently, it has been shown that a number of miRNAs play a major role in epigenetic processes by being involved in regulatory loops with the epigenetic enzymatic machinery.[15,16]

2.2. Histone modifications

The basic building block of chromatin is the nucleosome, composed of 147 bp of DNA wrapped around an octamer of histone proteins: two copies each of H2A, H2B, H3, and H4. Significant epigenetic control is exerted by dynamic histone modifications. The highly accessible N-terminal tails of histones are subject to posttranslational modifications, such as lysine acetylation (Ac), serine/threonine phosphorylation (P), lysine/arginine methylation (single methyl = Me, dimethyl = Me2, trimethyl = Me3), histone ubiquitination, and ADP-ribosylation. These modifications are especially prevalent at sites on histones H3 and H4 and contribute to the regulation of chromatin compaction. Histone modifications and the enzymes that catalyze these modifications have been the subject of numerous excellent reviews.[17–20]

The presence of electronegative acetyl or phospho groups decreases affinity between histone proteins and negatively charged DNA, contributing to an overall relaxation of chromatin and increased ability for transcriptional machinery to engage in active transcription (Fig. 2). For example, histone H3 serine 10 phosphorylation (PH3S10) serves as an indicator of active gene transcription.[21] However, the complete picture is not that simple. Histone methylation can have both activating and silencing impacts on gene expression; while histone H3 lysine 4 trimethylation (Me3H3K4) indicates active transcription, lysine 9 dimethylation (Me2H3K9) is an indicator of condensed and inactive chromatin. Furthermore, the interpretation of histone marks is not straightforward since genes can be in a "poised" state, presenting both active and repressive marks, and awaiting further cues. Such states may mark the potential for plasticity. Finally, some histone marks are known to participate in "crosstalk," for example, H3S10 phosphorylation and H3K14 acetylation (AcH3K14) are often concurrently occurring marks[22] and the combinatorial sum of DNA methylation and histone modifications marks can jointly affect recruitment and binding of chromatin-binding complexes (such as RNA Pol II).[23,24]

Specific enzymes are dedicated to imparting and removing histone modifications. For example, histone acetyltransferases (HATs) are responsible for the addition of histone acetyl marks, a process of transferring an acetyl group from acetyl Coenzyme A to histone tail lysine residues.[25] This is negatively regulated by histone deacetylases (HDACs). There are four classes of HDACs, zinc-dependent classes I, II, and IV, and the nicotinamide adenine dinucleotide (NAD)-dependent class III (sirtuins).[26] Importantly, HDACs have gained much consideration as pharmacological targets for inhibition. Indeed, epigenomic dysregulation of histone acetylation is a hallmark of cancer and HDAC inhibitors (HDACis) have received a lot of attention in cancer drug development.[27] Traditionally, HDACis are used to hyperacetylate histones and favor transcription of silenced tumor suppressor genes. Some HDACis are FDA approved in hematologic malignancies where they function as differentiation agents and are also used in combination with chemotherapy in solid tumors.[28] However, it is important to point out that their action on histones is nonspecific since they also deacetylate at lysine sites nonhistone targets, such as the tumor protein p53. Protein lysine acetylation and deacetylation are actually major regulators of cellular function.

Histone modifications are modulated by activation of intracellular transduction pathways such as MAPK and NFkB.[20,29,30] Specific signaling pathways target the enzymatic machinery devoted to modify chromatin, by either inducing phosphorylation of specific regulators or by modulating their interactions. For example, activation of the mitogen- and stress-activated protein kinase 1 (MSK1) via ERK/MAPK leads to the phosphorylation of H3S10, a mechanism important for synaptic activity-induced chromatin remodeling in neurons.[31,32]

2.3. DNA methylation

Methylation of cytosines in genomic DNA generates 5-methylcytosine (5mC) and is important for mediating repression of gene expression during mammalian development, which plays a key role in defining tissue-specific expression patterns. DNA methylation was, therefore, until recently regarded as fairly static, if not irreversible. However, it is now well accepted that DNA methylation is dynamically regulated in the adult nervous system and that this may be crucial for synaptic plasticity and memory formation.[33,34]

In most cell types, cytosine methylation occurs predominantly at CpG dinucleotides but it has just been shown that in neurons cytosine methylation can also occur at other dinucleotides such as CpA, CpC, or CpT.[35,36] While DNA methylation levels are very high in intergenic regions and repetitive sequences such as LINEs and SINEs (long and short intersped nuclear elements), CG-rich promoter sequences generally show low level of methylation. The functional consequence of DNA methylation is as complex as that of histone modifications and depends on the surrounding epigenome landscape and whether 5mC will go on recruiting transcriptional regulatory proteins or inhibit the binding of transcription factors. However, DNA methylation in 5′ regulatory regions usually silences promoter activity by interfering with transcription factor binding or by recruiting methylated DNA-binding proteins, such as MeCP2, that alter transcription efficiency by attracting silencing complexes.

DNA methylation is mediated by DNA methyltransferases (DNMTs). While DNMT1 is the maintenance DNMT that ensures that methylation patterns are copied faithfully throughout each cell division, DNMT3a and 3b regulate *de novo* DNA methylation in response to environmental factors, therefore encouraging gene silencing. Not only DNMT3a and 3b but also DNMT1 are expressed in postmitotic neurones and their expression has been shown to be dynamic in the mature nervous system,[37,38] suggesting that they could participate to activity-induced changes in gene expression.

The case of DNA demethylation is much more complex and remains somewhat to be fully elucidated. The first indication for DNA demethylation occurring in neurons was the finding of 5hmC in the brain,[39] the oxidized form of 5mC—a recognized intermediate of DNA demethylation. Soon after, the Ten-eleven translocation (Tet1–3) family of proteins was found to be necessary for 5hmC conversion and became intensely investigated as a regulator of memory formation.[40] Tet1 has been shown to play a critical role in activity-related demethylation in hippocampal neurons[41] and therefore activity-dependent gene expression. Other factors promoting demethylation have also emerged, in particular, the Gadd45 family implicated in DNA repair. Gadd45 genes are stress inducible and Gadd45b has been shown to be critical for the demethylation of promoters of plasticity-related genes such as *Bdnf*.[42,43] Finally, it has recently been suggested that the DNMTs themselves could catalyze the reverse reaction, demethylation.[44]

Drastic epigenomic dysregulation of DNA methylation is also a hallmark of cancer with aberrant hypermethylation and silencing of tumor suppressor genes commonly associated with the development of cancer.[45] During the last few decades, an increasing number of drugs targeting DNA methylation have been developed working toward better efficacy and stability and decreased toxicity.[46] However, as with HDACis, DNMT inhibitors (DNMTis) cannot act specifically on selected genes and induce both tumor suppressor and prometastatic genes.

While we have tried to present a clear picture of epigenetic marks and their role in controlling gene expression, we would like to reemphasize that the complete picture is very intricate and that elaborate crosstalk occurs between histone modifications, DNA methylation, and the genomic sequence. It is therefore important to be warned of overly simplistic interpretations of the meaning of epigenetic modifications.[24] Individual correlation between chromatin marks and translational activity are not enough to understand the potential of epigenetics for linking environmental signals with transcriptional regulation and neural function.

2.4. Technical advances in the investigation of the epigenome

The investigation of the epigenome would not be so advanced without the recent technical and bioinformatic developments for the generation and analysis of next-generation sequencing data. Whole genome can indeed now be sequenced at a fraction of the time required by capillary electrophoresis-based Sanger sequencing and new software and analysis algorithms are constantly being developed to address the need of specific applications.

The profiling of chromatin marks (such as histone modifications indicating active, silenced, or poised chromatin) is achieved by chromatin immunoprecipitation followed by sequencing (ChIP-seq). ChIP is a method used to selectively enrich a chromatin sample for DNA sequences bound by a particular protein. In this procedure, the proteins are crosslinked with DNA to preserve their *in vivo* association. The DNA is then sheared and DNA–protein complexes are pulled (isolation of DNA fragments via immunoprecipitation) using antibodies against diverse histone modifications. Commonly antibodies against H3K4me3 and H3K27me3 are used to identify transcriptionally active and silenced genes, respectively, while "poised" genes are marked by both H3K4me3 and H3K27me3. Finally, the crosslinks are reversed and the DNA is purified in preparation for sequencing.

Methylome profiling techniques usually rely on bisulphite conversion or employ a form of methylated DNA enrichment. Whole-genome bisulphite sequencing (WGBS) offers the most complete picture of the methylome but remains the most costly approach. Bisulphite treatment converts unmethylated cytosines to uracil while leaving methylated cytosines unchanged, and therefore allows for the discrimination between the two residues during sequencing. Recent developments in this field have facilitated the methylome analysis of very small amount of material such as nanogram quantities of genomic DNA.[47,48] It is important to point out that bisulfite treatment alone does not allow the distinction between 5mC and 5hmC or other oxidized cytosine modifications. These residues need to be protected by chemical modification (e.g., 5hmC can be glucosylated) to be quantified by sequencing[49] following bisulfite treatment.

A less expensive approach consists of selecting methylated DNA prior to the sequencing step. Methylated-DNA immunoprecipitation (MeDIP-Seq) is performed by immunoprecipitating methylated DNA with an antibody raised against 5mC. The presence or absence of a sequence gives an estimate of the methylation level in this area of the genome. It is worth noting that the analysis of CpG islands deserves particular attention. CpG islands are 300–3000 base pair stretches of DNA that are CG rich, often found within unmethylated regions of promoters, and play a critical role in regulating gene expression. CpG island shores, which are sequences up to 2 kb distant from the island themselves, also warrant some consideration since methylation is highly dynamic in island shores in disease states and correlates strongly with gene expression.[50]

Finally, it has recently become clear that the architecture of the nucleus and the spatial arrangement of the chromatin were key factors in the regulation of gene expression in relation to epigenetic marks and organization. Therefore, studying the structural properties and spatial organization of chromosomes is important when investigating the regulation of gene expression. Chromosome conformation capture (3C) investigates crosstalk between regulated genes and long range interactions between genes. When combined to next-generation sequencing, many interaction partners can be investigated to further develop our understanding of epigenetic networks.

3. EPIGENETIC REGULATION OF SYNAPTIC PLASTICITY IN LEARNING AND MEMORY

3.1. Introduction

Substrates for lifelong memory formation have long been under investigation, and indeed now include DNA methylation, histone modifications and the enzymes responsible for these epigenetic marks. We believe that in the relatively new field of pain epigenetics, it is essential to draw upon the extensive evidence showing that epigenetic processes are engaged in and necessary to memory formation. Striking similarities have been highlighted between the induction of central sensitization, the neural correlates of plasticity underlying the generation of long-term pain states, and early long-term potentiation (LTP) necessary to memory formation. Specifically, both involve molecules responsible for the posttranslational regulation of α-amino-3-hydroxy-5-methyl-4-isoxazolepropionic acid (AMPA) and *N*-methyl-D-aspartate receptor (NMDA) receptors, intracellular signaling cascades such as the ERK/MAPK pathway, and transcription of immediate early gene (IEG) programs such as *Zif268* and cAMP response element-binding protein (*CREB*).[51] Furthermore, it has been suggested that the molecular events underlying the synaptic plasticity observed in nociception and memory formation are evolutionarily conserved.[52]

Here, we highlight key studies identifying epigenetic mechanisms underlying the plasticity of memory formation and suggest that the pain field should learn from these studies. For extended interest, please see the excellent review articles cited here.[2,6,8,53–55]

3.2. Histones modifications and memory formation

It is well established that histone modifications are dynamic and altered during memory formation, but it is important to note that these marks vary spatiotemporally and by learning paradigm.[54] Relevant marks have been targeted to improve cognitive functions (Fig. 3).

3.2.1 Histone modifications

Pioneering studies by David Sweatt's group found that the transcriptionally permissive histone phosphorylation and acetylation marks on histone H3: PH3S10, AcH3K14, and the combined phospho-acetylated S10/K14 mark, were specifically increased in the hippocampus during the consolidation phase of contextual fear conditioning.[56,57] In another experiment, AcH4K5, along with PH3S10, AcH3K14 and the actively transcribed gene body mark Me3H3K36 were upregulated following (10 min–24 h) novel-object recognition learning in the hippocampus.[58] Using different paradigms, others have also reported an increase in hippocampal pan-acetyl H2B, and AcH4K12 during the consolidation phase of spatial memory formation using the Morris Water Maze.[59] Interestingly, AcH3K14, AcH4K5, and Me3H3K36 also increased with delayed expression in the prefrontal cortex,

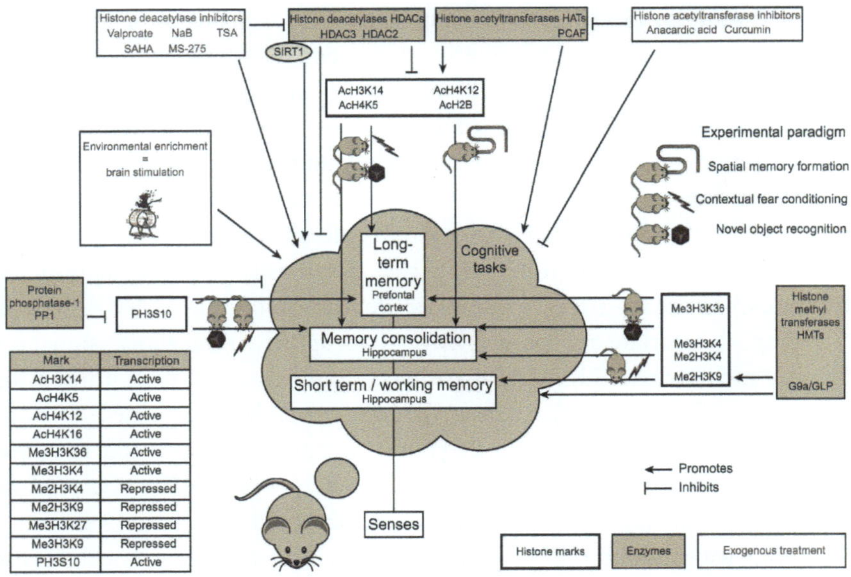

Figure 3 Histone modifications and their enzymes have been intensely investigated in the context of learning and memory. A number of histones marks are upregulated in the brain during cognitive tasks and are believed to promote memory formation. The role of histone acetyltransferases and deacetylases has been explored using both pharmacological and genetic tools.

persisting for up to 7 days, implicating an important role for these marks in remote memory formation.[58]

Additional studies using ChIP and ChIP-seq have localized the locus-specific increase of the aforementioned transcriptionally activating histone marks to the promoter regions of promemory formation genes such as *Bdnf*, *Zif268*, and *Homer1* following both contextual and cued fear conditioning, as well as novel-object recognition learning.[60–62] Finally, some histone modifications (specifically, PH3S10-AcH3K14, PH3S10) have been shown to be regulated by NMDA and ERK/MAPK signaling in memory induction, which could be crucial for the setting up of long-term pain states.[56,57,63,64]

3.2.2 Antagonizing histone-modifying enzymes to promote memory formation

It might be easy to correlate the appearance or loss of a number of histone marks with memory formation, however this does not imply that these marks are necessary to the full formation of memory. In order to assert the role of specific histone modifications in this process, it is important to antagonize these marks and observe the resulting behavioral effects. While largely unspecific, targeting the enzymes responsible for imparting or removing key histone modifications has been so far the preferred way forward. Since increases in histone acetylation have been observed in various brain areas and paradigms investigating memory formation, inhibiting HDACs has been the focus of numerous studies. In particular, the role of class I HDACs 2 and 3 has been the most extensively studied, using either genetic or pharmacological approaches.

3.2.2.1 Genetic modification of HDACs expression

By genetically overexpressing HDAC2 in neurons, Guan[65] and colleagues showed that HDAC2, but not HDAC1, was a negative regulator of memory formation. Furthermore, HDAC2 was upregulated in the hippocampi of mice exhibiting age-related memory deficits or experimentally induced neurodegeneration (CK-P25 mice) when compared with control animals.[66,67] In the latter case, HDAC2 was shown to associate with promoter regions for genes crucial to memory formation, indicating that they were silenced.[67] Finally, knockdown of HDAC2 with virally transfected shRNA in the hippocampus and PFC was also shown to improve memory deficits.[67] HDAC3 is also a crucial negative regulator of memory formation, and mice with HDAC3 disruption showed enhanced long-term memory.[68]

Contrastingly, Gao *et al.*[69] found that the class III HDAC sirtuin1 (SIRT1) had a positive association with contextual memory formation. SIRT1 was important in limiting miR134 expression, which forms a complex limiting CREB expression. Downregulation of SIRT1 leads to overexpression of miR134 and ultimately downregulation of CREB and *Bdnf*, a neurotrophin essential to synaptic plasticity.

3.2.2.2 Using HDAC inhibitors

In general, systemic or localized (i.e., hippocampal infusion) delivery of HDACis such as sodium butyrate (NaB), trichostatin A (TSA) and valproate (VPA) have been shown to enhance cognitive function when given during the consolidation phase shortly after training,[57,60–62,65,70–74] predominantly by increasing histone acetylation globally and at specific genes.

Moreover, HDACis have also been shown to recover deficits in memory formation. For example, chronic intraperitoneal (i.p.) NaB re-established access to long-term memory in memory-impaired CK-P25 neurodegenerating mice, acting to increase AcH3K14 and AcH4K5 and upregulate expression of synaptic proteins.[71] AcH4K12, a modification downregulated in aged (16-month old) mice exhibiting cognitive deficiencies was reinstated using hippocampal infusion of suberoyl anilide hydroxamic acid (SAHA), which restored cognitive abilities.[75] Interestingly, the hyperacetylating effects of NaB can be mimicked by environmental enrichment, a model that engages brain activity via physical environment, which begins to restore hippocampal and cortical acetylation levels and learning itself as early as 3 h following environmental enrichment in neurodegeneration-induced animals.[71]

It is important to note that while we would expect the use of HDACis to affect the expression of a very large number of genes, a gene array study has found that application of nonselective HDACis to cells upregulated only 10–15% of all genes analyzed.[76] This could indicate some specificity in genes targeted by HDACis but these results remain unexplained and need to be further investigated.

3.2.2.3 Other targets

Studies have also looked at HDAC counterpart enzymes, HATs. PCAF (the p300/CBP-associated factor complex) has HAT properties and is increased during memory consolidation.[59] Numerous studies by Ted Abel's group and others showed that CBP deficient mice displayed deficits in memory formation.[74,77–79] Other key studies worth mentioning have demonstrated a role for the G9a/GLP histone methyltransferase complex, which is responsible for the silencing histone methylation mark Me2H3K9, in the initiation of memory formation.[60,80] Finally, mice deficient for the protein phosphatase 1 (PP1), which is responsible for H3S10 dephosphorylation, exhibited enhanced long-term memory formation.[81,82]

3.3. DNA methylation and memory formation

Until recently, DNA methylation had been regarded as a fairly stable event leading to irreversible transcriptional silencing in the adult animal. However, the latest evidence has indicated that neuronal DNA methylation is dynamically regulated as a result of neural activity and that DNA demethylation regulates synaptic plasticity (Fig. 4). Moreover, since the methylation of DNA can be relatively persistent when compared to posttranslational histone modifications, it has often been suggested that DNA methylation was a likely environmental adaptation of the nervous system, such as memory formation, on a behavioral time scale.

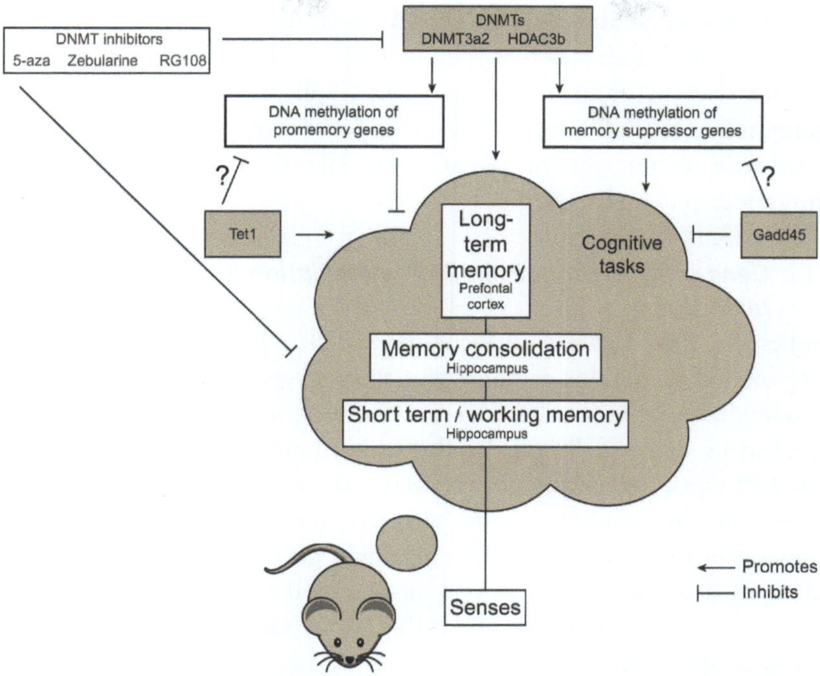

Figure 4 The role of DNA methylation in memory formation has been thoroughly investigated. Changes in DNA methylation in both promemory formation and memory suppressor genes have been reported in learning and memory paradigms. ?: unknown.

3.3.1 DNA methyltransferases sustain memory formation

Studies investigating the regulation of DNA methylation in neural processes have examined the expression of DNMTs. First of all, DNMT3a2 levels have been shown to be crucially indicative of cognitive ability and were decreased in animals displaying aging-impaired cognitive function.[83] DNMT3a2 is partly regulated by nuclear calcium signaling and restoration of DNMT3a2 levels (using recombinant adeno-associated virus) recovered deficits in cognitive function.[83] DNMT3a and 3b were upregulated in the hippocampus and the lateral amygdala shortly following memory formation in contextual fear conditioning.[84,85]

Subsequent studies have attempted to target the activity of DNMTs using the DNMTis 5-aza-cytidine (5-aza), zebularine, and RG108. As expected, inhibition of DNMT function was generally found to impair memory consolidation in contextual fear conditioning, reward learning, and remote memory paradigms.[61,84,86–88] It is important to note that the function of 5-aza and zebularine in postmitotic cells is debated as their mechanisms of action requires DNA replication whereby inactivation occurs because inhibitors trap the enzymes on the chromosome; RG108 however acts without requiring DNA replication. Thus, any studies attempting to block DNMT activity with 5-aza or zebularine must be interpreted with caution, and the role of nonneural cells in particular must also be considered.

3.3.2 Gene-specific changes in DNA methylation in learning and memory

Studies that have focused on locus-specific changes in DNA methylation have offered a complex picture. Not surprisingly, the methylation levels at individual plasticity-related genes crucial to memory formation were found to be dynamically regulated. For example, the memory suppressor gene *PP1* was hypermethylated and downregulated during memory consolidation process, indicating its targeted silencing.[84] Conversely, the promemory formation and LTP enhancing gene *reelin* was demethylated during memory formation, which co-occurred with an increase in gene expression. It has also been shown that DNA methylation was decreased at *Bdnf* exons I and IV in CA1 of the hippocampus which correlated with an increase in *Bdnf* mRNA after fear conditioning.[61]

Interestingly, while changes in DNA methylation at specific genes required for memory formation return to control within 24 h in the hippocampus, they can last at least 30 days in the cortex, a site important for long-term memory storage or remote memory. This parallels findings regarding histone modifications (see Section 3.2.1). For example, activity-induced DNA methylation of the memory suppressor gene *calcineurin*, was seen in the cortex 7 days after memory formation. Cortical infusion of DNMTs at 29 days disrupted memory recall, indicating that active maintenance of methylation (silencing of the memory suppressor gene) in the cortex was required for long-term memory.[88]

3.3.3 DNA demethylation and memory formation

Excitingly, Tet1 has recently been found to promote active DNA demethylation, and three key studies have shown its regulation of cognition using *Tet1* genetically modified mice, and silencing of *Tet1* with viral transfection of shRNA.[89–91] These studies showed that deregulation of *Tet1* impaired memory formation and memory extinction due to alterations in the expression levels of genes crucial to neural plasticity such as *Bdnf*, *Arc*, *Egr1*, *Fos*, *Homer1*, *Npas4*, and *Nr4a2*. Crucially for us, these IEGs are implicated in both memory formation and pain processing. *Tet1* knockouts exhibited abnormal hippocampal long-term depression and impaired memory extinction, although cued and contextual fear conditioning behavior at 24 h was normal.[90]

Other studies have focused on Gadd45, another candidate in activity-dependent DNA demethylation. However, the results are confusing and indicate an opposite regulation than that of Tet1. Indeed, Gadd45b was upregulated during memory consolidation but deletion of *Gadd45b* led to enhanced long-term memory formation.[43,92] It is possible that Gadd45b targets memory suppressor genes.[93] Interestingly, studies by Sultan *et al.*[92] as well as Rudenko *et al.*[90] found that *Tet1* and *Gadd45b* knockout mice displayed no deficits in acute nociception, but only as determined using hot plate, which measures thermal hypersensitivity.

3.4. The complexity of the interpretation of the epigenome

Before moving onto the field of pain, we would like to reiterate here that translating the epigenome is not straightforward. While we have tried in this section to be as clear as possible, we must remember the complexity of the crosstalks between epigenetic marks. To illustrate this, we would like to mention the work from Gupta *et al.*[60] In their study using paradigms of contextual and cued fear conditioning, this group reported that increased DNA methylation at the *Zif268* promoter region was associated with increased Me3H3K4 (a mark of active chromatin) and increased *Zif268* gene expression 30min post-fear conditioning training. These observations are in contrast to findings from developmental studies that suggest that DNA methylation is primarily associated with the repression of gene transcription. Moreover, in the same study, the authors found that Me3H3K4 (active) and Me2H3K9 (silencing) were both increased 24 h after fear conditioning training. We must therefore remain cautious when interpreting epigenetic marks, especially if the whole epigenetic landscape surrounding a gene of interest has not been fully identified.

4. EPIGENETIC MECHANISMS AND PAIN STATES

4.1. Introduction

In this section, we discuss the current evidence that epigenetic mechanisms play a role in determining the molecular and behavioral response to noxious stimulation. We have mentioned earlier on (Section 2.1) that some consider miRNA activity as epigenetic regulation. We have decided not to refer to such mechanisms but we would like to stress that few studies have suggested that changes in miRNA expression might contribute to pain sensation through translational regulation of pain-related genes. These studies have been reviewed elsewhere.[94,95]

Our purpose here is not to provide an exhaustive review of all the literature in the field since a number of recent reviews on the subject are available.[52,94,96–98] Our objective is rather to offer a perspective on the topic, by focusing on those studies we believe are key to unlocking regulation of nociception. We will first highlight studies that have looked at changes of the epigenome during the induction and persistent phase of pain states. We will then summarize studies looking at the behavioral and molecular effects of tentative treatments targeting epigenetic mechanisms since arguably the most enticing aspect of investigating these mechanisms in pain processing is the potential for reversibility. Our ultimate aim is to highlight the complexity of the findings so far, which makes targeting epigenetic mechanisms for the treatment of pain states a difficult task.

4.2. Injury induces changes to the epigenetic machinery and the epigenome

One clue that injury may act directly to regulate DNA methylation and histone modifications comes from the investigation of the expression of the epigenetic machinery. Few studies have indeed shown that a number of epigenetic regulators are subject to injury-induced changes in expression in the dorsal horn and DRGs: MeCP2,[99,100] the HAT, p300,[101,102] and several of the DNMTs and HDACs.[99,103–105] However, all studies looked at different pain models and there was no consensus in terms of time scale or direction of the changes (Fig. 5).

We also have some information concerning changes in the epigenome landscape, at different locations within the so-called pain pathways. So far, changes have been reported in the sciatic nerve, the DRGs, the dorsal horn as well as the brain, specifically the Nucleus Raphe Magnus (NRM) and the frontal cortex. Overall people have mainly focused on histone acetylation, whether at a global level or at gene-specific promoter areas (Fig. 5). While studies have mainly reported increased level in histone acetylation following injury, there is no agreement, with reports of both increase and decrease in histone acetylation in the promoter area of pronociceptive genes. This is important to bear in mind when considering targeting epigenetic mechanisms for pain relief. Another warning comes from reports of

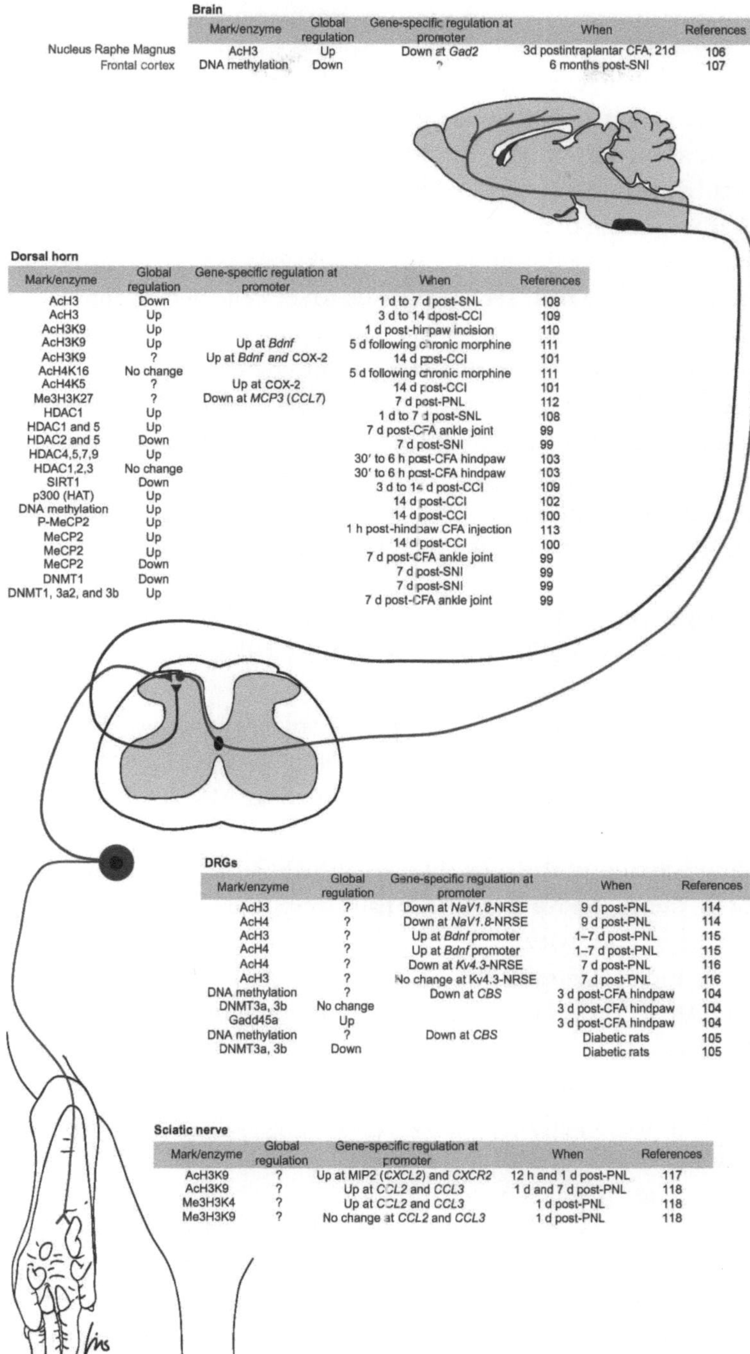

Brain

	Mark/enzyme	Global regulation	Gene-specific regulation at promoter	When	References
Nucleus Raphe Magnus	AcH3	Up	Down at *Gad2*	3d postintraplantar CFA, 21d	106
Frontal cortex	DNA methylation	Down	?	6 months post-SNI	107

Dorsal horn

Mark/enzyme	Global regulation	Gene-specific regulation at promoter	When	References
AcH3	Down		1 d to 7 d post-SNL	108
AcH3	Up		3 d to 14 dpost-CCI	109
AcH3K9	Up		1 d post-hindpaw incision	110
AcH3K9	Up	Up at *Bdnf*	5 d following chronic morphine	111
AcH3K9	?	Up at *Bdnf and* COX-2	14 d post-CCI	101
AcH4K16	No change		5 d following chronic morphine	111
AcH4K5	?	Up at COX-2	14 d post-CCI	101
Me3H3K27	?	Down at *MCP3* (*CCL7*)	7 d post-PNL	112
HDAC1	Up		1 d to 7 d post-SNL	108
HDAC1 and 5	Up		7 d post-CFA ankle joint	99
HDAC2 and 5	Down		7 d post-SNI	99
HDAC4,5,7,9	Up		30′ to 6 h post-CFA hindpaw	103
HDAC1,2,3	No change		30′ to 6 h post-CFA hindpaw	103
SIRT1	Down		3 d to 14 d post-CCI	109
p300 (HAT)	Up		14 d post-CCI	102
DNA methylation	Up		14 d post-CCI	100
P-MeCP2	Up		1 h post-hindpaw CFA injection	113
MeCP2	Up		14 d post-CCI	100
MeCP2	Up		7 d post-CFA ankle joint	99
MeCP2	Down		7 d post-SNI	99
DNMT1	Down		7 d post-SNI	99
DNMT1, 3a2, and 3b	Up		7 d post-CFA ankle joint	99

DRGs

Mark/enzyme	Global regulation	Gene-specific regulation at promoter	When	References
AcH3	?	Down at *NaV1.8*-NRSE	9 d post-PNL	114
AcH4	?	Down at *NaV1.8*-NRSE	9 d post-PNL	114
AcH3	?	Up at *Bdnf* promoter	1–7 d post-PNL	115
AcH4	?	Up at *Bdnf* promoter	1–7 d post-PNL	115
AcH4	?	Down at *Kv4.3*-NRSE	7 d post-PNL	116
AcH3	?	No change at Kv4.3-NRSE	7 d post-PNL	116
DNA methylation	?	Down at *CBS*	3 d post-CFA hindpaw	104
DNMT3a, 3b	No change		3 d post-CFA hindpaw	104
Gadd45a	Up		3 d post-CFA hindpaw	104
DNA methylation	?	Down at *CBS*	Diabetic rats	105
DNMT3a, 3b	Down		Diabetic rats	105

Sciatic nerve

Mark/enzyme	Global regulation	Gene-specific regulation at promoter	When	References
AcH3K9	?	Up at MIP2 (*CXCL2*) and *CXCR2*	12 h and 1 d post-PNL	117
AcH3K9	?	Up at *CCL2* and *CCL3*	1 d and 7 d post-PNL	118
Me3H3K4	?	Up at *CCL2* and *CCL3*	1 d post-PNL	118
Me3H3K9	?	No change at *CCL2* and *CCL3*	1 d post-PNL	118

Figure 5 Epigenetic marks and enzymes modified after injury. A number of studies have investigated changes in the epigenome and epigenetic enzymes following acute or long-term injuries. We have here summarized most of these studies and indicated where and when the changes were observed. ?: unknown.

gene-specific changes going the opposite direction than global changes; for example, Zhang *et al.*[106] found that global AcH3 was increased in the NRM but down at the promoter area of the *Gad2* gene, the gene encoding glutamic acid decarboxylase 65 (GAD65), the enzyme responsible for catalyzing the decarboxylation of glutamate to GABA and CO_2, following intraplantar Complete Freund's adjuvant (CFA).

Interestingly, we have recently started to see cell-specific approaches and there are reports of changes in epigenome within glia cells, a cell type which we know play an important role in nociceptive transmission.[112,118] Finally, other tissues may also bear epigenetic changes relevant to hypersensitivity. For example, increased DNA methylation of the secreted protein, acidic, rich in cysteine (*SPARC*) gene, a protein known to be involved with vertebrate disc degeneration, was found in discs and correlated with low expression of SPARC and behavioral manifestations of low back pain.[119]

4.3. Targeting histone modifications and DNA methylation to improve injury-induced hypersensitivity

4.3.1 Targeting HDACs and HATs

HDACis have been so far the most used epigenetic drugs in pain studies (Table 1). Contrastingly with the complex landscape described in Section 4.2, administration of HDACis such as MS-275 and SAHA has had overall positive outcome on pain behavior.

Administration of the class II HDACi SAHA has been shown to attenuate both the second phase of formalin response and intraplantar CFA-induced hypersensitivity.[103,106,120] However, effects have been assigned to various molecular mechanisms. Chiechio *et al.*[120] attributed the effect of the HDACis on formalin behavior to an upregulation of mGlu2 metabotropic glutamate receptors in the DRG and spinal cord. Pharmacological activation of mGlu2 is indeed known to produce analgesic effects in inflammatory and neuropathic pain states and the authors reported hyperacetylation of p65/RelA at lysine 310 and increased transcription of NFkB-regulated genes, which are known to include mGlu2. Zhang *et al.*[106] showed that NRM infusions of TSA and SAHA could improve injury-induced reduction of tail-flick latency and this was presumed to be the result of impaired presynaptic GABAergic synaptic function.[106] Drug-mediated attenuation of nociceptive behavior was indeed found to be the result of increased acetylation of histone H3 at the *Gad2* promoter.

HDACis have also been used to counteract the hypersensitivity seen in neuropathic pain models. Indeed, infusion of MS-275 via osmotic

Table 1 Targeting histone modifications and DNA methylation for the treatment of pain states using nonspecific inhibitors

Pain model(s)		Drug treatment	Behavior	Improved?	Proposed molecular mechanisms	Reference
Targeting histone modifications						
HDAC inhibitors						
Inflammatory	Formalin intraplantar	5 × preemptive subcutaneous injection; MS-275 3 mg/kg, SAHA 50 mg/kg	Formalin behavior attenuated with both drugs	Yes	Upregulation of p65/RelA acetylation in DRGs and spinal cord following injection of HDACis which correlated with mGlu2 upregulation. N.B.: Changes after injury not investigated	120
	CFA intraplantar	1 preemptive intrathecal injection; SAHA, TSA, LAQ, valproate, 4PB, MS-275; SAHA 25 μg	CFA-induced thermal hypersensitivity attenuated with all except MS-275 (class I)	Yes	Upregulation of AcH3K9/K18 in the spinal cord 30′ after IT SAHA, MS-275. N.B.: Changes after injury not investigated	103
	CFA intraplantar	4 × daily NRM infusion or 4 × daily intraperitoneal injection TSA 4 mg/kg, SAHA 40 mg/kg	NRM Infusion attenuated tail-flick latency behavior	Yes	Injury induced a global upregulation of AcH3 in the NRM and downregulation of AcH3 at Gad2 promoter. Drug treatment upregulated AcH3 at Gad2 promoter in the NRM which correlated with GAD65 upregulation	106
Neuropathic	d4T, L5 SNT, SNL	Intrathecal administration via osmotic minipump; MS-275, MGCD; 30 or 60 nmol/d	40–50% mechanical and thermal attenuation with MS-275	Yes	Upregulation of AcH3K9 (global and at specific gene promoters: Cacna2d1, Hdac1, MeCP2) in the spinal cord after HDACis administration. There were no changes in the DRGs. N.B.: Changes after injury not investigated	121

Continued

Table 1 Targeting histone modifications and DNA methylation for the treatment of pain states using nonspecific inhibitors—cont'd

Pain model(s)		Drug treatment	Behavior	Improved?	Proposed molecular mechanisms	Reference
	PNL	3× to 9× daily intraperitoneal injection; TSA 1 mg/kg, SAHA 5 mg/kg, VPA 200 mg/kg	TSA/VPA blocked hypoesthesia (mech/thermal/squeeze), SAHA reversed hypoesthesia	Yes	Downregulation of AcH3/H4 at Nav1.8 NRSE promoter in DRGs following PNL was reversed by HDACis, which reversed Nav1.8 mRNA downregulation	114
	L5/6 SNL	VPA, oral, daily for 21 days, 300 mg/5 ml/kg	Valproate attenuated SNL-induced mechanical hypersensitivity	Yes	Valproate prevented the SNL-induced downregulation of the glutamate transporters GLT-1 in the spinal cord. N.B.: Epigenetic changes were not investigated	122
Others	Incision and incisional priming	1× preemptive or 4× daily intraperitoneal injection; SAHA; 50 mg/kg	SAHA exacerbated the mechanical hypersensitivity seen postincision but no effect on thermal hypersensitivity	No	Global upregulation of AcH3K9 in the spinal cord postincision was exacerbated by SAHA and specific AcH3K9 at CXCR2/CXCL1 promoter was upregulated by SAHA. N.B.: Effect of incision on gene-specific acetylation changes not evaluated	110
	Opioid-induced hyperalgesia	Chronic administration; intraperitoneal injection; SAHA; 50 mg/kg	SAHA exacerbated morphine-induced mechanical, thermal hypersensitivity and physical dependence	No	Increase in global and at BDNF promoter AcH3K9 in the spinal cord (no change in AcH3K16), reduced HDAC activity and HAT activity unchanged after chronic morphine. N.B.: Molecular changes after drug treatment are not investigated.	111

Table 1 Targeting histone modifications and DNA methylation for the treatment of pain states using nonspecific inhibitors—cont'd

Pain model(s)		Drug treatment	Behavior	Improved?	Proposed molecular mechanisms	Reference
HAT inhibitors						
Neuropathic	PNL	30–100 μmol/kg intraperitoneal injection 30 min prior to surgery of AA	AA administered 30 min prior to PNL-reduced tactile allodynia and thermal hyperalgesia at Day 7 post-PNL	Yes	AA reduced the PNL-induced upregulation of AcH3K9 at MIP2 and CXCR2 promoter in the injured sciatic nerve, in neutrophils and macrophages	117
	PNL	30–100 μmol/kg intraperitoneal injection 30 min prior to surgery of AA	N.B.: Behavior was not investigated	Yes	AA reduced the PNL-induced upregulation of AcH3K9 and Me3H3K4 at CCL2 and CCL3 promoter in the injured sciatic nerve, in macrophages	118
	CCI	Curcumin; 7 × daily intraperitoneal injection post-CCI; 40–60 mg/kg	Curcumin delivered 7 days post-CCI for 7 days reduced CCI-induced thermal hyperalgesia and mechanical allodynia	Yes	Curcumin reduced CCI-induced AcH3K9 at BDNF promoter and both AcH3K9 and AcH4K5 at Cox-2 promoter in the spinal cord	101
Others	Incision and incisional priming	1 × preemptive or 4 × daily intraperitoneal injection; AA; 5 mg/kg intraperitoneally	AA 2 days before and 4 days after attenuated mechanical sensitivity postincision and incisional priming. No effect on thermal	Yes	Global upregulation of AcH3K9 in the spinal cord postincision. N.B.: Effect of AA not investigated at molecular level	110
	Opioid-induced hyperalgesia	Chronic administration; intraperitoneal injection; curcumin; 50 mg/kg	Curcumin reduced morphine-induced mechanical, thermal hypersensitivity and physical dependence	Yes	Increase in global and at BDNF promoter AcH3K9 in the spinal cord (no change in AcH3K16), reduced HDAC activity and HAT activity unchanged after chronic morphine. N.B.: Molecular changes after drug treatment not investigated	111
Targeting DNA methylation						
	CCI	5-azacytidine; IT catheter 10 μmol/d	5-aza attenuated CCI-induced mechanical and thermal hypersensitivity	Yes	CCI-induced spinal increase in global DNA Me was reduced by 5-aza	100

Various studies have investigated the effects of drugs targeting HDACs, HATs, and DNMTs in the treatment of pain states.

minipumps reduced by 50% both the thermal and mechanical hyperalgesia induced by neuropathic injury (clinically relevant d4T model, spinal nerve ligation (SNL), spinal nerve transection (SNT)).[121] MS-275, however, was unable to reverse existing neuropathic behavior. Neuropathic injury-induced c-fiber hypoesthesia was also prevented by HDACis, specifically i.p. TSA and VPA, and was reversed by SAHA.[114] While partial sciatic nerve ligation (PNL) induced the downregulation of *Nav1.8* in DRGs by binding of silencer element NRSF/REST to *Nav1.8* gene, TSA and VPA prevented the binding of NRSF by inducing the hyperacetylation of histone H3/H4 at the NRSE (binding sequence for NRSF) within the *Nav1.8* gene.[114,123] Using the same injury model, the authors reported that the NRSF/REST silencing mechanism also occurred within the *Kv4.3* gene in DRGs following a reduction of acetylation at NRSE.[116] Finally, PNL could also induce hyperacetylation of histone H3 and H4 at the exon I transcript of the *Bdnf* gene in primary afferent nociceptors, which might explain the upregulation of this pronociceptive gene in the development of neuropathic pain states. The authors, however, never fully investigated the behavioral consequences of targeting these latest epigenetic targets.[115]

Others, however, have reported that HDACis exacerbated the hypersensitivity induced by skin incision[110] and opioid use,[111] which was consistent with the spinal upregulation of global and pronociceptive genes acetylation seen in both models (Table 1). As expected, HAT inhibitors (anacardic acid (AA) and curcumin, respectively) attenuated the hypersensitivity seen in these models.

4.3.2 Targeting DNA methylation

A number of studies have investigated the dynamics of DNA methylation in the DNA sequence of pronociceptive genes. However, the interest in some of these studies might be limited considering the tissue analyzed (blood, keratinocytes, neural stem cells) and sometimes the absence of a pain paradigm (for review, see Bai *et al.*[94]). However, few studies are worth mentioning here. Lötsch and colleagues[124] have reported an increase in the level of DNA methylation of the μ-opioid receptor gene *OPRM1*, as well as global increase in DNA methylation at *LINE-1*, in blood cells of chronic opioid addicts. Increases in DNA methylation correlated with chronic pain scores but disappointingly did not have any influence on the expression level of the μ-opioid receptor. In another interesting study, Xu and colleagues correlated the upregulation of cystathionine-β-synthase (CBS) (both mRNA and protein levels) with a downregulation of DNA methylation in the promoter area in DRGs of rats with chemically induced diabetes or following injection of CFA in the hindpaw (3 days post-CFA). The authors then went on to show that CBS contributed to the hypersensitivity seen in both pain models. Surprisingly, in the CFA model, the change in DNA methylation was linked to an upregulation of the demethylating agent Gadd45a while in the diabetes model it was linked to a downregulation of DNMT3a and 3b.[104,105]

So far only one study has looked at antagonizing DNMTs for the treatment of pain states. The DNMT inhibitor, 5-aza delivered via *i.t.* catheter was found to attenuate both mechanical and thermal hyperalgesia seen after chronic constriction injury (CCI). The authors found that CCI caused an increase in both MeCP2 and global DNA methylation in the spinal cord, which was reversed by 5-aza.[100] Contrastingly, neuropathic (spared nerve injury—SNI) injury was found to cause a global decrease in DNA methylation in the prefrontal cortex and amygdala of mice 6 months postinjury.[107] These levels were reversed by environmental enrichment and correlated with levels of hyperalgesia. However, we must remember that the meaning of a global change in DNA methylation remains to be established; indeed, most methylation in CpGs occurs in repetitive sequences such as LINEs (see Section 2.3) while CpGs in promoter areas and CpG islands and their shores, which are more relevant to transcription, remains largely unmethylated and contribute only minimally to global changes in CpG methylation.

4.4. Epigenetic mechanisms: Pharmacological targets for the treatment of persistent pain states?

The picture emerging from the studies reviewed above is clearly one of a very complex scenario, with specific and dynamic epigenetic landscape patterns which would depend on the injury, and most probably the past experiences of each individual. Clinical observations indeed indicate that diseases or injuries of similar severity can cause a wide range of pain experiences across individuals. These observations do not come as a surprise and agree with genetic studies that have reported very precise patterns of gene expression following injury.[125,126] Therefore, it seems highly unlikely that HDACis and DNMTis, which are nonspecific drugs that affect the regulation of multiple genes and with numerous side effects,[127,128] will be considered for long-term treatment of chronic pain.

Another option to exploit epigenetic mechanisms to treat persistent pain states is to revert to manipulations at the single-gene level. Once changes in the epigenome have been identified in a gene-specific manner, genes could be targeted in order to investigate their contribution to the maintenance of pain states. This was the approach chosen by Geranton *et al.*[113] Indeed, the authors had found that a specific pattern of gene changes occurred in the dorsal horn over 7 days of inflammation, with an overall early increase in gene expression correlating with release of MeCP2 DNA binding. Knocking down a downstream target of MeCP2, the serum/glucocorticoid-regulated kinase *SGK1*, using siRNA, delayed the full expression of mechanical hypersensitivity.

Finally, a more state-of-the-art and subtle approach would be to modify the epigenome itself, in a gene-specific manner, using targeted genome engineering technologies such as designer DNA-binding proteins based on transcriptional activator-like effectors (TALENS) and zinc finger proteins (Fig. 6). Such artificial transcription factors are tailored to recognize

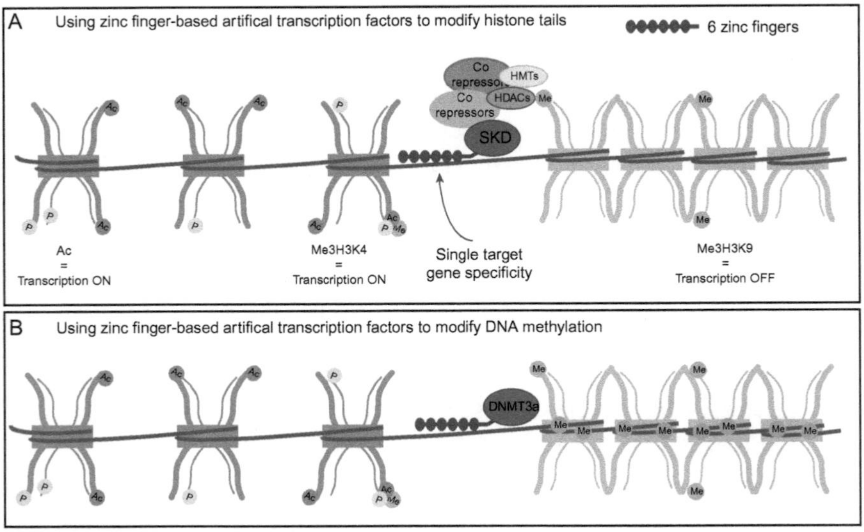

Figure 6 The epigenome can be modified at specific target sites using multimodular zinc finger proteins linked to epigenetic editing domains. These artificial transcription factors contain arrays of Cys2-His2 zinc finger domains, which specifically interact with 3 bp (triplet of recognition) of DNA. By combining a number of zing finger domains, multimodular proteins have been engineered to recognize specific sequences of promoters and regulate gene expression. 6 zinc finger proteins (18 bps targeted) represent optimized, state-of-the-art designs that regulate single target genes with high selectivity. (A) Zinc fingers linked to the Kruppel-associated box (SKD) can be used to change histone modifications at a specific target site. Upon recruitment of corepressors such as KRAB-associated protein 1 and nucleosome remodeling and deacetylase (NuRD), a repressive complex is assembled including HDACs, histone methyltransferases such as SETDB1 and the heterochromatin protein 1. This repressive complex could catalyze the formation of condensed chromatin by deacetylation of histones, demethylation of H3K4me3, and incorporation of H3K9me3.[129] (B) Zinc fingers linked to DNMT3a have been used previously to increase DNA methylation and silenced the oncogene SOX2.[130] (See the color plate.)

specific DNA sequences and can be engineered to generate tools for targeted genome modification.[129] They have been used successfully to modify the expression of tumor suppressor genes via increased DNA methylation.[130,131]

5. FUTURE DIRECTIONS

5.1. Long-lasting epigenetic modifications: The key to the transition from acute to chronic pain states?

It is well known that injuries of similar severity can cause a wide range of pain experiences across individuals, highlighting the complexity of pain processing. In particular, following similar injuries, some individuals will recover very quickly while others will go on suffering once the injury has resolved and there is currently no way to predict whether someone will develop chronic pain following injury. A number of studies have identified genes with polymorphisms associated with impaired pain sensitivity such as *SCN9A*, which encodes the alpha subunit of the sodium channel Nav1.7.[132] However, the response to injury cannot be simply linked to genetic variation since monozygotic twins can exhibit significantly different inflammatory and chronic pain phenotypes.[133] We strongly believe that since the response to injury is determined by complex interactions between the genome and the environment and these are governed by epigenetic mechanisms, such mechanisms could be the key to the transition from acute to chronic pain states.

A number of evidences support this idea. First, we have highlighted in this chapter how epigenetic mechanisms are engaged in the dorsal horn rapidly after noxious stimulation. Second, long-term changes in the epigenetic landscape are reported well into the maintenance phase of persistent pain states. Finally, it is known that in chronic pain the spinal cord nociceptive networks remain sensitized with long-lasting changes in gene expression and synaptic strength.[134] The similarity between LTP and spinal cord sensitization raises the possibility that these long-lasting changes resemble a form of learning. Chronic pain states could, therefore, be regarded as a form of nociceptive memory maintained by a specific epigenetic code. Interestingly, this theory could be investigated using hyperalgesic priming models.

Hyperalgesic priming models have been developed to investigate the transition from acute to chronic pain. In these models, animals who have suffered a minor injury remain in a long-lasting latent hyper-responsiveness to an inflammatory or surgical insult. In other words, these models produce a state of sensitization closely resembling clinical situations with increased risk of developing chronic pain. Long-lasting changes within primary afferent nociceptors have been reported and it has been suggested that the protein kinase Cε (PKCε) might be able to generate a pain memory in primary afferent nociceptors.[135] However, there is also increasing evidence that dorsal horn mechanisms are key to this primed state. A number of studies have indeed found that an atypical protein kinase C called protein kinase Mζ (PKMζ) might be involved in the maintenance of persistent nociceptive processing induced by inflammation in the rat dorsal horn.[136–138] Exploring the dynamics of the epigenome in the dorsal horn in the primed state of hyperalgesic priming models should therefore clarify the role of epigenetic mechanisms in maintaining a state of latent hypersensitivity.

5.2. Could an early life injury leave long-lasting epigenetic marks responsible for long-term changes in pain sensitivity?

As for the transition from acute to chronic pain, we believe that epigenetic mechanisms could contribute to the long-lasting nociceptive memory that develops following early life injury. Indeed, injury in early life results in alterations in sensory processing later on in life, but these persistent alterations in pain sensitivity vary depending on the type and severity of the injury and the age at which it occurs both in animals and humans.[139–143] We believe that a reason why priming mechanisms could differ across ages is because epigenetic modifications could be differentially engaged in the young and the mature nervous system by noxious stimulation. Indeed, the epigenome tends to change as we age; changes in both histone modifications and DNA methylation, specifically gene-specific and global hypermethylation and hypomethylation, have been reported with aging.[144–147] Some epigenetic actors are also expressed differently in young versus adult; e.g., the expression of DNMT1 and 3a decreases considerably between newborn and middle-age (23–50-years old) humans.[83,148] Finally, the most fascinating evidence comes from recent human studies on the gene *FKBP51*. Trauma-induced site-specific demethylation of the gene *FKBP51* was found to occur only if the trauma happened in childhood but not in adult life. This demethylation, which persisted into adulthood, could greatly increase the expression of FKBP51 after trauma in individuals with a specific *FKBP51* allele and enhanced the risk of developing posttraumatic stress disorder.[149] Consequently, we strongly believe that exploring the contribution of epigenetic mechanisms to injury-induced nociceptive memory in the context of early life challenges should be another focus of current pain research.

ACKNOWLEDGMENTS

This work was supported by the UK Medical Research Council New Investigator Research Grant G1100577. The authors would like to thank Fred Schwaller for his help with the illustration in Fig. 5.

REFERENCES

1. Waddington CH. Canalization of development and the inheritance of acquired characters. *Nature*. 1942;150(3811):563–565. http://dx.doi.org/10.1038/150563a0.
2. Borrelli E, Nestler EJ, Allis CD, Sassone-Corsi P. Decoding the epigenetic language of neuronal plasticity. *Neuron*. 2008;60(6):961–974. http://dx.doi.org/10.1016/j.neuron.2008.10.012.
3. Renthal W, Nestler EJ. Epigenetic mechanisms in drug addiction. *Trends Mol Med*. 2008;14(8):341–350. http://dx.doi.org/10.1016/j.molmed.2008.06.004.
4. Weaver ICG, Cervoni N, Champagne FA, et al. Epigenetic programming by maternal behavior. *Nat Neurosci*. 2004;7(8):847–854. http://dx.doi.org/10.1038/nn1276.
5. Lester BM, Tronick E, Nestler E, et al. Behavioral epigenetics: behavioral epigenetics. *Ann N Y Acad Sci*. 2011;1226(1):14–33. http://dx.doi.org/10.1111/j.1749-6632.2011.06037.x.
6. Day JJ, Sweatt JD. DNA methylation and memory formation. *Nat Neurosci*. 2010;13(11):1319–1323. http://dx.doi.org/10.1038/nn.2666.
7. Sweatt JD. The emerging field of neuroepigenetics. *Neuron*. 2013;80(3):624–632. http://dx.doi.org/10.1016/j.neuron.2013.10.023.
8. Day JJ, Sweatt JD. Epigenetic mechanisms in cognition. *Neuron*. 2011;70(5):813–829. http://dx.doi.org/10.1016/j.neuron.2011.05.019.
9. Akbarian S. The molecular pathology of schizophrenia—focus on histone and DNA modifications. *Brain Res Bull*. 2010;83(3–4):103–107. http://dx.doi.org/10.1016/j.brainresbull.2009.08.018.
10. Deutsch SI, Rosse RB, Mastropaolo J, Long KD, Gaskins BL. Epigenetic therapeutic strategies for the treatment of neuropsychiatric disorders: ready for prime time? *Clin Neuropharmacol*. 2008;31(2):104–119. http://dx.doi.org/10.1097/WNF.0b013e318067e255.
11. Qureshi IA, Mehler MF. Impact of nuclear organization and dynamics on epigenetic regulation in the central nervous system: implications for neurological disease states. *Ann N Y Acad Sci*. 2010;1204(Suppl):E20–E37. http://dx.doi.org/10.1111/j.1749-6632.2010.05718.x.
12. Cobb S, Guy J, Bird A. Reversibility of functional deficits in experimental models of Rett syndrome. *Biochem Soc Trans*. 2010;38(2):498. http://dx.doi.org/10.1042/BST0380498.
13. Landry CD, Kandel ER, Rajasethupathy P. New mechanisms in memory storage: piRNAs and epigenetics. *Trends Neurosci*. 2013;36(9):535–542. http://dx.doi.org/10.1016/j.tins.2013.05.004.
14. Rajasethupathy P, Antonov I, Sheridan R, et al. A role for neuronal piRNAs in the epigenetic control of memory-related synaptic plasticity. *Cell*. 2012;149(3):693–707. http://dx.doi.org/10.1016/j.cell.2012.02.057.
15. Denis H, Ndlovu MN, Fuks F. Regulation of mammalian DNA methyltransferases: a route to new mechanisms. *EMBO Rep*. 2011;12(7):647–656. http://dx.doi.org/10.1038/embor.2011.110.
16. Iorio MV, Piovan C, Croce CM. Interplay between microRNAs and the epigenetic machinery: an intricate network. *Biochim Biophys Acta*. 2010;1799(10–12):694–701. http://dx.doi.org/10.1016/j.bbagrm.2010.05.005.

17. Cheung P, Allis CD, Sassone-Corsi P. Signaling to chromatin through histone modifications. *Cell.* 2000;103(2):263–271.
18. Kouzarides T. Chromatin modifications and their function. *Cell.* 2007;128(4):693–705. http://dx.doi.org/10.1016/j.cell.2007.02.005.
19. Strahl BD, Allis CD. The language of covalent histone modifications. *Nature.* 2000;403(6765):41–45. http://dx.doi.org/10.1038/47412.
20. Zhou R, Gong A-Y, Chen D, Miller RE, Eischeid AN, Chen X-M. Histone deacetylases and NF-kB signaling coordinate expression of CX3CL1 in epithelial cells in response to microbial challenge by suppressing miR-424 and miR-503. In: Zilberstein D, ed. *PLoS One.* 2013;8((5):e65153. http://dx.doi.org/10.1371/journal.pone.0065153.
21. Crosio C, Heitz E, Allis CD, Borrelli E, Sassone-Corsi P. Chromatin remodeling and neuronal response: multiple signaling pathways induce specific histone H3 modifications and early gene expression in hippocampal neurons. *J Cell Sci.* 2003;116(Pt. 24):4905–4914. http://dx.doi.org/10.1242/jcs.00804.
22. Molina-Serrano D, Schiza V, Kirmizis A. Cross-talk among epigenetic modifications: lessons from histone arginine methylation. *Biochem Soc Trans.* 2013;41(3):751–759. http://dx.doi.org/10.1042/BST20130003.
23. Jenuwein T, Allis CD. Translating the histone code. *Science.* 2001;293(5532):1074–1080. http://dx.doi.org/10.1126/science.1063127.
24. Meaney MJ, Ferguson-Smith AC. Epigenetic regulation of the neural transcriptome: the meaning of the marks. *Nat Neurosci.* 2010;13(11):1313–1318. http://dx.doi.org/10.1038/nn1110-1313.
25. Lee KK, Workman JL. Histone acetyltransferase complexes: one size doesn't fit all. *Nat Rev Mol Cell Biol.* 2007;8(4):284–295. http://dx.doi.org/10.1038/nrm2145.
26. Haberland M, Montgomery RL, Olson EN. The many roles of histone deacetylases in development and physiology: implications for disease and therapy. *Nat Rev Genet.* 2009;10(1):32–42. http://dx.doi.org/10.1038/nrg2485.
27. Wee S, Dhanak D, Li H, et al. Targeting epigenetic regulators for cancer therapy: targeting epigenetic regulators for cancer therapy. *Ann N Y Acad Sci.* 2014;1309(1):30–36. http://dx.doi.org/10.1111/nyas.12356.
28. Beumer JH, Tawbi H. Role of histone deacetylases and their inhibitors in cancer biology and treatment. *Curr Clin Pharmacol.* 2010;5(3):196–208.
29. Riccio A. Dynamic epigenetic regulation in neurons: enzymes, stimuli and signaling pathways. *Nat Neurosci.* 2010;13(11):1330–1337. http://dx.doi.org/10.1038/nn.2671.
30. Suganuma T, Workman JL. MAP kinases and histone modification. *J Mol Cell Biol.* 2012;4(5):348–350. http://dx.doi.org/10.1093/jmcb/mjs043.
31. Arthur JSC, Fong AL, Dwyer JM, et al. Mitogen- and stress-activated protein kinase 1 mediates cAMP response element-binding protein phosphorylation and activation by neurotrophins. *J Neurosci.* 2004;24(18):4324–4332. http://dx.doi.org/10.1523/JNEUROSCI.5227-03.2004.
32. Soloaga A, Thomson S, Wiggin GR, et al. MSK2 and MSK1 mediate the mitogen- and stress-induced phosphorylation of histone H3 and HMG-14. *EMBO J.* 2003; 22(11):2788–2797. http://dx.doi.org/10.1093/emboj/cdg273.
33. Meaney MJ, Szyf M. Environmental programming of stress responses through DNA methylation: life at the interface between a dynamic environment and a fixed genome. *Dialogues Clin Neurosci.* 2005;7(2):103–123.
34. Roth TL, Sweatt JD. Regulation of chromatin structure in memory formation. *Curr Opin Neurobiol.* 2009;19(3):336–342. http://dx.doi.org/10.1016/j.conb.2009.05.011.
35. Lister R, Mukamel EA, Nery JR, et al. Global epigenomic reconfiguration during mammalian brain development. *Science.* 2013;341(6146):1237905. http://dx.doi.org/10.1126/science.1237905.

36. Varley KE, Gertz J, Bowling KM, et al. Dynamic DNA methylation across diverse human cell lines and tissues. *Genome Res.* 2013;23(3):555–567. http://dx.doi.org/10.1101/gr.147942.112.
37. Guo JU, Ma DK, Mo H, et al. Neuronal activity modifies the DNA methylation landscape in the adult brain. *Nat Neurosci.* 2011;14(10):1345–1351. http://dx.doi.org/10.1038/nn.2900.
38. Miller CA, Campbell SL, Sweatt JD. DNA methylation and histone acetylation work in concert to regulate memory formation and synaptic plasticity. *Neurobiol Learn Mem.* 2008;89(4):599–603. http://dx.doi.org/10.1016/j.nlm.2007.07.016.
39. Kriaucionis S, Heintz N. The nuclear DNA base 5-hydroxymethylcytosine is present in Purkinje neurons and the brain. *Science.* 2009;324(5929):929–930. http://dx.doi.org/10.1126/science.1169786.
40. Tahiliani M, Koh KP, Shen Y, et al. Conversion of 5-methylcytosine to 5-hydroxymethylcytosine in mammalian DNA by MLL partner TET1. *Science.* 2009;324(5929):930–935. http://dx.doi.org/10.1126/science.1170116.
41. Guo JU, Su Y, Zhong C, Ming G, Song H. Emerging roles of TET proteins and 5-hydroxymethylcytosines in active DNA demethylation and beyond. *Cell Cycle.* 2011;10(16):2662–2668.
42. Faigle R, Song H. Signaling mechanisms regulating adult neural stem cells and neurogenesis. *Biochim Biophys Acta.* 2013;1830(2):2435–2448. http://dx.doi.org/10.1016/j.bbagen.2012.09.002.
43. Ma DK, Jang M-H, Guo JU, et al. Neuronal activity-induced Gadd45b promotes epigenetic DNA demethylation and adult neurogenesis. *Science.* 2009;323(5917):1074–1077. http://dx.doi.org/10.1126/science.1166859.
44. Chen C-C, Wang K-Y, Shen C-KJ. DNA 5-methylcytosine demethylation activities of the mammalian DNA methyltransferases. *J Biol Chem.* 2013;288(13):9084–9091. http://dx.doi.org/10.1074/jbc.M112.445585.
45. Dong Y, Zhao H, Li H, Li X, Yang S. DNA methylation as an early diagnostic marker of cancer (review). *Biomed Rep.* 2014;2(3):326–330. http://dx.doi.org/10.3892/br.2014.237.
46. Yang X, Lay F, Han H, Jones PA. Targeting DNA methylation for epigenetic therapy. *Trends Pharmacol Sci.* 2010;31(11):536–546. http://dx.doi.org/10.1016/j.tips.2010.08.001.
47. Adey A, Shendure J. Ultra-low-input, tagmentation-based whole-genome bisulfite sequencing. *Genome Res.* 2012;22(6):1139–1143. http://dx.doi.org/10.1101/gr.136242.111.
48. Miura F, Enomoto Y, Dairiki R, Ito T. Amplification-free whole-genome bisulfite sequencing by post-bisulfite adaptor tagging. *Nucleic Acids Res.* 2012;40(17):e136. http://dx.doi.org/10.1093/nar/gks454.
49. Yu M, Hon GC, Szulwach KE, et al. Base-resolution analysis of 5-hydroxymethylcytosine in the mammalian genome. *Cell.* 2012;149(6):1368–1380. http://dx.doi.org/10.1016/j.cell.2012.04.027.
50. Irizarry RA, Ladd-Acosta C, Wen B, et al. The human colon cancer methylome shows similar hypo- and hypermethylation at conserved tissue-specific CpG island shores. *Nat Genet.* 2009;41(2):178–186. http://dx.doi.org/10.1038/ng.298.
51. Ji R-R, Kohno T, Moore KA, Woolf CJ. Central sensitization and LTP: do pain and memory share similar mechanisms? *Trends Neurosci.* 2003;26(12):696–705.
52. Rahn EJ, Guzman-Karlsson MC, David Sweatt J. Cellular, molecular, and epigenetic mechanisms in non-associative conditioning: implications for pain and memory. *Neurobiol Learn Mem.* 2013;105:133–150. http://dx.doi.org/10.1016/j.nlm.2013.06.008.

53. Barrett RM, Wood MA. Beyond transcription factors: the role of chromatin modifying enzymes in regulating transcription required for memory. *Learn Mem.* 2008; 15(7):460–467. http://dx.doi.org/10.1101/lm.917508.
54. Gräff J, Tsai L-H. Histone acetylation: molecular mnemonics on the chromatin. *Nat Rev Neurosci.* 2013;14(2):97–111. http://dx.doi.org/10.1038/nrn3427.
55. Peixoto L, Abel T. The role of histone acetylation in memory formation and cognitive impairments. *Neuropsychopharmacology.* 2013;38(1):62–76. http://dx.doi.org/10.1038/npp.2012.86.
56. Chwang WB, O'Riordan KJ, Levenson JM, Sweatt JD. ERK/MAPK regulates hippocampal histone phosphorylation following contextual fear conditioning. *Learn Mem.* 2006;13(3):322–328. http://dx.doi.org/10.1101/lm.152906.
57. Levenson JM, O'Riordan KJ, Brown KD, Trinh MA, Molfese DL, Sweatt JD. Regulation of histone acetylation during memory formation in the hippocampus. *J Biol Chem.* 2004;279(39):40545–40559. http://dx.doi.org/10.1074/jbc.M402229200.
58. Gräff J, Woldemichael BT, Berchtold D, Dewarrat G, Mansuy IM. Dynamic histone marks in the hippocampus and cortex facilitate memory consolidation. *Nat Commun.* 2012;3:991. http://dx.doi.org/10.1038/ncomms1997.
59. Bousiges O, de Vasconcelos AP, Neidl R, et al. Spatial memory consolidation is associated with induction of several lysine-acetyltransferase (histone acetyltransferase) expression levels and H2B/H4 acetylation-dependent transcriptional events in the Rat hippocampus. *Neuropsychopharmacology.* 2010;35(13):2521–2537. http://dx.doi.org/10.1038/npp.2010.117.
60. Gupta S, Kim SY, Artis S, et al. Histone methylation regulates memory formation. *J Neurosci.* 2010;30(10):3589–3599. http://dx.doi.org/10.1523/JNEUROSCI.3732-09.2010.
61. Lubin FD, Roth TL, Sweatt JD. Epigenetic regulation of BDNF gene transcription in the consolidation of fear memory. *J Neurosci.* 2008;28(42):10576–10586. http://dx.doi.org/10.1523/JNEUROSCI.1786-08.2008.
62. Mahan AL, Mou L, Shah N, Hu J-H, Worley PF, Ressler KJ. Epigenetic modulation of Homer1a transcription regulation in amygdala and hippocampus with Pavlovian fear conditioning. *J Neurosci.* 2012;32(13):4651–4659. http://dx.doi.org/10.1523/JNEUROSCI.3308-11.2012.
63. Chandramohan Y, Droste SK, Arthur JSC, Reul JMHM. The forced swimming-induced behavioural immobility response involves histone H3 phospho-acetylation and c-Fos induction in dentate gyrus granule neurons via activation of the *N*-methyl-D-aspartate/extracellular signal-regulated kinase/mitogen- and stress-activated kinase signalling pathway. *Eur J Neurosci.* 2008;27(10):2701–2713. http://dx.doi.org/10.1111/j.1460-9568.2008.06230.x.
64. Sarantis K, Antoniou K, Matsokis N, Angelatou F. Exposure to novel environment is characterized by an interaction of D1/NMDA receptors underlined by phosphorylation of the NMDA and AMPA receptor subunits and activation of ERK1/2 signaling, leading to epigenetic changes and gene expression in rat hippocampus. *Neurochem Int.* 2012;60(1):55–67. http://dx.doi.org/10.1016/j.neuint.2011.10.018.
65. Guan J-S, Haggarty SJ, Giacometti E, et al. HDAC2 negatively regulates memory formation and synaptic plasticity. *Nature.* 2009;459(7243):55–60. http://dx.doi.org/10.1038/nature07925.
66. Castellano JF, Fletcher BR, Kelley-Bell B, Kim DH, Gallagher M, Rapp PR. Age-related memory impairment is associated with disrupted multivariate epigenetic coordination in the hippocampus. Izquierdo I, ed. *PLoS One.* 2012;7(3):e33249. http://dx.doi.org/10.1371/journal.pone.0033249.
67. Gräff J, Rei D, Guan J-S, et al. An epigenetic blockade of cognitive functions in the neurodegenerating brain. *Nature.* 2012;483:222–226. http://dx.doi.org/10.1038/nature10849.

68. McQuown SC, Barrett RM, Matheos DP, et al. HDAC3 is a critical negative regulator of long-term memory formation. *J Neurosci.* 2011;31(2):764–774. http://dx.doi.org/10.1523/JNEUROSCI.5052-10.2011.
69. Gao J, Wang W-Y, Mao Y-W, et al. A novel pathway regulates memory and plasticity via SIRT1 and miR-134. *Nature.* 2010;466(7310):1105–1109. http://dx.doi.org/10.1038/nature09271.
70. Bredy TW, Barad M. The histone deacetylase inhibitor valproic acid enhances acquisition, extinction, and reconsolidation of conditioned fear. *Learn Mem.* 2008; 15(1):39–45. http://dx.doi.org/10.1101/lm.801108.
71. Fischer A, Sananbenesi F, Wang X, Dobbin M, Tsai L-H. Recovery of learning and memory is associated with chromatin remodelling. *Nature.* 2007;447(7141):178–182. http://dx.doi.org/10.1038/nature05772.
72. Roozendaal B, Hernandez A, Cabrera SM, et al. Membrane-associated glucocorticoid activity is necessary for modulation of long-term memory via chromatin modification. *J Neurosci.* 2010;30(14):5037–5046. http://dx.doi.org/10.1523/JNEUROSCI.5717-09.2010.
73. Stefanko DP, Barrett RM, Ly AR, Reolon GK, Wood MA. Modulation of long-term memory for object recognition via HDAC inhibition. *Proc Natl Acad Sci.* 2009;106(23):9447–9452. http://dx.doi.org/10.1073/pnas.0903964106.
74. Vecsey CG, Hawk JD, Lattal KM, et al. Histone deacetylase inhibitors enhance memory and synaptic plasticity via CREB:CBP-dependent transcriptional activation. *J Neurosci.* 2007;27(23):6128–6140. http://dx.doi.org/10.1523/JNEUROSCI.0296-07.2007.
75. Peleg S, Sananbenesi F, Zovoilis A, et al. Altered histone acetylation is associated with age-dependent memory impairment in mice. *Science.* 2010;328(5979):753–756. http://dx.doi.org/10.1126/science.1186088.
76. Tabuchi Y, Takasaki I, Doi T, Ishii Y, Sakai H, Kondo T. Genetic networks responsive to sodium butyrate in colonic epithelial cells. *FEBS Lett.* 2006;580(13):3035–3041. http://dx.doi.org/10.1016/j.febslet.2006.04.048.
77. Maddox SA, Watts CS, Doyère V, Schafe GE. A naturally-occurring histone acetyltransferase inhibitor derived from *Garcinia indica* impairs newly acquired and reactivated fear memories. *PLoS One.* 2013;8(1):e54463. http://dx.doi.org/10.1371/journal.pone.0054463.
78. Maddox SA, Watts CS, Schafe GE. p300/CBP histone acetyltransferase activity is required for newly acquired and reactivated fear memories in the lateral amygdala. *Learn Mem.* 2013;20(2):109–119. http://dx.doi.org/10.1101/lm.029157.112.
79. Oliveira AMM, Estévez MA, Hawk JD, Grimes S, Brindle PK, Abel T. Subregion-specific p300 conditional knock-out mice exhibit long-term memory impairments. *Learn Mem.* 2011;18(3):161–169. http://dx.doi.org/10.1101/lm.1939811.
80. Gupta-Agarwal S, Franklin AV, Deramus T, et al. G9a/GLP histone lysine dimethyltransferase complex activity in the hippocampus and the entorhinal cortex is required for gene activation and silencing during memory consolidation. *J Neurosci.* 2012;32(16):5440–5453. http://dx.doi.org/10.1523/JNEUROSCI.0147-12.2012.
81. Koshibu K, Gräff J, Mansuy IM. Nuclear protein phosphatase-1: an epigenetic regulator of fear memory and amygdala long-term potentiation. *Neuroscience.* 2011;173:30–36. http://dx.doi.org/10.1016/j.neuroscience.2010.11.023.
82. Koshibu K, Graff J, Beullens M, et al. Protein phosphatase 1 regulates the histone code for long-term memory. *J Neurosci.* 2009;29(41):13079–13089. http://dx.doi.org/10.1523/JNEUROSCI.3610-09.2009.
83. Oliveira AMM, Hemstedt TJ, Bading H. Rescue of aging-associated decline in Dnmt3a2 expression restores cognitive abilities. *Nat Neurosci.* 2012;15(8):1111–1113. http://dx.doi.org/10.1038/nn.3151.

84. Miller CA, Sweatt JD. Covalent modification of DNA regulates memory formation. *Neuron*. 2007;53(6):857–869. http://dx.doi.org/10.1016/j.neuron.2007.02.022.
85. Monsey MS, Ota KT, Akingbade IF, Hong ES, Schafe GE. Epigenetic alterations are critical for fear memory consolidation and synaptic plasticity in the lateral amygdala. *PLoS One*. 2011;6(5):e19958. http://dx.doi.org/10.1371/journal.pone.0019958.
86. Day JJ, Childs D, Guzman-Karlsson MC, et al. DNA methylation regulates associative reward learning. *Nat Neurosci*. 2013;16(10):1445–1452. http://dx.doi.org/10.1038/nn.3504.
87. Maddox SA, Watts CS, Schafe GE. DNA methyltransferase activity is required for memory-related neural plasticity in the lateral amygdala. *Neurobiol Learn Mem*. 2013;107:93–100. http://dx.doi.org/10.1016/j.nlm.2013.11.008.
88. Miller CA, Gavin CF, White JA, et al. Cortical DNA methylation maintains remote memory. *Nat Neurosci*. 2010;13(6):664–666. http://dx.doi.org/10.1038/nn.2560.
89. Kaas GA, Zhong C, Eason DE, et al. TET1 controls CNS 5-methylcytosine hydroxylation, active DNA demethylation, gene transcription, and memory formation. *Neuron*. 2013;79(6):1086–1093. http://dx.doi.org/10.1016/j.neuron.2013.08.032.
90. Rudenko A, Dawlaty MM, Seo J, et al. Tet1 is critical for neuronal activity-regulated gene expression and memory extinction. *Neuron*. 2013;79(6):1109–1122. http://dx.doi.org/10.1016/j.neuron.2013.08.003.
91. Zhang R-R, Cui Q-Y, Murai K, et al. Tet1 regulates adult Hippocampal neurogenesis and cognition. *Cell Stem Cell*. 2013;13(2):237–245. http://dx.doi.org/10.1016/j.stem.2013.05.006.
92. Sultan FA, Wang J, Tront J, Liebermann DA, Sweatt JD. Genetic deletion of gadd45b, a regulator of active DNA demethylation, enhances long-term memory and synaptic plasticity. *J Neurosci*. 2012;32(48):17059–17066. http://dx.doi.org/10.1523/JNEUROSCI.1747-12.2012.
93. Abel T. Memory suppressor genes: inhibitory constraints on the storage of long-term memory. *Science*. 1998;279(5349):338–341. http://dx.doi.org/10.1126/science.279.5349.338.
94. Bai G, Ren K, Dubner R. Epigenetic regulation of persistent pain. *Transl Res*. 2015;165:177–199. http://dx.doi.org/10.1016/j.trsl.2014.05.012.
95. Niederberger E, Kynast K, Lötsch J, Geisslinger G. MicroRNAs as new players in the pain game. *Pain*. 2011;152(7):1455–1458. http://dx.doi.org/10.1016/j.pain.2011.01.042.
96. Denk F, McMahon SB. Chronic pain: emerging evidence for the involvement of epigenetics. *Neuron*. 2012;73(3):435–444. http://dx.doi.org/10.1016/j.neuron.2012.01.012.
97. Géranton SM. Targeting epigenetic mechanisms for pain relief. *Curr Opin Pharmacol*. 2011;12:35–41. http://dx.doi.org/10.1016/j.coph.2011.10.012.
98. Stone LS, Szyf M. The emerging field of pain epigenetics. *Pain*. 2013;154(1):1–2. http://dx.doi.org/10.1016/j.pain.2012.10.016.
99. Tochiki KK, Cunningham J, Hunt SP, Géranton SM. The expression of spinal methyl-CpG-binding protein 2, DNA methyltransferases and histone deacetylases is modulated in persistent pain states. *Mol Pain*. 2012;8:14. http://dx.doi.org/10.1186/1744-8069-8-14.
100. Wang Y, Liu C, Guo Q-L, et al. Intrathecal 5-azacytidine inhibits global DNA methylation and methyl- CpG-binding protein 2 expression and alleviates neuropathic pain in rats following chronic constriction injury. *Brain Res*. 2011;1418:64–69. http://dx.doi.org/10.1016/j.brainres.2011.08.040.
101. Zhu X, Li Q, Chang R, et al. Curcumin alleviates neuropathic pain by inhibiting p300/CBP histone acetyltransferase activity-regulated expression of BDNF and Cox-2 in a rat model. In: Ma D, ed. *PLoS One*. 2014;9(3):e91303. http://dx.doi.org/10.1371/journal.pone.0091303.

102. Zhu X-Y, Huang C-S, Li Q, et al. p300 exerts an epigenetic role in chronic neuropathic pain through its acetyltransferase activity in rats following chronic constriction injury (CCI). *Mol Pain*. 2012;8(1):84. http://dx.doi.org/10.1186/1744-8069-8-84.
103. Bai G, Wei D, Zou S, Ren K, Dubner R. Inhibition of class II histone deacetylases in the spinal cord attenuates inflammatory hyperalgesia. *Mol Pain*. 2010;6:51. http://dx.doi.org/10.1186/1744-8069-6-51.
104. Qi F, Zhou Y, Xiao Y, et al. Promoter demethylation of cystathionine-β-synthetase gene contributes to inflammatory pain in rats. *Pain*. 2013;154(1):34–45. http://dx.doi.org/10.1016/j.pain.2012.07.031.
105. Zhang H-H, Hu J, Zhou Y-L, et al. Promoted interaction of nuclear factor-κB with demethylated cystathionine-β-synthetase gene contributes to gastric hypersensitivity in diabetic rats. *J Neurosci*. 2013;33(21):9028–9038. http://dx.doi.org/10.1523/JNEUROSCI.1068-13.2013.
106. Zhang Z, Cai Y-Q, Zou F, Bie B, Pan ZZ. Epigenetic suppression of GAD65 expression mediates persistent pain. *Nat Med*. 2011;17(11):1448–1455. http://dx.doi.org/10.1038/nm.2442.
107. Tajerian M, Alvarado S, Millecamps M, et al. Peripheral nerve injury is associated with chronic, reversible changes in global DNA methylation in the mouse prefrontal cortex. Taylor B, ed. *PLoS One*. 2013;8(1):e55259. http://dx.doi.org/10.1371/journal.pone.0055259.
108. Cherng C-H, Lee K-C, Chien C-C, et al. Baicalin ameliorates neuropathic pain by suppressing HDAC1 expression in the spinal cord of spinal nerve ligation rats. *J Formos Med Assoc*. 2013;113:513–520. http://dx.doi.org/10.1016/j.jfma.2013.04.007.
109. Yin Q, Lu F-F, Zhao Y, et al. Resveratrol facilitates pain attenuation in a rat model of neuropathic pain through the activation of spinal Sirt1. *Reg Anesth Pain Med*. 2013;38:93–99. http://dx.doi.org/10.1097/AAP.0b013e3182795b23.
110. Sun Y, Sahbaie P, Liang D-Y, et al. Epigenetic regulation of spinal CXCR2 signaling in incisional hypersensitivity in mice. *Anesthesiology*. 2013;119:1198–1208. http://dx.doi.org/10.1097/ALN.0b013e31829ce340.
111. Liang D-Y, Li X, Clark JD. Epigenetic regulation of opioid-induced hyperalgesia, dependence, and tolerance in mice. *J Pain*. 2013;14(1):36–47. http://dx.doi.org/10.1016/j.jpain.2012.10.005.
112. Imai S, Ikegami D, Yamashita A, et al. Epigenetic transcriptional activation of monocyte chemotactic protein 3 contributes to long-lasting neuropathic pain. *Brain J Neurol*. 2013;136(Pt. 3):828–843. http://dx.doi.org/10.1093/brain/aws330.
113. Géranton SM, Morenilla-Palao C, Hunt SP. A role for transcriptional repressor methyl-CpG-binding protein 2 and plasticity-related gene serum- and glucocorticoid-inducible kinase 1 in the induction of inflammatory pain states. *J Neurosci*. 2007;27(23):6163–6173. http://dx.doi.org/10.1523/JNEUROSCI.306-07.2007.
114. Matsushita Y, Araki K, Omotuyi Oi, Mukae T, Ueda H. HDAC inhibitors restore C-fibre sensitivity in experimental neuropathic pain model. *Br J Pharmacol*. 2013;170(5):991–998. http://dx.doi.org/10.1111/bph.12366.
115. Uchida H, Matsushita Y, Ueda H. Epigenetic regulation of BDNF expression in the primary sensory neurons after peripheral nerve injury: implications in the development of neuropathic pain. *Neuroscience*. 2013;240:147–154. http://dx.doi.org/10.1016/j.neuroscience.2013.02.053.
116. Uchida H, Sasaki K, Ma L, Ueda H. Neuron-restrictive silencer factor causes epigenetic silencing of Kv4.3 gene after peripheral nerve injury. *Neuroscience*. 2010;166(1):1–4. http://dx.doi.org/10.1016/j.neuroscience.2009.12.021.
117. Kiguchi N, Kobayashi Y, Maeda T, et al. Epigenetic augmentation of the macrophage inflammatory protein 2/C-X-C chemokine receptor type 2 axis through histone H3

acetylation in injured peripheral nerves elicits neuropathic pain. *J Pharmacol Exp Ther.* 2012;340(3):577–587. http://dx.doi.org/10.1124/jpet.111.187724.
118. Kiguchi N, Kobayashi Y, Saika F, Kishioka S. Epigenetic upregulation of CCL2 and CCL3 via histone modifications in infiltrating macrophages after peripheral nerve injury. *Cytokine.* 2013;64:666–672. http://dx.doi.org/10.1016/j.cyto.2013.09.019.
119. Tajerian M, Alvarado S, Millecamps M, et al. DNA methylation of SPARC and chronic low back pain. *Mol Pain.* 2011;7:65. http://dx.doi.org/10.1186/1744-8069-7-65.
120. Chiechio S, Zammataro M, Morales ME, et al. Epigenetic modulation of mGlu2 receptors by histone deacetylase inhibitors in the treatment of inflammatory pain. *Mol Pharmacol.* 2009;75(5):1014–1020. http://dx.doi.org/10.1124/mol.108.054346.
121. Denk F, Huang W, Sidders B, et al. HDAC inhibitors attenuate the development of hypersensitivity in models of neuropathic pain. *Pain.* 2013;154(9):1668–1679. http://dx.doi.org/10.1016/j.pain.2013.05.021.
122. Yoshizumi M, Eisenach JC, Hayashida K-I. Valproate prevents dysregulation of spinal glutamate and reduces the development of hypersensitivity in rats after peripheral nerve injury. *J Pain.* 2013;14(11):1485–1491. http://dx.doi.org/10.1016/j.jpain.2013.07.007.
123. Uchida H, Ma L, Ueda H. Epigenetic gene silencing underlies C-fiber dysfunctions in neuropathic pain. *J Neurosci.* 2010;30(13):4806–4814. http://dx.doi.org/10.1523/JNEUROSCI.5541-09.2010.
124. Doehring A, Oertel BG, Sittl R, Lötsch J. Chronic opioid use is associated with increased DNA methylation correlating with increased clinical pain. *Pain.* 2013;154(1):15–23. http://dx.doi.org/10.1016/j.pain.2012.06.011.
125. Lacroix-Fralish ML, Ledoux JB, Mogil JS. The pain genes database: an interactive web browser of pain-related transgenic knockout studies. *Pain.* 2007;131(1–2):3.e1–3.e4. http://dx.doi.org/10.1016/j.pain.2007.04.041.
126. Perkins JR, Lees J, Antunes-Martins A, et al. PainNetworks: a web-based resource for the visualisation of pain-related genes in the context of their network associations. *Pain.* 2013;154(12):2586.e1–2586.e12. http://dx.doi.org/10.1016/j.pain.2013.09.003.
127. Gravina GL, Festuccia C, Marampon F, et al. Biological rationale for the use of DNA methyltransferase inhibitors as new strategy for modulation of tumor response to chemotherapy and radiation. *Mol Cancer.* 2010;9(1):305. http://dx.doi.org/10.1186/1476-4598-9-305.
128. Subramanian S, Bates SE, Wright JJ, Espinoza-Delgado I, Piekarz RL. Clinical toxicities of histone deacetylase inhibitors. *Pharmaceuticals.* 2010;3(9):2751–2767. http://dx.doi.org/10.3390/ph3092751.
129. Stolzenburg S, Rots MG, Beltran AS, et al. Targeted silencing of the oncogenic transcription factor SOX2 in breast cancer. *Nucleic Acids Res.* 2012;40(14):6725–6740. http://dx.doi.org/10.1093/nar/gks360.
130. Rivenbark AG, Stolzenburg S, Beltran AS, et al. Epigenetic reprogramming of cancer cells via targeted DNA methylation. *Epigenetics.* 2012;7(4):350–360. http://dx.doi.org/10.4161/epi.19507.
131. Lara H, Wang Y, Beltran AS, et al. Targeting serous epithelial ovarian cancer with designer zinc finger transcription factors. *J Biol Chem.* 2012;287(35):29873–29886. http://dx.doi.org/10.1074/jbc.M112.360768.
132. Mogil JS. Pain genetics: past, present and future. *Trends Genet.* 2012;28(6):258–266. http://dx.doi.org/10.1016/j.tig.2012.02.004.
133. Nielsen CS, Knudsen GP, Steingrímsdóttir ÓA. Twin studies of pain. *Clin Genet.* 2012;82(4):331–340. http://dx.doi.org/10.1111/j.1399-0004.2012.01938.x.
134. Kuner R. Central mechanisms of pathological pain. *Nat Med.* 2010;16(11):1258–1266. http://dx.doi.org/10.1038/nm.2231.

135. Bogen O, Alessandri-Haber N, Chu C, Gear RW, Levine JD. Generation of a pain memory in the primary afferent nociceptor triggered by PKCε activation of CPEB. *J Neurosci.* 2012;32(6):2018–2026. http://dx.doi.org/10.1523/JNEUROSCI.5138-11.2012.
136. Asiedu MN, Tillu DV, Melemedjian OK, et al. Spinal protein kinase M ζ underlies the maintenance mechanism of persistent nociceptive sensitization. *J Neurosci.* 2011;31(18):6646–6653. http://dx.doi.org/10.1523/JNEUROSCI.6286-10.2011.
137. Marchand F, D'Mello R, Yip PK, et al. Specific involvement of atypical PKCζ/PKMζ in spinal persistent nociceptive processing following peripheral inflammation in rat. *Mol Pain.* 2011;7:86. http://dx.doi.org/10.1186/1744-8069-7-86.
138. Sacktor TC. Memory maintenance by PKMζ—an evolutionary perspective. *Mol Brain.* 2012;5:31. http://dx.doi.org/10.1186/1756-6606-5-31.
139. Beggs S, Currie G, Salter MW, Fitzgerald M, Walker SM. Priming of adult pain responses by neonatal pain experience: maintenance by central neuroimmune activity. *Brain J Neurol.* 2012;135(Pt. 2):404–417. http://dx.doi.org/10.1093/brain/awr288.
140. Fitzgerald M, Walker SM. Infant pain management: a developmental neurobiological approach. *Nat Clin Pract Neurol.* 2009;5(1):35–50. http://dx.doi.org/10.1038/ncpneuro0984.
141. Hermann C, Hohmeister J, Demirakça S, Zohsel K, Flor H. Long-term alteration of pain sensitivity in school-aged children with early pain experiences. *Pain.* 2006;125(3):278–285. http://dx.doi.org/10.1016/j.pain.2006.08.026.
142. Walker SM, Tochiki KK, Fitzgerald M. Hindpaw incision in early life increases the hyperalgesic response to repeat surgical injury: critical period and dependence on initial afferent activity. *Pain.* 2009;147(1–3):99–106. http://dx.doi.org/10.1016/j.pain.2009.08.017.
143. Wollgarten-Hadamek I, Hohmeister J, Demirakça S, Zohsel K, Flor H, Hermann C. Do burn injuries during infancy affect pain and sensory sensitivity in later childhood? *Pain.* 2009;141(1–2):165–172. http://dx.doi.org/10.1016/j.pain.2008.11.008.
144. D'Aquila P, Rose G, Bellizzi D, Passarino G. Epigenetics and aging. *Maturitas.* 2013;74(2):130–136. http://dx.doi.org/10.1016/j.maturitas.2012.11.005.
145. Rakyan VK, Down TA, Maslau S, et al. Human aging-associated DNA hypermethylation occurs preferentially at bivalent chromatin domains. *Genome Res.* 2010;20(4):434–439. http://dx.doi.org/10.1101/gr.103101.109.
146. Tammen SA, Friso S, Choi S-W. Epigenetics: the link between nature and nurture. *Mol Aspects Med.* 2013;34(4):753–764. http://dx.doi.org/10.1016/j.mam.2012.07.018.
147. Teschendorff AE, Menon U, Gentry-Maharaj A, et al. Age-dependent DNA methylation of genes that are suppressed in stem cells is a hallmark of cancer. *Genome Res.* 2010;20(4):440–446. http://dx.doi.org/10.1101/gr.103606.109.
148. Zhang Z, Deng C, Lu Q, Richardson B. Age-dependent DNA methylation changes in the ITGAL (CD11a) promoter. *Mech Ageing Dev.* 2002;123(9):1257–1268.
149. Klengel T, Mehta D, Anacker C, et al. Allele-specific FKBP5 DNA demethylation mediates gene-childhood trauma interactions. *Nat Neurosci.* 2013;16(1):33–41. http://dx.doi.org/10.1038/nn.3275.

CHAPTER SEVEN

Translational Control of Chronic Pain

Ohannes K. Melemedjian*^,1^, Arkady Khoutorsky†

*Department of Neural and Pain Sciences, School of Dentistry, University of Maryland, Baltimore, Maryland, USA

†Department of Biochemistry, Rosalind and Morris Goodman Cancer Research Centre, McGill University, Montréal, Quebec, Canada

[1]Corresponding author: e-mail address: ohannes@umaryland.edu

Contents

1. Introduction 186
2. Repair of Injured Tissue 186
3. Case for Axonal Translation 188
4. Regulation of Protein Synthesis 189
    4.1 Cap-dependent translation 189
    4.2 Signaling pathways that regulate translation 189
    4.3 eIF2 and the 43S preinitiation complex 192
    4.4 AMP-activated protein kinase and translational control 193
    4.5 Cap-dependent translation in the PNS and pain 194
    4.6 The role of the 3′ UTR and poly-A tail of mRNAs 198
    4.7 Cap-independent mechanism 198
5. Transport and Storage of mRNAs 199
6. Conclusions and Perspectives 200
7. Translational Control of Central Sensitization 201
    7.1 Peripheral inflammation- and injury-induced activation of mTOR in the spinal neurons contributes to the hyperalgesia state 203
    7.2 Translational control in opioid-induced hyperalgesia 204
    7.3 Translational control in spinal nonneuronal cells 204
    7.4 Reconsolidation-like processes in the pain pathway 205
    7.5 Molecular mechanisms underlying the effect of mTOR on spinal neuronal sensitization 205
8. Conclusions 206
References 207

Abstract

Pain is a crucial physiological response to injury and pathologies. The development and maintenance of pain requires the expression of novel genes. The expression of such genes occurs in highly regulated and orchestrated manner where protein translation provides an exquisite temporal and spatial fidelity within the axons and dendrites of

Progress in Molecular Biology and Translational Science, Volume 131
ISSN 1877-1173
http://dx.doi.org/10.1016/bs.pmbts.2014.11.006

neurons. Signaling pathways that regulate local translation are activated by cytokines, neurotrophic factors, or neurotransmitters, which are released either due to tissue damage or neuronal activity. In recent years, the ERK and mTOR pathways have been demonstrated to be central in regulating local translation in neurons of both the peripheral and central nervous systems in diverse models of chronic pain. The ERK and mTOR pathways converge onto the cap-dependent translational machinery that regulates genes essential for the development of nociceptive sensitization. Moreover, inhibition of these pathways has proved to be effective in normalizing the biochemical changes and the associated pain in various preclinical models.

1. INTRODUCTION

Pain and the accompanying nociceptive sensitization is a vital physiological response to injury. This protective response allows for organisms to escape the source of insult and guard against further damage, which is crucial for recuperation and survival.[1] However, maladaptive changes that occur in the peripheral and central nervous systems maintain the state of nociceptive sensitization which may be the underlying mechanism that sustains chronic pain conditions. Therefore, the elucidation of molecular mechanisms that lead to the development and maintenance of nociceptive sensitization is crucial for understanding chronic pain states. This chapter discusses the translational control mechanisms that sustain chronic pain states.

2. REPAIR OF INJURED TISSUE

In all tissue types, trauma is associated with the release of factors that are critical for the initiation of tissue repair and regeneration mechanisms. Moreover, factors released in response to injury promote nociceptive sensitization. Tissue injury disrupts blood vessels and releases plasma and the peripheral blood cells into the wound site. Early signals of tissue injury include the release of ATP and the exposure of collagen on tissue wall. This allows for the containment and hemostasis by initiating platelet activation and the formation of a platelet plug.[2] Platelet activation is accompanied with the release of signaling molecules that chemotactically recruit immune cells.[3] Mast cells contribute to the wound healing process by releasing histamine and serotonin that enhance blood vessel permeability and immune cell migration. The invading immune cells produce a variety of proinflammatory cytokines and growth factors that include interleukin-1 (IL-1) beta, tumor necrosis factor alpha, IL-6, and nerve growth factor (NGF).[4]

The body's reparative mechanisms display unique characteristics in various tissue types. For example, in collagenous soft tissue such as the skin, the extent of wound repair is dictated by depth of the injury. This is mainly due to the regenerative capability of the epidermis and limited capacity of the dermis to regenerate. Thus, deep tissue wound which causes loss of the dermis undergoes excessive extracellular matrix (ECM) production by fibroblasts leading to fibrous scar tissue formation. Alternatively, injured peripheral nerves display a unique series of events that can lead to regeneration and reinnervation of the target tissue.[5]

A nerve is comprised of a bundle of axons surrounded by their supporting Schwann cells. Schwann cells wrap around the axons of large diameter neurons where they produce myelin which insulates the axons from the ECM. This is important for rapid signal propagation. It should be noted that small diameter nociceptors are not myelinated. Each axon and its Schwann cells are enclosed by a connective tissue matrix called the endoneurium, subsequently grouped into bundles that are surrounded by another connective tissue matrix called the perineurium. A group of these bundles along with the vasculature comprise a nerve which in turn is encircled with the outer sheath, called epineurium.[5,6]

Regardless of the extent of the trauma to an axon, a successful regeneration hinges on the survival of the cell body of the neuron. The portion of the axon distal to the injury rapidly degenerates. Degeneration begins as Schwann cells stop making myelin proteins, and the axons begin to produce proteases that result in axonal fragmentation and the conversion of the surrounding myelin to droplet-shaped particulates. This process is termed Wallerian degeneration.[7] After several days, the axonal and myelin debris are cleared by Schwann cells and by the resident or invading macrophages. Proximal to the injury, Schwann cells proliferate and form interconnected cellular tubes, termed Bands of Büngner. These tubes act as conduits that allow the axons to regenerate. Within the conduit, Schwann cells release growth factors such as NGF and ECM proteins such as laminin and fibronectin.[7] Regenerating axons in turn respond to such signals by *de novo* synthesis of lipids and proteins that comprise the newly formed axons. In this chapter, we propose the coevolution of the nociceptive sensitization and injury repair mechanisms in the peripheral nervous system (PNS).

Injury without nociceptive sensitization bears a high evolutionary selective cost. A series of elegant predation studies in squid, *Doryteuthis pealei,* have provided the most compelling data in support of our proposal. Squid with injured arm demonstrate nociceptive sensitization and lower survival due to predation. However, squid with injured arm but lacking nociceptive sensitization due to the application of an anesthetic display a dramatically lower survival rate than squid with nociceptive sensitization.[8,9] Thus, biochemical signals that are generated during the wound repair and axon regeneration may also sensitize sensory neurons to allow for an organism to protect the injured area avoiding further damage.

3. CASE FOR AXONAL TRANSLATION

Sensory neurons have a unique cellular morphology (pseudo-unipolar), where a cell body emits a single axon that divides into two branches. One branch innervates the periphery and the other enters into the central nervous system (CNS). The length of the axon that innervates the periphery can exceed 1 m in humans.[10] Therefore, a typical cell body of sensory neuron with a diameter of 50 μm has a volume of around 65,000 μm^3, and a 120-cm axon has a volume of around 24,000,000 μm^3. Thus, the volume of the axon is 370 times larger than that of the cell body, and the length of the axon is 24,000 times longer than the diameter of the cell body. Local proteome homeostasis of such a large axonal volume requires the synthesis of vast amounts of protein in order to replenish damaged, degraded, or inactivated proteins.[11] Synthesizing a protein where and when it is needed confers several advantages over transporting it from the cell body. Axonal protein synthesis provides greatest spatial precision, as a protein is present only at a specific location along the length of the axon. Local synthesis also provides ultimate temporal precision, where protein is synthesized without delay.[12] The importance of temporal and spatial precision of local translation is highlighted in detection of guidance cues at the tip of a growing axon, termed the growth cone. The synthesis of the protein beta-actin closest to the gradient allows an axon to steer toward a trophic factor. Consequently, the growth cone is thought to be the nature's most sensitive sensor of chemical gradients detecting a difference as minute as 0.1%.[13] Local protein synthesis also eliminates the risk of modifying the structure and compromising function of a protein during transport. Lastly, local synthesis allows the production of multiple copies of proteins from a single mRNA molecule by multiple rounds of translation.[12] This eliminates the bioenergetic cost of the synthesis, transport, and localization of proteins to distal regions of an axon.

4. REGULATION OF PROTEIN SYNTHESIS

Translational control provides a powerful mechanism for gene regulation within the axons where the rate of mRNA translation dictates the abundance of proteins in a mammalian cell rather than the amount of mRNA present.[14] Translation is controlled through a variety of mechanisms that modulate the activity of translational machinery such as ribosomes, translation factors, and tRNA.[15]

4.1. Cap-dependent translation

The best-characterized regulatory mechanism is the phosphorylation events that regulate the function of the eukaryotic translation initiation factors. In eukaryotes, mRNAs that are transcribed within the nucleus have a 5′mRNA cap. Macromolecular complexes that assemble on the mRNA cap in order to initiate translation include the eukaryotic initiation factor 4F (eIF4F) and the 43S preinitiation complex.[15] These complexes are major targets of regulation for the type of translation known as cap-dependent translation. eIF4F is comprised of the cap-binding protein eIF4E,[16] the RNA helicase eIF4A, and the scaffolding protein eIF4G that binds to both eIF4E and eIF4A.[15] The assembly of the eIF4F complex on the mRNA cap is followed by the unwinding of secondary mRNA structures present in the 5′ UTR. This facilitates the recruitment of the 43S preinitiation complex and translation initiation.

The cap recognition step by eIF4E is the rate-limiting step for translation initiation, mainly because eIF4E is expressed in lowest levels compared to the other initiation factors. This makes the eIF4F complex a major site of regulation.[17] The best-characterized regulatory mechanism is mediated by the members of the eIF4E-binding proteins (4E-BP): 4E-BP1, 4E-BP2, and 4E-BP3.[18] The 4E-BPs bind to eIF4E through the same binding motif as eIF4G. Thus, 4E-BP and eIF4G compete for binding to eIF4E. 4E-BP is a negative regulator of translation where the hypo-phosphorylated form of this protein binds the eIF4E preventing the binding of eIF4G and the assembly of the eIF4F complex (Fig. 1).[17]

4.2. Signaling pathways that regulate translation

4E-BPs binding to eIF4E is inhibited by hierarchical phosphorylation, which promotes the formation of the eIF4F complex and the initiation of

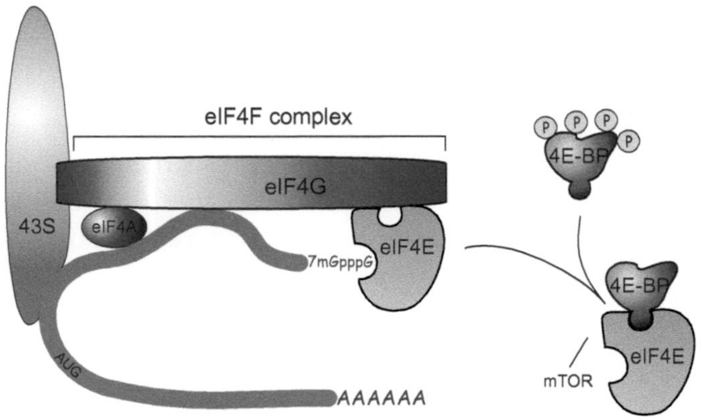

Figure 1 Regulation of translation initiation by eIF4F complex formation. The eukaryotic mRNA 5′ cap structure facilitates ribosome recruitment to the mRNA by the eIF4F complex, which consists of three subunits: (1) eIF4E, the cap-binding protein; (2) eIF4A, an RNA helicase that unwinds the secondary structure of the mRNA 5′ UTR; and (3) eIF4G, a large scaffolding protein that bridges the mRNA to the 43S preinitiation ribosomal complex. Once bound to the 5′ end of the mRNA, the 43S ribosomal complex traverses the 5′ UTR in a 5′→3′ direction, until it encounters the initiation codon (AUG) where translation begins. 4E-BP inhibits cap-dependent translation initiation by preventing the assembly of the eIF4F complex. 4E-BP and eIF4G compete for binding to the convex dorsal surface of eIF4E. Phosphorylation of 4E-BP by mTOR reduces its affinity for eIF4E and thus leads to eIF4F complex formation and translation initiation.

cap-dependent translation.[17] The phosphorylation of 4E-BPs is mainly regulated by the serine–threonine kinase mammalian/mechanistic target of rapamycin (mTOR) (Fig. 2).[19] The *TOR* genes were initially discovered in yeast following a forward genetic screen of a mutation that conferred resistance to the antimycotic agent rapamycin.[20] Various biochemical signals such as insulin, growth factors, nutrient, oxygen levels, and amino acids regulate the activity of the mTOR. mTOR in turn regulates cell growth, proliferation, and autophagy.[21] The mTOR catalytic subunit is part of multimeric complexes known as mTORC1 and mTORC2. Unlike mTORC2, mTORC1 kinase activity is inhibited by rapamycin.[19] In addition to promoting cap-dependent translation via 4E-BP phosphorylation, mTORC1 also phosphorylates p70 S6 kinase (S6K), which in turn phosphorylates the regulators of translation initiation such as S6, eIF4B, and programmed cell death 4 (PDCD4). mTORC1 also phosphorylates the eukaryotic elongation factor 2 kinase, which in turn phosphorylates the eukaryotic elongation factor 2, thus promoting the elongation phase of protein translation.[19]

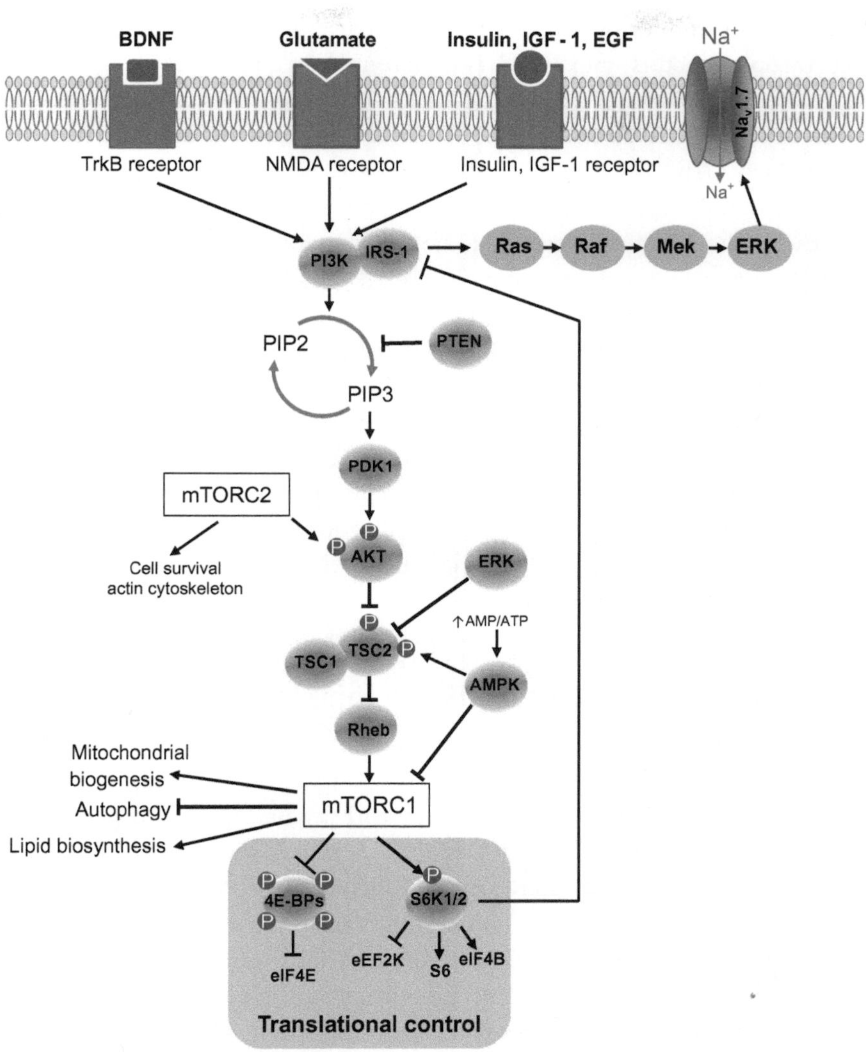

Figure 2 Regulation of translation initiation by mTOR signaling pathway. mTOR is a critical downstream target of the phosphatidylinositol-3 kinase (PI3K) signaling pathway. Insulin response element (IRS1) "transmits" signals from the TrkB, NMDA, insulin, and insulin-like growth factor-1 (IGF-1) receptors through PI3K and PI3K-dependent kinase 1 (PDK1), which phosphorylates Akt. mTOR forms two distinct protein complexes: mTORC1, which regulates translation, is sensitive to rapamycin and is defined by the scaffolding protein Raptor, and mTORC2, which is defined by Rictor and is generally rapamycin insensitive. mTORC1 phosphorylates 4E-BPs, S6K1, and S6K2. Akt can activate mTORC1 through phosphorylation and inhibition of the tuberous sclerosis complex (TSC). (See the color plate.)

The activation of cell-surface receptors of neurotransmitters, hormones, neurotrophic factors, and ECM components lead to the activation of the lipid kinase known as PI3K, which in turn activates PDK1 and mTORC2. These two kinases phosphorylate AKT on Thr308 and Ser473 residues, respectively. AKT in turn phosphorylates the tuberous sclerosis complex (TSC) 1 and 2, inactivating them. The inactivation of TSC1 and 2 in turn inactivate the small GTPases Ras homology enriched in brain (Rheb), which is a negative regulator of mTORC1.[19,21] These signaling events disinhibit mTORC1 leading to enhanced translation. In addition to the activation of the PI3K-AKT-mTORC1 pathway, the mitogen-activated protein (MAP) kinase pathway is known to regulate translation. The small GTPase Ras activates Raf which in turn activates Mek. Mek phosphorylates ERK leading to its activation. Similar to AKT, ERK phosphorylates TSC1 and 2 leading to their inhibition. ERK also phosphorylates the MAP kinase-interacting kinase (Mnk) 1 and 2. The phosphorylated form of Mnks binds to eIF4G leading to the phosphorylation of eIF4E,[21] an event that is associated with increased translation.[22]

4.3. eIF2 and the 43S preinitiation complex

The 43S preinitiation complex is comprised of 40S ribosomal subunit, eIFs distinct from eIF4F and the eIF2 ternary complex. This ternary complex includes eIF2, a GTPase that associates with a guanosine-5′ triphosphate (GTP), and the initiator Met-tRNAi^{Met}. A successful assembly of eIF4F is followed by the recruitment of the 43S preinitiation complex forming the 48S initiation complex.[23] This complex scans the 5′ UTR of mRNAs for the start codon, which triggers the hydrolysis of GTP to guanosine diphosphate (GDP) and inorganic phosphate by the eIF2. This allows for the recruitment of the 60S ribosomal subunit forming 80S ribosomal complex, which catalyzes the synthesis of proteins.

Cap-dependent translation is sustained by the replenishment of the GTP-bound eIF2 ternary complex. This process is mediated by eIF2B which functions as a guanine nucleotide exchange factor (GEF). eIF2 is another site of regulation of the cap-dependent translation. eIF2 is a multimeric complex comprised of alpha, beta, and gamma subunits. Phosphorylation of the α subunit of eIF2 at serine 51 converts eIF2 from a substrate to a competitive inhibitor of eIF2B. Since the amount of eIF2B is lower than eIF2, phosphorylation of only a small fraction of the eIF2α strongly inhibits eIF2B activity, thus preventing the exchange of GDP with GTP by eIF2B and leading to translation inhibition.[24] Diverse stress signals converge to phosphorylate eIF2α to integrate stress responses. The phosphorylation of eIF2α is mediated by several protein kinases such as the double-stranded RNA-activated protein kinase (PKR), the heme-regulated inhibitor kinase (HRI), the PKR-endoplasmic reticulum-related kinase (PERK), and the general control nonderepressible-2 (GCN2). These kinases are activated by double-stranded RNA during viral infection, iron deficiency, misfolded proteins in the endoplasmic reticulum, and amino acid deprivation, respectively, thus shutting down translation in response to various cellular stressors (Fig. 3).[25]

4.4. AMP-activated protein kinase and translational control

AMP-activated protein kinase (AMPK) is known as the "energy sensor" of a cell that plays a crucial role in the cellular energy homeostasis. AMPK is comprised of the alpha, beta, and gamma subunits.[26] The gamma subunit includes domains that confers AMPK sensitivity to AMP:ATP ratio. The binding of AMP molecules to certain domains in the gamma subunits exposes the catalytic domain found in the alpha subunit.[27] The exposure of the catalytic domain allows for the phosphorylation of AMPK at Thr172 residue by an upstream kinase complex AMPKK. AMPKK is comprised of STE-related adapter, mouse protein 25 (MO25), and LKB1.[28,29] Upon activation, AMPK increases energy levels of the cell by inhibiting

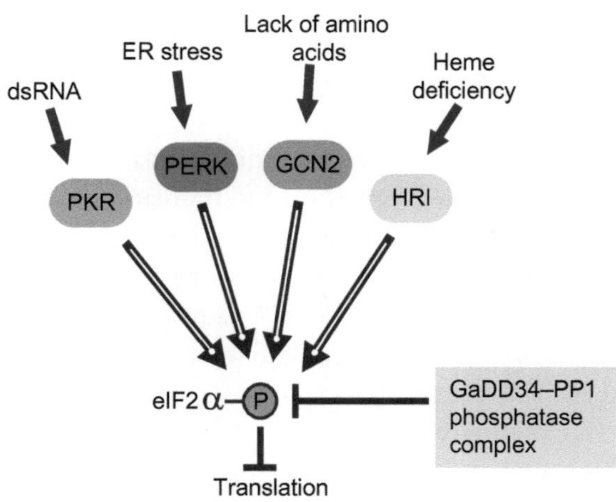

Figure 3 Cellular stressor activate eIF2α. Cellular stressor such as viral dsRNA, ER stress, amino acid, and heme deficiency activate eIF2α, which in turn suppresses translation. The Gadd34–PP1 phosphatase complex dephosphorylates eIF2α, thus inhibiting it.

energy-intensive processes such as protein synthesis and stimulating energy-producing processes such as fatty acid oxidation. AMPK can be activated by several compounds such as metformin, A769662, and AICAR.[30]

4.5. Cap-dependent translation in the PNS and pain

4.5.1 Algogens regulate cap-dependent translation

IL-6 and NGF are proinflammatory mediators that are released in response to tissue or nerve injury and are well-known sensitizers of nociceptors.[31–34] Many pathologies, associated with pain, demonstrate an increase in IL-6 levels,[33,35–38] and these pathological pain states depend on IL-6 receptor expression in sensory neurons.[31] IL-6 binds to IL-6R (gp80), which exists either as soluble or transmembrane protein, forming a composite ligand capable of binding to gp130 which is the signal-transducing component of the IL-6/IL-6R/gp130 complex. Similar to other cytokines, IL-6/IL-6R/gp130 recruits JAKs leading to the activation of STATs.[39,40] Moreover, IL-6 signals to the translational machinery in primary afferents leading to enhanced cap-dependent translation by activating the ERK-Mnk1-eIF4E pathway. Furthermore, IL-6-mediated nociceptive plasticity is crucially dependent on this pathway as it is blocked by inhibition of ERK and/or eIF4F complex formation. Thus, IL-6 controls translation in sensory neurons and induces mechanical allodynia via upregulation of nascent protein synthesis.[41]

NGF-mediated effects are central to inflammatory, surgical, visceral, and neuropathic pain.[35,42–46] The canonical NGF signal transduction pathway commences at the trkA receptor activating PI3 Kinase and AKT,[47] which in turn activate mTORC1 in sensory neurons through phosphorylation of Ser2448, a site directly phosphorylated by AKT.[41,48] Activation of mTORC1 inhibits the negative regulator of translation, 4E-BP leading to the enhanced eIF4F complex formation and translation initiation.[15] Similar to IL-6, NGF enhances nascent protein synthesis in sensory neurons. Therefore, NGF is a key regulator of translation in primary afferent neurons via stimulation of the PI3K-AKT-mTORC1 pathway.[41] Moreover, sustained mTOR activity is required to maintain the sensitivity of a subset of nociceptive A-type fibers.[49,50]

Cotreatment of primary afferent neurons with IL-6 and NGF results in the activation of both ERK and mTORC1 pathways. These signaling pathways converge onto the eIF4F complex resulting in its stabilization. The stabilization of the eIF4F complex leads to nascent protein synthesis in the axons of sensory neurons. IL-6- and NGF-mediated mechanical allodynia is completely blocked by inhibition of eIF4F complex formation with the small molecule 4EGI-1. Hence, IL-6 and NGF enhanced cap-dependent translation leads to prolonged mechanical hypersensitivity.[41]

4.5.2 Axonal synthesis and transport of CREB regulates nociception

The transcription factor cyclic-AMP response element-binding protein (CREB) has been identified as one of the proteins that is locally synthesized within axons and is critical to the development of nociceptive sensitization in response to IL-6 treatment.[51] Retrograde axonal trafficking of CREB in response to NGF has been demonstrated to be critical for the survival of developing primary sensory neurons.[52] In accordance with this, IL-6 injection into the paws of animals causes a rapid increase in CREB synthesis within the paw but not in the sciatic nerve. However, 2 h post injection, an increase in CREB in the sciatic nerve is observed, consistent with retrograde transport. Importantly, coinjection of IL-6 with the methionine analogue azidohomoalanine (AHA), to assess nascently synthesized proteins, revealed an increase in CREB containing AHA in the sciatic nerve 2 h post injection, indicating retrograde transport of nascently synthesized CREB within the paw. Behaviorally, blockade of retrograde transport by disruption of microtubules or inhibition of dynein prevents the development of mechanical allodynia, suggesting the critical role of axonal transport in IL-6-mediated sensitization. Moreover, intrathecal injection of cAMP response element (CRE) consensus sequence DNA oligonucleotides, which act as decoys for CREB DNA binding, prevents the development of IL-6-induced mechanical hypersensitivity. Consistent with previous studies in inflammatory models, intraplantar IL-6 enhances the expression of brain-derived neurotrophic factor (BDNF) in DRGs.[53,54] This effect was blocked by inhibition of retrograde axonal transport and by intrathecal CRE oligonucleotides.[51] Hence, these findings demonstrate that locally generated signals in the form of the nascent synthesis of a transcription factor within the peripheral axons of sensory neurons can mediate remote molecular effects within the cell bodies that manifest as mechanical hypersensitivity and hyperalgesic priming.

4.5.3 Peripheral nerve injury reorganizes the translational machinery

Investigations of the changes in translational regulatory mechanisms in response to peripheral nerve injury (PNI) have revealed profound changes in the peripheral nervous systems of rodents with neuropathic pain. PNI increases the expression of the proinflammatory mediators IL-6 and NGF in addition to the activity of kinases associated with translation control (e.g., AKT, mTORC1, and ERK) in the PNS. Moreover, injury increases the expression level of translational machinery (eIF4E, eIF4A, and eIF2α) and RNA processing and binding proteins (FMRP, Mov10, and rck/p54). Hence, these results demonstrate a fundamental reorganization of translational signaling pathways and machinery in the sciatic nerve induced by PNI. Notably, nerve injury is associated with a decrease in the phosphorylation of the eIF2 α consistent with sustained cap-dependent translation.[55] These reorganizations result in the stabilization of eIF4F complex on the 5′-mRNA cap and a profound enhancement of nascent protein synthesis in the injured sciatic nerves.[55]

4.5.4 mTORC1 inhibition and nociception

Inhibition of mTORC1 has emerged as an attractive therapeutic target because it is activated in pathologies underlying neuropathic pain at both peripheral and spinal levels.[50,55–57] Surprisingly, chronic treatment of patients with mTORC1 inhibitors such a rapamycin and its derivatives, termed rapalogues, are associated with increases in the incidence of pain,[58,59] including the possible development of a debilitating pain condition known as complex regional pain syndrome.[60–62] These clinical observations are corroborated in rodents where pharmacological (systemic rapalogue treatment) and genetic inhibition of mTORC1 causes pain. In primary sensory neurons, mTORC1 exerts its effect via phosphorylation of its two key substrates, S6K and 4E-BPs. When phosphorylated by mTORC1, S6K stimulates mRNA translation by phosphorylating proteins that control translation initiation and elongation.[63] As part of a negative feedback loop, S6K also inhibits the adaptor protein insulin receptor substrate 1 (IRS-1) by phosphorylating Ser636/639 residues, leading to inhibition of PI3K-AKT-mTOR and Ras-Raf-Mek-ERK signaling (Fig. 2). Pharmacological and genetic inhibition of mTORC1 function reduces the activity of S6K, leading to reduced IRS Ser636/639 phosphorylation, enhanced phosphorylation of IRS Tyr895 residue, and the recruitment of the adaptor protein Grb2, resulting in the activation of the of ERK pathway.[64,65] ERK in turn phosphorylates and modulates the gating properties of the voltage-gated sodium channel $Na_v1.7$. $Na_v1.7$ is a threshold channel that augments weak stimuli where ERK phosphorylation of specific residues on the intracellular L1 loop of $Na_v1.7$ mediates an increase in action potential firing and a decrease in latency to first action potential in sensory neurons.[66] Importantly, Nav1.7 is linked to human pain conditions wherein loss of function results in insensitivity to pain while gain of function causes erythromelalgia and paroxysmal extreme pain disorder.[67] Thus, direct mTORC1 inhibition by rapalogues may not present a viable therapeutic strategy due to development of mechanical allodynia via the feedback activation of the ERK pathway.[65]

4.5.5 AMPK as a therapeutic target

Protein synthesis is an energy-intensive process and, for this reason, an intricate system to control energy consumption has evolved in the form of the ubiquitous kinase, AMPK.[63] Activated AMPK blocks protein synthesis by inhibiting components of the mTOR and ERK signaling pathways.[63] Concurrent inhibition of multiple signaling pathways inherently prevents signaling crosstalk, which is commonly observed with the inhibition of a single kinase involved in these signaling cascades (e.g., inhibition of mTORC1 with rapamycin[64]). Phosphorylation of serine 636/639 and 789 residues is associated with decreased signaling from IRS. IRS S636/639 residues are phosphorylated by S6K[68,69] while the S789 site is phosphorylated by AMPK.[70] As mentioned above, treatment of primary sensory neurons and animals with rapalogues lead to the feedback activation of ERK and increased hyperexcitability of sensory neurons in part due to the modulation of the gating properties of $Na_v1.7$. In contrast, treatment of sensory neuron cultures with the AMPK activator enhances Ser789 phosphorylation and attenuated rapamycin-induced changes in Tyr895 phosphorylation of IRS-1. This increase in IRS-1 S789 phosphorylation and blockade of IRS-1 Y895 phosphorylation was associated with an AMPK-mediated block of ERK activation, the consequent hyperexcitability, and pain associated with rapalogue treatment.[65]

AMPK activation is associated with the inhibition of the signaling pathways that are activated in PNI-induced neuropathic pain. Thus, treatment of neuropathic animals with AMPK activators, metformin and A769662, reversed changes that are associated with PNI-induced neuropathic pain. The changes included normalization of translation regulation signaling pathways, eIF4F complex formation, nascent protein synthesis in injured nerves, and sodium channel-dependent excitability of sensory neurons, resulting in a resolution of neuropathic allodynia. The most dramatic finding is the potential disease modifying effect of AMPK activator metformin, where daily treatment with the drug for 7 days provides pain relief for at least 2 months. Therefore, injury-induced dysregulation of translation control may underlie the pathology of neuropathic pain and AMPK activation reverses these changes, providing a potential new therapeutic target.[55]

Further evidence on the therapeutic potential of AMPK activators has been demonstrated with the use of resveratrol. This AMPK activator was shown to be effective in inhibiting the postsurgical nociceptive sensitization as well as the prolonged sensitization caused by the intraplantar injection of IL-6. Furthermore, AMPK activators have been shown to be effective in inhibiting inflammatory pain and in preventing the development of chemotherapy-induced neuropathy.[71-73]

4.6. The role of the 3′ UTR and poly-A tail of mRNAs

The mRNA poly-A tail may also play a role in the cap-dependent translation. The cytoplasmic polyadenylation element (CPE) is a nucleotide sequence within the 3′ UTR of mRNAs which binds to the CPE-binding protein (CPEB). A highly conserved isoform of CPEB has been demonstrated to contain a prion-like domain in the N-terminus. Moreover, the expression of the Aplysia isoform of CPEB in yeast exhibits a crucial property of prion proteins where it causes heritable alterations in protein conformations.[74] Thus, it has been proposed that the long-lasting bistable prion-like properties of CPEB may contribute to long-term memory and chronification of pain.[75] CPEB may also act as a regulatory switch that inhibits translation in oocytes when it is associated with Maskin, which sequesters eIF4E, preventing the formation of the eIF4F complex. However, the phosphorylation of CPEB is thought to weaken the interaction of CPEB–Maskin–eIF4E allowing for the formation of the eIF4F complex. This process is proposed to be followed by the binding of the cleavage and polyadenylation specificity factor, thus promoting polyadenylation. The poly-A-binding protein is proposed to bind to both the eIF4G and the poly-A tail, resulting in the circularization of the mRNA which leads to enhance translation (Fig. 4).[76]

4.7. Cap-independent mechanism

A cap-independent mechanism of the translation was discovered in picornavirus RNAs which lack a 5′-mRNA cap.[77] Translation of these mRNA is mediated through an internal ribosome entry site (IRES). Several mRNAs that reside in dendrites contain IRES elements, however the extent and the importance of IRES-mediated translation in vertebrates is yet to be elucidated.[78]

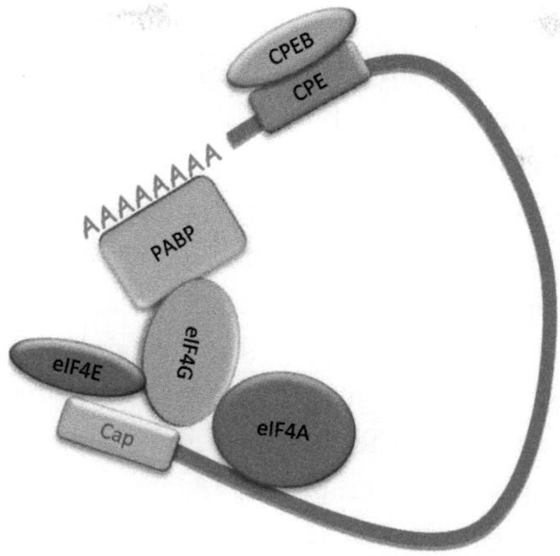

Figure 4 Poly-A-binding protein and mRNA circularization. PABP binds to both eIF4G and the poly-A tail of mRNA, which circularizes the mRNA. This circularization is proposed to prevent the translating ribosome from falling off the mRNA and is recycled to carry out several rounds of translation.

5. TRANSPORT AND STORAGE OF mRNAs

Cellular mRNAs associate with a group of proteins, termed RNA-binding proteins (RBPs). Ribonucleoprotein (mRNP) granules are composed of mRNAs, proteins, and regulatory RNAs where the composition of these granules determines whether an mRNA is transported, stored, or degraded.[79] One of the best described example of mRNA transport includes an RBP that belongs to the Pumilio family. This RBP specifically binds to mRNAs that are transported to mitochondria, ensuring localized expression of the required proteins.[80,81] Another key RBP that is involved in mRNA transport, localization,[82] and translation repression[83,84] is fragile X mental retardation protein (FMRP). Silencing of the gene that encodes for FMRP causes fragile X mental retardation in humans. Current evidence suggests that mGluR1-mediated activation of protein phosphatase 2A dephosphorylates FMRP causing its dissociation from the target mRNA, thus derepessing the translation of these mRNAs.[85] Crucially, the role of FMRP in nociceptive sensitization has been demonstrated in Fmr1 knockout mice. These mice show decreased nociceptive sensitization in multiple experimental paradigms demonstrating the critical role FMRP plays in sensitization of the nociceptive pathway.[86–88]

Another class of RNP granules include P bodies and stress granules. P bodies are thought to be important sites of mRNA metabolism in cells because they contain deadenylation and decapping enzymes. The shortening of the 3′ poly-A tail and/or removal of the 5′-mRNA cap leads to the rapid decay of mRNAs, effectively terminating their cellular life cycle.[89] P bodies are defined at the cytological level as aggregates of mRNA and protein composed of one of several accepted P body markers such as the translational repressor/decapping activator Rck/p54/dhh1 or the decapping enzyme Dcp2. Interestingly, P bodies are devoid of eukaryotic initiation factors (eIFs) with the exception of the 5′-mRNA cap-binding protein eIF4E.[90] In contrast, stress granules are cellular RNA granules composed of proteins including eIFs and mRNAs yet largely devoid of enzymes involved in mRNA decapping. These structures are induced by cellular stressors (e.g., starvation or arsenite exposure) in a wide variety of organisms and cell types and appear to play an important role in storing mRNAs stalled at translation initiation step when translation integrity may be compromised.[91] Decapping activators such as Rck, Scd6/Rap55, and Pat1 repress translation. This suggests that mRNA decapping/repression is preceded by (1) inhibition of translation, (2) exchange of the translation initiation factors for components of decapping/repression machinery, and (3) translational repression and possible storage or ultimate degradation of mRNA culminates this process. Cumulatively, this indicates that the 5′-mRNA cap is a site of direct competition between translation initiation factors and decapping/repression machinery.[92–95] Interestingly, in primary sensory neurons, P bodies are reciprocally controlled by factors that either stimulate (NGF and IL6[41]) or inhibit (AMPK activators[55,65,73]) the formation of eIF4F complex on the mRNA cap in sensory neurons.[96] Thus, the association of the components of the P bodies with the 5′-mRNA cap is inversely related to eIF4F complex formation in sensory neurons. Moreover, axons of primary sensory neurons display a robust regulation of P bodies in an apparent contrast to the CNS.

6. CONCLUSIONS AND PERSPECTIVES

Tissue injury is associated with the release of mediators that allow for regeneration to occur, which in turn is accompanied with *de novo* synthesis of lipids and proteins that are critical for the repair of damaged tissue. Moreover, many of the mediators that initiate regeneration also induce pain. This is likely due to the coevolution of the regeneration and pain mechanisms, since it may confer significant survival advantage. In chronic pain conditions such as PNI-induced neuropathic pain, the increase in protein synthesis in peripheral nerves as well as nociception is sustained through unknown mechanisms. Crucially, treatment with AMPK activators provide prolonged pain relief in neuropathic animals in conjunction with the normalization of protein synthesis in injured nerves, linking aberrant protein translation to pain. Thus, elucidating the mechanisms that sustain protein translation may provide crucial insights into the mechanisms that maintain chronic pain states.

7. TRANSLATIONAL CONTROL OF CENTRAL SENSITIZATION

Activation of sensory neurons in response to peripheral injury leads to a long-lasting increase in the excitability of the spinal cord circuit, resulting in a significant enhancement in the gain of the pain pathway. This activity-dependent increase in the excitability of the spinal circuit is known as "central sensitization." The central sensitization disproportionally amplifies the peripheral inputs, so that pain does no longer reflect the specific peripheral stimuli but rather the excitability state of the spinal nociceptive circuit. This results in a reduction in threshold for the induction of pain (allodynia), an increased responsiveness to noxious stimuli (hyperalgesia), and an expansion of the receptive field so that inputs from noninjured areas produce pain (secondary hyperalgesia). The increased responsiveness of the spinal circuit to peripheral stimulation promotes the recruitment of novel peripheral inputs to the nociceptive pathways that normally do not transmit pain, such as from the large low-threshold mechanoreceptor myelinated Aβ fibers. At the cellular level, the augmented responsiveness of the central circuit can be achieved via several mechanisms (a) an increase in presynaptic neurotransmitter release or postsynaptic response, (b) an increase in the excitability of spinal neurons via modulation of their membrane properties, (c) a reduction in the inhibition by modulation of an inhibitory synaptic transmission, and (d) the excitability state of spinal neurons is dependent not only on neuronal plasticity but is also modulated by the activity of astrocytes, microglia, and leukocytes.

Translational control plays an important role in regulating synaptic plasticity in the CNS. The long-lasting increase and decrease in the synaptic strength in mammals are known as long-term potentiation (LTP) and long-term depression (LTD), respectively.[97] LTP exhibits two temporally distinct phases, which are defined by their requirement for new gene expression: early-phase LTP (E-LTP) which does not depend on *de novo* protein synthesis, and which relies on the modification of preexisting components, whereas late-phase LTP (L-LTP) requires transcription of new sets of genes and synthesis of novel proteins. LTP is generally considered to be the cellular model for learning and memory, as both LTP and memory formation share similar molecular and cellular mechanisms. L-LTP and long-term memory can be impaired by inhibition of protein synthesis or the mTOR pathway (e.g., using rapamycin), supporting the critical role of translation control downstream of the mTOR pathway in regulating long-lasting synaptic plasticity. In the pain pathway, stimulation of nociceptors with electrical or natural activation increases synaptic efficiency of nociceptive neurons in the dorsal horn of the spinal cord and this synaptic potentiation lasts for many minutes to even hours.[98-100] Spinal LTP represents a particular component of the general phenomenon of central sensitization and contributes to the long-lasting increase in the gain of the nociceptive system. Tetanic stimulation of C-fibers leads to the long-lasting potentiation of field potentials in superficial spinal dorsal horn. Inhibition of protein synthesis in the spinal cord prior to titanic stimulation blocks the late phase of spinal LTP, whereas the E-LTP is not affected.[101] Together, these findings indicate that similar to hippocampal LTP, new protein synthesis is required for consolidation of spinal long-lasting synaptic potentiation.

Studies in mouse demonstrated that subcutaneous injection of formalin into the hindpaw induces biphasic neuronal and behavioral response.[102] The early-phase response (0–10 min) is caused predominantly by the peripheral C-fiber activation, whereas the late-phase response (10–55 min) is believed to be caused largely by functional changes in the dorsal horn of the spinal cord. Inhibition of general protein synthesis in the spinal cord using intrathecal administration of anisomycin, dramatically reduces pain behavior in the second phase of the formalin test, whereas the first phase is not affected.[103] Moreover, mTOR inhibition using intrathecal delivery of rapamycin reduces nociceptive behavior during the second phase of formalin test[88,104,105] and attenuates formalin-induced neuronal hyperexcitability in the wide dynamic range dorsal horn spinal neurons.[104] Additionally, intrathecal administration of rapamycin reduces the capsaicin-induced secondary mechanical hyperalgesia in the area surrounding the injection site, a response that is mediated via sensitization of the spinal cord neurons to the input from capsaicin-insensitive A-nociceptors.[49] These results support the role of *de novo* protein synthesis and its activation via mTOR in the regulation of peripheral injury-induced spinal synaptic plasticity that underlies central sensitization of dorsal horn neurons.

7.1. Peripheral inflammation- and injury-induced activation of mTOR in the spinal neurons contributes to the hyperalgesia state

In support for the role of translational control through mTOR pathway in central sensitization, previous studies showed that mTORC1 signaling components are present in the spinal neurons and are activated in several pain models. P-mTOR, P-S6, and P-4E-BP1/2 are present under basal conditions in lamina I/III projections neurons.[49,106,107] Moreover, there is a significant increase in the number of P-S6-expressing cells in the superficial dorsal horn following intraplantar capsaicin[49] and carrageenan injection[106]; and Western blot analysis showed that phosphorylation of mTOR, p70S6K, S6, and 4E-BP1 is increased in the dorsal horn in rodent models of peripheral inflammation[105,106,108,109] and bone-cancer-induced pain.[110] Consistent with these findings, activation of PI3K and AKT, two kinases acting upstream of mTOR, was elevated in the dorsal horn in models of peripheral inflammation.[105,111] The phosphorylation of mTOR pathway components (mTOR, p70S6K, and 4E-BP1) was found to be increased in a chronic constriction injury (CCI) neuropathic pain model in mice,[112] but not in spinal nerve ligation (SNL)[109] and spared nerve injury[49] models in rats, thus further studies will be required to confirm these findings.

In neurons, synaptic inputs activate mTOR via stimulation of NMDA and TrkB neurotrophin receptors, which signal to PI3K.[113] BDNF plays a critical role in the development of LTP and nociceptive sensitization. It has been demonstrated that the IL-6-induced nociceptive sensitization can be blocked by inhibiting spinal BDNF. Moreover, studies on spinal synaptoneurosomes (SNS) elucidated that BDNF activates the AKT and mTOR, resulting in the stabilization of eIF4F complex. These events lead to the nascent synthesis of proteins associated with establishment of LTP such as atypical PKCs and CAMKII alpha in spinal SNS.[51,54] Moreover, in a bone-cancer-induced pain model, mTOR activation requires activation of NMDA receptors.[110]

To study the functional role of mTOR in nociception at the level of the spinal cord, intrathecal administration of rapamycin and its derivative CCI-779 has been used. Intrathecal delivery of mTOR inhibitors does not affect acute responses to thermal and mechanical stimuli,[49,114] whereas it reduces nocifensive behaviors and hypersensitivity in the rodent models of peripheral inflammation (formalin, capsaicin, CFA, and carrageenan),[88,106,109] bone cancer-induced pain,[110] and neuropathic pain (SNL, CCI).[56,112,115]

All together, these studies clearly indicate that the mTOR pathway is activated in the spinal cord in different chronic pain conditions. However, since the intrathecal delivery of mTOR inhibitors can potentially inhibit the mTOR pathway in the central branches of sensory neurons, where it can locally regulate protein synthesis, or in the cell bodies, the behavioral effects cannot be exclusively attributed to the inhibition of central sensitization as the effect on primary nociceptors cannot be excluded. Thus, to further elucidate the roles of spinal mTOR signaling in mediating hyperalgesia states, mTOR pathway components should be genetically manipulated in different populations of spinal neurons.

7.2. Translational control in opioid-induced hyperalgesia

mTOR is involved in an additional nociceptive sensitization phenomenon, opioid-induced hyperalgesia. Opioids are used to treat pain; however, repeated administration of opioids may paradoxically sensitize patients to acute pain. Repeated intrathecal morphine administration activated mTOR and increased nascent protein synthesis in the dorsal horn of the spinal cord.[114] Moreover, the association of eIF4E with eIF4A was increased, whereas the association of 4E-BP1 with eIF4E was decreased, indicative of enhanced eIF4F complex formation and increased translation initiation. Behavioral studies in rats showed that morphine-induced tolerance and hyperalgesia can be significantly reduced by intrathecal rapamycin administered at the induction stage or when morphine tolerance and hyperalgesia are well established.[114] Morphine-induced tolerance and hyperalgesia are believed to result from adaptive neuroplastic changes in the nervous system. The involvement of mTOR in this form of spinal plasticity further demonstrates the important role of translational control in conditions associated with aberrant neuronal plasticity in the spinal cord.

7.3. Translational control in spinal nonneuronal cells

The role of translational control in spinal nonneuronal cells in response to peripheral inflammation or injury is not well characterized. Activation of spinal astrocytes in response to peripheral injury is termed reactive astrogliosis and has been reported to contribute to the increased sensitivity via enhanced expression of proinflammatory cytokines, neurotrophic factors,[116-118] and gliotransmitters such as glutamate and ATP. These glial mediators promote neuroinflammation and contribute to hypersensitivity states via activation of glial and immune cells and modulation of synaptic transmission in neurons. Reactive astrogliosis is characterized by astrocyte proliferation, hypertrophy, and enhanced expression of GFAP. Recent study showed that peripheral injury (CCI) activates astrocytes in the dorsal horn of the spinal cord, as assessed by enhanced GFAP expression, and this effect is sensitive to rapamycin.[115] Moreover, rapamycin inhibits an ADP-induced increase in ATP levels in cultured spinal astrocytes and an elevation of ATP in the cerebrospinal fluid of CCI rats. Together, these findings suggest that mTOR activity plays a prominent role in the activation of spinal astrocytes in neuropathic pain.

7.4. Reconsolidation-like processes in the pain pathway

An additional form of activity-dependent plasticity in the nervous system is reconsolidation. A newly formed memory remains in a dynamic or "labile" form for a short time, after which the memory trace is stored or consolidated into the physical organization of the brain. When a consolidated memory is retrieved or "reactivated", it becomes labile and needs to be restabilized in a protein synthesis-dependent manner, a process called reconsolidation.[119] Retrieval of consolidated memories coupled with inhibition of protein synthesis has been used to erase memories, therefore having important implications for the treatment of posttraumatic stress disorder. A recent study showed that similarly to memory reconsolidation, mechanical hyperalgesia, induced by capsaicin or CFA, can be made labile again by reactivation of the pain pathway and thus become sensitive to protein synthesis inhibition.[120] Moreover, the consolidated spinal LTP can be impaired by coupling the LTP induction stimulus to the inhibition of translation. These findings demonstrate the existence of reconsolidation phenomenon in the pain pathway and can potentially be used to erase the consolidated persistent pain.

7.5. Molecular mechanisms underlying the effect of mTOR on spinal neuronal sensitization

Molecular mechanisms underlying the inhibitory effects of intrathecal rapamycin in chronic pain models are not known. It is likely that the attenuation of LTP-like facilitation of synaptic transmission in the dorsal horn neurons is partially responsible for this inhibitory effect. Indeed, mTOR activity is associated with increased synaptic transmission and elevated ratio of excitatory to inhibitory (E/I) synaptic inputs. Enhancement of eIF4F complex formation by 4E-BP2 knockout[121,122] or eIF4E overexpression[123] increases both the excitatory and inhibitory miniature synaptic transmission in hippocampal pyramidal neurons. However, the E/I ratio is shifted in favor of excitation, supporting the idea that mTOR activity might increase the excitatory drive into the central spinal neurons. In hippocampus, 4E-BP2 regulates synaptic transmission, at least partially, via translational repression of mRNAs encoding neuroligins (NLGN1-4). Neuroligins are postsynaptic cell adhesion molecules that interact with presynaptic α-neurexins to regulate synapse formation, maturation, and functioning.[124] NLGN1 is expressed specifically at excitatory synapses, where it promotes excitatory synaptic transmission, whereas NLGN2 mainly regulates inhibitory synapses. NLGN3 is involved in the regulation of both excitatory and inhibitory synapses. Additionally, translation of mRNAs encoding for AMPA receptor subunits (GluA1 and GluA2) is repressed by 4E-BP2.[122] Accordingly, reduction of 4E-BP2 levels in hippocampal pyramidal neurons leads to facilitation of AMPA receptor-mediated synaptic transmission.[122]

Studies in hippocampal CA1 area showed that synaptic plasticity is regulated at the translational level via mTOR/4E-BP2 and eIF2α pathways.[125] The threshold for the induction of persistent synaptic potentiation is controlled by 4E-BP2 as in hippocampal slices prepared from 4E-BP2 knockout mice the L-LTP can be induced by a weak stimulation (one high-frequency stimulation), which in wild-type mice elicits E-LTP only.[122,126] Because 4E-BP1 and 4E-BP2 are the major mTOR downstream effectors, it is conceivable that mTOR might enhance excitatory synaptic transmission and stimulate synaptic plasticity via regulation of 4E-BP activity. However, the role of 4E-BPs in the pain pathway remains to be determined.

mTOR might also contribute to increased responsiveness of the spinal circuits by suppressing potassium channel $K_v1.1$ expression in dendrites,[127] which will lead to an increase in the excitability of dendritic membrane. $K_v1.1$ channel is expressed in spinal neurons;[128] however, the functional role of mTOR-dependent translational regulation of its expression in the pain pathway is for the moment unknown.

8. CONCLUSIONS

Translational control has emerged as an important regulator of activity-dependent plasticity in the pain pathway. Studies on mTOR, eIF4E, and AMPK have provided important insights into the mechanisms underlying hypersensitivity in primary nociceptors. Inhibition of mTOR activity alleviates chronic inflammatory and neuropathic pain; however, prolonged systemic treatment with mTOR inhibitors paradoxically leads to allodynia via disinhibition of the S6K1/IRS-1/ERK signaling. These data demonstrate that better strategies to control translation in the pain pathway are required. Despite the fact that most of the studies have focused on mTOR and 5' UTR-dependent regulation of protein synthesis, translational control at the 3' UTR by regulation of poly-A tail length is emerging as an important mechanism for synaptic plasticity and ongoing research will shed a new light on the role of this mode of translational regulation in the pain pathway. The molecular mechanisms by which translational upregulation sensitizes the pain pathway at the level of the spinal cord and most importantly, the identity of mRNAs whose aberrant translation leads to this sensitization remains poorly understood and further research is necessary to elucidate this.

REFERENCES

1. Woolf CJ, Walters ET. Common patterns of plasticity contributing to nociceptive sensitization in mammals and aplysia. *Trends Neurosci.* 1991;14:74–78.
2. Ruggeri ZM. Platelets in atherothrombosis. *Nat Med.* 2002;8:1227–1234.
3. Yager DR, Nwomeh BC. The proteolytic environment of chronic wounds. *Wound Repair Regen.* 1999;7:433–441.
4. Barrientos S, Stojadinovic O, Golinko MS, Brem H, Tomic-Canic M. Growth factors and cytokines in wound healing. *Wound Repair Regen.* 2008;16:585–601.
5. Lee SK, Wolfe SW. Peripheral nerve injury and repair. *J Am Acad Orthop Surg.* 2000;8:243–252.
6. Schmidt CE, Leach JB. Neural tissue engineering: strategies for repair and regeneration. *Annu Rev Biomed Eng.* 2003;5:293–347.
7. Gaudet AD, Popovich PG, Ramer MS. Wallerian degeneration: gaining perspective on inflammatory events after peripheral nerve injury. *J Neuroinflammation.* 2011;8:110.
8. Crook RJ, Dickson K, Hanlon RT, Walters ET. Nociceptive sensitization reduces predation risk. *Curr Biol.* 2014;24:1121–1125.
9. Price TJ, Dussor G. Evolution: the advantage of 'maladaptive' pain plasticity. *Curr Biol.* 2014;24:R384–R386.
10. Devor M. Unexplained peculiarities of the dorsal root ganglion. *Pain Suppl.* 1999;6: S27–S35.
11. Holt CE, Bullock SL. Subcellular mRNA localization in animal cells and why it matters. *Science.* 2009;326:1212–1216.
12. Jung H, Gkogkas CG, Sonenberg N, Holt CE. Remote control of gene function by local translation. *Cell.* 2014;157:26–40.
13. Rosoff WJ, Urbach JS, Esrick MA, McAllister RG, Richards LJ, Goodhill GJ. A new chemotaxis assay shows the extreme sensitivity of axons to molecular gradients. *Nat Neurosci.* 2004;7:678–682.
14. Schwanhausser B, Busse D, Li N, et al. Global quantification of mammalian gene expression control. *Nature.* 2011;473:337–342.

15. Sonenberg N, Hinnebusch AG. Regulation of translation initiation in eukaryotes: mechanisms and biological targets. *Cell.* 2009;136:731–745.
16. Sonenberg N, Rupprecht KM, Hecht SM, Shatkin AJ. Eukaryotic mRNA cap binding protein: purification by affinity chromatography on sepharose-coupled m7GDP. *Proc Natl Acad Sci U S A.* 1979;76:4345–4349.
17. Gingras AC, Raught B, Sonenberg N. eIF4 initiation factors: effectors of mRNA recruitment to ribosomes and regulators of translation. *Annu Rev Biochem.* 1999;68:913–963.
18. Pause A, Belsham GJ, Gingras AC, et al. Insulin-dependent stimulation of protein synthesis by phosphorylation of a regulator of 5'-cap function. *Nature.* 1994;371:762–767.
19. Laplante M, Sabatini DM. mTOR signaling in growth control and disease. *Cell.* 2012;149:274–293.
20. Heitman J, Movva NR, Hall MN. Targets for cell cycle arrest by the immunosuppressant rapamycin in yeast. *Science.* 1991;253:905–909.
21. Hay N, Sonenberg N. Upstream and downstream of mTOR. *Genes Dev.* 2004;18:1926–1945.
22. Buxade M, Parra-Palau JL, Proud CG. The Mnks: MAP kinase-interacting kinases (MAP kinase signal-integrating kinases). *Front Biosci.* 2008;13:5359–5373.
23. Hinnebusch AG, Lorsch JR. The mechanism of eukaryotic translation initiation: new insights and challenges. *Cold Spring Harb Perspect Biol.* 2012;4:a011544.
24. Mohammad-Qureshi SS, Jennings MD, Pavitt GD. Clues to the mechanism of action of eIF2B, the guanine-nucleotide-exchange factor for translation initiation. *Biochem Soc Trans.* 2008;36:658–664.
25. Costa-Mattioli M, Gobert D, Harding H, et al. Translational control of hippocampal synaptic plasticity and memory by the eIF2alpha kinase GCN2. *Nature.* 2005;436:1166–1173.
26. Stapleton D, Mitchelhill KI, Gao G, et al. Mammalian AMP-activated protein kinase subfamily. *J Biol Chem.* 1996;271:611–614.
27. Adams J, Chen ZP, Van Denderen BJ, et al. Intrasteric control of AMPK via the gamma1 subunit AMP allosteric regulatory site. *Protein Sci.* 2004;13:155–165.
28. Hawley SA, Davison M, Woods A, et al. Characterization of the AMP-activated protein kinase kinase from rat liver and identification of threonine 172 as the major site at which it phosphorylates AMP-activated protein kinase. *J Biol Chem.* 1996;271:27879–27887.
29. Shaw RJ, Kosmatka M, Bardeesy N, et al. The tumor suppressor LKB1 kinase directly activates AMP-activated kinase and regulates apoptosis in response to energy stress. *Proc Natl Acad Sci U S A.* 2004;101:3329–3335.
30. Price TJ, Dussor G. AMPK: an emerging target for modification of injury-induced pain plasticity. *Neurosci Lett.* 2013;557(pt A):9–18.
31. Andratsch M, Mair N, Constantin CE, et al. A key role for gp130 expressed on peripheral sensory nerves in pathological pain. *J Neurosci.* 2009;29:13473–13483.
32. Dina OA, Green PG, Levine JD. Role of interleukin-6 in chronic muscle hyperalgesic priming. *Neuroscience.* 2008;152:521–525.
33. Murphy PG, Ramer MS, Borthwick L, Gauldie J, Richardson PM, Bisby MA. Endogenous interleukin-6 contributes to hypersensitivity to cutaneous stimuli and changes in neuropeptides associated with chronic nerve constriction in mice. *Eur J Neurosci.* 1999;11:2243–2253.
34. Pezet S, McMahon SB. Neurotrophins: mediators and modulators of pain. *Annu Rev Neurosci.* 2006;29:507–538.
35. Anderson LC, Rao RD. Interleukin-6 and nerve growth factor levels in peripheral nerve and brainstem after trigeminal nerve injury in the rat. *Arch Oral Biol.* 2001;46:633–640.

36. Bolin LM, Verity AN, Silver JE, Shooter EM, Abrams JS. Interleukin-6 production by Schwann cells and induction in sciatic nerve injury. *J Neurochem.* 1995;64:850–858.
37. Clark JD, Shi X, Li X, et al. Morphine reduces local cytokine expression and neutrophil infiltration after incision. *Mol Pain.* 2007;3:28.
38. Nishimoto N, Miyasaka N, Yamamoto K, Kawai S, Takeuchi T, Azuma J. Long-term safety and efficacy of tocilizumab, an anti-IL-6 receptor monoclonal antibody, in monotherapy, in patients with rheumatoid arthritis (the STREAM study): evidence of safety and efficacy in a 5-year extension study. *Ann Rheum Dis.* 2009;68:1580–1584.
39. Boulanger MJ, Chow DC, Brevnova EE, Garcia KC. Hexameric structure and assembly of the interleukin-6/IL-6 alpha-receptor/gp130 complex. *Science.* 2003;300:2101–2104.
40. Kamimura D, Ishihara K, Hirano T. IL-6 signal transduction and its physiological roles: the signal orchestration model. *Rev Physiol Biochem Pharmacol.* 2003;149:1–38.
41. Melemedjian OK, Asiedu MN, Tillu DV, et al. IL-6- and NGF-induced rapid control of protein synthesis and nociceptive plasticity via convergent signaling to the eIF4F complex. *J Neurosci.* 2010;30:15113–15123.
42. Banik RK, Subieta AR, Wu C, Brennan TJ. Increased nerve growth factor after rat plantar incision contributes to guarding behavior and heat hyperalgesia. *Pain.* 2005;117:68–76.
43. Lewin GR, Rueff A, Mendell LM. Peripheral and central mechanisms of NGF-induced hyperalgesia. *Eur J Neurosci.* 1994;6:1903–1912.
44. McMahon SB, Bennett DL, Priestley JV, Shelton DL. The biological effects of endogenous nerve growth factor on adult sensory neurons revealed by a trkA-IgG fusion molecule. *Nat Med.* 1995;1:774–780.
45. Wild KD, Bian D, Zhu D, et al. Antibodies to nerve growth factor reverse established tactile allodynia in rodent models of neuropathic pain without tolerance. *J Pharmacol Exp Ther.* 2007;322:282–287.
46. Woolf CJ, Safieh-Garabedian B, Ma QP, Crilly P, Winter J. Nerve growth factor contributes to the generation of inflammatory sensory hypersensitivity. *Neuroscience.* 1994;62:327–331.
47. Kurt K, Karen M. Neurite growth promotion by nerve growth factor and insulin-like growth factor-1 in cultured adult sensory neurons: role of phosphoinositide 3-kinase and mitogen activated protein kinase. *J Neurosci Res.* 2001;63:486–499.
48. Nave BT, Ouwens M, Withers DJ, Alessi DR, Shepherd PR. Mammalian target of rapamycin is a direct target for protein kinase B: identification of a convergence point for opposing effects of insulin and amino-acid deficiency on protein translation. *Biochem J.* 1999;344(pt 2):427–431.
49. Geranton SM, Jimenez-Diaz L, Torsney C, et al. A rapamycin-sensitive signaling pathway is essential for the full expression of persistent pain states. *J Neurosci.* 2009;29:15017–15027.
50. Jimenez-Diaz L, Geranton SM, Passmore GM, et al. Local translation in primary afferent fibers regulates nociception. *PLoS One.* 2008;3:e1961.
51. Melemedjian OK, Tillu DV, Moy JK, et al. Local translation and retrograde axonal transport of CREB regulates IL-6-induced nociceptive plasticity. *Mol Pain.* 2014;10:45.
52. Cox LJ, Hengst U, Gurskaya NG, Lukyanov KA, Jaffrey SR. Intra-axonal translation and retrograde trafficking of CREB promotes neuronal survival. *Nat Cell Biol.* 2008;10:149–159.
53. Asiedu MN, Tillu DV, Melemedjian OK, et al. Spinal protein kinase M zeta underlies the maintenance mechanism of persistent nociceptive sensitization. *J Neurosci.* 2011;31:6646–6653.

54. Melemedjian OK, Tillu DV, Asiedu MN, et al. BDNF regulates atypical PKC at spinal synapses to initiate and maintain a centralized chronic pain state. *Mol Pain.* 2013;9:12.
55. Melemedjian OK, Asiedu MN, Tillu DV, et al. Targeting adenosine monophosphate-activated protein kinase (AMPK) in preclinical models reveals a potential mechanism for the treatment of neuropathic pain. *Mol Pain.* 2011;7:70.
56. Asante CO, Wallace VC, Dickenson AH. Mammalian target of rapamycin signaling in the spinal cord is required for neuronal plasticity and behavioral hypersensitivity associated with neuropathy in the rat. *J Pain.* 2010;11:1356–1367.
57. Obara I, Tochiki KK, Geranton SM, et al. Systemic inhibition of the mammalian target of rapamycin (mTOR) pathway reduces neuropathic pain in mice. *Pain.* 2011;152:2582–2595.
58. Budde K, Becker T, Arns W, et al. Everolimus-based, calcineurin-inhibitor-free regimen in recipients of de-novo kidney transplants: an open-label, randomised, controlled trial. *Lancet.* 2011;377:837–847.
59. McCormack FX, Inoue Y, Moss J, et al. Efficacy and safety of sirolimus in lymphangioleiomyomatosis. *N Engl J Med.* 2011;364(17):1595–1606.
60. Massard C, Fizazi K, Gross-Goupil M, Escudier B. Reflex sympathetic dystrophy in patients with metastatic renal cell carcinoma treated with Everolimus. *Investig New Drugs.* 2010;28:879–881.
61. Molina MG, Diekmann F, Burgos D, et al. Sympathetic dystrophy associated with sirolimus therapy. *Transplantation.* 2008;85:290–292, 210.1097/TP.1090b1013e3181601230.
62. Witzig TE, Geyer SM, Ghobrial I, et al. Phase II trial of single-agent temsirolimus (CCI-779) for relapsed mantle cell lymphoma. *J Clin Oncol.* 2005;23:5347–5356.
63. Zoncu R, Efeyan A, Sabatini DM. mTOR: from growth signal integration to cancer, diabetes and ageing. *Nat Rev Mol Cell Biol.* 2011;12:21–35.
64. Kinkade CW, Castillo-Martin M, Puzio-Kuter A, et al. Targeting AKT/mTOR and ERK MAPK signaling inhibits hormone-refractory prostate cancer in a preclinical mouse model. *J Clin Invest.* 2008;118:3051–3064.
65. Melemedjian OK, Khoutorsky A, Sorge RE, et al. mTORC1 inhibition induces pain via IRS-1-dependent feedback activation of ERK. *Pain.* 2013;154:1080–1091.
66. Stamboulian S, Choi JS, Ahn HS, et al. ERK1/2 mitogen-activated protein kinase phosphorylates sodium channel Na(v)1.7 and alters its gating properties. *J Neurosci.* 2010;30:1637–1647.
67. Dib-Hajj SD, Cummins TR, Black JA, Waxman SG. Sodium channels in normal and pathological pain. *Annu Rev Neurosci.* 2010;33:325–347.
68. Ozes ON, Akca H, Mayo LD, et al. A phosphatidylinositol 3-kinase/Akt/mTOR pathway mediates and PTEN antagonizes tumor necrosis factor inhibition of insulin signaling through insulin receptor substrate-1. *Proc Natl Acad Sci U S A.* 2001;98:4640–4645.
69. Um SH, Frigerio F, Watanabe M, et al. Absence of S6K1 protects against age- and diet-induced obesity while enhancing insulin sensitivity. *Nature.* 2004;431:200–205.
70. Tzatsos A, Tsichlis PN. Energy depletion inhibits phosphatidylinositol 3-kinase/Akt signaling and induces apoptosis via AMP-activated protein kinase-dependent phosphorylation of IRS-1 at Ser-794. *J Biol Chem.* 2007;282:18069–18082.
71. Mao-Ying QL, Kavelaars A, Krukowski K, et al. The anti-diabetic drug metformin protects against chemotherapy-induced peripheral neuropathy in a mouse model. *PLoS One.* 2014;9:e100701.
72. Russe OQ, Moser CV, Kynast KL, et al. Activation of the AMP-activated protein kinase reduces inflammatory nociception. *J Pain.* 2013;14:1330–1340.

73. Tillu DV, Melemedjian OK, Asiedu MN, et al. Resveratrol engages AMPK to attenuate ERK and mTOR signaling in sensory neurons and inhibits incision-induced acute and chronic pain. *Mol Pain.* 2012;8:5.
74. Shorter J, Lindquist S. Prions as adaptive conduits of memory and inheritance. *Nat Rev Genet.* 2005;6:435–450.
75. Bogen O, Alessandri-Haber N, Chu C, Gear RW, Levine JD. Generation of a pain memory in the primary afferent nociceptor triggered by PKCepsilon activation of CPEB. *J Neurosci.* 2012;32:2018–2026.
76. Mendez R, Richter JD. Translational control by CPEB: a means to the end. *Nat Rev Mol Cell Biol.* 2001;2:521–529.
77. Hellen CU, Sarnow P. Internal ribosome entry sites in eukaryotic mRNA molecules. *Genes Dev.* 2001;15:1593–1612.
78. Thompson SR. Tricks an IRES uses to enslave ribosomes. *Trends Microbiol.* 2012;20:558–566.
79. Erickson SL, Lykke-Andersen J. Cytoplasmic mRNP granules at a glance. *J Cell Sci.* 2011;124:293–297.
80. Garcia-Rodriguez LJ, Gay AC, Pon LA. Puf3p, a Pumilio family RNA binding protein, localizes to mitochondria and regulates mitochondrial biogenesis and motility in budding yeast. *J Cell Biol.* 2007;176:197–207.
81. Gerber AP, Herschlag D, Brown PO. Extensive association of functionally and cytotopically related mRNAs with Puf family RNA-binding proteins in yeast. *PLoS Biol.* 2004;2:E79.
82. Estes PS, O'Shea M, Clasen S, Zarnescu DC. Fragile X protein controls the efficacy of mRNA transport in Drosophila neurons. *Mol Cell Neurosci.* 2008;39:170–179.
83. Bassell GJ, Warren ST. Fragile X syndrome: loss of local mRNA regulation alters synaptic development and function. *Neuron.* 2008;60:201–214.
84. Napoli I, Mercaldo V, Boyl PP, et al. The fragile X syndrome protein represses activity-dependent translation through CYFIP1, a new 4E-BP. *Cell.* 2008;134:1042–1054.
85. Narayanan U, Nalavadi V, Nakamoto M, et al. FMRP phosphorylation reveals an immediate-early signaling pathway triggered by group I mGluR and mediated by PP2A. *J Neurosci.* 2007;27:14349–14357.
86. Melemedjian OK, Price TJ. Dendritic spine plasticity as an underlying mechanism of neuropathic pain: commentary on Tan et al. *Exp Neurol.* 2012;233:740–744.
87. Price TJ, Melemedjian OK. Fragile X mental retardation protein (FMRP) and the spinal sensory system. *Results Probl Cell Differ.* 2012;54:41–59.
88. Price TJ, Rashid MH, Millecamps M, Sanoja R, Entrena JM, Cervero F. Decreased nociceptive sensitization in mice lacking the fragile X mental retardation protein: role of mGluR1/5 and mTOR. *J Neurosci.* 2007;27:13958–13967.
89. Decker CJ, Parker R. P-bodies and stress granules: possible roles in the control of translation and mRNA degradation. *Cold Spring Harb Perspect Biol.* 2012;4:a012286.
90. Schwartz DC, Parker R. mRNA decapping in yeast requires dissociation of the cap binding protein, eukaryotic translation initiation factor 4E. *Mol Cell Biol.* 2000;20:7933–7942.
91. Balagopal V, Parker R. Polysomes, P bodies and stress granules: states and fates of eukaryotic mRNAs. *Curr Opin Cell Biol.* 2009;21:403–408.
92. Coller J, Parker R. General translational repression by activators of mRNA decapping. *Cell.* 2005;122:875–886.
93. Nissan T, Rajyaguru P, She M, Song H, Parker R. Decapping activators in Saccharomyces cerevisiae act by multiple mechanisms. *Mol Cell.* 2010;39:773–783.
94. Pilkington GR, Parker R. Pat1 contains distinct functional domains that promote P-body assembly and activation of decapping. *Mol Cell Biol.* 2008;28:1298–1312.

95. Tharun S, Parker R. Targeting an mRNA for decapping: displacement of translation factors and association of the Lsm1p-7p complex on deadenylated yeast mRNAs. *Mol Cell.* 2001;8:1075–1083.
96. Melemedjian OK, Mejia GL, Lepow TS, Zoph OK, Price TJ. Bidirectional regulation of P body formation mediated by eIF4F complex formation in sensory neurons. *Neurosci Lett.* 2014;563:169–174.
97. Bliss TV, Lomo T. Long-lasting potentiation of synaptic transmission in the dentate area of the anaesthetized rabbit following stimulation of the perforant path. *J Physiol.* 1973;232:331–356.
98. Sandkuhler J. Understanding LTP in pain pathways. *Mol Pain.* 2007;3:9.
99. Sandkuhler J, Benrath J, Brechtel C, Ruscheweyh R, Heinke B. Synaptic mechanisms of hyperalgesia. *Prog Brain Res.* 2000;129:81–100.
100. Woolf CJ, Salter MW. Neuronal plasticity: increasing the gain in pain. *Science.* 2000;288:1765–1769.
101. Hu NW, Zhang HM, Hu XD, et al. Protein synthesis inhibition blocks the late-phase LTP of C-fiber evoked field potentials in rat spinal dorsal horn. *J Neurophysiol.* 2003;89:2354–2359.
102. Dubuisson D, Dennis SG. The formalin test: a quantitative study of the analgesic effects of morphine, meperidine, and brain stem stimulation in rats and cats. *Pain.* 1977;4:161–174.
103. Kim SJ, Thomas KS, Calejesan AA, Zhuo M. Macromolecular synthesis contributes to nociceptive response to subcutaneous formalin injection in mice. *Neuropharmacology.* 1998;37:1091–1093.
104. Asante CO, Wallace VC, Dickenson AH. Formalin-induced behavioural hypersensitivity and neuronal hyperexcitability are mediated by rapid protein synthesis at the spinal level. *Mol Pain.* 2009;5:27.
105. Xu Q, Fitzsimmons B, Steinauer J, et al. Spinal phosphinositide 3-kinase-Akt-mammalian target of rapamycin signaling cascades in inflammation-induced hyperalgesia. *J Neurosci.* 2011;31:2113–2124.
106. Norsted Gregory E, Codeluppi S, Gregory JA, Steinauer J, Svensson CI. Mammalian target of rapamycin in spinal cord neurons mediates hypersensitivity induced by peripheral inflammation. *Neuroscience.* 2010;169:1392–1402.
107. Xu JT, Zhao X, Yaster M, Tao YX. Expression and distribution of mTOR, p70S6K, 4E-BP1, and their phosphorylated counterparts in rat dorsal root ganglion and spinal cord dorsal horn. *Brain Res.* 2010;1336:46–57.
108. Jiang F, Pang XY, Niu QS, Hua LM, Cheng M, Ji YH. Activation of mammalian target of rapamycin mediates rat pain-related responses induced by BmK I, a sodium channel-specific modulator. *Mol Pain.* 2013;9:50.
109. Liang L, Tao B, Fan L, Yaster M, Zhang Y, Tao YX. mTOR and its downstream pathway are activated in the dorsal root ganglion and spinal cord after peripheral inflammation, but not after nerve injury. *Brain Res.* 2013;1513:17–25.
110. Shih MH, Kao SC, Wang W, Yaster M, Tao YX. Spinal cord NMDA receptor-mediated activation of mammalian target of rapamycin is required for the development and maintenance of bone cancer-induced pain hypersensitivities in rats. *J Pain.* 2012;13:338–349.
111. Pezet S, Marchand F, D'Mello R, et al. Phosphatidylinositol 3-kinase is a key mediator of central sensitization in painful inflammatory conditions. *J Neurosci.* 2008;28:4261–4270.
112. Zhang W, Sun XF, Bo JH, et al. Activation of mTOR in the spinal cord is required for pain hypersensitivity induced by chronic constriction injury in mice. *Pharmacol Biochem Behav.* 2013;111:64–70.

113. Hoeffer CA, Klann E. mTOR signaling: at the crossroads of plasticity, memory and disease. *Trends Neurosci.* 2010;33:67–75.
114. Xu JT, Zhao JY, Zhao X, et al. Opioid receptor-triggered spinal mTORC1 activation contributes to morphine tolerance and hyperalgesia. *J Clin Invest.* 2014;124:592–603.
115. Cui J, He W, Yi B, et al. mTOR pathway is involved in ADP-evoked astrocyte activation and ATP release in the spinal dorsal horn in a rat neuropathic pain model. *Neuroscience.* 2014;275:395–403.
116. Cirillo G, Colangelo AM, Bianco MR, et al. BB14, a Nerve Growth Factor (NGF)-like peptide shown to be effective in reducing reactive astrogliosis and restoring synaptic homeostasis in a rat model of peripheral nerve injury. *Biotechnol Adv.* 2012;30:223–232.
117. Gwak YS, Hulsebosch CE. Remote astrocytic and microglial activation modulates neuronal hyperexcitability and below-level neuropathic pain after spinal injury in rat. *Neuroscience.* 2009;161:895–903.
118. Karimi-Abdolrezaee S, Billakanti R. Reactive astrogliosis after spinal cord injury-beneficial and detrimental effects. *Mol Neurobiol.* 2012;46:251–264.
119. Nader K, Schafe GE, Le Doux JE. Fear memories require protein synthesis in the amygdala for reconsolidation after retrieval. *Nature.* 2000;406:722–726.
120. Bonin RP, De Koninck Y. A spinal analog of memory reconsolidation enables reversal of hyperalgesia. *Nat Neurosci.* 2014;17:1043–1045.
121. Gkogkas CG, Khoutorsky A, Ran I, et al. Autism-related deficits via dysregulated eIF4E-dependent translational control. *Nature.* 2013;493:371–377.
122. Ran I, Gkogkas CG, Vasuta C, et al. Selective regulation of GluA subunit synthesis and AMPA receptor-mediated synaptic function and plasticity by the translation repressor 4E-BP2 in hippocampal pyramidal cells. *J Neurosci.* 2013;33:1872–1886.
123. Santini E, Huynh TN, MacAskill AF, et al. Exaggerated translation causes synaptic and behavioural aberrations associated with autism. *Nature.* 2013;493:411–415.
124. Sudhof TC. Neuroligins and neurexins link synaptic function to cognitive disease. *Nature.* 2008;455:903–911.
125. Costa-Mattioli M, Sossin WS, Klann E, Sonenberg N. Translational control of long-lasting synaptic plasticity and memory. *Neuron.* 2009;61:10–26.
126. Banko JL, Poulin F, Hou L, DeMaria CT, Sonenberg N, Klann E. The translation repressor 4E-BP2 is critical for eIF4F complex formation, synaptic plasticity, and memory in the hippocampus. *J Neurosci.* 2005;25:9581–9590.
127. Raab-Graham KF, Haddick PC, Jan YN, Jan LY. Activity- and mTOR-dependent suppression of Kv1.1 channel mRNA translation in dendrites. *Science.* 2006;314:144–148.
128. Karimi-Abdolrezaee S, Eftekharpour E, Fehlings MG. Temporal and spatial patterns of Kv1.1 and Kv1.2 protein and gene expression in spinal cord white matter after acute and chronic spinal cord injury in rats: implications for axonal pathophysiology after neurotrauma. *Eur J Neurosci.* 2004;19:577–589.

CHAPTER EIGHT

MicroRNA Biology and Pain

Marguerite K. McDonald, Seena K. Ajit[1]
Department of Pharmacology & Physiology, Drexel University College of Medicine, Philadelphia, Pennsylvania, USA
[1]Corresponding author: e-mail address: seena.ajit@drexelmed.edu

Contents

1. Introduction 216
2. History and Biogenesis 216
3. Genomic Location of miRNAs 218
4. miRNA Nomenclature 219
5. miRNA Target Determination 220
6. miRNA Detection, Quantification, and Functional Studies 221
7. Circulating miRNAs 224
8. miRNAs and Pain 227
9. Challenges Associated with miRNA Research 240
10. Conclusions 242
Acknowledgments 242
References 242

Abstract

Endogenously expressed small noncoding microRNAs (miRNAs) play an important role in posttranscriptionally regulating gene expression by binding to mRNAs with complementary sequences. miRNA–mRNA interactions allow for cellular flexibility to fine-tune gene expression by controlling translation in response to a multitude of signaling events. Disease states or perturbations in cellular homeostasis can lead to aberrant miRNA expression. The discovery of stable miRNAs in circulation generated enormous interest in exploring their utility as potential noninvasive biomarkers. Additionally, selectively inhibiting or supplementing an miRNA contributing to pathogenesis is being pursued as a therapeutic strategy for a variety of disorders. Studies from rodent models of pain and patients have now implicated a role for miRNAs in mediating various aspects of pain processing. These noncoding RNAs can provide mechanistic insights into the pathways modulated and could serve as therapeutic targets. Here, we discuss the challenges associated with miRNA research and the promises ahead in this vastly unexplored avenue in pain biology.

Progress in Molecular Biology and Translational Science, Volume 131
ISSN 1877-1173
http://dx.doi.org/10.1016/bs.pmbts.2014.11.015

1. INTRODUCTION

MicroRNAs (miRNAs) are single-stranded small noncoding RNAs that consist of approximately 22 nucleotides (nt) and play a crucial role in fine-tuning gene expression.[1] miRNAs regulate a wide range of biological processes, and dysregulation of individual or entire families of miRNAs is associated with the pathogenesis of an array of human diseases.[2–4] Modulation of gene expression is achieved by binding of the miRNA to one or more complementary sequences in the 3′ untranslated region (3′UTR) of specific-target mRNAs. This mRNA recognition is mediated by 2–7 nt usually found in the 5′ end of the miRNAs called the seed sequence. Binding of miRNA can result in either translational repression or mRNA degradation depending on the degree of complementarity between mRNA and miRNA. Because the required sequence complementarity is relatively short, a single miRNA can target many genes and mRNAs can harbor multiple miRNA-binding sites in their 3′UTR. Hence, it is estimated that ~60% of genes in humans are regulated by miRNAs.[5] It is now well established that miRNAs play key regulatory roles in nearly every aspect of biology, including physiological and pathological processes.

2. HISTORY AND BIOGENESIS

miRNAs, the fine-tuning regulators of the transcriptional milieu, are found in all plants and multicellular eukaryotes. In plants, miRNAs bind to mRNA targets with near complete complementarity, while many animal miRNAs show only partial complementarity to their cognate mRNAs.[6] In 1993, while studying larval developmental stages in *Caenorhabditis elegans*, a model organism for developmental genetics and neurobiology, the Ambros laboratory discovered regulation of the *lin-41* mRNA by *lin-4* miRNA.[7] The Ruvkun laboratory characterized the *lin-4:lin-41* RNA interaction by sequential nucleotide deletion and identified a strong complementary sequence in the 5′ region of the miRNA that could bind several stretches of sequence in the *lin-14* 3′UTR. This region of sequence complementarity of 6–8 bases in the 5′ portion of *lin-4* miRNA was eventually termed the seed sequence as computational studies revealed that many miRNAs were at least partially complementary to sequences in the untranslated regions of mRNAs.[1] In a study measuring the effects of increasing *lin-4* miRNA during development, a second target of *lin-4* was identified, *lin-14*, which showed decreased protein levels in parallel with *lin-4* expression.[8] Interestingly, *lin-14* mRNA levels remained constant, indicating that *lin-4* was able to translationally repress *lin-14* mRNA without causing its degradation. While *lin-4* has no human homolog, the second miRNA discovered, let-7 plays an important role in developmental timing in *C. elegans* and is found in humans and other higher species.[9]

miRNAs are transcribed as 150- to 200-nt precursor sequences (pri-miRNAs) that are capped and polyadenylated and reside in the nucleus in a double-stranded (ds) stem loop shape[1] (Fig. 1). RNA-mediated interference machinery facilitates miRNA processing by forming the

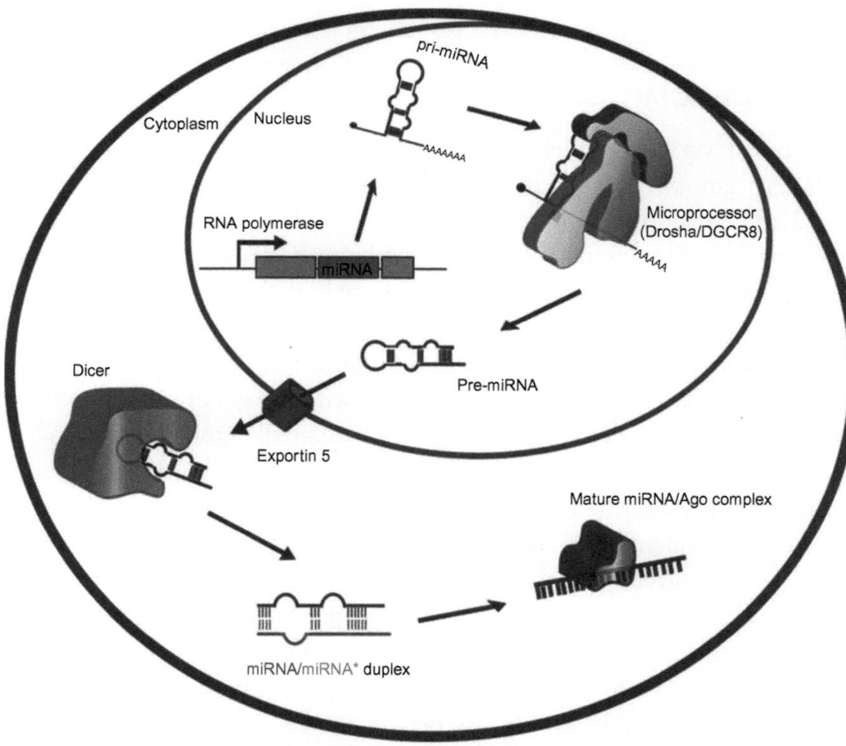

Figure 1 miRNA biogenesis. Primary miRNAs are transcribed from DNA as capped and polyadenylated hairpins that are processed by Drosha/DGCR8 nuclear complex into pre-miRNAs. After export from the nucleus by exportin 5, the pre-miRNA is recognized by DICER, an enzyme that produces short double-stranded intermediates with 5′ phosphates and 3′ overhangs. The miRNA encoded on the guide strand is loaded into argonaut (Ago), and the passenger strand is either degraded or loaded into Ago as a mature miRNA/Ago complex called an miRNP.

microprocessor. This complex contains two essential proteins, the dsRNA-binding protein DGCR8 (DiGeorge syndrome chromosome region) and the RNase III enzyme Drosha,[10] that process pri-miRNAs into 60- to 70-nt short hairpins with a phosphorylated 5′ end and a short overhang on the 3′ end that allows the hairpin to be recognized by the RNA-induced silencing complex (RISC) once exported by exportin 5 to the cytoplasm.[11] The RISC contains the essential RNA-binding proteins argonaut (Ago), the TAR RNA-binding protein, and the enzyme Dicer. Pre-miRNAs contain both the passenger strand and the guide strand that contains the mature miRNA. In a yet undiscovered mechanism, the RISC deciphers between the guide strand that will be loaded onto Ago to silence its cognate mRNA and the passenger strand, which will be destroyed by the "slicer" enzyme of the RISC.[1,12,13]

3. GENOMIC LOCATION OF miRNAs

Based on the sequences flanking miRNAs, genomic location of miRNAs can be broadly divided into intergenic, intronic, or exonic (Fig. 2). The majority of miRNAs are encoded in intergenic regions existing as independent transcription units, with miRNA-specific promoters. Intronic miRNAs processed from the introns of the host gene are transcribed from the same promoter as their host genes. These introns can be that of a protein-coding gene or within long noncoding RNAs. Synchronized expression of intronic miRNAs along with the genes harboring them has been widely observed.[14,15] Exonic miRNAs are far rarer and often overlap an exon and an intron of a noncoding gene. These miRNAs are also thought to be transcribed by their host gene promoter but their maturation is often independent of the host gene function. miRNAs can be transcribed in a cluster (polycistronic) with a shared promoter. Similar to single or monocistronic miRNAs, these clusters can originate from an intergenic or intronic location. miRNAs derived from short intronic hairpins are referred to as mirtrons. Such miRNA precursors bypass Drosha/microprocessor processing because these are already pre-miRNA with splice sites on either side. Both pre-miRNA and mirtron hairpins are exported from the nucleus by exportin-5 and cleaved by Dicer to generate ~22-nt RNA duplexes.[16] Interestingly, more than 50% of miRNA genes are located at fragile sites and cancer susceptibility loci,[17] and miRNAs are frequently amplified or downregulated in cancer. Thus the genomic location and transcription levels indicate tight integration of miRNAs in cellular regulatory circuits.

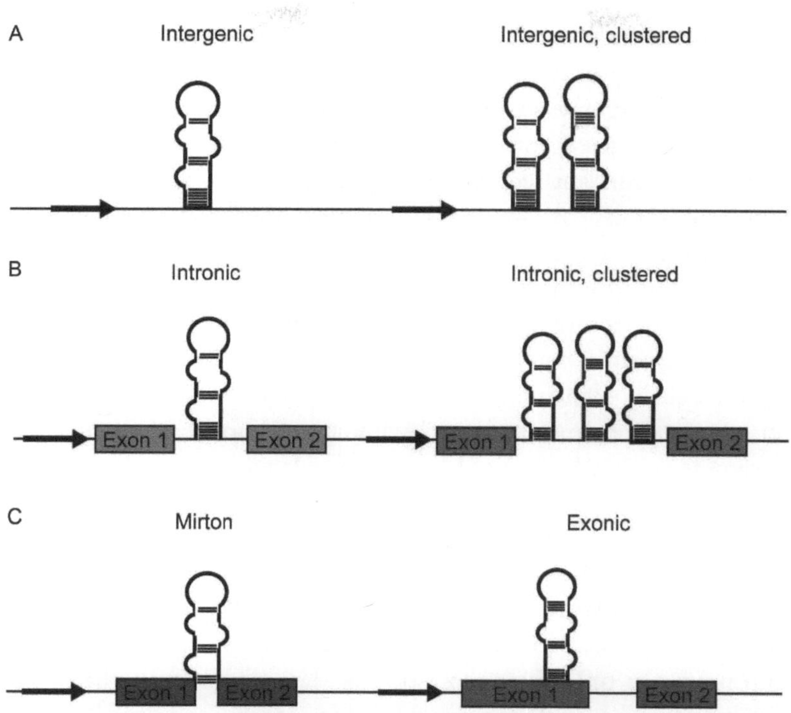

Figure 2 Genomic location of miRNAs. (A) Individual miRNAs or polycistronic miRNAs encoded in intergenic region are regulated by an miRNA-specific promoter. They are located in genomic regions distinct from where transcription for protein-coding genes occurs. (B) Individual or clustered miRNAs can be located between exon boundaries in intronic regions of DNA encoding for proteins or long noncoding RNAs. Typically, they are transcribed from the same promoter as their host genes. (C) miRNAs encoded in a short intron of a gene (mirtrons) or within a gene (exonic) are rare and often bypass nuclear processing.

4. miRNA NOMENCLATURE

Established in 2002, miRBase (formerly known as MicroRNA Registry[18]) is the online searchable database for all published miRNA sequences and associated annotations. According to miRBase (http://www.mirbase.org/), the numbering of miRNA genes is sequential, reflecting the order of their discovery.[19] miRNAs are named with the prefix "mir" followed by a dash and a number. For example, in hsa-miR-25, "hsa" represents the species ("hsa" is for *Homo sapiens*, "mmu" for *Mus musculus*, and "rno" for *Rattus norvegicus*). This is followed by "miR" (with capital "R") for mature miRNA or "mir" (with small "r") for pri-miRNA, pre-miRNA, or genomic locus. Two mature miRNAs are generated from either the 3′ or 5′ arm of the pre-miRNA hairpin sequence. Based on their relative abundance, one was designated with an asterisk following the name to indicate lower expression. Thus, the abundant form is referred to as miR, while the less abundant sequence produced from the opposite arm of the hairpin is called miR* (rni-miR-21* is a less abundant form compared with rno-miR-21). This convention was recently replaced by miRBase with rno-miR-25-5p (from the 5′ arm) and rno-miR-25-3p (from the 3′ arm). In addition to reflecting the origin from both arms of the predicted hairpin, this change accounts for the fact that both 5p and 3p versions may be expressed equally depending on the cell types.[20] miRNAs with nearly identical sequences are annotated with an additional lower case letter. For example, miR-181a would be closely related to miR-181b. Though transcribed from different loci or different regions in the genome, some pre-miRNAs can produce identical mature miRNAs. These are indicated with an additional dash-number suffix. Examples mentioned by miRBase are the pre-miRNAs hsa-mir-194-1 and hsa-mir-194-2.

Release 20 (June 2013) of miRBase contains 24,521 entries representing hairpin precursor miRNAs, expressing 30,424 mature miRNA products, in 206 species. The number reported for *H. sapiens* includes 1872 precursors and 2578 mature miRNAs. In addition to this comprehensive "encyclopedia" on miRNAs, other databases have been developed to consolidate information on verified mRNA targets of individual miRNAs (miRTarBase) and relationships between noncoding RNAs and disease states, including the mammalian noncoding RNA-disease database (MNDR) (www.rna-society.org/mndr/), the Human miRNA Disease Database (HMDD) (http://202.38.126.151/hmdd/mirna/md/), and miR2Disease (http://www.mir2disease.org/).[21–23]

5. miRNA TARGET DETERMINATION

The functional consequence of miRNA binding is defined by the genes targeted and the effect a miRNA has on regulating translation of the target mRNAs, thus influencing gene expression. A given mRNA can harbor either one or more miRNA-target binding sites in its 3′UTR. miRNAs regulate gene expression by inhibition of translation initiation, translation elongation, or by targeting mRNAs for degradation.[24,25] miRNAs predominantly bind to the 3′UTR of target mRNAs, often resulting in imperfect base heteroduplexes between the seed sequence and the mRNA sequence. However, the influence of the UTR sequence in providing the "sequence context" for miRNA–mRNA interactions has been observed. Reporter assays have demonstrated that an identical site can mediate repression in some UTRs but not in others.[26,27] Thus, seed matches are not always sufficient for repression, indicating that additional features of the UTR context are crucial to miRNA regulation and in determining the specificity of targeting. For example, AU-rich nucleotide composition near the site can be associated with weaker mRNA secondary structure in the vicinity of the site and thus increased accessibility to the seed sequence. Target locations are not evenly distributed throughout the whole UTR and there is a strong preference for targets to be located in close vicinity of the stop codon and the polyadenylation sites. This includes positioning within the 3′UTR at least 15 nt from the stop codon. mRNA enters the ribosome ~15 nt downstream of the decoding site and presumably, this will result in the removal of any silencing complex. This will block rebinding until the ribosome dissociated from mRNA. This interference by the ribosome could also explain why the 5′UTR and open reading frame harbor very few miRNA-target binding sites. Silencing complexes bound to these regions would be displaced by the translation machinery as it moves along from the cap-binding complex through the open reading frame. Genome-wide analyses indicate that more binding sites were selectively maintained near the ends than in the central region of the 3′UTR, especially for mRNAs with long (>1300 nt) 3′UTR.[1,28,29] Increased site accessibility or proximity to translation machinery could explain the increased efficacy of sites falling near the ends of long UTRs. One explanation that has been put forth for the presence of larger number of miRNA-binding sites near either ends of the 3′UTR and their increased efficacy is that, when mRNAs form the circularized structures with the poly(A) tail interacting with the 5′ cap, sites located in the middle of long 3′UTRs would be furthest from the translational machinery.[1]

6. miRNA DETECTION, QUANTIFICATION, AND FUNCTIONAL STUDIES

Several approaches are available to profile and quantify miRNA expression in tissues, cells, and bodily fluids. Because the expression of the pri- and pre-miRNA transcript does not correspond linearly to the expression level of mature miRNAs, only quantification of mature miRNAs will accurately indicate their presence and regulation.[30] However, if the goal is to identify miRNAs transcribed, different methods can be employed. Methods for genome-wide analysis of miRNA expression include microarrays, RNA sequencing, and quantitative PCR in array formats.[13] All three methods enable analysis of large numbers of miRNAs in parallel and can be used either for detection or for investigation of differential expression in diseased states. Data obtained from microarray require further confirmation and the study is limited to known miRNAs represented in the chip. Array-based qPCR technologies such as the Taqman low-density array (TLDA) and open array cards allow for profiling of ~750 miRNAs. Although the profiling is limited to a definite number of known miRNAs, use of TLDA can eliminate the qPCR confirmatory step. There is the additional option of custom designing TLDA and open array cards; hence these platforms are more flexible. With the advent of next-generation sequencing, total RNA or small RNA sequencing offers a complete overview of all RNAs including pri-miRNA and pre-miRNA. This method offers the advantage of potential identification of novel miRNAs and enables determination of various stages of biosynthesis for miRNAs of interest.

Once the miRNAs of interest have been identified, online databases listing miRNA sequences and annotations[20] for bioinformatically predicted[5] as well as experimentally validated[31] miRNA targets can be used to explore the functionality of miRNAs of interest. Computational methods of inferring miRNA functions have been invaluable in providing insights into miRNA function.[22] Integrating heterogeneous data sources and pathway analysis of individual miRNAs or disease-specific miRNA signatures has aided in inferring miRNA regulatory modules, thus advancing miRNA functional annotation.

Reporter assays are commonly used to confirm miRNA binding to its target mRNA. One approach to perform this assay is to clone the 3′UTR of interest harboring one or more of the miRNA-binding sites downstream of the luciferase reporter gene. A decrease in luciferase activity upon cotransfection of miRNA with the luciferase-3′UTR reporter plasmid indicates binding of the miRNA to the UTR which regulates the expression of luciferase gene. Mutating the miRNA-binding site in the UTR or utilizing an miRNA inhibitor to block miRNA function can be employed to further confirm the miRNA interaction with its target mRNA. Transfection of miRNAs into cells endogenously expressing target genes followed by qPCR and Western blot analysis will confirm whether the miRNA is mediating its effect via translational repression or mRNA degradation. Because individual mRNAs can harbor binding sites for different miRNAs, the resulting regulatory network is complex and can be functionally redundant. Pull-down assays for proteins involved in miRNA processing such as Ago have been used to identify target mRNA:miRNA interactions. Biochemical techniques such as high-throughput sequencing of RNA isolated by crosslinking immunoprecipitation[32] and photoactivatable-ribonucleoside-enhanced crosslinking and immunoprecipitation,[33] are commonly used to identify RNA sequences that bind to Ago proteins. Additional genome-wide approaches pursued for target identification include transcriptional profiling of specific miRNA-knockdown or miRNA-overexpressing cells and control cells. Though these studies are valuable in providing a global view of target gene expression changes, it is difficult to determine whether the resulting changes are mediated directly or indirectly by the miRNA of interest.

Both *in vitro* and *in vivo* inhibition or overexpression of miRNAs can be achieved using a variety of synthetic mimics, anti-miRNA oligonucleotide (AMO), expression constructs, and viral vectors. AMOs, originally called antagomirs, are synthetic DNA/LNA (locked nucleic acid) hybrids with complementarity to at least 15 nt of the endogenous miRNA[34] (see Text Box 1). LNAs, which incorporate stability by the addition of a 2′O methyl or 2′O to 4′C bond to a synthetic RNA, are also used to design miRNA mimics. A miRNA sponge is a plasmid-based technology encoding multiple miRNA-binding sites that are available to sequester the miRNA from the endogenous mRNA pool when the construct is expressed.[35] Either viral or nonviral gene transfer vectors can be used to deliver the miRNA sponge expression cassette into cells to inhibit miRNA function.[36] Genetic gain- and loss-of-function approaches to specific miRNAs or miRNA clusters have been successful. Overexpression, however, can lead to nonphysiological levels of miRNA and potentially to biologically irrelevant regulation of mRNAs. Genetic deletion of miRNAs present in introns or as a part of polycistronic transcript should be designed so that the mRNA of the gene-harboring intronic miRNAs or other miRNAs in the polycistronic cluster is not affected. The major limiting factor in genetic deletion studies is the lack of phenotypic effect. The redundancy in miRNA function coupled with compensation could hamper the elucidation of functional consequences of miRNA that was knocked out. miRNA-deficient embryonic stem cells[37] and conditional knockout mice[38] are freely available and can serve as a valuable resource in advancing our understanding of miRNA functions *in vivo*.

BOX 1

miRNA-based tools

Viral vectors: a viral-based approach to constitutively express a miRNA by transducing cells with viral particles encoding the miRNA. Several virus types are currently being used. These include adeno-associated virus (AAV), herpes simplex virus (HSV), or lentivirus (LV).

Anti-miRNA oligonucleotide (AMO): a synthetic oligonucleotide designed with high specificity and significant complementarity to its miRNA target to form a double-stranded duplex that blocks the miRNA function.

Antagomir: original AMO designed with 100% complementarity, containing RNA with 2′O methyl modification, phosphorothioate linkages, and 3′cholesterol to increase cellular uptake.

LNA (locked nucleic acid): a nucleic acid that contains at least one type of stabilizing element such as a 2′O methyl group added to the ribose, an extra bond between the 2′O and 4′C of the ribose ring, or phosphorothioate linkages between nts.

miRNA mimic: synthetic, double-stranded LNA-based oligonucleotide with a guide strand that is identical to the miRNA and a passenger strand containing stabilization and uptake elements such as cholesterol.

miRNA sponge: a vector-based approach using DNA-encoding multiple miRNA-binding sites downstream of a promoter that effectively reduces available miRNA when expressed in a cell.

7. CIRCULATING miRNAs

Recent identification of stable miRNAs in bodily fluids[39–41] generated a lot of interest in exploring their utility as novel noninvasive biomarkers. Cells employ multiple distinct processes to release miRNAs into circulation. miRNAs are found enclosed within membranous vesicles (exosomes, shedding vesicles, and apoptotic bodies), in association with high-density lipoprotein, and bound by RNA-binding proteins (Ago 2 or nucleophosmin 1) (Fig. 3). Thus, the term circulating miRNAs encompasses all miRNAs found in bodily fluids including those associated with RNA-binding proteins, cholesterols or encapsulated in exosomes.[42,43] Although it is still unclear whether all circulating miRNAs are Ago-bound and/or associated with exosomes, miRNAs found in plasma were reportedly resistant to endogenous RNases.[39,44] Extracellular vesicles are broadly classified

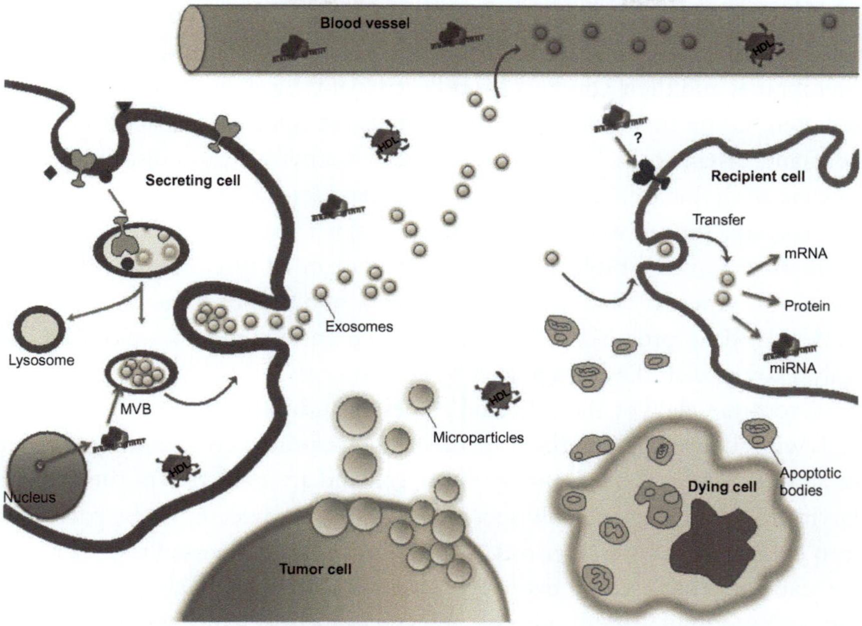

Figure 3 Circulating miRNAs and exosome biogenesis. Circulating miRNAs are secreted by cells in association with RNA-binding proteins such as Ago, high-density lipoproteins, or enclosed in microvesicles. Microvesicles include lipid-bound vesicles that are released during apoptosis (pink), shedding vesicles released by direct membrane budding (orange), or exosomes (blue) released from the multivesicular bodies (MVBs). Exosomes, the smallest of the microvesicles (30–100 nm), form within MVBs originating from the early endosomal pathway. Exosomes contain miRNAs, mRNAs, proteins, and lipids, the composition of which is reflective of the secreting cell. However, not all miRNAs present within the secreting cells are packaged into exosomes, and the mechanisms determining what is included in the exosomes for secretion is not well understood. After secretion, exosomes enter the blood stream and are transferred into recipient cells, releasing functional contents. Mechanisms similar to receptor-ligand mediated recognition have been postulated to guide exosomal uptake by recipient cells. It has been shown that extracellular Ago-bound miRNAs influence cellular function through receptor binding and hypothesized that HDL-associated miRNAs are secreted and taken up in a receptor-dependent mechanism. (See the color plate.)

into shedding microvesicles, exosomes, and apoptotic bodies based on their distinct mode of biogenesis. Exosomes are secreted microvesicles (30–100 nm in diameter) that form from the early endosomal pathway and their membranes originate from invaginations of the cell membrane (Fig. 3). Intraluminal budding results in multivesicular bodies, and the fusion of these multivesicular bodies with the plasma membrane results in the release of the exosomes into the extracellular space.[45,46] Thus, exosomal cell membranes and their contents are reflective of the cell from which they are secreted. Upon their release from cells, these vesicles can impact both adjacent and distant cells.[47,48] Exosomes carry numerous proteins, lipids, and nucleic acids, including mRNA, miRNA, and long noncoding RNAs.[49,50] Vesiclepedia (http://www.microvesicles.org) is a manually curated compendium that contains molecular data identified in all classes of extracellular vesicles[51] whereas ExoCarta (http://www.exocarta.org), a database for molecular data (proteins, RNA, and lipids) identified in exosomes, catalogs only exosomal studies as reported by the authors.[52]

Exosomes deliver these biomolecules including miRNAs to recipient cells with functional targeting capabilities.[49] Exosomes contain only a subset of all miRNAs present in the secreting cell and are reflective of the cellular environment and state. How specific miRNAs are selected to be packaged into exosomes and whether mRNAs and miRNAs are sorted together into the exosomes are areas of extensive investigation, but it is clear that RNAs are not randomly loaded into exosomes. Specific mechanisms exist for their recognition, packaging, and secretion.[53] Specific proteins such as hnRNPA2B1, along with *cis*-acting elements in the RNA sequence, control the sorting of RNAs into exosomes.[53] Additionally, exosome release is inhibited by small molecules that inactivate neutral sphingomyelinase 2, nSMase2, an essential enzyme of the ceramide pathway.[54]

Defining the source and destination of secreted miRNAs has been challenging. Their presence in circulation indicates that they can be transported long distance, and the communication mediated by exosomes will be defined by the contents. Exosomes can deliver their cargo by fusing with the cell membrane, thereby directly releasing the contents into the cytoplasm of the recipient cells. Classical adhesion molecules involved in cell–cell interactions such as integrins and intercellular adhesion molecules, along with exosomal-membrane-rich molecules such as TIM-binding phosphatidylserines, carbohydrate/lectin receptors, and heparan sulfateproteoglycans, have been implicated in mediating the binding of exosomes to the surface of recipient cells. In addition to direct fusion, exosomes can be internalized by endocytic pathways employing clathrin–dynamin–caveolae-dependent endocytosis, pinocytosis, or phagocytosis.[53]

Though secreted miRNAs can be transported between distant locations via circulation, how an miRNA taken up by a cell might achieve a sufficient concentration to inhibit its targets is not known. It is unclear whether the differentially expressed miRNAs participate in the disease process or simply serve as markers of disease progression. Though circulating miRNAs are ubiquitous in all body fluids, they exhibit fluid-specific expression profile signatures that can be used to distinguish individuals with disease from healthy controls.[39,55] Thus, there is increased interest in exploring the utility of the biomolecular signature of exosomes, including miRNA profiles, in tumor biology; cardiovascular, immunological, and neurological disorders; and stem cell biology, both as clinical biomarkers and therapeutic tools.[56,57]

A recent review focusing on circulating miRNAs as potential biomarkers of human cancers highlighted approximately 100 miRNAs that may allow earlier diagnosis and represent potential therapeutic avenues.[56] Additionally, miRNA signatures in cerebrospinal fluid might be useful for diagnosing brain diseases such as Alzheimer's disease.[58,59] A study profiling circulating miRNAs after nerve transplantation indicated that three miRNAs were specifically upregulated in the whole blood and in the serum of immunosuppressed animals that received an allograft. Nerve transplants are commonly supplemented with immunosuppressive drugs to enhance reinnervation through the allograft. These miRNAs are not expressed in the nerve graft but are found circulating in the blood, which makes them potential biomarkers for monitoring the immunosuppression status of an allograft.[60] Circulating miRNAs have been utilized as a measure of the quality of stored blood, as an addition to noninvasive prenatal diagnostics, and as a means of monitoring treatment efficacy after myocardial infarction or infection.[56] Thus there is immense potential in unraveling the role of circulating miRNAs in both normal physiology and pathogenic conditions.

8. miRNAs AND PAIN

The critical role played by miRNAs in the development and pathophysiology of the nervous system is now well established.[61] In recent years, interest has increased in deciphering the role of miRNAs in modulating pain and analgesia, in defining the importance of miRNAs in pain circuitries, and in the cognitive, emotional, and behavioral components involved in pain.[62] Animal models have been developed to mimic persistent pain states, facilitating translational research on various types of pain.[63,64] Although animal models of pain have limitations, it is clear that much can be learned about the mechanisms and maintenance of chronic pain from existing models. In conjunction with human studies utilizing circulating biomolecules that are altered in bodily fluids, these models have been instrumental in elucidating the role of miRNAs in pain states.

Specific regions of the nervous system express different levels of miRNAs that are known to regulate neuronal activity and modulate translation in dendrites.[65] A handful of miRNAs with known function in neuronal activity are also dysregulated in pain, although the mechanisms mediated by interactions between neuronal miRNAs and identified pain targets are not completely elucidated.[65] Distinct expression patterns of miRNAs in mammalian development as well as tissue-specific expression of miRNAs in disease states suggest that miRNAs play a role in differentiation and maintenance of cell types. In a study investigating the role of miRNAs in peripheral pain pathways, Dicer was conditionally knocked out in neurons that express Nav1.8 using the Cre-loxP system.[66] Dicer is the cytoplasmic ribonuclease that is a component of RISC and generates small ds miRNAs from hairpin loop precursors. Nav1.8 is a sensory neuron-specific voltage-gated sodium channel that is known to play a critical role in inflammatory pain, cold and noxious mechanosensation. Unlike in postmitotic Purkinje cells where Dicer is necessary for survival,[67] dorsal root ganglia (DRG) neurons maintained normal function in the absence of this enzyme. Dicer deletion resulted in deficits in inflammatory but not acute pain. Inflammatory agents including carrageenan, formalin, or complete Freund's adjuvant (CFA) did not alter the pain thresholds in these Dicer-null mice. At the cellular level, inflammatory mediators failed to enhance the excitability of Nav1.8 sensory neurons from null mice. They also observed that nociceptor-specific genes were transcriptionally downregulated, while other genes were either expressed at normal levels or upregulated. Downregulation of nociceptor-associated transcripts including Nav1.8, P2xr3, and CamKII could have reduced functional protein, thus altering the thresholds of activation of sensory neurons. Microarray and deep sequencing revealed that ablation of Dicer in Nav1.8 neurons resulted in a significant reduction or loss of more than 60 miRNAs. Since deletion of Dicer can affect all miRNAs in the Nav1.8 neurons, it was difficult to identify the miRNAs that specifically contributed to pain. However, this study showed that miRNAs are crucial for altering pain thresholds after inflammatory stimuli.[66] Another study using genetic ablation of Dicer showed impaired sciatic nerve regeneration and delayed functional recovery after sciatic nerve crush.[68]

Measuring changes in miRNA levels in rodent models of pain is another commonly used approach pursued for delineating the molecular mechanisms that underlie pain states. Recent reviews have cataloged the studies linking miRNAs to various pain conditions.[62,65,69–72] Profiling studies have identified tissue-specific deregulated expression of miRNAs in DRG and spinal cord after peripheral inflammatory stimulus,[73,74] nerve injury,[74–80] and cancer pain.[81] Additionally, differential regulation of individual miRNAs has been observed in specific regions of the brain in acute and chronic pain models.[82–84] The few studies that link specific miRNA changes to a pain condition suggest that temporal and spatial regulation of intracellular miRNA expression occurs at specific time points that correspond to early or late phases of pain conditions.[69,71,70] miRNAs altered in animal models could serve as an efficient strategy to elucidate the pathophysiology of pain, and miRNA changes specific to certain models or pain conditions could provide insights unique to the model being studied.

The natural follow-up from profiling studies is to measure the target mRNA levels in affected tissue after inducing pain. Our miRNA expression profiling study of DRG in spinal nerve ligation (SNL) model of neuropathic pain in rats 4 weeks after surgery identified significant alterations in expression of 63 miRNAs. We used the same RNA for global transcriptional profiling and observed a general upregulation of mRNAs; however, a strict inverse correlation between the differentially regulated miRNAs that were mostly downregulated, and their predicted mRNA targets was not observed. It is not surprising that a strict inverse correlation between the differentially regulated miRNAs and their predicted mRNA targets was lacking and a number of factors could have contributed to this. A large-scale proteomics study performed to investigate the influence of specific miRNAs on protein levels revealed that a number of targets were predominantly regulated by translational repression.[85] Thus, miRNAs could be mediating regulatory effect on mRNAs by suppression of translation and mRNA levels would remain unchanged.

A number of transcription factors were perturbed by SNL in DRG and could exert their influence on gene expression following SNL. In addition, several transcription factors are known to be regulated by miRNAs.[86,87] Though the classic switch interactions where miRNA induction results in the downregulation of preexisting mRNA targets is well known, other types of interactions such as fine-tuning and neutral mode of action are also possible.[1] In a "tuning" target–miRNA interaction, the target gene remains functional because the expression is not eliminated by miRNA.[88] Thus, in its fine-tuning role, a miRNA could adjust protein output that allows for customized expression in a particular cell type, in response to a signaling cue. Neutral miRNA–target mRNA interactions are the one with no particular consequence to the cell because they are neither advantageous nor do they cause any adverse outcome.[1,88] The complexity of the miRNA regulatory networks, with many miRNAs affecting many target transcripts makes it challenging to extract individual miRNA/target transcript interactions. Prediction of miRNA-target sites is an evolving field, and any errors in prediction of miRNA-target sites would also undermine our ability to observe a correlation between miRNA and mRNA expression. Detailed mechanistic studies of individual genes and the miRNAs targeting them are required to classify the mode of miRNA regulation in pain.

Deep sequencing was used in another study to investigate alterations of miRNA expression following rat sciatic nerve injury in DRG and the proximal stumps of the nerve. They identified 201 and 225 known miRNAs with significant expression variance at five time points after injury, in DRG and nerve stump, respectively.[77] Sequencing of miRNAs in primary sensory neurons of L4 and L5 DRG (from two strains of rats that differed in their response to peripheral nerve injury) 3 days after SNL showed only three abundant (rno-miR-30d-5p, rno-miR-125b-5p) and one moderately expressed (rno-miR-379-5p) miRNA to be differentially regulated,[80] suggesting temporal regulation of miRNAs postinjury.

Several studies have now successfully established negative correlation between miRNAs and target mRNA levels.[89–93] Some have successfully provided the proof-of concept for a role of specific miRNAs in influencing behavior using rodent models[79,94–96] (Fig. 4). miRNA-induced attenuation or exacerbation of pain phenotype in rodent models of inflammatory, neuropathic, and cancer pain[79,81,94–97] has been demonstrated. miR-103 was shown to simultaneously regulate the three subunits forming the L-type calcium channel (Cav1.2 LTC). Knockdown of miR-103 in naive rats induced hypersensitivity to pain. All three Cav1.2 LTC subunit mRNAs were overexpressed with a concomitant decrease in miR-103 in the dorsal horn after SNL injury. Intrathecal applications of miR-103 repressed Cav1.2 LTC upregulation and attenuated mechanical hypersensitivity and cold allodynia in SNL rats.[94]

The let-7 family of miRNAs has been implicated in pain processing, specifically in opioid tolerance owing to its upregulation after morphine treatment and consequential downregulation of mu-opioid receptors.[98,99] A recent study demonstrated that miR-7a dysfunction underlies maintenance of neuropathic pain through regulation of DRG neuronal excitability.[79] Decrease of miR-7a in injured DRG was associated with neuropathic pain. Adeno-associated viral vector was used for local induction of miR-7a

miRNA	Model	miRNA	mRNA	Therapeutic agent	Injection site	Effect	Reference
miR-124	Formalin	Decrease (SC)	MeCP2 increases	miR-124 mimic	Tail vein	Decreased nociception, downregulated MeCP2	Kynast *et al.*[69]
				miR-124 AMO	Tail vein	Increased nociception, upregulation of MeCP2	
	Carrageenan	N.D.		miR-124 mimic	Intrathecal	Attenuated hyperalgesia	Willeman *et al.*[95]
	SNI	Decrease (SC)			Intrathecal	Prevented development of mechanical allodynia	
miR-183	SNL	Decrease (DRG)	Nav1.3 increases BDNF increases	LV-miR-183	Intrathecal	Downregulated Nav1.3 and BDNF Attenuation of SNL-induced mechanical allodynia	Lin *et al.* (2014)[145]
miR-96	CCI	Decrease (DRG)	Nav1.3 increase	miR-96 mimic	Intrathecal	Suppressed CCI-induced NAv1.3 expression Attenuated mechanical and thermal hyperalgesia	Chen *et al.*[89]
miR-21	SNL	Increase (DRG)		miR-21 AMO	Intrathecal	Partially attenuated mechanical and thermal hyperalgesia	Sakai *et al.*[109]
miR-103	SNL	Decrease (SC)	Cav1.2 increases	miR-103 mimic	Intrathecal	Reduced cold and mechanical allodynia	Favereaux *et al.* (2011)[94]
				miR-103 AMO	Intrathecal	Induced mechanical hypersensitivity	
miR-7a	SNL	Decrease (DRG)	Scn2b increases	AAV-anti-miR-7a	DRG	Alleviated late phase mechanical and thermal hyperalgesia and decreased levels of Scn2b	Sakai *et al.*[79]
	Naive	N.D.		AAV-anti-miR-7a		Causes allodynia and hyperalgesia	
miR-844	Carrageenan	N.D.	ASIC3 increases	HSV-miR-844	Gastrocnemius muscle	Reduced muscle and paw mechanical hyperalgesia	Walder *et al.*[113]
miR-1a-3p	Bone metastatic cancer pain	Increase (DRG)	Clcn3 decreases	miR-1a-3p AMO	Intrathecal	Reduced mechanical hypersensitivity	Bali *et al.*[81]
miR-34c-5p				miR-34c-5p AMO		Prevented mechanical hypersensitivity	
miR-483-3p		Decrease (DRG)		miR-483-3p mimic		Lesser mechanical hypersensitivity	

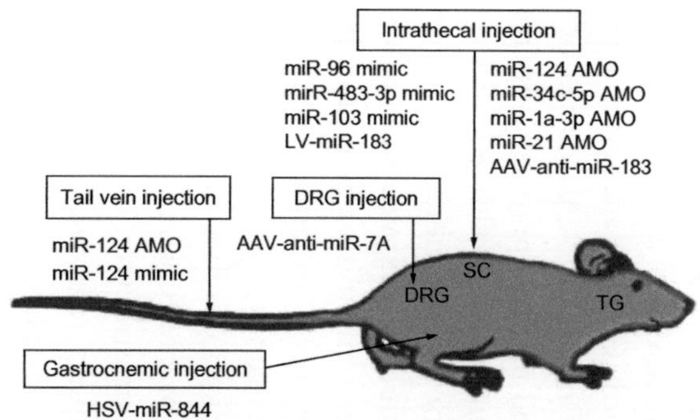

Figure 4 miRNA administration in rodent models of pain. Specific miRNAs that are altered in animal model of neuropathic, inflammatory, and cancer pain have been identified as potential therapeutic targets and are summarized in the table. Schematic indicates therapeutic delivery strategies that have been successful at reversing the pain phenotype including viral vectors (adeno-associated virus (AAV), herpes simplex virus (HSV), or lentivirus (LV)), anti-miRNA oligonucleotides (AMO), and stabilized locked nucleic acid (LNA) mimics. *Figure modified from http://www.doyourownpestcontrol.com/rodent_cdc.jpg.*

in DRG and this alleviated neuropathic pain in SNL model rats. Naïve rats developed both mechanical allodynia and thermal hyperalgesia when miR-7 expression was blocked. miR-7a was ineffective in reversing inflammatory pain and these results imply a role for miR-7a in neuropathic pain. The beta2 subunit of voltage-gated sodium channels is crucial for the cell surface expression, and it is targeted by miR-7a. Downregulation of beta subunit protein by miR-7a suppressed hyperexcitability of nociceptive DRG neurons demonstrating that miR-7a downregulation is causally involved in maintenance of neuropathic pain through regulation of neuronal excitability.[79]

While earlier studies revealed the role of miR-21 in tumorigenesis,[100,101] it was upregulated in spinal cord in response to spinal cord injury.[102–104] Although astrocytes are not the only cells to express miR-21 in the injured spinal cord, GFAP+ astrocytes adjacent to the lesion area expressed high levels of miR-21.[105] Astrocytes becomes reactive following spinal cord injury and initial hypertrophy (increase in volume due to the enlargement of its component) followed by hyperplasia (increase in number) around the injury site contributing to progression of the glial scar. Upregulation of the inhibitory extracellular matrix components is known to contribute to glial scar, and it is a significant barrier for axonal regeneration. However, glial scar serves to repair the blood–brain barrier and limit infiltration of inflammatory cells and cellular degeneration.[106] The role of miR- 21 in astrogliosis was investigated using transgenic mice that overexpress either miR-21 or a miRNA sponge (see Text Box 1) designed to inhibit miR-21 function in astrocytes. Overexpression of miR-21 attenuated the beneficial hypertrophic response to injury and miR-21 sponge augmented the hypertrophic phenotype.[105] Inhibition of miR-21 decreased scar formation without interfering with blood–brain barrier repair. These results suggest miR-21 can be a potential target for manipulating gliosis and enhancing functional outcome after spinal cord injury.[61] Peripherally, Sakai and Suzuki[107] characterized miR-21 in DRG neurons after SNL injury. They found increased expression of miR-21 in the injured DRG and were able to reverse the pain phenotype associated with this injury by administering an inhibitor to decrease miR-21.

Inflammatory pain models have specific miRNA signatures and responses to miRNA therapy. For example, miR-124 is enriched in central nervous system (CNS) tissues and has low expression in microglia and peripheral macrophages.[108–110] In a rodent model of inflammatory pain induced by formalin, animals had measurably less endogenous miR-124 in spinal cord and delivery of miR-124 mimics (see Text Box 1) attenuated hyperalgesia.[96] In another study, delivery of miR-124 mimics reversed persistent hyperalgesia induced by carrageenan and prevented development of mechanical allodynia in SNI model.[95] Delivery of miR-124 analogs in multiple inflammatory pain models consistently reduced thermal hyperalgesia suggesting that increasing miR-124 can be a therapeutic strategy for the treatment of pain.

A miRNA microarray profiling study identified downregulation of miR-219 in spinal cord from CFA model. miR-219 negatively regulates NMDA receptors and calcium/calmodulin-dependent protein kinase II γ (CaMKII γ), two target genes that are known to be important in central sensitization.[111] CFA-induced inflammatory pain reduced miR-219 expression and increased the expression of CaMKII γ. Overexpression of miR-219 prevented and reversed CFA-induced thermal hyperalgesia, mechanical allodynia and spinal neuronal sensitization. Knockdown of miR-219 by miR-219-sponge produced thermal hyperalgesia and mechanical allodynia and increased Fos protein expression in naive mice suggesting that miR-219 contributes to the modulation of chronic pain and the increase in CaMKIIγ expression mediates miR-219 downregulation-induced pain behavior. This study is also novel for highlighting how two different epigenetic mechanisms, DNA methylation and noncoding RNAs can collectively contribute to pain state. Methylation of cytosine bases found in cytosine-guanine dinucleotides are predominantly present in enriched regions of the DNA referred to as CpG islands. DNA methylation is an important epigenetic modification and this study demonstrated that chronic inflammatory pain can induce hypermethylation of CpG islands in the miR-219 promoter. Methylation of cytosine is usually associated with stable transcriptional repression when present in promoters. A demethylation agent increased miR-219 and decreased CaMKII expression, resulting in alleviation of pain in CFA model.[97] An earlier report had linked DNA methylation to miRNA modulation in neuropathic pain. A drastic decrease in the expression of miR-200b and miR-429 along with increase in DNA methyltransferase 3a (Dnmt3a) was observed in nucleus accumbens neurons, 7 days after SNL. Dnmt3a, commonly referred to as *de novo* methyltransferases, is one of the two enzymes responsible for establishing methylation patterns rather than maintaining the existing methylation patterns. Future studies no doubt will unravel additional convergence of multiple epigenetic mechanisms in mediating pain.

In a first genome-wide miRNA profiling study investigating the development and maintenance of tumor-mediated chronic pain using a mouse model of metastatic bone cancer pain, dysregulation of 57 miRNAs in sensory neurons corresponding to areas affected by tumor was observed in the DRG. Six candidate miRNAs (three upregulated and three downregulated) were chosen for *in vivo* functional analysis, and intrathecal delivery of miRNA inhibitors or mimics into lumbar DRGs were undertaken to demonstrate the efficacy of miRNA manipulation *in vivo*. Intrathecal application of inhibitors showed that inhibiting the tumor-induced upregulation of miR-1a-3p or miR-34c-5p, but not of miRNA-544- 3p, in sensory neurons, markedly attenuated tumor-mediated hyperalgesia. Reversing the decrease of miR-483-3p, but not of miR-291b-5p, with miRNA mimics attenuated tumor-mediated hyperalgesia but increase in miR-370-3p in DRGs led to exaggerated tumor mediated hyperalgesia. They further investigated the impact of miRNA-1a-3p inhibition on the expression of target genes using a nanostring device which digitally quantifies a wide set of mRNA transcripts, and confirmed that chloride channel 3 (Clcn3) was a target for miRNA-1a-3p. Though there are several ways by which Clcn3-mediated Cl^- flux could affect sensory neuron function, the exact mechanism underlying how decreased Clcn3 levels cause hyperalgesia is unknown. In addition to confirming the inverse correlation in expression and miRNA binding to the 3′UTR, siRNAs (Table 1) designed against Clcn3 mRNA were used to demonstrate that a knockdown of Clcn3 expression in DRG was associated with exaggerated mechanical hypersensitivity. Thus, downregulating expression of Clcn3 in the DRG had the same functional effect as inhibition of miR-1a-3p, again indicating the *in vivo* significance of miR-1a-3p mediated regulation of Clnc3 as a key modulator of nociceptive hypersensitivity associated with cancer pain.[81]

Utility of miRNAs in reversing hyperalgesia associated with diabetic neuropathy and muscle inflammation has been explored using viral vectors expressing miRNA mimics. Increase in voltage-gated sodium channel isoforms Nav1.7 and Nav1.3 protein in the DRG of rats with streptozotocin-induced diabetes was reversed using a nonreplicating herpes simplex virus-based vector expressing a miRNA against Nav alpha subunits.[112] Artificial miRNAs to downregulate acid-sensing ion channel 3 (ASIC3) expression was delivered into the muscle using a herpes simplex viral vector and this resulted in reduction of both muscle and paw mechanical hyperalgesia in mouse model of carrageenan-induced muscle inflammation.[113]

While a link between miRNA therapeutics and pain has been established in animal models, miRNA alterations in human pain conditions are being

Table 1 Comparison of miRNA and siRNA

miRNA	**siRNA**
~21-bp RNA	Similar length
Derived from longer ~70-bp imperfectly base-paired hairpin	Derived from longer, perfectly complementary double-stranded RNA
Precursor of endogenous origin	Precursors of endogenous or exogenous origin
Nuclear processing by Drosha and secondary processing in the cytoplasm by Dicer	Processed in the cytoplasm by Dicer
Translational repression or mRNA degradation	Degradation of mRNA by RNAi
Heterosilencing; i.e., target remote loci	Autosilencing; i.e., target the same locus
miRNA can regulate several target RNAs	Gene (family) specific silencing
Highly conserved among species	Rarely conserved
Only common feature of the targets are the short sequences complementary to 6–7 bases seed sequence of the miRNA	High complementarity to target mRNA (often 100%)
Clinical use	
Drug targets or drug agents themselves	Valuable laboratory tool
Expression levels (potential diagnostic or biomarker tools)	Several siRNAs in clinical trials as therapeutic agents

Small interfering RNAs (siRNAs) repress target mRNAs by binding with 100% complementarity to target mRNAs, forming a double-stranded (dsRNA) motif that is degraded by the RNA-induced silencing complex (RISC). An association of noncoding RNA, mRNA, and proteins including Ago2, make up the RISC and guide the process of RNA interference (RNAi).[114] Both miRNAs and siRNAs are bound by Ago2, a small RNA scaffold that has endonucleolytic activity, but the smaller region of complementarity between miRNAs and their target mRNA ensures that not all miRNA:mRNA interactions leads to mRNA degradation.[115] These two types of noncoding RNAs are similar in size and function, but have important distinctions in the mechanisms employed to regulate endogenous mRNA translation.

explored for their utility as biomarkers as well as therapeutics. Profiling studies in cancer have yielded many specific alterations in miRNAs that are being developed as biomarkers and explored as potential therapeutics, but translating these types of studies to neurological conditions proves more difficult because of the difficulty in obtaining healthy and diseased tissue from each individual.[116] Screening, diagnosis, and prognosis of various human diseases through miRNA profiling has capitalized on the stability of miRNAs in all bodily fluids and their aberrant expression often correlate with disease severity or progression.[117,118] Altered miRNA expression profiles have been reported for immune disorders such as rheumatoid arthritis[119] and systemic lupus erythematosus[120] as well as for pain conditions such as irritable bowel syndrome,[121] chronic bladder syndrome,[122] and endometriosis.[123] Cerebrospinal fluid from fibromyalgia patients showed differential expression of nine miRNAs and miR-145-5p correlated positively with pain ratings based on the fibromyalgia impact questionnaire[124] (Fig. 5).

Analysis of miRNA signature in whole blood from patients with complex regional pain syndrome (CRPS),[125] a severe and multifactorial

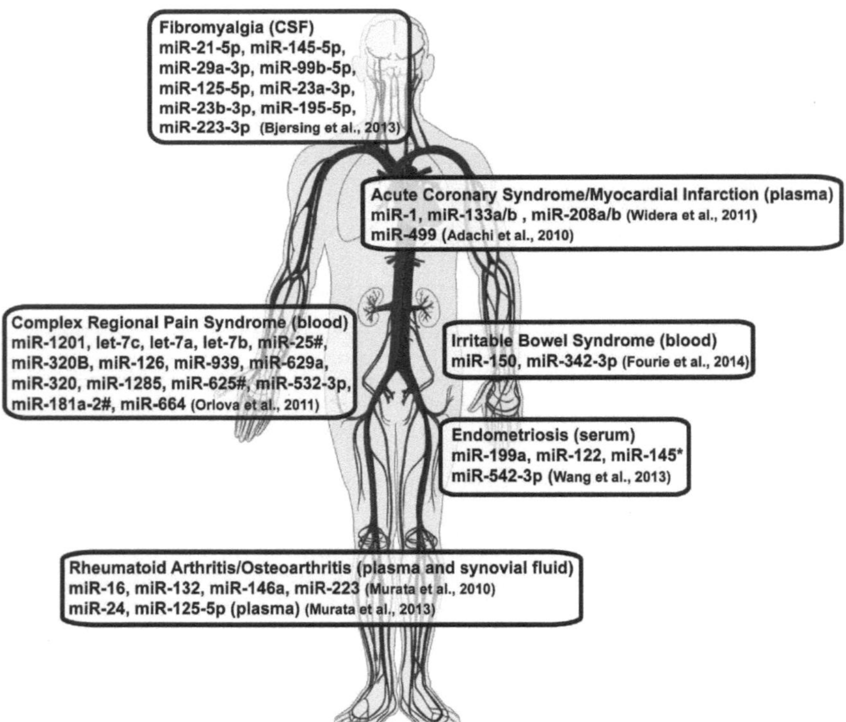

Figure 5 Circulating miRNAs altered in painful disorders. Schematic showing altered miRNAs in bodily fluids of patients with painful conditions.[124–129,121,130] Altered miRNAs are potential biomarkers for disease progression and diagnosis as well as therapeutic targets for disease intervention. Alterations in circulating miRNAs were useful for stratifying patients with CRPS, a multifactorial neuropathic condition with complex pathophysiology. *Figure modified from http://en.wikipedia.org/wiki/File:Circulatory_System_en.svg.*

neuropathic pain condition characterized by dysregulated immune responses, demonstrated the utility of circulating miRNAs in patient stratification. Clustering of patients based on miRNA signature revealed additional miRNAs and inflammatory markers that were not significant when the whole population were analyzed together. These stratifications and identification of subset specific biomarkers are especially beneficial for disorders such as CRPS where the patients present with a variety of clinical symptoms. This study was also the first to show that circulating miRNAs can serve as potential biomarkers in pain.[125]

Pathological pain is correlated with cognitive and mood disorders.[131] In our efforts to obtain additional insights into comorbid conditions commonly associated with CRPS, we performed correlation analyses of all miRNAs detected in whole blood with clinical parameters. Strong correlations were observed between miRNAs and comorbidities such as migraine, high blood pressure, cholesterol, thyroid disease, and use of narcotics and antiepileptic medications.[125] For example, hsa-miR-339-5p was correlated with narcotics use in CRPS patients. A recent study showed that miR-339 can downregulate mu-opioid receptor at the posttranscriptional level in response to opioid treatment.[132] Opioid analgesics fentanyl and morphine act mainly on mu-opioid receptors. Fentanyl is currently the most widely used synthetic opioid in clinical practice and it is up to 100 times more potent than morphine. miR-339-3p was upregulated in the hippocampus of mice chronically treated with morphine and fentanyl. miR-339-3p inhibited the production of mu-opioid receptor protein by destabilizing the mRNA.[132] This shows that miRNAs associated with clinical parameters may provide insights on mechanisms by linking it to observations in animal models and *vice versa*.

To investigate the functional consequences of circulating miRNAs in pain, we recently characterized exosomes derived from mouse RAW 264.7 macrophage cell line and demonstrated that macrophages primed with the inflammatory stimulus lipopolysaccharide (LPS) release exosomes containing a distinct miRNA signature. Next-generation sequencing of exosomal RNA showed the reads that mapped to miRNAs were more abundant after LPS stimulation. Exosomes from LPS-stimulated cells showed a dose-dependent NF-κB activation compared with exosomes from control cells, confirming that exosomes were functional in recipient cells. To determine the effect of exosomes in mediating pain hypersensitivity, exosomes were directly injected into CFA-treated paws. Although exosomes derived from both unstimulated and LPS-stimulated cells reduced

CFA-induced thermal hyperalgesia, those from stimulated cells also attenuated paw swelling and showed an analgesic response at earlier time points after delivery.[133] The absence of pain and swelling in saline-treated paws after exosome injection demonstrates that exosomal delivery does not induce a hypersensitivity response. This combined with the fact that exosomes can be loaded with nucleic acids or drugs suggests a potential role for exosomes in pain therapy. Exosomes may provide an avenue to regulate inflammatory response and reduce chronic inflammatory pain by modulating dysregulated signaling pathways. Exosomes derived from a patient and reintroduced after loading with the molecule of interest should be better tolerated by the immune system.[134–136] Further investigations of exosomal biology and other microvesicle-mediated transport will no doubt provide insights on how regulation of cellular processes can be mediated via circulation. Although little is known about the mechanisms of recognition and uptake of circulating miRNAs, it was recently demonstrated that hematopoietic cells, which include the cells of the immune system, release mRNA and miRNAs that are taken up by the neurons.[137] Using transgenic mice expressing Cre-recombinase under the hematopoietic-specific promoter in a Cre reporter background, Purkinje neurons expressed the reporter. Having ruled out cell fusion, they identified Cre-recombinase mRNA, but not the protein in the extracellular vesicles. Induction of peripheral inflammation increased the number of Cre-recombined neurons demonstrating that mRNA can be transferred from immune cells in blood to neurons in CNS via extracellular vesicles. Cre-recombined Purkinje neurons contained three miRNAs not found in nonrecombined neurons and the same three miRNAs were also present in exosomes isolated from blood of mice 24 h after induction of peritonitis.[137] Thus, exosomes provide yet another mode of regulation of the nervous system mediated by the immune system.

miRNAs have been reported to act as agonists of Toll-like receptors (TLR) and activate downstream signaling pathway in target cells.[138] Additionally, miRNAs have been shown to act extracellularly by directly interacting with cell-surface receptors; miR-21, miR-29a, and let-7b act as agonists of TLR7 and 8, influencing downstream signaling pathway in target cells.[138] Another recent study showed that secreted miRNAs may represent a new class of pain mediators; miRNA let-7b induced rapid inward currents and action potentials in DRG neurons coexpressing TLR7 and TRPA1 receptors. miRNA-induced neuronal activation required miRNAs with a GUUGUGU motif indicating that GU-rich core is critical. DRG neurons from mice lacking either TLR7 or TRPA1 did not respond to let-7b. In earlier studies describing actions of let-7b at TLR7, myeloid differentiation factor 88 (Myd88), an adapter protein associated with TLR7 was critical for the canonical signaling of most TLRs including TLR7. In *Myd88*-deficient DRG neurons, let-7b-induced inward currents were largely intact, although the amplitude of currents was slightly reduced compared to that of wild-type neurons. TLR7 and TRPA1 colocalized with let-7b, both on the cell surface and inside cells and let-7 activated TLR7 and TRPA1 in HEK293 cells. Expression of TRPA1 alone was not effective confirming that TLR7 is essential for mediating the let-7b-induced currents in HEK293 cells. Interestingly, let-7b was more potent than the synthetic TLR7 agonists loxoribine and imiquimod in inducing inward currents in DRG neurons. Single-channel recordings in both DRG neurons and HEK293 cells confirmed that let-7b could bind TLR7 on the extracellular surface to induce single-channel activities of TRPA1. Spontaneous pain was induced upon intraplantar injection of let-7b, and an inhibitor of let-7b reduced formalin-induced TRAPA1 currents and spontaneous pain.[139] A single intraplantar injection of let-7b produced mechanical allodynia that was dependent on TLR7, Myd88 and TrpA1 activation in a dose-dependent manner that was abolished in Tlr7 knockout mice and reduced in Trpa1 and Myd88 knockout mice. Pretreatment with an inhibitor of let-7b reduced formalin-induced inflammatory pain in wild-type animals. These studies demonstrated that extracellular let-7b is required and sufficient for inducing inflammatory pain via the activation of TLR7 and TRPA1. Additionally, let-7b is highly enriched in DRG tissues and endogenous let-7b can be released from DRG neurons in an activity-dependent manner.[139] These studies indicate an unconventional role for miRNAs that could potentially act in an autocrine fashion on nociceptors, or mediate novel signaling mechanisms.

Studies using rodent models of pain highlighting the effectiveness of synthetic miRNAs in reducing hyperalgesia have been beneficial in providing the proof-of-principle for miRNA-based therapy. Different animal models have also been used to determine tissue-specific expression changes under various acute and chronic pain conditions. The difficulty in obtaining human tissue samples of interest has led to the analysis of circulating miRNA in patients with chronic pain disorders. For miRNAs and their mRNA targets conserved across species, observations in patients can be validated in rodents. Though miRNA therapeutics for pain is still unrealized, other diseases have benefited and are forging ahead with clinical trials. Miraversen has been proven to be an effective therapeutic for chronic hepatitis C.[140] This is not surprising because liver is relatively easy to be targeted for therapy; when administered in the absence of a carrier, miRNA oligos are taken up by the liver and kidney and rapidly excreted in urine.[141] miRNAs found in plasma and blood samples have emerged as potential biomarkers for pain. Although pain is the common denominator in a variety of disorders, the underlying neurobiological mechanisms differ between pain conditions and circulating miRNAs may provide insight into unique miRNA signatures that could be exploited for diagnostic uses as well as therapeutics. Identifying several miRNAs as biomarkers, rather than relying on one specific molecule or parameter, may increase the chances of successful treatment in an extremely heterogeneous group of patients suffering from pain. Exploring the basis for the altered expression of such biomarkers, their relationship to disease progression, and utility in predicting disease progression and clinical outcome will undoubtedly be beneficial in unraveling the significance of miRNAs in the context of pain.

9. CHALLENGES ASSOCIATED WITH miRNA RESEARCH

Lack of overt phenotype in loss-of-function studies has been an impediment in the miRNA field. Though the initial identification of miRNAs in *C. elegans* was accomplished using classic genetic screens, the vast majority of deletion and inhibition studies lacked significant phenotypes. The targets of *lin-4* and *let-7* miRNAs were identified based on discrete developmental timing events, but identifying mRNA targets of miRNAs that are ubiquitously expressed in adult tissues is more challenging. The task of determining valid miRNA:mRNA interactions requires multiple bioinformatics prediction algorithms, individual validation of binding, and translational suppression in conjunction with forward genetic techniques to uncover new potential targets.[5] With miRNAs acting as a rheostat, the influence of miRNAs under a normal or homeostatic state is often described as modest. It has been proposed that under stress or disease states, the functions of miRNAs are more pronounced.[3]

The majority of the 25 miRNA knockout mice that have been examined thus far lack severe defects in embryonic development.[3] This is not surprising considering that miRNAs are fine tuners of gene regulation. With multiple target binding sites within a gene, and the presence of several miRNA family members with the same seed sequence, compensatory mechanisms can easily override any functional perturbations resulting from loss or inhibition of an individual miRNA. It has been proposed that investigating miRNA signaling under conditions of cellular stress or injury perturbing cellular homeostasis may be required for unraveling the role of miRNAs. Several potential mechanisms by which miRNAs mediate stress signaling have been proposed. An increase or decrease in expression of a miRNA can act as a critical intermediate in modulating signaling pathways. A miRNA can serve as a negative or positive feedback loop to amplify or dampen a signal. miRNAs can also have a buffering role by targeting both positive and negative regulators in a signal transduction pathway, thus enabling a cell or tissue to return to a homeostatic state.[3] An example for negative regulation has been demonstrated for miR-146a. This miRNA is induced in THP-1 monocytic cell line in response to LPS and proinflammatory cytokines. Promoter analysis of miR-146a showed that its induction in response to LPS, TNFα, and IL-1β was dependent on NF-κB. One of the branches of TLR4 signaling cascade is initiated when adaptor protein MyD88 is recruited to the receptor. MyD88 serves as a bridge between TLR4- and IL-1 receptor-associated kinase 1 (IRAK1) that then recruits TNF receptor-associated factor 6 (TRAF6) into the complex. miR-146a can target the UTRs of TRAF6 and IRAK1 genes. Thus, miR-146a which is activated by NF-κB, negatively feeds back on this signaling cascade by targeting TRAF6 and IRAK1, the two upstream activators of the pathway.[142] NF-κB pathway also has been shown to employ a miRNA-mediated positive feedback circuit. Lin28B-dependent downregulation of let-7 causes upregulation of IL-6 which is a target of let-7, and this result in further stimulation of NF-κB.[143] Applying these emerging principles of miRNA regulation of stress signaling pathways to our understanding of the roles of miRNAs in disease states including pain, will be beneficial in elucidating their role.

Another challenging aspect of miRNA research is the absence of high-throughput biological approaches to identify miRNA targets. Elucidation of functional consequences of aberrant miRNA expression in a disease state is dependent on understanding the target genes being regulated in a physiologically relevant context. To describe the mechanistic basis of miRNA function in the context of a single downstream target is an oversimplification of the mode of action of miRNAs.[4] Thus, a combination of systems biology approaches with functional studies of individual-target genes will help capture global perturbations affecting signaling cascades.

Though significant progress has been made, substantial challenges remain in the delivery and design of effective synthetic RNAs. Success of antimiR or antagomiR therapy will depend on achieving meaningful down regulation of the targeted miRNA. Anti-miRs accumulate predominantly in liver and kidneys. Hence, it is not surprising that the first miRNA-based therapeutic to reach the clinic is against a liver-specific miRNA.[144] As with any good pharmacological modulator, cell permeability, stability, and rate of excretion will be crucial for a successful therapeutic outcome.[30]

10. CONCLUSIONS

It is thought that each miRNA can bind and repress multiple mRNAs, which ultimately affects protein levels that can be crucial in regulating multiple physiological pathways. Induction of extensive miRNA expression changes in chronic pain conditions indicates their involvement in mediating underlying mechanisms of chronic pain, ranging from neuronal hyperexcitability and neuroinflammation to possibly altered higher brain function.[70] miRNAs represent an emerging new tool for patient management, with a wide range of potential clinical applications, including diagnosis, prognosis, and prediction of treatment efficiency. Larger patient cohorts will be needed to reach definitive conclusions regarding the prognostic power of extracellular miRNAs in pain. miRNAs also have potential therapeutic applications, but more effort is needed before miRNAs are translated from research tools to clinical use. Investigating the role of circulating miRNAs and exosome-mediated information transfer and employing systems biology approaches will undoubtedly advance our understanding of miRNA biology under conditions of homeostasis and disease in the context of pain.

ACKNOWLEDGMENTS

We acknowledge grant support from the Rita Allen Foundation and National Institutes of Health (1R21NS082991-01) to S. A.

REFERENCES

1. Bartel DP. MicroRNAs: target recognition and regulatory functions. *Cell.* 2009;136:215–233.
2. Esteller M. Non-coding RNAs in human disease. *Nat Rev Genet.* 2011;12:861–874.
3. Mendell JT, Olson EN. MicroRNAs in stress signaling and human disease. *Cell.* 2012;148:1172–1187.
4. Olson EN. MicroRNAs as therapeutic targets and biomarkers of cardiovascular disease. *Sci Transl Med.* 2014;6, 239 ps3.
5. Friedman RC, Farh KK, Burge CB, Bartel DP. Most mammalian mRNAs are conserved targets of microRNAs. *Genome Res.* 2009;19:92–105.

6. Ambros V. The functions of animal microRNAs. *Nature*. 2004;431:350–355.
7. Lee RC, Feinbaum RL, Ambros V. The C. elegans heterochronic gene lin-4 encodes small RNAs with antisense complementarity to lin-14. *Cell*. 1993;75:843–854.
8. Olsen PH, Ambros V. The lin-4 regulatory RNA controls developmental timing in Caenorhabditis elegans by blocking LIN-14 protein synthesis after the initiation of translation. *Dev Biol*. 1999;216:671–680.
9. Pasquinelli AE, Reinhart BJ, Slack F, et al. Conservation of the sequence and temporal expression of let-7 heterochronic regulatory RNA. *Nature*. 2000;408:86–89.
10. Gregory RI, Yan KP, Amuthan G, et al. The microprocessor complex mediates the genesis of microRNAs. *Nature*. 2004;432:235–240.
11. Bernstein E, Caudy AA, Hammond SM, Hannon GJ. Role for a bidentate ribonuclease in the initiation step of RNA interference. *Nature*. 2001;409:363–366.
12. Gregory R, Chendrimada T, Shiekhattar R. MicroRNA biogenesis: isolation and characterization of the microprocessor complex. In: Ying S-Y, ed. *MicroRNA Protocols*. New York: Humana Press; 2006:33–47.
13. Djuranovic S, Nahvi A, Green R. A parsimonious model for gene regulation by miRNAs. *Science*. 2011;331:550–553.
14. Baskerville S, Bartel DP. Microarray profiling of microRNAs reveals frequent coexpression with neighboring miRNAs and host genes. *RNA*. 2005;11:241–247.
15. Zhu Y, Lu Y, Zhang Q, et al. MicroRNA-26a/b and their host genes cooperate to inhibit the G1/S transition by activating the pRb protein. *Nucleic Acids Res*. 2012;40:4615–4625.
16. Olena AF, Patton JG. Genomic organization of microRNAs. *J Cell Physiol*. 2009;222:540–545.
17. Calin GA, Sevignani C, Dumitru CD, et al. Human microRNA genes are frequently located at fragile sites and genomic regions involved in cancers. *Proc Natl Acad Sci USA*. 2004;101:2999–3004.
18. Griffiths-Jones S. The microRNA registry. *Nucleic Acids Res*. 2004;32:D109–D111.
19. Griffiths-Jones S, Grocock RJ, van Dongen S, Bateman A, Enright AJ. miRBase: microRNA sequences, targets and gene nomenclature. *Nucleic Acids Res*. 2006;34: D140–D144.
20. Kozomara A, Griffiths-Jones S. miRBase: integrating microRNA annotation and deep-sequencing data. *Nucleic Acids Res*. 2011;39:D152–D157.
21. Hsieh WJ, Lin FM, Huang HD, Wang H. Investigating microRNA-target interaction-supported tissues in human cancer tissues based on miRNA and target gene expression profiling. *PLoS One*. 2014;9:e95697.
22. Liu B, Li J, Cairns MJ. Identifying miRNAs, targets and functions. *Brief Bioinform*. 2014;15:1–19.
23. Jiang Q, Wang Y, Hao Y, et al. miR2Disease: a manually curated database for microRNA deregulation in human disease. *Nucleic Acids Res*. 2009;37:D98–D104.
24. Filipowicz W, Bhattacharyya SN, Sonenberg N. Mechanisms of post-transcriptional regulation by microRNAs: are the answers in sight? *Nat Rev Genet*. 2008;9:102–114.
25. Krol J, Loedige I, Filipowicz W. The widespread regulation of microRNA biogenesis, function and decay. *Nat Rev Genet*. 2010;11:597–610.
26. Brennecke J, Stark A, Russell RB, Cohen SM. Principles of microRNA-target recognition. *PLoS Biol*. 2005;3:e85.
27. Farh KK-H, Grimson A, Jan C, et al. The widespread impact of mammalian MicroRNAs on mRNA repression and evolution. *Science*. 2005;310:1817–1821.
28. Grimson A, Farh KK, Johnston WK, Garrett-Engele P, Lim LP, Bartel DP. MicroRNA targeting specificity in mammals: determinants beyond seed pairing. *Mol Cell*. 2007;27:91–105.

29. Majoros WH, Ohler U. Spatial preferences of microRNA targets in 3′ untranslated regions. *BMC Genomics*. 2007;8:152.
30. van Rooij E. The art of microRNA research. *Circ Res*. 2011;108:219–234.
31. Hsu S-D, Lin F-M, Wu W-Y, et al. miRTarBase: a database curates experimentally validated microRNA–target interactions. *Nucleic Acids Res*. 2011;39:D163–D169.
32. Chi SW, Zang JB, Mele A, Darnell RB. Argonaute HITS-CLIP decodes microRNA-mRNA interaction maps. *Nature*. 2009;460:479–486.
33. Hafner M, Landthaler M, Burger L, et al. Transcriptome-wide identification of RNA-binding protein and MicroRNA target sites by PAR-CLIP. *Cell*. 2010;141:129–141.
34. Lennox KA, Owczarzy R, Thomas DM, Walder JA, Behlke MA. Improved performance of anti-miRNA oligonucleotides using a novel non-nucleotide modifier. *Mol Ther Nucleic acids*. 2013;2:e117.
35. Ebert MS, Neilson JR, Sharp PA. MicroRNA sponges: competitive inhibitors of small RNAs in mammalian cells. *Nat Methods*. 2007;4:721–726.
36. Tay FC, Lim JK, Zhu H, Hin LC, Wang S. Using artificial microRNA sponges to achieve microRNA loss-of-function in cancer cells. *Adv Drug Deliv Rev*. 2015;81: 117–127.
37. Prosser HM, Koike-Yusa H, Cooper JD, Law FC, Bradley A. A resource of vectors and ES cells for targeted deletion of microRNAs in mice. *Nat Biotechnol*. 2011;29:840–845.
38. Park Chong Y, Jeker Lukas T, Carver-Moore K, et al. A resource for the conditional ablation of microRNAs in the mouse. *Cell Rep*. 2012;1:385–391.
39. Mitchell PS, Parkin RK, Kroh EM, et al. Circulating microRNAs as stable blood-based markers for cancer detection. *Proc Natl Acad Sci USA*. 2008;105:10513–10518.
40. Chen X, Ba Y, Ma L, et al. Characterization of microRNAs in serum: a novel class of biomarkers for diagnosis of cancer and other diseases. *Cell Res*. 2008;18:997–1006.
41. Gilad S, Meiri E, Yogev Y, et al. Serum microRNAs are promising novel biomarkers. *PLoS One*. 2008;3:e3148.
42. Grasedieck S, Sorrentino A, Langer C, et al. Circulating microRNAs in hematological diseases: principles, challenges, and perspectives. *Blood*. 2013;121:4977–4984.
43. Boon RA, Vickers KC. Intercellular transport of microRNAs. *Arterioscler Thromb Vasc Biol*. 2013;33:186–192.
44. Turchinovich A, Weiz L, Langheinz A, Burwinkel B. Characterization of extracellular circulating microRNA. *Nucleic Acids Res*. 2011;39:7223–7233.
45. Kowal J, Tkach M, Théry C. Biogenesis and secretion of exosomes. *Curr Opin Cell Biol*. 2014;29:116–125.
46. Robbins PD, Morelli AE. Regulation of immune responses by extracellular vesicles. *Nat Rev Immunol*. 2014;14:195–208.
47. Stoorvogel W. Functional transfer of microRNA by exosomes. *Blood*. 2012;119:646–648.
48. Thery C, Ostrowski M, Segura E. Membrane vesicles as conveyors of immune responses. *Nat Rev Immunol*. 2009;9:581–593.
49. Valadi H, Ekstrom K, Bossios A, Sjostrand M, Lee JJ, Lotvall JO. Exosome-mediated transfer of mRNAs and microRNAs is a novel mechanism of genetic exchange between cells. *Nat Cell Biol*. 2007;9:654–659.
50. Gyorgy B, Szabo TG, Pasztoi M, et al. Membrane vesicles, current state-of-the-art: emerging role of extracellular vesicles. *Cell Mol Life Sci*. 2011;68:2667–2688.
51. Kalra H, Simpson RJ, Ji H, et al. Vesiclepedia: a compendium for extracellular vesicles with continuous community annotation. *PLoS Biol*. 2012;10:e1001450.
52. Mathivanan S, Fahner CJ, Reid GE, Simpson RJ. ExoCarta 2012: database of exosomal proteins, RNA and lipids. *Nucleic Acids Res*. 2012;40:D1241–D1244.
53. Villarroya-Beltri C, Baixauli F, Gutierrez-Vazquez C, Sanchez-Madrid F, Mittelbrunn M. Sorting it out: regulation of exosome loading. *Semin Cancer Biol*. 2014;28:3–13.

54. Vickers KC, Remaley AT. Lipid-based carriers of microRNAs and intercellular communication. *Curr Opin Lipidol.* 2012;23:91–97.
55. Weber JA, Baxter DH, Zhang S, et al. The microRNA spectrum in 12 body fluids. *Clin Chem.* 2010;56:1733–1741.
56. Weiland M, Gao X-H, Zhou L, Mi Q-S. Small RNAs have a large impact: circulating microRNAs as biomarkers for human diseases. *RNA Biol.* 2012;9:850–859.
57. Ajit SK. Circulating microRNAs as biomarkers, therapeutic targets, and signaling molecules. *Sensors (Basel).* 2012;12:3359–3369.
58. Cogswell JP, Ward J, Taylor IA, et al. Identification of miRNA changes in Alzheimer's disease brain and CSF yields putative biomarkers and insights into disease pathways. *J Alzheimers Dis.* 2008;14:27–41.
59. Machida A, Ohkubo T, Yokota T. Circulating microRNAs in the cerebrospinal fluid of patients with brain diseases. In: Kosaka N, ed. *Circulating MicroRNAs.* New York: Humana Press; 2013:203–209.
60. Rau C-S, Jeng J, Jeng S-F, et al. Entrapment neuropathy results in different microRNA expression patterns from denervation injury in rats. *BMC Musculoskelet Disord.* 2010;11:181.
61. Bhalala OG, Srikanth M, Kessler JA. The emerging roles of microRNAs in CNS injuries. *Nat Rev Neurol.* 2013;9:328–339.
62. Kress M, Hüttenhofer A, Landry M, et al. microRNAs in nociceptive circuits as predictors of future clinical applications. *Front Mol Neurosci.* 2013;6(33):1–11.
63. Mogil JS, Davis KD, Derbyshire SW. The necessity of animal models in pain research. *Pain.* 2010;151:12–17.
64. Gregory NS, Harris AL, Robinson CR, Dougherty PM, Fuchs PN, Sluka KA. An overview of animal models of pain: disease models and outcome measures. *J Pain.* 2013;14:1255–1269.
65. Elramah S, Landry M, Favereaux A. MicroRNAs regulate neuronal plasticity and are involved in pain mechanisms. *Front Cell Neurosci.* 2014;8:31.
66. Zhao J, Lee MC, Momin A, et al. Small RNAs control sodium channel expression, nociceptor excitability, and pain thresholds. *J Neurosci.* 2010;30:10860–10871.
67. Schaefer A, O'Carroll D, Tan CL, et al. Cerebellar neurodegeneration in the absence of microRNAs. *J Exp Med.* 2007;204:1553–1558.
68. Wu D, Raafat A, Pak E, Clemens S, Murashov AK. Dicer-microRNA pathway is critical for peripheral nerve regeneration and functional recovery *in vivo* and regenerative axonogenesis *in vitro*. *Exp Neurol.* 2012;233:555–565.
69. Kynast KL, Russe OQ, Geisslinger G, Niederberger E. Novel findings in pain processing pathways: implications for miRNAs as future therapeutic targets. *Expert Rev Neurother.* 2013;13:515–525.
70. Sakai A, Suzuki H. Emerging roles of microRNAs in chronic pain. *Neurochem Int.* 2014;77:58–67.
71. Bali KK, Kuner R. Noncoding RNAs: key molecules in understanding and treating pain. *Trends Mol Med.* 2014;20:437–448.
72. Tan P-H, Pao Y-Y, Cheng J-K, Hung K-C, Liu C-C. MicroRNA-based therapy in pain medicine: current progress and future prospects. *Acta Anaesthesiol Taiwan.* 2013;51:171–176.
73. Bai G, Ambalavanar R, Wei D, Dessem D. Downregulation of selective microRNAs in trigeminal ganglion neurons following inflammatory muscle pain. *Mol Pain.* 2007;3:15.
74. Kusuda R, Cadetti F, Ravanelli MI, et al. Differential expression of microRNAs in mouse pain models. *Mol Pain.* 2011;7:17.
75. Aldrich BT, Frakes EP, Kasuya J, Hammond DL, Kitamoto T. Changes in expression of sensory organ-specific microRNAs in rat dorsal root ganglia in association with

mechanical hypersensitivity induced by spinal nerve ligation. *Neuroscience*. 2009;164:711–723.
76. Strickland IT, Richards L, Holmes FE, Wynick D, Uney JB, Wong L-F. Axotomy-induced miR-21 promotes axon growth in adult dorsal root ganglion neurons. *PLoS One*. 2011;6:e23423.
77. Yu B, Zhou S, Wang Y, Ding G, Ding F, Gu X. Profile of MicroRNAs following rat sciatic nerve injury by deep sequencing: implication for mechanisms of nerve regeneration. *PLoS One*. 2011;6:e24612.
78. von Schack D, Agostino MJ, Murray BS, et al. Dynamic changes in the microRNA expression profile reveal multiple regulatory mechanisms in the spinal nerve ligation model of neuropathic pain. *PLoS One*. 2011;6:e17670.
79. Sakai A, Saitow F, Miyake N, Miyake K, Shimada T, Suzuki H. miR-7a alleviates the maintenance of neuropathic pain through regulation of neuronal excitability. *Brain*. 2013;136:2738–2750.
80. Bali K, Hackenberg M, Lubin A, Kuner R, Devor M. Sources of individual variability: miRNAs that predispose to neuropathic pain identified using genome-wide sequencing. *Mol Pain*. 2014;10:22.
81. Bali KK, Selvaraj D, Satagopam VP, Lu J, Schneider R, Kuner R. Genome-wide identification and functional analyses of microRNA signatures associated with cancer pain. *EMBO Mol Med*. 2013;5:1740–1758.
82. Poh KW, Yeo JF, Ong WY. MicroRNA changes in the mouse prefrontal cortex after inflammatory pain. *Eur J Pain*. 2011;15:801, e1-12.
83. Imai S, Saeki M, Yanase M, et al. Change in microRNAs associated with neuronal adaptive responses in the nucleus accumbens under neuropathic pain. *J Neurosci*. 2011;31:15294–15299.
84. Arai M, Genda Y, Ishikawa M, Shunsuke T, Okabe T, Sakamoto A. The miRNA and mRNA changes in rat hippocampi after chronic constriction injury. *Pain Med*. 2013;14:720–729.
85. Selbach M, Schwanhausser B, Thierfelder N, Fang Z, Khanin R, Rajewsky N. Widespread changes in protein synthesis induced by microRNAs. *Nature*. 2008;455:58–63.
86. Shalgi R, Lieber D, Oren M, Pilpel Y. Global and local architecture of the mammalian microRNA–transcription factor regulatory network. *PLoS Comput Biol*. 2007;3:e131.
87. Hobert O. Gene regulation by transcription factors and microRNAs. *Science*. 2008;319:1785–1786.
88. Flynt AS, Lai EC. Biological principles of microRNA-mediated regulation: shared themes amid diversity. *Nat Rev Genet*. 2008;9:831–842.
89. Chen H-P, Zhou W, Kang L-M, et al. Intrathecal miR-96 inhibits Nav1.3 expression and alleviates neuropathic pain in rat following chronic construction injury. *Neurochem Res*. 2014;39:76–83.
90. Sengupta JN, Pochiraju S, Kannampalli P, et al. MicroRNA-mediated GABAAα-1 receptor subunit down-regulation in adult spinal cord following neonatal cystitis-induced chronic visceral pain in rats. *Pain*. 2013;154:59–70.
91. Im YB, Jee MK, Choi JI, Cho HT, Kwon OH, Kang SK. Molecular targeting of NOX4 for neuropathic pain after traumatic injury of the spinal cord. *Cell Death Dis*. 2012;3:e426.
92. Sun Y, Li XQ, Sahbaie P, et al. miR-203 regulates nociceptive sensitization after incision by controlling phospholipase A2 activating protein expression. *Anesthesiology*. 2012;117:626–638.
93. Ni J, Gao Y, Gong S, Guo S, Hisamitsu T, Jiang X. Regulation of mu-opioid type 1 receptors by microRNA134 in dorsal root ganglion neurons following peripheral inflammation. *Eur J Pain*. 2013;17:313–323.

94. Favereaux A, Thoumine O, Bouali-Benazzouz R, et al. Bidirectional integrative regulation of Cav1.2 calcium channel by microRNA miR-103: role in pain. *EMBO J.* 2011;30:3830–3841.
95. Willemen H, Huo X-J, Mao-Ying Q-L, Zijlstra J, Heijnen C, Kavelaars A. MicroRNA-124 as a novel treatment for persistent hyperalgesia. *J Neuroinflammation.* 2012;9:143.
96. Kynast KL, Russe OQ, Möser CV, Geisslinger G, Niederberger E. Modulation of central nervous system -specific microRNA-124a alters the inflammatory response in the formalin test in mice. *PAIN.* 2013;154(3):368–376.
97. Pan Z, Zhu L-J, Li Y-Q, et al. Epigenetic modification of spinal miR-219 expression regulates chronic inflammation pain by targeting CaMKIIγ. *J Neurosci.* 2014;34: 9476–9483.
98. He Y, Yang C, Kirkmire CM, Wang ZJ. Regulation of opioid tolerance by let-7 family microRNA targeting the mu opioid receptor. *J Neurosci.* 2010;30:10251–10258.
99. He Y, Wang ZJ. Let-7 microRNAs and opioid tolerance. *Front Genet.* 2012;3(110):1–7.
100. Si ML, Zhu S, Wu H, Lu Z, Wu F, Mo YY. miR-21-mediated tumor growth. *Oncogene.* 2007;26:2799–2803.
101. Chan JA, Krichevsky AM, Kosik KS. MicroRNA-21 is an antiapoptotic factor in human glioblastoma cells. *Cancer Res.* 2005;65:6029–6033.
102. Liu N-K, Wang X-F, Lu Q-B, Xu X-M. Altered microRNA expression following traumatic spinal cord injury. *Exp Neurol.* 2009;219:424–429.
103. Yunta M, Nieto-Díaz M, Esteban FJ, et al. MicroRNA dysregulation in the spinal cord following traumatic injury. *PLoS One.* 2012;7:e34534.
104. Nieto-Díaz M, Esteban FJ, Reigada D, et al. MicroRNA dysregulation in spinal cord injury: causes, consequences and therapeutics. *Front Cell Neurosci.* 2014;8.
105. Bhalala OG, Pan L, Sahni V, et al. microRNA-21 regulates astrocytic response following spinal cord injury. *J Neurosci.* 2012;32:17935–17947.
106. Silver J, Miller JH. Regeneration beyond the glial scar. *Nat Rev Neurosci.* 2004;5:146–156.
107. Sakai A, Suzuki H. Nerve injury-induced upregulation of miR-21 in the primary sensory neurons contributes to neuropathic pain in rats. *Biochem Biophys Res Commun.* 2013;435:176–181.
108. Veremeyko T, Siddiqui S, Sotnikov I, Yung A, Ponomarev ED. IL-4/IL-13-dependent and independent expression of miR-124 and its contribution to M2 phenotype of monocytic cells in normal conditions and during allergic inflammation. *PLoS One.* 2013;8:e81774.
109. Mishima T, Mizuguchi Y, Kawahigashi Y, Takizawa T. RT-PCR-based analysis of microRNA (miR-1 and -124) expression in mouse CNS. *Brain Res.* 2007; 1131:37–43.
110. Ponomarev ED, Veremeyko T, Barteneva N, Krichevsky AM, Weiner HL. MicroRNA-124 promotes microglia quiescence and suppresses EAE by deactivating macrophages via the C/EBP-alpha-PU.1 pathway. *Nat Med.* 2011;17:64–70.
111. Kocerha J, Faghihi MA, Lopez-Toledano MA, et al. MicroRNA-219 modulates NMDA receptor-mediated neurobehavioral dysfunction. *Proc Natl Acad Sci USA.* 2009;106:3507–3512.
112. Chattopadhyay M, Zhou Z, Hao S, Mata M, Fink D. Reduction of voltage gated sodium channel protein in DRG by vector mediated miRNA reduces pain in rats with painful diabetic neuropathy. *Mol Pain.* 2012;8:17.
113. Walder RY, Gautam M, Wilson SP, Benson CJ, Sluka KA. Selective targeting of ASIC3 using artificial miRNAs inhibits primary and secondary hyperalgesia after muscle inflammation. *Pain.* 2011;152:2348–2356.
114. Sen GL, Blau HM. A brief history of RNAi: the silence of the genes. *FASEB J.* 2006;20:1293–1299.

115. Carthew RW, Sontheimer EJ. Origins and mechanisms of miRNAs and siRNAs. *Cell*. 2009;136:642–655.
116. Hébert SS, De Strooper B. Alterations of the microRNA network cause neurodegenerative disease. *Trends Neurosci*. 2009;32:199–206.
117. Siddeek B, Inoubli L, Lakhdari N, et al. MicroRNAs as potential biomarkers in diseases and toxicology. *Mutat Res Genet Toxicol Environ Mutagen*. 2014;764–765:46–57.
118. De Guire V, Robitaille R, Tetreault N, et al. Circulating miRNAs as sensitive and specific biomarkers for the diagnosis and monitoring of human diseases: promises and challenges. *Clin Biochem*. 2013;46:846–860.
119. Ammari M, Jorgensen C, Apparailly F. Impact of microRNAs on the understanding and treatment of rheumatoid arthritis. *Curr Opin Rheumatol*. 2013;25:225–233.
120. Shen N, Liang D, Tang Y, de Vries N, Tak PP. MicroRNAs—novel regulators of systemic lupus erythematosus pathogenesis. *Nat Rev Rheumatol*. 2012;8:701–709.
121. Fourie NH, Peace RM, Abey SK, et al. Elevated circulating miR-150 and miR-342-3p in patients with irritable bowel syndrome. *Exp Mol Pathol*. 2014;96:422–425.
122. Gheinani A, Burkhard F, Monastyrskaya K. Deciphering microRNA code in pain and inflammation: lessons from bladder pain syndrome. *Cell Mol Life Sci*. 2013;70:3773–3789.
123. Ohlsson Teague EMC, Print CG, Hull ML. The role of microRNAs in endometriosis and associated reproductive conditions. *Hum Reprod Update*. 2010;16:142–165.
124. Bjersing JL, Lundborg C, Bokarewa MI, Mannerkorpi K. Profile of cerebrospinal microRNAs in fibromyalgia. *PLoS One*. 2013;8:e78762.
125. Orlova IA, Alexander GM, Qureshi RA, et al. MicroRNA modulation in complex regional pain syndrome. *J Transl Med*. 2011;9:195.
126. Murata K, Yoshitomi H, Tanida S, et al. Plasma and synovial fluid microRNAs as potential biomarkers of rheumatoid arthritis and osteoarthritis. *Arthritis Res Ther*. 2010;12:R86.
127. Murata K, Furu M, Yoshitomi H, et al. Comprehensive microRNA analysis identifies miR-24 and miR-125a-5p as plasma biomarkers for rheumatoid arthritis. *PLoS One*. 2013;8:e69118.
128. Adachi T, Nakanishi M, Otsuka Y, et al. Plasma microRNA 499 as a biomarker of acute myocardial infarction. *Clin Chem*. 2010;56:1183–1185.
129. Widera C, Gupta SK, Lorenzen JM, et al. Diagnostic and prognostic impact of six circulating microRNAs in acute coronary syndrome. *J Mol Cell Cardiol*. 2011;51:872–875.
130. Wang WT, Zhao YN, Han BW, Hong SJ, Chen YQ. Circulating microRNAs identified in a genome-wide serum microRNA expression analysis as noninvasive biomarkers for endometriosis. *J Clin Endocrinol Metab*. 2013;98:281–289.
131. McWilliams LA, Cox BJ, Enns MW. Mood and anxiety disorders associated with chronic pain: an examination in a nationally representative sample. *Pain*. 2003;106:127–133.
132. Wu Q, Hwang CK, Zheng H, et al. MicroRNA 339 down-regulates μ-opioid receptor at the post-transcriptional level in response to opioid treatment. *FASEB J*. 2013;27:522–535.
133. McDonald MK, Tian Y, Qureshi RA, et al. Functional significance of macrophage-derived exosomes in inflammation and pain. *Pain*. 2014;155(8):1527–1539.
134. Morse MA, Garst J, Osada T, et al. A phase I study of dexosome immunotherapy in patients with advanced non-small cell lung cancer. *J Transl Med*. 2005;3:9.
135. Zitvogel L, Regnault A, Lozier A, et al. Eradication of established murine tumors using a novel cell-free vaccine: dendritic cell-derived exosomes. *Nat Med*. 1998;4:594–600.

136. Viaud S, Ploix S, Lapierre V, et al. Updated technology to produce highly immunogenic dendritic cell-derived exosomes of clinical grade: a critical role of interferon-gamma. *J Immunother.* 2011;34:65–75.
137. Ridder K, Keller S, Dams M, et al. Extracellular vesicle-mediated transfer of genetic information between the hematopoietic system and the brain in response to inflammation. *PLoS Biol.* 2014;12:e1001874.
138. Fabbri M, Paone A, Calore F, Galli R, Croce CM. A new role for microRNAs, as ligands of toll-like receptors. *RNA Biol.* 2013;10:169–174.
139. Park CK, Xu ZZ, Berta T, et al. Extracellular microRNAs activate nociceptor neurons to elicit pain via TLR7 and TRPA1. *Neuron.* 2014;82:47–54.
140. Janssen HLA, Reesink HW, Lawitz EJ, et al. Treatment of HCV infection by targeting microRNA. *N Eng J Med.* 2013;368:1685–1694.
141. Li Z, Rana TM. Therapeutic targeting of microRNAs: current status and future challenges. *Nat Rev Drug Discov.* 2014;13:622–638.
142. Taganov KD, Boldin MP, Chang KJ, Baltimore D. NF-kappaB-dependent induction of microRNA miR-146, an inhibitor targeted to signaling proteins of innate immune responses. *Proc Natl Acad Sci USA.* 2006;103:12481–12486.
143. Iliopoulos D, Hirsch HA, Struhl K. An epigenetic switch involving NF-kappaB, Lin28, Let-7 MicroRNA, and IL6 links inflammation to cell transformation. *Cell.* 2009; 139:693–706.
144. Lindow M, Kauppinen S. Discovering the first microRNA-targeted drug. *J Cell Biol.* 2012;199:407–412.
145. Lin C-R, Chen K-H, Yang C-H, Huang H-W, Sheen-Chen S-M. Intrathecal miR-183 delivery suppresses mechanical allodynia in mononeuropathic rats. *Eur J Neurosci.* 2014;39(10):1682–1689.

CHAPTER NINE

Role of Extracellular Damage-Associated Molecular Pattern Molecules (DAMPs) as Mediators of Persistent Pain

Jungo Kato, Camilla I. Svensson[1]
Department of Physiology and Pharmacology, Karolinska Institutet, Stockholm, Sweden
[1]Corresponding author: e-mail address: camilla.svensson@ki.se

Contents

1. Introduction 252
2. High-Mobility Group Box 1 254
   2.1 Structure, localization, and release mechanisms 254
   2.2 Roles as an inflammatory mediator 255
   2.3 Emerging roles of HMGB1 as a nociceptive molecule 256
   2.4 Molecular mechanisms underlying the nociceptive action of HMGB1 260
3. S100 Proteins 262
   3.1 S100A8 and S100A9 262
   3.2 S100B 263
4. Heat-Shock Proteins 264
   4.1 HSP27, HSP70, and HSP90 265
5. microRNA 266
   5.1 Lethal-7 miRNA 266
6. Purine Metabolites 267
   6.1 ATP 267
   6.2 Uric acid 268
7. Summary and Perspective 268
References 269

Abstract

Damage-associated molecular pattern molecules (DAMPs) are endogenous molecules that are constitutively expressed and released upon tissue damage, resulting in activation of the immune system. In the absence of injury or infection, DAMPs play important intracellular roles. However, once released subsequent to cell damage or cell stress, DAMPs promote activation of innate immune cells and recruitment and activation of antigen-presenting cells engaged in host defense and tissue repair. This process involves pattern recognition receptors, such as the Toll-like receptors (TLRs) and

Progress in Molecular Biology and Translational Science, Volume 131
ISSN 1877-1173
http://dx.doi.org/10.1016/bs.pmbts.2014.11.014

receptor for advanced glycation end products (RAGE). Several of the TLRs and RAGE have been implicated to play key roles not only in the detection of injury but also in pain signaling. Pain-like behavior is reduced in TLR2- and TLR4-deficient mice, and after injection of TLR2-, TLR4-, and RAGE antagonists in experimental models of nerve injury, arthritis, and bone cancer pain. This suggests that the pathological processes in these models are associated with release of endogenous TLR and RAGE ligands, and further that DAMPs play an important role in persistent pain. There is now a rapidly growing list of DAMPs in the literature and here we give an overview of DAMPs that have been associated with nociceptive signaling.

1. INTRODUCTION

Pathogen-associated molecular pattern molecules (PAMPs), for example, lipopolysaccharide (LPS), are a diverse set of microbial molecules that share a number of different general "patterns," or structures, that alert immune cells to destroy intruding pathogens.[1,2] It is well established that PAMPs are recognized by pattern recognition receptors (PRRs), such as Toll-like receptors (TLRs), nucleotide-binding oligomerization domain-like receptors, and the mannose receptor expressed on innate immune cells.[1] About 20 years ago, it was postulated by Matzinger that the immune system not only responds to pathogens but also senses signals arising from necrosis of healthy cells in response to inflammation, ischemia, or hypoxia, in the absence of microbial components.[3,4] This hypothesis is based on the observation that the immune system is activated by so-called misplaced self-molecules released from damaged or stressed cells such that a sterile inflammation is generated. These endogenous signals are referred to as alarmins, or danger-associated molecular pattern molecules (DAMPs).[1,5] Importantly, emerging data show that certain PRRs are also expressed on nociceptors, indicating that enhanced activity in the nociceptive system in response to danger signals is part of the machinery that is activated to protect the host. Dysfunction, or exaggerated activation of this fine-tuned system, may therefore be a part of the pathology leading to persistent pain. Understanding how and which DAMPs modulate activity in the sensory nervous system is crucial in order to delineate the underlying mechanisms of various types of chronic pain.

Under physiological conditions, most DAMPs play important intracellular roles, but subsequent to injury- or disease-induced cellular damage and cell death, they can be released to the extracellular space and exert a completely different function. DAMPs comprise structurally diverse and evolutionarily unrelated multifunctional endogenous molecules that communicate to the host that cellular injury has occurred and promote early innate and adaptive immune responses important for mobilization of repair mechanisms. In addition to being released from necrotic cells, and in certain conditions from apoptotic cells, DAMP molecules are also actively released to the extracellular space by leukocytes and other cells in response to microbial components, cytokines, or cell stress.[1] DAMP molecules lack leader signal sequences (required for transport through the classical secretory reticulum—Golgi pathway), thus secretion during cell activation represents a nonclassical process and vesicular exocytosis regulated by caspase-1 and inflammasomes has been suggested.[6] The effects of DAMPs are mediated through PRRs, such as TLR2, TLR4, TLR7, the receptor for advanced glycation end products (RAGE), and Siglec (Table 1), often in company by cytokines, chemokines, and other inflammatory mediators.[5,11,31] Emerging evidence suggests that TLRs, RAGE, and their associated signaling

Table 1 DAMPs and associated PRRs

DAMPs	Receptors	References
Nuclear/cytosolic components		
HMGB1	TLR2, TLR4, TLR9, CD24, RAGE, Siglec, Mac-1	7–11
S100 proteins	RAGE, TLR4	12,13
HSPs	TLR2, TLR4, CD40, CD91	14–17
Histone	TLR9	18
RNA	TLR3, TLR7	19,20
DNA	TLR9	21
Purine metabolites (ATP, uric acid)	P1, P2X and P2Y receptors, NLRP3, TLR2, TLR4	22–24
Beta-amyloid	RAGE, NLRP3	12,25
Extracellular matrix proteins		
Proteoglycans (e.g., hyaluronic acid)	TLR2, TLR4, NLRP3	26,27
Heparan sulfate	TLR4	28
Fibrinogen	TLR4	29
Fibronectin	TLR4	30

components, contribute to pain hypersensitivity. For example, pain-like behavior is reduced in TLR2- and TLR4-deficient mice and after injection of TLR2, TLR4, and RAGE antagonists in experimental models of nerve injury, arthritis, and bone cancer pain.[32–35] These types of models are not associated with pathogenic infection, which suggest that the injury and inflammatory-driven processes in these models generate endogenous TLR and RAGE stimulating DAMPs, and that those DAMPs play an important role in pathological pain. DAMPs drive expression of cytokines, which in turn induce expression ofDAMPs. Thus, signaling events mediated by these types of molecules can create a feed-forward cycle, which may also contribute to the potentiation and/or prolongation of pain hypersensitivity.

As our understanding of the multifaceted roles of DAMPs evolves, so does the number of nominal DAMP family members (Table 1). The majority of DAMPs are nuclear and cytosolic proteins and nucleotides, including high-mobility group box 1 (HMGB1),[36,37] heat-shock proteins (HSPs),[38,39] histones,[18] microRNA (miRNA),[40] mitochondrial RNA and DNA,[41,42] and ATP[43,44] (Table 1). However, not all DAMPs are located intracellularly, they can also be generated by proteases that degrade extracellular matrix and thereby generate DAMPs like hyaluronic acid,[45] fibrinogen,[29] and fibronectin[30] (Table 1). Here, we give an overview of DAMPs that have been associated with nociceptive signaling, with a specific emphasis on HMGB1.

2. High-Mobility Group Box 1

2.1. Structure, localization, and release mechanisms

HMGB1 is a nuclear nonhistone DNA-binding protein that is ubiquitously expressed in most cells types. As a nuclear molecule, HMGB1 regulates a number of key housekeeping events, such as DNA replication, repair, recombination, and transcription through exerting effects on the chromosomal architecture.[46] HMGB1 is a 25-kDa protein containing two positively charged DNA-binding domains (HMG boxes A and B) and a negatively charged C-tail. The primary HMGB1 amino acid sequence is 99% identical in all mammals. Truncation of the full-length HMGB1 demonstrates that the extracellular cytokine activity resides within the B box. Although the mechanism is still unresolved, the truncated HMGB1 A box protein antagonizes HMGB1 activities.[47,48]

While HMGB1 is predominantly located in the nucleus in most cells at baseline, biologically active HMGB1 can be transported to the cytosol, plasma membrane, and extracellular space in response to a variety of stimuli, such as pathogen invasion, tissue injury, and inflammation.[36] HMGB1 can be released via two major pathways, which differ in terms of inducing events, molecular mechanisms, release kinetics, and downstream signaling responses. Passive release of HMGB1 from dead, dying, or injured cells is rapid and associated with increased cell permeability,[37,49] while active secretion is a slower process, involving membrane receptor interaction with extracellular products (ligands), and subsequent intracellular signal transduction.[50,51] Active secretion of HMGB1 can be induced in leukocytes, endothelial cells, neurons, astrocytes, and microglia[52] following stimulation by a number of factors, including LPS, IL-1, TNF,[51] interferons,[53–55] and neuropeptide Y.[56] In activated immune cells, HMGB1 is acetylated and translocated to cytoplasmic vesicles. Exocytosis of HMGB1 via nonclassical, dsRNA-activated protein kinase R and inflammasome controlled secretory pathways have been proposed as a route of the active release.[50,54,57] HMGB1 release is also regulated by other posttranslational modifications, including methylation[58] and phosphorylation by calcium/calmodulin-dependent protein kinase and PKC.[59–61] However, although the details of how immune cells actively release HMGB1 is starting to emerge, the mechanism of how other cells, including sensory neurons and spinal glial cells, regulate HMGB1 release in response to different signals remains largely unknown.

2.2. Roles as an inflammatory mediator

Once released to the extracellular space, HMGB1 has a broad repertoire of immunological functions that includes activation of immune cells and induction of cytokine production, chemotaxis, cell proliferation, and cell differentiation.[36] Intriguingly, it has also been shown that HMGB1 can directly activate the nociceptive system through several different mechanisms (as discussed below). HMGB1 acts as a ligand for TLR2, TLR4, TLR9, RAGE, CD24, Siglec-10 (Siglec-G in mice), and Mac-1 [macrophage-1 antigen (complement receptor consisting of CD11b and CD18)][7–11] and has diverse proinflammatory roles through interactions with these different receptors. For example, HMGB1 signaling through RAGE mediates chemotaxis, proliferation, and differentiation of immune cells, while TLR4 signaling is a strong inducer of cytokine production.

Accumulating evidence suggests that the functional diversity of HMGB1 is partly dependent on its redox state. HMGB1 has three cysteines at amino acid positions 23, 45, and 106.[62,63] When these three cysteines are all reduced (all-thiol HMGB1), HMGB1 acts as a RAGE ligand[64] and potentiates chemotactic activity via CXCR4, by forming a heterocomplex with the CXCL12 chemokine.[62] When C23 and C45 are engaged in a disulfide bridge and the cysteine in position C106 is in the reduced thiol form (disulfide HMGB1), HMGB1 is stabilized in a conformation in which it functions as a cytokine-inducing TLR4 ligand.[63,65] When HMGB1 is terminally oxidized to contain sulfonyl groups on all cysteines (sulfonyl HMGB1), it is thought to be nonactive as this form has not yet been associated with any *in vivo* functions (Fig. 1).

In addition to direct receptor activation, HMGB1 can form complexes with different ligands (e.g., LPS, IL-1, bacterial DNA, CXCL12, and viral RNA) and potentiate the activities of these specific ligands[66–68] (Fig. 2). The exact mechanism underlying this synergy is not known, although dependence on partner ligand's receptors, but not on HMGB1's own receptors, has been demonstrated.[67]

2.3. Emerging roles of HMGB1 as a nociceptive molecule

2.3.1 Evidence for nociceptive properties of HMGB1 from behavioral studies

More than 10 years ago, Chacur and colleagues applied HMGB1 onto the sciatic nerve via a preimplanted indwelling perisciatic catheter and found that this induced dose-dependent reduction in mechanical thresholds.[69] O'Connor and colleagues injected HMGB1 intrathecally and observed a pronounced reduction in the response threshold to Von Frey stimulation[70] in male rats. Similar pronociceptive actions of HMGB1 were later made after application of HMGB1 to rat sciatic nerves exposed by a blunt dissection[71] and intrathecal (i.t.) injection of HMGB1 to male and female mice.

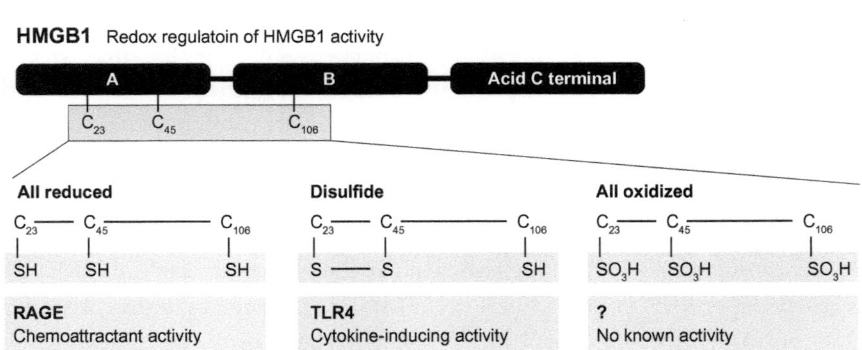

Figure 1 The structure and function of HMGB1.

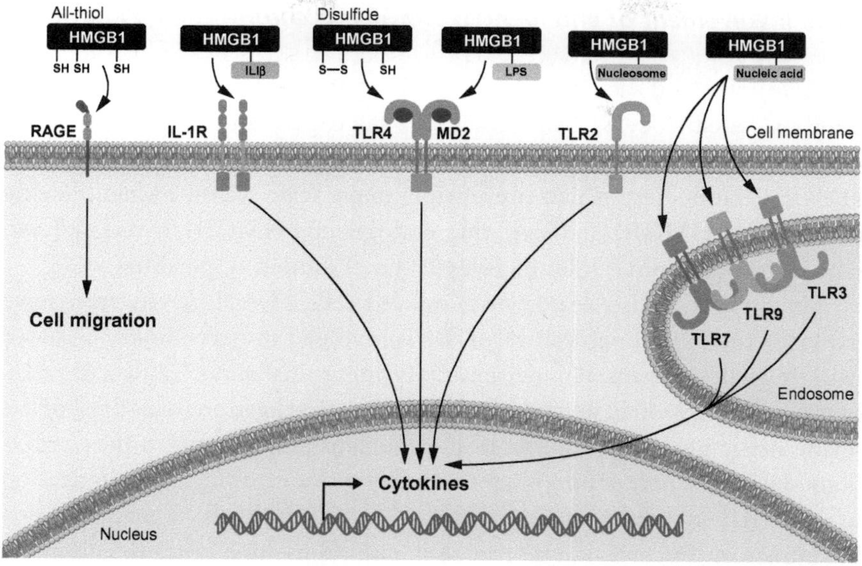

Figure 2 HMGB1 activates multiple signaling pathways. HMGB1 interaction with TLRs and RAGE induces cytokine production, migration, proliferation, and differentiation. When HMGB1 is in the all-thiol form it activates RAGE and when in the disulfide form it activates TLR4. The HMGB1 redox dependence in interactions with the other receptors is presently unknown. In addition to direct receptor interactions, HMGB1 can form complexes with different ligands and thereby enhance the responses induced by the bound ligand. (See the color plate.)

In addition, subcutaneous injection of HMGB1 to the plantar side of the paw of rats[72] and intraarticular injection of disulfide HMGB1 to the ankle joint of mice (N. Agalave and C.I. Svensson, unpublished observations) drives mechanical and thermal hypersensitivity. Interestingly, the spinal pronociceptive property of HMGB1 is dependent on its redox state. In naïve spinal cord, HMGB1 induces mechanical hypersensitivity only when it is in the disulfide TLR4 activating, but not the all-thiol or sulfonyl non-TLR4-binding forms. Moreover, i.t. injection of disulfide HMGB1, but not all-thiol HMGB1, induced increased levels of mRNA of GFAP, CD11b, TNF, IL-1, and MCP-1, typically found during microglia and astrocyte activation. The nociceptive, proinflammatory, and glial cell activating effects of disulfide HMGB1 was abolished in TLR4, but not TLR2- or RAGE-deficient mice, indicating that disulfide HMGB1 mediates processes associated with spinal sensitization via TLR4.[73] Taken together, these behavioral data demonstrate that HMGB1 can exert nociceptive actions both in the peripheral and in the central nervous systems, when applied exogenously.

2.3.2 Involvement of endogenous HMGB1 in chronic pain: Peripheral actions

In naïve mice and rats, nuclear HMGB1 immunoreactivity is detected in neurons, satellite cells, Schwann cells, microglia, and astrocytes.[71,73–75] Thus, HMGB1 could be released from any of these cells. Studies using HMGB1 antibodies, which presumably exert their action by neutralizing extracellular HMGB1, indicate that endogenous HMGB1 is released and served pronociceptive roles in several animal models of persistent pain.

Since HMGB1 is released from injured cells, several investigators have explored the hypothesis that HMGB1 is involved in nerve injury-induced pain. Indeed, mechanical hypersensitivity after spinal nerve ligation (ligation of the L4 and L5 spinal nerves), partial sciatic nerve ligation (one-third of the sciatic nerve ligated), or application of nucleus pulposus on to nerve roots (a model of disc herniation) is attenuated by systemic or local administration of HMGB1 neutralizing antibody.[71,75,76] Glycyrrhizin, a natural anti-inflammatory and antiviral triterpene in clinical use that binds to and interfere with HMGB1 function,[77] has also been used in studies of neuropathic pain and found to attenuate mechanical hypersensitivity induced by ligation of the tibial nerve after systemic administration[74,78] (Table 2).

Nerve injury changes HMGB1 gene and protein expression levels: in dorsal root ganglia (DRG), increased levels of HMGB1 mRNA were observed after spinal nerve ligation[71] and higher HMGB1 protein levels were detected after tibial nerve injury.[74] Immunohistochemistry showed that the increased levels of HMGB1 found in DRG whole cell homogenates, were associated with a translocation from the nucleus to the cytoplasm in DRG neurons,[71,74] which can be interpreted as a movement of HMGB1 toward release. As elevated HMGB1 expression levels were found in satellite cells and Schwann cells subsequent to spinal nerve ligation,[71] HMGB1 in nonneuronal cells may also be important in the pathogenesis following nerve injury.

In addition to neuropathic pain, two recent studies have also shown that HMGB1 contributes to the hypernociception in conditions of more inflammatory character. Tanaka and colleagues reported that mechanical hypersensitivity, induced by intraplantar injection of LPS to the paw, was attenuated by systemic administration of thrombomodulin (an endothelial anticoagulant known to sequester HMGB1 and promotes its degradation[82,83]) and anti-HMGB1 neutralizing antibodies.[72] In another study, they also showed that thrombomodulin and anti-HMGB1 neutralizing antibody reduced the visceral pain caused by cyclophosphamide-induced cystitis[79]

Table 2 Effects of blocking the actions of HGMB1 in different experimental models of pain

HMGB1 inhibitor (mechanism of action)	Pain model	Route of administration	Effects	References
Thrombomodulin (promotes degradation)	HMGB1 i.pl./LPS i.pl.	i.p.	Reduced mechanical hypersensitivity	72
	Cyclophosphamide-induced bladder pain	i.p.	Reduced bladder pain-like behavior and referred hyperalgesia	79
Anti-HMGB1 antibodies (neutralization)	Spinal nerve ligation (SNL)	Catheter[a]	Reduced mechanical allodynia	71
	Partial sciatic nerve ligation (PSNL)	i.v.	Ameliorated tactile hypersensitivity	75
	Application of autologous nucleus pulposus onto nerve roots	i.p.	Reduced mechanical hypersensitivity	76
	LPS i.pl.	i.p.	Reduced mechanical hypersensitivity	72
	Cyclophosphamide-induced bladder pain	i.p.	Reduced bladder-pain like behavior and referred hyperalgesia	79
	Bone cancer pain (intra-tibial injection of cancer cells)	i.t.	Reduced mechanical allodynia	80
	Diabetic pain (db/db mice)	i.t.	Inhibited mechanical allodynia	81
	Rheumatoid arthritis (CAIA model)	i.t.	Reversed mechanical hypersensitivity	73
Glycyrrhizin (binds to HMG boxes)	Tibial nerve injury (TNI)	i.p.	Reversed mechanical hyperalgesia	74
A box (unknown)	Rheumatoid arthritis (CAIA model)	i.t.	Reversed mechanical hypersensitivity	73

[a]Via indwelling catheter placed at the ligation site. *Abbreviations*: i.pl., intraplantar; i.p., intraperitoneal; i.t., intrathecal; i.v., intravenous.

(Table 2). These findings provide new evidence that HMGB1 may play an important role in chronic inflammatory pain as well as neuropathic pain.

2.3.3 Involvement of endogenous HMGB1 in chronic pain: Central actions

As mentioned above, HMGB1 is also ubiquitously expressed in the central nervous system. HMGB1 promotes neurite outgrowth during development[84] but shows limited expression in the regions of neurogenesis in the adult brain.[85] While HMGB1 is downregulated in neurons of the aged brain, it is upregulated in astrocytes, suggesting that HMGB1 expression during aging is differentially regulated between neurons and astrocytes.[86] Moreover, recent studies have shown that spinally released HMGB1 potentiates neuroinflammation and contributes to the development of several neurodegenerative diseases, such as multiple sclerosis and amyotrophic lateral sclerosis.[87] In analogy, since neuroinflammation is a key mechanism underlying the spinal sensitization, spinal HMGB1 has been attracting attention as a candidate mediator of persistent pathological pain. Recent studies have shown that blocking the spinal action of HMGB1 by i.t. injection of anti-HMGB1 antibodies, reversed pain-like behavior in experimental models of bone cancer, diabetic neuropathy, and rheumatoid arthritis (Table 2).[73,80,81]

In the bone cancer model (induced by inoculation of carcinoma cells in the tibia), HMGB1 protein levels were elevated in lumbar spinal cord whole cell homogenates. In addition to the antihyperalgesic effect, i.t. injection of neutralizing antibodies reduced cancer-induced IL-1β protein expression in the spinal cord.[80] Similarly, in *db/db* mice, used as a model of diabetic painful neuropathy, blocking the actions of spinal HMGB1 also prevented spinal upregulation of IL-1β, IL-6, TNF, MCP-1 mRNA, and GFAP protein expression, associated with increased glial cell reactivity.[81] Finally, in the collagen antibody-induced arthritis model, it was found that spinal HMGB1 mRNA was increased, as well as HMGB1 protein levels in the lumbar spinal extranuclear fraction. In addition to the effect of neutralizing HMGB1 antibodies, the arthritis-induced mechanical hypersensitivity was also reversed by i.t. injection of recombinant A box[73] (Table 2). Taken together, these reports point to the crucial role of spinal HMGB1 in spinal cord sensitization, underlying the development of different types of chronic pain.

2.4. Molecular mechanisms underlying the nociceptive action of HMGB1

If assumed that the neutralizing HMGB1 antibodies do not cross plasma membranes, the fact that systemic and i.t. injection of anti-HMGB1 antibodies have antinociceptive effects, point to a presence of HMGB1 in the extracellular space in the periphery, and also in the spinal cord, in a number of animal models of persistent pain. Presence of HMGB1 in the cytoplasm has been detected both in DRG and in the spinal cord, and this is often used as an indicator of HMGB1 moving from the nucleus to the extracellular space. This is further supported by *in vitro* studies: It has been shown that F11 cells, a DRG neuron/neruoblastoma hybrid cell line, release HMGB1 in response to depolarization,[74] indicating that neuronal HMGB1 release may occur in an activity-dependent manner. In addition, HMGB1 has been detected in the cell culture media and cytosol of BV2 cells, a microglia cell line, in response to LPS stimulation[88–90] and in media from human astrocyte cultures in response to IL-1β.[91]

Once present in the extracellular space, HMGB1 can interact with both neurons and nonneuronal cells, to mediate or facilitate pain signal transmission. Using ratiometric calcium imaging and patch-clamp electrophysiology, White and colleagues showed that HMGB1 drives calcium mobilization and increases the excitability of primary DRG neurons.[74] A majority of the cells in which HMGB1-induced calcium mobilization also responded to capsaicin. Furthermore, both all-thiol HMGB1 and disulfide HGMB1 evoke calcium transients in primary DRG neurons. Patch-clamp recordings showed that all-thiol HMGB1 induces neuronal excitability via RAGE, and that its ability to enhance the neuronal excitability was significantly reduced in the presence of a RAGE antibody (11E6), but not small molecule TLR2- or TLR4 antagonists.[78] Thus, in contrast to the spinal *in vivo* actions, where the disulfide, but not the all-thiol form of HMGB1 had an effect of nociceptive signaling, current data suggest that both disulfide and all-thiol HMGB1 have direct effects on peripheral neuronal excitability in these *in vitro* assays.

Both microglia and astrocytes express TLRs and RAGE, and HMGB1 stimulation of primary microglia cell cultures induce increased mRNA levels of TNF, IL-1β, COX-2,[88] and release of IL-6.[92] Furthermore, cultured astrocytes exposed to HMGB1 produce inflammatory molecules, such as COX-2, chemokines, and MMP-9.[93] Thus, HMGB1 can stimulate microglia and astrocytes to produce factors that are not only associated with a heighted inflammatory state but also known to drive spinal sensitization. Though the exact details of the cellular source of spinal HMGB1 and the cell(s) responsible for driving the nociceptive actions of spinal HMGB1 remains to be elucidated, the studies mentioned above consistently point in the same direction; indicating HMGB1 as a proinflammatory factor that regulates pain processing.

3. S100 PROTEINS

There is emerging evidence that the S100 proteins play important roles in the pathogenesis of various inflammatory, metabolic, and neoplastic disorders.[94–98] S100 proteins are a family of more than 20 different small (9–13 kDa) proteins that are characterized by the presence of two calcium-binding EF-hand motifs, connected by variable hinge regions that are often associated with biological activity.[98,99] They are produced as monomers and spontaneously form homodimers, heterodimers, and oligomers.[100–103] Different S100 proteins are expressed in specific tissues and cells, which indicate a functional complexity and diversification. Intracellularly, they are involved in regulating essential functions, including calcium- and zinc-dependent signaling, cell cycle regulation, cell growth, migration, cytoskeletal interactions, protein phosphorylation, and regulation of transcriptional factors.[104] Besides the known intracellular functions, some of the S100 proteins, namely S100A8, S100A9, S100A12, and S100B, can also be secreted to the extracellular space in response to cell damage or activation and act as DAMPs.[103,105,106] S100A8 and S100A9 are present in neutrophils, monocytes, and myeloid progenitors and can be induced in keratinocytes during inflammation.[107–110] S100A12 expression, on the other hand, is restricted to neutrophils. There is 60–80% homology between human, rat, and mouse S100A8/A9, but S100A12 is not expressed in rodents.[111–113] S100B is predominantly expressed in the CNS and is especially abundant in astrocytes.[114,115] Like other DAMP molecules, the S100 proteins lack secretion signals, and the mechanisms of release remain somewhat unclear, but an energy- and tubulin-dependent active release process, which requires activation of PKC, has been suggested.[104,116] Other proposed mechanisms include vesicular S100A8/S100A9 release in neutrophil extracellular traps, triggered by the production of ROS from dying neutrophils.[117] Interestingly, both S100A9 and S100B have been proposed to participate in pain signal transmission, but with opposing actions.

3.1. S100A8 and S100A9

S100A8 and S100A9 proteins are secreted in a heterodimeric form by activated immune cells, such as monocytes, granulocytes, and neutrophils.[118] S100A8 and S100A9 proteins are also referred to as calgranulin A and B, and MRP-8 and MRP-14, respectively. While most reports described how the extracellular S100A8/A9 protein complex can trigger immune responses by activation of RAGE and TLR4,[119] anti-inflammatory effects of the S100A8/A9 complex has been noted in a model of adjuvant-induced arthritis model.[120,121] In addition, S100A9, but not S100A8, has anti-nociceptive properties. Intraperitoneal (i.p.) injection of S100A9, but not S100A8, reduces the number of abdominal contortions induced by i.p. injection of 0.6% acetic acid.[122] Treatment of animals with a monoclonal S100A9 antibody reinstated the writhing response reduced by i.p. injection of glycogen in the acetic acid model. I.p. injection of carrageenan also reduced the number of acetic acid-induced abdominal contractions, and i.p. injection of S100A9 monoclonal antibodies reversed the antinociceptive effect of carrageenan.[123] Although the mechanism by which S100A9 induces antinociception has not been completely established, it has been proposed that the C-terminal domain of S100A9 has a direct inhibitory effect on the voltage-gated N-type calcium channel expressed in DRG neurons.[124] A synthetic peptide identical to the C-terminus of the murine S100A9 protein (mS100A9p) has been used to explore the antinociceptive effects of S100A9. While i.p. injection of mS100A9p did not alter tail flick and hot plate responses in naïve mice, it reduced the number of acetic acid-induced abdominal contractions, and i.p. and i.t. injection of mS100A9p reversed mechanical and thermal hypersensitivity in the carrageenan and the chronic constriction-induced (CCI) nerve injury models.[125,126] Mechanical hypersensitivity evoked by activation of protease-activated receptor type 2 (PAR_2) by intraplantar injection of trypsin, or a synthetic PAR_2 ligand, was blocked by intraplantar injection of mS100A9p.[124] Moreover, mS100A9p inhibited PAR_2, bradykinin- and capsaicin-induced calcium mobilization in primary DRG cell cultures, implicating that S100A9 can act directly on sensory neurons. These results indicate that sensory neurons are another important cellular target for the effects of the S100A9 protein in the context of inflammation, and also suggest an anti-nociceptive role for S100A9, independent of its effects on inflammation.

3.2. S100B

S100B is widely expressed in astrocytes, certain neuronal populations, Schwann cells, myeloid-derived cells, and a few other cell types.[127] S100B is released in the extracellular space in response to glutamate,[128] serotonin,[129] TNF,[130] IL-1β,[131] beta-amyloid peptides,[132] and lysophosphatidic acid.[133] Extracellular S100B exerts both paracrine and autocrine effects on neurons and glia[100,134–136] through RAGE, and possibly other receptors. The effect of S100B is concentration dependent and can be either trophic or toxic.[134,137–139] In humans, high levels of extracellular S100B have been detected under various clinical conditions, for example, brain trauma, ischemia and neurodegenerative, and inflammatory and psychiatric diseases.[140,141] A recent study showed that S100B levels were higher in serum from fibromyalgia patients with lower pressure pain thresholds.[142]

Although extracellular levels of S100B have not been examined in experimental models of pain, there are indications that peripheral painful pathology can change expression levels of S100B in the spinal cord. S100B mRNA levels increase in the lumbar spinal cord after L5 spinal nerve transection[35] and CFA-induced paw inflammation.[34] S100B protein levels in the spinal cord are increased after spinal nerve ligation in rats[143] and after mammary metastasis tumor-induced bone cancer in rats.[143] It has also been found in human postmortem material that S100B is elevated in the spinal dorsal horn of HIV patients with neuropathic pain, but not in pain-free HIV patents.[144] A functional role of S100B in nociceptive signaling has been investigated using S100B deficient and overexpressing mice. While S100B knockout mice did not develop mechanical hypersensitivity after spinal nerve transection, S100B overexpressing mice displayed significantly decreased tactile threshold responses in comparison to wild-type mice,[35] indicating an active role of S100B in pain processing.

Elevated levels of S100B have also been detected in brain regions associated with pain signaling including the trigeminal subnucleus caudalis after tooth pulp injury[145] and anterior cingulate cortex subsequent to injection of formalin to the paw of rats.[146] It is possible that centrally released S100B induces neuronal hyperexcitability through the interaction with RAGE, or other S100B receptors, expressed in neurons, though the downstream signaling with respect to S100B/RAGE in neurons has not been extensively studied. Apart from the central nervous system, S100B release from Schwann cells following peripheral nerve damage has been shown, which promotes the recruitment of macrophages and the migration of Schwann cells, thereby facilitating the repair of injured peripheral nerves.[147] However, the involvement of Schwann cell-derived S100B in the pathogenesis of neuropathic pain is not clear.

4. HEAT-SHOCK PROTEINS

HSPs were originally identified as a family of proteins induced by heat shock, but later turned out to be induced also by other cellular stresses.[148–150]

HSPs include a broad spectrum of molecules ranging from 10 to 170 kDa, with various intracellular localizations, and classified into subfamilies according to their molecular weights.[151] In normal healthy cells, HSPs are ubiquitously expressed, intracellular protein chaperones that guide newly synthesized polypeptide chains to prevent aggregation and misfolding. In addition to the HSPs' roles as intracellular chaperones, the function of extracellularly released HSPs has been gaining intense interest. HSPs are passively released extracellulary from injured cells or dying necrotic cells.[16] They are also actively secreted to the extracellular milieu upon cell stress or activation[152] and have been shown to function as DAMPs to produce a robust innate immune response with production and release of cytokines and chemokines. In addition, complexes between extracellular HSPs and antigens can form, and are taken up by antigen-presenting cells and presented to MHCI to activate T cells through cross-presentation.[153,154]

The status of HSPs as DAMPs was questioned by evidence showing that the proinflammatory effects of HSPs could be attributed to bacterial contaminants, obtained during the HSP expression process using prokaryotic cells.[155–157] However, the proinflammatory roles of HSPs have been reestablished by later studies using purified HSPs prepared by bacteria-free expression systems. In those studies, mammalian HSP60 and 70 have been shown to activate immune cells through TLR2 and TLR4.[15,158,159]

4.1. HSP27, HSP70, and HSP90

HSPs are upregulated in axons of injured neurons and are likely to be released to the extracellular space.[160,161] Following peripheral axotomy, the mRNA and protein levels of HSP27 are upregulated in all sizes of DRG neurons, transported antegradely, and accumulated in the spinal cord.[160] Increased expression of HSP70 following axotomy has also been reported and this upregulation was associated with improved survival of the damaged neuron.[162] HSP70 is also upregulated in the dental pulp following pulpal trauma or inflammation.[163,164] Extensive upregulation of HSP70 in the CNS has also been seen during the hypersensitive phase of morphine withdrawal.[165] Although an association between upregulation of HSP27 and HSP70 and the development of neuropathic pain is yet to be established, a role for HSPs in neuropathic pain is supported by studies in which treatment with the two structurally different HSP90 inhibitors, geldanamycin (s.c.), and 17-DMAG (i.t.) reversed CCI mechanical hypersensitivity.[166] Furthermore, the work by Hutchinson and colleagues suggests that HSP90 can function as a cofactor, contributing to the development of morphine-induced hyperalgesia, and i.t. LPS-induced mechanical hypersensitivity in a TLR4-dependent manner.[166] Inhibition of HSP90, on the other hand, has been shown to reverse hypoalgesia in diabetic mice through upregulation of HSP70.[167] Previous work has also shown that upregulation of HSP70 decreases phosphorylation of JNK. An interaction of HSP70 with JNK is possible in the context of diabetic neuropathy, since JNK activity increases in diabetic nerves[168] and contributes to features of neuropathic pain.[169] Thus, upregulating HSP70 may be an effective treatment for both insensate and painful neuropathy. Although some controversy still remains with regards to the role of HSPs as DAMPs, the proinflammatory and pronociceptive roles of HSPs deserve close attention. In addition, analogously to the complex-forming ability of HMGB1, the potential synergistic effects of HSPs as cofactors with other TRL ligands is an interesting topic for future pain research.

5. microRNA

miRNAs are endogenous, small, nonprotein-coding RNAs, of around 22 nucleotides in length, that mediate cellular responses through posttranscriptional regulation of mRNAs.[169a] In humans, up to one-third of the protein-coding genes are predicted to be potential miRNA targets.[170] Interestingly, several groups have shown that miRNAs can be secreted and detected in, e.g., blood, saliva, and urine, and that different levels of specific circulating miRNAs differentiate patients with cancer from healthy subjects.[171,172] Increased levels of some miRNAs have also been detected in the cerebrospinal fluid of Alzheimer's disease patients.[40] The seminal discovery that miRNAs can be released by mast cells and taken up by another mast cell, showed that exosome-released miRNAs represented a previously unknown mechanism of cross-talk and genetic exchange between cells.[173] Recently, it has also been demonstrated that miRNAs may act as DAMPs, via activation of murine TLR7 and human TLR8, to trigger innate immune responses in uninfected or undamaged cells.[174,175]

5.1. Lethal-7 miRNA

The TLR7-binding capacity of miRNAs is not a general feature, but is restricted to miRNAs that contain a specific GU-rich motif (GUUGUGU). The miRNA lethal-7 (let-7), originally discovered as a key developmental regulator in the nematode,[176] and its associated family, have such motifs and possess binding affinity to TLR7.[40,177] Noteworthy, let-7 miRNA is abundant in the brain and has been implicated in neurogenesis. Furthermore, let-7 miRNA induces TLR7-mediated Myd88-dependent neuronal apoptosis in healthy neurons and is released from dying neurons.[40] TLR7 is expressed on small-diameter nociceptive DRG neurons.[178] In the recent work by Park and colleagues, it was shown that let-7b miRNA induces TLR7-dependent, MyD88-independent, inward currents, and action potentials in small-diameter nociceptive neurons (transient receptor potential A1, TRPA1).[20] Notably, unlike let-7-triggered apoptosis, the precise sequence of the GU motif of let-7b was critical for the nociceptive neuron activation. This action was coupled to a TLR7-dependent triggering of activation of the cation channel TRPA1. Addition of formalin to cultured DRG cells, to mimic a nociceptive stimulus, increased let-7b release. Furthermore, intraplantar injection of let-7b miRNA in mice produced a TLR7-dependent mechanical hypersensitivity, which was prevented by coinjection with a TRPA1 antagonist. Taken together, this study suggests that let-7b–TLR7 interactions coordinate pain processing within the DRG by influencing the activity of TRPA1. Thus, in addition to other evidence showing the involvement of dysregulated miRNA in the posttranscriptional modulation of pain-related genes,[179] the unique nociceptive mechanism of extracellularly released let-7b has shed light onto the role of miRNA as a nociceptive DAMP molecule.

6. PURINE METABOLITES

Purine metabolites, such as ATP and uric acid, may also be considered as DAMPs, as they are released from damaged cells and exert DAMP-like functions in the extracellular environment, including activation of nociceptive neurons.

6.1. ATP

ATP is abundantly present in the cytoplasm in any types of cells, and the extracellular concentration of ATP is usually kept negligible under healthy condition. However, large amounts of ATP can be passively, or actively, released in response to cellular stress or damage, such as trauma, oxidants, and pathogen invasion.[180–182] Although not generally presented as a DAMP, the role of ATP as a proinflammatory and pronociceptive factor through the activation of purinergic receptors, expressed both on immune cells and nociceptive neurons, has been extensively studied and reviewed elsewhere.[183–186]

6.2. Uric acid

An elevated uric acid plasma concentration has long been known to cause gout, after precipitation and formation of urate crystals in joints and soft tissues. However, recent evidence suggests that uric acid can be released also from dying cells and function as a painful endogenous DAMP.[187–189] The active moiety, monosodium urate crystals (MSU), forms in the extracellular space, and promotes inflammation through the upregulation of IL-1β via NALP3 inflammasome/caspase 1 activation.[22,190,191] It has also been demonstrated that MSU can cause hypernociception without inflammatory activation.[192] In an MSU intraarticular injection model, sensitization of TRPV1 expressed in nociceptive neurons has been shown to participate in MSU-induced hypernociception.[193,194] Taken together, extracellular purine metabolites, such as ATP and uric acid, play an important role in the development of sterile inflammation in response to endogenous danger and induce hypernociception through both inflammation-dependent and -independent pathways.

7. SUMMARY AND PERSPECTIVE

The ability to experience pain and develop pain sensitivity in response to tissue and cell damage is an important component that promotes healing and survival. The pain experience serves as a warning and activator of plasticity mechanisms, such as learning and memory, which limits the use of the injured body part by evoking behavioral strategies to avoid further or repeated involvement, or signals that something is "wrong." In this light, it is not surprising that DAMPs, in addition to their other roles in the extracellular environment after injury and sterile inflammation, can induce and maintain pain transmission. Considering this role in acute and protective pain, exaggerated and/or prolonged release of DAMPs, and excessive or *de novo* expression of PRRs on nociceptors, are likely an important part of the pathology underlying persistent pain. Indeed, there are reports suggesting such roles for HMGB1, S100B, Hsp90, let-7 miRNA, ATP, and uric acid.

Dysregulation of inflammation and immune reactions is the basis for the pathophysiological process of many disorders. Cytokines are key mediators of many inflammatory reactions, and in conditions where beneficial effects are thought to be achieved by downtuning the activity of these factors, cytokines have received significant attention as potential points of intervention. TNF was recognized as a therapeutic target in the 1990s, which leads to a paradigm shift with impressive clinical benefits, most notable in patients with RA.[195] Unfortunately, cytokine blockade is not always effective, and a significant proportion of RA patients,[196] patients with acute inflammatory disorder, such as trauma-induced systemic inflammatory response syndrome and sepsis[197] are not responding to TNF blockade. The identification of DAMPs as crucial mediators of the inflammatory processes in these disorders, suggests that DAMPs, PRRs and associated intracellular signaling pathways, can be alternative or complementary targets. Indeed, there are a growing number of reports indicating that many of the DAMP molecules can directly or indirectly activate the nociceptive system and play an important role in the pathogenesis of chronic pain. However, compared to the immune system, the DAMP/PRR signaling in the nociceptive system remains poorly characterized and leaves much room for further studies, such as detailed downstream signaling, expression patterns of PRRs, underlying release mechanisms of DAMP molecules, and so on. The DAMP molecules that have been reviewed here based on their identified roles in pain transmission, most importantly HMGB1, S100B, Hsp90, and let-7 miRNA, may just represent the tip of an iceberg. Advancement of our understanding of which DAMP molecules, and how the DAMPs modulate activity in the sensory nervous system, may shed light onto the pathophysiology of chronic pain and lead to the development of novel treatments against refractory chronic pain.

REFERENCES

1. Bianchi ME. DAMPs, PAMPs and alarmins: all we need to know about danger. *J Leukoc Biol*. 2007;81:1–5.
2. Lotze MT, Zeh HJ, Rubartelli A, et al. The grateful dead: damage-associated molecular pattern molecules and reduction/oxidation regulate immunity. *Immunol Rev*. 2007;220:60–81.
3. Matzinger P. Tolerance, danger, and the extended family. *Annu Rev Immunol*. 1994;12:991–1045.
4. Matzinger P. The danger model: a renewed sense of self. *Science*. 2002;296:301–305.
5. Chan JK, Roth J, Oppenheim JJ, et al. Alarmins: awaiting a clinical response. *J Clin Invest*. 2012;122:2711–2719.
6. Keller M, Ruegg A, Werner S, Beer HD. Active caspase-1 is a regulator of unconventional protein secretion. *Cell*. 2008;132:818–831.
7. Hori O, Brett J, Slattery T, et al. The receptor for advanced glycation end products (RAGE) is a cellular binding site for amphoterin. Mediation of neurite outgrowth and co-expression of rage and amphoterin in the developing nervous system. *J Biol Chem*. 1995;270:25752–25761.

8. Park JS, Svetkauskaite D, He Q, et al. Involvement of Toll-like receptors 2 and 4 in cellular activation by high mobility group box 1 protein. *J Biol Chem*. 2004;279: 7370–7377.
9. Ivanov S, Dragoi AM, Wang X, et al. A novel role for HMGB1 in TLR9-mediated inflammatory responses to CpG-DNA. *Blood*. 2007;110:1970–1981.
10. Orlova VV, Choi EY, Xie C, et al. A novel pathway of HMGB1-mediated inflammatory cell recruitment that requires Mac-1-integrin. *EMBO J*. 2007;26:1129–1139.
11. Chen GY, Tang J, Zheng P, Liu Y. CD24 and Siglec-10 selectively repress tissue damage-induced immune responses. *Science*. 2009;323:1722–1725.
12. Yan SD, Chen X, Fu J, et al. RAGE and amyloid-beta peptide neurotoxicity in Alzheimer's disease. *Nature*. 1996;382:685–691.
13. Ehrchen JM, Sunderkotter C, Foell D, Vogl T, Roth J. The endogenous Toll-like receptor 4 agonist S100A8/S100A9 (calprotectin) as innate amplifier of infection, autoimmunity, and cancer. *J Leukoc Biol*. 2009;86:557–566.
14. Vabulas RM, Ahmad-Nejad P, da Costa C, et al. Endocytosed HSP60s use Toll-like receptor 2 (TLR2) and TLR4 to activate the Toll/interleukin-1 receptor signaling pathway in innate immune cells. *J Biol Chem*. 2001;276:31332–31339.
15. Zanin-Zhorov A, Nussbaum G, Franitza S, Cohen IR, Lider O. T cells respond to heat shock protein 60 via TLR2: activation of adhesion and inhibition of chemokine receptors. *FASEB J*. 2003;17:1567–1569.
16. Basu S, Binder RJ, Suto R, Anderson KM, Srivastava PK. Necrotic but not apoptotic cell death releases heat shock proteins, which deliver a partial maturation signal to dendritic cells and activate the NF-kappa B pathway. *Int Immunol*. 2000;12:1539–1546.
17. Becker T, Hartl FU, Wieland F. CD40, an extracellular receptor for binding and uptake of Hsp70–peptide complexes. *J Cell Biol*. 2002;158:1277–1285.
18. Huang H, Evankovich J, Yan W, et al. Endogenous histones function as alarmins in sterile inflammatory liver injury through Toll-like receptor 9 in mice. *Hepatology*. 2011;54:999–1008.
19. Kariko K, Ni H, Capodici J, Lamphier M, Weissman D. mRNA is an endogenous ligand for Toll-like receptor 3. *J Biol Chem*. 2004;279:12542–12550.
20. Park CK, Xu ZZ, Berta T, et al. Extracellular microRNAs activate nociceptor neurons to elicit pain via TLR7 and TRPA1. *Neuron*. 2014;82:47–54.
21. Lamphier MS, Sirois CM, Verma A, Golenbock DT, Latz E. TLR9 and the recognition of self and non-self nucleic acids. *Ann N Y Acad Sci*. 2006;1082:31–43.
22. Martinon F, Mayor A, Tschopp J. The inflammasomes: guardians of the body. *Annu Rev Immunol*. 2009;27:229–265.
23. Bours MJ, Swennen EL, Di Virgilio F, Cronstein BN, Dagnelie PC. Adenosine 5′-triphosphate and adenosine as endogenous signaling molecules in immunity and inflammation. *Pharmacol Ther*. 2006;112:358–404.
24. Liu-Bryan R, Scott P, Sydlaske A, Rose DM, Terkeltaub R. Innate immunity conferred by Toll-like receptors 2 and 4 and myeloid differentiation factor 88 expression is pivotal to monosodium urate monohydrate crystal-induced inflammation. *Arthritis Rheum*. 2005;52:2936–2946.
25. Halle A, Hornung V, Petzold GC, et al. The NALP3 inflammasome is involved in the innate immune response to amyloid-beta. *Nat Immunol*. 2008;9:857–865.
26. Jiang D, Liang J, Fan J, et al. Regulation of lung injury and repair by Toll-like receptors and hyaluronan. *Nat Med*. 2005;11:1173–1179.
27. Yamasaki K, Muto J, Taylor KR, et al. NLRP3/cryopyrin is necessary for interleukin-1beta (IL-1beta) release in response to hyaluronan, an endogenous trigger of inflammation in response to injury. *J Biol Chem*. 2009;284:12762–12771.

28. Johnson GB, Brunn GJ, Kodaira Y, Platt JL. Receptor-mediated monitoring of tissue well-being via detection of soluble heparan sulfate by Toll-like receptor 4. *J Immunol.* 2002;168:5233–5239.
29. Smiley ST, King JA, Hancock WW. Fibrinogen stimulates macrophage chemokine secretion through Toll-like receptor 4. *J Immunol.* 2001;167:2887–2894.
30. Gondokaryono SP, Ushio H, Niyonsaba F, et al. The extra domain A of fibronectin stimulates murine mast cells via Toll-like receptor 4. *J Leukoc Biol.* 2007;82:657–665.
31. Chiba S, Baghdadi M, Akiba H, et al. Tumor-infiltrating DCs suppress nucleic acid-mediated innate immune responses through interactions between the receptor TIM-3 and the alarmin HMGB1. *Nat Immunol.* 2012;13:832–842.
32. Cao L, Tanga FY, Deleo JA. The contributing role of CD14 in Toll-like receptor 4 dependent neuropathic pain. *Neuroscience.* 2009;158:896–903.
33. Kim D, Kim MA, Cho IH, et al. A critical role of Toll-like receptor 2 in nerve injury-induced spinal cord glial cell activation and pain hypersensitivity. *J Biol Chem.* 2007;282:14975–14983.
34. Raghavendra V, Tanga FY, DeLeo JA. Complete Freunds adjuvant-induced peripheral inflammation evokes glial activation and proinflammatory cytokine expression in the CNS. *Eur J Neurosci.* 2004;20:467–473.
35. Tanga FY, Nutile-McMenemy N, DeLeo JA. The CNS role of Toll-like receptor 4 in innate neuroimmunity and painful neuropathy. *Proc Natl Acad Sci USA.* 2005;102: 5856–5861.
36. Andersson U, Tracey KJ. HMGB1 is a therapeutic target for sterile inflammation and infection. *Annu Rev Immunol.* 2011;29:139–162.
37. Scaffidi P, Misteli T, Bianchi ME. Release of chromatin protein HMGB1 by necrotic cells triggers inflammation. *Nature.* 2002;418:191–195.
38. Asea A, Rehli M, Kabingu E, et al. Novel signal transduction pathway utilized by extracellular HSP70: role of Toll-like receptor (TLR) 2 and TLR4. *J Biol Chem.* 2002;277:15028–15034.
39. Ohashi K, Burkart V, Flohe S, Kolb H. Cutting edge: heat shock protein 60 is a putative endogenous ligand of the Toll-like receptor-4 complex. *J Immunol.* 2000;164:558–561.
40. Lehmann SM, Kruger C, Park B, et al. An unconventional role for miRNA: let-7 activates Toll-like receptor 7 and causes neurodegeneration. *Nat Neurosci.* 2012;15:827–835.
41. Kariko K, Buckstein M, Ni H, Weissman D. Suppression of RNA recognition by Toll-like receptors: the impact of nucleoside modification and the evolutionary origin of RNA. *Immunity.* 2005;23:165–175.
42. Zhang Q, Itagaki K, Hauser CJ. Mitochondrial DNA is released by shock and activates neutrophils via p38 map kinase. *Shock.* 2010;34:55–59.
43. Mariathasan S, Weiss DS, Newton K, et al. Cryopyrin activates the inflammasome in response to toxins and ATP. *Nature.* 2006;440:228–232.
44. Trautmann A. Extracellular ATP in the immune system: more than just a "danger signal" *Sci Signal.* 2009;2:pe6.
45. Termeer C, Benedix F, Sleeman J, et al. Oligosaccharides of Hyaluronan activate dendritic cells via Toll-like receptor 4. *J Exp Med.* 2002;195:99–111.
46. Ueda T, Yoshida M. HMGB proteins and transcriptional regulation. *Biochim Biophys Acta.* 2010;1799:114–118.
47. Andersson U, Erlandsson-Harris H, Yang H, Tracey KJ. HMGB1 as a DNA-binding cytokine. *J Leukoc Biol.* 2002;72:1084–1091.
48. Yang H, Ochani M, Li J, et al. Reversing established sepsis with antagonists of endogenous high-mobility group box 1. *Proc Natl Acad Sci USA.* 2004;101:296–301.

49. Tsung A, Klune JR, Zhang X, et al. HMGB1 release induced by liver ischemia involves Toll-like receptor 4 dependent reactive oxygen species production and calcium-mediated signaling. *J Exp Med.* 2007;204:2913–2923.
50. Gardella S, Andrei C, Ferrera D, et al. The nuclear protein HMGB1 is secreted by monocytes via a non-classical, vesicle-mediated secretory pathway. *EMBO Rep.* 2002;3:995–1001.
51. Wang H, Bloom O, Zhang M, et al. HMG-1 as a late mediator of endotoxin lethality in mice. *Science.* 1999;285:248–251.
52. Kang R, Chen R, Zhang Q, et al. HMGB1 in health and disease. *Mol Aspects Med.* 2014;40c:1–116.
53. Jiang W, Pisetsky DS. The role of IFN-alpha and nitric oxide in the release of HMGB1 by RAW 264.7 cells stimulated with polyinosinic-polycytidylic acid or lipopolysaccharide. *J Immunol.* 2006;177:3337–3343.
54. Lu B, Antoine DJ, Kwan K, et al. JAK/STAT1 signaling promotes HMGB1 hyperacetylation and nuclear translocation. *Proc Natl Acad Sci USA.* 2014;111:3068–3073.
55. Rendon-Mitchell B, Ochani M, Li J, et al. IFN-gamma induces high mobility group box 1 protein release partly through a TNF-dependent mechanism. *J Immunol.* 2003;170:3890–3897.
56. Zhou JR, Zhang LD, Wei HF, et al. Neuropeptide Y induces secretion of high-mobility group box 1 protein in mouse macrophage via PKC/ERK dependent pathway. *J Neuroimmunol.* 2013;260:55–59.
57. Lu B, Nakamura T, Inouye K, et al. Novel role of PKR in inflammasome activation and HMGB1 release. *Nature.* 2012;488:670–674.
58. Ito I, Fukazawa J, Yoshida M. Post-translational methylation of high mobility group box 1 (HMGB1) causes its cytoplasmic localization in neutrophils. *J Biol Chem.* 2007;282:16336–16344.
59. Ma L, Kim SJ, Oh KI. Calcium/calmodulin-dependent protein kinase is involved in the release of high mobility group box 1 via the interferon-beta signaling pathway. *Immune Netw.* 2012;12:148–154.
60. Oh YJ, Youn JH, Ji Y, et al. HMGB1 is phosphorylated by classical protein kinase C and is secreted by a calcium-dependent mechanism. *J Immunol.* 2009;182:5800–5809.
61. Zhang X, Wheeler D, Tang Y, et al. Calcium/calmodulin-dependent protein kinase (CaMK) IV mediates nucleocytoplasmic shuttling and release of HMGB1 during lipopolysaccharide stimulation of macrophages. *J Immunol.* 2008;181:5015–5023.
62. Venereau E, Casalgrandi M, Schiraldi M, et al. Mutually exclusive redox forms of HMGB1 promote cell recruitment or proinflammatory cytokine release. *J Exp Med.* 2012;209:1519–1528.
63. Yang H, Lundback P, Ottosson L, et al. Redox modification of cysteine residues regulates the cytokine activity of high mobility group box-1 (HMGB1). *Mol Med.* 2012;18:250–259.
64. Huttunen HJ, Fages C, Kuja-Panula J, Ridley AJ, Rauvala H. Receptor for advanced glycation end products-binding COOH-terminal motif of amphoterin inhibits invasive migration and metastasis. *Cancer Res.* 2002;62:4805–4811.
65. Yang H, Hreggvidsdottir HS, Palmblad K, et al. A critical cysteine is required for HMGB1 binding to Toll-like receptor 4 and activation of macrophage cytokine release. *Proc Natl Acad Sci USA.* 2010;107:11942–11947.
66. Campana L, Bosurgi L, Bianchi ME, Manfredi AA, Rovere-Querini P. Requirement of HMGB1 for stromal cell-derived factor-1/CXCL12-dependent migration of macrophages and dendritic cells. *J Leukoc Biol.* 2009;86:609–615.
67. Hreggvidsdottir HS, Lundberg AM, Aveberger AC, Klevenvall L, Andersson U, Harris HE. High mobility group box protein 1 (HMGB1)-partner molecule complexes

enhance cytokine production by signaling through the partner molecule receptor. *Mol Med*. 2012;18:224–230.
68. Sha Y, Zmijewski J, Xu Z, Abraham E. HMGB1 develops enhanced proinflammatory activity by binding to cytokines. *J Immunol*. 2008;180:2531–2537.
69. Chacur M, Milligan ED, Gazda LS, et al. A new model of sciatic inflammatory neuritis (SIN): induction of unilateral and bilateral mechanical allodynia following acute unilateral peri-sciatic immune activation in rats. *Pain*. 2001;94:231–244.
70. O'Connor KA, Hansen MK, Rachal Pugh C, et al. Further characterization of high mobility group box 1 (HMGB1) as a proinflammatory cytokine: central nervous system effects. *Cytokine*. 2003;24:254–265.
71. Shibasaki M, Sasaki M, Miura M, et al. Induction of high mobility group box-1 in dorsal root ganglion contributes to pain hypersensitivity after peripheral nerve injury. *Pain*. 2010;149:514–521.
72. Tanaka J, Seki Y, Ishikura H, et al. Recombinant human soluble thrombomodulin prevents peripheral HMGB1-dependent hyperalgesia in rats. *Br J Pharmacol*. 2013; 170:1233–1241.
73. Agalave NM, Larsson M, Abdelmoaty S, et al. Spinal HMGB1 induces TLR4-mediated long-lasting hypersensitivity and glial activation and regulates pain-like behavior in experimental arthritis. *Pain*. 2014;155:1802–1813.
74. Feldman P, Due MR, Ripsch MS, Khanna R, White FA. The persistent release of HMGB1 contributes to tactile hyperalgesia in a rodent model of neuropathic pain. *J Neuroinflammation*. 2012;9:180.
75. Nakamura Y, Morioka N, Abe H, et al. Neuropathic pain in rats with a partial sciatic nerve ligation is alleviated by intravenous injection of monoclonal antibody to high mobility group box-1. *PLoS One*. 2013;8:e73640.
76. Otoshi K, Kikuchi S, Kato K, Sekiguchi M, Konno S. Anti-HMGB1 neutralization antibody improves pain-related behavior induced by application of autologous nucleus pulposus onto nerve roots in rats. *Spine (Phila Pa 1976)*. 2011;36:E692–E698.
77. Mollica L, De Marchis F, Spitaleri A, et al. Glycyrrhizin binds to high-mobility group box 1 protein and inhibits its cytokine activities. *Chem Biol*. 2007;14:431–441.
78. Allette YM, Due MR, Wilson SM, et al. Identification of a functional interaction of HMGB1 with Receptor for Advanced Glycation End-products in a model of neuropathic pain. *Brain Behav Immun*. 2014;42:169–177.
79. Tanaka J, Yamaguchi K, Ishikura H, et al. Bladder pain relief by HMGB1 neutralization and soluble thrombomodulin in mice with cyclophosphamide-induced cystitis. *Neuropharmacology*. 2014;79:112–118.
80. Tong W, Wang W, Huang J, Ren N, Wu SX, Li YQ. Spinal high-mobility group box 1 contributes to mechanical allodynia in a rat model of bone cancer pain. *Biochem Biophys Res Commun*. 2010;395:572–576.
81. Ren PC, Zhang Y, Zhang XD, et al. High-mobility group box 1 contributes to mechanical allodynia and spinal astrocytic activation in a mouse model of type 2 diabetes. *Brain Res Bull*. 2012;88:332–337.
82. Abeyama K, Stern DM, Ito Y, et al. The N-terminal domain of thrombomodulin sequesters high-mobility group-B1 protein, a novel antiinflammatory mechanism. *J Clin Invest*. 2005;115:1267–1274.
83. Ito T, Kawahara K, Okamoto K, et al. Proteolytic cleavage of high mobility group box 1 protein by thrombin–thrombomodulin complexes. *Arterioscler Thromb Vasc Biol*. 2008;28:1825–1830.
84. Rauvala H, Pihlaskari R. Isolation and some characteristics of an adhesive factor of brain that enhances neurite outgrowth in central neurons. *J Biol Chem*. 1987;262: 16625–16635.

85. Guazzi S, Strangio A, Franzi AT, Bianchi ME. HMGB1, an architectural chromatin protein and extracellular signalling factor, has a spatially and temporally restricted expression pattern in mouse brain. *Gene Expr Patterns*. 2003;3:29–33.
86. Enokido Y, Yoshitake A, Ito H, Okazawa H. Age-dependent change of HMGB1 and DNA double-strand break accumulation in mouse brain. *Biochem Biophys Res Commun*. 2008;376:128–133.
87. Fang P, Schachner M, Shen YQ. HMGB1 in development and diseases of the central nervous system. *Mol Neurobiol*. 2012;45:499–506.
88. Kim ID, Lee JK. HMGB1-binding heptamer confers anti-inflammatory effects in primary microglia culture. *Exp Neurobiol*. 2013;22:301–307.
89. Kim SW, Jin Y, Shin JH, et al. Glycyrrhizic acid affords robust neuroprotection in the postischemic brain via anti-inflammatory effect by inhibiting HMGB1 phosphorylation and secretion. *Neurobiol Dis*. 2012;46:147–156.
90. Shin JH, Lee HK, Lee HB, Jin Y, Lee JK. Ethyl pyruvate inhibits HMGB1 phosphorylation and secretion in activated microglia and in the postischemic brain. *Neurosci Lett*. 2014;558:159–163.
91. Zurolo E, Iyer A, Maroso M, et al. Activation of Toll-like receptor, RAGE and HMGB1 signalling in malformations of cortical development. *Brain*. 2011;134:1015–1032.
92. Laird MD, Shields JS, Sukumari-Ramesh S, et al. High mobility group box protein-1 promotes cerebral edema after traumatic brain injury via activation of Toll-like receptor 4. *Glia*. 2014;62:26–38.
93. Pedrazzi M, Patrone M, Passalacqua M, et al. Selective proinflammatory activation of astrocytes by high-mobility group box 1 protein signaling. *J Immunol*. 2007;179: 8525–8532.
94. Grevers LC, de Vries TJ, Vogl T, et al. S100A8 enhances osteoclastic bone resorption in vitro through activation of Toll-like receptor 4: implications for bone destruction in murine antigen-induced arthritis. *Arthritis Rheum*. 2011;63:1365–1375.
95. Hiratsuka S, Watanabe A, Aburatani H, Maru Y. Tumour-mediated upregulation of chemoattractants and recruitment of myeloid cells predetermines lung metastasis. *Nat Cell Biol*. 2006;8:1369–1375.
96. Hoyaux D, Decaestecker C, Heizmann CW, et al. S100 proteins in Corpora amylacea from normal human brain. *Brain Res*. 2000;867:280–288.
97. Salama I, Malone PS, Mihaimeed F, Jones JL. A review of the S100 proteins in cancer. *Eur J Surg Oncol*. 2008;34:357–364.
98. van Lent PL, Grevers L, Blom AB, et al. Myeloid-related proteins S100A8/S100A9 regulate joint inflammation and cartilage destruction during antigen-induced arthritis. *Ann Rheum Dis*. 2008;67:1750–1758.
99. Yan WX, Armishaw C, Goyette J, et al. Mast cell and monocyte recruitment by S100A12 and its hinge domain. *J Biol Chem*. 2008;283:13035–13043.
100. Donato R. S100: a multigenic family of calcium-modulated proteins of the EF-hand type with intracellular and extracellular functional roles. *Int J Biochem Cell Biol*. 2001;33:637–668.
101. Heizmann CW, Fritz G, Schafer BW. S100 proteins: structure, functions and pathology. *Front Biosci*. 2002;7:d1356–d1368.
102. Pietzsch J. S100 proteins in health and disease. *Amino Acids*. 2011;41:755–760.
103. Roth J, Vogl T, Sorg C, Sunderkotter C. Phagocyte-specific S100 proteins: a novel group of proinflammatory molecules. *Trends Immunol*. 2003;24:155–158.
104. Donato R, Cannon BR, Sorci G, et al. Functions of S100 proteins. *Curr Mol Med*. 2013;13:24–57.
105. Foell D, Wittkowski H, Vogl T, Roth J. S100 proteins expressed in phagocytes: a novel group of damage-associated molecular pattern molecules. *J Leukoc Biol*. 2007;81:28–37.

106. Roth J, Goebeler M, Sorg C. S100A8 and S100A9 in inflammatory diseases. *Lancet.* 2001;357:1041.
107. Gebhardt C, Nemeth J, Angel P, Hess J. S100A8 and S100A9 in inflammation and cancer. *Biochem Pharmacol.* 2006;72:1622–1631.
108. Hogg N, Allen C, Edgeworth J. Monoclonal antibody 5.5 reacts with p8,14, a myeloid molecule associated with some vascular endothelium. *Eur J Immunol.* 1989;19: 1053–1061.
109. Wilkinson MM, Busuttil A, Hayward C, Brock DJ, Dorin JR, Van Heyningen V. Expression pattern of two related cystic fibrosis-associated calcium-binding proteins in normal and abnormal tissues. *J Cell Sci.* 1988;91(pt 2):221–230.
110. Zwadlo G, Bruggen J, Gerhards G, Schlegel R, Sorg C. Two calcium-binding proteins associated with specific stages of myeloid cell differentiation are expressed by subsets of macrophages in inflammatory tissues. *Clin Exp Immunol.* 1988;72:510–515.
111. Fuellen G, Foell D, Nacken W, Sorg C, Kerkhoff C. Absence of S100A12 in mouse: implications for RAGE–S100A12 interaction. *Trends Immunol.* 2003;24:622–624.
112. Imamichi T, Uchida I, Wahl SM, McCartney-Francis N. Expression and cloning of migration inhibitory factor-related protein (MRP)8 and MRP14 in arthritis-susceptible rats. *Biochem Biophys Res Commun.* 1993;194:819–825.
113. Yang Z, Tao T, Raftery MJ, Youssef P, Di Girolamo N, Geczy CL. Proinflammatory properties of the human S100 protein S100A12. *J Leukoc Biol.* 2001;69:986–994.
114. Fukumoto H, Kakihana M, Suno M. Characterization of C6-10A glioma cells highly responsive to beta-adrenergic receptor agonist-induced NGF synthesis/secretion. *Glia.* 1994;12:151–160.
115. Van Eldik LJ, Zimmer DB. Secretion of S-100 from rat C6 glioma cells. *Brain Res.* 1987;436:367–370.
116. Rammes A, Roth J, Goebeler M, Klempt M, Hartmann M, Sorg C. Myeloid-related protein (MRP) 8 and MRP14, calcium-binding proteins of the S100 family, are secreted by activated monocytes via a novel, tubulin-dependent pathway. *J Biol Chem.* 1997;272:9496–9502.
117. Urban CF, Ermert D, Schmid M, et al. Neutrophil extracellular traps contain calprotectin, a cytosolic protein complex involved in host defense against Candida albicans. *PLoS Pathog.* 2009;5:e1000639.
118. Altwegg LA, Neidhart M, Hersberger M, et al. Myeloid-related protein 8/14 complex is released by monocytes and granulocytes at the site of coronary occlusion: a novel, early, and sensitive marker of acute coronary syndromes. *Eur Heart J.* 2007;28:941–948.
119. Markowitz J, Carson 3rd WE. Review of S100A9 biology and its role in cancer. *Biochim Biophys Acta.* 2013;1835:100–109.
120. Brun JG, Haland G, Haga HJ, Fagerhol MK, Jonsson R. Effects of calprotectin in avridine-induced arthritis. *APMIS.* 1995;103:233–240.
121. Ikemoto M, Murayama H, Itoh H, Totani M, Fujita M. Intrinsic function of S100A8/A9 complex as an anti-inflammatory protein in liver injury induced by lipopolysaccharide in rats. *Clin Chim Acta.* 2007;376:197–204.
122. Giorgi R, Pagano RL, Dias MA, Aguiar-Passeti T, Sorg C, Mariano M. Antinociceptive effect of the calcium-binding protein MRP-14 and the role played by neutrophils on the control of inflammatory pain. *J Leukoc Biol.* 1998;64:214–220.
123. Pagano RL, Dias MA, Dale CS, Giorgi R. Neutrophils and the calcium-binding protein MRP-14 mediate carrageenan-induced antinociception in mice. *Mediators Inflamm.* 2002;11:203–210.
124. Dale CS, Altier C, Cenac N, et al. Analgesic properties of S100A9 C-terminal domain: a mechanism dependent on calcium channel inhibition. *Fundam Clin Pharmacol.* 2009;23:427–438.

125. Dale CS, Cenac N, Britto LR, et al. The C-terminus of murine S100A9 protein inhibits hyperalgesia induced by the agonist peptide of protease-activated receptor 2 (PAR_2). *Br J Pharmacol*. 2006;149:374–384.
126. Paccola CC, Gutierrez VP, Longo I, Juliano L, Juliano MA, Giorgi R. Antinociceptive effect of the C-terminus of murine S100A9 protein on experimental neuropathic pain. *Peptides*. 2008;29:1806–1814.
127. Donato R, Sorci G, Riuzzi F, et al. S100B's double life: intracellular regulator and extracellular signal. *Biochim Biophys Acta*. 2009;1793:1008–1022.
128. Ciccarelli R, Di Iorio P, Bruno V, et al. Activation of A(1) adenosine or mGlu3 metabotropic glutamate receptors enhances the release of nerve growth factor and S-100beta protein from cultured astrocytes. *Glia*. 1999;27:275–281.
129. Whitaker-Azmitia PM, Murphy R, Azmitia EC. Stimulation of astroglial 5-HT1A receptors releases the serotonergic growth factor, protein S-100, and alters astroglial morphology. *Brain Res*. 1990;528:155–158.
130. Edwards MM, Robinson SR. TNF alpha affects the expression of GFAP and S100B: implications for Alzheimer's disease. *J Neural Transm*. 2006;113:1709–1715.
131. de Souza DF, Leite MC, Quincozes-Santos A, et al. S100B secretion is stimulated by IL-1beta in glial cultures and hippocampal slices of rats: likely involvement of MAPK pathway. *J Neuroimmunol*. 2009;206:52–57.
132. Pena LA, Brecher CW, Marshak DR. beta-Amyloid regulates gene expression of glial trophic substance S100 beta in C6 glioma and primary astrocyte cultures. *Brain Res Mol Brain Res*. 1995;34:118–126.
133. Pinto SS, Gottfried C, Mendez A, et al. Immunocontent and secretion of S100B in astrocyte cultures from different brain regions in relation to morphology. *FEBS Lett*. 2000;486:203–207.
134. Donato R. Intracellular and extracellular roles of S100 proteins. *Microsc Res Tech*. 2003;60:540–551.
135. Ponath G, Schettler C, Kaestner F, et al. Autocrine S100B effects on astrocytes are mediated via RAGE. *J Neuroimmunol*. 2007;184:214–222.
136. Zimmer DB, Cornwall EH, Landar A, Song W. The S100 protein family: history, function, and expression. *Brain Res Bull*. 1995;37:417–429.
137. Dietrich MO, Tort AB, Schaf DV, et al. Increase in serum S100B protein level after a swimming race. *Can J Appl Physiol*. 2003;28:710–716.
138. Lam AG, Koppal T, Akama KT, et al. Mechanism of glial activation by S100B: involvement of the transcription factor NFkappaB. *Neurobiol Aging*. 2001;22:765–772.
139. Van Eldik LJ, Wainwright MS. The Janus face of glial-derived S100B: beneficial and detrimental functions in the brain. *Restor Neurol Neurosci*. 2003;21:97–108.
140. Rothermundt M, Peters M, Prehn JH, Arolt V. S100B in brain damage and neurodegeneration. *Microsc Res Tech*. 2003;60:614–632.
141. Rothermundt M, Ponath G, Glaser T, Hetzel G, Arolt V. S100B serum levels and long-term improvement of negative symptoms in patients with schizophrenia. *Neuropsychopharmacology*. 2004;29:1004–1011.
142. Zanette SA, Dussan-Sarria JA, Souza A, Deitos A, Torres IL, Caumo W. Higher serum S100B and BDNF levels are correlated with a lower pressure-pain threshold in fibromyalgia. *Mol Pain*. 2014;10:46.
143. Ducourneau VR, Dolique T, Hachem-Delaunay S, et al. Cancer pain is not necessarily correlated with spinal overexpression of reactive glia markers. *Pain*. 2014;155:275–291.
144. Shi Y, Gelman BB, Lisinicchia JG, Tang SJ. Chronic-pain-associated astrocytic reaction in the spinal cord dorsal horn of human immunodeficiency virus-infected patients. *J Neurosci*. 2012;32:10833–10840.

145. Canzobre MC, Rios H. Pulpar tooth injury induces plastic changes in S100B positive astroglial cells in the trigeminal subnucleus caudalis. *Neurosci Lett.* 2010;470:71–75.
146. Lu Y, Zhu L, Gao YJ. Pain-related aversion induces astrocytic reaction and proinflammatory cytokine expression in the anterior cingulate cortex in rats. *Brain Res Bull.* 2011;84:178–182.
147. Sbai O, Devi TS, Melone MA, et al. RAGE-TXNIP axis is required for S100B-promoted Schwann cell migration, fibronectin expression and cytokine secretion. *J Cell Sci.* 2010;123:4332–4339.
148. Feder ME, Hofmann GE. Heat-shock proteins, molecular chaperones, and the stress response: evolutionary and ecological physiology. *Annu Rev Physiol.* 1999;61:243–282.
149. Sedger L, Ruby J. Heat shock response to vaccinia virus infection. *J Virol.* 1994;68:4685–4689.
150. Zugel U, Kaufmann SH. Immune response against heat shock proteins in infectious diseases. *Immunobiology.* 1999;201:22–35.
151. Ritossa F. Discovery of the heat shock response. *Cell Stress Chaperon.* 1996;1:97–98.
152. Mambula SS, Calderwood SK. Heat shock protein 70 is secreted from tumor cells by a nonclassical pathway involving lysosomal endosomes. *J Immunol.* 2006;177:7849–7857.
153. Li Z, Menoret A, Srivastava P. Roles of heat-shock proteins in antigen presentation and cross-presentation. *Curr Opin Immunol.* 2002;14:45–51.
154. Tsan MF, Gao B. Heat shock proteins and immune system. *J Leukoc Biol.* 2009;85:905–910.
155. Gao B, Tsan MF. Recombinant human heat shock protein 60 does not induce the release of tumor necrosis factor alpha from murine macrophages. *J Biol Chem.* 2003;278:22523–22529.
156. Gao B, Tsan MF. Endotoxin contamination in recombinant human heat shock protein 70 (Hsp70) preparation is responsible for the induction of tumor necrosis factor alpha release by murine macrophages. *J Biol Chem.* 2003;278:174–179.
157. Gao B, Tsan MF. Induction of cytokines by heat shock proteins and endotoxin in murine macrophages. *Biochem Biophys Res Commun.* 2004;317:1149–1154.
158. Zanin-Zhorov A, Bruck R, Tal G, et al. Heat shock protein 60 inhibits Th1-mediated hepatitis model via innate regulation of Th1/Th2 transcription factors and cytokines. *J Immunol.* 2005;174:3227–3236.
159. Zanin-Zhorov A, Tal G, Shivtiel S, et al. Heat shock protein 60 activates cytokine-associated negative regulator suppressor of cytokine signaling 3 in T cells: effects on signaling, chemotaxis, and inflammation. *J Immunol.* 2005;175:276–285.
160. Costigan M, Mannion RJ, Kendall G, et al. Heat shock protein 27: developmental regulation and expression after peripheral nerve injury. *J Neurosci.* 1998;18:5891–5900.
161. Osterloh A, Breloer M. Heat shock proteins: linking danger and pathogen recognition. *Med Microbiol Immunol.* 2008;197:1–8.
162. Tidwell JL, Houenou LJ, Tytell M. Administration of Hsp70 in vivo inhibits motor and sensory neuron degeneration. *Cell Stress Chaperon.* 2004;9:88–98.
163. Ohara K, Shimizu K, Matsuura S, et al. Toll-like receptor 4 signaling in trigeminal ganglion neurons contributes tongue-referred pain associated with tooth pulp inflammation. *J Neuroinflammation.* 2013;10:139.
164. Pileggi R, Holland GR. The expression of heat shock protein 70 in the dental pulp following trauma. *Dent Traumatol.* 2009;25:426–428.
165. Sharma HS, Ali SF, Patnaik R, Zimmermann-Meinzingen S, Sharma A, Muresanu DF. Cerebrolysin attenuates heat shock protein (HSP 72 KD) expression in the rat spinal cord following morphine dependence and withdrawal: possible new therapy for pain management. *Curr Neuropharmacol.* 2011;9:223–235.

166. Hutchinson MR, Ramos KM, Loram LC, et al. Evidence for a role of heat shock protein-90 in Toll like receptor 4 mediated pain enhancement in rats. *Neuroscience*. 2009;164:1821–1832.
167. Urban MJ, Li C, Yu C, et al. Inhibiting heat-shock protein 90 reverses sensory hypoalgesia in diabetic mice. *ASN Neuro*. 2010;2:e00040.
168. Fernyhough P, Gallagher A, Averill SA, et al. Aberrant neurofilament phosphorylation in sensory neurons of rats with diabetic neuropathy. *Diabetes*. 1999;48:881–889.
169. Gao YJ, Ji RR. Activation of JNK pathway in persistent pain. *Neurosci Lett*. 2008;437:180–183.
169a. Bartel DP. MicroRNAs: genomics, biogenesis, mechanism, and function. *Cell*. 2004;116:281–297.
170. Lim LP, Lau NC, Garrett-Engele P, et al. Microarray analysis shows that some microRNAs downregulate large numbers of target mRNAs. *Nature*. 2005;433:769–773.
171. Braconi C, Henry JC, Kogure T, Schmittgen T, Patel T. The role of microRNAs in human liver cancers. *Semin Oncol*. 2011;38:752–763.
172. Mitchell PS, Parkin RK, Kroh EM, et al. Circulating microRNAs as stable blood-based markers for cancer detection. *Proc Natl Acad Sci USA*. 2008;105:10513–10518.
173. Valadi H, Ekstrom K, Bossios A, Sjostrand M, Lee JJ, Lotvall JO. Exosome-mediated transfer of mRNAs and microRNAs is a novel mechanism of genetic exchange between cells. *Nat Cell Biol*. 2007;9:654–659.
174. Chen X, Liang H, Zhang J, Zen K, Zhang CY. Secreted microRNAs: a new form of intercellular communication. *Trends Cell Biol*. 2012;22:125–132.
175. Fabbri M, Paone A, Calore F, et al. MicroRNAs bind to Toll-like receptors to induce prometastatic inflammatory response. *Proc Natl Acad Sci USA*. 2012;109:E2110–E2116.
176. Reinhart BJ, Slack FJ, Basson M, et al. The 21-nucleotide let-7 RNA regulates developmental timing in Caenorhabditis elegans. *Nature*. 2000;403:901–906.
177. Heil F, Hemmi H, Hochrein H, et al. Species-specific recognition of single-stranded RNA via Toll-like receptor 7 and 8. *Science*. 2004;303:1526–1529.
178. Liu T, Xu ZZ, Park CK, Berta T, Ji RR. Toll-like receptor 7 mediates pruritus. *Nat Neurosci*. 2010;13:1460–1462.
179. Andersen HH, Duroux M, Gazerani P. MicroRNAs as modulators and biomarkers of inflammatory and neuropathic pain conditions. *Neurobiol Dis*. 2014;71c:159–168.
180. Burnstock G. Pathophysiology and therapeutic potential of purinergic signaling. *Pharmacol Rev*. 2006;58:58–86.
181. Ferrari D, Chiozzi P, Falzoni S, Hanau S, Di Virgilio F. Purinergic modulation of interleukin-1 beta release from microglial cells stimulated with bacterial endotoxin. *J Exp Med*. 1997;185:579–582.
182. Wang X, Arcuino G, Takano T, et al. P2X7 receptor inhibition improves recovery after spinal cord injury. *Nat Med*. 2004;10:821–827.
183. Burnstock G. Purinergic mechanisms and pain—an update. *Eur J Pharmacol*. 2013;716:24–40.
184. Chiu IM, von Hehn CA, Woolf CJ. Neurogenic inflammation and the peripheral nervous system in host defense and immunopathology. *Nat Neurosci*. 2012;15:1063–1067.
185. Molliver DC, Cook SP, Carlsten JA, Wright DE, McCleskey EW. ATP and UTP excite sensory neurons and induce CREB phosphorylation through the metabotropic receptor, P2Y2. *Eur J Neurosci*. 2002;16:1850–1860.
186. Sanada M, Yasuda H, Omatsu-Kanbe M, et al. Increase in intracellular Ca(2+) and calcitonin gene-related peptide release through metabotropic P2Y receptors in rat dorsal root ganglion neurons. *Neuroscience*. 2002;111:413–422.
187. Chen GY, Nunez G. Sterile inflammation: sensing and reacting to damage. *Nat Rev Immunol*. 2010;10:826–837.

188. Kono H, Chen CJ, Ontiveros F, Rock KL. Uric acid promotes an acute inflammatory response to sterile cell death in mice. *J Clin Invest.* 2010;120:1939–1949.
189. McCarty DJ, Hollander JL. Identification of urate crystals in gouty synovial fluid. *Ann Intern Med.* 1961;54:452–460.
190. Liu-Bryan R. Intracellular innate immunity in gouty arthritis: role of NALP3 inflammasome. *Immunol Cell Biol.* 2010;88:20–23.
191. Martinon F, Petrilli V, Mayor A, Tardivel A, Tschopp J. Gout-associated uric acid crystals activate the NALP3 inflammasome. *Nature.* 2006;440:237–241.
192. Horowitz MD, Abbey L, Sirota DK, Spiera H. Intraarticular noninflammatory free urate suspension (urate milk) in 3 patients with painful joints. *J Rheumatol.* 1990;17:712–714.
193. Hoffmeister C, Trevisan G, Rossato MF, de Oliveira SM, Gomez MV, Ferreira J. Role of TRPV1 in nociception and edema induced by monosodium urate crystals in rats. *Pain.* 2011;152:1777–1788.
194. Hoffmeister C, Silva MA, Rossato MF, et al. Participation of the TRPV1 receptor in the development of acute gout attacks. *Rheumatology (Oxford).* 2014;53:240–249.
195. Taylor PC, Feldmann M. Anti-TNF biologic agents: still the therapy of choice for rheumatoid arthritis. *Nat Rev Rheumatol.* 2009;5:578–582.
196. Feldmann M, Maini RN. Lasker Clinical Medical Research Award. TNF defined as a therapeutic target for rheumatoid arthritis and other autoimmune diseases. *Nat Med.* 2003;9:1245–1250.
197. Reinhart K, Karzai W. Anti-tumor necrosis factor therapy in sepsis: update on clinical trials and lessons learned. *Crit Care Med.* 2001;29:S121–S125.

CHAPTER TEN

mGluRs Head to Toe in Pain

Benedict J. Kolber[1]
Department of Biological Sciences, Duquesne University, Pittsburgh, Pennsylvania, USA
[1]Corresponding author: e-mail address: kolberb@duq.edu

Contents

1. Introduction 282
2. Metabotropic Receptors 282
   2.1 Metabotropic glutamate receptors 283
3. Analysis of mGluRs at Each Level of the Pain Neuraxis 288
   3.1 Periphery (nociceptor ending and DRG cell body) 289
   3.2 Spinal cord (secondary sensory neurons and interneurons) 294
   3.3 Brainstem 300
   3.4 Brain 305
   3.5 Conventional knockout and systemic effects 312
   3.6 mGluR drug development 313
4. Conclusion 316
References 316

Abstract

Metabotropic glutamate receptors (mGluRs) and their role in modulating pain throughout the peripheral and central nervous system are the focus of this chapter. Because these receptors are so prolifically involved in pain signaling throughout the neuraxis, we will use them as a vehicle to explore the totality of the neuraxis. These diverse receptors can increase or decrease pain depending on the subtype of receptor involved and anatomical location of activity. We will cover the basic molecular structure and function of mGluRs and then evaluate the role of different mGluRs at each level of the pain neuraxis. Similar to the functional anatomy involved in the processing of exogenous noxious stimuli, we will start from the peripheral nociceptive terminal and end in higher brain centers that are involved in the cognitive and emotional components of pain. We will conclude by examining the cutting-edge technology involved in the development of mGluR agents for the treatment of pain.

Progress in Molecular Biology and Translational Science, Volume 131
ISSN 1877-1173
http://dx.doi.org/10.1016/bs.pmbts.2014.12.003

1. INTRODUCTION

Since pain is subjective, understating pain in nonhumans depends on behavioral changes (which themselves may be mediated through neurons expressing and responding to glutamate through mGluRs) or on measurement of the underlying phenomena (EPSPs, IPSPs) thought to underlie the neuronal communication of noxious stimuli. Experimentally, we induce pain by selective application of inflammatory agents (e.g., formalin) or injury to some part of the peripheral (e.g., sciatic nerve) or central nervous system (e.g., spinal cord). Behavioral output is measured by monitoring spontaneous behavior (e.g., lifting, licking, biting of affected area), evaluating social interactions and conditioning, or by monitoring responses to experimenter-delivered stimuli. These stimuli include mechanical von Frey filaments or heat/cold stimuli. mGluRs have been evaluated at each stop along the pain neuraxis.

These diverse receptors can increase or decrease pain depending on the subtype of receptor involved and anatomical location of activity. We will cover the basic molecular structure and function of mGluRs and then evaluate the role of different mGluRs at each level of the pain neuraxis. Similar to the functional anatomy involved in the processing of exogenous noxious stimuli, we will start from the peripheral nociceptive terminal and end in higher brain centers that are involved in the cognitive and emotional components of pain. We will conclude by examining the cutting-edge technology involved in the development of mGluR agents for the treatment of pain.

2. METABOTROPIC RECEPTORS

Metabotropic G-protein-coupled receptors (GPCRs) can mediate both quick and long-lasting signaling changes through interaction with G-proteins. G-proteins are a trimeric signaling complex that consist of alpha, beta, and gamma subunits. At baseline most G-proteins are inactive. An inactive G-protein is converted to an active form when bound GDP is exchanged for GTP in the alpha subunit of the G-protein. The G-protein then slowly hydrolyzes the GTP to GDP and then becomes inactive; in this way, the G-protein has a sort of ingenious internal timer. Depending on the type of receptor and ligand, activation of a G-protein can activate or inhibit cell-signaling cascades and, for excitable neurons, can ultimately activate or inhibit the cell itself. GPCRs can be found on the soma, dendrites, or axons of neurons and can modulate glial function. The GPCRs are targets for roughly one-third of all medications, and they are prolifically involved in both acute and chronic pain. A better understanding of GPCRs and their role in pain may be the key to unlocking the mystery of the complex transition from acute to chronic pain. Well-studied GPCRs in pain include the opioid receptors, serotonin receptors, ATP receptors, neurokinin receptor, corticotropin-releasing factor receptors, glutamate receptors, and countless others. Given the prominence of glutamate at many nociceptive synapses, it is not surprising that G-protein-coupled glutamate receptors (better known as mGluRs) have received considerable attention from academic and industrial scientists.

2.1 Metabotropic glutamate receptors

mGluRs (synonyms: GRM#, mGluR#, mGlu#, where "#" stands for subtype number) are found throughout the peripheral and central nervous system where they fine-tune signaling through NMDA and AMPA ionotropic glutamate receptors (iGluRs).[1] mGluRs belong to class C GPCRs, which are characterized by a large N-terminal extracellular domain, seven transmembrane domains, and highly variable C-terminal intracellular domains (reviewed in Refs. 1,2) (Fig. 1). The N-terminal extracellular domains of three mGluRs have been crystalized.[3–5] These crystallized domains contain several important motifs for mGluR function; first, they contain the Venus flytrap domain, which is thought to be critical for ligand binding (i.e., glutamate), exogenous modulation (by orthostatic agonist/antagonists; Table 1), and constitutive receptor dimerization; second, they contain a cysteine-rich domain that propagates conformation changes due to ligand binding to the transmembrane, or heptahelical, domains.[2] The transmembrane domains are responsible for activation of the intracellular G-protein and are thought to contain binding sites for multiple allosteric modulators of mGluRs (Table 1).[2,6] The various cytoplasmic C-terminal domains of mGluRs contain binding sites for a variety of intracellular proteins, sites for posttranslational modification, and are a major area of splice variant expression.[7] Eight distinct mGluRs (mGluR1, mGluR2, mGluR3, mGluR4, mGluR5, mGluR6, mGluR7, mGluR8) belonging to three groups (Group I, Group II, Group III) have been identified based on sequence homology[1] (Fig. 1).

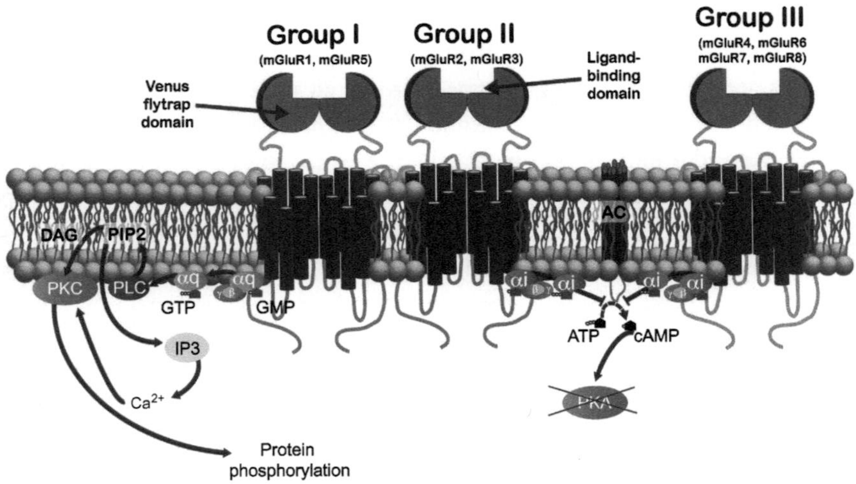

Figure 1 Schematic illustration of mGluR dimers from Group I, II, and III receptors. All mGluRs contain large extracellular domains known as Venus flytrap domains. These domains contain the binding site for glutamate and other agents. Group I mGluRs (mGluR1 and mGluR5) couple to the G-protein $G_{\alpha q}$. $G_{\alpha q}$ activates phospholipase C (PLC) which hydrolyzes phospholipid phosphatidylinositol 4,5-bisphosphate ($PI(4,5)P_2$) to generate DAG and IP3 to increase intracellular calcium and activate PKC. Both Group II (mGluR2 and mGluR3) and Group III (mGluR4, mGluR6, mGluR7, mGluR8) mGluRs couple to the G-protein $G_{\alpha i}$. Activated $G_{\alpha i}$ inhibits the conversion of ATP to cAMP by adenylyl cyclase (AC), which ultimately lowers protein kinase A (PKA) activation. (See the color plate.)

2.1.1 Group I mGluRs (mGluR1 and mGluR5)

Group I mGluRs include mGluR1 and mGluR5. mGluR1 is associated with four main splice variants (mGluR1a, b, c, d) and mGluR5 with two variants (mGluR5a, b).[2] Group I mGluRs are most commonly associated with the G_q/G_{11} subtype of G-protein, which activates phospholipase C (PLC) (Fig. 1). PLC hydrolyzes phospholipid phosphatidylinositol 4,5-bisphosphate ($PI(4,5)P_2$) to generate diacylglycerol (DAG) and inositol 1,4,5-triphosphate (IP3). This pathway leads to increased intracellular calcium and activation of protein kinase C (PKC). A number of studies[2] including those studying the role of Group I mGluRs in pain[8,9] have shown that these receptors can modulate other downstream signaling cascades and G-proteins including $G_{i/o}$ and G_s. Group I mGluRs are associated with increased activation of the mitogen-activated protein kinase pathway,[10] protein kinase A (PKA),[8] potassium channel $Kv_{4.2}$,[10] and regulation by Homer proteins,[11] all of which are associated with amplification of pain signaling. These mGluRs (Group I) are expressed on neurons throughout the pain

Table 1 Targets and names of mGluR pharmacological agents listed in text

mGluR group	Target	Compound; chemical name	Activity
Nonspecific	Group I/II > Group III	(1*S*,3*R*)-ACPD; (1*S*,3*R*)-1-aminocyclopentane-1,3-dicarboxylic acid	Antagonist
	Group II > Group III > Group I	LY341495; [2(*S*)-2-amino-2-(1*S*,2*S*)-2-carboxycycloprop-1-yl]-3-xanth-9-yl) propanoic acid	Antagonist
	mGluR1/5	(*S*)-3,5-DHPG; (*S*)-3,5-dihydroxyphenylglycine	Agonist
	mGluR1 > mGluR5	AIDA; 1-aminoindan-1,5-dicarboxylic acid	Antagonist
	mGluRI	LY367385; (*S*)-(+)-a-amino-4-carboxy-2-methylbenzeneacetic acid	Antagonist
Group I	mGluRI	4-CPG; (*S*)-4-carboxyphenylglycine	Antagonist
	mGluRI	CPCCPEt; 7-(hydroxyimino)cyclopropa[*b*]chromen-1a-carboxylate ethyl ester	NAM
	mGluR5	CHPG; (*RS*)-2-chloro-5-hydroxyphenylglycine	Agonist
	mGluR5	MPEP; 2-methyl-6-(phenylethynyl)pyridine	NAM
	mGluR5	MTEP; 3-[(2-methyl-1,3-thiazol-4-yl)ethynyl]pyridine	NAM
	mGluR5	Fenobam; 1-(3-chlorophenyl)-3-(3-methyl-5-oxo-4H-imidazol-2-yl) urea	NAM
	mGluR2/3	LY354740; (1*S*,2*S*,5*R*,6*S*)-2-aminobicyclo[3.1.0]hexane-2,6-dicarboxylic acid	Agonist
	mGluR2/3	DCG-IV; (2*S*,10*R*,20*R*,30*R*)-2-(20,30-dicarboxycyclopropyl) glycine	Agonist

Continued

Table 1 Targets and names of mGluR pharmacological agents listed in text—cont'd

mGluR group	Target	Compound; chemical name	Activity
Group II	mGluR2/3	L-CCG-I; (2*S*,1′*S*,2′*S*)-2-(carboxycyclopropyl)glycine	Agonist
	mGluR2/3	APDC; (2R,4R)-4-aminopyrrolidine-2,4-dicarboxylate	Agonist
	mGluR2/3	LY404039; (−)-(1 *R*,4*S*,5*S*,6*S*)-4-amino-2-sulfonylbicyclo[3.1.0] hexane-4,6-dicarboxylic acid	Agonist
	mGluR2/3 > group III	EGLU; (2*S*)-alpha-ethylglutamic acid	Antagonist
	mGluR4,6,8 > mGluR7	LAP4; L-2-amino-4-phosphonobutyric acid	Agonist
	mGluR8 > mGluR4/6	DCPG; (*S*)-3,4-dicarboxyphenylglycine	Agonist
	Group III	L-SOP; L-serine-*O*-phosphate	Agonist
	Group III	MAP4; (*S*)-2-amino-2-methyl-4-phosphonobutanoic acid	Antagonist
Group III	Group III	CPPG; (*RS*)-a-cyclopropyl-4-phosphonophenylglycine	Antagonist
	Group III	MSOP; (*RS*)-a-methylserine-*O*-phosphate	Antagonist
	Group III	UBP112; α-methyl-3-methyl-4-phosphonophenylglycine	Antagonist
	mGluR4	PHCCC; *N*-phenyl-7-(hydroxyimino)cyclopropa[b]chromen-1a-carboxamide	PAM
	mGluR7	AMN082; *N*,*N*′-dibenzhydrylethane-1,2-diamine dihydrochloride	Agonist

This is a list of the main pharmacological agents that have been used to understand the role of mGluRs in pain and are referred to in the text. PAM, positive allosteric modulator; NAM, negative allosteric modulator.

neuraxis, including in the peripheral nociceptors (including DRG),[12–16] spinal cord neurons,[17,18] periaqueductal gray (PAG),[19] thalamus,[20] amygdala,[21] and prefrontal cortex (PFC)[22] (Fig. 2). Activation of Group I mGluRs usually, although not always, increases pain or pain-like responses.

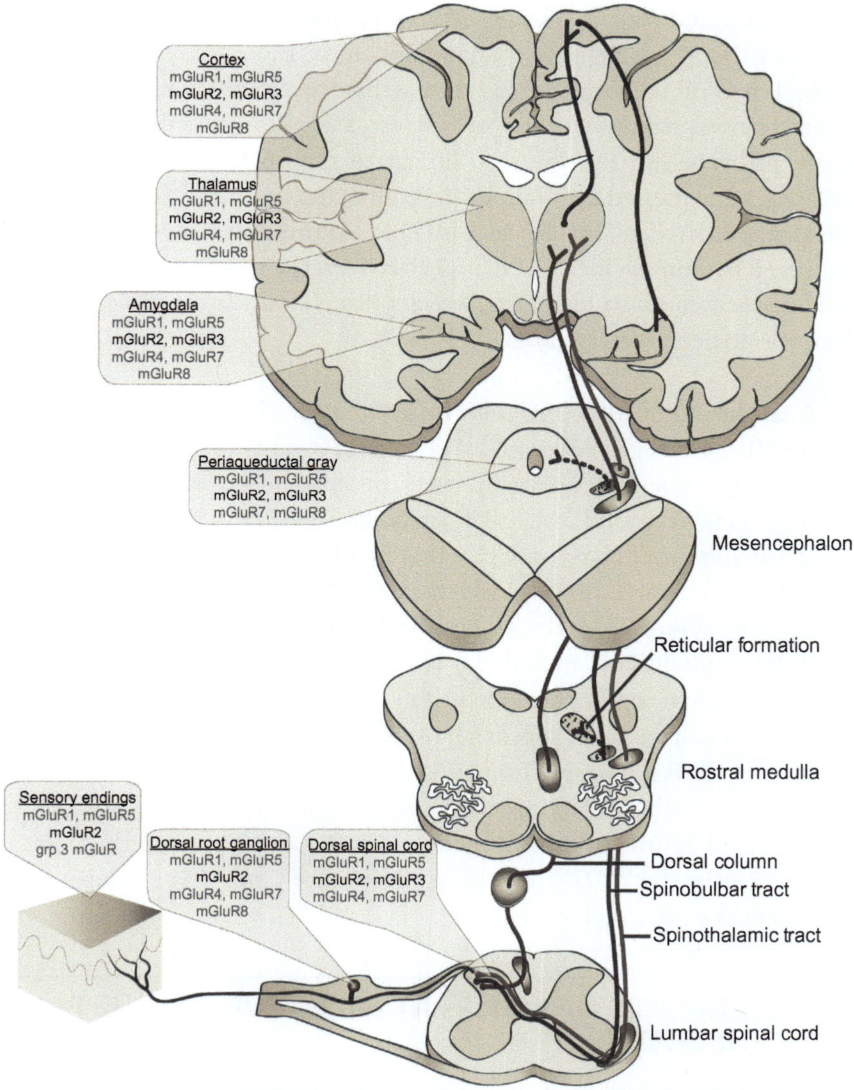

Figure 2 Illustration of mGluR expression at different levels of the pain neuraxis. mGluRs are expressed throughout the nervous system including at all major sites involved in nociception and pain. *Image adapted from original art by Maria Elena Morales and previous publication,[23] used with permission.*

2.1.2 Group II mGluRs (mGluR2 and mGluR3)

Group II mGluRs include mGluR2 and mGluR3. There are no known splice variants of mGluR2. mGluR3, by contrast, has four splice variants: the full-length protein, GRM3Δ2 (missing exon 2), GRM3Δ4 (missing exon 4), and GRM3Δ2Δ3 (missing exons 2 and 3).[24] Group II mGluRs are most commonly associated with the G_i/G_o G-protein, which inhibits adenylyl cyclase and PKA (Fig. 1). A number of studies including those studying the role of Group II mGluRs in pain have shown that these receptors can modulate other downstream signaling cascades to activate the mitogen-activated protein kinase pathway and PI3 kinase.[9] Similar to Group I mGluRs, Group II mGluRs are expressed throughout the pain neuraxis[17,20,25–28] (Fig. 2). Specific delineation of mGluR2 versus mGluR3 expression, however, has been hampered by nonspecificity of staining between the receptors in many antibody histochemical studies.[13,29] Activation of Group II mGluRs typically dampens nociceptive signaling.

2.1.3 Group III mGluRs (mGluR4, mGluR6, mGluR7, and mGluR8)

Group III mGluRs include mGluR4, mGluR6, mGluR7, and mGluR8. mGluR4 has one splice variant (taste mGluR4) that is found in taste neurons. mGluR6 and mGluR8 are each associated with three main splice variants (mGluR6a, mGluR6b, mGluR6c; mGluR8a, mGluR8b, mGluR8c).[30,31] mGluR7 is associated with five variants (mGluR7a, mGluR7b, mGluR7c, mGluR7d, mGluR7e).[32] Three Group III variants (mGluR6b, mGluR6c, mGluR8c) terminate within the N-terminus and may be soluble receptors for glutamate.[30,31] None of the minor Group III splice variants have been specifically studied in the context of pain. Like Group II mGluRs, Group III mGluRs are most commonly associated with the G_i/G_o, which inhibits adenylyl cyclase and PKA (Fig. 1). Similar to Group I and II mGluRs, Group III mGluRs are expressed throughout the pain neuraxis[13,18,33–35] (Fig. 2). Activation of Group III mGluRs typically dampens nociceptive signaling.

3. ANALYSIS OF mGluRs AT EACH LEVEL OF THE PAIN NEURAXIS

mGluRs, as indicated above, are expressed throughout the pain neuraxis on both presynaptic and postsynaptic neurons and on glial cells. In the following sections, the role of each mGluR group will be described at each individual section of the pain neuraxis. This organizational layout is meant to facilitate your understanding of the complex function that glutamate signaling serves in pain. It is likely that glutamate signals through multiple groups of mGluRs at the same synapse in the same "pain" state and the dynamic regulation of each mGluR may determine whether pain is increased (e.g., more Group I signaling) or decreased (e.g., more Group II and III signaling) by glutamate. In some anatomical sites, activation of a specific mGluR such as mGluR1 may increase pain (e.g., spinal cord), while in other sites, activation of this same receptor may decrease pain (e.g., PAG). Many of the studies on pain are done with systemic application of agents which work through mGluRs; combining these studies with those using local application is one way to improve our understanding of where each mGluR plays a specific role. Such an approach is logically appealing, but these agents are seldom as clean (in their target effects) as reported and the physiology is not always as simple as logic would make it seem. Because pain is subjective, understating pain in nonhumans depends on behavioral changes (which themselves may be mediated though neurons expressing and responding to mGluRs) or on measurement of the underlying phenomena (EPSPs, IPSPs) thought to underlie the neuronal communication of pain. In the following sections, we will describe the roles of mGluRs at the level of the peripheral nervous system, the spinal cord, the brainstem, and the brain. Finally, evidence from systemic experiments (e.g., conventional animal knockout approaches) will be evaluated.

3.1 Periphery (nociceptor ending and DRG cell body)

One of the more interesting questions about the role of mGluRs in pain is the extent to which receptors from different families interact and/or are coexpressed on the same cells. In the peripheral nervous system, a coexpression analysis has been only partially completed for a subset of receptors (mGluR1, mGluR2/3, mGluR8).[13] Nonetheless, two important findings have come out of this analysis. First, the number of cells expressing the inhibitory Group II (mGluR2/3; >50% of DRG neurons) and Group III (mGluR8; ~80% of DRG neurons) receptors is considerably higher than cells expressing excitatory Group I (mGluR1; >6%) receptors. Second, there is more overlap (in percent) between Group I-positive cells and either Group II or Group III cells (~50% of mGluR1 cells are double labeled) than between cells expressing Group II and Group III cells (~30% of cells are mGluR2/3 and mGluR8 double labeled). The abundance of inhibitory Group II and III mGluRs on small diameter mGluR1-positive neurons suggests that the push and pull of glutamate within a single nociceptive cell could have a significant impact on DRG activity and ultimately on pain (Table 2). Keep these data in mind when thinking about the physiological and behavioral consequences of peripheral mGluR signaling described below.

3.1.1 Group I activity in the periphery

Below, we outline evidence that mGluR1 and mGluR5 function in the sensory endings of nociceptors and in the DRG cell bodies themselves. Both mGluR1 and mGluR5 are expressed in unmyelinated and myelinated axons in the periphery of rats and mice[12,14–16] (Fig. 3). Peripheral activation of mGluR1 and mGluR5 with the Group I agonist DHPG (Table 1) causes mechanical allodynia that is blocked with peripheral application of the mGluR1 > mGluR5 antagonist AIDA (Table 1)[16] or the mGluR5

Table 2 Summary of behavioral effects of mGluR activation in the periphery

		Acute effects (naive animals)			Postinjury		
		Spontaneous	Mechanical	Thermal	Spontaneous	Mechanical	Thermal
	mGluR1/5	no change in pain	increase in pain	increase in pain	not tested	not tested	not tested
Group I	mGluR1	not tested	increase in pain / no change in pain	increase in pain	increase in pain	no change in pain	not tested
	mGluR5	not tested	increase in pain	increase in pain	increase in pain	increase in pain / no change in pain	increase in pain
Group II	mGluR2/3	no change in pain	no change in pain	no change in pain	decrease in pain	decrease in pain	decrease in pain
Group III	mGluR4/7/8	no change in pain	no change in pain	no change in pain	decrease in pain	not tested	decrease in pain

Legend: increase in pain; decrease in pain; no change in pain; not tested

Specifically, effects are indicated by the color (different gray shades) of the circle. Red (black in the print version) means that activation of the receptor(s) increases pain-like responses, while blue (medium gray in the print version) indicates a reduction in pain-like responses with activation of the indicated receptor(s). A light gray circle represents no change in pain upon activation, and a white circle represents a situation that has not been specifically addressed yet in the literature. In situations with conflicting data (i.e., one study shows an increase in pain and another shows a decrease), circles are half colored (different gray shades). Finally, all circles are representative of the effect of activating the receptor(s), but the actual experimental data may be from a study using antagonists to analyze the function of the receptor. In this table, "postinjury" includes any manipulation to the animal that induces or mimics an injury. Injury can include intraplantar injections of formalin, carrageenan, or capsaicin along with injury models such as chronic constriction injury and spinal cord injury. Spontaneous tests include standard spontaneous assays (e.g., formalin test) as well as newer tests using conditioned place preference unless the conditioning was done with an experimenter-guided mechanical or thermal stimulus.

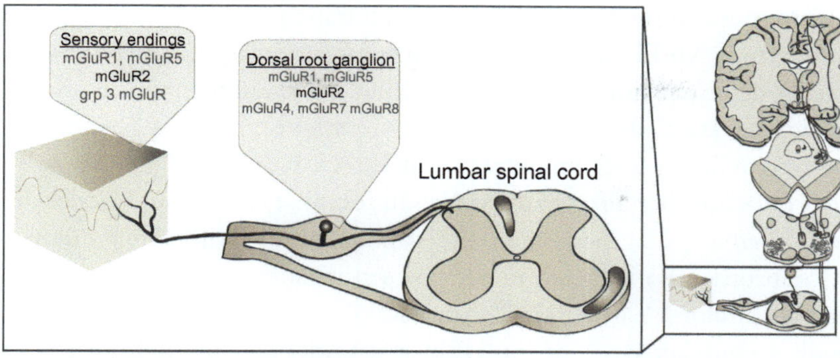

Figure 3 Illustration of mGluR expression in the periphery. mGluRs are expressed in the sensory endings, DRG cell bodies, and/or presynaptic areas in primary sensory neurons. *Image adapted from original art by Maria Elena Morales, used with permission.*

antagonist MPEP but not with the mGluR1 antagonist 4-CPG (Table 1).[15] DHPG also induces thermal hyperalgesia that is blocked with mGluR1 antagonists CPCCOEt/LY367385 (Table 1) or the mGluR5 antagonist MPEP.[12] Intraplantar activation of mGluR5 with the selective agonist CHPG (Table 1) induces mechanical hypersensitivity in naïve animals.[15] In addition, both mGluR1 and mGluR5 at peripheral terminals are involved in spontaneous behavior following intraplantar inflammation.[12,36] Intraplantar blockade of mGluR1 or mGluR5 reduces spontaneous formalin behavior[12]; application of AIDA (mGluR1 > mGluR5 antagonist) decreases spontaneous behavior following intraplantar bee venom application.[36] Interestingly, in the same study, intraplantar AIDA did not reduce or reverse bee venom-induced mechanical hypersensitivity,[36] suggesting that mechanical nocifensive behavior in this pain model is mediated through other receptors, possibly mGluR5. Following intraplantar inflammation with carrageenan, the mGluR5 antagonist MPEP reduces injury-induced mechanical hypersensitivity.[15] In three different models of peripheral nerve injury (total sciatic nerve ligation, partial sciatic nerve ligation, L5 spinal nerve ligation), expression of mGluR5 increases in the injured nerve and DRG neurons.[14] However, mGluR5 inhibition with orally delivered MPEP reversed thermal hypersensitivity in the L5 model only and had no significant effects on mechanical hypersensitivity in any model.[14] As MPEP was delivered systemically, it is unclear whether this effect of MPEP was truly acting in the periphery or whether the effect was central.

In a model of colorectal visceral pain, local peripheral inhibition of mGluR5 can reduce activation of mechanically sensitive nociceptors in the gastrointestinal tract.[37] In bladder pain, intraperitoneal (IP) inhibition of mGluR5 using MPEP or the negative allosteric modulator (NAM) fenobam (Table 1) decreases nociceptive responses in naïve animals and in animals with a urinary tract infection.[38] This antinociceptive effect may be mediated by central mGluR5 since inhibition of mGluR5 with MPEP directly in the bladder does not alter afferent discharges in response to noxious bladder distention *in vitro*.[39] The peripheral effects of mGluR5, studied via cultured DRG neurons, are mediated, in part, by sensitization of transient receptor potential vanilloid 1 (TRPV1).[8] Interestingly, these effects are mediated through a complex series of intracellular cascades and multiple receptors. This pathway is dependent on PLC, DAG, and PKA but not on PKC.[8] Although there are some disagreements in the data, overall activation of mGluR1 and mGluR5 tends to increase pain-like behavior, and inhibition of these Group I mGluRs can reverse some postinjury pain-like changes (Table 2). Studies using peripherally restricted knockout approaches for mGluR1 or 5 would help fine-tune our knowledge of Group I mGluRs in the periphery.

3.1.2 Group II activity in the periphery

Using a nonspecific antibody for mGluR2/3, there is evidence for expression of mGluR2 and/or mGluR3 in DRG soma and presynaptically in the dorsal horn of the spinal cord[26,28,27] (Fig. 3). mRNA for mGluR3 has not been found in DRG, suggesting that earlier immunostaining was due to mGluR2.[40] A large proportion of DRG neurons that express mGluR2/3 binds the isolectin B4 (IB4)[26] and coexpress TRPV1.[41] Subcutaneous injections of Group II agonists do not alter baseline mechanical[15,42] or thermal sensitivity.[43] In the context of prostaglandin E2- or carrageenan-induced peripheral inflammation, however, the Group II agonist APDC (Table 1) reverses mechanical[42] and thermal hyperalgesia.[43] Intraplantar delivery of the Group II agonist APDC is able to reduce bee-venom-induced spontaneous behavior and mechanical hypersensitivity,[36] and APDC is able to reduce subcutaneous IL-1beta-induced mechanical allodynia.[44] Furthermore, capsaicin-induced spontaneous behaviors are attenuated with intraplantar APDC.[41] Intraplantar blockade of Group II and III mGluRs increases capsaicin-induced spontaneous behaviors, suggesting that during noxious stimulation, there is ongoing antinociceptive activity of Group II and/or Group III signaling in the periphery.[45] The natural product L-acetylcarnitine (LAC), a dietary supplement, reduces allodynia caused by chronic constriction injury.[29] This analgesic effect of LAC is blocked by the systemically active mGluR2/3 antagonist LY 341495 (Table 1); LAC acts to increase mRNA expression of mGluR2 but not mGluR3 in the spinal cord[29] and in cultured DRG neurons.[46,47] In the spinal cord, mGluR2 is likely acting presynaptically to reduce excitatory neurotransmission in the dorsal horn.[48,49]

Peripheral mGluR2/3 effects have been shown to be cAMP dependent.[43] In cultured DRG neurons, activation of adenylyl cyclase with forskolin resulted in an increase in TTX-resistant current that was blocked by co-application of mGluR2/3 agonist APDC.[50] These data suggest that mGluR2/3 in the periphery is likely coupled to G_i. Corresponding to behavioral findings, Group II agonists reduce capsaicin-induced TRPV1-dependent excitability in a skin-nerve preparation.[41] Overall, activation of Group II mGluRs in the periphery tends to decrease behavioral (Table 2) and electrophysiological manifestations of pain. Although mGluR2 is likely a more predominant player in peripheral pain, future studies with more selective mGluR2 versus mGluR3 antibodies or pharmacological agents will allow for a more complete and discriminative understanding of Group II mGluRs in the peripheral nervous system.

3.1.3 Group III activity in the periphery

Of the Group III mGluRs, the presence of mGlu4, mGlu7, and mGluR8 has been detected in nociceptive primary afferent neurons (Fig. 3). mGluR4 has been found in DRG soma[35] and at presynaptic sites in the spinal cord.[33] This presynaptic staining is presumed to come from primary afferent neurons and spinal interneurons.[33] Similarly, mGluR7 expression has been confirmed in the DRG[35] and primary afferent presynaptic sites in the spinal cord.[34,51] mGluR8 is widely expressed in DRG soma[13] but has not been detected in the spinal cord (on presynaptic or postsynaptic sites).[18] In DRG, approximately 25% of mGluR8-positive cells were co-labeled for TRPV1.[52] Although the functional behavioral role of Group III mGluRs has been studied extensively at the level of the spinal cord, only recently there have been any reports showing peripheral (e.g., intradermal) manipulation of Group III mGluRs in pain. In naïve animals, activation of Group III mGluRs with L-AP4 (Table 1) does not alter baseline mechanical[15] or thermal sensitivity[52] and inhibition of Group III mGluRs with the antagonist UBP1112 (Table 1) does not induce any spontaneous behaviors.[52] In the context of capsaicin injection, two similar studies demonstrated that intraplantar delivery of L-AP4 reduced capsaicin-induced spontaneous behaviors[45] and thermal hyperalgesia.[52] Antagonism of Group II/III or just

Group III mGluRs enhanced capsaicin-induced pain-like behaviors, and this enhancement was reversed with L-AP4.[45,52] The Group III agonist L-AP4 can reduce forskolin-induced thermal hyperalgesia.[52] Thus, the ability of Group III mGluRs to inhibit TRPV1-induced neuronal activation at the sensory terminal may be through a G_i cAMP-dependent mechanism. Group II/III antagonism in the paw causes thermal hypersensitivity when co-applied with the exogenous ligand L-glutamic acid hydrochloride (GLU) even though GLU does not cause thermal hyperalgesia on its own.[45] Intraplantar delivery of the Group III agonist L-AP4 reduced bee-venom-induced spontaneous nocifensive behavior but did not reduce bee-venom-induced mechanical hypersensitivity.[36] Overall, these data suggest that signaling via Group III mGluRs during peripheral nociceptor activation attenuates thermal pain-like responses.

In an intact skin-nerve preparation, skin-applied capsaicin induces an increase in single unit activity in the nerve.[45] This increase is enhanced with Group III inhibition and attenuated with Group III activation. Corresponding studies from *in vitro* DRG neurons demonstrate an increase in capsaicin-induced Ca^{2+} activity with co-application of a Group II/III antagonist but no effect on Ca^{2+} activity with Group II/III antagonist alone.[45] Application of the Group II/III antagonist alone (no capsaicin) does not change single unit activity in the skin-nerve preparation. Overall, these data suggest that Group III mGluRs are recruited after nociceptor activation (Table 2). The possible role such recruitment has in chronic pain remains unexplored.

3.2 Spinal cord (secondary sensory neurons and interneurons)

The role of mGluRs in nociceptive processing in the spinal cord has been intensely studied. As the first site of central processing of nociceptive signals, mGluRs in the spinal cord play a critical modulatory role in acute nociception and in the transition from acute to chronic pain (Table 3). Primary afferents use glutamate as the primary excitatory neurotransmitter at the afferent-dorsal horn neuron synapse. There are of course many challenges and caveats to studying mGluRs in the spinal cord. These include the presence of presynaptic receptors (e.g., mGluR4/7) on peripheral neurons and the diversity of cell types in the spinal cord (local interneurons, ascending neurons, glia). In the context of this complexity, a picture has emerged during the last 20 years that firmly cements the importance of various mGluRs in the spinal cord processing of pain. In particular, the dynamic modulation of mGluR signaling

Table 3 Summary of behavioral effects of mGluR activation in the spinal cord

		Acute effects (naive animals)			Postinjury		
		Spontaneous	Mechanical	Thermal	Spontaneous	Mechanical	Thermal
Group I	mGluR1/5	increase in pain	increase in pain	increase in pain	increase in pain	increase in pain	no change in pain
	mGluR1	increase in pain	not tested	not tested	increase in pain	increase in pain	increase / decrease in pain
	mGluR5	increase in pain	not tested	increase in pain	increase in pain	increase in pain	increase in pain
Group II	mGluR2/3	increase in pain	increase in pain	not tested	increase / decrease in pain	increase / decrease in pain	decrease in pain
Group III	mGluR4/7	no change in pain	decrease / no change in pain	decrease / no change in pain	decrease in pain	decrease in pain	decrease in pain

Legend: increase in pain; decrease in pain; no change in pain; not tested

Specifically, effects are indicated by the color (different gray shades) of the circle. Red (black in the print version) means that activation of the receptor(s) increases pain-like responses, while blue (medium gray in the print version) indicates a reduction in pain-like responses with activation of the indicated receptor(s). A light gray circle represents no change in pain upon activation, and a white circle represents a situation that has not been specifically addressed yet in the literature. In situations with conflicting data (i.e., one study shows an increase in pain and another shows a decrease), circles are half colored (different gray shades). Finally, all circles are representative of the effect of activating the receptor(s), but the actual experimental data may be from a study using antagonists to analyze the function of the receptor. In this table, "postinjury" includes any manipulation to the animal that induces or mimics an injury. Injury can include intraplantar injections of formalin, carrageenan, or capsaicin along with injury models such as chronic constriction injury and spinal cord injury. Spontaneous tests include standard spontaneous assays (e.g., formalin test) as well as newer tests using conditioned place preference unless the conditioning was done with an experimenter-guided mechanical or thermal stimulus.

appears to play an important role in the development of central sensitization.[53] In the spinal cord, there is evidence for mRNA expression of mGluR1, 5, 2, 3, 4, and 7 but not 6 or 8[18,25] (Fig. 4).

3.2.1 Group I activity in the spinal cord

mGluR1 and mGluR5 have been found to be expressed in a variety of cell types in the dorsal horn of the spinal cord[18] (Fig. 4). This includes both GABAergic interneurons and ascending neurons and glial cells.[27] mGluR1 expression is concentrated in lamina I and II but diffuse mGluR1 staining is found throughout the dorsal horn.[17] mGluR5 is found perisynaptically or postsynaptically on synapses between unmyelinated afferents and spinal cord neurons in lamina I and II.[27,54,55] Similar to the role of Group I mGluRs in

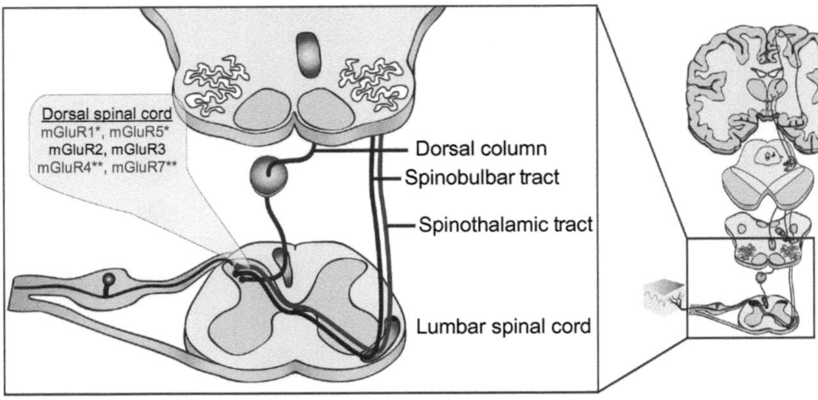

Figure 4 Illustration of mGluR expression in the spinal cord. mGluRs are expressed both presynaptically and/or postsynaptically in the dorsal horn of the spinal cord. *, postsynaptic only (in spinal interneuron or on dorsal horn projection neuron); **, presynaptic only (on primary sensory neuron). *Image adapted from original art by Maria Elena Morales, used with permission.*

the periphery (nociceptors or DRG), activation of mGluR1 and mGluR5 in the dorsal spinal cord tends to increase pain-like behaviors and neuronal excitability. Intrathecal delivery of the Group I agonist DHPG induces mechanical and thermal (hot or cold) hypersensitivity[56,57] as well as spontaneous nocifensive behavior in naïve animals.[58,59] Both mGluR1 and mGluR5 inhibition in the spinal cord can reduce the spontaneous effects of DHPG.[59] Intrathecal DHPG also enhances nocifensive responses to peripheral formalin[60] or complete Freund's adjuvant (CFA) inflammation.[61] Intrathecal inhibition of mGluR5 with MPEP, inhibition of mGluR1 with CPCCOEt, or antisense knockdown of mGluR1 (~40%) reduces pain-like behavior in the formalin test.[59,62]

Suggesting a functional difference in spinal Group I mGluRs after injury, mGluR1 but not mGluR5 expression is increased in the spinal cord following spinal cord injury.[63] Antagonism of mGluR1 but not mGluR5 in the spinal cord reduces mechanical allodynia after this spinal injury, while antagonism of mGluR1 and mGluR5 has opposite effects on thermal hyperalgesia.[64] Inhibition of mGluR1 increases thermal hyperalgesia, while inhibition of mGluR5 decreases thermal hyperalgesia following this spinal injury. However, in other neuropathic models (sciatic nerve injury), mGluR5 expression is increased in lamina II neurons and modulates post-injury thermal hyperalgesia.[14] In another model of nerve injury, chronic constriction injury, intrathecal inhibition of both mGluR1 (with AIDA) and mGluR5 (with MPEP) reduces mechanical hyperalgesia, and inhibition of mGluR1 reduces cold hyperalgesia.[65] These data from injury models suggest two possibilities. First, distinct injuries may recruit different receptors as part of the long-term changes that accompany the injury. This interpretation would suggest that the efficacy of specific Group I analgesic agents for use in humans would depend in part on the type of pain condition or injury. Alternatively, the differences in expression and function between mGluR1 and mGluR5 after injury may be due to more subtle experimental differences in the studies described above. In this context, caution must be used in interpreting whether one Group I mGluR dominates over the other in chronic pain states.

As postsynaptic modulators, activation of Group I mGluRs tends to depolarize dorsal horn neurons.[10] Both mGluR1 and mGluR5 in the spinal cord are capable of phosphorylating extracellular-signal-regulated kinases 1 and 2 (ERK1/2), which is necessary for spontaneous nocifensive behaviors after intraplantar formalin[59] or CFA[61] inflammation. Subsequent studies identified the potassium leak channel $Kv_{4.2}$ as a critical target downstream of mGluR1/5 and ERK1/2.[10,66] Activation of mGluR1/5 in dorsal horn neurons leads to activation of ERK1/2, which then phosphorylates (and closes) $Kv_{4.2}$. In $Kv_{4.2}$ knockout mice, DHPG does not increase excitability of spinal cord neurons *in vitro* and induces only mild spontaneous nocifensive behavior compared to wild-type mice.[10] Interestingly, the connection between Group I mGluRs and $Kv_{4.2}$ seems to be mediated by mGluR5 only as the mGluR1 antagonist LY367385 does not change DHPG-induced ERK1/2 activation or $Kv_{4.2}$ A-type currents. In this circumstance, the phosphorylation of ERK1/2 by mGluR5 could be PKC and/or PKA dependent,[66] although future studies will need to assess this directly. The intracellular signaling cascade associated with mGluR1 in the spinal cord is still unknown, although the behavioral results described above suggest a role in increasing excitability in dorsal horn neurons. Supporting this, the Group I agonist DHPG enhances excitability of primate dorsal horn neurons during noxious mechanical stimulation under naïve conditions and following peripheral capsaicin stimulation.[67] This facilitation was blocked by the mGluR1 antagonist CPCCOEt, and the effect was actually reversed (i.e., inhibition rather than facilitation of neurons) at high doses of DHPG. The net result of mGluR1 and mGluR5 modulation is to depolarize the neuron, leading to hyperexcitability and long-term potentiation.[53] However, there is also evidence for spinal Group I mGluR activation leading to long-term depression (LTD),[68] which may explain some studies showing that high levels of the Group I mGluR agonist, DHPG, inhibit spinal neuronal responses to noxious stimulation.[67] Group I mGluR LTD is driven by IP3 production leading to increased intracellular calcium and activation of protein phosphatase I by calcineurin.[68]

Activation of mGluR1 and mGluR5 in the spinal cord tends to increase all manner of pain-like behavior, but there are some intriguing differences in the receptors based on the type of injury (inflammatory vs. neuropathic) or site of injury (peripheral nerve vs. spinal cord), which suggest that there may be subtle but potentially interesting differences between mGluR1 and mGluR5 in the spinal cord (Table 3).

3.2.2 Group II activity in the spinal cord

Although Group II expression is thought to be mainly presynaptic from primary afferents in the spinal cord, there is evidence using nonspecific mGluR2/3 antibodies of postsynaptic staining in the inner part of lamina II[17,27] and on astrocytes[17] (Fig. 4). Group II expression in the spinal cord is decreased following spinal cord injury.[63] Following intradermal CFA inflammation, mGluR2 and mGluR3 mRNA is increased,[69] and following peripheral UV radiation-induced inflammation, mGluR3 mRNA is increased.[25] In the UV radiation study, no change in mRNA for mGluR1, 2, 4, 5, and 7 was found. Intrathecal activation of Group II mGluRs with ACPD (Table 1) induces spontaneous nociceptive behaviors[58] and facilitates formalin pain behavior,[60] but these data are difficult to interpret because of the poor specificity of ACPD for mGluR2/3 (Table 1). More recent data with a selective Group II mGluR agonist, DCG-IV (Table 1), show that mGluR2/3 activation in the spinal cord induces hyperalgesia in paw-pressure nociception in naïve rats.[70] Selective mGluR2/3 inhibition dose-dependently reduces CFA-induced mechanical but not thermal hyperalgesia.[69] In contrast to the pronociceptive data for mGluR2/3 in naïve animals and after peripheral inflammation, mGluR2/3 seems to be antinociceptive in neuropathic models. Following spinal nerve ligation (Chung method), intrathecal DCG-IV decreases injury-induced mechanical hypersensitivity and injury-induced paw-pressure nociception.[70] Pre-treatment with the selective Group II agonist APDC (not ACPD) reduced chronic constriction injury-induced mechanical and cold hyperalgesia.[65] Finally, intrathecal APDC can dose-dependently decrease spontaneous nociceptive behaviors driven by intrathecal delivery of the mGluR1/5 agonist DHPG.[71] Overall, activation of spinal Group II mGluRs reduces pain-like responses after neuropathic injury but can, in other circumstances, increase pain-like responses in naïve or inflammation-injured animals (Table 3). This apparent paradox may be explained by distinct populations of Group II mGluRs in the spinal cord (e.g., pre- vs. postsynaptic).

At the physiological level, Group II receptor agonists depress dorsal horn excitability via a presynaptic mechanism.[49] The Group II antagonist EGLU (Table 1) had a greater inhibitory effect on polysynaptic EPSCs in spinal nerve-ligated animals compared to control animals.[70] Intrathecal Group II blockade suppressed monosynaptic postsynaptic potentials via a presynaptic mechanism.[49] A role of Group II mGluRs in the spinal cord seems to be in reducing GABAergic interneuron activity during increased peripheral nociceptive input.[72] This reduction in GABAergic tone would serve to disinhibit projection neurons and may be a component of the pronociceptive role that Group II mGluRs play in mechanical pressure pain described above.[70]

3.2.3 Group III activity in the spinal cord

There is evidence for expression of mGluR4 and mGluR7 in the spinal cord although both receptors appear to be primarily expressed on the afferent terminals of primary afferent neurons in the superficial laminae of the spinal cord[33,34] (Fig. 4). On a functional level, intrathecal injection of the Group III agonist LAP4 does not induce any spontaneous nociceptive behaviors[58] and does not alter baseline thermal or mechanical sensitivity.[73] Inhibition of spinal Group III mGluRs with MAP4, however, induces mechanical and thermal hyperalgesia in naïve animals.[73] These differential effects of activating Group III mGluRs (no effect) versus inhibiting Group III mGluRs (pronociceptive) in naïve animals may be due to cell-type-specific (glutamatergic vs. GABAergic cells) activity of Group III mGluRs. Specifically, pharmacological activation of Group III mGluRs may produce a net-zero effect by simultaneously reducing both primary sensory neuron excitatory and interneuron inhibitory inputs on dorsal horn neurons. In contrast, inhibition of Group III mGluRs may specifically facilitate glutamate release at the central synapse by reducing ongoing Group III autoinhibition of primary sensory neurons, which would increase pain facilitation. For this hypothesis to be true, it would suggest that presynaptic Group III mGluRs on interneurons are not occupied when the primary sensory neuron is largely quiet (i.e., glutamate release is low).

In fact, Group III mGluR activation has an effect in the context of injury when primary sensory neurons are excited. Pretreatment of the spinal cord with the Group III agonist LAP4 reduces chronic constriction injury-induced mechanical and cold hyperalgesia[65] and L5/L6 spinal nerve ligation-induced mechanical allodynia.[73] Again, the difference in LAP4 between naïve and injured animals may be due to the increase in primary sensory neuron glutamatergic drive and decrease in spinal cord interneuron inhibitory drive that occurs after neuropathic injury.[74,75] Thus, after injury, LAP4 is able to induce antinociception by reducing excitatory drive on ascending dorsal horn neurons.

Activation of Group III mGluRs by LAP4 in the spinal cord reduces evoked activity to mechanical stimulation (brush, press, pinch) in control conditions and following capsaicin in anesthetized macaques,[76] but LAP4 only reduces evoked activity after injury in rats.[73] These effects after injury were likely mediated by Group III mGluRs in the spinal cord reducing glutamatergic drive on dorsal horn neurons. Activation of the Group III mGluRs with LAP4 suppresses both monosynaptic and polysynaptic EPSPs in spinal cord slices.[49] As mentioned above, a role of Group III mGluRs in the spinal cord seems to be in reducing GABAergic tone during increased peripheral nociceptive input.[49,72] Capsaicin applied to primary nerve endings in a spinal cord preparation inhibits IPSCs. This effect of capsaicin can be blocked by treatment with CPPG, a Group III mGluR antagonist (Table 1).[72] Overall, activation of Group III mGluRs in the spinal cord inhibits nociceptors and interneurons. The determinative factors as to whether activation of Group III mGluRs fails to alter pain-like responses or decrease pain-like responses (Table 3) likely include the state of the primary sensory nociceptor (normal or sensitized), state of the GABAergic interneurons (normal or desensitized), and the expression and localization of mGluRs in these cells. One of the biggest open questions is whether the expression or subcellular localization of these receptors changes on either primary afferents or interneurons. Dynamic changes following injury may explain the complex and confusing activity of Group III mGluRs in the spinal cord.

3.3 Brainstem

As mentioned above, a tremendous amount of research has focused on the processing and modulation of nociceptive information at anatomic loci in the brainstem (e.g., rostral ventral medulla (RVM), PAG, parabrachial nucleus (PB)) and brain (e.g., thalamus, anterior cingulate cortex (ACC), PFC, and amygdala). Not surprisingly, mGluRs have been identified at the expression and functional level in nearly all higher centers in the pain neuraxis[77,78] (Fig. 2). In the following sections, we will review these data separately for the brainstem (e.g., PAG; Fig. 5) and nociceptive brain areas (Section 3.4; e.g., thalamus, ACC, and amygdala; Fig. 6).

The ascending projections of dorsal horn neurons project primarily to the PAG/PB (via spinobulbar projections) and thalamus (via spinothalamic

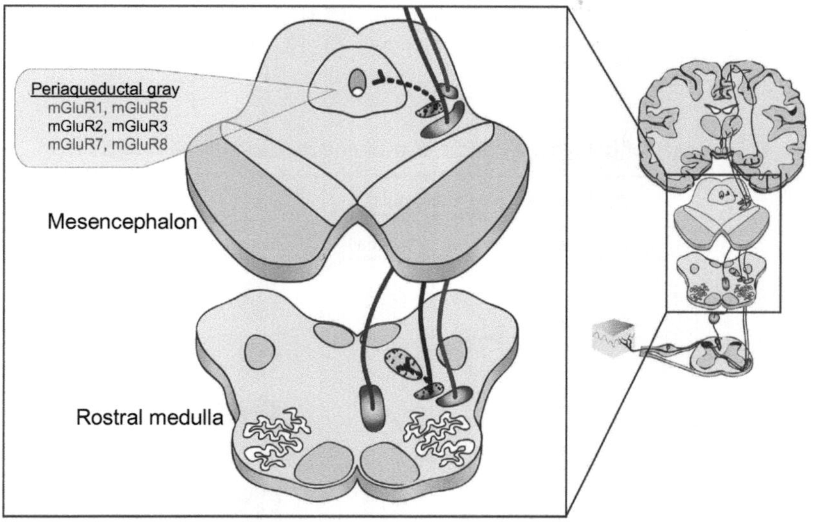

Figure 5 Illustration of mGluR expression in the PAG. mGluRs are expressed in the PAG as part of both the ascending and descending components of the pain neuraxis. *Image adapted from original art by Maria Elena Morales, used with permission.*

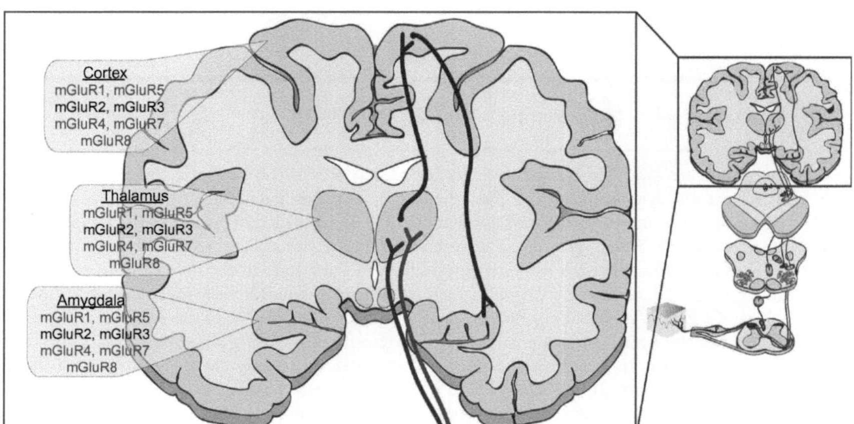

Figure 6 Illustration of mGluR expression in the brain. mGluRs expressed in the thalamus, cortex, and amygdala modulate ascending and descending components of nociceptive information. *Image adapted from original art by Maria Elena Morales, used with permission.*

tract). Modulation at these levels can influence nociceptive input as well as descending output to modulate pain. The behavioral roles of mGluRs in the PAG can be opposite of those found in the periphery, spinal cord, or other areas. For example, the activation of Group I mGluRs (which are pronociceptive in the periphery) causes antinociception, and activation of Group II and III mGluRs (which are antinociceptive in the periphery) causes increased nociception in the PAG (Table 4).

Table 4 Summary of behavioral effects of mGluR activation in the brainstem

		Acute effects (naive animals)			Postinjury		
		Spontaneous	Mechanical	Thermal	Spontaneous	Mechanical	Thermal
	mGluR1/5	not tested	not tested	decrease in pain	decrease in pain	not tested	not tested
Group I	mGluR1	not tested	not tested	not tested	decrease in pain	not tested	no change in pain
	mGluR5	not tested	not tested	not tested	increase in pain / decrease in pain	not tested	decrease in pain
Group II	mGluR2/3	not tested	not tested	increase in pain / no change in pain	decrease in pain	not tested	decrease in pain
	mGluR4/7/8	not tested	not tested	increase in pain / decrease in pain	increase in pain	not tested	decrease in pain
Group III	mGluR7	no change in pain	not tested	increase in pain	not tested	not tested	not tested
	mGluR8	no change in pain	not tested	decrease in pain	decrease in pain	decrease in pain	decrease in pain

Legend: ● increase in pain; ● decrease in pain; ● no change in pain; ⊘ not tested

Specifically, effects are indicated by the color (different gray shades) of the circle. Red (black in the print version) means that activation of the receptor(s) increases pain-like responses, while blue (medium gray in the print version) indicates a reduction in pain-like responses with activation of the indicated receptor(s). A light gray circle represents no change in pain upon activation, and a white circle represents a situation that has not been specifically addressed yet in the literature. In situations with conflicting data (i.e., one study shows an increase in pain and another shows a decrease), circles are half colored (different gray shades). Finally, all circles are representative of the effect of activating the receptor(s), but the actual experimental data may be from a study using antagonists to analyze the function of the receptor. In this table, "postinjury" includes any manipulation to the animal that induces or mimics an injury. Injury can include intraplantar injections of formalin, carrageenan, or capsaicin along with injury models such as chronic constriction injury and spinal cord injury. Spontaneous tests include standard spontaneous assays (e.g., formalin test) as well as newer tests using conditioned place preference unless the conditioning was done with an experimenter-guided mechanical or thermal stimulus.

3.3.1 Group I activity in the brainstem

At the level of the PAG, mGluR5 receptors seem to play a more prominent role in modulating pain compared to mGluR1 (Table 4). mGluR5 is expressed on dendrites and cell bodies of PAG neurons[19] (Fig. 5). The mGluR1/5 agonist DHPG in the PAG induces antinociceptive behavior in the hot plate test.[78] Low doses of the mGluR5 antagonist MPEP in the dorsal PAG increase nociceptive behavior in the first phase of the formalin test, while higher doses of MPEP decrease nociceptive behavior in the second phase of the formalin test.[79] The antinociceptive effect of high dose MPEP may be related to nonspecific effects of MPEP because DHPG (mGluR1/5 agonist) also decreases nociceptive behavior in the second phase of the formalin test.[80] In addition to mGluR5, the antinociceptive effects of DHPG in the formalin test are also driven by mGluR1 activity.[80]

PAG-directed MPEP but not the mGluR1 antagonist CPCCOEt blocked the antinociceptive effects of cannabinoid agonists in the thermal Hargreaves test.[81] Further suggesting an overall antinociceptive role of mGluR5 in the PAG, the mGluR5 agonist CHPG reduced thermal sensitivity in the Hargreaves assay.[81] One proposed mechanism of mGluR5-induced analgesia in the PAG is via presynaptic inhibition of GABAergic cells. In this scenario, activity-dependent increases in nearby glutamate release "spill over" to GABAergic neurons where presynaptic mGluR5 (on the GABAergic cells) causes circuit disinhibition. In other words, after mGluR5 activation, the subsequent reduction in GABA release disinhibits descending analgesic neurons. There is also evidence from PAG slice recordings for a postsynaptic excitatory role for mGluR1 and mGluR5.[82] Inhibition of mGluR5, but not mGluR1, in the PAG is able to block analgesia induced by PAG delivery of capsaicin.[83] In this scenario, activation of TRPV1 increases glutamate release, which activates mGluR5. mGluR activation leads to PLC-beta activation and yields DAG, which is then converted to the endocannabinoid 2-arachidonoylglycerol (2-AG). 2-AG then activates CB1 receptors to presynaptically reduce GABA release onto PAG analgesic projection neurons. A similar role of mGluR5 in the analgesic effects of intra-PAG neurotensin has also been postulated,[84] suggesting that multiple types of inputs in the PAG may all use mGluR5 as an intermediate in PAG-induced antinociception. This retrograde inhibition of neurons is similar to endocannabinoid and mGluR5-mediated LTD in the hippocampus.[85]

3.3.2 Group II activity in the brainstem

Depending on the pain assay used, there is evidence for both a pronociceptive and an antinociceptive action of Group II mGluRs in the PAG (Table 4). Delivery of the Group II agonist L-CCG-I (Table 1) to the PAG increased pain-like responses in the hot plate test[78] and in the Hargreaves plantar test.[81] In contrast to this pronociceptive role of Group II mGluRs in the PAG, CCG-I reduced spontaneous behavior in formalin test[80] and inhibition of Group II mGluRs with EGLU was able to block the thermal analgesic effects of a cannabinoid agonist.[81] When given alone, EGLU does not alter baseline thermal sensitivity.[81] Although Group II receptors are thought to act presynaptically in the PAG, there is evidence that a small number of PAG neurons respond to postsynaptic Group II mGluR activation.[82] Presynaptic activation of Group II mGluRs with the agonist, DCG-IV, reduced GABA-evoked IPSCs.[86] This regional diversity of Group II mGluRs may explain somewhat contradictory behavioral results during PAG Group II mGluR activation.

3.3.3 Group III activity in the brainstem

Similar to the dual nature of Group II mGluRs in the brainstem, activation of the Group III mGluRs in the PAG can be pro- or antinociceptive[80] (Table 4). Activation of Group III mGluRs with the agonist L-SOP (Table 1) induces thermal hypersensitivity[78] and increases pain behavior in the second phase of the formalin test.[80] Inhibition of Group III mGluRs with the antagonist MSOP (Table 1) reduced pain-like responses in the second phase of the formalin test.[79] However, MSOP reversed the thermal analgesic effects of a cannabinoid agonist and induced thermal hyperalgesia when used alone,[81] suggesting that Group III mGluRs can also be antinociceptive. In fact, further analyses of Group III mGluRs using receptor-specific agents have revealed some interesting subtleties for the role of these receptors at the level of the PAG. Intra-PAG delivery of the mGluR7 agonist, AMN082 (Table 1), increased pain-like behavior in the hot plate and Hargreaves plantar test.[87,88] Activation of mGluR8 with DCPG (Table 1) decreased pain-like behavior in the hot plate,[88] reduced spontaneous formalin behavior, temporarily blocked carrageenan-induced thermal and mechanical hyperalgesia, and reduced mechanical and thermal hyperalgesia 3 days after chronic constriction injury.[87] Corresponding to the behavioral effects of mGluR7 and mGluR8 in the PAG, activation of these receptors had opposing effects on glutamate release in the PAG and contrasting effects on the RVM ON and RVM OFF cells. Molecularly, activation of mGluR7 in the PAG causes a decrease in glutamate with a corresponding activation of RVM ON cells and inhibition of RVM OFF cells.[88] mGluR8, by contrast, in the PAG causes an increase in glutamate[89] with an opposite effect on RVM ON and OFF cells.[88] Presynaptic activation of all Group III mGluRs with the agonist, L-AP4, reduced GABA-evoked IPSCs.[86]

3.4 Brain

Due to the complex and subjective nature of chronic pain along with the high incident of affective comorbidities in chronic pain patients,[90] many studies have evaluated the role of mGluRs in modulating the sensory, cognitive, and emotional components of pain in higher brain areas (Fig. 6) such as the thalamus (Table 5), amygdala (Table 6), and PFC and ACC (Table 7).

Table 5 Summary of behavioral effects of mGluR activation in the thalamus

		Acute effects (naive animals)			Postinjury		
		Spontaneous	Mechanical	Thermal	Spontaneous	Mechanical	Thermal
	mGluR1/5	no change in pain	not tested	not tested	increase in pain	not tested	not tested
Group I	mGluR1	not tested	not tested	not tested	increase in pain	not tested	not tested
	mGluR5	not tested	not tested	not tested	no change in pain	not tested	not tested
Group II	mGluR2/3	not tested	not tested	not tested	not tested	increase in pain	not tested
Group III	mGluR4/7/8	not tested	not tested	not tested	not tested	not tested	not tested

● increase in pain ● decrease in pain ● no change in pain ⊘ not tested

Specifically, effects are indicated by the color (different gray shades) of the circle. Red (black in the print version) means that activation of the receptor(s) increases pain-like responses, while blue (medium gray in the print version) indicates a reduction in pain-like responses with activation of the indicated receptor(s). A light gray circle represents no change in pain upon activation, and a white circle represents a situation that has not been specifically addressed yet in the literature. In situations with conflicting data (i.e., one study shows an increase in pain and another shows a decrease), circles are half colored (different gray shades). Finally, all circles are representative of the effect of activating the receptor(s), but the actual experimental data may be from a study using antagonists to analyze the function of the receptor. In this table, "postinjury" includes any manipulation to the animal that induces or mimics an injury. Injury can include intraplantar injections of formalin, carrageenan, or capsaicin along with injury models such as chronic constriction injury and spinal cord injury. Spontaneous tests include standard spontaneous assays (e.g., formalin test) as well as newer tests using conditioned place preference unless the conditioning was done with an experimenter-guided mechanical or thermal stimulus.

Table 6 Summary of behavioral effects of mGluR activation in the amygdala

		Acute effects (naive animals)			Postinjury		
		Spontaneous	Mechanical	Thermal	Spontaneous	Mechanical	Thermal
CeA							
Group I	mGluR1/5	no change in pain	increase in pain	not tested	increase in pain	not tested	not tested
	mGluR1	not tested	not tested	not tested	increase in pain	increase in pain	not tested
	mGluR5	no change in pain	increase in pain / no change in pain	not tested	no change in pain	increase in pain	not tested
Group II	mGluR2/3	not tested	not tested	not tested	not tested	not tested	not tested
Group III	mGluR7	not tested	increase in pain	not tested	not tested	no change in pain	not tested
	mGluR8	not tested	no change in pain	no change in pain	not tested	decrease in pain	decrease in pain
BLA							
Group I	mGluR1	not tested	not tested	not tested	not tested	increase in pain	not tested
	mGluR5	not tested	not tested	not tested	decrease in pain	no change in pain	not tested
Group II	mGluR2/3	not tested	not tested	not tested	not tested	not tested	not tested
Group III	mGluR4/7/8	not tested	not tested	not tested	not tested	not tested	not tested

Legend: increase in pain (black circle); decrease in pain (dark gray circle); no change in pain (light gray circle); not tested (white circle with slash)

Specifically, effects are indicated by the color (different gray shades) of the circle. Red (black in the print version) means that activation of the receptor(s) increases pain-like responses, while blue (medium gray in the print version) indicates a reduction in pain-like responses with activation of the indicated receptor(s). A light gray circle represents no change in pain upon activation, and a white circle represents a situation that has not been specifically addressed yet in the literature. In situations with conflicting data (i.e., one study shows an increase in pain and another shows a decrease), circles are half colored (different gray shades). Finally, all circles are representative of the effect of activating the receptor(s), but the actual experimental data may be from a study using antagonists to analyze the function of the receptor. In this table, "postinjury" includes any manipulation to the animal that induces or mimics an injury. Injury can include intraplantar injections of formalin, carrageenan, or capsaicin along with injury models such as chronic constriction injury and spinal cord injury. Spontaneous tests include standard spontaneous assays (e.g., formalin test) as well as newer tests using conditioned place preference unless the conditioning was done with an experimenter-guided mechanical or thermal stimulus. CeA, central nucleus of the amygdala; BLA, basolateral nucleus of the amygdala.

Table 7 Summary of behavioral effects of mGluR activation in the cortex

			Acute effects (naive animals)			Postinjury		
			Spontaneous	Mechanical	Thermal	Spontaneous	Mechanical	Thermal
PFC								
	Group I	mGluR1/5	Increase in pain	Not tested	Not tested	Not tested	Not tested	Not tested
	Group II	mGluR2/3	Not tested	Not tested	Not tested	Not tested	Not tested	Not tested
	Group III	mGluR4/7/8	Not tested	Not tested	Not tested	Not tested	Not tested	Not tested
ACC								
	Group I	mGluR1/5	No change in pain	Not tested	Increase in pain	Not tested	Not tested	Not tested
	Group II	mGluR2/3	Not tested	Not tested	Increase in pain	Not tested	Not tested	Not tested
	Group III	mGluR4/7/8	Not tested	Not tested	Not tested	Not tested	Not tested	Not tested

Legend: (black circle) increase in pain; (medium gray circle) decrease in pain; (light gray circle) no change in pain; (slashed circle) not tested

Specifically, effects are indicated by the color (different gray shades) of the circle. Red (black in the print version) means that activation of the receptor(s) increases pain-like responses, while blue (medium gray in the print version) indicates a reduction in pain-like responses with activation of the indicated receptor(s). A light gray circle represents no change in pain upon activation, and a white circle represents a situation that has not been specifically addressed yet in the literature. In situations with conflicting data (i.e., one study shows an increase in pain and another shows a decrease), circles are half colored. Finally, all circles are representative of the effect of activating the receptor(s), but the actual experimental data may be from a study using antagonists to analyze the function of the receptor. In this table, "postinjury" includes any manipulation to the animal that induces or mimics an injury. Injury can include intraplantar injections of formalin, carrageenan, or capsaicin along with injury models such as chronic constriction injury and spinal cord injury. Spontaneous tests include standard spontaneous assays (e.g., formalin test) as well as newer tests using conditioned place preference unless the conditioning was done with an experimenter-guided mechanical or thermal stimulus. ACC, anterior cingulate cortex; PFC, prefrontal cortex.

3.4.1 Group I activity in cortical and subcortical areas

In the thalamus, Group I mGluRs are widely expressed with prominent expression of mGluR1 in the ventroposterolateral (VPL) and ventroposteromedial (VPM) nuclei[20] (Fig. 6). Chronic constriction injury induces increases in expression of both mGluR1 and mGluR5 in the thalamus.[11] Administration of the Group I agonist DHPG in the VPL enhanced pain responses in both the first and second phases of the formalin test; this DHPG enhancement of the second phase of the test was dependent on PLC-beta.[91] However, DHPG alone in the VPL (i.e., no formalin) did not induce any spontaneous pain-like behaviors. Furthermore, VPL administration of mGluR1 antagonists (CPCCOEt or AIDA) but not an mGluR5 antagonist (MPEP) reduced pain-like behavior in the second phase of the formalin test. Consistent with a local role in the thalamus for mGluR1 but not mGluR5, it was shown that systemic (IV) but not local thalamic administration of the mGluR5 antagonist MPEP reduced mechanical noxious stimulation-induced excitability of VPL neurons.[92] However, another study found that thermal-induced excitability of thalamic neurons was attenuated by local thalamic administration of both mGluR1 and mGluR5 antagonists.[93]

In the last decade, the amygdala has been a site of intense study of Group I mGluRs in the modulation of pain (Table 6). In particular, distinct roles of Group I mGluRs in the central nucleus of the amygdala (CeA) and basolateral nucleus of the amygdala (BLA) have been described (Fig. 6). The CeA receives direct nociceptive information from the spinal cord and indirect signals from the thalamus and PB.[94] mGluR1 and mGluR5 expression is increased in the CeA following peripheral injury.[21] Activation of mGluR1/5 with DHPG is sufficient to increase somatic and visceral pain-like responses in naïve animals[95–97] via mGluR5 receptors but induces aversion in a pain-related conditioned aversion task acting primarily through mGluR1 receptors.[98] Inhibition of mGluR5 does not alter spontaneous formalin behavior or somatic sensitivity in naïve animals,[96] but the mGluR5 antagonist, MPEP, does decrease responses to acute noxious visceral bladder distention.[95] In the context of spared nerve injury, the mGluR1/5 agonist DHPG facilitated and the mGluR1 antagonist CPCCOEt inhibited pain-related conditioned aversion.[98] Inhibition of mGluR5 through amygdala-specific conditional deletion of mGluR5 did not alter spontaneous formalin behavior but did decrease formalin-induced mechanical hypersensitivity.[96] In an arthritis model of pain, mGluR1 or mGluR5 antagonists had inhibitory effects on ultrasonic or audible vocalizations, respectively, to noxious mechanical stimulation, while mGluR1 but not mGluR5 inhibition reversed arthritis-induced mechanical hypersensitivity.[99] There is evidence that the role of mGluR1/5 in the modulation of pain may be lateralized to the right amygdala. Specifically, inhibition of mGluR5 in the right but not left amygdala with MPEP decreased formalin-induced mechanical hypersensitivity.[96] These data are consistent with reports of lateralization of ERK1/2 activation in the amygdala during injury[100] and physiological data showing that right amygdala neurons have larger nociceptive receptive fields before and after injury compared to left amygdala neurons.[101] Using a mixture of pharmacological and electrophysiological data, the molecular signaling pathway of Group I mGluRs in the amygdala has emerged.

It appears that mGluR1 signals primarily presynaptically (in GABAergic cells) and mGluR5 signals postsynaptically (in CeA neurons).[21] A group of neurons located in the lateral capsular division of the CeA (CeLC) respond to noxious stimuli and show an increase of excitatory transmission relative to inhibitory transmission following arthritis induction.[102] Under acute noxious (i.e., uninjured) circumstances, only mGluR5 antagonists can block nociceptive responses in the CeLC.[103] After injury, by contrast, both mGluR5 and mGluR1 antagonists can reduce the injury-induced hyperexcitability.[21,103] mGluR1 appears to act presynaptically to inhibit GABAergic neurons whose cell bodies lie in the intercalated cell mass outside the CeA.[102] This activity disinhibits neurons within the CeLC. On the other hand, mGluR5 increases excitability postsynaptically on CeA neurons.[102] mGluR5 effects are dependent on IP3 (but not PKC) leading to reactive oxygen species (ROS) production and activation of ERK1/2 and PKA.[96,97] mGluR1 action in the CeA is dependent on Homer1a[104] signaling and ROS production.[105]

In the BLA, inhibition of mGluR1 (with CPCCOEt) but not mGluR5 (with MPEP) reduces mechanical allodynia associated with an arthritis model.[106] The role of Group I mGluRs in the BLA is related to role of these receptors in the PFC and plays a significant role in the modulation of decision making during pain.[107] After painful injury, medial PFC neurons are inhibited and this is thought to disrupt normal decision making. Activation of mGluR1/5 with DHPG in the BLA depresses PFC neurons through a BLA mGluR1-dependent mechanism.[106] Activation with DHPG in the PFC but not the ACC can impair decision making as is seen in chronic pain states.[22] Chronic pain impairs decision making through a physiological inhibition of PFC pyramidal neurons.[107] The physiological inhibition can be mimicked in naïve animals using the mGluR1/5 agonist DHPG directed at the PFC and can be reversed with an mGluR1 antagonist (LY367385) but not an mGluR5 antagonist (MPEP) in the PFC.[108] Chronic constriction injury induces increases in expression of mGluR1 along with Homer scaffold proteins in the PFC.[11]

Activation of Group I and Group II mGluRs in the ACC with ACPD facilitates pain-like behavior in the tail-flick reflex.[109] Following painful peripheral amputation, LTD in the ACC is impaired.[110] This impairment of LTD is that the ACC is coupled with findings showing an enhancement of excitatory synaptic activity in the ACC after injury. Interestingly, the loss of LTD is related to signaling via mGluR1. Although activation of mGluR1 on its own cannot induce LTD in amputated animals, mGluR1 activation can prime the system for LFS (low-frequency stimulation)-induced LTD. DHPG can prime LFS–LTD after injury, and this effect is blocked with the mGluR1 antagonist, LY367385, but not an mGluR5 antagonist.[110]

3.4.2 Group II activity in cortical and subcortical areas

Moderate Group II mGluR staining is found in presynaptic and postsynaptic sites throughout the brain.[111] Although lack of specific antibodies has limited the specific characterization of mGluR2 versus mGluR3, evidence from mGluR2 knockout animals (i.e., mGluR3 staining only) has shown that both receptors are expressed at high levels throughout brain areas involved in pain processing (Fig. 6). Group II mGluRs are expressed at low levels in the nociceptive components of the thalamus[20,28] including the reticular thalamus. Inhibition of Group II mGluRs in the reticular thalamus with the antagonist, EGLU, reduces mechanical allodynia in an arthritic model.[112] No other studies to date have probed the behavioral (Table 5), physiological, or molecular implications of Group II mGluRs in the thalamus leaving this as an important area for future study.[20] Nonetheless, the available data suggest that Group II mGluRs in the thalamus serve a pronociceptive role during injury. It has been hypothesized that these receptors inhibit GABAergic cells in the thalamus.[112,113] Thus, inhibition of Group II mGluRs with EGLU would serve to disinhibit GABAergic cells causing a reduction in the excitability of thalamic projection neurons and a reduction in allodynia.

In the CeA, the Group II agonist LY354740 reduces EPSCs in normal animals via a presynaptic mechanism; this effect is enhanced in slices from animals with arthritis.[114] Extracellular recordings from the CeA in injured animals show that Group II activation can reduce responses to noxious stimulation of the knee in arthritis.[115] The effects of the agonist are enhanced in injured animals relative to normal animals but only for noxious stimulation. Inhibitory effects of Group II activation during innocuous stimulation do not differ based on the pain state. As Group II mGluRs are classically thought to be G_i coupled, modulation of PKA via Group II mGluR signaling in the CeA may counterbalance mGluR1 signaling, which is PKA dependent in the CeA.[97]

Following arthritis-like injury, mGluR3 mRNA is increased throughout the cortex with noticeably strong increases in the ACC.[116] As mentioned above, the dietary supplement LAC decreases mechanical allodynia through a mechanism involving mGluR2 receptors. LAC treatment also selectively increases expression of Group II protein in the cerebral cortex.[117] This may be related to emotion and cognition during pain. Increased expression of Group II mGluRs with LAC could serve to inhibit excitatory cortical areas and would thereby lessen the impact of noxious stimuli on cognition and emotion. Otherwise, no other behavioral (Table 7) or electrophysiological studies have evaluated Group II mGluRs in the ACC or other cortical areas in the context of pain.

3.4.3 Group III activity in cortical and subcortical areas

mGluR4, mGluR7, and mGluR8 are expressed in the thalamus, amygdala, frontal cortex, and ACC[35] (Fig. 6). Similar to the Group II mGluRs, Group III mGluRs are expressed at low levels in the nociceptive components of the thalamus.[20,118] mGluR4 and mGluR7 predominate in the VPL/VPM with mGluR7 and mGluR8 also expressed in the reticular thalamic nucleus.[20] No study to date has systematically studied the behavioral (Table 5), physiological, or molecular implications of Group III mGluRs in the thalamus.

mGluR8 expression is increased in the CeA following arthritis injury in animals.[119] mGluR7 and mGluR8 in the CeA have differential effects on pain-like behavior depending on the pain state of the animal (Table 6). In naïve animals, mGluR7 (AMN082) but not mGluR8 (DCPG) agonists in the CeA increase pain-like responses to noxious mechanical stimulation.[119,120] After arthritis injury, by contrast, mGluR8 but not mGluR7 activation actually decreases pain-like responses to mechanical stimulation[120] and mGluR8 activation decreases responses to thermal stimulation after intraplantar carrageenan.[119] These opposing roles for mGluR7 and mGluR8 in pain are similar to the effects of mGluR7 versus mGluR8 in the PAG (see above Section 3.3.3). Somewhat surprisingly, this antinociceptive effect of pharmacological activation of mGluR8 in the CeA is coupled to an increase in CeA glutamate and decrease in GABA.[119] This association is surprising because a similar effect on neurotransmitters has been postulated for the opposite behavioral effects of Group I mGluR activation in the CeA. In extracellular recordings from the CeA in injured animals, Group III activation reduces responses to noxious stimulation of the knee in arthritis,[115] but unlike Group II mGluRs, activation of Group III mGluRs causes greater inhibition of CeA neurons to innocuous stimulation after injury. In slices from arthritic animals, mGluR8 activation reduces evoked EPSCs, while mGluR7 activation with AMN082 increases glutamatergic signaling by presynaptic inhibition of GABAergic neurons.[121] It should be noted that the evolving role of Group III mGluRs in the CeA is missing important data on other members of this mGluR family in the amygdala. Group III mRNA is increased in arthritic rats in the ACC, but no studies to date have analyzed these receptors in pain-like behavior (Table 7) or physiology.[116]

3.5 Conventional knockout and systemic effects

A number of scientific studies have evaluated the role of mGluRs in pain at an organismal level. These include studies looking at whole-animal genetic "knockouts" of *mGluR* genes or systemic delivery of pharmacological agents to an animal. (*Note*: All of the mGluRs have been genetically knocked-out out, but only a few have been evaluated for pain-like changes.) The drawback of these systemic studies, of course, is that by treating the entire animal, region-specific information is hard to obtain. Any developmental compensation for the knockout may also give rise to a different phenotype than other functional studies reveal. On the other hand, studies of the systemic effects of blocking or activating different mGluRs can be useful in predicting outcomes in future clinical trials with these or new agents, including revealing important side effects of mGluR activation or inhibition outside of the pain system. In particular, as we have seen, there are some areas of the pain neuraxis where the "typical" roles of the mGluRs seem to reverse. For example, while mGluR1 and mGluR5 are pronociceptive throughout most parts of the body, they seem to be antinociceptive in the PAG. Thus, a systemically delivered antagonist could reveal which one of these roles (i.e., pro- vs. antinociceptive) might predominate during treatment of a pain patient. What follows are a few examples showing that systemic manipulation of mGluRs tends to yield effects in pain that are in line with the most well-described role of a particular receptor. mGluR5 appears *in vivo* to be primarily pronociceptive, and inhibition of mGluR5 systemically tends to decrease pain. This suggests that the antinociceptive role of mGluR5 in the PAG may not be of serious clinical consequence in the context of chronic pain or injury with systemic treatments.

During noxious stimulation of a rat's paw, IV administration of the mGluR5 antagonist MPEP but not the mGluR1 antagonist AIDA reduces excitatory responses in VPL thalamic neurons.[92] Conventional knockouts (i.e., whole body deletion from conception) for mGluR5 have a complex phenotype that includes learning and memory abnormalities[122] and overall hypermobility.[123] Although a complete analysis of these mice in the context of pain is ongoing, published data show that conventional knockout of mGluR5 leads to a reduction in both the first and second phases of the spontaneous formalin test, a small amount of baseline mechanical hyposensitivity, a partial reduction in formalin-induced mechanical hypersensitivity,[96] and reduced visceromotor responses to noxious bladder distention.[38] This physiological effect was similar to reduction in visceromotor responses with systemic application of the mGluR5 antagonist, fenobam.[38] Recently, we found that systemic delivery of either fenobam- or MPEP-induced conditioned place preference in spared nerve injured male and female mice but not in sham-operated mice.[124] These data suggest that inhibition of mGluR5 may be effective in providing relief from spontaneous negative stimulation after nerve injury. Furthermore, the observation that sham-operated mice did not show preference for fenobam or MPEP suggests that unlike morphine, mGluR5-inhibiting drugs do not provide positive reinforcement in the absence of injury.

Most of the observed phenotype in mGluR2 and mGluR3 conventional knockouts supports a more prominent role of mGluR2 rather than mGluR3 in pain.[29] mGluR2 knockout mice exhibit exaggerated behavior in the second phase of the formalin test, while mGluR3KO mice are not different from wild-type littermates.[29] Furthermore, the established analgesic effects of the nonspecific mGluR2/mGluR3 agonist LY354740 (Table 1) are blocked in mGluR2 knockouts only.[29] Systemic (IP) delivery of three different Group II mGluR agonists reduced spontaneous formalin behavior and mechanical allodynia in nerve-ligated rats.[125]

Systemic activation of mGluR8 reduced the second phase of formalin test and mechanical/thermal hyperalgesia after carrageenan injection.[87] This effect on the formalin test was reversed with intra-PAG delivery of the Group III antagonist, MSOP.

3.6 mGluR drug development

As described above, mGluRs impact nearly every anatomical stop along the pain neuraxis. G_q-coupled Group I mGluRs tend to increase pain (in naïve and injured conditions) and are likely involved in pain chronicity. In contrast, G_i-coupled Group II and III mGluRs are typically analgesic after injury. As such, there is considerable interest in targeting the mGluR system to treat chronic pain in humans. This quest is furthered by the fact that mGluR signaling is also heavily involved in other conditions such as schizophrenia, depression, anxiety, epilepsy, and fragile-X syndrome. The obvious advantage of targeting mGluRs as opposed to iGluRs (e.g., NMDA receptors) is that mGluRs primarily have a modulatory role; thus, mGluR modulation would be less likely to affect normal functioning. This fact is particularly true for positive allosteric modulators (PAMs) and NAMs, which due to their nonligand-binding site interaction with receptors will only have an effect when glutamate is bound to the receptor (i.e., if the ligand is not there, the drug will have minimal effect). Clinical trials for specific mGluR agents have been limited to schizophrenia and fragile-X syndrome. However, preclinical modeling of mGluR agents for human use in pain is well underway.

3.6.1 Group I mGluR drug development

One of the more promising drugs for testing in human pain patients is the NAM fenobam. This drug was originally developed by McNeil Laboratories in the 1970s as an anxiolytic agent.[126] Taken through phase II clinical trials, it was eventually dropped because of low anxiolytic effects in patients and psychostimulant side effects.[127] Interest in fenobam has been renewed in the last 10 years, however, when it was shown to be a potent antagonist for mGluR5.[126] In animal models, fenobam reduces pain in the formalin test.[128] The specificity of the agent for mGluR5 seems better compared to the prototypical mGluR5 antagonist, MPEP. Specifically, when MPEP or fenobam was given to mGluR5 conventional knockout animals, only MPEP reduced spontaneous formalin behavior below the knockout vehicle control mice.[128] Consistent with its original development, fenobam also reduces anxiety-like behavior in mice,[123] a property that may be advantageous in pain clinical trials. The major side effect of fenobam appears to include hypermobility in a number of locomotor assays. However, data from our lab suggest that mice quickly develop tolerance to the locomotor but not to the analgesic effects of the drug.[124] To date, one modern study has used fenobam in human patients. In a study of fragile-X syndrome (characterized in part by hyperactivity of mGluR5), a single dose of fenobam did not produce any adverse events and improved prepulse inhibition is a subset of patients.[129]

In addition to specific agents for mGluR1 or mGluR5, there is considerable interest in agents that target Group I mGluRs and other receptors at the same time. A recent study described the development of MMG22 (Table 1), which is an mGluR5 antagonist and mu opioid receptor

(MOR) agonist.[130] Such an agent could provide improved benefits to patients by simultaneous reducing mGluR5 and increasing MOR signaling to induce analgesia. One of the more exciting tenets of such a strategy is that the mGluR5 antagonist could reduce the dose of MOR agonist needed, which may reduce unwanted side effects of mu agonists including addiction and tolerance. Of additional interest of this specific agent is that it likely targets mGluR5–MOR heteromers, which have been identified previously in culture models only.[131] In this new study, MMG22 reduced mechanical allodynia in lipopolysaccharide-injected animals.[130]

3.6.2 Group II mGluR drug development

As described above, activation of Group II mGluRs at many points in the pain neuraxis is analgesic. Changes in Group II mGluRs (expression and/or function) are suspected in chronic conditions and in the normal healing process. To date, no clinical trials in pain have been reported for specific Group II mGluR agents. A phase II double-blind study of the mGluR2/3 agonist LY404039 (Table 1) in schizophrenia demonstrated that the compound was well tolerated and had antipsychotic effects compared to placebo in patients.[132,133] The antipsychostimulant effects of Group II mGluR activation is in contrast to the early trials for mGluR5 inhibition in humans, which showed psychostimulant effects in a double-blind placebo-controlled study.[127] Although there are more data supporting the role of Group I mGluRs in pain compared to Group II mGluRs, it is worth considering additional focus on the Group II mGluRs in the treatment of pain in humans. LAC, the dietary supplement, has been reported to reduce pain in humans[134] through an action that may depend on activation of mGluR2.[117] Recent evidence suggests that mGluR2 may heterodimerize with the serotonin receptor 5-HT$_{2A}$.[135] 5-HT$_{2A}$ has been shown to dynamically alter pain based in part on condition and site of modulation (peripheral vs. central nervous system).[136] As such, the existence of an mGluR2/5-HT$_{2A}$ heterodimer (similar to the story described above for mGluR5 and MOR) may prove to be advantageous for treatment of pain conditions.

3.6.3 Group III mGluR drug development

As the most diverse group of mGluRs, there is considerable interest in developing agents with specificity for different Group III mGluRs for the treatment of pain. Future clinical studies are likely to investigate targeting of Group III mGluRs to reduce pain through activation of the receptors.

4. CONCLUSION

With a distribution that ranges from the nociceptive sensory terminal, DRG, spinal cord, descending pain modulatory centers all the way to higher cognitive processing centers, mGluRs have been shown to be important in the normal and pathological processing of noxious information. The diversity of actions of the mGluRs as they excite or inhibit cells through pre-, post-, or perisynaptic mechanisms on the soma, dendrites, or axons of primary, secondary, and tertiary nociceptive neurons suggests that there are still many undiscovered or poorly understood roles for the mGluRs in the pain neuraxis. In general, Group I mGluRs (mGluR1 and mGluR5) play a pronociceptive role through excitatory modulation at pre- or postsynaptic sites. Group II (mGluR2 and mGluR3) and Group III mGluRs (mGluR4, mGluR6, mGluR7, mGluR8), by contrast, tend to reduce pain through a presynaptic mechanism. Given this distribution, it may be that an ideal drug would be one that acts as an antagonist at Group I mGluRs and an agonist at Group II/III mGluRs. The signaling pathways involved in the modulation of pain by mGluRs have only been studied in the periphery and spinal cord. Even with the paucity of data on the subject, it is clear that a given receptor can act through both canonical and noncanonical signaling cascades to induce an effect. The future of mGluRs in pain will rest on the ability of basic scientists and clinicians to transition from the fragmentary evidence presented above into clinically relevant and meaningful treatment paradigms.

REFERENCES

1. Gereau RW, Swanson G. *The Glutamate Receptors*. Totowa, NJ: Humana Press; 2008.
2. Niswender CM, Conn PJ. Metabotropic glutamate receptors: physiology, pharmacology, and disease. *Annu Rev Pharmacol Toxicol*. 2010;50:295–322. http://dx.doi.org/10.1146/annurev.pharmtox.011008.145533.
3. Kunishima N, Shimada Y, Tsuji Y, et al. Structural basis of glutamate recognition by a dimeric metabotropic glutamate receptor. *Nature*. 2000;407(6807):971–977. http://dx.doi.org/10.1038/35039564.
4. Muto T, Tsuchiya D, Morikawa K, Jingami H. Structures of the extracellular regions of the group II/III metabotropic glutamate receptors. *Proc Natl Acad Sci USA*. 2007;104(10):3759–3764. http://dx.doi.org/10.1073/pnas.0611577104.
5. Tsuchiya D, Kunishima N, Kamiya N, Jingami H, Morikawa K. Structural views of the ligand-binding cores of a metabotropic glutamate receptor complexed with an antagonist and both glutamate and Gd3+. *Proc Natl Acad Sci USA*. 2002;99(5):2660–2665. http://dx.doi.org/10.1073/pnas.052708599.
6. Goudet C, Gaven F, Kniazeff J, et al. Heptahelical domain of metabotropic glutamate receptor 5 behaves like rhodopsin-like receptors. *Proc Natl Acad Sci USA*. 2004;101(1):378–383. http://dx.doi.org/10.1073/pnas.0304699101.

7. Enz R. Structure of metabotropic glutamate receptor C-terminal domains in contact with interacting proteins. *Front Mol Neurosci*. 2012;5:52. http://dx.doi.org/10.3389/fnmol.2012.00052.
8. Hu HJ, Bhave G, Gereau RW. Prostaglandin and protein kinase A-dependent modulation of vanilloid receptor function by metabotropic glutamate receptor 5: potential mechanism for thermal hyperalgesia. *J Neurosci*. 2002;22(17):7444–7452, doi: 22/17/7444 [pii].
9. Neugebauer V. Metabotropic glutamate receptors—important modulators of nociception and pain behavior. *Pain*. 2002;98(1–2):1–8, doi: S0304395902001409 [pii].
10. Hu HJ, Alter BJ, Carrasquillo Y, Qiu CS, Gereau RW. Metabotropic glutamate receptor 5 modulates nociceptive plasticity via extracellular signal-regulated kinase-Kv4.2 signaling in spinal cord dorsal horn neurons. *J Neurosci*. 2007;27(48):13181–13191. http://dx.doi.org/10.1523/JNEUROSCI.0269-07.2007, 27/48/13181 [pii].
11. Obara I, Goulding SP, Hu JH, Klugmann M, Worley PF, Szumlinski KK. Nerve injury-induced changes in Homer/glutamate receptor signaling contribute to the development and maintenance of neuropathic pain. *Pain*. 2013;154(10):1932–1945. http://dx.doi.org/10.1016/j.pain.2013.03.035.
12. Bhave G, Karim F, Carlton SM, Gereau RW. Peripheral group I metabotropic glutamate receptors modulate nociception in mice. *Nat Neurosci*. 2001;4(4):417–423.
13. Carlton SM, Hargett GL. Colocalization of metabotropic glutamate receptors in rat dorsal root ganglion cells. *J Comp Neurol*. 2007;501(5):780–789. http://dx.doi.org/10.1002/cne.21285.
14. Hudson LJ, Bevan S, McNair K, et al. Metabotropic glutamate receptor 5 upregulation in A-fibers after spinal nerve injury: 2-methyl-6-(phenylethynyl)-pyridine (MPEP) reverses the induced thermal hyperalgesia. *J Neurosci*. 2002;22(7):2660–2668, doi: 20026229 22/7/2660 [pii].
15. Walker K, Reeve A, Bowes M, et al. mGlu5 receptors and nociceptive function II. mGlu5 receptors functionally expressed on peripheral sensory neurones mediate inflammatory hyperalgesia. *Neuropharmacology*. 2001;40(1):10–19, doi: S0028390800001143 [pii].
16. Zhou S, Komak S, Du J, Carlton SM. Metabotropic glutamate 1alpha receptors on peripheral primary afferent fibers: their role in nociception. *Brain Res*. 2001;913(1): 18–26, doi: S0006-8993(01)02747-0 [pii].
17. Tang FR, Sim MK. Pre- and/or post-synaptic localisation of metabotropic glutamate receptor 1alpha (mGluR1alpha) and 2/3 (mGluR2/3) in the rat spinal cord. *Neurosci Res*. 1999;34(2):73–78.
18. Valerio A, Paterlini M, Boifava M, Memo M, Spano P. Metabotropic glutamate receptor mRNA expression in rat spinal cord. *Neuroreport*. 1997;8(12):2695–2699.
19. Azkue JJ, Knopfel T, Kuhn R, Mateos JM, Grandes P. Distribution of the metabotropic glutamate receptor subtype mGluR5 in rat midbrain periaqueductal grey and relationship with ascending spinofugal afferents. *Neurosci Lett*. 1997;228(1):1–4.
20. Lourenco Neto F, Schadrack J, Berthele A, Zieglgansberger W, Tolle TR, Castro-Lopes JM. Differential distribution of metabotropic glutamate receptor subtype mRNAs in the thalamus of the rat. *Brain Res*. 2000;854(1–2):93–105.
21. Neugebauer V, Li W, Bird GC, Bhave G, Gereau RW. Synaptic plasticity in the amygdala in a model of arthritic pain: differential roles of metabotropic glutamate receptors 1 and 5. *J Neurosci*. 2003;23(1):52–63.
22. Sun H, Neugebauer V. mGluR1, but not mGluR5, activates feed-forward inhibition in the medial prefrontal cortex to impair decision making. *J Neurophysiol*. 2011;106(2):960–973. http://dx.doi.org/10.1152/jn.00762.2010.
23. Gereau RW, Cavallone LF. Mechanisms of pain transmission and transduction. In: Evers AS, Maze M, Kharasch ED, eds. *Anesthetic Pharmacology*. 2nd ed. Cambridge: Cambridge University Press; 2013:227–247.

24. Sartorius LJ, Nagappan G, Lipska BK, et al. Alternative splicing of human metabotropic glutamate receptor 3. *J Neurochem*. 2006;96(4):1139–1148. http://dx.doi.org/10.1111/j.1471-4159.2005.03609.x.
25. Boxall SJ, Berthele A, Laurie DJ, et al. Enhanced expression of metabotropic glutamate receptor 3 messenger RNA in the rat spinal cord during ultraviolet irradiation induced peripheral inflammation. *Neuroscience*. 1998;82(2):591–602.
26. Carlton SM, Hargett GL, Coggeshall RE. Localization of metabotropic glutamate receptors 2/3 on primary afferent axons in the rat. *Neuroscience*. 2001;105(4):957–969.
27. Jia H, Rustioni A, Valtschanoff JG. Metabotropic glutamate receptors in superficial laminae of the rat dorsal horn. *J Comp Neurol*. 1999;410(4):627–642.
28. Tamaru Y, Nomura S, Mizuno N, Shigemoto R. Distribution of metabotropic glutamate receptor mGluR3 in the mouse CNS: differential location relative to pre- and postsynaptic sites. *Neuroscience*. 2001;106(3):481–503.
29. Zammataro M, Chiechio S, Montana MC, et al. mGlu2 metabotropic glutamate receptors restrain inflammatory pain and mediate the analgesic activity of dual mGlu2/mGlu3 receptor agonists. *Mol Pain*. 2011;7:6. http://dx.doi.org/10.1186/1744-8069-7-6.
30. Malherbe P, Kratzeisen C, Lundstrom K, Richards JG, Faull RL, Mutel V. Cloning and functional expression of alternative spliced variants of the human metabotropic glutamate receptor 8. *Brain Res Mol Brain Res*. 1999;67(2):201–210.
31. Valerio A, Ferraboli S, Paterlini M, Spano P, Barlati S. Identification of novel alternatively-spliced mRNA isoforms of metabotropic glutamate receptor 6 gene in rat and human retina. *Gene*. 2001;262(1–2):99–106.
32. Corti C, Restituito S, Rimland JM, et al. Cloning and characterization of alternative mRNA forms for the rat metabotropic glutamate receptors mGluR7 and mGluR8. *Eur J Neurosci*. 1998;10(12):3629–3641.
33. Azkue JJ, Murga M, Fernandez-Capetillo O, et al. Immunoreactivity for the group III metabotropic glutamate receptor subtype mGluR4a in the superficial laminae of the rat spinal dorsal horn. *J Comp Neurol*. 2001;430(4):448–457.
34. Li H, Ohishi H, Kinoshita A, Shigemoto R, Nomura S, Mizuno N. Localization of a metabotropic glutamate receptor, mGluR7, in axon terminals of presumed nociceptive, primary afferent fibers in the superficial layers of the spinal dorsal horn: an electron microscope study in the rat. *Neurosci Lett*. 1997;223(3):153–156.
35. Ohishi H, Akazawa C, Shigemoto R, Nakanishi S, Mizuno N. Distributions of the mRNAs for L-2-amino-4-phosphonobutyrate-sensitive metabotropic glutamate receptors, mGluR4 and mGluR7, in the rat brain. *J Comp Neurol*. 1995;360(4):555–570. http://dx.doi.org/10.1002/cne.903600402.
36. Chen HS, Qu F, He X, Kang SM, Liao D, Lu SJ. Differential roles of peripheral metabotropic glutamate receptors in bee venom-induced nociception and inflammation in conscious rats. *J Pain*. 2010;11(4):321–329. http://dx.doi.org/10.1016/j.jpain.2009.07.013.
37. Lindstrom E, Brusberg M, Hughes PA, et al. Involvement of metabotropic glutamate 5 receptor in visceral pain. *Pain*. 2008;137(2):295–305. http://dx.doi.org/10.1016/j.pain.2007.09.008, doi: S0304-3959(07)00517-9 [pii].
38. Crock LW, Stemler KM, Song DG, et al. Metabotropic glutamate receptor 5 (mGluR5) regulates bladder nociception. *Mol Pain*. 2012;8:20. http://dx.doi.org/10.1186/1744-8069-8-20.
39. Hu Y, Dong L, Sun B, et al. The role of metabotropic glutamate receptor mGlu5 in control of micturition and bladder nociception. *Neurosci Lett*. 2009;450(1):12–17. http://dx.doi.org/10.1016/j.neulet.2008.11.026, S0304-3940(08)01576-0 [pii].
40. Ohishi H, Shigemoto R, Nakanishi S, Mizuno N. Distribution of the mRNA for a metabotropic glutamate receptor (mGluR3) in the rat brain: an in situ hybridization study. *J Comp Neurol*. 1993;335(2):252–266. http://dx.doi.org/10.1002/cne.903350209.

41. Carlton SM, Du J, Zhou S. Group II metabotropic glutamate receptor activation on peripheral nociceptors modulates TRPV1 function. *Brain Res.* 2009;1248:86–95. http://dx.doi.org/10.1016/j.brainres.2008.10.066.
42. Yang D, Gereau RW. Peripheral group II metabotropic glutamate receptors mediate endogenous anti-allodynia in inflammation. *Pain.* 2003;106(3):411–417.
43. Yang D, Gereau RW. Peripheral group II metabotropic glutamate receptors (mGluR2/3) regulate prostaglandin E2-mediated sensitization of capsaicin responses and thermal nociception. *J Neurosci.* 2002;22(15):6388–6393, doi: 20026660.
44. Jung CY, Lee SY, Choi HS, et al. Participation of peripheral group I and II metabotropic glutamate receptors in the development or maintenance of IL-1beta-induced mechanical allodynia in the orofacial area of conscious rats. *Neurosci Lett.* 2006;409(3):173–178. http://dx.doi.org/10.1016/j.neulet.2006.09.043.
45. Carlton SM, Zhou S, Govea R, Du J. Group II/III metabotropic glutamate receptors exert endogenous activity-dependent modulation of TRPV1 receptors on peripheral nociceptors. *J Neurosci.* 2011;31(36):12727–12737. http://dx.doi.org/10.1523/JNEUROSCI.6558-10.2011.
46. Chiechio S, Copani A, De Petris L, Morales ME, Nicoletti F, Gereau RW. Transcriptional regulation of metabotropic glutamate receptor 2/3 expression by the NF-kappaB pathway in primary dorsal root ganglia neurons: a possible mechanism for the analgesic effect of L-acetylcarnitine. *Mol Pain.* 2006;2:20. http://dx.doi.org/10.1186/1744-8069-2-20.
47. Chiechio S, Copani A, Melchiorri D, et al. Metabotropic receptors as targets for drugs of potential use in the treatment of neuropathic pain. *J Endocrinol Invest.* 2004;27(6 suppl):171–176.
48. Chiechio S, Copani A, Gereau RW, Nicoletti F. Acetyl-L-carnitine in neuropathic pain: experimental data. *CNS Drugs.* 2007;21(suppl 1):31–38, discussion 45–36.
49. Gerber G, Zhong J, Youn D, Randic M. Group II and group III metabotropic glutamate receptor agonists depress synaptic transmission in the rat spinal cord dorsal horn. *Neuroscience.* 2000;100(2):393–406.
50. Yang D, Gereau RW. Group II metabotropic glutamate receptors inhibit cAMP-dependent protein kinase-mediated enhancemednt of tetrodotoxin-resistant sodium currents in mouse dorsal root ganglion neurons. *Neurosci Lett.* 2004;357(3):159–162. http://dx.doi.org/10.1016/j.neulet.2003.11.074.
51. Ohishi H, Nomura S, Ding YQ, et al. Presynaptic localization of a metabotropic glutamate receptor, mGluR7, in the primary afferent neurons: an immunohistochemical study in the rat. *Neurosci Lett.* 1995;202(1–2):85–88.
52. Govea RM, Zhou S, Carlton SM. Group III metabotropic glutamate receptors and transient receptor potential vanilloid 1 co-localize and interact on nociceptors. *Neuroscience.* 2012;217:130–139. http://dx.doi.org/10.1016/j.neuroscience.2012.05.014.
53. Gerber G, Youn DH, Hsu CH, Isaev D, Randic M. Spinal dorsal horn synaptic plasticity: involvement of group I metabotropic glutamate receptors. *Prog Brain Res.* 2000;129:115–134. http://dx.doi.org/10.1016/s0079-6123(00)29009-2.
54. Valerio A, Rizzonelli P, Paterlini M, et al. mGluR5 metabotropic glutamate receptor distribution in rat and human spinal cord: a developmental study. *Neurosci Res.* 1997;28(1):49–57, doi: S0168-0102(97)01175-9 [pii].
55. Vidnyanszky Z, Hamori J, Negyessy L, et al. Cellular and subcellular localization of the mGluR5a metabotropic glutamate receptor in rat spinal cord. *Neuroreport.* 1994;6(1):209–213.
56. Fisher K, Coderre TJ. Hyperalgesia and allodynia induced by intrathecal (RS)-dihydroxyphenylglycine in rats. *Neuroreport.* 1998;9(6):1169–1172.
57. Hama AT. Acute activation of the spinal cord metabotropic glutamate subtype-5 receptor leads to cold hypersensitivity in the rat. *Neuropharmacology.* 2003;44(4):423–430.

58. Fisher K, Coderre TJ. Comparison of nociceptive effects produced by intrathecal administration of mGluR agonists. *Neuroreport.* 1996;7(15–17):2743–2747.
59. Karim F, Wang CC, Gereau RW. Metabotropic glutamate receptor subtypes 1 and 5 are activators of extracellular signal-regulated kinase signaling required for inflammatory pain in mice. *J Neurosci.* 2001;21(11):3771–3779.
60. Fisher K, Coderre TJ. The contribution of metabotropic glutamate receptors (mGluRs) to formalin-induced nociception. *Pain.* 1996;68(2–3):255–263.
61. Adwanikar H, Karim F, Gereau RW. Inflammation persistently enhances nocifensive behaviors mediated by spinal group I mGluRs through sustained ERK activation. *Pain.* 2004;111(1–2):125–135.
62. Noda K, Anzai T, Ogata M, Akita H, Ogura T, Saji M. Antisense knockdown of spinal-mGluR1 reduces the sustained phase of formalin-induced nociceptive responses. *Brain Res.* 2003;987(2):194–200.
63. Mills CD, Fullwood SD, Hulsebosch CE. Changes in metabotropic glutamate receptor expression following spinal cord injury. *Exp Neurol.* 2001;170(2):244–257. http://dx.doi.org/10.1006/exnr.2001.7721.
64. Mills CD, Johnson KM, Hulsebosch CE. Group I metabotropic glutamate receptors in spinal cord injury: roles in neuroprotection and the development of chronic central pain. *J Neurotrauma.* 2002;19(1):23–42. http://dx.doi.org/10.1089/089771502753460213.
65. Fisher K, Lefebvre C, Coderre TJ. Antinociceptive effects following intrathecal pretreatment with selective metabotropic glutamate receptor compounds in a rat model of neuropathic pain. *Pharmacol Biochem Behav.* 2002;73(2):411–418, doi: S0091305702008328 [pii].
66. Hu HJ, Glauner KS, Gereau RW. ERK integrates PKA and PKC signaling in superficial dorsal horn neurons. I. Modulation of A-type K+ currents. *J Neurophysiol.* 2003;90(3):1671–1679.
67. Neugebauer V, Chen PS, Willis WD. Role of metabotropic glutamate receptor subtype mGluR1 in brief nociception and central sensitization of primate STT cells. *J Neurophysiol.* 1999;82(1):272–282.
68. Heinke B, Sandkuhler J. Signal transduction pathways of group I metabotropic glutamate receptor-induced long-term depression at sensory spinal synapses. *Pain.* 2005;118(1–2):145–154. http://dx.doi.org/10.1016/j.pain.2005.08.004.
69. Zhang T, Zhang J, Shi J, Feng Y, Sun ZS, Li H. Antinociceptive synergistic effect of spinal mGluR2/3 antagonist and glial cells inhibitor on peripheral inflammation-induced mechanical hypersensitivity. *Brain Res Bull.* 2009;79(3–4):219–223. http://dx.doi.org/10.1016/j.brainresbull.2009.01.009.
70. Zhou HY, Chen SR, Chen H, Pan HL. Functional plasticity of group II metabotropic glutamate receptors in regulating spinal excitatory and inhibitory synaptic input in neuropathic pain. *J Pharmacol Exp Ther.* 2011;336(1):254–264. http://dx.doi.org/10.1124/jpet.110.173112.
71. Lefebvre C, Fisher K, Cahill CM, Coderre TJ. Evidence that DHPG-induced nociception depends on glutamate release from primary afferent C-fibres. *Neuroreport.* 2000;11(8):1631–1635.
72. Zhou HY, Zhang HM, Chen SR, Pan HL. Increased nociceptive input rapidly modulates spinal GABAergic transmission through endogenously released glutamate. *J Neurophysiol.* 2007;97(1):871–882. http://dx.doi.org/10.1152/jn.00964.2006.
73. Chen SR, Pan HL. Distinct roles of group III metabotropic glutamate receptors in control of nociception and dorsal horn neurons in normal and nerve-injured Rats. *J Pharmacol Exp Ther.* 2005;312(1):120–126. http://dx.doi.org/10.1124/jpet.104.073817.

74. Harris JA, Corsi M, Quartaroli M, Arban R, Bentivoglio M. Upregulation of spinal glutamate receptors in chronic pain. *Neuroscience*. 1996;74(1):7–12.
75. Moore KA, Kohno T, Karchewski LA, Scholz J, Baba H, Woolf CJ. Partial peripheral nerve injury promotes a selective loss of GABAergic inhibition in the superficial dorsal horn of the spinal cord. *J Neurosci*. 2002;22(15):6724–6731, doi: 20026611.
76. Neugebauer V, Chen PS, Willis WD. Groups II and III metabotropic glutamate receptors differentially modulate brief and prolonged nociception in primate STT cells. *J Neurophysiol*. 2000;84(6):2998–3009.
77. Catania MV, De Socarraz H, Penney JB, Young AB. Metabotropic glutamate receptor heterogeneity in rat brain. *Mol Pharmacol*. 1994;45(4):626–636.
78. Maione S, Marabese I, Leyva J, Palazzo E, de Novellis V, Rossi F. Characterisation of mGluRs which modulate nociception in the PAG of the mouse. *Neuropharmacology*. 1998;37(12):1475–1483.
79. Berrino L, Oliva P, Rossi F, Palazzo E, Nobili B, Maione S. Interaction between metabotropic and NMDA glutamate receptors in the periaqueductal grey pain modulatory system. *Naunyn Schmiedebergs Arch Pharmacol*. 2001;364(5):437–443. http://dx.doi.org/10.1007/s002100100477.
80. Maione S, Oliva P, Marabese I, et al. Periaqueductal gray matter metabotropic glutamate receptors modulate formalin-induced nociception. *Pain*. 2000;85(1–2):183–189.
81. Palazzo E, Marabese I, de Novellis V, et al. Metabotropic and NMDA glutamate receptors participate in the cannabinoid-induced antinociception. *Neuropharmacology*. 2001;40(3):319–326.
82. Wilson-Poe AR, Mitchell VA, Vaughan CW. Postsynaptic mGluR mediated excitation of neurons in midbrain periaqueductal grey. *Neuropharmacology*. 2013;66:348–354. http://dx.doi.org/10.1016/j.neuropharm.2012.06.057.
83. Liao HT, Lee HJ, Ho YC, Chiou LC. Capsaicin in the periaqueductal gray induces analgesia via metabotropic glutamate receptor-mediated endocannabinoid retrograde disinhibition. *Br J Pharmacol*. 2011;163(2):330–345. http://dx.doi.org/10.1111/j.1476-5381.2011.01214.x.
84. Mitchell VA, Kawahara H, Vaughan CW. Neurotensin inhibition of GABAergic transmission via mGluR-induced endocannabinoid signalling in rat periaqueductal grey. *J Physiol*. 2009;587(pt 11):2511–2520. http://dx.doi.org/10.1113/jphysiol.2008.167429.
85. Izumi Y, Zorumski CF. NMDA receptors, mGluR5, and endocannabinoids are involved in a cascade leading to hippocampal long-term depression. *Neuropsychopharmacology*. 2012;37(3):609–617. http://dx.doi.org/10.1038/npp.2011.243.
86. Drew GM, Vaughan CW. Multiple metabotropic glutamate receptor subtypes modulate GABAergic neurotransmission in rat periaqueductal grey neurons in vitro. *Neuropharmacology*. 2004;46(7):927–934. http://dx.doi.org/10.1016/j.neuropharm.2004.01.015.
87. Marabese I, de Novellis V, Palazzo E, et al. Effects of (S)-3,4-DCPG, an mGlu8 receptor agonist, on inflammatory and neuropathic pain in mice. *Neuropharmacology*. 2007;52(2):253–262. http://dx.doi.org/10.1016/j.neuropharm.2006.04.006.
88. Marabese I, Rossi F, Palazzo E, et al. Periaqueductal gray metabotropic glutamate receptor subtype 7 and 8 mediate opposite effects on amino acid release, rostral ventromedial medulla cell activities, and thermal nociception. *J Neurophysiol*. 2007;98(1):43–53. http://dx.doi.org/10.1152/jn.00356.2007.
89. Marabese I, de Novellis V, Palazzo E, et al. Differential roles of mGlu8 receptors in the regulation of glutamate and gamma-aminobutyric acid release at periaqueductal grey level. *Neuropharmacology*. 2005;49(suppl 1):157–166. http://dx.doi.org/10.1016/j.neuropharm.2005.02.006.

90. Bair MJ, Robinson RL, Katon W, Kroenke K. Depression and pain comorbidity: a literature review. *Arch Intern Med.* 2003;163(20):2433–2445. http://dx.doi.org/10.1001/archinte.163.20.2433.
91. Miyata M, Kashiwadani H, Fukaya M, et al. Role of thalamic phospholipase C[beta]4 mediated by metabotropic glutamate receptor type 1 in inflammatory pain. *J Neurosci.* 2003;23(22):8098–8108.
92. Bordi F, Ugolini A. Involvement of mGluR(5) on acute nociceptive transmission. *Brain Res.* 2000;871(2):223–233.
93. Salt TE, Binns KE. Contributions of mGlu1 and mGlu5 receptors to interactions with N-methyl-D-aspartate receptor-mediated responses and nociceptive sensory responses of rat thalamic neurons. *Neuroscience.* 2000;100(2):375–380.
94. Neugebauer V, Li W, Bird GC, Han JS. The amygdala and persistent pain. *Neuroscientist.* 2004;10(3):221–234.
95. Crock LW, Kolber BJ, Morgan CD, et al. Central amygdala metabotropic glutamate receptor 5 in the modulation of visceral pain. *J Neurosci.* 2012;32(41):14217–14226. http://dx.doi.org/10.1523/JNEUROSCI.1473-12.2012.
96. Kolber BJ, Montana MC, Carrasquillo Y, et al. Activation of metabotropic glutamate receptor 5 in the amygdala modulates pain-like behavior. *J Neurosci.* 2010;30(24):8203–8213. http://dx.doi.org/10.1523/JNEUROSCI.1216-10.2010.
97. Li Z, Ji G, Neugebauer V. Mitochondrial reactive oxygen species are activated by mGluR5 through IP3 and activate ERK and PKA to increase excitability of amygdala neurons and pain behavior. *J Neurosci.* 2011;31(3):1114–1127.
98. Ansah OB, Goncalves L, Almeida A, Pertovaara A. Enhanced pronociception by amygdaloid group I metabotropic glutamate receptors in nerve-injured animals. *Exp Neurol.* 2009;216(1):66–74.
99. Han JS, Neugebauer V. mGluR1 and mGluR5 antagonists in the amygdala inhibit different components of audible and ultrasonic vocalizations in a model of arthritic pain. *Pain.* 2005;113(1–2):211–222.
100. Carrasquillo Y, Gereau RW. Hemispheric lateralization of a molecular signal for pain modulation in the amygdala. *Mol Pain.* 2008;4:24.
101. Ji G, Neugebauer V. Hemispheric lateralization of pain processing by amygdala neurons. *J Neurophysiol.* 2009;102(4):2253–2264.
102. Ren W, Neugebauer V. Pain-related increase of excitatory transmission and decrease of inhibitory transmission in the central nucleus of the amygdala are mediated by mGluR1. *Mol Pain.* 2010;6:93. http://dx.doi.org/10.1186/1744-8069-6-93.
103. Li W, Neugebauer V. Differential roles of mGluR1 and mGluR5 in brief and prolonged nociceptive processing in central amygdala neurons. *J Neurophysiol.* 2004;91(1):13–24.
104. Tappe-Theodor A, Fu Y, Kuner R, Neugebauer V. Homer1a signaling in the amygdala counteracts pain-related synaptic plasticity, mGluR1 function and pain behaviors. *Mol Pain.* 2011;7:38. http://dx.doi.org/10.1186/1744-8069-7-38.
105. Ji G, Neugebauer V. Reactive oxygen species are involved in group I mGluR-mediated facilitation of nociceptive processing in amygdala neurons. *J Neurophysiol.* 2010;104(1):218–229.
106. Luongo L, de Novellis V, Gatta L, et al. Role of metabotropic glutamate receptor 1 in the basolateral amygdala-driven prefrontal cortical deactivation in inflammatory pain in the rat. *Neuropharmacology.* 2013;66:317–329. http://dx.doi.org/10.1016/j.neuropharm.2012.05.047.
107. Ji G, Sun H, Fu Y, et al. Cognitive impairment in pain through amygdala-driven prefrontal cortical deactivation. *J Neurosci.* 2010;30(15):5451–5464. http://dx.doi.org/10.1523/JNEUROSCI.0225-10.2010.

108. Ji G, Neugebauer V. Pain-related deactivation of medial prefrontal cortical neurons involves mGluR1 and GABA(A) receptors. *J Neurophysiol.* 2011;106(5):2642–2652. http://dx.doi.org/10.1152/jn.00461.2011.
109. Calejesan AA, Kim SJ, Zhuo M. Descending facilitatory modulation of a behavioral nociceptive response by stimulation in the adult rat anterior cingulate cortex. *Eur J Pain.* 2000;4(1):83–96. http://dx.doi.org/10.1053/eujp.1999.0158.
110. Kang SJ, Liu MG, Chen T, et al. Plasticity of metabotropic glutamate receptor-dependent long-term depression in the anterior cingulate cortex after amputation. *J Neurosci.* 2012;32(33):11318–11329. http://dx.doi.org/10.1523/JNEUROSCI.0146-12.2012.
111. Petralia RS, Wang YX, Niedzielski AS, Wenthold RJ. The metabotropic glutamate receptors, mGluR2 and mGluR3, show unique postsynaptic, presynaptic and glial localizations. *Neuroscience.* 1996;71(4):949–976.
112. Neto FL, Castro-Lopes JM. Antinociceptive effect of a group II metabotropic glutamate receptor antagonist in the thalamus of monoarthritic rats. *Neurosci Lett.* 2000;296(1):25–28.
113. Govindaiah G, Cox CL. Metabotropic glutamate receptors differentially regulate GABAergic inhibition in thalamus. *J Neurosci.* 2006;26(52):13443–13453. http://dx.doi.org/10.1523/JNEUROSCI.3578-06.2006.
114. Han JS, Fu Y, Bird GC, Neugebauer V. Enhanced group II mGluR-mediated inhibition of pain-related synaptic plasticity in the amygdala. *Mol Pain.* 2006;2:18. http://dx.doi.org/10.1186/1744-8069-2-18.
115. Li W, Neugebauer V. Differential changes of group II and group III mGluR function in central amygdala neurons in a model of arthritic pain. *J Neurophysiol.* 2006;96(4):1803–1815. http://dx.doi.org/10.1152/jn.00495.2006.
116. Neto FL, Schadrack J, Platzer S, Zieglgansberger W, Tolle TR, Castro-Lopes JM. Up-regulation of metabotropic glutamate receptor 3 mRNA expression in the cerebral cortex of monoarthritic rats. *J Neurosci Res.* 2001;63(4):356–367.
117. Chiechio S, Caricasole A, Barletta E, et al. L-Acetylcarnitine induces analgesia by selectively up-regulating mGlu2 metabotropic glutamate receptors. *Mol Pharmacol.* 2002;61(5):989–996.
118. Bradley SR, Rees HD, Yi H, Levey AI, Conn PJ. Distribution and developmental regulation of metabotropic glutamate receptor 7a in rat brain. *J Neurochem.* 1998;71(2):636–645.
119. Palazzo E, Marabese I, Soukupova M, et al. Metabotropic glutamate receptor subtype 8 in the amygdala modulates thermal threshold, neurotransmitter release, and rostral ventromedial medulla cell activity in inflammatory pain. *J Neurosci.* 2011;31(12):4687–4697. http://dx.doi.org/10.1523/JNEUROSCI.2938-10.2011.
120. Palazzo E, Fu Y, Ji G, Maione S, Neugebauer V. Group III mGluR7 and mGluR8 in the amygdala differentially modulate nocifensive and affective pain behaviors. *Neuropharmacology.* 2008;55(4):537–545. http://dx.doi.org/10.1016/j.neuropharm.2008.05.007.
121. Ren W, Palazzo E, Maione S, Neugebauer V. Differential effects of mGluR7 and mGluR8 activation on pain-related synaptic activity in the amygdala. *Neuropharmacology* 2011;61(8):1334–1344. http://dx.doi.org/10.1016/j.neuropharm.2011.08.006.
122. Lu YM, Jia Z, Janus C, et al. Mice lacking metabotropic glutamate receptor 5 show impaired learning and reduced CA1 long-term potentiation (LTP) but normal CA3 LTP. *J Neurosci.* 1997;17(13):5196–5205.
123. Montana MC, Conrardy BA, Cavallone LF, et al. Metabotropic glutamate receptor 5 antagonism with fenobam: examination of analgesic tolerance and side effect profile in mice. *Anesthesiology.* 2011;115(6):1239–1250. http://dx.doi.org/10.1097/ALN.0b013e318238c051.

124. Lax NC, George DC, Ignatz C, Kolber BJ. The mGluR5 antagonist fenobam induces analgesic conditioned place preference in mice with spared nerve injury. *PLoS One.* 2014;9(7):e103524. http://dx.doi.org/10.1371/journal.pone.0103524.
125. Simmons RM, Webster AA, Kalra AB, Iyengar S. Group II mGluR receptor agonists are effective in persistent and neuropathic pain models in rats. *Pharmacol Biochem Behav.* 2002;73(2):419–427.
126. Porter RH, Jaeschke G, Spooren W, et al. Fenobam: a clinically validated non-benzodiazepine anxiolytic is a potent, selective, and noncompetitive mGlu5 receptor antagonist with inverse agonist activity. *J Pharmacol Exp Ther.* 2005;315(2):711–721. http://dx.doi.org/10.1124/jpet.105.089839, jpet.105.089839 [pii].
127. Friedmann CTH, Davis LJ, Ciccone PE, Rubin RT. Phase II double-blind controlled study of a new anxiolytic, fenobam (McN-3377) vs placebo. *Curr Ther Res.* 1980;27(2):144–151.
128. Montana MC, Cavallone LF, Stubbert KK, Stefanescu AD, Kharasch ED, Gereau RW. The metabotropic glutamate receptor subtype 5 antagonist fenobam is analgesic and has improved in vivo selectivity compared with the prototypical antagonist 2-methyl-6-(phenylethynyl)-pyridine. *J Pharmacol Exp Ther.* 2009;330(3):834–843.
129. Berry-Kravis E, Hessl D, Coffey S, et al. A pilot open label, single dose trial of fenobam in adults with fragile X syndrome. *J Med Genet.* 2009;46(4):266–271. http://dx.doi.org/10.1136/jmg.2008.063701.
130. Akgun E, Javed MI, Lunzer MM, Smeester BA, Beitz AJ, Portoghese PS. Ligands that interact with putative MOR-mGluR5 heteromer in mice with inflammatory pain produce potent antinociception. *Proc Natl Acad Sci USA.* 2013;110(28):11595–11599. http://dx.doi.org/10.1073/pnas.1305461110.
131. Schroder H, Wu DF, Seifert A, et al. Allosteric modulation of metabotropic glutamate receptor 5 affects phosphorylation, internalization, and desensitization of the micro-opioid receptor. *Neuropharmacology.* 2009;56(4):768–778. http://dx.doi.org/10.1016/j.neuropharm.2008.12.010.
132. Mezler M, Geneste H, Gault L, Marek GJ. LY-2140023, a prodrug of the group II metabotropic glutamate receptor agonist LY-404039 for the potential treatment of schizophrenia. *Curr Opin Investig Drugs.* 2010;11(7):833–845.
133. Patil ST, Zhang L, Martenyi F, et al. Activation of mGlu2/3 receptors as a new approach to treat schizophrenia: a randomized Phase 2 clinical trial. *Nat Med.* 2007;13(9):1102–1107. http://dx.doi.org/10.1038/nm1632.
134. Chiechio S, Copani A, Nicoletti F, Gereau RW. L-Acetylcarnitine: a proposed therapeutic agent for painful peripheral neuropathies. *Curr Neuropharmacol.* 2006;4(3):233–237.
135. Gonzalez-Maeso J, Ang RL, Yuen T, et al. Identification of a serotonin/glutamate receptor complex implicated in psychosis. *Nature.* 2008;452(7183):93–97. http://dx.doi.org/10.1038/nature06612.
136. Sommer C. Is serotonin hyperalgesic or analgesic? *Curr Pain Headache Rep.* 2006;10(2):101–106.

CHAPTER ELEVEN

Nonneuronal Central Mechanisms of Pain: Glia and Immune Response

E. Alfonso Romero-Sandoval[1], Sarah Sweitzer

Department of Pharmaceutical and Administrative Sciences, Presbyterian College School of Pharmacy, Clinton, South Carolina, USA

[1]Corresponding author: e-mail address: asandoval@presby.edu

Contents

1. Introduction 326
2. The Central Multipartite Synapse in the Mechanisms of Chronic Pain 326
3. Glial Activation or Reactivity: A Change in Cellular Phenotype in Pain States 330
4. Triggers of the Change of Glial Phenotype in Pain States 333
5. Drugs with Glial Modulation Capabilities and Efficacy in Preclinical Models of Pain 336
6. What It Is Known of Glial Cells in Humans with Pain Conditions? 338
7. Glial Cells: Human Versus Rodents and Other Species 342
8. Clinical Data Using Glial Modulating Agents 343
9. Conclusions 345
Acknowledgments 349
References 349

Abstract

The role of central glial cells in the mechanisms underlying pain has been intensively studied in the last two decades. Most studies on glia and pain focused on the potential detrimental role of glial cells following noxious stimulus/insults manifested as an "activation" or a "reactive" state (increase in glial marker expression and production of proinflammatory/nociceptive molecules). Therefore, "activated" or "reactive" glial cells became a target for the future generation of drugs to treat chronic pain. Several glial modulators that reduce the activation of glial cells have shown great efficacy in multiple animal (rodents mostly) models of pain (acute, subacute, chronic, inflammatory, neuropathic, surgical, etc.). These encouraging findings inspired clinical trials that have been completed in the last 5 years. Unfortunately, all clinical trials with these glial modulators have failed to demonstrate efficacy for the treatment of pain. New lines of investigation and elegant experimental designs are shedding light on alternative glial functions, which demonstrate that "glial reactivity" is not necessarily deleterious in some pathological conditions. New strategies to validate findings through our current animal models are necessary to enhance the translational value of our preclinical studies.

Progress in Molecular Biology and Translational Science, Volume 131
ISSN 1877-1173
http://dx.doi.org/10.1016/bs.pmbts.2014.11.007

Also, more studies using human subjects would enhance our understanding of glial cells in the context of pain. This chapter explores the available literature to objectively ponder the potential role of glial cells in human pain conditions.

1. INTRODUCTION

Chronic pain research in the previous two decades has identified specific cellular and molecular changes in the spinal cord and brain that are proposed to be critical in the transition from acute to persistent pain states. Glial cells, which were once functionally classified as merely structural support, have been postulated to be key players in chronic pain states. In animal models, reactive glia have been shown to modulate neuronal responsiveness to painful sensory stimuli. Preclinical models of neuropathic, inflammatory, and cancer pain have highlighted the importance of reactive glia in the transition from acute to chronic pain and suggested the potential clinical utility of glial modulators in the treatment of chronic pain states. Unfortunately, this promise has not translated to clinical breakthroughs and in fact, it has been quite disappointing and calls for a reassessment of our basic science to clinical therapy pipeline as it pertains to the study of glia in pain.

2. THE CENTRAL MULTIPARTITE SYNAPSE IN THE MECHANISMS OF CHRONIC PAIN

The three major groups that comprise glia in the central nervous system (CNS) are microglia, astrocytes, and oligodendrocytes. Since astrocytes and microglia are the types of nonneuronal cells that have been more studied in the last two decades in the context of pain, this chapter focuses on these cells. Throughout the CNS, astrocytes and microglia are integral to neuronal survival and homeostasis. Astrocytes are diverse in nature, have multiple functions that varie even depending on their anatomical location in the CNS. They are very territorial (in adulthood), serve as the conduit between the endothelial cells of the blood–brain barrier (BBB) and neurons, maintain the extracellular ionic balance and neurotransmitter levels in the synapse, release neurotrophic factors, influence synaptogenesis, and communicate in a coordinated manner via gap junctions and calcium waves.[1] In contrast, microglia function as the "macrophages" of the CNS under pathological conditions. Microglial cells have a vital role in immune surveillance and orchestrate the immune response in the CNS; they serve as sensors for threats and are the first cells to respond to injury. They help to maintain the integrity of neuronal synapses and possess neuroprotective functions (kill exogenous pathogens, phagocyte dead neurons, remove excess of neuro-transmitters, etc.).[2,3]

Both central microglia[4] and astrocytes[5,6] interact with neurons in the CNS. Thus, the presynaptic and postsynaptic neurons, astrocytes and microglia constitute the tetrapartite synapse in the brain and spinal cord, especially under pathological conditions.[7] Following infection or injury to the CNS, both astrocytes and microglia undergo a rapid and stereotypical "reactivity."[3,8] In fact, major peripheral damage, such as sciatic nerve ligation, or rather minor tissue damage, such as a paw incision, promote over-expression of cellular markers and changes in glial morphology in the dorsal horn of the spinal cord (L4–L5) in rats.[9] It is believed that a similar tetrapartite synapse between pre- and postsynaptic neurons and "reactive" glial cells is an integral signaling unit in spinal cord (and likely in some brain areas) following peripheral insults that result in pain states (Fig. 1).[7,10] For example, in rodent models of neuropathic pain, spinal astrocytes modulate the synaptic cleft level of glutamate via glutamate transporters and contribute to increase spinal cord neuronal activity.[11–14]

The BBB helps to isolate the CNS from peripheral cells and large molecules. However, under peripheral or CNS damage, the BBB becomes permeable to peripheral leukocytes. The best-studied leukocytes in preclinical models of persistent pain are lymphocytes and monocyte/macrophages. Peripheral lymphocytes and monocytes gain access to the spinal cord following peripheral nerve injury in different models of neuropathic pain, and they contribute to the induction of nociceptive behaviors.[15–17] These studies suggested a functional interaction of peripheral lymphocytes with CNS cells. It is likely that lymphocytes can modulate the tetrapartite synapse in the spinal cord under neuropathic conditions. In fact, these studies demonstrated an enhancement of microglia reactivity[16] or astrocytes[15] in association with the lymphocyte infiltration, which suggests a spinal cord lymphocyte–glia interaction in preclinical models of neuropathic pain.

In models of peripheral nerve injury and persistent neuropathic pain, the spinal cord vasculature is not structurally compromised, which suggests that peripheral leukocytes infiltration likely results from an increase in the permeability of the BBB. Endothelial cells and astrocytes form and control the integrity and permeability of the BBB.[18–20] In fact, astrocytes are pivotal in modulating lymphocyte trafficking into the CNS.[20] The direct interaction between lymphocytes and astrocytes has already been

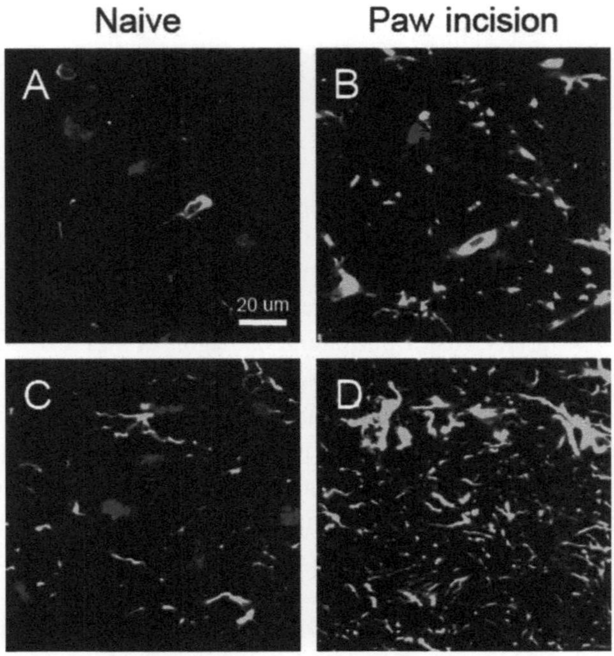

Figure 1 Physical interaction between glial cells and neurons in the rat spinal cord. This figure displays confocal microscopy images of dorsal horn (superficial laminae) spinal cord of rats. (A) Immunolabeling for Iba1 (green microglial marker) and NeuN (red neuronal nuclear marker) in a naïve rat. (B) Immunolabeling for Iba1 (green microglial marker) and NeuN (red neuronal nuclear marker) in a rat undergone paw incision (postoperative day 3); note the apparent increase in the physical interaction with neurons compared to naïve conditions. (C) Immunolabeling for GFAP (green astrocytic marker) and NeuN (red neuronal nuclear marker) in a naïve rat. (D) Immunolabeling for GFAP (green astrocytic marker) and NeuN (red neuronal nuclear marker) in a rat undergone paw incision (postoperative day 3); note the apparent loss of territoriality of astrocytes, which appears to increase the physical interaction with neurons. (See the color plate.)

demonstrated, specifically for clearance of virally infected brain astrocytes.[21,22] At this structural level, perivascular microglia, which are instrumental in the communication between the periphery and the CNS, also play a role in neuroinflammatory conditions, including rodent models of pain[17,23] (Fig. 2).

In summary, glial cells interact with presynaptic and postsynaptic neurons, with infiltrated leukocytes and perivascular cells, and with endothelial cells in the microvasculature of the BBB (Fig. 3). These multipartite

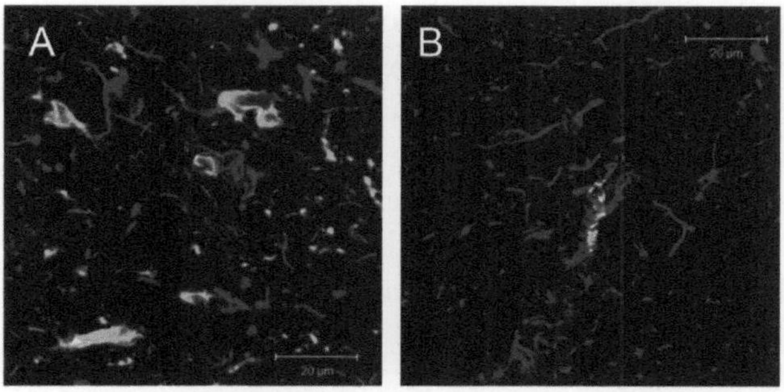

Figure 2 Physical interactions between microglia and astrocytes in the rat spinal cord. This figure displays confocal microscopy images of dorsal horn (superficial laminae) spinal cord of rats. (A) Immunolabeling of parenchymal microglia (Iba1, green) and astrocytes (GFAP, red) in a rat undergone paw incision (postoperative day 3). A "reactive" glial morphology is observed. (B) Immunolabeling of a perivascular microglial cell (ED2, green) and astrocytes (GFAP, red). Note the intimate interaction of astrocytes and a perivascular cell in what appears to be a capillary. (See the color plate.)

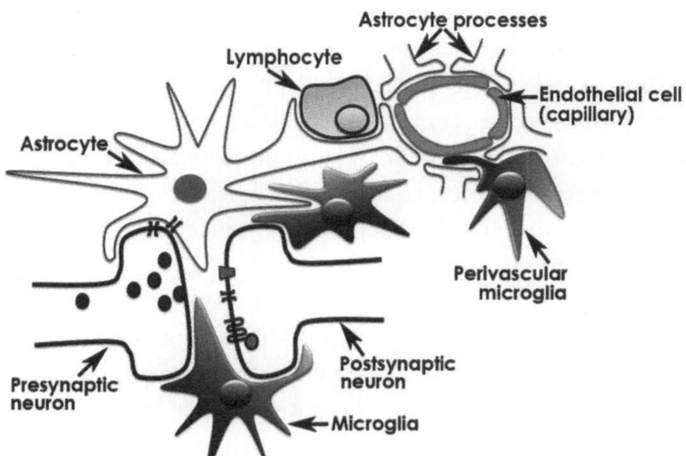

Figure 3 Cellular interactions of glial cells in the spinal cord and brain. Microglia and astrocytes physically and molecularly interact with presynaptic and postsynaptic neurons, perivascular cells, endothelial cells, and infiltrating lymphocytes under pathological conditions, like peripheral nerve injury (demonstrated in animal models).

interactions reflect the complexity of the pathophysiological mechanisms that underlie the initiation and maintenance of neuropathic pain and challenge the classic neuronal centered notion. The discovery of the involvement of nonneuronal cell players in the pathophysiology of chronic pain identified glial cells as a novel target to disrupt the aberrant interactions of the multipartite synapse in neuropathic pain states. Thus, several compounds with glial modulation capabilities were discovered, developed, and tested in multiple preclinical models of pain that had demonstrated glial cell "activation" or "reactivity" following injury. These studies were the foundation for clinical trials that have been mostly disappointing. This lack of translation from the bench to bedside has brought into question the real role of glial cells in human chronic pain conditions.

3. GLIAL ACTIVATION OR REACTIVITY: A CHANGE IN CELLULAR PHENOTYPE IN PAIN STATES

The terms "activation" and "reactivity" have been used to describe a particular state of glial cells following a nonphysiological stimulus (nerve injury, peripheral soft tissue damage, joint inflammation, bone pathology, infections, neurotoxic chemotherapy, etc.). However, as the glial field has matured, a new perspective on what constitutes activation and reactivity has emerged. The terminology "glial activation" in the context of pain has been particularly characterized mainly by changes in cell morphology and expression of cell surface molecules. Even though glial cells under physiological conditions express certain molecules in a modest degree and possess a rather ramified (nonhypertrophic) morphology, glial cells are certainly not "inactive." Therefore, the term "activation" is not accurate when referring to a functional state of glial cells that has been induced by pathophysiological (harmful or threatening) conditions.

Similarly, glial cells react to their environment, either physiologic or pathologic, thus the term "reactive glia" could be misleading in some instances. For example, microglia react to a homeostatic environment to modulate synapse formation and function (under physiological conditions).[24–26] Astrocytes also react under homeostatic conditions by interacting with neuronal synapses and blood vessels.[5,27,28] Thus, glial cells are also "activated" and "reactive" under physiological states of the CNS.

However, in several preclinical models of pain (nonphysiological conditions), glial cells adopt a differential phenotype characterized by expression of multiple surface markers of immunocompetence. For example, the expression of spinal cord microglial CCR3/CD11b, major histocompatibility complex II, and ionized calcium-binding adaptor molecule-1 (Iba-1) is enhanced in several preclinical pain models. The immunostaining of these markers allows the visualization of cellular morphology, which in pain models is characterized by hypertrophy, shorter, and thicker branches, and in some cases ameboid shapes (Fig. 4). Microglial cells also enhance the expression of other molecules (enzymes, receptors, etc.) that are known to play a role in nociceptive signaling such as cyclooxygenase-1, 2; Toll-like receptor-4; purinergic receptors (P2X4, P2X7, P2Y6, and P2Y12); CXCCR1, CCR2, etc.[29] In addition, microglia demonstrate phagocytic capabilities in some preclinical models of pain.[30] Under pathological conditions, microglia also undergo higher rates of proliferation[31] such as observed in preclinical chronic pain models.[32]

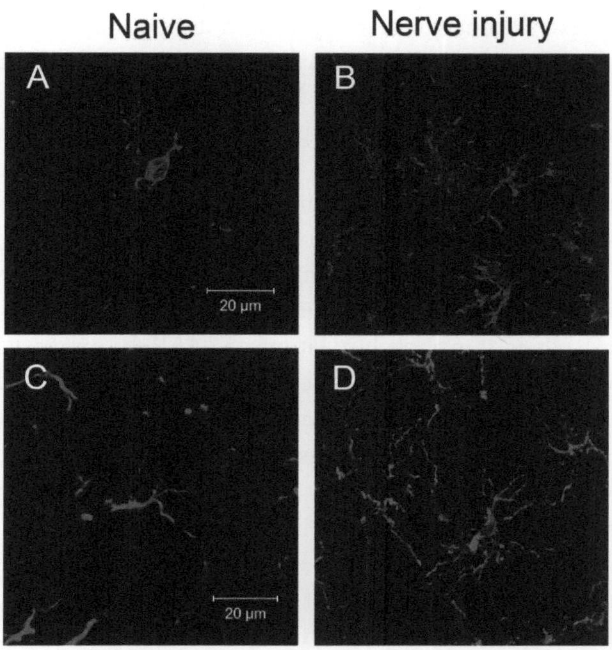

Figure 4 Glial morphology in spinal cord under naïve and peripheral nerve injury conditions in the rat. This figure displays confocal microscopy images of dorsal horn (superficial laminae) spinal cord of rats. (A) Immunolabeling for Iba1 (red microglial marker) in a naïve rat; note the defined cell body and thin processes. (B) Immunolabeling for Iba1 (red microglial marker) in a rat undergone L5 nerve transection (postoperative day 4); note the increased Iba1 expression, thicker processes, larger cell bodies, and cell confluence. (C) Immunolabeling for GFAP (green astrocytic marker) in a naïve rat; note the small cell body and defined processes with virtually no overlap among astrocyte territories. D. Immunolabeling for GFAP (green astrocytic marker) in a rat undergone L5 nerve transection (postoperative day 4); note the increased GFAP expression, larger and thicker processes and cell bodies, loss of territoriality of astrocytes.

The expression of spinal cord astrocytic S-100b and glial fibrillary acidic protein (GFAP) is typically increased in multiple preclinical pain models. The immunolabeling of these markers also allows visualization of the morphology of astrocytes and the transition from ramified to hypertrophic[33] (Fig. 4). Even though the staining of astrocytes with these markers represent only ~15% of the entire cell, their expression in pain models demonstrate an enhancement of cell volume, thicker branches, and a loss of their otherwise well-defined nonoverlapping territoriality.[29,33,34] Even though cell proliferation also enhances in astrocytes following peripheral nerve injury that models persistent neuropathic pain,[35] this occurs to a lesser extent than microglial proliferation.

Nevertheless, the changes that represent a distinct functional phenotype in glial cells under pain conditions may be better represented by the expression of intracellular molecules that drive the production of extracellular effectors, such as mitogen activator protein kinases (MAPKs) or NF-κB, both are pivotal for the production of proinflammatory effectors.[36] It is important to note that these molecular signaling pathways are not exclusive of glial cells. In fact, they are pivotal for the proper innate immune response and participate in all types of inflammatory processes.[37] Perhaps, the most important intracellular change that determines a "pronociceptive" phenotype in glial cells is the phosphorylation of MAPKs: extracellular signal-regulated kinase 1 and 2 (ERK-1 and ERK-2, respectively), p38, and c-Jun N-terminal kinases (JNK). Interestingly, MAPKs are phosphorylated in both neurons and glia following insults that produce nociceptive behaviors. Inhibition of spinal cord MAPKs reduces pain-related behaviors in different preclinical models of acute and chronic pain.[38,39] Thus, p-p38 in microglia,[36,40–43] p-JNK in astrocytes,[44–47], and p-ERK in neurons[48–50] play a major role in the induction of pain-related behaviors in preclinical models. Interestingly, two p-p38 inhibitors have shown efficacy in acute postsurgical pain and chronic neuropathic pain in clinical trials.[51,52] However, these studies did not conduct any analysis that determines whether these compounds exerted their actions into the CNS and directly on glial cells.

The signaling pathways of MAPKs drive the production of cytokines and chemokines, such as tumor necrosis factor alpha (TNF), interleukin (IL)-1 beta (IL-1), IL-6, etc., by microglia[38]; and IL-1, calcium signaling, and monocyte chemoattractant protein-1 (MCP-1) in astrocytes.[44,53–56] These major proinflammatory cytokines drive neuronal sensitization in the spinal cord, thus contributing to chronic pain behaviors in models of pain.[55,57–61]

Furthermore, these cytokines also induce the phosphorylation of spinal p38,[41,59,62] JNK,[44,55] and ERK[50] in glia and neurons in spinal cord. Similarly, MAPKs (i.e., p-p38) in glial cells (particularly microglia) can also activate NF-κB signaling, which contributes to further produce proinflammatory cytokines.[36] The direct activation of neurons by cytokines (i.e., TNF) also promotes the phosphorylation of ERK in neurons, which leads to *N*-methyl-D-aspartate (NMDA) receptor hyperactivity and neuronal sensitization in the spinal cord.[48–50] These reverberant processes reinforce the MAPK–cytokine–MAPK vicious cycle that is believed to underlie the perpetuation of central neuronal sensitization and constitutes the primordial mechanism of maintenance of chronic pain.

In addition to the role of glial cells in enhancing excitatory signaling at the multipartite synapse in the spinal cord, there is evidence for a role in reducing the endogenous inhibitory signaling pathways. Following nerve injury, microglia release brain-derived neurotrophic factor. Brain-derived neurotrophic factor reduces inhibitory GABAergic signaling,* resulting in a net increase in the neuronal excitatory milieu in the spinal cord.[63] Thus, at a multipartite synapse, glial cells may induce central neuronal sensitization by both increasing pronociceptive signaling and inhibiting antinociceptive pathways.

4. TRIGGERS OF THE CHANGE OF GLIAL PHENOTYPE IN PAIN STATES

The exact mechanism by which glial cells adopt a pronociceptive cellular phenotype following peripheral or central tissue damage is not fully understood. However, some molecules have been identified as major contributors in rodent models: substance P, glutamate, adenosine triphosphate (ATP), fractalkine, or activators/agonists of TLR-4 receptors (yet unknown endogenous ligand). Also, it is known that glial responses in the spinal cord increase in parallel to the degree of peripheral tissue (i.e., nerve) damage[64]; thus, mild nerve damage produces a mild glial reaction, and a robust nerve damage produces a robust glial reaction.

Peripheral tissue damage induces nociceptor excitability, which is believed to be the critical step in inducing central changes, including glial responses. Activated nociceptors release glutamate and substance P, which in turn stimulate second-order neurons in the spinal cord. The use of local anesthetics or spinal NMDA receptor blockers following peripheral nerve injury in rodent models of pain attenuates some responses in microglia and astrocytes.[65–67] The neuronal activity-induced release of substance P in the spinal cord can directly activate NK1 receptors on microglia[68] and astrocytes.[69,70] However, the biological relevance of substance P and glutamate release on glial reactions in the context of pain seems to be very limited, since the attenuation of glial reactions induced by peripheral anesthetics does not always parallel to attenuation in pain-related behaviors in animal models of neuropathic pain.[65] Whether neuronal glutamate is playing a role in glial responses is difficult to elucidate since glutamate can be released from both neuronal and glial cells. Along the same lines, whether substance P's effects on glial changes have a biological significance on pain is questionable since NK1 knockout mice display similar pain-related behaviors as wild-type mice following peripheral nerve injury.[71] These findings highlight the complexity of understanding not just the role of glial cells in chronic pain but also in understanding the factors that modulate glial cell reactivity following peripheral nerve injury.

Purinergic receptor expression and activation in microglia following nerve injury has been well studied and suggested as a trigger for microglial changes. Microglia express purinergic receptors especially under a pathologic "reactive" state or following peripheral nerve injury. The ligand of these receptors, ATP, is released by neurons (and also by glial cells) and has been postulated to induce peripheral nerve injury-induced glial changes. For example, microglial expression of P2X4,[72] P2X7,[73,74] or P2Y12 receptors[75] is enhanced following peripheral nerve injury, and the blockade of these receptors reduces pain-related behaviors in models of neuropathic pain. However, whether the antinociceptive effects of purinergic receptor blockers are due to actions on microglia is not completely clear, since microglia still react and adopt a pronociceptive/inflammatory phenotype to peripheral nerve injury (or ATP application *in vitro*) in the absence of P2X4 expression (rodents or cells lacking, knockout, P2X4 receptors).[76] It remains to be determined whether neuronal ATP triggers microglial reactions, since ATP may be released from both neurons and glia following peripheral nerve injury.

Several chemokines and chemokine receptors have been well studied in the pursuit of understanding the role of glia in pain pathology and the identification of potential new therapeutic targets to treat chronic pain. Fractalkine is a chemokine that is produced by nociceptors (neurons) and binds CX3CR1 receptor that might be expressed only on microglial cells.[77] Following peripheral nerve injury, fractalkine is released from neurons in response to microglial signaling via cathepsin S,[78] which through activation of CX3CR1 receptors induces the phosphorylation of p38 in microglia.[79] The subsequent effect is the production of proinflammatory/nociceptive cytokines, such as TNF, IL-1, or IL-6. In fact, IL-6 induces microglial CX3CR1 expression via p-p38,[41] and this might contribute to establish a reverberant circuit that could amplify and maintain central neuronal sensitization and chronic pain. Another chemokine that produces similar effects in rodent neuropathic pain models is CCL2, also known as monocyte chemotactic protein-1, MCP-1. CCL2/MCP-1 is released by neurons and astrocytes and acts on CCR2 receptors that are expressed in microglia and neurons.[44,80] An antibody against CCL2 administered intrathecally in rats reduces pain-related behaviors and the expression of microglial markers in a model of neuropathic pian.[80]

One additional signaling molecule that has been identified to play a role in inducing glial responses is TLR-4. Signaling molecules triggered by TLR-4, which regulate innate immunity, contribute to glial reactions following nerve damage.[81] Bacterial lipopolysaccharide is the better-known TLR-4 agonist (i.e., pathogen-associated molecular patterns), while endogenous TLR-4 agonists have also been described and called danger-associated molecular patterns.[82] TLR-4 is expressed in microglia,[81] astrocytes,[83] and nociceptors/neurons.[84,85] Mice lacking TLR-4 that have undergone peripheral nerve injury or induced arthritis display reduced glial responses and pain-related behaviors. The main signaling pathway that is activated in microglia through TLR-4 is the activation of p38, which results in the production of proinflammatory factors such as cytokines, prostaglandins, BDNF, and nitric oxide.[29,81,86] In astrocytes, TLR-4 activation triggers NF-κB, MAPKs (i.e., JNK), and Jak1/Stat1 signaling pathways, all of them involved in the effective production of proinflammatory/nociceptive effectors, such as CCL2.[82,83] In transient receptor potential vanilloid-1 (TRPV1)-expressing nociceptive neurons, TLR-4 activation produces the release of calcitonin gene-related peptide (CGRP, a major pronociceptive effector), TRPV1 neuronal sensitization, and intracellular calcium accumulation.[87] Curiously, it has been shown that spinal cord TLR-4 plays a role in nociception under inflammatory or neuropathic conditions in male mice, but not in female mice.[88] Based on rodent studies, other TLR receptors (TLR-2, TLR-3, TLR-7, and TLR-9) are also present in microglia, astrocytes, and neurons, and all of them trigger signaling pathways that govern the production of proinflammatory/nociceptive products that can also contribute to central sensitization under chronic pain conditions.[82] TLR-4 antagonists have demonstrated success in preclinical neuropathic models.[89] Even though TLR-4 antagonists might have a place for the treatment of sepsis in humans,[90] targeting TLR-4 for the treatment of pain states could result in a risky strategy due to the multiple implications of TLR-4 in immune responses that protect from cancer or infections.

5. DRUGS WITH GLIAL MODULATION CAPABILITIES AND EFFICACY IN PRECLINICAL MODELS OF PAIN

Numerous glial modulating compounds have been examined in a wide variety of preclinical pain models (from acute to persistent, including nociceptive, inflammatory, and neuropathic). Fluorocitrate, a glial metabolic inhibitor, was one of the first glial modulators to demonstrate a reduction of pain-related behaviors when administered intrathecally in models of inflammatory pain.[91,92] Subsequently, three glial modulators were extensively tested in multiple preclinical pain models: minocycline,[93,94] propentofylline,[95] and ibudilast.[96] These compounds reduced pain-related behaviors in virtually all preclinical pain models examined. Interestingly, these pain models (acute, inflammatory, visceral, chronic, neuropathic, metabolic, chemotherapy-induced, etc.) produced spinal glial "reactivity," a terminology that refers to a specific pattern of cellular and molecular changes in glial cells following peripheral or central insults in pain models/conditions.

Minocycline is a tetracycline glial modulator that targets preferentially (but not exclusively) microglia. When minocycline was administered systemically or intrathecally prior to nerve injury in rats, it prevented the development of allodynia. In contrast, when minocycline was given 5–7 days after peripheral nerve injury, it did not reduce established hypersensitivity.[97,98] However, minocycline reduced microglial reactivity when given before or days after neuronal damage,[98] suggesting that minocycline's antinociceptive effects may not be related to changes in glial reactivity. Accordingly, minocycline administered systemically did not reduce postoperative pain-related behaviors in rats, but it does reduce the expression of CD11b, a marker of glial "reactivity."[99] Minocycline reduced pain-related behaviors in several preclinical rodent pain models: formalin,[100] carrageenan inflammation,[101] taxol/chemotherapy-induced neuropathy,[102] burn pain,[103] diabetic neuropathy,[104] visceral pain,[105,106] and spinal cord injury.[107] At the molecular or functional level, minocycline inhibits spinal cord microglial p-p38[108] and neuronal p-ERK,[105] reduces microglial motility and expression of Kv1.3 channels *in vitro*,[109] and enhances endocannabinoid levels in the spinal cord.[110] Interestingly, minocycline failed to reduce microglia and astrocyte reactivity, neuronal loss, and clinical progression in a Batten disease model in the sheep. This disease is characterized by neurodegeneration due to neuroinflammation (believed to be driven by microglia). In this sheep model of a chronic inflammatory neurodegenerative disease, minocycline was given orally for 11 months and reached detectable levels in serum and cerebrospinal fluid (CSF).[111] It remains to be determined whether species differences exist in microglial function or microglial drug targets between rodents and nonrodents. The answer to this question is extremely relevant, since it can help determine the translational potential of our current animal models of glial function, glial targets, and pain.

Propentofylline is a methylxanthine compound that inhibits phosphodiesterase and reduces glial "reactivity" in neuroinflammatory conditions in rodents. Even though the effects on glial cells are the most studied actions of propentofylline, a direct effect of propentofylline on neurons cannot be ruled out.[112,113] Systemic or intrathecal administration of propentofylline before, days after, or weeks after peripheral nerve injury in rodents prevented and treated neuropathic pain-related behaviors and reduced spinal cord microglia and astrocyte reactivity.[114–116] Propentofylline reduced pain-related behaviors induced by vincristine in a rat model of chemotherapy-induced neuropathy[117] in parallel with a reduction of microglia and astrocyte reactivity. Propentofylline reduced morphine-induced antinociceptive tolerance/hypersensitivity and morphine-induced glial reactivity in rats.[118] At the molecular and functional level, propentofylline reduces CD11b, GFAP,[119] TNF, IL-1, and IL-6,[114,120] p38 and ERK phosphorylation,[121] blocks the presynaptic downregulation of glutamic acid decarboxylase 65,[113] and reduces spinal cord neuronal sensitization interpreted as neuronal "windup" in electrophysiological experiments.[112]

Ibudilast, a phosphodiesterase inhibitor, was originally described to prevent oligodendrocyte toxicity induced by kainate[122] and to inhibit *in vitro* TNF release in both microglia and astrocytes.[123] Ibudilast also has direct potent effects on neurons, since it protects neurons from glutamate toxicity in a concentration-dependent manner.[124] Ibudilast inhibits microglial production of IL-12, reactive oxygen species, nitric oxide, IL-1, and IL-6 and is neuroprotective in *in vitro* neuronal/microglial cultures.[125] Ibudilast reduces the production of TNF and IL-1 induced by the HIV protein Tat in human microglia *in vitro*.[126] *In vivo*, ibudilast administered systemically reduced allodynia in several rodent models of peripheral or spinal cord injury neuropathic pain, paclitaxel-induced neuropathy, and morphine-induced antinociceptive tolerance.[127,128]

These studies formed a strong foundation of preclinical support for the critical role of glia in numerous pain models and the promise of glial modulators for the treatment of chronic pain (a condition in which microglia cells are implicated) in humans. Of course, it is always possible that the modulation of a single type of cell (i.e., microglia) might elicit a compensatory response by the other players at the multipartite synapse and the net effect would be of no change at a behavioral level. Or at a molecular level, it could be postulated that the modulation of glial function could be physiologically nonsignificant given the vast myriad of molecular signaling occurring in the multipartite synapse. Nevertheless, based on the foundation of positive studies with multiple glial modulating compounds across numerous rodent models of pain, the modulation of glial cell function is an attractive therapeutic strategy for the treatment of pain. For best translation of these preclinical findings to the clinic, one must ask what is the evidence is for glial contributions to human disease processes, especially pain conditions.

6. WHAT IT IS KNOWN OF GLIAL CELLS IN HUMANS WITH PAIN CONDITIONS?

The detection of changes in glial phenotype in patients with pain conditions has been attempted via measurements of glial marker levels in CSF and serum, or through imaging techniques of radiolabeled glial markers. Tissue extraction to determine glial reactions via histological assessments in patients with chronic pain is obviously not a viable alternative. Therefore, the access to glial markers in CSF and serum, or through imaging approaches, is an attractive possibility.

YKL-40, a 40-kDa heparin- and chitin-binding glycoprotein also known as human cartilage glycoprotein 39 (HC-gp39), is a molecule that is involved in inflammatory processes and is present in "reactive" microglia and astrocytes. YKL-40 is elevated in CSF of patients with multiple sclerosis (a neuroimmunologic condition that is presented with chronic pain) in comparison to multiple sclerosis patients on immunosuppressant treatment or healthy controls.[129,130] Whether the source of CSF YKL-40 is peripheral (leukocytes) or central (glial cells) could not be determined in these studies, since serum YKL-40 was comparable in all three groups.[129]

A reactive state of glial cells in patients with brain trauma is supported by the fact that microglial cells express high levels of proinflammatory markers, such as MHC class II (CD/43, histology).[131] Additionally, in patients with traumatic brain injury, GFAP (measured indirectly through GFAP-breakdown products) in serum is elevated implying a change in astrocytic phenotype[131] but similar limitations exist in determining whether the source of GFAP is peripheral or central. GFAP is expressed mostly by astrocytes (CNS) and in a lesser extent by satellite cells (at the dorsal root ganglia level). The detection of GFAP in serum in patients with brain trauma might be related to the physical disruption of the BBB that would allow the diffusion of GFAP from the CNS to the periphery. Nonetheless, GFAP (measured indirectly through GFAP-breakdown products) levels in serum are elevated in patients with peripheral orthopedic/traumatic conditions in the absence of CNS trauma and without physical damage of the BBB.[132] These findings indicate that in pain conditions GFAP may diffuse through the BBB due to changes in permeability. Together, these studies suggest that a potential change in astrocyte phenotype could be determined by serum levels of GFAP in patients with neuroimmune conditions, and possibly in chronic pain states. Interestingly, mice lacking GFAP (knockouts) display a reduced duration of pain-related behaviors following peripheral nerve injury, and the knockdown of GFAP (by antisense oligonucleotide) reduces pain-related behavior in this model of neuropathic pain.[133] Based on these findings, the investigation of the association of spinal cord (i.e., in CSF) levels of GFAP and chronic pain intensity or duration (ideally, in the presence of glial modulators) in human subjects is warranted and feasible.

An alternative to monitor changes in glial marker expression in patients with pain conditions is the use of positron emission tomography (PET) and radiolabeling glial markers. PK11195 is an isoquinoline that binds to the peripheral benzodiazepine receptor. This receptor is widely expressed in the periphery but expressed in very low levels under normal states in the CNS. Even though (R)-PK11195 has been found to bind to astrocytes and microglia, some animal studies suggest that under neuroimmune pathological conditions (R)-PK11195 binding correlates better with reactive microglia than with reactive astrocytes.[134] The labeling of (R)-PK11195 with carbon-11 ([11C]-(R)-PK11195) allows its visualization using PET imaging. This *in vivo* imaging approach has been used to indirectly determine changes in brain microglia. Patients with peripheral nerve injury,[135,136] or multiple sclerosis,[137] conditions that result in chronic pain, show elevated binding of [11C]-(R)-PK11195 in the brain. Similarly, patients suffering chronic fatigue syndrome/myalgic encephalomyelitis show higher [11C]-(R)-PK11195 binding in the cingulate cortex and thalamus, which positively correlated with pain scores.[138]

This *in vivo* technique could be a way to identify specific patients and specific pain conditions displaying changes in microglia. The use of this technology would allow studies that test glial modulators to determine the real biological and clinical significance of glial changes in pain conditions in human subjects. However, this technique has its limitations and may not be suitable for longitudinal monitoring proposes, due to high levels of intrasubject variability of [11C]-(R)-PK11195 binding and its likely binding to peripheral benzodiazepine receptor under normal nonpathological conditions.[139]

Changes in the CSF levels of proinflammatory factors, presumably produced by glial cells, are another indirect approach that has been used to study the role of glial cells in patients with chronic pain. For example, IL-8 (a proinflammatory chemokine, also known as CXCL8), but not IL-1, is elevated in CSF in patients with fibromyalgia (a chronic pain condition). The central origin of the IL-8 elevation is supported by the low levels of this cytokine in plasma in these patients.[140] Similarly, cerebrospinal levels of IL-8 are elevated in patients with osteoarthritis and arthroplastic surgery, another chronic pain condition, in comparison to serum levels or CSF levels of control subjects.[141] The levels of IL-8 are also elevated in CSF from patients with postherpetic neuralgia, and CSF IL-8 elevated levels are predictors of developing neuropathic pain following herpes zoster skin rash.[142] Furthermore, IL-8 levels could be reduced with intrathecal treatment with methylprednisolone which paralleled improvement in pain scores.[142,143] Since microglia is the major source of cytokines in the CNS in humans,[144] it is reasonable to think that intrathecal methylprednisolone reduces CSF IL-8 by acting on spinal microglial cells.

Studies using postmortem tissue assessments from HIV-positive patients with chronic pain showed that microglia did not display changes in cellular markers (CD11b and Iba-1), while astrocytes displayed increased expression of S100b and GFAP in the dorsal horn of the spinal cord when compared to tissues from HIV-positive patients without chronic pain.[145] In addition, in these patients, TNF and IL-1 (proinflammatory/nociceptive cytokines), p-ERK and p-p38 (MAPKs), c-fos (marker of neuronal activation) and p-CREB (marker of neuronal sensitization) were overexpressed in the dorsal horn of the spinal cord.[145] These data argue for a glial proinflammatory (probably astrocyte rather than microglial) and neuronal sensitizing spinal cord phenotype in HIV patients with chronic pain. These spinal cord changes are associated with the activation of neuronal signaling pathways that are involved in synaptic plasticity (i.e., the glycoprotein Wnt)[146] and with HIV gp120 protein.[147]

GFAP is also elevated in CSF in about 50% of patients with complex regional pain syndrome (CRPS). These patients also have high levels of IL-6, IL-1, TNF, MCP-1, glutamate, calcium, and NO metabolites (proinflammatory/nociceptive factors) and low levels of IL-4 or IL-10 (anti-inflammatory/nociceptive cytokine).[148,149] Curiously, in these patients, there is a positive correlation of higher levels of GFAP with higher levels of MCP-1, strongly suggesting an astrocyte origin of MCP-1.[148] Since the major source of cytokines in the CNS is microglia, it seems that these cells are likely playing a role in this pain condition. However, due to the lack of concordant serum measurements in these studies,[148] a peripheral source (leukocytes) cannot be ruled out. Accordingly, in a patient (case report) with long-lasting CRPS, GFAP (astrocytic marker) and CD-68 (microglial and monocyte marker) expression is enhanced in dorsal horn spinal cord (postmortem tissue), and astrocytes and microglia display large cell bodies (hypertrophic morphology) not only at the original site of the injury (more prominent) but also along the entire spinal cord.[150] These data suggest the potential involvement of glial cells in the central pathological mechanisms of CRPS. Also, these findings identify CSF levels of glial markers, such as GFAP as potentially useful biomarkers to determine glial reactivity in correlation with CSF levels of inflammatory molecules (i.e., cytokines), to assess a neuroimmune process in patients with chronic pain conditions.

Expression of the proinflammatory/nociceptive cytokine IL-1 in CSF positively correlated with pain intensity ratings in patients with post-traumatic neuralgia or painful neuropathy.[151] Lower levels of IL-1 in serum in these patients argue for a central source (likely glial cells) of this cytokine.[151] In these patients, IL-10 levels in the periphery (plasma) were also evaluated. In fact, the peripheral contribution of IL-10 (plasma levels) was negatively correlated with pain intensity ratings, suggesting that this also play a major role in these neuropathic pain patients.[151]

In other chronic conditions such as rheumatoid arthritis (RA) that manifests with chronic pain, fatigue, and sleep disturbances, IL-1 is elevated in CSF (10 times higher than serum levels). However in patients with RA, these high levels of IL-1 did not correlate to pain intensity and sleep disturbances, but it did correlate to fatigue scores.[152] These data together indicate that in humans the central production of proinflammatory cytokines (detected in CSF), such as IL-1, is not sufficient to diagnose or monitor the progression or resolution of chronic pain. This is intriguing since the *in vitro* addition of CSF from patients suffering postherpetic neuralgia to human microglia cultures (immortalized, nonprimary cell line) induces IL-6 production.[153] Interestingly, CSF from patients with noneffective steroid treatment (dexamethasone, no pain improvement) induced higher levels of microglial IL-6 than CSF from patients with effective (pain improvement) steroid therapy.[153] Furthermore, *in vitro* incubation of microglial with methylprednisolone produced a greater reduction of IL-6 in cells with effectively treated patient CSF versus ineffectively treated patients CSF.[153] All these clinical findings strongly suggest that other centrally produced molecules independent of proinflammatory cytokines likely determine pain intensity or therapy efficacy in patients with chronic pain. More intriguing, these molecules also may condition glial responses that can be associated or not with pain or therapeutic efficacy. The identification of such factors is an exciting and vast area that warrants future research.

7. GLIAL CELLS: HUMAN VERSUS RODENTS AND OTHER SPECIES

Glial cells in rodents, which are the most widespread preclinical model organism for the study of glial cell functions in pain, are vastly different from nonhuman primate and human glial cells. For example, human astrocytes are 1.75- and 2.6-fold larger in diameter and more complex than chimpanzee and mice astrocytes.[154] Human astrocytes compare to mouse astrocytes, express 10 times more GFAP, conduct calcium waves four times faster, display a more complex structure, exist in more diverse types than mouse astrocytes, and reside in brain areas in which mouse astrocytes do not reside.[154,155] Similarly, human microglia display a larger diversity of cellular morphology than mouse microglia, even though these species possess similar size microglial cells.[156] Functionally, human microglia are significantly less responsive than rat microglia when stimulated with LPS and also respond differentially to the glial modulator, propentofylline, when compared to rat microglia.[157] These findings indicate that glial function and pharmacological effects of glial modulators observed in rodent models or rodent cells are not directly translatable to human conditions or human cells. Therefore, these differences between mice and humans in terms of glial cell functions and pharmacological responses should be taken into consideration before moving glial modulators from preclinical studies to clinical trials.

8. CLINICAL DATA USING GLIAL MODULATING AGENTS

One of the first studies using a glial modulator in human subject studies (double-blind, multinational, randomized, and placebo-controlled) was performed in patients with multiple-system-atrophy Parkinson type that received 1-year therapy with minocycline. Interestingly, this study evaluated clinical outcomes (motor function scores and parkinsonism symptoms for progression rate), and [11C](R)-PK11195-PET was performed in very small subgroups.[158] This study failed to show a clinical effect of minocycline, despite a strong trend ($p=0.07$) in [11C](R)-PK11195 binding decrease in the minocycline group ($n=3$) compared to the placebo group ($n=5$).[158] Despite the limitations of this clinical trial, this study strongly suggests that the microglia-modulating effects of minocycline are preserved in humans. Furthermore, it shows that microglia has a limited significance in neurological diseases affecting motor function (i.e., parkinsonism).

In another study using nonhuman primates (rhesus macaques) infected with simian immunodeficiency virus (the equivalent of HIV in humans), minocycline (daily for 4 weeks) reduced viral load in CSF and serum and reduced the expression of brain GFAP and Iba-1 (histology).[159] These studies indicate that minocycline seems to be a suitable drug to test the clinical relevance of microglia modulation in the context of pain in primates.

Three drugs have been studied in patients with chronic pain conditions: minocycline, propentofylline, and AZD2423 (Table 1).

In a multicenter, double-blind, placebo-controlled, randomized clinical trial, minocycline given (100 mg twice daily, orally) for 8 days (from the day of the surgery) failed to show any beneficial effect on the intensity and the incidences of leg or back pain in patients undergone lumbar radiculopathy secondary to disk disease in the acute (days after surgery) and chronic phase (3 months after surgery).[160]

It is very likely that glial cells play a more relevant role in rodents than in humans, since differential effects of minocycline have been reported in other neurological states among species. For example, minocycline seems to have detrimental effects in patients with amyotrophic lateral sclerosis, but some minocycline's beneficial effects have seen in pilot clinical studies in patients with multiple sclerosis.[93] Even though these studies suggest a differential role of microglia in different neurological diseases, they also may represent a warning call, similar to the studies in pain patients, since these findings also

Table 1 Description of clinical trials using glial modulator drugs

Drug	Condition	Type of clinical trial	Dose and duration of treatment	Primary end point	Pain treatment efficacy	Glial response marker used	Drug levels and function assessment	Sample number (drug–placebo groups)	References
Minocycline	Lumbar radiculopathy secondary to disk disease	Randomized, double-blind, placebo-controlled	100 mg P.O. bid—8 days starting the day before surgery	Leg or back pain at rest vs. baseline	None (1 day and 3 months after surgery)	None used	None	43 (drug)–42 (placebo)	Martinez *et al.*[160]
Propentofylline	Postherpetic neuralgia	Randomized, double-blind, placebo-controlled, parallel group	300 mg P.O. tid—4 weeks starting at least 6 months after rash	Pain intensity associated with skin rash vs. baseline	None (for 4 weeks)	None used	Plasma levels of drug were detected	66 (drug)–75 (placebo)	Landry *et al.*[157]
AZD2423 (CCL2 antagonist)	Posttraumatic neuralgia	Randomized, double-blind, placebo-controlled, parallel group, multicenter	20 or 150 mg P.O. once a day—28 days starting at least 3 months (and up to 5 years) after trauma or surgery	Pain intensity vs. baseline	None (first 5 days of treatment, and last 5 days of treatment)	None used	Plasma levels of drug were detected. Reduction of mean blood monocyte counts in drug groups was observed	48 (20 mg), 41 (150 mg)–44 (placebo)	Kalliomaki *et al.*[161]

point out the lack of translation from animal models to human neurological conditions (i.e., ALS[162]).

Propentofylline, another glial modulator, has been studied in post-herpetic neuralgia patients (randomized, double-blind, placebo-controlled, parallel group clinical trial). Propentofylline was given orally (300 mg, three times a day) for 4 weeks and this regimen failed to show any pain relief or reduction in rescue medication (acetaminophen).[157] This study supports the notion of the potential limited role of glial cells in chronic pain conditions.

AZD2423 is a CCR2 antagonist that was studied in patients with chronic pain, posttraumatic peripheral nerve injury neuralgia (randomized, double-blind, placebo-controlled, multicenter clinical trial). AZD2423 given orally (20 or 150 mg, once daily) for 4 weeks failed to reduce pain in these neuropathic patients.[161] This compound targets CCR2, which is expressed in the CNS in neurons and microglia (but also in the periphery). Pharmacologically, targeting two systems should produce a more efficacious or potent effect, especially when the systems possess differential functions. Therefore, it is revealing that AZD2423 was not efficacious in neuropathic pain patients. These data also underline the potential limited clinical relevance of CCR2 and the neuronal–microglial interactions in neuropathic pain conditions in humans.

These clinical trials using three different glial modulators consistently failed in translation from rodent/animal models to human pain conditions. Together, all these studies should stimulate discussion in the pain community around three points: (1) that perhaps minocycline, propentofylline, and AZD2423 are not effective in treating neuropathic pain in humans in contrast to preclinical studies, (2) that the role of microglia in pain conditions might be limited and possibly is not significant in humans when compared to the robust response in rodent preclinical models, and (3) that perhaps there is low translational value of preclinical rodent models that strongly demonstrated an instrumental role of microglia in pain conditions.

It seems that glial neuroimmune mechanisms in chronic pain patients are more complex than the mechanisms observed in rodent models, and this complexity may reflect a potential limited clinical significance of glial function in the context of pain.

9. CONCLUSIONS

The nonneuronal central mechanisms of pain are mostly orchestrated by glial cells not only in the spinal cord but also in superior CNS areas. These mechanisms involve immune response-like processes that affect other non-neuronal cells from the periphery, the BBB, and the CNS vasculature. These processes ultimately influence and modify the physical and functional integrity of neurons. Virtually, all the studies that explore the neuronal and glial interactions in the context of pain have been focused on the pronociceptive or detrimental effects of glial cells on neurons using mostly rodent models. These studies have not translated to a clinical reduction in pain. The differential functional profile of glial cells in rodents and primates or humans, or the differential targets of glial modulators (drugs) in rodent and primates or humans could explain at least partially the lack of translation of preclinical studies. Certainly, more studies exploring these potential differences are warranted. The validation of glial targets or functions in human tissues or human conditions before moving forward to clinical trials could potentially better predict the translational value of glial modulators. Based on the current clinical evidence (Table 2), it seems that GFAP is a common molecule in rodents and humans in terms of glial "reactivity" (namely, astrocytic reactivity) and its association with pain states. Another molecule that is produced by microglia in humans and seems to be related to pain conditions is IL-8. Curiously, IL-8 has not been studied extensively in rodent models of pain. However, it seems that spinal cord IL-8 in the rat plays similar roles to those in humans in the context of pain and pharmacologic pain control.[163] For a biomarker perspective, it is interesting that both GFAP and IL-8 are molecules that can be measured in CSF in humans. In the case of GFAP that is produced mostly in the CNS, it is important to highlight that can be also measured in serum, an advantageous feature for a CNS biomarker. Future studies using human subjects or tissues should include these potential biomarkers to prove the role of glial cells in the pathophysiology of pain conditions and the mechanism of actions of glial modulators.

As an alternative, a different view of glial cells and their role in the mechanisms of pain could also provide valuable novel information. Glial cells have been largely seen as detrimental and "bad" in the context of pain.[92] However, targeting glial cells to block the production of proinflammatory factors has resulted in a failed strategy, and this paradigm should probably change. Glial cells do not possess only deleterious functions under pathological conditions. Indeed, microglia and astrocyte cells have been described to have beneficial and neuroprotective actions in the spinal cord following spinal cord injury and using very elegant experiments. The ablation of reactive astrocytes in a mice model of spinal cord injury results in larger tissue degeneration.[164] Similarly, the activation of TLR-2 enhances microglia reactivity

Table 2 Description of glial markers studied in human subjects under pain conditions

Glial marker	Condition	Main source	Levels—CNS condition group:CNS control group	Levels—periphery condition group: periphery control group	Levels—CNS condition group: periphery condition group	Major limitation	Type of specimen	Refs.
GFAP	TBI	Astrocytes (CNS)	Not investigated	Periph TBI > Periph control	Not investigated	No CSF levels studied	Serum	Papa *et al.*[132]
GFAP	Orthopedic trauma (OT, non-TBI)	Astrocytes (CNS)	Not investigated	Periph OT > Periph control	Not investigated	No CSF levels studied	Serum	Papa *et al.*[132]
GFAP and S100b	HIV+ with chronic pain	Astrocytes (CNS)	CNS HIV > CNS control	Not investigated	Not investigated	Limited sample ($n=5$)	Spinal cord (Western blot)	Shi *et al.*[145]
GFAP	CRPS	Astrocytes (CNS)	CNS CRPS (50% of patients) > CNS control	Not investigated	Not investigated	50% cases were negative ($n=11/22$)	CSF	Alexander *et al.*[148]
GFAP and CD68	CRPS (case report)	Astrocytes and microglia respectively (CNS)	CNS CRPS > CNS control (both markers)	Not investigated	Not investigated	Only one CRPS case vs. four controls	Spinal cord (histology)	Del Valle *et al.*[150]

Continued

Table 2 Description of glial markers studied in human subjects under pain conditions—cont'd

Glial marker	Condition	Main source	Levels—CNS condition group:CNS control group	Levels—periphery condition group: periphery control group	Levels—CNS condition group: periphery condition group	Major limitation	Type of specimen	Refs.
CD11b	HIV + with chronic pain	Microglia (CNS) and monocytes (periphery)	CNS HIV = CNS control	Not investigated	Not investigated	Limited sample ($n=5$)	Spinal cord (Western blot)	Shi *et al.*[145]
IL-8	Fibromyalgia (FM)	Microglia (CNS) and monocytes (periphery)	CNS FM > CNS control	Periph FM > Periph control	CNS FM > Periph FM	Limited sample ($n=15$)	Serum and CSF	Kadetoff *et al.*[140]
IL-8	Osteoarthritis and arthroplastic surgery (OA)	Microglia (CNS) and monocytes (periphery)	CNS OA > CNS control	Periph OA > Periph control	CNS OA > Periph OA	Limited sample ($n=20$)	Serum and CSF	Lundborg *et al.*[141]
IL-8	Postherpetic neuralgia (PHN)	Microglia (CNS) and monocytes (periphery)	CNS PHN > CNS control	Not investigated	Not investigated	No plasma levels studied	CSF	Kotani *et al.*[142]

following spinal cord injury, which results in a reduction of tissue degeneration in mice.[165] These studies strongly suggest that the inhibition of the "harmful" signals of glial cells can also reduce glial capabilities to repair tissue or induce homeostasis. Therefore, the potentiation of the protective functions of glial cells could be an alternative strategy to provide pain relief by targeting glial cells. In accordance to this hypothesis, it has been shown that the induction of spinal ED2/CD163[166] or spinal phosphatases that deactivate MAPKs, MKP-1, or MKP-3[157,167] prevent or treat pain-related behaviors in rat models of neuropathic pain. Nevertheless, all these studies on alternative glial functions or molecules that induce an alternative phenotype in glial cells should be validated in human tissues or human pain conditions before concluding that these are valid targets to be tested in clinical trials.

In conclusion, it is clear that glial cells play a role in the nonneuronal mechanisms of pain in preclinical models. However, these preclinical foundations have failed to demonstrate that the modulation of glial functions in the context of pain results in clinical benefits for pain patients. The question of whether glial cells are a valid target to treat human pain conditions is far from conclusive and uncovers more questions than answers. Once again a failure in translating from bench to bedside in the pain research community should engage researchers in discussion of the validity and clinical predictive value of preclinical models and a need to clarify the impact of species differences on translation.

ACKNOWLEDGMENTS

This chapter is dedicated to Dr. Joyce DeLeo for her scientific and personal mentorship and friendship for both authors. She taught these authors several important lessons on being a good scientist and mentor. Perhaps the most important lessons were to be able to question the scientific dogma with integrity and to be autocritical before being critical to the work or others.

REFERENCES

1. Freeman MR. Specification and morphogenesis of astrocytes. *Science*. 2010;330(6005): 774–778.
2. Graeber MB. Changing face of microglia. *Science*. 2010;330(6005):783–788.
3. Graeber MB, Streit WJ. Microglia: biology and pathology. *Acta Neuropathol*. 2010;119(1):89–105.
4. Dustin ML. A dynamic view of the immunological synapse. *Semin Immunol*. 2005;17(6):400–410.
5. Panatier A, Theodosis DT, Mothet JP, et al. Glia-derived D-serine controls NMDA receptor activity and synaptic memory. *Cell*. 2006;125(4):775–784.
6. Pascual O, Casper KB, Kubera C, et al. Astrocytic purinergic signaling coordinates synaptic networks. *Science*. 2005;310(5745):113–116.

7. De Leo JA, Tawfik VL, LaCroix-Fralish ML. The tetrapartite synapse: path to CNS sensitization and chronic pain. *Pain.* 2006;122(1–2):17–21.
8. Nedergaard M, Ransom B, Goldman SA. New roles for astrocytes: redefining the functional architecture of the brain. *Trends Neurosci.* 2003;26(10):523–530.
9. Romero-Sandoval A, Chai N, Nutile-McMenemy N, Deleo JA. A comparison of spinal Iba1 and GFAP expression in rodent models of acute and chronic pain. *Brain Res.* 2008;1219:116–126.
10. Romero-Sandoval EA, Horvath RJ, DeLeo JA. Neuroimmune interactions and pain: focus on glial-modulating targets. *Curr Opin Investig Drugs.* 2008;9(7):726–734.
11. Maeda S, Kawamoto A, Yatani Y, Shirakawa H, Nakagawa T, Kaneko S. Gene transfer of GLT-1, a glial glutamate transporter, into the spinal cord by recombinant adenovirus attenuates inflammatory and neuropathic pain in rats. *Mol Pain.* 2008;4:65.
12. Tawfik VL, Lacroix-Fralish ML, Bercury KK, Nutile-McMenemy N, Harris BT, Deleo JA. Induction of astrocyte differentiation by propentofylline increases glutamate transporter expression in vitro: heterogeneity of the quiescent phenotype. *Glia.* 2006;54(3):193–203.
13. Tawfik VL, Regan MR, Haenggeli C, et al. Propentofylline-induced astrocyte modulation leads to alterations in glial glutamate promoter activation following spinal nerve transection. *Neuroscience.* 2008;152(4):1086–1092.
14. Xin WJ, Weng HR, Dougherty PM. Plasticity in expression of the glutamate transporters GLT-1 and GLAST in spinal dorsal horn glial cells following partial sciatic nerve ligation. *Mol Pain.* 2009;5:15.
15. Cao L, DeLeo JA. CNS-infiltrating $CD4^+$ T lymphocytes contribute to murine spinal nerve transection-induced neuropathic pain. *Eur J Immunol.* 2008;38(2):448–458.
16. Costigan M, Moss A, Latremoliere A, et al. T-cell infiltration and signaling in the adult dorsal spinal cord is a major contributor to neuropathic pain-like hypersensitivity. *J Neurosci.* 2009;29(46):14415–14422.
17. Sweitzer SM, Hickey WF, Rutkowski MD, Pahl JL, DeLeo JA. Focal peripheral nerve injury induces leukocyte trafficking into the central nervous system: potential relationship to neuropathic pain. *Pain.* 2002;100(1–2):163–170.
18. Abbott NJ, Ronnback L, Hansson E. Astrocyte-endothelial interactions at the blood–brain barrier. *Nat Rev Neurosci.* 2006;7(1):41–53.
19. Alvarez JI, Dodelet-Devillers A, Kebir H, et al. The Hedgehog pathway promotes blood–brain barrier integrity and CNS immune quiescence. *Science.* 2011;334(6063):1727–1731.
20. Bush TG, Puvanachandra N, Horner CH, et al. Leukocyte infiltration, neuronal degeneration, and neurite outgrowth after ablation of scar-forming, reactive astrocytes in adult transgenic mice. *Neuron.* 1999;23(2):297–308.
21. Barcia C, Sanderson NS, Barrett RJ, et al. T cells' immunological synapses induce polarization of brain astrocytes in vivo and in vitro: a novel astrocyte response mechanism to cellular injury. *PLoS ONE.* 2008;3(8):e2977.
22. Barcia C, Thomas CE, Curtin JF, et al. In vivo mature immunological synapses forming SMACs mediate clearance of virally infected astrocytes from the brain. *J Exp Med.* 2006;203(9):2095–2107.
23. Bauer J, Huitinga I, Zhao W, Lassmann H, Hickey WF, Dijkstra CD. The role of macrophages, perivascular cells, and microglial cells in the pathogenesis of experimental autoimmune encephalomyelitis. *Glia.* 1995;15(4):437–446.
24. Paolicelli RC, Bolasco G, Pagani F, et al. Synaptic pruning by microglia is necessary for normal brain development. *Science.* 2011;333(6048):1456–1458.
25. Stephan AH, Barres BA, Stevens B. The complement system: an unexpected role in synaptic pruning during development and disease. *Annu Rev Neurosci.* 2012;35:369–389.

26. Tremblay ME, Stevens B, Sierra A, Wake H, Bessis A, Nimmerjahn A. The role of microglia in the healthy brain. *J Neurosci*. 2011;31(45):16064–16069.
27. Iadecola C, Nedergaard M. Glial regulation of the cerebral microvasculature. *Nat Neurosci*. 2007;10(11):1369–1376.
28. Wang F, Smith NA, Xu Q, et al. Astrocytes modulate neural network activity by Ca^{2+}-dependent uptake of extracellular K^{+}. *Sci Signal*. 2012;5(218):ra26.
29. Ji RR, Berta T, Nedergaard M. Glia and pain: is chronic pain a gliopathy? *Pain*. 2013;154(suppl 1):S10–S28.
30. Calvo M, Bennett DL. The mechanisms of microgliosis and pain following peripheral nerve injury. *Exp Neurol*. 2012;234(2):271–282.
31. Ajami B, Bennett JL, Krieger C, Tetzlaff W, Rossi FM. Local self-renewal can sustain CNS microglia maintenance and function throughout adult life. *Nat Neurosci*. 2007;10(12):1538–1543.
32. Liu L, Rudin M, Kozlova EN. Glial cell proliferation in the spinal cord after dorsal rhizotomy or sciatic nerve transection in the adult rat. *Exp Brain Res*. 2000;131(1):64–73.
33. Bushong EA, Martone ME, Jones YZ, Ellisman MH. Protoplasmic astrocytes in CA1 stratum radiatum occupy separate anatomical domains. *J Neurosci*. 2002;22(1):183–192.
34. Ogata K, Kosaka T. Structural and quantitative analysis of astrocytes in the mouse hippocampus. *Neuroscience*. 2002;113(1):221–233.
35. Echeverry S, Shi XQ, Zhang J. Characterization of cell proliferation in rat spinal cord following peripheral nerve injury and the relationship with neuropathic pain. *Pain*. 2008;135(1–2):37–47.
36. Ji RR, Suter MR. p38 MAPK, microglial signaling, and neuropathic pain. *Mol Pain*. 2007;3:33.
37. Arthur JS, Ley SC. Mitogen-activated protein kinases in innate immunity. *Nat Rev Immunol*. 2013;13(9):679–692.
38. Ji RR, Gereau RW, Malcangio M, Strichartz GR. MAP kinase and pain. *Brain Res Rev*. 2009;60(1):135–148.
39. Sweitzer SM, Peters MC, Ma JY, et al. Peripheral and central p38 MAPK mediates capsaicin-induced hyperalgesia. *Pain*. 2004;111(3):278–285.
40. Jin SX, Zhuang ZY, Woolf CJ, Ji RR. p38 mitogen-activated protein kinase is activated after a spinal nerve ligation in spinal cord microglia and dorsal root ganglion neurons and contributes to the generation of neuropathic pain. *J Neurosci*. 2003;23(10):4017–4022.
41. Lee KM, Jeon SM, Cho HJ. Interleukin-6 induces microglial CX3CR1 expression in the spinal cord after peripheral nerve injury through the activation of p38 MAPK. *Eur J Pain*. 2010;14(7):682.e1–682.e12.
42. Svensson CI, Marsala M, Westerlund A, et al. Activation of p38 mitogen-activated protein kinase in spinal microglia is a critical link in inflammation-induced spinal pain processing. *J Neurochem*. 2003;86(6):1534–1544.
43. Wen YR, Suter MR, Ji RR, et al. Activation of p38 mitogen-activated protein kinase in spinal microglia contributes to incision-induced mechanical allodynia. *Anesthesiology*. 2009;110(1):155–165.
44. Gao YJ, Zhang L, Samad OA, et al. JNK-induced MCP-1 production in spinal cord astrocytes contributes to central sensitization and neuropathic pain. *J Neurosci*. 2009;29(13):4096–4108.
45. Ji RR, Kawasaki Y, Zhuang ZY, Wen YR, Decosterd I. Possible role of spinal astrocytes in maintaining chronic pain sensitization: review of current evidence with focus on bFGF/JNK pathway. *Neuron Glia Biol*. 2006;2(4):259–269.
46. Katsura H, Obata K, Miyoshi K, et al. Transforming growth factor-activated kinase 1 induced in spinal astrocytes contributes to mechanical hypersensitivity after nerve injury. *Glia*. 2008;56(7):723–733.

47. Zhuang ZY, Wen YR, Zhang DR, et al. A peptide c-Jun N-terminal kinase (JNK) inhibitor blocks mechanical allodynia after spinal nerve ligation: respective roles of JNK activation in primary sensory neurons and spinal astrocytes for neuropathic pain development and maintenance. *J Neurosci.* 2006;26(13):3551–3560.
48. Kawasaki Y, Kohno T, Zhuang ZY, et al. Ionotropic and metabotropic receptors, protein kinase A, protein kinase C, and Src contribute to C-fiber-induced ERK activation and cAMP response element-binding protein phosphorylation in dorsal horn neurons, leading to central sensitization. *J Neurosci.* 2004;24(38):8310–8321.
49. Kawasaki Y, Xu ZZ, Wang X, et al. Distinct roles of matrix metalloproteases in the early- and late-phase development of neuropathic pain. *Nat Med.* 2008;14(3):331–336.
50. Xu ZZ, Zhang L, Liu T, et al. Resolvins RvE1 and RvD1 attenuate inflammatory pain via central and peripheral actions. *Nat Med.* 2010;16(5):592–597 [591p following 597].
51. Anand P, Shenoy R, Palmer JE, et al. Clinical trial of the p38 MAP kinase inhibitor dilmapimod in neuropathic pain following nerve injury. *Eur J Pain.* 2011;15(10): 1040–1048.
52. Tong SE, Daniels SE, Black P, Chang S, Protter A, Desjardins PJ. Novel p38alpha mitogen-activated protein kinase inhibitor shows analgesic efficacy in acute postsurgical dental pain. *J Clin Pharmacol.* 2012;52(5):717–728.
53. Berta T, Liu YC, Xu ZZ, Ji RR. Tissue plasminogen activator contributes to morphine tolerance and induces mechanical allodynia via astrocytic IL-1beta and ERK signaling in the spinal cord of mice. *Neuroscience.* 2013;247:376–385.
54. Gao K, Wang CR, Jiang F, et al. Traumatic scratch injury in astrocytes triggers calcium influx to activate the JNK/c-Jun/AP-1 pathway and switch on GFAP expression. *Glia.* 2013;61(12):2063–2077.
55. Gao YJ, Zhang L, Ji RR. Spinal injection of TNF-alpha-activated astrocytes produces persistent pain symptom mechanical allodynia by releasing monocyte chemoattractant protein-1. *Glia.* 2010;58(15):1871–1880.
56. Takahashi K, Watanabe M, Suekawa Y, et al. IL-1beta in the trigeminal subnucleus caudalis contributes to extra-territorial allodynia/hyperalgesia following a trigeminal nerve injury. *Eur J Pain.* 2011;15(5):467.e1–467.e14.
57. Kawasaki Y, Zhang L, Cheng JK, Ji RR. Cytokine mechanisms of central sensitization: distinct and overlapping role of interleukin-1beta, interleukin-6, and tumor necrosis factor-alpha in regulating synaptic and neuronal activity in the superficial spinal cord. *J Neurosci.* 2008;28(20):5189–5194.
58. Sung CS, Wen ZH, Chang WK, et al. Inhibition of p38 mitogen-activated protein kinase attenuates interleukin-1beta-induced thermal hyperalgesia and inducible nitric oxide synthase expression in the spinal cord. *J Neurochem.* 2005;94(3):742–752.
59. Svensson CI, Schafers M, Jones TL, Powell H, Sorkin LS. Spinal blockade of TNF blocks spinal nerve ligation-induced increases in spinal P-p38. *Neurosci Lett.* 2005;379(3):209–213.
60. Vazquez E, Kahlenbach J, Segond von Banchet G, Konig C, Schaible HG, Ebersberger A. Spinal interleukin-6 is an amplifier of arthritic pain in the rat. *Arthritis Rheum.* 2012;64(7):2233–2242.
61. Zhang L, Berta T, Xu ZZ, Liu T, Park JY, Ji RR. TNF-alpha contributes to spinal cord synaptic plasticity and inflammatory pain: distinct role of TNF receptor subtypes 1 and 2. *Pain.* 2011;152(2):419–427.
62. Willemen HL, Eijkelkamp N, Wang H, et al. Microglial/macrophage GRK2 determines duration of peripheral IL-1beta-induced hyperalgesia: contribution of spinal cord CX3CR1, p38 and IL-1 signaling. *Pain.* 2010;150(3):550–560.
63. Coull JA, Beggs S, Boudreau D, et al. BDNF from microglia causes the shift in neuronal anion gradient underlying neuropathic pain. *Nature.* 2005;438(7070):1017–1021.

64. Sweitzer SM, Colburn RW, Rutkowski M, DeLeo JA. Acute peripheral inflammation induces moderate glial activation and spinal IL-1beta expression that correlates with pain behavior in the rat. *Brain Res.* 1999;829(1–2):209–221.
65. Colburn RW, DeLeo JA, Rickman AJ, Yeager MP, Kwon P, Hickey WF. Dissociation of microglial activation and neuropathic pain behaviors following peripheral nerve injury in the rat. *J Neuroimmunol.* 1997;79(2):163–175.
66. Garrison CJ, Dougherty PM, Carlton SM. GFAP expression in lumbar spinal cord of naive and neuropathic rats treated with MK-801. *Exp Neurol.* 1994;129(2):237–243.
67. Lin SC, Yeh JH, Chen CL, Chou SH, Tsai YJ. Effects of local lidocaine treatment before and after median nerve injury on mechanical hypersensitivity and microglia activation in rat cuneate nucleus. *Eur J Pain.* 2011;15(4):359–367.
68. Rasley A, Bost KL, Olson JK, Miller SD, Marriott I. Expression of functional NK-1 receptors in murine microglia. *Glia.* 2002;37(3):258–267.
69. Guo W, Wang H, Watanabe M, et al. Glial-cytokine-neuronal interactions underlying the mechanisms of persistent pain. *J Neurosci.* 2007;27(22):6006–6018.
70. Palma C, Minghetti L, Astolfi M, et al. Functional characterization of substance P receptors on cultured human spinal cord astrocytes: synergism of substance P with cytokines in inducing interleukin-6 and prostaglandin E2 production. *Glia.* 1997;21(2):183–193.
71. Martinez-Caro L, Laird JM. Allodynia and hyperalgesia evoked by sciatic mononeuropathy in NKI receptor knockout mice. *Neuroreport.* 2000;11(6):1213–1217.
72. Tsuda M, Shigemoto-Mogami Y, Koizumi S, et al. P2X4 receptors induced in spinal microglia gate tactile allodynia after nerve injury. *Nature.* 2003;424(6950):778–783.
73. Chu YX, Zhang Y, Zhang YQ, Zhao ZQ. Involvement of microglial P2X7 receptors and downstream signaling pathways in long-term potentiation of spinal nociceptive responses. *Brain Behav Immun.* 2010;24(7):1176–1189.
74. Clark AK, Staniland AA, Marchand F, Kaan TK, McMahon SB, Malcangio M. P2X7-dependent release of interleukin-1beta and nociception in the spinal cord following lipopolysaccharide. *J Neurosci.* 2010;30(2):573–582.
75. Tozaki-Saitoh H, Tsuda M, Miyata H, Ueda K, Kohsaka S, Inoue K. P2Y12 receptors in spinal microglia are required for neuropathic pain after peripheral nerve injury. *J Neurosci.* 2008;28(19):4949–4956.
76. Ulmann L, Hatcher JP, Hughes JP, et al. Up-regulation of P2X4 receptors in spinal microglia after peripheral nerve injury mediates BDNF release and neuropathic pain. *J Neurosci.* 2008;28(44):11263–11268.
77. Verge GM, Milligan ED, Maier SF, Watkins LR, Naeve GS, Foster AC. Fractalkine (CX3CL1) and fractalkine receptor (CX3CR1) distribution in spinal cord and dorsal root ganglia under basal and neuropathic pain conditions. *Eur J Neurosci.* 2004;20(5):1150–1160.
78. Clark AK, Yip PK, Malcangio M. The liberation of fractalkine in the dorsal horn requires microglial cathepsin S. *J Neurosci.* 2009;29(21):6945–6954.
79. Zhuang ZY, Kawasaki Y, Tan PH, Wen YR, Huang J, Ji RR. Role of the CX3CR1/p38 MAPK pathway in spinal microglia for the development of neuropathic pain following nerve injury-induced cleavage of fractalkine. *Brain Behav Immun.* 2007;21(5):642–651.
80. Thacker MA, Clark AK, Bishop T, et al. CCL2 is a key mediator of microglia activation in neuropathic pain states. *Eur J Pain.* 2009;13(3):263–272.
81. Tanga FY, Nutile-McMenemy N, DeLeo JA. The CNS role of Toll-like receptor 4 in innate neuroimmunity and painful neuropathy. *Proc Natl Acad Sci U S A.* 2005;102(16):5856–5861.
82. Liu T, Gao YJ, Ji RR. Emerging role of Toll-like receptors in the control of pain and itch. *Neurosci Bull.* 2012;28(2):131–144.

83. Gorina R, Font-Nieves M, Marquez-Kisinousky L, Santalucia T, Planas AM. Astrocyte TLR4 activation induces a proinflammatory environment through the interplay between MyD88-dependent NFkappaB signaling, MAPK, and Jak1/Stat1 pathways. *Glia*. 2011;59(2):242–255.
84. Ferraz CC, Henry MA, Hargreaves KM, Diogenes A. Lipopolysaccharide from *Porphyromonas gingivalis* sensitizes capsaicin-sensitive nociceptors. *J Endod*. 2011;37(1): 45–48.
85. Wadachi R, Hargreaves KM. Trigeminal nociceptors express TLR-4 and CD14: a mechanism for pain due to infection. *J Dent Res*. 2006;85(1):49–53.
86. Matsui T, Svensson CI, Hirata Y, Mizobata K, Hua XY, Yaksh TL. Release of prostaglandin E(2) and nitric oxide from spinal microglia is dependent on activation of p38 mitogen-activated protein kinase. *Anesth Analg*. 2010;111(2):554–560.
87. Diogenes A, Ferraz CC, Akopian AN, Henry MA, Hargreaves KM. LPS sensitizes TRPV1 via activation of TLR4 in trigeminal sensory neurons. *J Dent Res*. 2011;90(6):759–764.
88. Sorge RE, LaCroix-Fralish ML, Tuttle AH, et al. Spinal cord Toll-like receptor 4 mediates inflammatory and neuropathic hypersensitivity in male but not female mice. *J Neurosci*. 2011;31(43):15450–15454.
89. Bettoni I, Comelli F, Rossini C, et al. Glial TLR4 receptor as new target to treat neuropathic pain: efficacy of a new receptor antagonist in a model of peripheral nerve injury in mice. *Glia*. 2008;56(12):1312–1319.
90. Opal SM, Laterre PF, Francois B, et al. Effect of eritoran, an antagonist of MD2-TLR4, on mortality in patients with severe sepsis: the ACCESS randomized trial. *JAMA*. 2013;309(11):1154–1162.
91. Meller ST, Dykstra C, Grzybycki D, Murphy S, Gebhart GF. The possible role of glia in nociceptive processing and hyperalgesia in the spinal cord of the rat. *Neuropharmacology*. 1994;33(11):1471–1478.
92. Watkins LR, Hutchinson MR, Ledeboer A, et al. Norman Cousins Lecture. Glia as the "bad guys": implications for improving clinical pain control and the clinical utility of opioids. *Brain Behav Immun*. 2007;21(2):131–146.
93. Chen X, Ma X, Jiang Y, Pi R, Liu Y, Ma L. The prospects of minocycline in multiple sclerosis. *J Neuroimmunol*. 2011;235(1–2):1–8.
94. Mika J, Zychowska M, Popiolek-Barczyk K, Rojewska E, Przewlocka B. Importance of glial activation in neuropathic pain. *Eur J Pharmacol*. 2013;716(1–3):106–119.
95. Sweitzer S, De Leo J. Propentofylline: glial modulation, neuroprotection, and alleviation of chronic pain. *Handb Exp Pharmacol*. 2011;200:235–250.
96. Ledeboer A, Hutchinson MR, Watkins LR, Johnson KW. Ibudilast (AV-411). A new class therapeutic candidate for neuropathic pain and opioid withdrawal syndromes. *Expert Opin Investig Drugs*. 2007;16(7):935–950.
97. Ledeboer A, Sloane EM, Milligan ED, et al. Minocycline attenuates mechanical allodynia and proinflammatory cytokine expression in rat models of pain facilitation. *Pain*. 2005;115(1–2):71–83.
98. Raghavendra V, Tanga F, DeLeo JA. Inhibition of microglial activation attenuates the development but not existing hypersensitivity in a rat model of neuropathy. *J Pharmacol Exp Ther*. 2003;306(2):624–630.
99. Ito N, Obata H, Saito S. Spinal microglial expression and mechanical hypersensitivity in a postoperative pain model: comparison with a neuropathic pain model. *Anesthesiology*. 2009;111(3):640–648.
100. Cho IH, Chung YM, Park CK, et al. Systemic administration of minocycline inhibits formalin-induced inflammatory pain in rat. *Brain Res*. 2006;1072(1):208–214.
101. Bastos LF, Merlo LA, Rocha LT, Coelho MM. Characterization of the antinociceptive and anti-inflammatory activities of doxycycline and minocycline in different experimental models. *Eur J Pharmacol*. 2007;576(1–3):171–179.

102. Cata JP, Weng HR, Dougherty PM. The effects of thalidomide and minocycline on taxol-induced hyperalgesia in rats. *Brain Res*. 2008;1229:100–110.
103. Chang YW, Waxman SG. Minocycline attenuates mechanical allodynia and central sensitization following peripheral second-degree burn injury. *J Pain*. 2010;11(11): 1146–1154.
104. Pabreja K, Dua K, Sharma S, Padi SS, Kulkarni SK. Minocycline attenuates the development of diabetic neuropathic pain: possible anti-inflammatory and anti-oxidant mechanisms. *Eur J Pharmacol*. 2011;661(1–3):15–21.
105. Cho IH, Lee MJ, Jang M, Gwak NG, Lee KY, Jung HS. Minocycline markedly reduces acute visceral nociception via inhibiting neuronal ERK phosphorylation. *Mol Pain*. 2012;8:13.
106. Kannampalli P, Pochiraju S, Bruckert M, Shaker R, Banerjee B, Sengupta JN. Analgesic effect of minocycline in rat model of inflammation-induced visceral pain. *Eur J Pharmacol*. 2014;727:87–98.
107. Hains BC, Waxman SG. Activated microglia contribute to the maintenance of chronic pain after spinal cord injury. *J Neurosci*. 2006;26(16):4308–4317.
108. Hua XY, Svensson CI, Matsui T, Fitzsimmons B, Yaksh TL, Webb M. Intrathecal minocycline attenuates peripheral inflammation-induced hyperalgesia by inhibiting p38 MAPK in spinal microglia. *Eur J Neurosci*. 2005;22(10):2431–2440.
109. Nutile-McMenemy N, Elfenbein A, Deleo JA. Minocycline decreases in vitro microglial motility, beta1-integrin, and Kv1.3 channel expression. *J Neurochem*. 2007;103(5): 2035–2046.
110. Guasti L, Richardson D, Jhaveri M, et al. Minocycline treatment inhibits microglial activation and alters spinal levels of endocannabinoids in a rat model of neuropathic pain. *Mol Pain*. 2009;5:35.
111. Kay GW, Palmer DN. Chronic oral administration of minocycline to sheep with ovine CLN6 neuronal ceroid lipofuscinosis maintains pharmacological concentrations in the brain but does not suppress neuroinflammation or disease progression. *J Neuroinflammation*. 2013;10:97.
112. Arriagada O, Constandil L, Hernandez A, Barra R, Soto-Moyano R, Laurido C. Effects of interleukin-1beta on spinal cord nociceptive transmission in intact and propentofylline-treated rats. *Int J Neurosci*. 2007;117(5):617–625.
113. Gwak YS, Crown ED, Unabia GC, Hulsebosch CE. Propentofylline attenuates allodynia, glial activation and modulates GABAergic tone after spinal cord injury in the rat. *Pain*. 2008;138(2):410–422.
114. Sweitzer S, Martin D, DeLeo JA. Intrathecal interleukin-1 receptor antagonist in combination with soluble tumor necrosis factor receptor exhibits an anti-allodynic action in a rat model of neuropathic pain. *Neuroscience*. 2001;103(2):529–539.
115. Sweitzer SM, Schubert P, DeLeo JA. Propentofylline, a glial modulating agent, exhibits antiallodynic properties in a rat model of neuropathic pain. *J Pharmacol Exp Ther*. 2001;297(3):1210–1217.
116. Tawfik VL, Nutile-McMenemy N, Lacroix-Fralish ML, Deleo JA. Efficacy of propentofylline, a glial modulating agent, on existing mechanical allodynia following peripheral nerve injury. *Brain Behav Immun*. 2007;21(2):238–246.
117. Sweitzer SM, Pahl JL, DeLeo JA. Propentofylline attenuates vincristine-induced peripheral neuropathy in the rat. *Neurosci Lett*. 2006;400(3):258–261.
118. Raghavendra V, Tanga FY, DeLeo JA. Attenuation of morphine tolerance, withdrawal-induced hyperalgesia, and associated spinal inflammatory immune responses by propentofylline in rats. *Neuropsychopharmacology*. 2004;29(2):327–334.
119. Raghavendra V, Tanga F, Rutkowski MD, DeLeo JA. Anti-hyperalgesic and morphine-sparing actions of propentofylline following peripheral nerve injury in rats: mechanistic implications of spinal glia and proinflammatory cytokines. *Pain*. 2003;104(3):655–664.

120. Arruda JL, Sweitzer S, Rutkowski MD, DeLeo JA. Intrathecal anti-IL-6 antibody and IgG attenuates peripheral nerve injury-induced mechanical allodynia in the rat: possible immune modulation in neuropathic pain. *Brain Res.* 2000;879(1–2): 216–225.
121. Garry EM, Delaney A, Blackburn-Munro G, et al. Activation of p38 and p42/44 MAP kinase in neuropathic pain: involvement of VPAC2 and NK2 receptors and mediation by spinal glia. *Mol Cell Neurosci.* 2005;30(4):523–537.
122. Yoshioka A, Shimizu Y, Hirose G, Kitasato H, Pleasure D. Cyclic AMP-elevating agents prevent oligodendroglial excitotoxicity. *J Neurochem.* 1998;70(6):2416–2423.
123. Suzumura A, Ito A, Yoshikawa M, Sawada M. Ibudilast suppresses TNFalpha production by glial cells functioning mainly as type III phosphodiesterase inhibitor in the CNS. *Brain Res.* 1999;837(1–2):203–212.
124. Tominaga Y, Nakamura Y, Tsuji K, Shibata T, Kataoka K. Ibudilast protects against neuronal damage induced by glutamate in cultured hippocampal neurons. *Clin Exp Pharmacol Physiol.* 1996;23(6–7):519–523.
125. Mizuno T, Kurotani T, Komatsu Y, et al. Neuroprotective role of phosphodiesterase inhibitor ibudilast on neuronal cell death induced by activated microglia. *Neuropharmacology.* 2004;46(3):404–411.
126. Kiebala M, Maggirwar SB. Ibudilast, a pharmacologic phosphodiesterase inhibitor, prevents human immunodeficiency virus-1 Tat-mediated activation of microglial cells. *PLoS ONE.* 2011;6(4):e18633.
127. Ellis A, Wieseler J, Favret J, et al. Systemic administration of propentofylline, ibudilast, and (+)-naltrexone each reverses mechanical allodynia in a novel rat model of central neuropathic pain. *J Pain.* 2014;15(4):407–421.
128. Ledeboer A, Liu T, Shumilla JA, et al. The glial modulatory drug AV411 attenuates mechanical allodynia in rat models of neuropathic pain. *Neuron Glia Biol.* 2006;2(4):279–291.
129. Malmestrom C, Axelsson M, Lycke J, Zetterberg H, Blennow K, Olsson B. CSF levels of YKL-40 are increased in MS and replaces with immunosuppressive treatment. *J Neuroimmunol.* 2014;269(1–2):87–89.
130. Stoop MP, Singh V, Stingl C, et al. Effects of natalizumab treatment on the cerebrospinal fluid proteome of multiple sclerosis patients. *J Proteome Res.* 2013;12(3): 1101–1107.
131. Kou Z, Vandevord PJ. Traumatic white matter injury and glial activation: from basic science to clinics. *Glia.* 2014;62(11):1831–1855.
132. Papa L, Lewis LM, Falk JL, et al. Elevated levels of serum glial fibrillary acidic protein breakdown products in mild and moderate traumatic brain injury are associated with intracranial lesions and neurosurgical intervention. *Ann Emerg Med.* 2012;59(6):471–483.
133. Kim DS, Figueroa KW, Li KW, Boroujerdi A, Yolo T, Luo ZD. Profiling of dynamically changed gene expression in dorsal root ganglia post peripheral nerve injury and a critical role of injury-induced glial fibrillary acidic protein in maintenance of pain behaviors [corrected]. *Pain.* 2009;143(1–2):114–122.
134. Banati RB. Visualising microglial activation in vivo. *Glia.* 2002;40(2):206–217.
135. Banati RB. Brain plasticity and microglia: is transsynaptic glial activation in the thalamus after limb denervation linked to cortical plasticity and central sensitisation? *J Physiol Paris.* 2002;96(3–4):289–299.
136. Banati RB, Cagnin A, Brooks DJ, et al. Long-term trans-synaptic glial responses in the human thalamus after peripheral nerve injury. *Neuroreport.* 2001;12(16):3439–3442.
137. Giannetti P, Politis M, Su P, et al. Microglia activation in multiple sclerosis black holes predicts outcome in progressive patients: an in vivo [(11)C](R)-PK11195-PET pilot study. *Neurobiol Dis.* 2014;65:203–210.

138. Nakatomi Y, Mizuno K, Ishii A, et al. Neuroinflammation in patients with chronic fatigue syndrome/myalgic encephalomyelitis: an 11C-(R)-PK11195 PET study. *J Nucl Med*. 2014;55(6):945–950.
139. Jucaite A, Cselenyi Z, Arvidsson A, et al. Kinetic analysis and test-retest variability of the radioligand [11C](R)-PK11195 binding to TSPO in the human brain—a PET study in control subjects. *EJNMMI Res*. 2012;2:15.
140. Kadetoff D, Lampa J, Westman M, Andersson M, Kosek E. Evidence of central inflammation in fibromyalgia-increased cerebrospinal fluid interleukin-8 levels. *J Neuroimmunol*. 2012;242(1–2):33–38.
141. Lundborg C, Hahn-Zoric M, Biber B, Hansson E. Glial cell line-derived neurotrophic factor is increased in cerebrospinal fluid but decreased in blood during long-term pain. *J Neuroimmunol*. 2010;220(1–2):108–113.
142. Kotani N, Kudo R, Sakurai Y, et al. Cerebrospinal fluid interleukin 8 concentrations and the subsequent development of postherpetic neuralgia. *Am J Med*. 2004;116(5):318–324.
143. Kikuchi A, Kotani N, Sato T, Takamura K, Sakai I, Matsuki A. Comparative therapeutic evaluation of intrathecal versus epidural methylprednisolone for long-term analgesia in patients with intractable postherpetic neuralgia. *Reg Anesth Pain Med*. 1999;24(4):287–293.
144. D'Aversa TG, Eugenin EA, Berman JW. CD40-CD40 ligand interactions in human microglia induce CXCL8 (interleukin-8) secretion by a mechanism dependent on activation of ERK1/2 and nuclear translocation of nuclear factor-kappaB (NFkappaB) and activator protein-1 (AP-1). *J Neurosci Res*. 2008;86(3):630–639.
145. Shi Y, Gelman BB, Lisinicchia JG, Tang SJ. Chronic-pain-associated astrocytic reaction in the spinal cord dorsal horn of human immunodeficiency virus-infected patients. *J Neurosci*. 2012;32(32):10833–10840.
146. Shi Y, Shu J, Gelman BB, Lisinicchia JG, Tang SJ. Wnt signaling in the pathogenesis of human HIV-associated pain syndromes. *J Neuroimmune Pharmacol*. 2013;8(4): 956–964.
147. Yuan SB, Shi Y, Chen J, et al. Gp120 in the pathogenesis of human immunodeficiency virus-associated pain. *Ann Neurol*. 2014;75(6):837–850.
148. Alexander GM, Perreault MJ, Reichenberger ER, Schwartzman RJ. Changes in immune and glial markers in the CSF of patients with Complex Regional Pain Syndrome. *Brain Behav Immun*. 2007;21(5):668–676.
149. Alexander GM, van Rijn MA, van Hilten JJ, Perreault MJ, Schwartzman RJ. Changes in cerebrospinal fluid levels of pro-inflammatory cytokines in CRPS. *Pain*. 2005;116(3):213–219.
150. Del Valle L, Schwartzman RJ, Alexander G. Spinal cord histopathological alterations in a patient with longstanding complex regional pain syndrome. *Brain Behav Immun*. 2009;23(1):85–91.
151. Backonja MM, Coe CL, Muller DA, Schell K. Altered cytokine levels in the blood and cerebrospinal fluid of chronic pain patients. *J Neuroimmunol*. 2008;195(1–2):157–163.
152. Lampa J, Westman M, Kadetoff D, et al. Peripheral inflammatory disease associated with centrally activated IL-1 system in humans and mice. *Proc Natl Acad Sci U S A*. 2012;109(31):12728–12733.
153. Tay AS, Liu EH, Lee TL, et al. Cerebrospinal fluid of postherpetic neuralgia patients induced interleukin-6 release in human glial cell-line T98G. *Neurochem Int*. 2013;63(5):517–521.
154. Oberheim NA, Takano T, Han X, et al. Uniquely hominid features of adult human astrocytes. *J Neurosci*. 2009;29(10):3276–3287.
155. Oberheim NA, Wang X, Goldman S, Nedergaard M. Astrocytic complexity distinguishes the human brain. *Trends Neurosci*. 2006;29(10):547–553.

156. Torres-Platas SG, Comeau S, Rachalski A, et al. Morphometric characterization of microglial phenotypes in human cerebral cortex. *J Neuroinflammation*. 2014;11:12.
157. Landry RP, Jacobs VL, Romero-Sandoval EA, DeLeo JA. Propentofylline, a CNS glial modulator does not decrease pain in post-herpetic neuralgia patients: in vitro evidence for differential responses in human and rodent microglia and macrophages. *Exp Neurol*. 2012;234(2):340–350.
158. Dodel R, Spottke A, Gerhard A, et al. Minocycline 1-year therapy in multiple-system-atrophy: effect on clinical symptoms and [(11)C] (R)-PK11195 PET (MEMSA-trial). *Mov Disord*. 2010;25(1):97–107.
159. Ratai EM, Bombardier JP, Joo CG, et al. Proton magnetic resonance spectroscopy reveals neuroprotection by oral minocycline in a nonhuman primate model of accelerated NeuroAIDS. *PLoS ONE*. 2010;5(5):e10523.
160. Martinez V, Szekely B, Lemarie J, et al. The efficacy of a glial inhibitor, minocycline, for preventing persistent pain after lumbar discectomy: a randomized, double-blind, controlled study. *Pain*. 2013;154(8):1197–1203.
161. Kalliomaki J, Attal N, Jonzon B, et al. A randomized, double-blind, placebo-controlled trial of a chemokine receptor 2 (CCR2) antagonist in posttraumatic neuralgia. *Pain*. 2013;154(5):761–767.
162. Gordon PH, Moore DH, Miller RG, et al. Efficacy of minocycline in patients with amyotrophic lateral sclerosis: a phase III randomised trial. *Lancet Neurol*. 2007;6(12):1045–1053.
163. Kim SJ, Park SM, Cho YW, et al. Changes in expression of mRNA for interleukin-8 and effects of interleukin-8 receptor inhibitor in the spinal dorsal horn in a rat model of lumbar disc herniation. *Spine*. 2011;36(25):2139–2146.
164. Faulkner JR, Herrmann JE, Woo MJ, Tansey KE, Doan NB, Sofroniew MV. Reactive astrocytes protect tissue and preserve function after spinal cord injury. *J Neurosci*. 2004;24(9):2143–2155.
165. Stirling DP, Cummins K, Mishra M, Teo W, Yong VW, Stys P. Toll-like receptor 2-mediated alternative activation of microglia is protective after spinal cord injury. *Brain J Neurol*. 2014;137(pt 3):707–723.
166. Romero-Sandoval A, Nutile-McMenemy N, DeLeo JA. Spinal microglial and perivascular cell cannabinoid receptor type 2 activation reduces behavioral hypersensitivity without tolerance after peripheral nerve injury. *Anesthesiology*. 2008;108(4):722–734.
167. Ndong C, Landry RP, DeLeo JA, Romero-Sandoval EA. Mitogen activated protein kinase phosphatase-1 prevents the development of tactile sensitivity in a rodent model of neuropathic pain. *Mol Pain*. 2012;8:34.

CHAPTER TWELVE

Synaptic Inhibition and Disinhibition in the Spinal Dorsal Horn

Steven A. Prescott[*,†,1]
[*]Neurosciences and Mental Health, The Hospital for Sick Children, Toronto, Ontario, Canada
[†]Department of Physiology, University of Toronto, Toronto, Ontario, Canada
[1]Corresponding author: e-mail address: steve.prescott@utoronto.ca

Contents

1. Introduction 360
2. Synthesis and Handling of Inhibitory Neurotransmitters 360
3. GABA and Glycine Receptors 362
4. Driving Force and Ion Flux 364
5. Regulation of Intracellular Anion Concentrations 366
6. Mechanisms of Inhibition 369
7. Inhibitory Interneurons and Dorsal Horn Circuitry 370
8. Pathological Changes in Inhibition 372
9. Implications for Therapeutic Interventions 374
10. Summary 375
References 376

Abstract

Nociceptive signals originating in the periphery must be transmitted to the brain to evoke pain. Rather than being conveyed unchanged, those signals undergo extensive processing in the spinal dorsal horn. Synaptic inhibition plays a crucial role in that processing. On the one hand, neuropathy and inflammation are associated with reduced spinal inhibition; on the other hand, the hypersensitivity associated with inflammatory and neuropathic pain can be reproduced by blocking inhibition at the spinal level. To understand the consequences of disinhibition and how to therapeutically reverse it, one must understand how synaptic inhibition normally operates. To that end, this chapter will discuss the structure and function of $GABA_A$ and glycine receptors together with the role of associated molecules involved in transmitter handling and chloride regulation. Mechanisms by which inhibition modulates cellular excitability will be described. The chapter will end with discussion of how inhibition goes awry under pathological conditions and what the implications are for the treatment of resulting pain.

Progress in Molecular Biology and Translational Science, Volume 131
ISSN 1877-1173
http://dx.doi.org/10.1016/bs.pmbts.2014.11.008

1. INTRODUCTION

Synaptic inhibition plays a crucial role in processing neural information. In the context of somatosensation and pain, the importance of inhibition was brought to the fore by Melzack and Wall.[1] According to their Gate Control Theory, low-threshold inputs conveyed by large-diameter fibers and high-threshold inputs conveyed by small-diameter fibers converge on spinal projection neurons that relay sensory information to higher brain centers; however, the activation of those projection neurons depends not only on monosynaptic input from low- and high-threshold afferents, but also on the disynaptic inhibition engaged by low-threshold afferents. Many of the mechanistic details have been disproven over the last 50 years but, in the process, Gate Control Theory's main conceptual advance—that inhibition plays a pivotal role in pain processing—has been firmly established and its mechanistic basis clarified. The time is ripe to exploit this increased mechanistic understanding to precisely modulate inhibition in order to relieve persistent pain without compromising other sensory functions.

2. SYNTHESIS AND HANDLING OF INHIBITORY NEUROTRANSMITTERS

Fast synaptic inhibition is mediated through γ-aminobutyric acid (GABA) or glycine. GABA is the sole inhibitory neurotransmitter in many parts of the nervous system but glycine is used extensively in the spinal dorsal horn. Slow modulation mediated through other transmitters, although inhibitory in nature, will not be discussed.

The synthesis, packaging, and reuptake of GABA and glycine depend on the coordinated activities of several molecules (Fig. 1). GABA is synthesized from glutamate by the enzyme glutamic acid decarboxylase (GAD). There are two isoforms of GAD known as GAD65 and GAD67 (genes *gad2* and *gad1*, respectively). GABA is loaded into presynaptic vesicles by the vesicular GABA transporter (VGAT) (gene *slc32a1*). VGAT is also responsible for loading glycine into presynaptic vesicles[2] and thus also goes by the name vesicular inhibitory amino acid transporter (VIAAT). Unlike GABA, which is synthesized exclusively in neurons that express one or the other isoform of GAD, glycine is an amino acid present in all cells. One might therefore suspect that presynaptic vesicles in GABAergic neurons will invariably contain glycine in addition to GABA, but this is not the case. Because VGAT has a

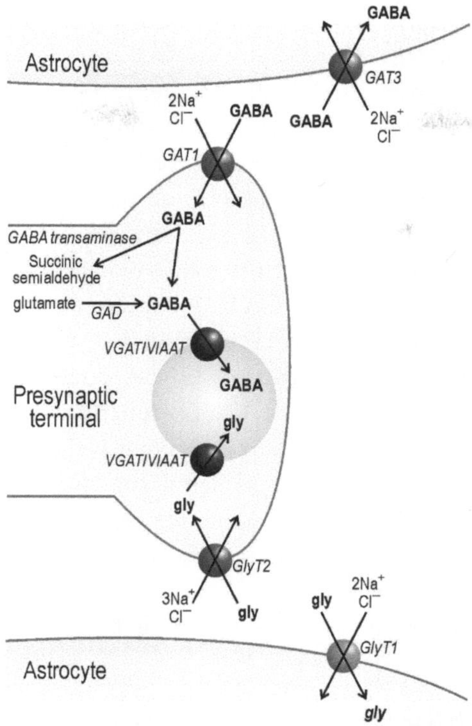

Figure 1 Summary of the processes involved in the handing of GABA and glycine. *GAD*, glutamic acid decarboxylase; *VGAT/VIAAT*, vesicular GABA transporter/vesicular inhibitory amino acid transporter; *GAT*, GABA transporter; *GlyT*, glycine transporter.

relatively low affinity for glycine, glycine only reaches appreciable intravesicular levels if cytoplasmic levels of glycine are high.[3] High cytoplasmic glycine levels are reached via the actions of the glycine transporter GlyT2 (gene *slc6a5*), which is present in the cell membrane of presynaptic terminals of inhibitory neurons. Another glycine transporter, GlyT1 (gene *slc6a9*), exists but is present on astrocytes and near excitatory synapses.[4] GlyT1 has been shown to be important for regulating the strength of tonic glycinergic inhibition in the spinal dorsal horn.[5]

GABAergic and glycinergic neurons both require VGAT to load their vesicles with transmitters. If a neuron has both GAD and GlyT2, then VGAT can load vesicles with appreciable amounts of both GABA and glycine, meaning that both transmitters are coreleased upon fusion of the vesicle with the cell membrane during synaptic release. Corelease of GABA and glycine occurs in the spinal dorsal horn,[6] consistent with immunohistochemical data showing colocalization of GABA and glycine in a subset of dorsal horn neurons.[7,8] Corelease does not, however, guarantee that both transmitters will mediate an inhibitory effect since codetection depends on the expression of both GABA and glycine receptors.[9]

There are also several GABA transporters present in the cell membrane that load GABA back into the cell after vesicular release. These include GAT1, (gene *slc6a1*), GAT2 (gene *slc6a13*), and GAT3 (gene *slc6a11*). Apart from recycling GABA for reuse after it has been released, GABA transporters are important for reducing extracellular GABA levels and thereby terminating the effects of released GABA. GAT1 is expressed predominantly in presynaptic terminals and appears to be the first line of defense in regulating extracellular GABA levels; however, if GABA uptake by GAT1 is overwhelmed (consequent to reduced GAT1 function or increased GABA release), then GAT3, which is expressed exclusively in astrocytes, comes into play.[10] Though not widely appreciated, the glycine and GABA transporters also load chloride and sodium into the cell when taking up transmitter (which has implications for intracellular chloride levels; see Section 5) and they can reverse their direction.[11] If not loaded into a vesicle, GABA can be broken down to succinic semialdehyde by GABA transaminase.

3. GABA AND GLYCINE RECEPTORS

Upon release, GABA and/or glycine molecules diffuse through the extracellular space and must bind to an appropriate receptor to mediate their effect. These receptors are present inside and outside the synaptic cleft, on the postsynaptic cell and presynaptic terminal, as well as on nonneuronal cells such as astrocytes. Fast synaptic inhibition is mediated through ionotropic receptors, which include various forms of $GABA_A$ and glycine receptors. Slower inhibition is mediated through metabotropic $GABA_B$ receptors, which will be discussed only briefly.

$GABA_A$ receptors are composed of five subunits, each with four transmembrane domains. There are a total of 19 different subunits with which to assemble the pentameric receptor: α1–α6, β1–β3, γ1–γ3, δ, ε, π, θ, and ρ1–ρ3.[12–14] The subunits are not, however, combined randomly; instead, the typical $GABA_A$ receptor comprises two α subunits, two β subunits, and one γ subunit arranged in a ring structure. This typical composition gives a receptor that has a rather low affinity for binding GABA and is responsible for phasic inhibition, i.e., brief inhibition occurring immediately after vesicular release of transmitter. Indeed, most $GABA_A$ receptors are clustered at the synapse through interaction of their γ2 subunit with gephyrin.[15] If the γ subunit is replaced with a δ subunit, the resulting receptor is not clustered at synapses yet has a much higher affinity for GABA, so much so that it tends to be activated by the ambient levels of GABA present in the extracellular space, thus producing tonic inhibition.[16] Tonic GABAergic (and glycinergic) inhibition is evident in some dorsal horn neurons.[5] The unmasking of GABAergic transmission by benzodiazepines, which increase the probability of the central pore opening when GABA binds the receptor, also argues that extrasynaptic $GABA_A$ receptors are present.[17] GABA binds at the interface between the α and β subunits, leading to a structural change that opens a central transmembrane pore that is permeable to anions. Chloride and bicarbonate are the two most important anion species under physiological conditions.[18] The net current passing through the pore depends on the relative flux of chloride and bicarbonate, which depends on their electrochemical gradients (see Section 4).

Apart from dictating the affinity of the receptor for its ligand, the subunit composition affects the susceptibility of that receptor to various modulators including neurosteroids, alcohol, anesthetics, barbiturates, and benzodiazepines.[12,14] There is not necessarily a simple one-to-one link between modulator sensitivity and subunit expression. For instance, the γ2 subunit is necessary for benzodiazepine to bind with high affinity (while γ1 and γ3 subunits allow lower affinity binding) but the identity of the adjacent α subunit is important for the effect of that binding.[19] $GABA_A$ receptors can be selectively blocked by bicuculline or gabazine.

Glycine receptors are, like $GABA_A$ receptors, composed of five subunits but the subunit variety is much less: α1–α4 and β.[14,20,21] They comprise two α subunits and three β subunits, and are distinguished by their blockade by strychnine. Glycine receptor function is also subject to modulation[22] but has received less attention than $GABA_A$ receptors in this regard. The subunits forming glycine and $GABA_A$ receptors are encoded by distinct genes but are part of a larger superfamily that also includes nicotinic acetylcholine receptors and type-3 serotonin receptors, all of which form pentameric channels and have a disulfide bridge between two cysteine residues in the amino terminal.

On a practical note, inhibitory synapses are often identified by immunostaining for gephyrin. This scaffolding protein, analogous to PSD-95 in excitatory synapses, associates with many but not all glycine and $GABA_A$ receptors.[23] Gephyrin constitutes a good marker for inhibitory synapses considering that staining for any one $GABA_A$ or glycine receptor subunit would overlook receptors that do not include that subunit. Of course gephyrin staining does not reveal extrasynaptic receptors. Recent work has shown that the terminals of both peptidergic and nonpeptidergic C fibers lack gephyrin.[24] This suggests a more diffuse $GABA_A$ receptor distribution although gephyrin-independent clustering of α2 GABA receptor subunits is possible[25] and α2 subunits are predominant presynaptically.[26] Selective deletion of the α2 GABA receptor subunit from nociceptive primary afferents significantly reduced the antihyperalgesic effects of spinal diazepam.[26] The potentiating effect of diazepam on primary afferent depolarization (PAD) and presynaptic inhibition were also lost, all of which argues that benzodiazepines act largely through a presynaptic mechanism. This is especially notable given that selective activation of $GABA_A$ receptors containing α2 and/or α3 is effective against neuropathic and inflammatory pain.[27] Lorenzo *et al.*[24] showed that the α1 glycine receptor subunit was absent from those terminals, arguing against the involvement of glycine in presynaptic inhibition and pointing instead to GABA, consistent with the above results and with other physiological evidence.[28]

$GABA_B$ receptors are metabotropic receptors whose activation leads, amongst other things, to activation of potassium channels via G-proteins.[29] Consequently, the inhibition mediated by $GABA_B$ receptors is slow and associated with a reversal potential near −90 mV, the reversal potential for potassium. This differs from the reversal potential associated with $GABA_A$ and glycine receptors, which depends on chloride and bicarbonate (see Section 4). $GABA_B$ receptors can also be present presynaptically. These receptors comprise only two principal subunits although additional auxiliary subunits are important for its function and modulation.

4. DRIVING FORCE AND ION FLUX

Ligand binding causes the central pore of $GABA_A$ and glycine receptors to open. Once afforded passage across the membrane, chloride and bicarbonate anions move down their respective electrochemical gradients, into or out of the cell. The electrochemical gradient can be described as a driving force, which corresponds to the difference between the membrane potential and the reversal potential for a given ion. Driving forces control the efficacy of GABAergic and glycinergic inhibition (irrespective of receptor subunit composition), are disrupted in various pathological conditions (Section 8), and are a valuable target for therapeutic intervention (Section 9). It is, therefore, imperative that one understand their biophysical basis.

Reversal potentials are calculated by the Nernst equation, which takes into account intracellular and extracellular ion concentrations. When more than one ion species are involved, the relative permeability of each species must be factored in using the Goldman–Hodgkin–Katz (GHK) equation.[30] With the relative permeability of chloride and bicarbonate being 4:1,[31] the GHK equation is

$$E_{\text{anion}} = 10^3 \frac{RT}{F} \ln\left(\frac{4[\text{Cl}^-]_\text{i} + [\text{HCO}_3{}^-]_\text{i}}{4[\text{Cl}^-]_\text{o} + [\text{HCO}_3{}^-]_\text{o}}\right),$$

where R is the gas constant, T is absolute temperature, and F is Faraday's constant, and where subscripts i and o designate the anion concentration inside and outside the cell, respectively.

Because of the contribution of bicarbonate, the net reversal potential for the mixed-anion conductance (which is written here as E_{anion} but is commonly referred to as E_{GABA}) is several millivolts less negative than E_{Cl}. In other words, for typical conditions in adult central neurons, where $[\text{HCO}_3{}^-]_\text{i} = 12\,\text{mM}$, $[\text{HCO}_3{}^-]_\text{o} = 25\,\text{mM}$, $[\text{Cl}^-]_\text{i} = 6\,\text{mM}$, and $[\text{Cl}^-]_\text{o} = 120\,\text{mM}$, E_{HCO_3} is about −20 mV and E_{Cl} is about −80 mV, which combines to give E_{anion} of about −70 mV. Under these conditions, chloride flows into the cell whereas bicarbonate flows out, but the former predominates resulting in a net outward (depolarizing) current; despite a net *influx* of negative charge, the net current is referred to as *outward* because the direction of current is defined by which way a positive test charge would move. If the chloride gradient were to collapse, net current would become inward when bicarbonate efflux exceeds chloride influx.[32] This is one way in which GABA can cause depolarization. Note that this mechanism does not involve chloride efflux. GABA can cause depolarization via chloride efflux but this requires that chloride is actively maintained at a high intracellular concentrations, as occurs in immature neurons and in primary afferents because of their expression of sodium–potassium–chloride cotransporter 1 (NKCC1; Fig. 2; Section 5).

Although often overlooked, it should be clear from the above discussion that transmembrane ion flux leads to changes in the intracellular and extracellular concentrations of that ion, or at least this would occur were it not for countermeasures that regulate ion concentration. Whereas ions that might accumulate outside the cell tend to dissipate via diffusion into a relatively large extracellular volume, ions accumulating inside the cell, like chloride, do not have that option. Chloride can diffuse within the cytosol, moving from an area of high chloride concentration (such as a dendrite with a high surface area to volume ratio) to an area of lower chloride concentration (such

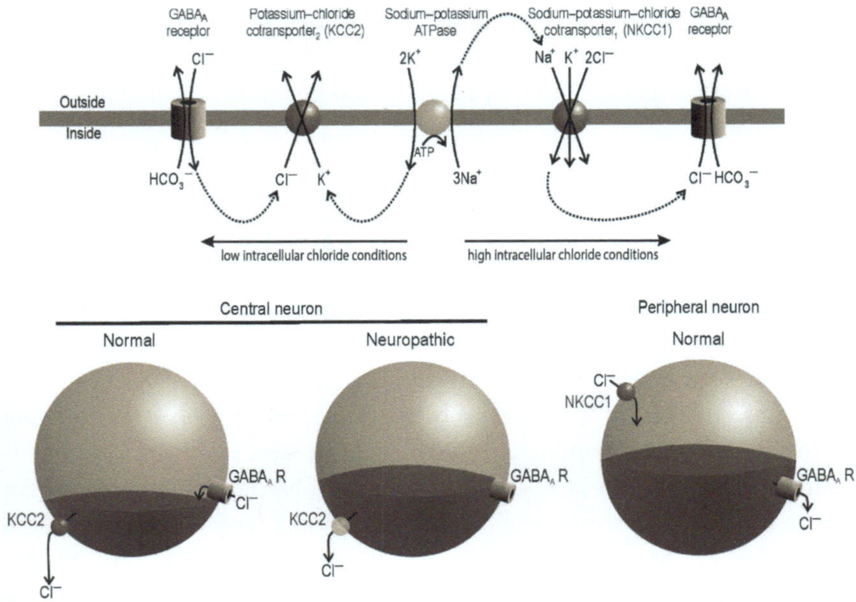

Figure 2 Summary of mechanisms involved in chloride regulation. In normal adult central neurons, KCC2 expression is high, resulting in low intracellular chloride levels that insure chloride influx via activated GABA$_A$ and glycine receptors (bottom left). After nerve injury, KCC2 is downregulated, resulting in elevated intracellular chloride levels and reduced chloride influx via GABA$_A$ and glycine receptors (bottom middle); under these conditions, depolarization occurs on the basis of bicarbonate efflux. In peripheral neurons, NKCC1 expression is high and KCC2 is absent, resulting in high intracellular chloride levels that result in chloride efflux via GABA$_A$ and glycine receptors (bottom right). *Top panel adapted from Price* et al.[33]

as the soma, which has a lower surface area to volume ratio). This intracellular diffusion helps mitigate short-term local concentration changes[34] but the long-term global regulation of chloride depends on chloride being extruded from the cytosol, as described in the next section.

5. REGULATION OF INTRACELLULAR ANION CONCENTRATIONS

Chloride regulation depends on the coordination of several processes (Fig. 2). Certain chloride leak channels have been suggested to reduce intracellular chloride concentration by acting as one-way valves. This idea stems from the observation that chloride channels like ClC-2 (gene *clcn2*) are more permeable to chloride exiting the cell than to chloride entering the cell.[35] Regardless of this differential permeability, which is referred to as rectification, the direction of chloride flux still depends on the chloride driving force. This means chloride will rarely if ever have the opportunity to exit the cell via ClC-2 because the chloride driving force is almost always in the opposite direction. Because the "valve" is imperfect, ClC-2 channels actually let chloride leak into the cell.[36]

The failure of channels to let chloride out of the cell highlights the need for different ion transport mechanisms that can move chloride against its gradient.[37] Cotransporters, or symporters, move two or more ion species in the same direction across the cell membrane; chloride can move against its gradient by piggybacking another ion that moves down its gradient. Exchangers, or antiporters, do effectively the same thing but by coupling the movement of ion species that flow in opposite directions across the membrane. The main chloride extruder in neurons is the potassium-chloride cotransporter 2 (KCC2) (gene *slc12a5*). KCC2 lets chloride piggyback potassium ions that flow down their gradient and out of the cell. The process is electroneutral because of the 1:1 stoichiometry of chloride and potassium. The process is not active insofar as it does not directly involve hydrolysis of ATP (and therefore it should not be referred to as pumping); instead, the process is secondarily active since KCC2 relies on the potassium gradient that is maintained by the sodium–potassium ATPase, which pumps potassium into the cell.

The sodium-potassium-chloride-cotransporter 1 or NKCC1 (gene *slc12a2*) is another important contributor to neuronal chloride homeostasis. NKCC1 harnesses the sodium gradient to move potassium and chloride into the cell, thus resulting in a high intracellular chloride concentration. This is of course the opposite to how KCC2 affects chloride. The relative expression of NKCC1 and KCC2 thus dictates the intracellular chloride concentration, notwithstanding the effects of chloride load through various chloride channels including activated $GABA_A$ and glycine channels. Several points should be noted. First, NKCC1 is strongly expressed early in development while KCC2 is only weakly expressed, but a developmental switch occurs that leads to the inverse pattern in adulthood.[38,39] In the rat spinal dorsal horn, E_{anion} appears to reach its mature value around 2 weeks after birth,[40] but full chloride extrusion capacity is not reached until 3–4 weeks after birth[41]; in other words, chloride loads more readily overwhelm KCC2-mediated chloride extrusion in young neurons. Second, the developmental switch does not occur in primary afferent neurons, which means NKCC1 levels remain high, resulting in high intracellular chloride concentrations in those cells.[42,43] Third, NKCC1 and KCC2 are not expressed uniformly within even a single neuron, which can lead to high intracellular chloride in one compartment (such as the axon initial segment) and low intracellular chloride in other compartments (such as the soma and dendrites).[44,45] And lastly, the normal adult levels of KCC2 expression can be pathologically altered (Section 8).

Some mention must be made about how electrophysiological recordings are conducted since this can (deliberately or inadvertently) lead to changes in intracellular chloride concentration. With the whole cell patch clamp technique, the cell membrane is ruptured to gain electrical access to the cell after sealing the patch pipette to the cell membrane; consequently, the cytosol is dialyzed with the pipette solution. The pipette solution is often designed to have a chloride concentration approximating the natural intracellular level, but sometimes it deliberately has a high chloride concentration in order to increase the chloride driving force (e.g., to facilitate the detection of small inhibitory postsynaptic currents). Either approach is acceptable depending on the question being asked. But, in both cases, dialyzing the cell means that intracellular chloride is effectively clamped at or near the chloride level in the pipette solution, which is obviously not appropriate for measuring the natural chloride level in the cell. This problem can be solved by using the perforated patch technique.[46] That said, dialyzing the cell can be used to test extrusion capacity by determining whether the intracellular chloride concentration equilibrates with the pipette concentration or if the cell manages to maintain a lower level because of its extrusion mechanisms.[47,48] Moreover, in voltage clamp, membrane potential is abruptly changed and held at arbitrarily chosen values, which can lead to very unnatural chloride driving forces. As explained by Ratté and Prescott,[36] such experimental details must be carefully considered to avoid misinterpretations.

As already mentioned, bicarbonate flows out through activated $GABA_A$ and glycine receptors. The likelihood of bicarbonate efflux causing extracellular accumulation is low given the relatively unrestricted diffusion of bicarbonate in the extracellular space, but bicarbonate efflux can deplete intracellular bicarbonate levels and cause a drop in pH.[49] However, this tends not to occur under normal conditions because intracellular bicarbonate is replenished by the conversion of carbon dioxide and water to bicarbonate and protons by the enzyme carbonic anhydrase; as a gas, carbon dioxide freely diffuses across the cell membrane. Intracellular bicarbonate can be depleted (and its efflux thus curtailed) through blockade of carbonic anhydrase by acetazolamide,[32] which can in fact have analgesic effects (Section 9). Regulation of pH implicates other chemical reactions and transport mechanisms, and bicarbonate itself can be shuttled across the cell membrane in exchange for chloride.[50]

6. MECHANISMS OF INHIBITION

Based on their relative permeability to chloride and bicarbonate, $GABA_A$ (and glycine) receptors are normally associated with a reversal potential of around −70 mV (see Section 4), which is near the resting membrane potential of most neurons. Therefore, activating $GABA_A$ receptors on a neuron at rest will not cause any overt hyperpolarization because there is no driving force. $GABA_A$ receptors are thus said to mediate "shunting" inhibition because their activation reduces, or shunts, the depolarization caused by concurrent excitatory input.[51] However, the term *shunting* is often misinterpreted to mean a nonspecific reduction in input resistance. Shunting is not literally caused by cations that enter through glutamate receptors leaving through $GABA_A$ receptors (which are selectively permeable to anions). Instead, the depolarization caused by sodium influx via glutamate receptors leads to increased chloride driving force and, in turn, increased chloride influx that counterbalances the cation current. Thus, even shunting inhibition depends on the chloride gradient and imposes a chloride load.

If E_{anion} were to shift to −60 mV, activation of $GABA_A$ (or glycine) receptors in a resting cell would cause depolarization. This might be considered excitatory but it is usually insufficient to evoke spikes. Importantly, if the neuron were to receive glutamatergic excitation causing depolarization beyond −60 mV, then chloride driving force would be inverted and $GABA_A$ receptor activation would once again act to mitigate depolarization rather than contributing to it. The resulting inhibition would be less effective in reducing depolarization than if E_{anion} was at its normal value of −70 mV, but the point is that $GABA_A$ receptor activation is still inhibitory, just less inhibitory. Even if E_{anion} were to undergo a depolarizing shift large enough to allow $GABA_A$ receptor activation to evoke spiking, $GABA_A$ receptor activation may continue to have a net inhibitory effect under conditions in which the neuron receives strong glutamatergic input; in other words, the neuron would have spiked more in response to glutamatergic input alone.

The above discussion, while pertinent for understanding pathological disinhibition (see Section 8), is also relevant for appreciating how presynaptic inhibition works. As explained in Section 5, primary afferent neurons express NKCC1 but no KCC2. Because chloride is actively loaded into the cell, primary afferents normally have E_{anion} values near −35 mV.[43,52] Under these conditions, $GABA_A$ receptor activation will trigger efflux of both bicarbonate and chloride, leading to PAD. Although depolarization is typically equated with excitation, PAD reduces spike propagation and synaptic transmission in the affected axon, and is therefore inhibitory. There are several possible mechanisms through which such inhibition could occur including the inactivation of sodium channels, shunting (which could compromise action potential propagation), or via effects on calcium channels.[28,53–55] The precise mechanism remains uncertain and might well include multiple mechanisms whose proportional contribution differs between fiber types. PAD can, under certain pathological conditions, become excitatory and produce a dorsal root reflex (see Section 8).

As explained in Section 3, $GABA_B$ receptor activation opens potassium channels via G proteins. Because E_K is -90 mV or so, this form of inhibition causes gross hyperpolarization as opposed to "shunting." $GABA_B$ receptors are also present presynaptically on GABAergic terminals and have been shown to reduce GABA/glycine corelease from spinal dorsal horn neurons.[56] This example of presynaptic inhibition differs from PAD, and instead involves G protein-dependent inhibition of calcium channels. From this, one can appreciate that blocking $GABA_B$ receptors could enhance fast synaptic inhibition. That said, Iyadomi *et al.*[57] observed that baclofen, a $GABA_B$ receptor agonist, reduced the frequency of miniature inhibitory and excitatory postsynaptic currents at around the same dose, leading the authors to suggest that the analgesic effects of higher dose baclofen occur through a postsynaptic mechanism.

7. INHIBITORY INTERNEURONS AND DORSAL HORN CIRCUITRY

Dorsal horn circuitry will be discussed here only briefly since it has been the focus of several recent reviews.[58–61] Our current understanding is summarized in Fig. 3. Most inhibition originates from local inhibitory interneurons, which comprise around 1/3 of all neurons in the superficial dorsal horn based on immunocytochemistry.[62] The remaining neurons are mostly excitatory interneurons, with projection neurons constituting as little as 5% of the total population.[63] There is also some evidence for descending fibers from the rostral ventral medulla that release GABA and glycine[64,65] but descending modulation is predominantly monoaminergic.

Identification of inhibitory neurons has traditionally relied on immunostaining for GAD or GABA, which when combined with electrophysiology and intracellular labeling, has revealed important yet imperfect correlations

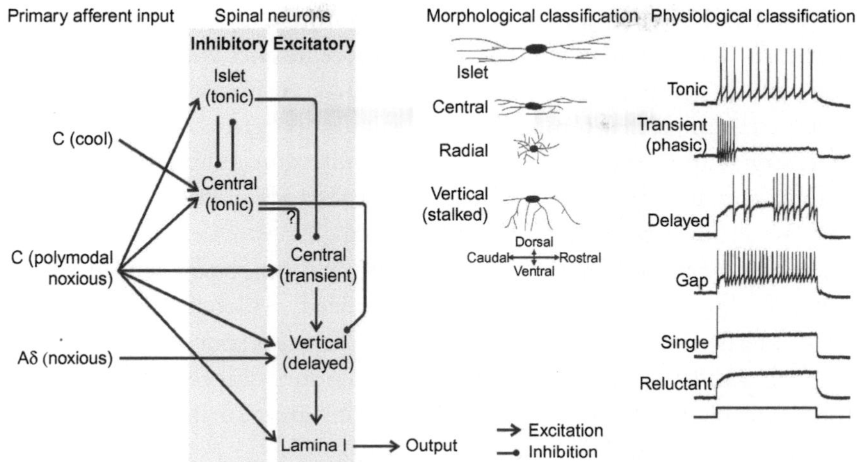

Figure 3 Circuit diagram depicting the connectivity between defined cell types in the superficial dorsal horn. Basis for morphological and physiological classification is illustrated on the right. *Adapted from Prescott and Ratté.*[58]

between cellular properties. Tonic-spiking islet and central cells tend to be inhibitory whereas other morphologies and spiking patterns (many of which are connected with the A-type potassium current) tend to be associated with excitatory neurons.[58,66] This is largely consistent with the results of paired recordings.[67–71] Various transgenic mouse line in which inhibitory neurons are labeled with GFP have also been used to characterize inhibitory neurons in the spinal dorsal horn.[5,7,71–78] Identification and systematic classification of the full set of different inhibitory and excitatory cell types awaits careful genetic dissection.[14,79]

Some notable observations emerge from Fig. 3. On the one hand, inhibitory interneurons inhibit each other[71,78] as well as various types of excitatory interneurons; on the other hand, excitatory interneurons do not appear to connect to inhibitory interneurons. It is difficult to prove that those connections do not exist, but the lack of paired recordings demonstrating such connections suggests that they are sparse. Instead, it would appear that inhibitory neurons rely on inputs from primary afferents or descending fibers for their activation. Some inhibitory neurons receive low-threshold input from Aβ fibers,[72] consistent with *in vivo* patch clamp recording showing that light touch increases the rate of inhibitory postsynaptic currents.[80] It is also notable that inhibitory interneurons experience strong tonic inhibition. Takazawa and MacDermott[5] showed that tonic inhibition at the lamina II/III border was predominantly mediated by glycine receptors whereas $GABA_A$ receptors were largely responsible for tonic inhibition in lamina IIo, with the separation becoming more distinct with development. Other work demonstrated only tonic GABAergic inhibition[81] but both studies established that tonic inhibition is quite powerful. It is not clear if excitatory neurons experience similar tonic inhibition, although this is obviously important. Figure 3 also shows that excitatory interneurons connect with one another, forming polysynaptic chains. Transmission through those polysynaptic chains is under tight inhibitory control, as evidenced by the consequences of disinhibition (Section 8). Indeed, evidence points to excitatory interneurons being crucial for the development of neuropathic pain,[82,83] and it appears equally true that inhibitory control of those excitatory neurons is key. Inhibitory interneurons are also implicated in the control of itch.[84,85]

8. PATHOLOGICAL CHANGES IN INHIBITION

Normal pain processing can be disrupted by countless forms of maladaptive plasticity that occur within the spinal dorsal horn.[86,87] We will focus on changes involving fast inhibition (see also Refs. 14,88). The importance of spinal inhibition for pain processing was firmly established by intrathecal injection of $GABA_A$ and/or glycine receptor blockers like bicuculline and strychnine.[89–98] These studies showed that pharmacological reduction of inhibition can acutely reproduce features of persistent neuropathic and inflammatory pain such as mechanical allodynia. Such experiments suggest but do not demonstrate that inhibition is reduced in these pain conditions. In that regard, demonstration that nerve injury occluded the effects of pharmacologically induced disinhibition[99] is notable. Still other studies have shown reduced immunoreactivity for GAD or GABA,[100–104] reduced synaptosomal GABA,[105] reduced GABA release[106] or physiological evidence of less inhibition.[107,108] Moore *et al.*[103] argued that there was a selective loss of inhibitory neurons, although other studies have provided evidence to the contrary.[62,109] Furthermore, Schoffnegger *et al.*[110] showed that the excitability of inhibitory neurons was not altered by nerve injury. Notwithstanding evidence that inhibitory neurons are intact and functioning properly, if reduced GABAergic transmission were the underlying cause of disinhibition, it stands to reason that replacing the lost transmitter[91,99,111,112] including through transplantation of GABAergic cells[113–115] would reverse the problem. Similarly, one could aim to enhance the inhibition produced by endogenously released GABA by applying allosteric modulators of GABA receptors (Section 9).

Although activating or modulating $GABA_A$ receptors can be therapeutic, there is little evidence suggesting that the receptors themselves are downregulated or their subunit composition changed. But, given the numerous ways in which $GABA_A$ receptor function can be altered,[116] many changes occurring in the spinal dorsal horn may still be undiscovered. Nonetheless, evidence points to reduction in GABA release rather than to changes in the receptors. By comparison, glycinergic transmission can also be reduced by prostaglandin E_2 activating protein kinase A, which in turn phosphorylates the glycine receptor α3 subunit to reduce glycine receptor conductance.[117,118] The same modulation is not involved in neuropathic pain.[119,120] In terms of the glycine transmission, blocking GlyT1 (but not GlyT2) can reduce allodynia[121] although this does not demonstrate the glycine release is reduced in neuropathic pain. That said, given the overlap in neurons releasing both GABA and glycine, a reduction in GABAergic cells (see above) may imply a reduction in glycinergic neurons.

But a completely different form of disinhibition exists that does not directly involve changes in GABA, glycine, or any of their receptors. Coull *et al.*[122] discovered that nerve injury triggers downregulation of KCC2 in spinal neurons, leading to dysregulation of their intracellular chloride levels and a depolarizing shift in E_{anion}. Other studies have found similar downregulation of KCC2 after inflammation, but no alteration in NKCC1 expression.[123,124] Subsequent work has linked the downregulation of KCC2 in neuropathic pain to BDNF release from activated microglia[125,126] and the events leading up to this have now been worked out in detail.[127] Interestingly, morphine-induced hyperalgesia seems to involve a similar disruption of chloride regulation.[47] Although sufficient downregulation of KCC2 could make GABA (and glycine) paradoxically excitatory, Coull *et al.*[122] showed after nerve injury that GABA evoked spiking in very few cells, the message being that chloride dysregulation can make GABA less inhibitory without making it paradoxically excitatory. This is important because it predicts that GABA agonist should retain some efficacy in reducing pain rather than exacerbating pain (see Section 9).

Downregulation of KCC2 should leave presynaptic inhibition intact given that primary afferents do not express KCC2 in the first place. Presynaptic inhibition could be altered by changes in NKCC1, but its expression appears to be relatively stable (see above). That said, membrane trafficking and phosphorylation of NKCC1 is affected by painful stimulation[128] and

E_{anion} does undergo a depolarizing shift after inflammation.[129] This is notable because intense nociceptor activation can enhance PAD to the point that depolarization (that is usually inhibitory) manages to evoke spikes in the primary afferent terminals, leading to what is referred to as a dorsal root reflex[130] that may in turn cause neurogenic inflammation, central sensitization, and hypersensitivity (for detailed discussion see Ref. 33). Separate from a change in chloride regulation, inflammation has been found to potentiate high affinity $GABA_A$ receptors in primary afferents, albeit in larger fibers.[131]

Despite much focus on the molecular mechanisms of disinhibition and its consequences for cellular excitability, relatively little attention has been given to the effects on network function yet this is obviously important; for instance, if an inhibitory neuron were disinhibited, it may end up providing stronger inhibition to downstream excitatory neurons and have a net inhibitory effect on the network. *In vivo* recordings have established that disinhibition unmasks low-threshold input to normally nociceptive-specific lamina I neurons.[126,132] This is consistent with accumulating evidence from *in vitro* experiments showing that disinhibition ungates polysynaptic circuits that relay low-threshold inputs from lamina IIi toward lamina I.[67,92,133–135] These findings highlight the importance of changes occurring at the network level insofar as disinhibition would not cause lamina I neurons to become responsive to low threshold input, presumably resulting in mechanical allodynia, if the network was organized differently. As our understanding of dorsal horn circuitry becomes clearer (Section 7), there will be greater opportunity to investigate these issues.

9. IMPLICATIONS FOR THERAPEUTIC INTERVENTIONS

Numerous articles have reviewed ongoing efforts and prospects for therapies targeting different aspects of synaptic inhibition: restoring normal chloride regulation,[136–138] enhancing GABAergic transmission[139–142] or enhancing glycinergic transmission[143,144] not to mention related issues such as microglia and inflammation. Rather than repeat the details here, a few simple observations will be made to highlight the practical utility of the information conveyed earlier in this chapter.

Putting aside how disinhibition affects neural networks and whether inhibition is reduced pre- or postsynaptically, disinhibitory mechanisms can be roughly subdivided into two categories: impaired synaptic transmission versus impaired inhibitory effect (i.e., reduced inhibition despite normal receptor activation). The first results from reduced transmitter release and/or receptor function whereas the latter results from chloride dysregulation. The distinction is important because the former mechanism implies that disinhibition can be reversed or at least mitigated by enhancing transmitter release and/or receptor activation. Disinhibition caused by chloride dysregulation does not imply that benzodiazepines or other means of augmenting transmission will necessarily exacerbate pain, but it does predict that such treatments will be less effective compared with their efficacy against pain arising from impaired synaptic transmission.[34,145] On the other hand, enhancing KCC2 function is not predicted to[34] and nor does it[48] enhance inhibition unless inhibition has been compromised by chloride dysregulation. This discussion highlights that when trying to reduce pain caused by disinhibition, one must identify the precise disinhibitory mechanism and choose therapy accordingly.

That said, it would be naïve to think that pathological changes occur in isolation of one another; on the contrary, nerve injury and/or inflammation are liable to evoke multiple different pathological changes as well as trigger endogenous compensatory changes.[146] This makes it very difficult to predict, for example, how effective benzodiazepines will be against a certain type of pain (for detailed discussions, see Refs. 14,33,111,147). But by the same token, one need not target a single disinhibitory mechanism; indeed, combination pharmacotherapy can be a useful approach toward pathological pain.[148] Most research in this area is largely empirical, but consider how understanding disinhibitory mechanisms can inform the choice of combinations. For instance, considering that reducing bicarbonate efflux through blockade of carbonic anhydrase by acetazolamide would cause a hyperpolarizing shift in E_{anion}, thereby restoring the inhibitory efficacy of $GABA_A$ receptor activation, it stands to reason that acetazolamide and benzodiazepines will have synergistic effects, which indeed they do.[149,150] In contrast, combining acetazolamide with a chloride extrusion enhancer like CLP257[48] is redundant and liable to give subadditive effects.

10. SUMMARY

Synaptic inhibition plays a crucial role in regulating the flow of sensory information through the spinal cord. This involves not only two different transmitters, but also a multitude of different receptor subtypes. As if that were not complicated enough, the efficacy of receptor activation depends intimately on chloride (and bicarbonate) regulation, which itself differs between pre- and postsynaptic loci. As we understand more about how dorsal horn circuits are organized, we will be able to understand more precisely how changes in inhibition impact network function. When Melzack and Wall formulated their Gate Control Theory, none of these details were known. The challenge now is to make sense of the details, fill in the gaps, and use the resulting knowledge to clinical benefit.

REFERENCES

1. Melzack R, Wall PD. Pain mechanisms: a new theory. *Science*. 1965;150(699):971–979.
2. Wojcik SM, Katsurabayashi S, Guillemin I, et al. A shared vesicular carrier allows synaptic corelease of GABA and glycine. *Neuron*. 2006;50(4):575–587.
3. Aubrey KR, Rossi FM, Ruivo R, et al. The transporters GlyT2 and VIAAT cooperate to determine the vesicular glycinergic phenotype. *J Neurosci*. 2007;27(23):6273–6281.
4. Zafra F, Aragon C, Olivares L, Danbolt NC, Gimenez C, Storm-Mathisen J. Glycine transporters are differentially expressed among CNS cells. *J Neurosci*. 1995;15(5 Pt 2): 3952–3969.
5. Takazawa T, MacDermott AB. Glycinergic and GABAergic tonic inhibition fine tune inhibitory control in regionally distinct subpopulations of dorsal horn neurons. *J Physiol*. 2010;588(Pt 14):2571–2587.
6. Jonas P, Bischofberger J, Sandkuhler J. Corelease of two fast neurotransmitters at a central synapse. *Science*. 1998;281(5375):419–424.
7. Heinke B, Ruscheweyh R, Forsthuber L, Wunderbaldinger G, Sandkuhler J. Physiological, neurochemical and morphological properties of a subgroup of GABAergic spinal lamina II neurones identified by expression of green fluorescent protein in mice. *J Physiol*. 2004;560(Pt 1):249–266.
8. Todd AJ, Sullivan AC. Light microscope study of the coexistence of GABA-like and glycine-like immunoreactivities in the spinal cord of the rat. *J Comp Neurol*. 1990;296(3):496–505.
9. Keller AF, Coull JA, Chery N, Poisbeau P, De Koninck Y. Region-specific developmental specialization of GABA-glycine cosynapses in laminas I-II of the rat spinal dorsal horn. *J Neurosci*. 2001;21(20):7871–7880.
10. Kersante F, Rowley SC, Pavlov I, et al. A functional role for both aminobutyric acid (GABA) transporter-1 and GABA transporter-3 in the modulation of extracellular GABA and GABAergic tonic conductances in the rat hippocampus. *J Physiol*. 2013;591(Pt 10):2429–2441.
11. Richerson GB, Wu Y. Dynamic equilibrium of neurotransmitter transporters: not just for reuptake anymore. *J Neurophysiol*. 2003;90(3):1363–1374.
12. D'Hulst C, Atack JR, Kooy RF. The complexity of the $GABA_A$ receptor shapes unique pharmacological profiles. *Drug Discov Today*. 2009;14(17–18):866–875.
13. Olsen RW, Sieghart W. $GABA_A$ receptors: subtypes provide diversity of function and pharmacology. *Neuropharmacology*. 2009;56(1):141–148.
14. Zeilhofer HU, Wildner H, Yevenes GE. Fast synaptic inhibition in spinal sensory processing and pain control. *Physiol Rev*. 2012;92(1):193–235.
15. Essrich C, Lorez M, Benson JA, Fritschy JM, Luscher B. Postsynaptic clustering of major GABAA receptor subtypes requires the gamma 2 subunit and gephyrin. *Nat Neurosci*. 1998;1(7):563–571.
16. Belelli D, Harrison NL, Maguire J, Macdonald RL, Walker MC, Cope DW. Extrasynaptic GABAA receptors: form, pharmacology, and function. *J Neurosci*. 2009;29(41):12757–12763.

17. Chéry N, De Koninck Y. Junctional versus extrajunctional glycine and $GABA_A$ receptor-mediated IPSCs in identified lamina I neurons of the adult rat spinal cord. *J Neurosci*. 1999;19(17):7342–7355.
18. Bormann J, Hamill OP, Sakmann B. Mechanism of anion permeation through channels gated by glycine and gamma-aminobutyric acid in mouse cultured spinal neurones. *J Physiol*. 1987;385:243–286.
19. Pritchett DB, Sontheimer H, Shivers BD, et al. Importance of a novel GABAA receptor subunit for benzodiazepine pharmacology. *Nature*. 1989;338(6216):582–585.
20. Dutertre S, Becker CM, Betz H. Inhibitory glycine receptors: an update. *J Biol Chem*. 2012;287(48):40216–40223.
21. Lynch JW. Native glycine receptor subtypes and their physiological roles. *Neuropharmacology*. 2009;56(1):303–309.
22. Laube B, Maksay G, Schemm R, Betz H. Modulation of glycine receptor function: a novel approach for therapeutic intervention at inhibitory synapses? *Trends Pharmacol Sci*. 2002;23(11):519–527.
23. Tyagarajan SK, Fritschy JM. Gephyrin: a master regulator of neuronal function? *Nat Rev Neurosci*. 2014;15(3):141–156.
24. Lorenzo LE, Godin AG, Wang F, et al. Gephyrin clusters are absent from small diameter primary afferent terminals despite the presence of $GABA_A$ receptors. *J Neurosci*. 2014;34(24):8300–8317.
25. Levi S, Logan SM, Tovar KR, Craig AM. Gephyrin is critical for glycine receptor clustering but not for the formation of functional GABAergic synapses in hippocampal neurons. *J Neurosci*. 2004;24(1):207–217.
26. Witschi R, Punnakkal P, Paul J, et al. Presynaptic alpha2-$GABA_A$ receptors in primary afferent depolarization and spinal pain control. *J Neurosci*. 2011;31(22):8134–8142.
27. Knabl J, Witschi R, Hosl K, et al. Reversal of pathological pain through specific spinal $GABA_A$ receptor subtypes. *Nature*. 2008;451(7176):330–334.
28. Willis WD. John Eccles' studies of spinal cord presynaptic inhibition. *Prog Neurobiol*. 2006;78(3–5):189–214.
29. Gassmann M, Bettler B. Regulation of neuronal $GABA_B$ receptor functions by subunit composition. *Nat Rev Neurosci*. 2012;13(6):380–394.
30. Hille B. *Ion Channels of Excitable Membranes*. Sunderland, MA: Sinauer Associates Inc.; 2001.
31. Voipio J, Pasternack M, Rydqvist B, Kaila K. Effect of gamma-aminobutyric acid on intracellular pH in the crayfish stretch-receptor neurone. *J Exp Biol*. 1991;156:349–360.
32. Staley KJ, Soldo BL, Proctor WR. Ionic mechanisms of neuronal excitation by inhibitory $GABA_A$ receptors. *Science*. 1995;269(5226):977–981.
33. Price TJ, Cervero F, Gold MS, Hammond DL, Prescott SA. Chloride regulation in the pain pathway. *Brain Res Rev*. 2009;60(1):149–170.
34. Doyon N, Prescott SA, Castonguay A, Godin AG, Kroger H, De Koninck Y. Efficacy of synaptic inhibition depends on multiple dynamically interacting mechanisms implicated in chloride homeostasis. *PLoS Comput Biol*. 2011;7:e1002149.
35. Duran C, Thompson CH, Xiao Q, Hartzell HC. Chloride channels: often enigmatic, rarely predictable. *Annu Rev Physiol*. 2010;72:95–121.
36. Ratté S, Prescott SA. ClC-2 channels regulate neuronal excitability, not intracellular chloride levels. *J Neurosci*. 2011;31(44):15838–15843.
37. Payne JA, Rivera C, Voipio J, Kaila K. Cation-chloride co-transporters in neuronal communication, development and trauma. *Trends Neurosci*. 2003;26(4):199–206.
38. Ben Ari Y. Excitatory actions of gaba during development: the nature of the nurture. *Nat Rev Neurosci*. 2002;3(9):728–739.
39. Rivera C, Voipio J, Payne JA, et al. The K^+/Cl^- co-transporter KCC2 renders GABA hyperpolarizing during neuronal maturation. *Nature*. 1999;397(6716):251–255.

40. Baccei ML, Fitzgerald M. Development of GABAergic and glycinergic transmission in the neonatal rat dorsal horn. *J Neurosci.* 2004;24(20):4749–4757.
41. Cordero-Erausquin M, Coull JA, Boudreau D, Rolland M, De Koninck Y. Differential maturation of GABA action and anion reversal potential in spinal lamina I neurons: impact of chloride extrusion capacity. *J Neurosci.* 2005;25(42):9613–9623.
42. Price TJ, Hargreaves KM, Cervero F. Protein expression and mRNA cellular distribution of the NKCC1 cotransporter in the dorsal root and trigeminal ganglia of the rat. *Brain Res.* 2006;1112(1):146–158.
43. Sung KW, Kirby M, McDonald MP, Lovinger DM, Delpire E. Abnormal $GABA_A$ receptor-mediated currents in dorsal root ganglion neurons isolated from Na-K-2Cl cotransporter null mice. *J Neurosci.* 2000;20(20):7531–7538.
44. Khirug S, Yamada J, Afzalov R, Voipio J, Khiroug L, Kaila K. GABAergic depolarization of the axon initial segment in cortical principal neurons is caused by the Na-K-2Cl cotransporter NKCC1. *J Neurosci.* 2008;28(18):4635–4639.
45. Szabadics J, Varga C, Molnar G, Olah S, Barzo P, Tamas G. Excitatory effect of GABAergic axo-axonic cells in cortical microcircuits. *Science.* 2006;311(5758):233–235.
46. Akaike N. Gramicidin perforated patch recording and intracellular chloride activity in excitable cells. *Prog Biophys Mol Biol.* 1996;65(3):251–264.
47. Ferrini F, Trang T, Mattioli TA, et al. Morphine hyperalgesia gated through microglia-mediated disruption of neuronal Cl^- homeostasis. *Nat Neurosci.* 2013;16(2):183–192.
48. Gagnon M, Bergeron MJ, Lavertu G, et al. Chloride extrusion enhancers as novel therapeutics for neurological diseases. *Nat Med.* 2013;19(11):1524–1528.
49. Kaila K, Voipio J. Postsynaptic fall in intracellular pH induced by GABA-activated bicarbonate conductance. *Nature.* 1987;330(6144):163–165.
50. Alper SL. Molecular physiology of SLC4 anion exchangers. *Exp Physiol.* 2006;91(1):153–161.
51. Eccles JC. *The Physiology of Synapses.* Berlin: Springer-Verlag; 1964.
52. Rocha-Gonzalez HI, Mao S, Alvarez-Leefmans FJ. Na^+, K^+, $2Cl^-$ cotransport and intracellular chloride regulation in rat primary sensory neurons: thermodynamic and kinetic aspects. *J Neurophysiol.* 2008;100(1):169–184.
53. Graham B, Redman S. A simulation of action potentials in synaptic boutons during presynaptic inhibition. *J Neurophysiol.* 1994;71(2):538–549.
54. Kullmann DM, Ruiz A, Rusakov DM, Scott R, Semyanov A, Walker MC. Presynaptic, extrasynaptic and axonal $GABA_A$ receptors in the CNS: where and why? *Prog Biophys Mol Biol.* 2005;87(1):33–46.
55. MacDermott AB, Role LW, Siegelbaum SA. Presynaptic ionotropic receptors and the control of transmitter release. *Annu Rev Neurosci.* 1999;22:443–485.
56. Chéry N, De Koninck Y. $GABA_B$ receptors are the first target of released GABA at lamina I inhibitory synapses in the adult rat spinal cord. *J Neurophysiol.* 2000;84(2):1006–1011.
57. Iyadomi M, Iyadomi I, Kumamoto E, Tomokuni K, Yoshimura M. Presynaptic inhibition by baclofen of miniature EPSCs and IPSCs in substantia gelatinosa neurons of the adult rat spinal dorsal horn. *Pain.* 2000;85(3):385–393.
58. Prescott SA, Ratté S. Pain processing by spinal microcircuits: afferent combinatorics. *Curr Opin Neurobiol.* 2012;22(4):631–639.
59. Ribeiro-da-Silva A, De Koninck Y. Morphological and neurochemical organization of the spinal dorsal horn. In: Basbaum AI, Bushnell MC, eds. *Science of Pain.* New York, NY: Elsevier; 2008.
60. Todd AJ. Neuronal circuitry for pain processing in the dorsal horn. *Nat Rev Neurosci.* 2010;11(12):823–836.
61. Todd AJ, Koerber HR. Neuroanatomical substrates of spinal nociception. In: McMahon SB, Koltzenburg M, Tracey I, Turk DC, eds. *Wall and Melzack's Textbook of Pain.* Philadelphia, PA: Elsevier; 2013:77–93.

62. Polgar E, Hughes DI, Riddell JS, Maxwell DJ, Puskar Z, Todd AJ. Selective loss of spinal GABAergic or glycinergic neurons is not necessary for development of thermal hyperalgesia in the chronic constriction injury model of neuropathic pain. *Pain.* 2003;104(1–2):229–239.
63. Spike RC, Puskar Z, Andrew D, Todd AJ. A quantitative and morphological study of projection neurons in lamina I of the rat lumbar spinal cord. *Eur J Neurosci.* 2003;18(9):2433–2448.
64. Antal M, Petko M, Polgar E, Heizmann CW, Storm-Mathisen J. Direct evidence of an extensive GABAergic innervation of the spinal dorsal horn by fibres descending from the rostral ventromedial medulla. *Neuroscience.* 1996;73(2):509–518.
65. Kato G, Yasaka T, Katafuchi T, et al. Direct GABAergic and glycinergic inhibition of the substantia gelatinosa from the rostral ventromedial medulla revealed by in vivo patch-clamp analysis in rats. *J Neurosci.* 2006;26(6):1787–1794.
66. Yasaka T, Tiong SY, Hughes DI, Riddell JS, Todd AJ. Populations of inhibitory and excitatory interneurons in lamina II of the adult rat spinal dorsal horn revealed by a combined electrophysiological and anatomical approach. *Pain.* 2010;151(2):475–488.
67. Lu Y, Dong H, Gao Y, et al. A feed-forward spinal cord glycinergic neural circuit gates mechanical allodynia. *J Clin Invest.* 2013;123(9):4050–4062.
68. Lu Y, Perl ER. A specific inhibitory pathway between substantia gelatinosa neurons receiving direct C-fiber input. *J Neurosci.* 2003;23(25):8752–8758.
69. Lu Y, Perl ER. Modular organization of excitatory circuits between neurons of the spinal superficial dorsal horn (laminae I and II). *J Neurosci.* 2005;25(15):3900–3907.
70. Lu Y, Perl ER. Selective action of noradrenaline and serotonin on neurones of the spinal superficial dorsal horn in the rat. *J Physiol.* 2007;582(Pt 1):127–136.
71. Zheng J, Lu Y, Perl ER. Inhibitory neurones of the spinal substantia gelatinosa mediate interaction of signals from primary afferents. *J Physiol.* 2010;588(Pt 12):2065–2075.
72. Daniele CA, MacDermott AB. Low-threshold primary afferent drive onto GABAergic interneurons in the superficial dorsal horn of the mouse. *J Neurosci.* 2009;29(3): 686–695.
73. Dougherty KJ, Sawchuk MA, Hochman S. Properties of mouse spinal lamina I GABAergic interneurons. *J Neurophysiol.* 2005;94(5):3221–3227.
74. Gassner M, Ruscheweyh R, Sandkuhler J. Direct excitation of spinal GABAergic interneurons by noradrenaline. *Pain.* 2009;145(1–2):204–210.
75. Hantman AW, van den Pol AN, Perl ER. Morphological and physiological features of a set of spinal substantia gelatinosa neurons defined by green fluorescent protein expression. *J Neurosci.* 2004;24(4):836–842.
76. Hu HJ, Gereau RW. Metabotropic glutamate receptor 5 regulates excitability and Kv4.2-containing K channels primarily in excitatory neurons of the spinal dorsal horn. *J Neurophysiol.* 2011;105(6):3010–3021.
77. Iwagaki N, Garzillo F, Polgar E, Riddell JS, Todd AJ. Neurochemical characterisation of lamina II inhibitory interneurons that express GFP in the PrP-GFP mouse. *Mol Pain.* 2013;9:56.
78. Labrakakis C, Lorenzo LE, Bories C, Ribeiro-da-Silva A, De Koninck Y. Inhibitory coupling between inhibitory interneurons in the spinal cord dorsal horn. *Mol Pain.* 2009;5:24.
79. Prescott SA, Ma Q, De Koninck Y. Normal and abnormal coding of somatosensory stimuli causing pain. *Nat Neurosci.* 2014;17(2):183–191.
80. Narikawa K, Furue H, Kumamoto E, Yoshimura M. In vivo patch-clamp analysis of IPSCs evoked in rat substantia gelatinosa neurons by cutaneous mechanical stimulation. *J Neurophysiol.* 2000;84(4):2171–2174.
81. Ataka T, Gu JG. Relationship between tonic inhibitory currents and phasic inhibitory activity in the spinal cord lamina II region of adult mice. *Mol Pain.* 2006;2:36.

82. Hu HJ, Carrasquillo Y, Karim F, et al. The kv4.2 potassium channel subunit is required for pain plasticity. *Neuron*. 2006;50(1):89–100.
83. Wang X, Zhang J, Eberhart D, et al. Excitatory superficial dorsal horn interneurons are functionally heterogeneous and required for the full behavioral expression of pain and itch. *Neuron*. 2013;78(2):312–324.
84. Kardon AP, Polgar E, Hachisuka J, et al. Dynorphin acts as a neuromodulator to inhibit itch in the dorsal horn of the spinal cord. *Neuron*. 2014;82(3):573–586.
85. Ross SE, Mardinly AR, McCord AE, et al. Loss of inhibitory interneurons in the dorsal spinal cord and elevated itch in Bhlhb5 mutant mice. *Neuron*. 2010;65(6):886–898.
86. Sandkuhler J. Models and mechanisms of hyperalgesia and allodynia. *Physiol Rev*. 2009;89(2):707–758.
87. Sandkuhler J. Spinal cord plasticity and pain. In: McMahon SB, Koltzenburg M, Tracey I, Turk DC, eds. *Wall and Melzack's Textbook of Pain*. Philadelphia, PA: Elsevier; 2013:94–100.
88. Bardoni R, Takazawa T, Tong CK, Choudhury P, Scherrer G, Macdermott AB. Pre- and postsynaptic inhibitory control in the spinal cord dorsal horn. *Ann N Y Acad Sci*. 2013;1279:90–96.
89. Beyer C, Roberts LA, Komisaruk BR. Hyperalgesia induced by altered glycinergic activity at the spinal cord. *Life Sci*. 1985;37(9):875–882.
90. Loomis CW, Khandwala H, Osmond G, Hefferan MP. Coadministration of intrathecal strychnine and bicuculline effects synergistic allodynia in the rat: an isobolographic analysis. *J Pharmacol Exp Ther*. 2001;296(3):756–761.
91. Malan TP, Mata HP, Porreca F. Spinal $GABA_A$ and $GABA_B$ receptor pharmacology in a rat model of neuropathic pain. *Anesthesiology*. 2002;96(5):1161–1167.
92. Miraucourt LS, Moisset X, Dallel R, Voisin DL. Glycine inhibitory dysfunction induces a selectively dynamic, morphine-resistant, and neurokinin 1 receptor-independent mechanical allodynia. *J Neurosci*. 2009;29(8):2519–2527.
93. Sherman SE, Loomis CW. Morphine insensitive allodynia is produced by intrathecal strychnine in the lightly anesthetized rat. *Pain*. 1994;56(1):17–29.
94. Sherman SE, Loomis CW. Strychnine-sensitive modulation is selective for non-noxious somatosensory input in the spinal cord of the rat. *Pain*. 1996;66(2–3):321–330.
95. Sivilotti L, Woolf CJ. The contribution of GABA(A) and glycine receptors to central sensitization: disinhibition and touch-evoked allodynia in the spinal cord. *J Neurophysiol*. 1994;72(1):169–179.
96. Sorkin LS, Puig S. Neuronal model of tactile allodynia produced by spinal strychnine: effects of excitatory amino acid receptor antagonists and a mu-opiate receptor agonist. *Pain*. 1996;68(2–3):283–292.
97. Sorkin LS, Puig S, Jones DL. Spinal bicuculline produces hypersensitivity of dorsal horn neurons: effects of excitatory amino acid antagonists. *Pain*. 1998;77(2):181–190.
98. Yaksh TL. Behavioral and autonomic correlates of the tactile evoked allodynia produced by spinal glycine inhibition: effects of modulatory receptor systems and excitatory amino acid antagonists. *Pain*. 1989;37(1):111–123.
99. Hwang JH, Yaksh TL. The effect of spinal GABA receptor agonists on tactile allodynia in a surgically-induced neuropathic pain model in the rat. *Pain*. 1997;70(1):15–22.
100. Castro-Lopes JM, Tavares I, Coimbra A. GABA decreases in the spinal cord dorsal horn after peripheral neurectomy. *Brain Res*. 1993;620(2):287–291.
101. Eaton MJ, Plunkett JA, Karmally S, Martinez MA, Montanez K. Changes in GAD- and GABA- immunoreactivity in the spinal dorsal horn after peripheral nerve injury and promotion of recovery by lumbar transplant of immortalized serotonergic precursors. *J Chem Neuroanat*. 1998;16(1):57–72.
102. Ibuki T, Hama AT, Wang XT, Pappas GD, Sagen J. Loss of GABA-immunoreactivity in the spinal dorsal horn of rats with peripheral nerve injury and promotion of recovery by adrenal medullary grafts. *Neuroscience*. 1997;76(3):845–858.

103. Moore KA, Kohno T, Karchewski LA, Scholz J, Baba H, Woolf CJ. Partial peripheral nerve injury promotes a selective loss of GABAergic inhibition in the superficial dorsal horn of the spinal cord. *J Neurosci.* 2002;22(15):6724–6731.
104. Ralston DD, Behbehani M, Sehlhorst SC, Meng XW, Ralston HJ. Decreased GABA immunoreactivity in rat dorsal horn is correlated with pain behavior: a light and electron microscope study. In: Jensen TS, Turner JA, Wiesenfeld-Hallin Z, eds. *Proceedings of the Eighth World Congress on Pain*. Seattle, WA: IASP Press; 1997:547–560.
105. Somers DL, Clemente FR. Dorsal horn synaptosomal content of aspartate, glutamate, glycine and GABA are differentially altered following chronic constriction injury to the rat sciatic nerve. *Neurosci Lett.* 2002;323(3):171–174.
106. Stiller CO, Cui JG, O'Connor WT, Brodin E, Meyerson BA, Linderoth B. Release of gamma-aminobutyric acid in the dorsal horn and suppression of tactile allodynia by spinal cord stimulation in mononeuropathic rats. *Neurosurgery*. 1996;39(2):367–374.
107. Lever I, Cunningham J, Grist J, Yip PK, Malcangio M. Release of BDNF and GABA in the dorsal horn of neuropathic rats. *Eur J Neurosci.* 2003;18(5):1169–1174.
108. Woolf CJ, Wall PD. Chronic peripheral nerve section diminishes the primary afferent A-fibre mediated inhibition of rat dorsal horn neurones. *Brain Res.* 1982;242(1):77–85.
109. Polgar E, Todd AJ. Tactile allodynia can occur in the spared nerve injury model in the rat without selective loss of GABA or GABA(A) receptors from synapses in laminae I-II of the ipsilateral spinal dorsal horn. *Neuroscience*. 2008;156(1):193–202.
110. Schoffnegger D, Heinke B, Sommer C, Sandkuhler J. Physiological properties of spinal lamina II GABAergic neurons in mice following peripheral nerve injury. *J Physiol.* 2006;577(Pt 3):869–878.
111. Anseloni VC, Gold MS. Inflammation-induced shift in the valence of spinal $GABA_A$ receptor-mediated modulation of nociception in the adult rat. *J Pain.* 2008;9(8):732–738.
112. Rode F, Jensen DG, Blackburn-Munro G, Bjerrum OJ. Centrally-mediated antinociceptive actions of GABA(A) receptor agonists in the rat spared nerve injury model of neuropathic pain. *Eur J Pharmacol.* 2005;516(2):131–138.
113. Braz JM, Sharif-Naeini R, Vogt D, et al. Forebrain GABAergic neuron precursors integrate into adult spinal cord and reduce injury-induced neuropathic pain. *Neuron.* 2012;74(4):663–675.
114. Eaton MJ, Plunkett JA, Martinez MA, et al. Transplants of neuronal cells bioengineered to synthesize GABA alleviate chronic neuropathic pain. *Cell Transplant.* 1999;8(1):87–101.
115. Stubley LA, Martinez MA, Karmally S, Lopez T, Cejas P, Eaton MJ. Only early intervention with gamma-aminobutyric acid cell therapy is able to reverse neuropathic pain after partial nerve injury. *J Neurotrauma.* 2001;18(4):471–477.
116. Kittler JT, Moss SJ. Modulation of GABAA receptor activity by phosphorylation and receptor trafficking: implications for the efficacy of synaptic inhibition. *Curr Opin Neurobiol.* 2003;13(3):341–347.
117. Ahmadi S, Lippross S, Neuhuber WL, Zeilhofer HU. PGE2 selectively blocks inhibitory glycinergic neurotransmission onto rat superficial dorsal horn neurons. *Nat Neurosci.* 2002;5(1):34–40.
118. Harvey RJ, Depner UB, Wassle H, et al. GlyR alpha3: an essential target for spinal PGE2-mediated inflammatory pain sensitization. *Science.* 2004;304(5672):884–887.
119. Harvey VL, Caley A, Muller UC, Harvey RJ, Dickenson AH. A selective role for alpha3 subunit glycine receptors in inflammatory pain. *Front Mol Neurosci.* 2009;2:14.
120. Hösl K, Reinold H, Harvey RJ, Müller U, Narumiya S, Zeilhofer HU. Spinal prostaglandin E receptors of the EP2 subtype and the glycine receptor alpha3 subunit, which mediate central inflammatory hyperalgesia, do not contribute to pain after peripheral nerve injury or formalin injection. *Pain.* 2006;126(1–3):46–53.

121. Nishikawa Y, Sasaki A, Kuraishi Y. Blockade of glycine transporter (GlyT) 2, but not GlyT1, ameliorates dynamic and static mechanical allodynia in mice with herpetic or postherpetic pain. *J Pharmacol Sci*. 2010;112(3):352–360.
122. Coull JA, Boudreau D, Bachand K, et al. Trans-synaptic shift in anion gradient in spinal lamina I neurons as a mechanism of neuropathic pain. *Nature*. 2003;424(6951): 938–942.
123. Nomura H, Sakai A, Nagano M, Umino M, Suzuki H. Expression changes of cation chloride cotransporters in the rat spinal cord following intraplantar formalin. *Neurosci. Res*. 2006;56(4):435–440.
124. Zhang W, Liu LY, Xu TL. Reduced potassium-chloride co-transporter expression in spinal cord dorsal horn neurons contributes to inflammatory pain hypersensitivity in rats. *Neuroscience*. 2008;152(2):502–510.
125. Coull JAM, Beggs S, Boudreau D, et al. BDNF from microglia causes the shift in neuronal anion gradient underlying neuropathic pain. *Nature*. 2005;438:1017–1021.
126. Keller AF, Beggs S, Salter MW, De Koninck Y. Transformation of the output of spinal lamina I neurons after nerve injury and microglia stimulation underlying neuropathic pain. *Mol Pain*. 2007;3:27.
127. Beggs S, Trang T, Salter MW. P2X4R+ microglia drive neuropathic pain. *Nat Neurosci*. 2012;15(8):1068–1073.
128. Galan A, Cervero F. Painful stimuli induce in vivo phosphorylation and membrane mobilization of mouse spinal cord NKCC1 co-transporter. *Neuroscience*. 2005;133(1):245–252.
129. Zhu Y, Lu SG, Gold MS. Persistent inflammation increases GABA-induced depolarization of rat cutaneous dorsal root ganglion neurons in vitro. *Neuroscience*. 2012;220:330–340.
130. Willis Jr WD. Dorsal root potentials and dorsal root reflexes: a double-edged sword. *Exp Brain Res*. 1999;124(4):395–421.
131. Lee KY, Gold MS. Inflammatory mediators potentiate high affinity $GABA_A$ currents in rat dorsal root ganglion neurons. *Neurosci Lett*. 2012;518(2):128–132.
132. Lavertu G, Cote SL, De Koninck Y. Enhancing K-Cl co-transport restores normal spinothalamic sensory coding in a neuropathic pain model. *Brain*. 2014;137(Pt 3): 724–738.
133. Baba H, Ji RR, Kohno T, et al. Removal of GABAergic inhibition facilitates polysynaptic A fiber-mediated excitatory transmission to the superficial spinal dorsal horn. *Mol Cell Neurosci*. 2003;24(3):818–830.
134. Miraucourt LS, Dallel R, Voisin DL. Glycine inhibitory dysfunction turns touch into pain through PKCgamma interneurons. *PLoS One*. 2007;2(11):e1116.
135. Torsney C, MacDermott AB. Disinhibition opens the gate to pathological pain signaling in superficial neurokinin 1 receptor-expressing neurons in rat spinal cord. *J Neurosci*. 2006;26(6):1833–1843.
136. Bonin RP, De Koninck Y. Restoring ionotropic inhibition as an analgesic strategy. *Neurosci Lett*. 2013;557(Pt A):43–51.
137. De Koninck Y. Altered chloride homeostasis in neurological disorders: a new target. *Curr Opin Pharmacol*. 2007;7(1):93–99.
138. Kahle KT, Khanna A, Clapham DE, Woolf CJ. Therapeutic restoration of spinal inhibition via druggable enhancement of potassium-chloride cotransporter KCC2-mediated chloride extrusion in peripheral neuropathic pain. *JAMA Neurol*. 2014;71(5): 640–645.
139. McCarson KE, Enna SJ. GABA pharmacology: the search for analgesics. *Neurochem Res*. 2014;39(10):1948–1963.
140. Munro G, Hansen RR, Mirza NR. $GABA_A$ receptor modulation: potential to deliver novel pain medicines? *Eur J Pharmacol*. 2013;716(1–3):17–23.

141. Rudolph U, Knoflach F. Beyond classical benzodiazepines: novel therapeutic potential of $GABA_A$ receptor subtypes. *Nat Rev Drug Discov*. 2011;10(9):685–697.
142. Zeilhofer HU, Mohler H, Di Lio A. GABAergic analgesia: new insights from mutant mice and subtype-selective agonists. *Trends Pharmacol Sci*. 2009;30(8):397–402.
143. Harvey RJ, Yee BK. Glycine transporters as novel therapeutic targets in schizophrenia, alcohol dependence and pain. *Nat Rev Drug Discov*. 2013;12(11):866–885.
144. Vandenberg RJ, Ryan RM, Carland JE, Imlach WL, Christie MJ. Glycine transport inhibitors for the treatment of pain. *Trends Pharmacol Sci*. 2014;35(8):423–430.
145. Prescott SA, Sejnowski TJ, De Koninck Y. Reduction of anion reversal potential subverts the inhibitory control of firing rate in spinal lamina I neurons: towards a biophysical basis for neuropathic pain. *Mol Pain*. 2006;2:32.
146. Ratté S, Zhu Y, Lee KY, Prescott SA. Criticality and degeneracy in injury-induced changes in primary afferent excitability and the implications for neuropathic pain. *eLife*. 2014;3:02370.
147. Asiedu MN, Mejia G, Ossipov MK, Malan TP, Kaila K, Price TJ. Modulation of spinal GABAergic analgesia by inhibition of chloride extrusion capacity in mice. *J Pain*. 2012;13(6):546–554.
148. Chaparro LE, Wiffen PJ, Moore RA, Gilron I. Combination pharmacotherapy for the treatment of neuropathic pain in adults. *Cochrane Database Syst Rev*. 2012;7, CD008943.
149. Asiedu M, Ossipov MH, Kaila K, Price TJ. Acetazolamide and midazolam act synergistically to inhibit neuropathic pain. *Pain*. 2009;148(2):302–308.
150. Asiedu MN, Mejia GL, Hubner CA, Kaila K, Price TJ. Inhibition of carbonic anhydrase augments GABAA receptor-mediated analgesia via a spinal mechanism of action. *J Pain*. 2014;15(4):395–406.

CHAPTER THIRTEEN

Dendritic Spine Dysgenesis in Neuropathic Pain

Andrew Michael Tan[1]

Department of Neurology and Center for Neuroscience and Regeneration Research, Yale University School of Medicine, New Haven, Connecticut, USA
Rehabilitation Research Center, Veterans Affairs Connecticut Healthcare System, West Haven, Connecticut, USA
Hopkins School, New Haven, Connecticut, USA
[1]Corresponding author: e-mail address: Andrew.Tan@yale.edu

Contents

1. A Brief History 386
2. Dendritic Spine Morphology 387
3. Dendritic Spines in Synaptic Function 389
4. Dendritic Spines in Pathology 390
5. Dendritic Spines in Neuropathic Pain 391
6. Spinal Cord Injury 392
7. Peripheral Nerve Injury 394
8. Diabetes Mellitus 394
9. Thermal Burn Injury 395
10. Molecular Control of Dendritic Spine Dysgenesis 396
11. Spinal Memory Mechanism in Neuropathic Pain 400
12. Final Perspectives 401
References 402

Abstract

The failure of neuropathic pain to abate even years after trauma suggests that adverse changes to synaptic function must exist in a chronic pathological state in nociceptive pathways. The chronicity of neuropathic pain therefore underscores the importance of understanding the contribution of dendritic spines—micron-sized postsynaptic structures that represent modifiable sites of synaptic contact. Historically, dendritic spines have been of great interest to the learning and memory field. More recent evidence points to the exciting implication that abnormal dendritic spine structure following disease or injury may represent a "molecular memory" for maintaining chronic pain. Dendritic spine dysgenesis in dorsal horn neurons contributes to nociceptive hyperexcitability associated with neuropathic pain, as demonstrated in multiple pain models, i.e., spinal cord injury, peripheral nerve injury, diabetic neuropathy, and thermal burn injury. Because of the relationship between dendritic spine structure and neuronal function, a thorough investigation of dendritic spine behavior in the spinal cord is a unique opportunity to better understand the mechanisms of sensory dysfunction after injury or

Progress in Molecular Biology and Translational Science, Volume 131
ISSN 1877-1173
http://dx.doi.org/10.1016/bs.pmbts.2014.12.001

disease. At a conceptual level, a spinal memory mechanism that engages dendritic spine remodeling would also contribute to a broad range of intractable neurological conditions. Molecules involved in regulating dendritic spine plasticity may offer novel targets for the development of effective and durable therapies for neurological disease.

1. A BRIEF HISTORY

Ramon y Cajal first described dendritic spines located on neurons in the cerebellum in the late nineteenth century.[1,2] It was apparent by his writings that he found beauty in neuronal architecture and the power of the Golgi-staining method for visualizing fine morphological processes. From stained images of Purkinje cells in the cerebellum of birds, Cajal described the surface of these neurons as having thorns or "espina" (short spines).[3] Subsequent drawings demonstrated that dendritic spines had a variety of morphological features that differed across cell types. His publications and drawings were the first to describe and name these microscopic structures as "dendritic spines" (Fig. 1).

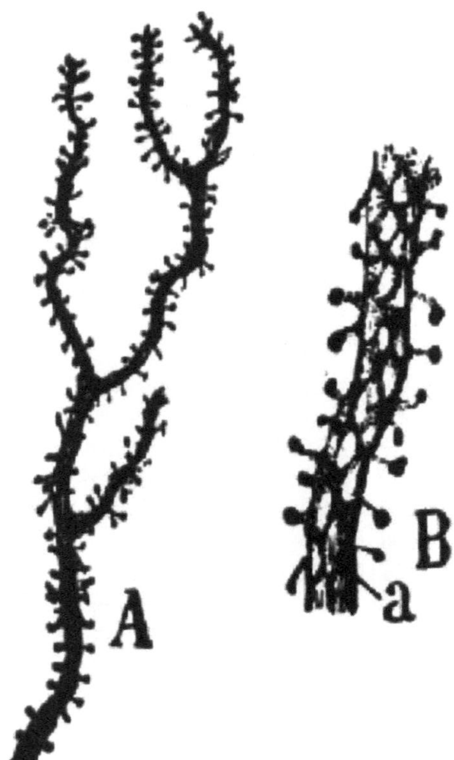

Figure 1 Original drawing by Cajal showing a low (A) magnification and a high (B) magnification of dendritic spines on the dendritic branch of a cerebellar Purkinje cell.[4]

Cajal's impressive speculations during his time now form the basis for much of what neuroscientists study today. He argued that dendritic spines would be a main contact point between axons and dendrites.[1,5] He proposed that dendritic spines would be involved in the process of learning and that the physical characteristics of dendritic spines would contribute to neuronal activity.[1,5] Although his early work drew criticism with arguments that dendritic spines were mere artifacts of the staining methods he used, Cajal counter-argued that dendritic spines were real anomalies on neurons because (1) spines could be visualized using different staining methods, including Golgi, Cox, or Ehrlich methylene-blue stains; (2) dendritic spines appeared in the same manner and in similar areas of the nervous system; (3) spines were always absent in the regions of axons, soma, or proximal dendritic regions, e.g., close to cell bodies; (4) dendritic spines did not resemble the crystal deposit artifacts observed when viewed at high magnification; and (5) dendritic spine necks are visibly distinguishable from the dendritic arbor. Indeed, after more than 100 years of study, dendritic spines have been more than confirmed as real structures; dendritic spines continue to inform and pique curiosity.

2. DENDRITIC SPINE MORPHOLOGY

Dendritic spines are microscopic protrusions located on dendritic branches of neurons[6] (Fig. 2). Dendritic spines represent sites of synaptic contact and primarily comprised of polymerized actin or filamentous actin. Dendritic spines are found throughout the nervous system, represent sites of synaptic contact, and are primarily associated with neurons that receive convergent input. "Spiny" neurons include hippocampal neurons, pyramidal neurons in the motor cortex, Purkinje neurons in the cerebellum, and more recently characterized, dorsal horn sensory neurons in the spinal cord.[7–11] More than 90% of excitatory synapses in the central nervous system (CNS) occur upon dendritic spines.[12] The architecture of dendritic spines is variable, but estimates suggest that the thin spine neck, which connects the spine to the main dendritic branch, has lengths between 0.04 and 1 μm and has "door knob"-shaped head structures with diameters that between 0.5 and 2 μm.[13] Dendritic spine density depends upon the type of neuron profiled and the method for visualization and analysis,[14,15] with estimates ranging from 1 to 10 spines per micron length of dendritic branch. Because of the large variations in shape, investigators have categories dendritic spines using simple visual characteristics. Common terms for spine shapes include "mushroom," "thin or filopodia," or "stubby"[16,17] (Fig. 3).

Figure 2 Photomicrograph of the apical dendritic branch of a pyramidal neuron in the motor cortex (lamina V). Dendritic spines, which vary in shape and size, populate the entire length of the dendritic branch. Scale bar = 10 μm.

Figure 3 Common descriptors for dendritic spines include stubby, thin, or mushroom shapes. Dendritic spine measurements of spine head diameter and spine length ("a" and "b" with red arrows, respectively).

3. DENDRITIC SPINES IN SYNAPTIC FUNCTION

Dendritic spines have been of greatest interest to the field of learning and memory. The most well-known concept for a neural basis for learning and memory is the phenomenon of long-term potentiation (LTP).[18] LTP is defined as a localized increase in synaptic strength between two neurons following coincident and synchronous activity. Following LTP induction *in vitro*, increased synaptic efficacy can last for hours. *In vivo*, LTP can persist for days, an observation which has supported a synaptic basis for neural circuit maintenance and long-term memory.

Most LTP studies have been performed in the cortex or hippocampus and can be divided into an early and late phase. Early LTP is not dependent on protein synthesis and occurs rapidly and lasts for up to 60 min following induction, as shown in hippocampal slice experiments. Early LTP involves the posttranslational modification of excitatory glutamatergic AMPA receptors, and the trafficking and membrane insertion of AMPA receptors.[19] LTP induction protocols include high-frequency electrical tetanus stimulation ($\sim$100 Hz) and acute application of exogenous glutamate. Both protocols increase the responsiveness of the postsynaptic cell from subsequent stimulation of the presynaptic cell.[20–22]

The transition from early to late LTP involves an activation of downstream kinases, such as PKA, PKC, and the extracellular signal-related kinase, a member of the family of mitogen-activated protein kinases (MAPKs). Activation of these molecular pathways ultimately leads to gene transcription and protein synthesis, which are necessary for late LTP.[23–27] Actin polymerization and stabilization within dendritic spine structures are also important for enhanced synaptic efficacy. Inhibition of actin polymerization with latriculins or cytochalasins blocks late LTP and disrupts dendritic spine maturation.[28] The neurotrophin BDNF is involved in some forms of translation-dependent long-term memory and has also been shown to contribute to the development of dendritic spines.[29]

Calcium influx via activated NMDA receptors and an increased level of intracellular calcium is required for LTP induction, since blockade of calcium signaling prevents synaptic potentiation.[30,31] Interestingly, three properties of dendritic spines may contribute to regulating calcium signaling.[32,33] First, the narrow dimension of the spine neck acts as a diffusion barrier to Ca^{2+} influx (with the synapse located at the spine head). Second, the spine neck increases longitudinal electrical resistance during a synaptic transmission event, thereby amplifying and isolating the effect of local membrane depolarization of the spine synapse. Third, although the spine head is partially isolated from the dendritic branch, bidirectional current flow can occur. This electrically couples the spine to the rest of the dendritic arbor.[33] In an elegant study, activated CAMKII (or Ca^{2+}/calmodulin-dependent protein kinases II) within a single dendritic spine accompanied LTP induction.[34,35] CAMKII activation directly resulted from postsynaptic activation of NMDA receptors and L-type voltage-sensitive calcium channels, demonstrating that calcium signaling within the dendritic spine activated CAMKII. LTP across a single dendritic spine also led to its increase in volume.[35,36]

Indeed, dendritic spine morphology has an integral role in regulating synaptic function.[37] Larger mushroom-shaped dendritic spines are associated with stronger synapses and memory formation.[38,39] The dendritic spine neck region acts as a diffusion barrier against passive movement of cations and other molecules, which can directly affect local function. By changing shape, dendritic spines can open and close their "diffusion" gates in response to activity.[40] Dendritic spine morphology is therefore a key factor in regulating intracellular function at synapses. For example, by slowing calcium diffusion, dendritic spines contribute to the local accumulation of calcium and likewise maintain elevated calcium for longer periods of time.[41–43] Together, this compartmental model for dendritic spine function allows neurons to isolate and integrate information at a discrete synaptic level. A "gating property" for spines would be expected to grant incredible computational power to a single neuron.[41,44–48]

4. DENDRITIC SPINES IN PATHOLOGY

Dendritic spine dysgenesis (or malformation) is associated with disease. Abnormal dendritic spine structures have been observed in epilepsy, schizophrenia, dementia, stroke, and chronic substance abuse.[49] In human patients and animal models of mental retardation, dendritic spines appear smaller and thinner (i.e., less mature spine morphologies).[50–52] Because mental retardation may be traced to genetic defects in scaffolding and structural proteins that maintain synaptic structure,[53] abnormal dendritic spine structure may indicate synaptic dysfunction. In mice lacking a copy of the shank-1 gene, which encodes a crucial scaffolding protein in spines, dendritic spines appear stunted. These mice had deficits in memory function compared with wild-type mice.[54] In a model of Fragile X mental retardation, mice with Fmr1 protein knockout show impaired LTP[55] and malformed dendritic spines due to dysfunctional Rac1 signaling.[56] Taken together, a large body of evidence strongly supports associations between abnormal dendritic spine structure and pathology.

Synaptic potentiation in spinal cord nociceptive circuits is a well-known phenomenon that partially explains central sensitization associated with neuropathic pain.[57] Low- and high-frequency electrical stimulation as well as inflammation can induce LTP between primary c-fiber afferents and nociceptive neurons in the superficial dorsal horn (i.e., lamina II) and wide-dynamic range (WDR) neurons located in the intermediate zone.[58–60] Injury-induced increases in glutamate signaling, release of neuromodulator substance-P, and neurotrophic factors (e.g., neurotrophin-3 (NT-3) and BDNF) contribute to central sensitization.[61–63] In addition, glutamate ligand binding with AMPA and NMDA receptors, substance-P binding with neurokinin 1 receptor, and neurotrophin binding with tyrosine kinase receptors (e.g., TrkB/C) increase intracellular calcium concentration and trigger synaptic potentiation. Thus, emerging evidence strongly suggest that nervous system injury or disease can adversely modify synaptic function by engaging synaptic-based memory mechanisms in the spinal cord.[64]

5. DENDRITIC SPINES IN NEUROPATHIC PAIN

The chronicity of neuropathic pain underscores the importance of understanding the contribution of dendritic spines. Multiple factors contribute to hyperexcitability of nociceptive neurons associated with neuropathic pain. These factors include chronic nerve damage, e.g., amputation or diabetic neuropathy, CNS inflammation, dysregulation of potassium-chloride cotransporter 2 (KCC2) activity, loss of inhibitory inputs, e.g., reduced GABAergic signaling, and altered growth factor and sodium channel expression in dorsal root ganglia (DRG) or spinal cord neurons.[65–72] A synaptic model similar to that involved in long-term memory storage has also been proposed to explain the persistent, intractable nature of neuropathic pain.[65] Previous studies in multiple pain models, i.e., spinal cord injury, peripheral nerve injury, diabetic neuropathy, and burn injury, demonstrate that dendritic spine dysgenesis can abnormally change the electrical properties of dorsal horn neurons in a manner that reflects nociceptive hyperexcitability associated with neuropathic pain[73] (Fig. 4). Because of the link between dendritic spine structure and neuronal function,[74] a thorough investigation of dendritic spine behavior in the spinal cord is a unique opportunity to better understand the mechanisms of sensory dysfunction after injury or disease.

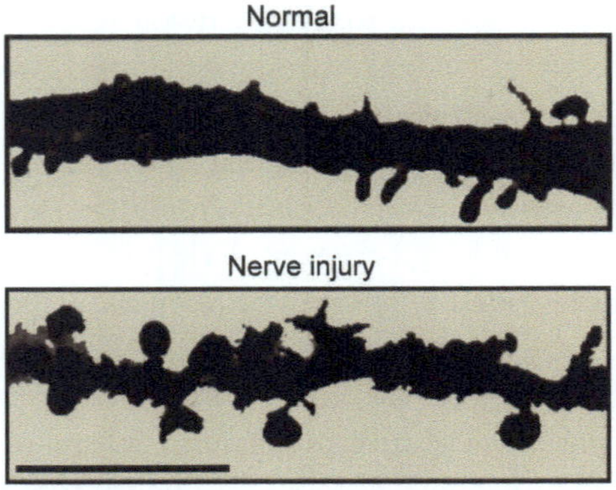

Figure 4 Dendritic branches from a dorsal horn neuron in a normal condition (top panel) and 10 days after peripheral nerve injury (bottom panel) demonstrate qualitative differences in dendritic spine morphology, including changes in dendritic spine size, shape, and number. Scale bar = 10 μm.

6. SPINAL CORD INJURY

Spinal cord injury (SCI) often leads to reactive neuronal plasticity, which can partially explain the development of intractable chronic pain.[75] Dendritic spine remodeling after SCI has been largely unexplored. Levels of postsynaptic density protein (PSD-95), a postsynaptic marker localized at excitatory glutamatergic synapses,[76] and neuronal PDZ protein, which demarcates sites of protein complex interaction and is associated with AMPA and NMDA receptors, increase more than 20% in the injured spinal cord compared with uninjured animals,[77,78] demonstrating that synaptic plasticity occurs following injury. Although dendritic spine morphologies change after SCI in the motor cortex,[79,80] the functional role for these dendritic spine alterations is not known.

To investigate the contribution of dendritic spine remodeling in neuropathic pain, a previous study analyzed dendritic spine morphometry and functional influence on dorsal horn neurons in adult rats 1 month after SCI.[81] The authors profiled dendritic spines located on WDR neurons, which relay multiple primary afferent sensory inputs through the spinothalamic pathway.[82,83] After SCI, dendritic spine density increased along dendritic branches, altered their spatial distribution (relocalizing closer to the cell body), and changed shape, increasing in length and head diameter. Such changes in dendritic spine elements have been shown to directly influence postsynaptic neuronal function.[84] In agreement, abnormal physiological and behavioral effects accompanied abnormal dendritic spine morphologies after SCI. WDR neurons exhibited hyperexcitability in response to evoked innocuous and noxious stimulation in electrophysiological recordings taken from animals with SCI. WDR units had increased evoked-action potential firing rates that were 50–100% greater than units in uninjured, normal animals.[81] Notably, the electrical properties of WDR units *in vivo* were analogous to predictions derived from computer simulations, which demonstrated that dendritic spine structure alone could increase excitability of postsynaptic neurons[73] (Fig. 5). Taken together, these findings revealed that SCI leads to abnormal dendritic spine remodeling that is accompanied by electrical signs and behavioral symptoms of neuropathic pain.

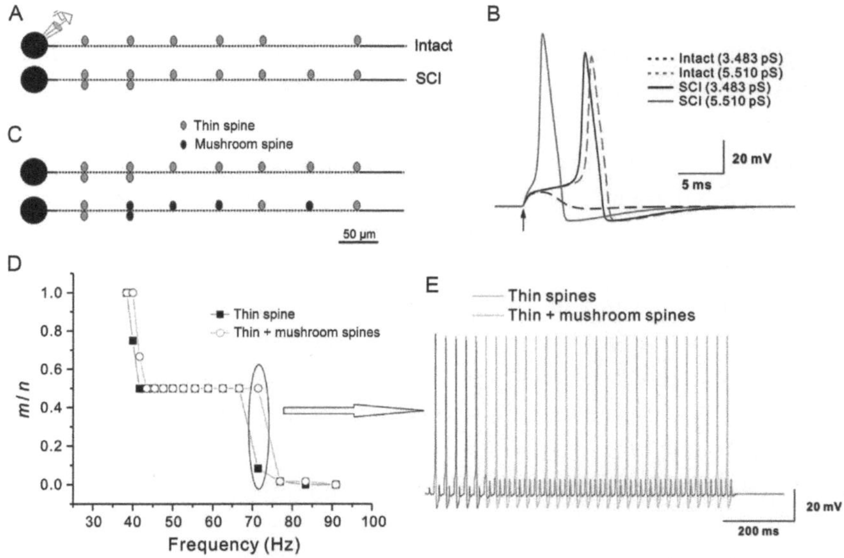

Figure 5 Increased spine density and altered spine distribution can contribute to hyperexcitable neuronal output. (A) A simple neuronal model of intact (upper panel) and SCI (lower panel) dorsal horn neurons based upon primary anatomical data. (B) AMPA-like excitatory stimulation on spines for the SCI neuron model (solid line) produced single somatic action potentials whose activation threshold was less than the uninjured, control neuron. (C) Five thin spines converted into mushroom spines in the SCI model (upper to lower panel) to mimic observations *in vivo*. (D) The response-to-stimulus ratio (*m*/*n*) is plotted as a function of the frequency. The SCI neuron produced more action potentials at higher frequency. (E) Action potentials simulated at 71.4 Hz.[73]

7. PERIPHERAL NERVE INJURY

Primary afferent plasticity contributes to pain after peripheral nerve injury, but it is less clear if plasticity occurs within spinal nociceptive circuits, postsynaptically.[85–87] In adult rats, ipsilateral WDR neurons exhibited increased dendritic spine density and altered spine distribution 10 days following a chronic constriction injury of the sciatic nerve. Dendritic spine length and head diameter also increased. These postinjury dendritic spine profiles were accompanied by WDR unit hyperexcitability, as demonstrated by hyperresponsiveness to evoked low- and high-threshold stimuli applied to cutaneous receptive fields. Nerve-injured animals with abnormal dendritic spines also exhibited behavioral evidence of neuropathic mechanical allodynia and heat hyperalgesia.[85] Interestingly, in FMR1 knockout mice in which dendritic spine development and maturation is impaired,[88,89] peripheral nerve injury fails to increase nociceptive sensitization to the same extent as control wild-type mice.[90] The finding of dendritic spine remodeling after nerve injury suggests the existence of a putative peripheral nervous system (PNS)-to-CNS signal that chronically and adversely changes the central sensory circuit system. Potential candidates for such a transneuronal signal may include remote neuro-immune signals, e.g., cytokine release.[91,92] Microglial inflammation within the spinal cord occurs after peripheral nerve injury, and activated microglia can contribute to central sensitization.[92–94] Microglia, as well as astrocytes, contribute to dendritic spine plasticity and function. Using high-resolution electron microscopy, microglial processes have been observed to directly contact synaptic spine structures, suggesting a direct interaction between microglia and spine synapses.[95,96] In the hippocampus, time-lapse imaging has demonstrated that astrocytes rapidly extend and retract fine processes to interact with dendritic spines[97] and these astrocyte–spine interactions appear coordinated and more stable at sites of larger, more mature dendritic spines.[98,99] Overall, glial activity may also have an important role in dendritic spine remodeling associated with injury.[85,92,96,100,101]

8. DIABETES MELLITUS

The mechanisms underlying diabetic neuropathic pain are extraordinarily elusive due to the complexity of metabolic and systemic diseases. Peripheral neuropathy due to secondary vascular disease,[102] dysfunctional KCC2 activity,[103] and sodium channel misexpression in DRG neurons[68] have been proposed as potential explanations for diabetic neuropathic pain. In addition, a few studies have shown that maladaptive plasticity within nociceptive circuits in the spinal cord could also contribute to mechanisms underlying diabetic neuropathic pain.[104,105] In regard to dendritic spine reorganization, a study has demonstrated that abnormal dendritic spines appear on WDR neurons following streptozotocin (STZ)-induced diabetic neuropathic pain.[106] STZ is a target-specific toxin for insulin-producing beta cells of the pancreas, and systemic administration in rodents significantly increases blood glucose levels, producing a hyperglycemic model of type 1 diabetes.[107] About 2 months after STZ-induced hyperglycemia, dendritic spine density decreases in the cortex[108] and parietal cortex.[104] Reduced dendritic spine number in these hyperglycemic animals accompanied impaired memory performance in a water maze test. Although most diabetic pain studies implicate PNS mechanisms in pain development, changes within the CNS may further account for neuropathic pain in diabetes.[109,110]

Similar to SCI and peripheral nerve injury, in animals with diabetic neuropathic pain, abnormal spine profiles accompanied WDR unit hyperexcitability and mechanical allodynia.[73] Interestingly, because tactile allodynia develops slowly following STZ-induced diabetes, a time window exists (in the first week after STZ induction) when animals are hyperglycemic, but lack detectable pain phenotype. Within this timeframe, hyperglycemic animals without pain also had near-normal spine profiles. This temporal relationship between the development of neuropathic pain and abnormal spines revealed two exciting implications: first, the presence of maladaptive dendritic spines can predict the manifestation of neuropathic pain in hyperglycemic conditions. Second, a therapeutic window is available for intervention in early diabetes that could prevent the establishment of intractable neuropathic pain by targeting dendritic spine reorganization.[106] Taken together, diabetes can induce dendritic spine remodeling in the spinal cord, which suggests that metabolic diseases in general can trigger abnormal synaptic plasticity associated with pathological nociception.[57] However, it remains to be seen whether dendritic spine changes in diabetic neuropathic pain are the direct result of hyperglycemia, secondary inflammation, or changes in presynaptic elements.

9. THERMAL BURN INJURY

Thermal injury is one of the least studied and poorly understood types of injuries that often produce chronic or neuropathic pain conditions. In humans and animal models with mild–moderate burn injury, primary hyperalgesia (which arises within the cutaneous area of the burn injury) is followed by a progressive expansion of secondary tactile allodynia to uninjured areas.[111,112] Inflammatory responses within the spinal cord and activation of p38 MAPK in spinal microglia have been implicated in the pathogenesis of burn-injury pain.[111,113,114] Pain after burn may also develop from abnormalities in newly regenerated nerve endings following damage and poor reinnervation of scar tissue.[115,116] In a study by Hill and colleagues, skin incision alone was able to induce expression of plasticity and regeneration-associated genes in DRG, including activating transcription factor 3, galanin, GAP-43, and Gadd45a. Thus, skin damage, in general, through a variety of pathways may contribute to subsequent central sensitization.[117,118]

The chronicity of neuropathic pain after severe burn injuries emphasizes the importance of understanding contribution of dendritic spine dysgenesis.[49,85,119] Indeed, maladaptive plasticity in the CNS has been offered as an explanation for the notorious opioid resistance observed in burn patients.[120] In a study utilizing adult rats, a unilateral second-degree burn injury on the hind paw resulted in abnormal dendritic spine remodeling on ipsilateral WDR neurons in the spinal cord dorsal horn.[119] Similar to other models of neuropathic pain, 4 weeks after burn injury, dendritic spines on WDR neurons increased in density, redistributed along dendrites, and exhibited changes in morphological structure, increasing in length and head diameter. These abnormal structural profiles were accompanied by an expansion of cutaneous receptive fields, increased WDR neuron excitability, and the development of mechanical allodynia. While a time course investigating the progression from acute inflammatory pain to more chronic pain after burn injury was not performed, the findings strongly suggest a relationship between the presence of abnormal dendritic spines and abnormal pain after skin burn injury.

10. MOLECULAR CONTROL OF DENDRITIC SPINE DYSGENESIS

Of the molecular signaling pathways governing dendritic spine dynamics, the Rac1 small molecule (~21 kDa) kinase is well studied.[121,122] Rac1 is a soluble intracellular protein in the family of Rho GTPases and involved in a broad array of cellular functions, including cell growth, cytoskeletal rearrangement, intracellular protein trafficking, and the activation of a host of downstream protein kinases.[123] Rac1 activity regulates dendritic spine morphology through actin cytoskeletal organization and also promotes clustering of glutamatergic AMPA receptors in dendritic spines.[84] Rac1 exists in a dynamic equilibrium between a GTP-bound activated state and a GDP-bound inactivated state. Constitutively activated Rac1 increases the rate of dendritic spine turnover, density, stability, and spine volume. Conversely, dominant negative expression of Rac1 (mutant RacN17) decreases spine density and inhibits spine maturation in the hippocampus[121,124] (Fig. 6). As mentioned earlier, mice with Fmr1 protein knockout exhibit malformed dendritic spines due to dysfunctional Rac1 signaling.[56] In animal models of injury or stress, Rac1 mRNA expression and activation increases.[125,126] Specifically, Rac1 mRNA expression increases after SCI and remains elevated for up to 3 months.[126]

In studies investigating the contribution of dendritic spine remodeling in injury- and disease-induced neuropathic pain, a Rac1-specific inhibitor, NSC23766, has proven to be a useful and powerful tool.[9,127] NSC23766 has been shown to block guanine exchange factors, Trio and Tiam1, without interfering with closely related GTPases cdc42 or RhoA-GEF and does not influence the binding of Rac1 to its downstream effector PAK1.[127] In primary neuron cultures, treatment of spinal cord dorsal horn neurons with NSC23766 effectively decreases the density of dendritic spines compared with untreated neurons.[85] Moreover, intrathecal administration of NSC23766 in a model of SCI reduces protein levels of postsynaptic density marker PSD-95 without affecting levels of presynaptic terminal-associated

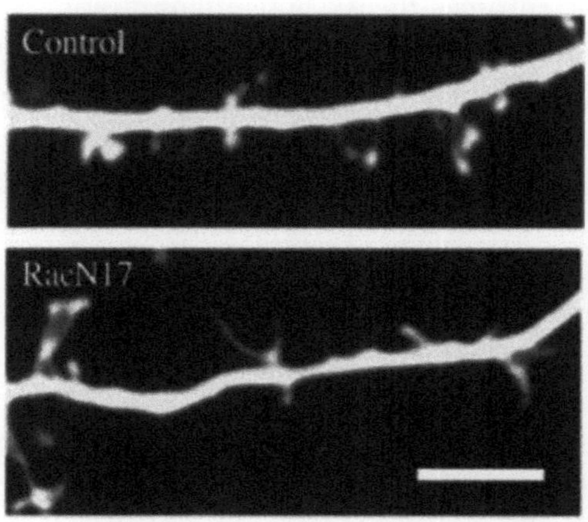

Figure 6 Dendritic spines in control or RacN17 (dominant negative)-transfected neurons. The neurons were transfected with GFP for visualization. Reduced Rac1 function disrupts spine formation and stabilization. *Adapted from Ref. 121.*

protein, synaptophysin.[81,128] This result suggests that NSC23766 administration *in vivo* may have preferential effects on postsynaptic elements and function. A treatment regimen over a 3-day period with NSC23766 restored near-normal dendritic spine profiles on WDR neurons 1 month after SCI; this included a reduction of injury-induced changes in spine length and spine head size, as well as a reduction in spine density. NSC23766 treatment returned mechanical and heat pain thresholds closer to values comparable with uninjured controls. *In vivo* electrophysiological recordings demonstrated that NSC23766 treatment reduces SCI-induced WDR hyperexcitability.[81] Taken together, these data demonstrate that Rac1-regulated dendritic spine remodeling contributes to neuropathic pain after SCI, and suggest that targeting the Rac1 pathway may have utility in other conditions of neuropathic pain.

In support of this notion, follow-up studies demonstrated that intrathecal administration of NSC23766 after peripheral nerve injury also reduces dendritic spine density on WDR neurons and returns the distribution of mature (i.e., mushroom-shaped) spines closer to those of uninjured controls.[85] Pharmacological disruption of nerve injury-induced abnormal spine profiles decreased WDR neuron hyperexcitability and restored mechanical nociceptive thresholds. However, it was noted in this study that NSC23766 treatment failed to restore heat pain thresholds, which revealed that Rac1 inhibition has a disparate action on mechanical and thermal pain modalities in peripheral nerve injury. As an explanation for the disparity in drug effect, it was argued that the geometry of mushroom-shaped spines can preferentially enhance the synaptic transmission of high-frequency tactile input.[73,129] Therefore, the reduction of mushroom spine morphology by NSC23766 treatment would mitigate tactile allodynia. Thermal nociception would continue to transmit with more fidelity, because thermal inputs travel at lower frequency along unmyelinated or thinly myelinated C- or Aδ-afferents.[130] The presence or absence of aberrant dendritic spines on different spinal cord neurons could further explain discrepancies in neuropathic pain responses. Additionally, it is worth noting that alternative mechanisms that underlie thermal pain transmission, e.g., TRPV1-expressing neurons,[131,132] may not be affected by pharmacological Rac1 inhibition. It is not known whether dendritic spine profiles change on other neuron cell types in the spinal cord, which may include low-threshold mechanoreceptive, thermal-, or nociceptive-specific neurons, e.g., neurons in lamina II.

Previous work has suggested that inhibitors of RhoA/Rho kinases are promising candidates for the treatment of painful diabetic neuropathies, as demonstrated by experiments in STZ-induced diabetic mice.[133,134] In a model of STZ-induced diabetic rats, administration of the Rac1-inhibitor NSC23766 can also alleviate neuropathic tactile allodynia and mitigate the hyperexcitability of WDR neurons. Treatment with NSC23766 restores close-to-normal dendritic spine profiles in association with reduced behavioral and electrophysiological signs of neuropathic allodynia.[106] Similarly, in a model of burn injury, NSC23766 treatment also attenuated dendritic spine dysgenesis, which was accompanied by decreased mechanical allodynia and reduced the hyperexcitability of WDR neurons.[119] In this study, however, NSC23766 disruption of dendritic spines only partially restored mechanical pain thresholds, which suggests that other factors contribute to a pain phenotype after burn injury.[111,113,119]

The mode of action of Rac1 inhibition in neuropathic pain is not firmly understood. NSC23766 treatment in SCI animals alleviates neuropathic pain without significantly affecting sensory–motor coordination,[81] which suggests that NSC23766 treatments do not appear to have a detectable behavioral effect on the spinal cord motor system, at least with the utilized drug dosages. After injury or disease insults to the CNS, activated microglia can contribute to the initiation and maintenance of neuropathic pain.[100] Rac1 activity regulates microglial lamellipodia formation and membrane ruffling.[135] After peripheral nerve injury, increased immunoreactivity for microglia in the dorsal horn was not reduced by NSC23766 treatment at dosages that were effective for reducing neuropathic pain.[85] Thus, it currently seems that the analgesic effect of Rac1 inhibition is due to an action that does not reduce microglial activity in the spinal cord. Furthermore, NSC23766 treatment in sham animals did not affect dendritic spine morphologies and did not change electrical or behavioral outcomes compared with untreated sham animals. This demonstrated that although NSC23766 treatment can potentially affect dendritic spine structure in normal animals, its action does significantly not affect normal function. NSC23766 is capable of directly disrupting Rac1 regulation of dendritic spine morphology when administered to isolated neurons *in vitro*[85,127,136]; but systemic administration of the drug *in vivo* is likely to have off-target effects, a limitation of many conventional pharmacological agents, which can often lead to adverse side effects. The emerging availability of genomic tools, e.g., viral-mediated gene manipulation or transgenic mice, offers promising opportunities to explore the molecular underpinnings of dendritic spine remodeling in neuropathic pain and other neuropathology.

11. SPINAL MEMORY MECHANISM IN NEUROPATHIC PAIN

Emerging evidence has begun to reveal a chain of events leading from injury or disease to the activation of a putative spinal memory mechanism involving dendritic spine reorganization (Fig. 7). According to this model, mechanical injury such as SCI or acutely injured nerves leads to chronic transmission of uncontrolled and spontaneous electrical activity.[137] Inflammatory cells, e.g., microglia, release their internal stores of glutamate, which

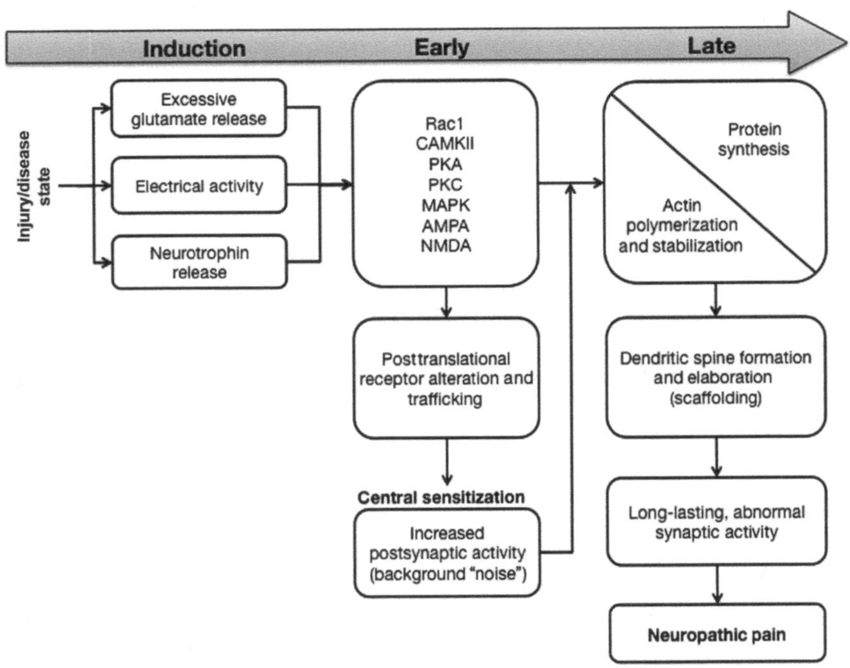

Figure 7 Dendritic spine dysgenesis may involve an array of events and molecular pathways that engage a putative spinal memory mechanism in the spinal cord. Because of the strong correlation between spine structure and synaptic function, maladaptive dendritic spine remodeling could directly contribute to long-lasting, abnormal synaptic activity accompanying neuropathic pain states after injury or disease. *Adapted from Ref. 85.*

can bind and activate AMPA and NMDA receptors, further leading to neuronal depolarization and excessive calcium signaling. As noted earlier, neurotrophins in the environment such as BDNF and NT-3 can activate intracellular signaling pathways that can adversely disrupt neuronal function.[138] NT-3 is known to contribute to LTP-like synaptic enhancement of transmission in the spinal cord.[139] BDNF is necessary for transcription-dependent, late-phase LTP.[140] Activation of a host of other molecular pathways, e.g., Rac1, CAMKII, PKC, and PKA, may further contribute to stimulating glutamate signaling. Evidence has also shown that increased expression of NMDA receptor subunits NR1 and NR2, and AMPA receptor subunit, GluR1, can amplify excitatory glutamate neurotransmission. Collectively, these spinal cord molecular events converge upon pathways that are similar to synaptic-based models of learning and memory in the cortex. Central sensitization is produced by chronic potentiation of spinal nociceptive pathways. Empirically, this is evident by hyperresponsiveness to stimulation and increased background "noise."[57] The failure of neuropathic pain to abate even years after trauma suggests that adverse changes to synaptic function must exist in a chronic pathological state in nociceptive pathways. Both protein synthesis and actin reorganization are required for synaptic reorganization and stabilization. As a result, dendritic spine formation and maturation contribute to the long-term maintenance of neuropathic pain.[38,43]

12. FINAL PERSPECTIVES

The mechanistic link between dendritic spine remodeling and neuronal function in the spinal cord has potential ramifications beyond chronic pain phenomena. Spasticity and hyperreflexia are common after SCI, multiple sclerosis, and other diseases. While altered cation channel expression[141] and disinhibition[72] increase the excitability of spinal reflex circuits, a spinal memory mechanism may also be a factor in dysfunctional motor control.[142,143] At a conceptual level, a spinal memory mechanism that engages dendritic spine remodeling could contribute to a broad range of intractable neurological conditions. Taken together, the observation of dendritic spine remodeling in the spinal cord after injury or disease provides a novel avenue for investigating chronic disease (e.g., pain), and molecules implicated in cortical memory processes involving dendritic spines are promising targets for more effective and durable therapies.

REFERENCES

1. Garcia-Lopez P, Garcia-Marin V, Freire M. The discovery of dendritic spines by Cajal in 1888 and its relevance in the present neuroscience. *Prog Neurobiol.* 2007;83:110–130.
2. Ramon y Cajal S. Estructura de los centros nerviosos de las aves. *Rev Trimest Histol Normal Patol.* 1888;1:1–10.
3. Ramon y Cajal S. Las espinas colaterales de las células del cerebro teñidas por el azul de metileno. *Rev Trimest Micrograf Madrid.* 1896;1:123–136.
4. Ramon y Cajal S. *La Textura del Sistema Nerviosa del Hombre y los Vertebrados.* 1st ed. Madrid: Moya; 1899, re-edited.
5. Yuste R, Bonhoeffer T. Morphological changes in dendritic spines associated with long-term synaptic plasticity. *Annu Rev Neurosci.* 2001;24:1071–1089.
6. Calabrese B, Wilson MS, Halpain S. Development and regulation of dendritic spine synapses. *Physiology (Bethesda).* 2006;21:38–47.
7. Gazzaley A, Kay S, Benson DL. Dendritic spine plasticity in hippocampus. *Neuroscience.* 2002;111:853–862.
8. Strata P, Morando L, Bravin M, Rossi F. Dendritic spine density in Purkinje cells. *Trends Neurosci.* 2000;23:198.
9. Tan AM, Waxman SG. Spinal cord injury, dendritic spine remodeling, and spinal memory mechanisms. *Exp Neurol.* 2012;235:142–151.
10. Thomas CC, Combe CL, Dyar KA, Inglis FM. Modest alterations in patterns of motor neuron dendrite morphology in the Fmr1 knockout mouse model for fragile X. *Int J Dev Neurosci.* 2008;26:805–811.
11. Wolpaw JR. Adaptive plasticity in the spinal stretch reflex: an accessible substrate of memory? *Cell Mol Neurobiol.* 1985;5:147–165.
12. Kirov SA, Sorra KE, Harris KM. Slices have more synapses than perfusion-fixed hippocampus from both young and mature rats. *J Neurosci.* 1999;19:2876–2886.
13. Harris KM, Kater SB. Dendritic spines: cellular specializations imparting both stability and flexibility to synaptic function. *Annu Rev Neurosci.* 1994;17:341–371.
14. Feldman ML, Peters A. A technique for estimating total spine numbers on Golgi-impregnated dendrites. *J Comp Neurol.* 1979;188:527–542.
15. Koh IY, Lindquist WB, Zito K, Nimchinsky EA, Svoboda K. An image analysis algorithm for dendritic spines. *Neural Comput.* 2002;14:1283–1310.
16. Alvarez VA, Sabatini BL. Anatomical and physiological plasticity of dendritic spines. *Annu Rev Neurosci.* 2007;30:79–97.
17. Bonhoeffer T, Yuste R. Spine motility. Phenomenology, mechanisms, and function. *Neuron.* 2002;35:1019–1027.
18. Malenka RC. The long-term potential of LTP. *Nat Rev Neurosci.* 2003;4:923–926.
19. Frey S, Frey JU. 'Synaptic tagging' and 'cross-tagging' and related associative reinforcement processes of functional plasticity as the cellular basis for memory formation. *Prog Brain Res.* 2008;169:117–143.
20. Collin C, Miyaguchi K, Segal M. Dendritic spine density and LTP induction in cultured hippocampal slices. *J Neurophysiol.* 1997;77:1614–1623.
21. Emptage N, Bliss TV, Fine A. Single synaptic events evoke NMDA receptor-mediated release of calcium from internal stores in hippocampal dendritic spines. *Neuron.* 1999;22:115–124.
22. Korkotian E, Segal M. Morphological constraints on calcium dependent glutamate receptor trafficking into individual dendritic spine. *Cell Calcium.* 2007;42:41–57.
23. Ackermann M, Matus A. Activity-induced targeting of profilin and stabilization of dendritic spine morphology. *Nat Neurosci.* 2003;6:1194–1200.
24. Bramham CR. Local protein synthesis, actin dynamics, and LTP consolidation. *Curr Opin Neurobiol.* 2008;18:524–531.

25. Impey S, Obrietan K, Wong ST, et al. Cross talk between ERK and PKA is required for Ca^{2+} stimulation of CREB-dependent transcription and ERK nuclear translocation. *Neuron*. 1998;21:869–883.
26. Matsuzaki M, Honkura N, Ellis-Davies GC, Kasai H. Structural basis of long-term potentiation in single dendritic spines. *Nature*. 2004;429:761–766.
27. Tanaka J, Horiike Y, Matsuzaki M, Miyazaki T, Ellis-Davies GC, Kasai H. Protein synthesis and neurotrophin-dependent structural plasticity of single dendritic spines. *Science*. 2008;319:1683–1687.
28. Fischer M, Kaech S, Knutti D, Matus A. Rapid actin-based plasticity in dendritic spines. *Neuron*. 1998;20:847–854.
29. Lu Y, Christian K, Lu B. BDNF: a key regulator for protein synthesis-dependent LTP and long-term memory? *Neurobiol Learn Mem*. 2008;89:312–323.
30. Appleby VJ, Correa SA, Duckworth JK, et al. LTP in hippocampal neurons is associated with a CaMKII-mediated increase in GluA1 surface expression. *J Neurochem*. 2011;116:530–543.
31. Otmakhov N, Tao-Cheng JH, Carpenter S, et al. Persistent accumulation of calcium/calmodulin-dependent protein kinase II in dendritic spines after induction of NMDA receptor-dependent chemical long-term potentiation. *J Neurosci*. 2004;24:9324–9331.
32. Nevian T, Sakmann B. Spine Ca^{2+} signaling in spike-timing-dependent plasticity. *J Neurosci*. 2006;26:11001–11013.
33. Wickens J. Electrically coupled but chemically isolated synapses: dendritic spines and calcium in a rule for synaptic modification. *Prog Neurobiol*. 1988;31:507–528.
34. Fagnou DD, Tuchek JM. The biochemistry of learning and memory. *Mol Cell Biochem*. 1995;149–150:279–286.
35. Lee SJ, Escobedo-Lozoya Y, Szatmari EM, Yasuda R. Activation of CaMKII in single dendritic spines during long-term potentiation. *Nature*. 2009;458:299–304.
36. Lee SJ, Yasuda R. Spatiotemporal regulation of signaling in and out of dendritic spines: CaMKII and Ras. *Open Neurosci J*. 2009;3:117–127.
37. Fifkova E. A possible mechanism of morphometric changes in dendritic spines induced by stimulation. *Cell Mol Neurobiol*. 1985;5:47–63.
38. Bourne J, Harris KM. Do thin spines learn to be mushroom spines that remember? *Curr Opin Neurobiol*. 2007;17:381–386.
39. Harris KM, Jensen FE, Tsao B. Three-dimensional structure of dendritic spines and synapses in rat hippocampus (CA1) at postnatal day 15 and adult ages: implications for the maturation of synaptic physiology and long-term potentiation. *J Neurosci*. 1992;12:2685–2705.
40. Yuste R, Majewska A, Holthoff K. From form to function: calcium compartmentalization in dendritic spines. *Nat Neurosci*. 2000;3:653–659.
41. Araya R, Eisenthal KB, Yuste R. Dendritic spines linearize the summation of excitatory potentials. *Proc Natl Acad Sci USA*. 2006;103:18799–18804.
42. Majewska A, Brown E, Ross J, Yuste R. Mechanisms of calcium decay kinetics in hippocampal spines: role of spine calcium pumps and calcium diffusion through the spine neck in biochemical compartmentalization. *J Neurosci*. 2000;20:1722–1734.
43. Majewska A, Tashiro A, Yuste R. Regulation of spine calcium dynamics by rapid spine motility. *J Neurosci*. 2000;20:8262–8268.
44. Carter AG, Soler-Llavina GJ, Sabatini BL. Timing and location of synaptic inputs determine modes of subthreshold integration in striatal medium spiny neurons. *J Neurosci*. 2007;27:8967–8977.
45. Jaslove SW. The integrative properties of spiny distal dendrites. *Neuroscience*. 1992;47:495–519.
46. Koester HJ, Sakmann B. Calcium dynamics in single spines during coincident pre- and postsynaptic activity depend on relative timing of back-propagating action potentials

and subthreshold excitatory postsynaptic potentials. *Proc Natl Acad Sci USA*. 1998;95:9596–9601.
47. Martiel JL, Mouchet P, Boissier MD. Modeling the integrative properties of dendrites: application to the striatal spiny neuron. *Synapse*. 1994;16:269–279.
48. Yuste R, Urban R. Dendritic spines and linear networks. *J Physiol Paris*. 2004;98:479–486.
49. Halpain S, Spencer K, Graber S. Dynamics and pathology of dendritic spines. *Prog Brain Res*. 2005;147:29–37.
50. Bardoni B, Schenck A, Mandel JL. The fragile X mental retardation protein. *Brain Res Bull*. 2001;56:375–382.
51. Carlisle HJ, Kennedy MB. Spine architecture and synaptic plasticity. *Trends Neurosci*. 2005;28:182–187.
52. Chechlacz M, Gleeson JG. Is mental retardation a defect of synapse structure and function? *Pediatr Neurol*. 2003;29:11–17.
53. Huber KM. The fragile X-cerebellum connection. *Trends Neurosci*. 2006;29:183–185.
54. Hung AY, Futai K, Sala C, et al. Smaller dendritic spines, weaker synaptic transmission, but enhanced spatial learning in mice lacking Shank1. *J Neurosci*. 2008;28:1697–1708.
55. Wilson BM, Cox CL. Absence of metabotropic glutamate receptor-mediated plasticity in the neocortex of fragile X mice. *Proc Natl Acad Sci USA*. 2007;104:2454–2459.
56. Chen LY, Rex CS, Babayan AH, et al. Physiological activation of synaptic Rac>PAK (p-21 activated kinase) signaling is defective in a mouse model of fragile X syndrome. *J Neurosci*. 2010;30:10977–10984.
57. Ji RR, Woolf CJ. Neuronal plasticity and signal transduction in nociceptive neurons: implications for the initiation and maintenance of pathological pain. *Neurobiol Dis*. 2001;8:1–10.
58. Ikeda H, Asai T, Murase K. Robust changes of afferent-induced excitation in the rat spinal dorsal horn after conditioning high-frequency stimulation. *J Neurophysiol*. 2000;83:2412–2420.
59. Sandkuhler J. Understanding LTP in pain pathways. *Mol Pain*. 2007;3:9.
60. Svendsen F, Rygh LJ, Hole K, Tjolsen A. Dorsal horn NMDA receptor function is changed after peripheral inflammation. *Pain*. 1999;83:517–523.
61. Chen Z, Muscoli C, Doyle T, et al. NMDA-receptor activation and nitroxidative regulation of the glutamatergic pathway during nociceptive processing. *Pain*. 2010;149:100–106.
62. Mendell LM, Albers KM, Davis BM. Neurotrophins, nociceptors, and pain. *Microsc Res Tech*. 1999;45:252–261.
63. Zhou LJ, Zhong Y, Ren WJ, Li YY, Zhang T, Liu XG. BDNF induces late-phase LTP of C-fiber evoked field potentials in rat spinal dorsal horn. *Exp Neurol*. 2008;212:507–514.
64. Ikeda H, Stark J, Fischer H, et al. Synaptic amplifier of inflammatory pain in the spinal dorsal horn. *Science*. 2006;312:1659–1662.
65. Besson JM. The neurobiology of pain. *Lancet*. 1999;353:1610–1615.
66. Boulton A. What causes neuropathic pain? *J Diabetes Complications*. 1992;6:58–63.
67. Craner MJ, Klein JP, Black JA, Waxman SG. Preferential expression of IGF-I in small DRG neurons and down-regulation following injury. *Neuroreport*. 2002; 13:1649–1652.
68. Craner MJ, Klein JP, Renganathan M, Black JA, Waxman SG. Changes of sodium channel expression in experimental painful diabetic neuropathy. *Ann Neurol*. 2002;52:786–792.
69. Dib-Hajj SD, Cummins TR, Black JA, Waxman SG. From genes to pain: Na v 1.7 and human pain disorders. *Trends Neurosci*. 2007;30:555–563.

70. Drew GM, Siddall PJ, Duggan AW. Mechanical allodynia following contusion injury of the rat spinal cord is associated with loss of GABAergic inhibition in the dorsal horn. *Pain.* 2004;109:379–388.
71. Hasbargen T, Ahmed MM, Miranpuri G, et al. Role of NKCC1 and KCC2 in the development of chronic neuropathic pain following spinal cord injury. *Ann N Y Acad Sci.* 2010;1198:168–172.
72. Price TJ, Cervero F, Gold MS, Hammond DL, Prescott SA. Chloride regulation in the pain pathway. *Brain Res Rev.* 2009;60:149–170.
73. Tan AM, Choi JS, Waxman SG, Hains BC. Dendritic spine remodeling after spinal cord injury alters neuronal signal processing. *J Neurophysiol.* 2009;102:2396–2409.
74. Segev I, Rall W. Excitable dendrites and spines: earlier theoretical insights elucidate recent direct observations. *Trends Neurosci.* 1998;21:453–460.
75. Costigan M, Scholz J, Woolf CJ. Neuropathic pain: a maladaptive response of the nervous system to damage. *Annu Rev Neurosci.* 2009;32:1–32.
76. Chetkovich DM, Bunn RC, Kuo SH, Kawasaki Y, Kohwi M, Bredt DS. Postsynaptic targeting of alternative postsynaptic density-95 isoforms by distinct mechanisms. *J Neurosci.* 2002;22:6415–6425.
77. El-Husseini AE, Schnell E, Chetkovich DM, Nicoll RA, Bredt DS. PSD-95 involvement in maturation of excitatory synapses. *Science.* 2000;290:1364–1368.
78. Sharma K, Fong DK, Craig AM. Postsynaptic protein mobility in dendritic spines: long-term regulation by synaptic NMDA receptor activation. *Mol Cell Neurosci.* 2006;31:702–712.
79. Kim BG, Dai HN, McAtee M, Vicini S, Bregman BS. Remodeling of synaptic structures in the motor cortex following spinal cord injury. *Exp Neurol.* 2006;198:401–415.
80. Kim E, Ko J. Molecular organization and assembly of the postsynaptic density of excitatory brain synapses. *Results Probl Cell Differ.* 2006;43:1–23.
81. Tan AM, Stamboulian S, Chang YW, et al. Neuropathic pain memory is maintained by Rac1-regulated dendritic spine remodeling after spinal cord injury. *J Neurosci.* 2008;28:13173–13183.
82. Mendell LM. Physiological properties of unmyelinated fiber projection to the spinal cord. *Exp Neurol.* 1966;16:316–332.
83. Woolf CJ, King AE. Physiology and morphology of multireceptive neurons with C-afferent fiber inputs in the deep dorsal horn of the rat lumbar spinal cord. *J Neurophysiol.* 1987;58:460–479.
84. Yuste R, Bonhoeffer T. Genesis of dendritic spines: insights from ultrastructural and imaging studies. *Nat Rev Neurosci.* 2004;5:24–34.
85. Tan AM, Chang YW, Zhao P, Hains BC, Waxman SG. Rac1-regulated dendritic spine remodeling contributes to neuropathic pain after peripheral nerve injury. *Exp Neurol.* 2011;232:222–233.
86. Woolf CJ, Salter MW. Neuronal plasticity: increasing the gain in pain. *Science.* 2000;288:1765–1769.
87. Woolf CJ, Shortland P, Coggeshall RE. Peripheral nerve injury triggers central sprouting of myelinated afferents. *Nature.* 1992;355:75–78.
88. Comery TA, Harris JB, Willems PJ, et al. Abnormal dendritic spines in fragile X knockout mice: maturation and pruning deficits. *Proc Natl Acad Sci USA.* 1997;94:5401–5404.
89. Mineur YS, Sluyter F, de Wit S, Oostra BA, Crusio WE. Behavioral and neuroanatomical characterization of the Fmr1 knockout mouse. *Hippocampus.* 2002;12:39–46.
90. Price TJ, Rashid MH, Millecamps M, Sanoja R, Entrena JM, Cervero F. Decreased nociceptive sensitization in mice lacking the fragile X mental retardation protein: role of mGluR1/5 and mTOR. *J Neurosci.* 2007;27:13958–13967.

91. Saab CY, Hains BC. Remote neuroimmune signaling: a long-range mechanism of nociceptive network plasticity. *Trends Neurosci.* 2009;32:110–117.
92. Saab CY, Waxman SG, Hains BC. Alarm or curse? The pain of neuroinflammation. *Brain Res Rev.* 2008;58:226–235.
93. Hains BC, Saab CY, Klein JP, Craner MJ, Waxman SG. Altered sodium channel expression in second-order spinal sensory neurons contributes to pain after peripheral nerve injury. *J Neurosci.* 2004;24:4832–4839.
94. Tan AM, Zhao P, Waxman SG, Hains BC. Early microglial inhibition preemptively mitigates chronic pain development after experimental spinal cord injury. *J Rehabil Res Dev.* 2009;46:123–133.
95. Tremblay ME, Lowery RL, Majewska AK. Microglial interactions with synapses are modulated by visual experience. *PLoS Biol.* 2010;8:e1000527.
96. Tremblay ME, Majewska AK. A role for microglia in synaptic plasticity? *Commun Integr Biol.* 2011;4:220–222.
97. Haber M, Zhou L, Murai KK. Cooperative astrocyte and dendritic spine dynamics at hippocampal excitatory synapses. *J Neurosci.* 2006;26:8881–8891.
98. Klein R. Bidirectional modulation of synaptic functions by Eph/ephrin signaling. *Nat Neurosci.* 2009;12:15–20.
99. Nishida H, Okabe S. Direct astrocytic contacts regulate local maturation of dendritic spines. *J Neurosci.* 2007;27:331–340.
100. Scholz J, Woolf CJ. The neuropathic pain triad: neurons, immune cells and glia. *Nat Neurosci.* 2007;10:1361–1368.
101. Zhao P, Waxman SG, Hains BC. Modulation of thalamic nociceptive processing after spinal cord injury through remote activation of thalamic microglia by cysteine cysteine chemokine ligand 21. *J Neurosci.* 2007;27:8893–8902.
102. Beisswenger PJ. Neuropathic and vascular complications occurring in diabetes. *Postgrad Med.* 1976;59:169–174.
103. Morgado C, Pereira-Terra P, Cruz CD, Tavares I. Minocycline completely reverses mechanical hyperalgesia in diabetic rats through microglia-induced changes in the expression of the potassium chloride co-transporter 2 (KCC2) at the spinal cord. *Diabetes Obes Metab.* 2011;13:150–159.
104. Malone JI, Hanna S, Saporta S, et al. Hyperglycemia not hypoglycemia alters neuronal dendrites and impairs spatial memory. *Pediatr Diabetes.* 2008;9:531–539.
105. Yamano T, Shimada M, Fujizeki Y, Kawasaki H, Onaga A. Quantitative synaptic changes on Purkinje cell dendritic spines of rats born from streptozotocin-induced diabetic mothers. *Brain Dev.* 1986;8:269–273.
106. Tan AM, Samad OA, Fischer TZ, Zhao P, Persson AK, Waxman SG. Maladaptive dendritic spine remodeling contributes to diabetic neuropathic pain. *J Neurosci.* 2012;32:6795–6807.
107. Morrow TJ. Animal models of painful diabetic neuropathy: the STZ rat model. *Curr Protoc Neurosci.* 2004;Chapter 9:Unit 9.18.
108. Joghataie MT, Roghani M, Jalali MR, Baluchnejadmojarad T, Sharayeli M. Dendritic spine changes in medial prefrontal cortex of male diabetic rats using Golgi-impregnation method. *Arch Iran Med.* 2007;10:54–58.
109. Fischer TZ, Tan AM, Waxman SG. Thalamic neuron hyperexcitability and enlarged receptive fields in the STZ model of diabetic pain. *Brain Res.* 2009;1268:154–161.
110. Fischer TZ, Waxman SG. Neuropathic pain in diabetes—evidence for a central mechanism. *Nat Rev Neurol.* 2010;6:462–466.
111. Chang YW, Tan A, Saab C, Waxman S. Unilateral focal burn injury is followed by long-lasting bilateral allodynia and neuronal hyperexcitability in spinal cord dorsal horn. *J Pain.* 2010;11:119–130.
112. Fischer TZ, Waxman SG. Extraterritorial temperature pain threshold abnormalities in subjects with healed thermal injury. *J Rehabil Res Dev.* 2012;49:515–522.

113. Chang YW, Waxman SG. Minocycline attenuates mechanical allodynia and central sensitization following peripheral second-degree burn injury. *J Pain.* 2010;11:1146–1154.
114. Tal M, Bennett GJ. Extra-territorial pain in rats with a peripheral mononeuropathy: mechano-hyperalgesia and mechano-allodynia in the territory of an uninjured nerve. *Pain.* 1994;57:375–382.
115. Atchison NE, Osgood PF, Carr DB, Szyfelbein SK. Pain during burn dressing change in children: relationship to burn area, depth and analgesic regimens. *Pain.* 1991;47:41–45.
116. Choiniere M, Melzack R, Papillon J. Pain and paresthesia in patients with healed burns: an exploratory study. *J Pain Symptom Manage.* 1991;6:437–444.
117. Hill CE, Harrison BJ, Rau KK, et al. Skin incision induces expression of axonal regeneration-related genes in adult rat spinal sensory neurons. *J Pain.* 2010; 11:1066–1073.
118. Stewart LS, Martin WJ. Influence of postoperative analgesics on the development of neuropathic pain in rats. *Comp Med.* 2003;53:29–36.
119. Tan AM, Samad OA, Liu S, Bandaru S, Zhao P, Waxman SG. Burn injury-induced mechanical allodynia is maintained by Rac1-regulated dendritic spine dysgenesis. *Exp Neurol.* 2013;248:509–519.
120. Latarjet J, Choinere M. Pain in burn patients. *Burns.* 1995;21:344–348.
121. Tashiro A, Yuste R. Regulation of dendritic spine motility and stability by Rac1 and Rho kinase: evidence for two forms of spine motility. *Mol Cell Neurosci.* 2004;26:429–440.
122. Tashiro A, Yuste R. Role of Rho GTPases in the morphogenesis and motility of dendritic spines. *Methods Enzymol.* 2008;439:285–302.
123. Ridley AJ. Rho GTPases and actin dynamics in membrane protrusions and vesicle trafficking. *Trends Cell Biol.* 2006;16:522–529.
124. Nakayama AY, Luo L. Intracellular signaling pathways that regulate dendritic spine morphogenesis. *Hippocampus.* 2000;10:582–586.
125. Dubreuil CI, Winton MJ, McKerracher L. Rho activation patterns after spinal cord injury and the role of activated Rho in apoptosis in the central nervous system. *J Cell Biol.* 2003;162:233–243.
126. Erschbamer MK, Hofstetter CP, Olson L. RhoA, RhoB, RhoC, Rac1, Cdc42, and Tc10 mRNA levels in spinal cord, sensory ganglia, and corticospinal tract neurons and long-lasting specific changes following spinal cord injury. *J Comp Neurol.* 2005;484:224–233.
127. Gao Y, Dickerson JB, Guo F, Zheng J, Zheng Y. Rational design and characterization of a Rac GTPase-specific small molecule inhibitor. *Proc Natl Acad Sci USA.* 2004;101:7618–7623.
128. Masliah E, Fagan AM, Terry RD, DeTeresa R, Mallory M, Gage FH. Reactive synaptogenesis assessed by synaptophysin immunoreactivity is associated with GAP-43 in the dentate gyrus of the adult rat. *Exp Neurol.* 1991;113:131–142.
129. Timofeeva Y, Lord GJ, Coombes S. Spatio-temporal filtering properties of a dendritic cable with active spines: a modeling study in the spike-diffuse-spike framework. *J Comput Neurosci.* 2006;21:293–306.
130. Ossipov MH, Bian D, Malan Jr TP, Lai J, Porreca F. Lack of involvement of capsaicin-sensitive primary afferents in nerve-ligation injury induced tactile allodynia in rats. *Pain.* 1999;79:127–133.
131. Pabbidi RM, Cao DS, Parihar A, Pauza ME, Premkumar LS. Direct role of streptozotocin in inducing thermal hyperalgesia by enhanced expression of transient receptor potential vanilloid 1 in sensory neurons. *Mol Pharmacol.* 2008;73:995–1004.
132. Wilson-Gerwing TD, Dmyterko MV, Zochodne DW, Johnston JM, Verge VM. Neurotrophin-3 suppresses thermal hyperalgesia associated with neuropathic pain

and attenuates transient receptor potential vanilloid receptor-1 expression in adult sensory neurons. *J Neurosci.* 2005;25:758–767.
133. Ohsawa M, Aasato M, Hayashi SS, Kamei J. RhoA/Rho kinase pathway contributes to the pathogenesis of thermal hyperalgesia in diabetic mice. *Pain.* 2011;152:114–122.
134. Ohsawa M, Kamei J. RhoA/Rho kinase signaling in the spinal cord and diabetic painful neuropathy. *Eur J Pharmacol.* 2010;644:1–4.
135. Ohsawa K, Imai Y, Kanazawa H, Sasaki Y, Kohsaka S. Involvement of Iba1 in membrane ruffling and phagocytosis of macrophages/microglia. *J Cell Sci.* 2000;113(Pt 17):3073–3084.
136. Akbar H, Cancelas J, Williams DA, Zheng J, Zheng Y. Rational design and applications of a Rac GTPase-specific small molecule inhibitor. *Methods Enzymol.* 2006; 406:554–565.
137. Hulsebosch CE. Recent advances in pathophysiology and treatment of spinal cord injury. *Adv Physiol Educ.* 2002;26:238–255.
138. Ramer MS. Endogenous neurotrophins and plasticity following spinal deafferentation. *Exp Neurol.* 2010;235:70–77.
139. Arvanov VL, Seebach BS, Mendell LM. NT-3 evokes an LTP-like facilitation of AMPA/kainate receptor-mediated synaptic transmission in the neonatal rat spinal cord. *J Neurophysiol.* 2000;84:752–758.
140. Messaoudi E, Ying SW, Kanhema T, Croll SD, Bramham CR. Brain-derived neurotrophic factor triggers transcription-dependent, late phase long-term potentiation in vivo. *J Neurosci.* 2002;22:7453–7461.
141. Vinay L, Jean-Xavier C. Plasticity of spinal cord locomotor networks and contribution of cation-chloride cotransporters. *Brain Res Rev.* 2008;57:103–110.
142. Chen Y, Chen XY, Jakeman LB, Chen L, Stokes BT, Wolpaw JR. Operant conditioning of H-reflex can correct a locomotor abnormality after spinal cord injury in rats. *J Neurosci.* 2006;26:12537–12543.
143. Wolpaw JR. The complex structure of a simple memory. *Trends Neurosci.* 1997;20:588–594.

CHAPTER FOURTEEN

Commonalities Between Pain and Memory Mechanisms and Their Meaning for Understanding Chronic Pain

Theodore J. Price[1], Kufreobong E. Inyang
Behavioral and Brain Sciences, The University of Texas at Dallas, Richardson, Texas, USA
[1]Corresponding author: e-mail address: theodore.price@utdallas.edu

Contents

1. Introduction: Pain Plasticity and "Pain Memory" 410
2. The Adaptive Nature of Pain Plasticity 412
3. The Peripheral Nociceptor as the Locus of "Pain Memory," Focus on Translation Control 413
   3.1 Local translation is a key mediator of nociceptor priming 416
4. "Pain Memory" in the Spinal Dorsal Horn 420
   4.1 LTP in the spinal dorsal horn, a neurophysiological correlate of pain memory? 420
   4.2 Atypical PKCs and BDNF 422
   4.3 Reconsolidation of pain memory 425
5. Clinical Implications of Pain Memory 427
6. Conclusions 428
Acknowledgments 428
References 428

Abstract

Pain sensing neurons in the periphery (called nociceptors) and the central neurons that receive their projections show remarkable plasticity following injury. This plasticity results in amplification of pain signaling that is now understood to be crucial for the recovery and survival of organisms following injury. These same plasticity mechanisms may drive a transition to a nonadaptive chronic pain state if they fail to resolve following the termination of the healing process. Remarkable advances have been achieved in the past two decades in understanding the molecular mechanisms that underlie pain plasticity following injury. The mechanisms bear a striking resemblance to molecular mechanisms involved in learning and memory processes in other brain regions, including the hippocampus and cerebral cortex. Here those mechanisms, their commonalities and subtle differences, will be highlighted and their role in causing chronic pain will be discussed. Arising from these data is the striking argument that chronic pain is a disease of

Progress in Molecular Biology and Translational Science, Volume 131
ISSN 1877-1173
http://dx.doi.org/10.1016/bs.pmbts.2014.11.010

the nervous system, which distinguishes this phenomena from acute pain that is frequently a symptom alerting the organism to injury. This argument has important implications for the development of disease modifying therapeutics.

1. INTRODUCTION: PAIN PLASTICITY AND "PAIN MEMORY"

A core feature of all nervous systems is an ability to adapt to sensory information. This adaptive process is referred to as plasticity and the study of neuronal plasticity has led to some of the most exciting advances in modern biological research. The pain system, comprised of peripheral neurons responsible for detecting, damaging, or potentially damaging peripheral stimuli, called nociceptors, and the neurons of the CNS that receive direct or indirect inputs from these neurons, rapidly change upon injury. In almost all studied cases this adaptation results in an amplification of signaling.[1] This pain amplification is thought to underlie some important psychophysical aspects of pain such as an enhanced response to a normally noxious stimulus (hyperalgesia) and a noxious response to a normally innocuous stimulus (allodynia, Ref. 2). Plasticity may also lead to changes in nociceptors or other neurons in the pain pathway that cause them to fire action potentials without any direct stimulation (ectopic activity) or fire continuously following stimulation (after discharge) both of which likely contribute to what is commonly called spontaneous pain that is a common feature of chronic neuropathic pain.[3,4] While all of these states can exist acutely following an injury, they are also prominent features of chronic pain disorders.

On the most general level, plasticity in the pain system occurs at two locations, the primary afferent nociceptor[5] and at synapses receiving nociceptive input throughout the CNS.[6–8] Preclinical models of acute and chronic inflammatory pain as well as models of neuropathic pain have revealed a plethora of molecular targets that have advanced our understanding of how chronic pain develops as well as revealing important potential therapeutic intervention points. These experimental studies have also revealed a striking similarity in mechanisms underlying pain amplification and learning and memory in areas of the brain such as hippocampus and cerebral cortex.[6,9,10] These findings have given rise to the idea that a "pain memory" is encoded within the nervous system and that reversing this pain memory may be the key to terminate chronic pain disorders.[5,11,12] In other words, reversing plasticity in pain circuits may provide the opportunity to permanently alleviate chronic pain.

While the term "pain memory" has been used in a variety of forms for decades, the first specific uses in the scientific literature, to our knowledge, can be attributed to Ronald Melzack, one of the experimental pioneers widely credited with advancing pain science into the modern age of neuroscience. In 1979, Dennis and Melzack described a series of experiments where painful irritation of the rat paw prior to a denervation injury led to an exacerbation of neuropathic pain.[13] They hypothesized that this preirritation led to the generation of a "pain memory" that could not be repressed by descending pain modulation centers, due to the subsequent nerve injury, and therefore persisted unabated after the nerve injury. Subsequently, in 1990, Katz and Melzack presented this same term in the context of phantom limb pain,[14] pain arising in a limb that has been amputated. They postulated that this sort of pain occurs due to the "memory" of pain that was caused by damage to the limb that was subsequently amputated. Since many amputations occur due to injury to a limb that is irreparable, this could explain this common clinical finding in amputees. While this idea is now widely disputed in the pain field,[15] it remains an area of intense investigation. Very recently, clinical evidence was published indicating that this "pain memory" causing phantom limb pain is very likely to reside in the peripheral nervous system.[16] These investigators found that local nerve block applied to the dorsal root ganglion (DRG) innervating the limb with phantom pain led to an immediate resolution of pain in 31 out of 31 patients. Strikingly, this finding is directly paralleled by preclinical (a term used to describe animal model experiments in the neuroscience area) work from Coderre and Melzack in 1987, where they concluded that neuropathic "pain memory" almost certainly resides in the peripheral nervous system.[17]

Since the first experiments into "pain memory" in the Melzack lab at McGill University, many animal models of acute and chronic pain have been developed. For the purposes of the present chapter we will focus on a particular model developed in the late 1990s and early 2000s by Jon Levine and colleagues called "hyperalgesic priming" (for extensive review see Ref. 12). This model provides unique insight into plasticity in the nociceptive system because it allows for molecular dissection of pain states in two distinct phases. These models involve a priming stimulus, aimed at causing an acute sensitization of peripheral nociceptors and their central inputs. Next, the initial sensitization is allowed to resolve and a second, normally subthreshold, stimulus is delivered. This second stimulus, which has only a transient effect in naïve animals, leads to a prolonged state of nociceptive hypersensitivity that allows for investigation of molecular mechanisms that define the primed nociceptor and/or the primed nociceptive system. Here we will describe how models of hyperalgesic priming give unique insight into how acute pain states lead to reorganization of molecular machinery throughout the pain system rendering animals, and almost certainly humans, susceptible to prolonged pain states provoked by insults that would have little effect in unprimed individuals. This primed state, therefore, represents a kind of "pain memory." Hence, our goals in this chapter will be to highlight mechanisms underlying this "pain memory." This will include three major themes (1) molecular signaling in the peripheral nociceptor and their parallels to memory mechanisms (local translation), (2) mechanisms controlling plasticity of synapses in the spinal dorsal horn and their relation to memory mechanisms (long-term potentiation), and (3) signaling mechanisms in the spinal dorsal horn that parallel findings from the learning and memory area (atypical PKCs (aPKC) and brain derived neurotrophic factor (BDNF)).

2. THE ADAPTIVE NATURE OF PAIN PLASTICITY

Before delving into molecular mechanisms of pain plasticity and pain memory, it is useful to first consider the evolutionary relevance of nociception and nociceptive plasticity. First, nearly all organisms with nervous systems have sensory neurons that can be considered nociceptors. That is, they have sensory detectors that are capable of sensing damaging chemicals, temperatures, or tissue insult and whose action can lead to avoidance of real or potential damage to the organism. Investigators interested in nociceptor biology have employed model organisms such as Drosophila melanogaster (fruit flies), Caenorhabditis elegans (transparent nematodes), and Aplysia californica (sea hares) due to the ease of manipulation of their genome, their transparency (for imaging purposes), rapid life cycle, or their stereotypic behaviors and large, easily accessible neurons. Some of the most relevant work has been done using Aplysia. The gill withdrawal reflex in these animals has been studied in great detail and involves a simple circuit made up of a sensory neuron, an interneuron, and a motor neuron. Plasticity in this circuit has been widely studied as a simple model for learning and memory[18] and molecular mechanisms discovered in these neurons led to the awarding of the Nobel Prize for Physiology or Medicine to Eric Kandel in 2000.[19] Importantly, the sensory neuron that plays a critical role in this reflex shares many properties with nociceptors and has been extensively investigated as a model for nociceptive plasticity.[1,20,21] Studying the response of this neuron to injury led to the first demonstration of local translation contributing to the increased excitability of nociceptors after injury.[22] Hence, there is strong evidence that nociceptive plasticity is evolutionarily ancient and that the basic mechanisms of this plasticity are conserved across a broad variety of organisms.[1,23] This has profound implications for understanding the crucial nature of nociceptive plasticity for organisms to survive in their natural environments and also suggests that mechanisms of neuronal plasticity in general may have first evolved in the nociceptive system.

Does nociceptive plasticity provide an adaptive advantage to organisms? On the surface, the answer to this question seems obvious. Of course it does. It is paramount for animals to protect an injured area until the healing process has run its course. This idea, however, is difficult to demonstrate experimentally. In an elegant set of experiments, Crook and colleagues used squid to address this question directly.[24] When the squid arm is injured it causes nociceptive neurons that innervate the squid arm to become sensitized. The sensitization results in reduced thresholds for mechanical stimulation similar to observations of hyperalgesia in rodents and humans. Local anesthetic applied at the time of squid arm injury completely blocks nociceptor sensitization and amplified nociceptive behavioral responses. Hence, it is possible to experimentally injury squid and have two distinct outcomes: squid with injury and hyperalgesia and squid with identical injury but no hyperalgesia. When these squid are exposed to a natural predator a remarkable difference in behavior is observed. Squid with injury and hyperalgesia orient to the predator more quickly and are able to escape attack. On the other hand, squid with injury and no hyperalgesia fail to mount this response to the predator resulting in markedly increased predation. Hence, hyperalgesia, or pain amplification, holds a distinct survival advantage, placing pain plasticity in a new evolutionary light. Moreover, these findings highlight the evolutionarily ancient nature of plasticity in peripheral nociceptors as a mechanism to drive persistent pain.[24,25]

3. THE PERIPHERAL NOCICEPTOR AS THE LOCUS OF "PAIN MEMORY," FOCUS ON TRANSLATION CONTROL

In 1982 Steward and Levy published a landmark study describing the preferential localization of ribosomes at the base of dendritic spines.[26] While the potential importance of dendritic spines for nervous system function had been postulated since the time of Santiago Ramon y Cajal, this finding came less than a decade after the original description of LTP by Bliss and Lomo in 1973.[27] At the time of the work of Steward and Levy, the mechanisms underlying LTP were still poorly understood but their finding led to a striking conclusion, the localization of ribosomes at the base of dendritic spines places the translation machinery at the perfect location to control local changes that regulate postsynaptic plasticity. What has followed has been a veritable explosion of work on translation control and its contribution to synaptic plasticity and LTP.[28] While other aspects of LTP will be covered in more detail below, it is now known that LTP decays if local translation is blocked and signaling molecules that control activity-dependent plasticity are enriched in and around dendritic spines. Moreover, RNA binding proteins anchor mRNAs at dendritic spines and play critical roles in brain function, for instance, they are mutated in several neurodevelopmental disorders (e.g., fragile X mental retardation protein (FMRP) and fragile X syndrome, Ref. 29). Therefore, local, activity-dependent control of gene expression at the level of translation is at the very foundation of plasticity in the brain, including the most widely studied learning and memory mechanism, LTP.[30]

Why is local translation control so important for neuronal plasticity? Here it is important to remember the basic anatomy/organization of neurons in relation to gene expression demands and the correct sorting of protein localization given this complexity. Individual neurons can have tens of thousands of synapses and plasticity can occur at any of these individual synapses depending on afferent input to the neuron. This creates a difficult problem if changes in gene expression are required for the full expression of plasticity, as is now well accepted. If new proteins required for this plasticity were all contributed from the cell body, the neuron would need a mechanism in place to correctly sort all of these plasticity-related proteins to their correct location. A more parsimonious solution is for the neuron to traffic mRNAs to dendritic spines and hold them in a translationally dormant state until an appropriate signal is received (e.g., strong synaptic input). Changes in gene expression can then be achieved locally via activity-dependent translation. Overwhelming evidence, from a broad variety of brain regions, including the dorsal horn of the spinal cord, supports the preceding statements.[11,28,30–32]

But which mRNAs are trafficked to these distal sites in neurons? This has likewise been an area of intense investigation. Investigators have focused on finding mRNAs that associate with mRNA binding proteins, such as FMRP[33,34] but these experiments have been technically demanding and have led to disparate results depending on the technique. Another approach has been to use multichambered devices where neuronal extensions, be they dendrites or axons, can extend into an isolated chamber where these neuronal components can be isolated and mRNA can be extracted[35,36] (Fig. 1). While these approaches have identified important candidate mRNAs it has only been recently that bioinformatics approaches have led to insights into the species of mRNAs that are found at distal sites in neurons and whether these mRNAs differ from those that are preferentially translated in the neuronal soma.[37] These investigations found several distinguishing factors in the proteins encoded by these distally localized mRNAs, most prominently an enrichment of sites for posttranslational modification (e.g., phosphorylation) and increased intrinsic disorder. These protein properties indicate that the protein is highly prone to changes in conformation upon interaction with other proteins or signaling factors.[38] Consistent with this, proteins encoded by distally localized mRNAs have more protein–protein interaction partners and a broader interaction network. They also show large changes in protein expression after cellular stimulation, likely because their mRNAs are targets for signaling mechanisms to the translation machinery, and the proteins have shorter half-lives, consistent with tight temporal control of distally localized signaling processes that are integrated by translation control.[37]This study reached several conclusions yielding important insights into distally localized mRNAs and their localized translation: (1) local translation of these genes likely minimizes off-target

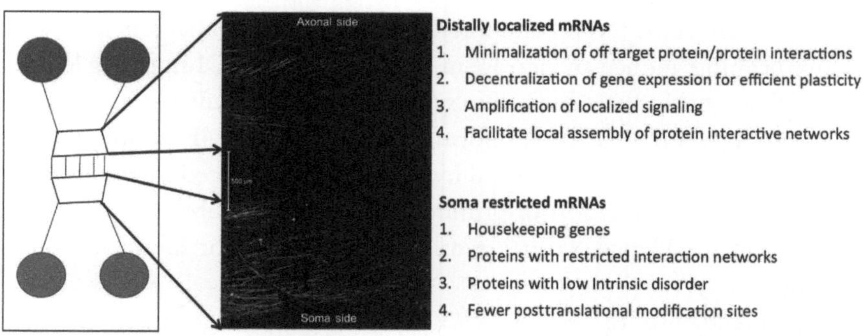

Figure 1 *Use of microfluidic devices to elucidate properties of distally localized mRNAs.* The left panel shows a schematic of a microfluidic device while the middle panel shows an immunocytochemical image of DRG neurons in culture labeled with βIII-tubulin staining. DRG somas are found on the bottom side of the chamber and extend axons through the microfluidic barrier where they then elaborate extensive axonal arobrations on the axonal side. Properties of distally (e.g., axonal) localized mRNAs versus those found restricted to cell bodies are listed on the right.

protein–protein interactions that might otherwise be expected amongst proteins that have wide interaction networks, (2) it decentralizes protein expression for efficient processing of plasticity-related signaling, (3) it permits strong amplification of spatially localized signaling, an important property of proteins with high intrinsic disorder and (4) it facilitates the organization of localized assemblies such as the postsynaptic density (Fig. 1). Interestingly, these features share a variety of similarities with mRNAs and proteins that are highly regulated in cancer and the upstream mechanisms that are thought to regulate these genes are likewise shared.[39,40]

3.1. Local translation is a key mediator of nociceptor priming

In hyperalgesic priming models, there is now clear evidence that persistent plasticity in peripheral nociceptors is critical to the initiation and maintenance of the primed state.[12] A broad variety of signaling mechanisms are altered in this state including switches in kinase and G protein coupled receptor (GPCR) signaling cascades[41–45] but a key feature of this form of plasticity is changes in gene expression regulated at the level of translation. Translation can be controlled, in an activity-dependent fashion, by extracellular factors signaling via kinase cascades offering rapid, locally mediated control of gene expression. Two important kinases for translation control are the mechanistic target of rapamycin complex 1 (mTORC1) and extracellular signal regulated kinase (ERK, Ref. 46). Both of these kinases signal to proteins that bind to the 5′ cap structure of mRNAs. In sensory neurons, nerve growth factor (NGF) and interleukin 6 (IL-6), two factors known to induce priming, induce an increase in ERK and mTORC1 signaling leading to a local, axonal increase in protein synthesis.[47,48] Blockade of these kinases, or blockade of eIF4F complex formation with the eIF4F inhibitor compound 4EGI1, inhibits mechanical hypersensitivity induced by these factors and abrogates precipitation of priming by a normally subthreshold stimulus[48,49] (Fig. 2). Hence, axonal translation is required for the induction of priming.

One mechanism to decrease ERK and mTORC1 signaling is via stimulation of adenosine monophosphate activated protein kinase (AMPK). AMPK is a widely expressed kinase well known to inhibit mTORC1[50,51] and ERK signaling[52,53] (Fig. 3). In sensory neurons, AMPK activation with pharmacological stimulators (for review see Ref. 54) leads to decreased ERK and mTORC1 activity,[55,56] decreased eIF4F complex formation[55,56] and inhibition of axonal protein synthesis.[47] AMPK activators also decrease

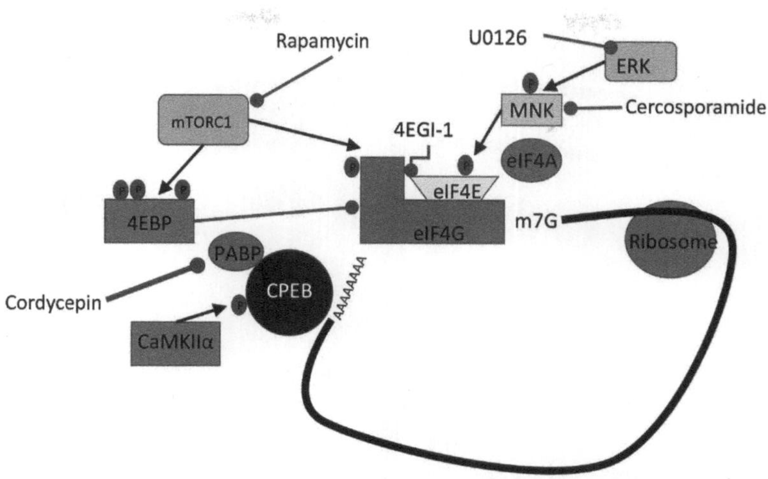

Figure 2 *Translational control pathways involved in hyperalgesic priming.* mTORC1 phosphorylates 4EBPs, negative regulators of eIF4F formation. This results in its dissociation from eIF4E, allowing the binding of eIF4E to eIF4G. Phosphorylation of eIF4E (via ERK/MNK1/2) or eIF4G (via mTORC1) enhances the formation of the eIF4F complex, promoting translation. Phosphorylation of CPEB by CamKIIα enhances translation efficiency by increasing the length of the poly A tail in mRNAs containing a CPE sequence. Taken together, eIF4F complex formation enhances cap-dependent translation, which is necessary for the induction of priming via translational control of gene expression in sensory afferents.

peripheral nerve injury- and inflammation-induced mechanical hyperalgesia[55,57] suggesting an important role for this kinase in peripheral pain plasticity across pain models. In the context of hyperalgesic priming, AMPK activation decreases mechanical hypersensitivity caused by incision or IL-6 exposure and completely blocks the development of priming when given locally around the time of incision.[56]

The regulation of translation via 5′ cap binding proteins (the eIFs) and their upstream kinases clearly comprise an important mechanism for the priming of nociceptors. Translation is also regulated by RNA binding proteins that bind to either 5′ or 3′ untranslated regions (UTRs) of mRNAs. For instance, the fragile X mental retardation protein (FMRP) is a key RNA binding protein regulating plasticity in the PNS and CNS.[29] As such, FMRP knockout mice fail to sensitize in several preclinical pain models[58,59] and these mice also have deficits in priming induced by NGF and IL-6.[49] Cytoplasmic polyadenylation element binding protein (CPEB) binds preferentially to mRNAs containing a CPE sequence in their 3′ UTR near the polyadenylation sequence. These mRNAs contain short poly A tails and

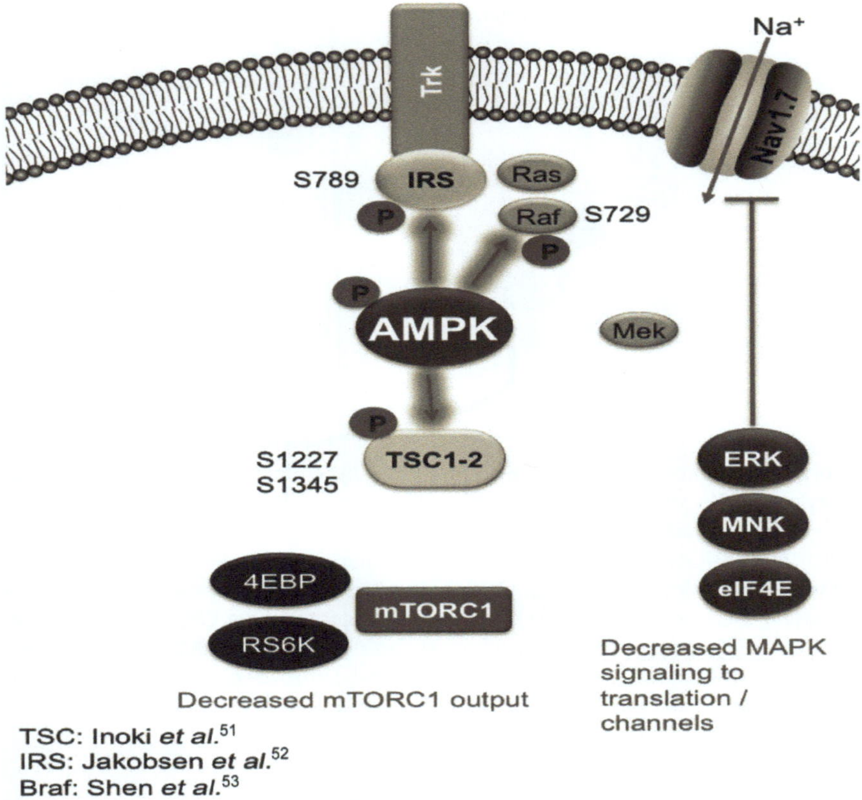

Figure 3 *AMPK activation pathway.* AMPK activation phosphorylates TSCs at Ser 1227 and 1345 leading to the inhibition of mTORC1. This is shown in the figure as an uncoupling of mTORC1 from Trk signaling via phosphorylation of TSC1/2 by AMPK. AMPK activation also phosphorylates Braf (Raf) at Ser 729 leading to inhibition of ERK signaling. Again, this is shown in the figure by an uncoupling of Raf/Mek signaling to ERK, MNK, and eIF4E via phosphorylation of Raf at Ser 729. Finally, AMPK phosphorylates IRS1 at Ser 789 leading to further inhibition of tyrosine kinase receptor signaling. (See the color plate.)

CPEB acts to enhance poly A tail length leading to enhanced translation efficiency in an activity-dependent fashion.[60] This process is linked to LTP in the CNS[61] and plays a central role in nociceptor priming[41,43,62] as evidenced by inhibition of the initiation of priming via CPEB knockdown in the DRG.[41] CPEB is phosphorylated by the aurora family kinases and by Ca^{2+}/calmodulin activated protein kinase II α (CaMKIIα, Ref. 63). Importantly, in priming induced by peripheral inflammation, CPEB may act downstream of CaMKIIα to initiate and maintain a primed state.[62] Since CPEB is thought to have prion-like properties that are linked to its role in memory maintenance,[64,65] these findings highlight a potential role for CPEB in creating a permanently primed state in peripheral nociceptors. This could occur due to the self-perpetuating, prion-like properties of activated CPEB.

Which mRNAs are locally translated in the setting of hyperalgesic priming? One candidate is the cyclic AMP response element binding protein (CREB) transcription factor.[66] Interestingly, CREB was one of the first memory genes identified using the aplysia gill reflex model system.[67,68] Subsequently, a broad variety of studies have shown a key role for CREB in memory throughout the brain and spinal cord.[69] More recently CREB mRNA was identified in developing DRG axons where its local translation is regulated by NGF to control the survival of developing neurons.[70] In adult DRG axons, stimulation with IL-6 leads to local, nascent synthesis of CREB that is then retrogradely transported to the DRG nucleus where it regulates changes in transcription that are crucial for the establishment of hyperalgesic priming. One gene that is regulated by this retrograde signaling transcription factor is BDNF.[66]

While it is clear that translation regulation is required to initiate a primed state in the periphery, an important question is whether continuous local translation is required to maintain priming once it has been established. An experimental paradigm to test this translation dependency is to induce priming with a locally administered stimulus and then allow the initial mechanical hypersensitivity to resolve. Then, prior to injection of the stimulus to precipitate the second episode of mechanical hypersensitivity in primed animals, translation inhibitors can be administered locally to test whether continuous translation is required to express a primed state.[49,62] In this regard, following injection of IL-6 and resolution of mechanical hypersensitivity in mice, injection of anisomycin or rapamycin (at doses that block the initiation of priming) fail to reverse a primed state when given 2 days prior to PGE_2 injection.[49] In contrast, in rats, injection of carrageenan causes priming that is disrupted both at the time of carrageenan injection and during the maintenance phase by either the mTORC1 inhibitor rapamycin or the polyadenylation inhibitor cordycepin.[71] Similar effects with rapamycin and cordycepin are observed in rats primed with paw injection of CaMKIIα. Since CaMKIIα phosphorylates CPEB and CPEB regulates CaMKIIα translation this raises the intriguing possibility that CaMKIIα/CPEB signaling could represent a positive feedback mechanism to maintain pain memory in the peripheral nociceptor.[62] Hence, while there are conflicting results in different models, it is possible that brief disruption of local translation in primed nociceptors is capable of resolving a "pain memory" stored in the peripheral nociceptor.

4. "PAIN MEMORY" IN THE SPINAL DORSAL HORN

4.1. LTP in the spinal dorsal horn, a neurophysiological correlate of pain memory?

LTP has been described at synapses throughout the CNS and is widely believed to be a core mechanism of plasticity for the nervous system. In the hippocampus LTP occurs during learning and its persistence is correlated with memory acquisition and consolidation.[72] Likewise, learning and LTP induce changes in dendritic spines in the hippocampus and cortex and these changes in spine shape are thought to be critical for the maintenance of potentiation of postsynaptic responses.[73] Interestingly, changes in spine shape also occur in the spinal dorsal horn after injury providing a potential structural change associated with pain memory.[74–76] Likewise, and similar to memory circuits, LTP can be observed in synapses activated by C-fiber afferent activity.[9,10] Key sites for nociceptor activity-dependent LTP are the outer lamina of the spinal dorsal horn[77] where projection neurons expressing the neurokinin receptor type 1 (NK1, substance P receptor) are found and in the deep dorsal horn where many wide dynamic range neurons (neurons that receive both Aβ- and C-fiber input) reside. LTP at these synapses shares many molecular and electrophysiological mechanisms with hippocampal and cortical LTP,[9,10] with a notable and very important distinction. Whereas low-frequency afferent stimulation causes long term depression (LTD) at most synapses in the brain, low-frequency stimulation of C-fibers, which largely matches their normal firing frequency, is sufficient to evoke LTP at a subset of dorsal horn neurons receiving direct C-fiber input.[77]

LTP has been equated with an important term used to describe activity-dependent plasticity in the dorsal horn: central sensitization. This topic has been covered by recent reviews[7] and editorial comments[78,79] and it is still controversial as to whether spinal LTP and central sensitization are parallel processes. What is less controversial is that LTP correlates well with hyperalgesia and, from a neurophysiological perspective, provides a parsimonious explanation for this form of pain amplification, at least at lamina I synapses in the dorsal horn. This is because homosynaptic LTP is observed at these synapses and hyperalgesia, an enhanced response to a normally noxious stimulus, can be explained by a monosynaptic amplification of C-fiber input onto projection neurons.[9,10] Allodynia, a noxious response to a normally innocuous stimulus, is more difficult to explain in terms of LTP because it would require a heterosynaptic form of plasticity. This is because Aβ-fibers that are stimulated by light touch do not have a monosynaptic connection to lamina I projection neurons. Interestingly, heterosynaptic LTP has been described at GABAergic synapses in the dorsal horn[80] but a potential role for this form of plasticity in allodynia has not been established.

Critically, LTP can be induced by natural stimulation of C-fibers with algogens such as capsaicin and formalin.[77] This form of LTP is consolidated into late-LTP and shares mechanisms with hippocampal and cortical LTP.[9,10,77,81,82] Once spinal dorsal horn LTP reaches the late-phase it does not readily reverse over a several hour time course. However, hyperalgesia induced by both capsaicin and formalin eventually reverses after several days. Does this mean that late phase spinal LTP eventually decays? One possibility is that endogenous analgesic mechanisms mask the hyperalgesic state that would otherwise be evident as a result of a persistent form of LTP. Hyperalgesic priming models may be capable of revealing such a mechanism. In this scenario, the precipitation of hyperalgesia in primed animals would override these endogenous inhibitory mechanisms leading to the reemergence of a hyperalgesic state revealed by a normally subthreshold stimulus. One candidate for endogenous analgesia overriding late-phase LTP and hyperalgesia is the endogenous opioid system. This system is robust in the dorsal horn with interneurons capable of releasing peptides that act on μ opioid receptors (MORs) expressed throughout the dorsal horn,[83,84] including presynaptic nociceptor nerve endings.[85,86] In support of this idea, infusion of MOR inverse agonists immediately precipitates a reinstatement of hyperalgesia in animals that have been primed with an inflammatory stimulus that is known to induce spinal LTP after the resolution of hyperalgesia.[87] This effect is absent in sham animals and is analogous to precipitation of hyperalgesia in primed animals with a subthreshold peripheral stimulus. What governs this effect? Peripheral inflammation, and presumably other nociceptive stimuli, induces a change in spinal MORs such that they now acquire constitutive activity (signaling through G proteins in the absence of agonist). This MOR constitutive activity then causes a tonic inhibition of pain signaling that masks a hyperalgesic state that would otherwise persist following the initial insult.[87]

These findings have several important implications for understanding central mechanisms governing hyperalgesic priming. They provide an elegant solution to why initial hyperalgesia resolves despite the persistence of a primed state and the potential durability of spinal LTP. This evidence also provides links to priming and late-LTP maintenance that potentially solve questions stated above. Opioid-dependent mechanisms play an important role in regulating spinal LTP. While there is evidence that high dose opioids can stimulate LTP at certain synapses after their abrupt removal,[81] there is likewise evidence that MOR activation can resolve even late-LTP at spinal synapses.[82] Based on this, it is possible that the initial priming stimulus, or induction of LTP, leads to late-LTP consolidation but this is subsequently resolved by endogenous opioid-mediated mechanisms. Does this mean that the previous establishment of late-LTP at central synapses causes a drop in threshold for establishment of subsequent LTP? If the mechanisms governing the MOR-dependent reversal of spinal late-LTP are constitutively expressed, as appears to be the case,[87] then this may lead to a tonic reversal of late-LTP with underlying mechanisms (e.g., aPKC and BDNF/trkB signaling, discussed below) still in place. While this idea obviously requires experimental testing, it could represent an important mechanism linking changes in peripheral sensitivity to CNS plasticity responsible for the maintenance of priming. Reversing these mechanisms could lead to revolutionary new therapeutics with disease modifying effects on chronic pain.

4.2. Atypical PKCs and BDNF

Early-LTP requires the activation of CaMKIIα, PKA, and conventional PKC leading to the phosphorylation of AMPA receptors.[88] Early-LTP also leads to changes in gene expression that occur both at the level of transcription and translation. These changes in gene expression are needed for the consolidation of early-LTP into late-LTP.[30] Mechanisms involved in the maintenance of late-LTP have been more difficult to clearly elucidate but are thought to involve an atypical PKC (aPKC) isoform called PKMζ.[32] Late-LTP can be reversed by inhibition of aPKCs with a peptide-based, pseudo substrate inhibitor called ZIP.[89] Intrathecal injection of ZIP either at the time of priming induction or following the resolution of the initial hyperalgesia leads to a complete reversal of hyperalgesic priming.[49,90] ZIP also reverses established pain states that have become dependent on central plasticity following sustained afferent input.[91] These findings are consistent with a role for PKMζ in the maintenance of late-LTP, memory retention, and the maintenance of a chronic pain state. On the other hand, recent experiments using genetic models to dissect the role of PKMζ in late-LTP and memory maintenance have called the specificity of ZIP and the role of PKMζ in these effects into question.[92,93] It remains to be seen if

PKMζ plays a specific role in the maintenance of hyperalgesic priming in the dorsal horn of the spinal cord (for review on this topic see Ref. 11).

An important component of the proposed role of PKMζ in LTP and memory is the trafficking of AMPA receptors to synaptic sites leading to a persistent augmentation of postsynaptic glutamate-mediated signaling.[32] This trafficking can be disrupted with a peptide called pep2m that blocks AMPA receptor association with trafficking molecules.[94] Similar to experiments in other CNS regions, intrathecal injection of pep2m disrupts the maintenance of hyperalgesic priming[49] suggesting that aPKC-mediated regulation of AMPA receptor trafficking may play a central role in chronic pain states. This is consistent with a wide variety of experimental findings indicating that AMPA receptor trafficking plays a central role in mediating pain plasticity induced by peripheral injury[95] and that an increase in AMPA receptors at the postsynaptic density is required for LTP consolidation and maintenance.

As mentioned above, while it is clear that ZIP is capable of permanently reversing a primed state in a variety of experimental models,[49,90,91] the molecular mechanisms engaged by ZIP are less clear based on evidence from transgenic mice.[92,93] One possibility is that aPKC isoforms play a redundant role in synaptic plasticity and therefore another aPKC, PKCλ, may be involved in maintenance mechanisms of hyperalgesic priming.[11] This isoform is also inhibited by ZIP,[90,93] therefore, this provides an explanation for the discrepancy between pharmacological effects of ZIP and findings from mice lacking aPKCs derived from the *Prckz* locus (PKMζ and PKCζ). If this were the case, upstream mechanisms that regulate all aPKCs isoforms would represent alternative targets to reverse hyperalgesic priming. A candidate molecule fitting this description is BDNF.

BDNF is well recognized as an important mediator of pain plasticity. BDNF is expressed by DRG neurons and released in the spinal dorsal horn,[96] where it can act on pre- and postsynaptic trkB receptors to regulate plasticity of presynaptic afferent fibers[97] and postsynaptic dorsal horn neurons.[98–100] BDNF expression increases following peripheral injury,[101] nociceptor-specific knockout of BDNF leads to a profound reduction in many forms of injury-induced pain plasticity[102] and microglial BDNF expression is increased by nerve injury.[103] BDNF is also a key factor in LTP. In hippocampus, BDNF is required for the induction of LTP and dendritic-expressed BDNF is postulated to play an autocrine role in maintenance of late phase LTP.[104] Likewise, BDNF is sufficient to induce LTP in dorsal horn neurons[105] linking BDNF-induced pain plasticity to memory-like mechanisms that may be involved in the maintenance of hyperalgesic priming. Indeed, intrathecal injection of compounds that interfere with BDNF action, blocks hyperalgesia induced by priming agents and prevents the precipitation of a primed state by subsequent stimulation. Significantly, interruption of BDNF/trkB signaling after the establishment of a primed state leads to a resolution of priming[90] suggesting a key role of BDNF/trkB signaling in the maintenance of a primed state. At spinal synapses, BDNF induces phosphorylation and translation of the two major aPKC isoforms found in the CNS, PKMζ and PKCλ indicating a link between BDNF/trkB and aPKCs[90] (Fig. 4). Therefore, these findings point to BDNF/trkB

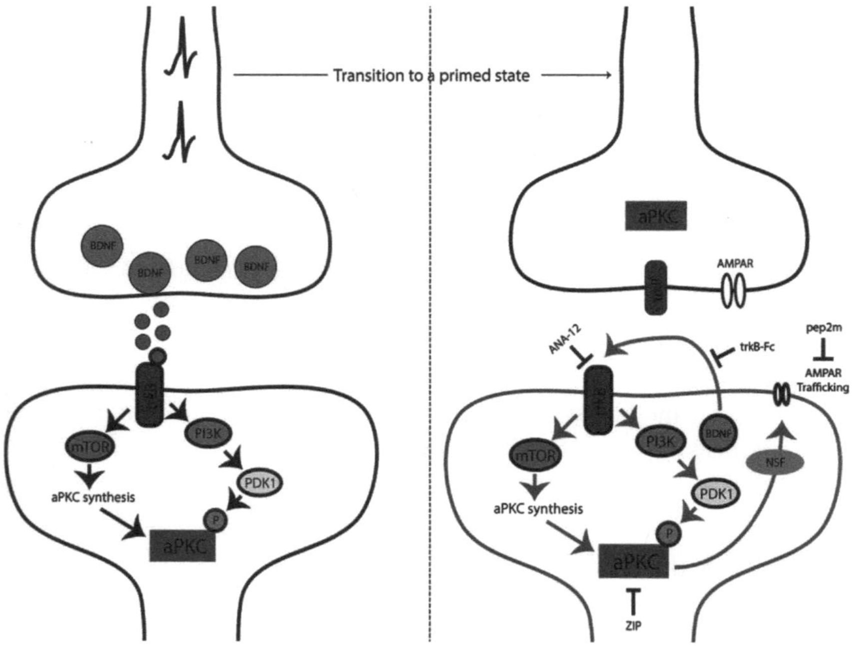

Figure 4 *The role of aPKCs and BDNF in hyperalgesic priming initiation and maintenance.* Nociceptor activation leads to spinal BDNF release and a postsynaptic mTORC1-dependent translation of aPKC protein. These newly synthesized aPKCs are then phosphorylated by PDK1. Increased levels and phosphorylation of aPKCs are thought to be involved in initiating priming. Once priming is established (right panel), increased aPKC protein and phosphorylation leads to a constitutive increase in AMPAR trafficking to the postsynaptic membrane. This appears to be regulated by BDNF signaling via trkB with BDNF potentially being released from postsynaptic dendrites in the maintenance stage of priming. Presynaptic trkB may also be activated by increased BDNF action in primed animals. Once established, hyperalgesic priming can be permanently reversed by inhibition of aPKCs with ZIP, disruption of AMPAR trafficking with pep2M or via inhibition of trkB/BDNF signaling with ANA-12 or trkB-Fc, respectively. (See the color plate.)

signaling as a therapeutic target for the reversal of established chronic pain states. Therapeutics aimed at this signaling axis could lead to disease modification in chronic pain patients.

4.3. Reconsolidation of pain memory

The mechanisms discussed above provide a foundation for understanding spinal mechanisms that encode a memory trace for pain. These mechanisms primarily tie into the concept that spinal dorsal horn plasticity shares molecular, neurophysiological, and structural similarities with plasticity in other regions of the brain that are involved in learning and memory. Do spinal memory traces share other features with hippocampal or cortical memory mechanisms? An important concept in the learning and memory literature is the idea of reconsolidation. Reconsolidation is a memory updating mechanism during which a memory trace can become labile and susceptible to revision with behavioral intervention and/or pharmacological manipulation.[106] Classical reconsolidation experiments involve fear conditioning. Here an animal is trained to fear a neutral stimulus through pairing with a foot shock. This creates a long-lasting memory that can be revealed by freezing induced by presentation of the neutral stimulus that has been paired with shock. Infusion of a protein synthesis inhibitor at the time of training blocks the acquisition of the fear memory, however, after the memory has consolidated (e.g., within 6 h after training) protein synthesis inhibition no longer alters the memory trace, so long as the protein synthesis inhibitor is given outside of the training context. However, if the animal is placed back in the training context and exposed to the neutral stimulus to trigger recall of the memory, infusion of a protein synthesis inhibitor will again lead to a reversal of fear conditioning.[107] Hence, during reconsolidation, a memory trace is made labile and can be reversed by pharmacological mechanisms that were only effective prior to consolidation of the long-term memory. Importantly, these pharmacological mechanisms share strong similarities with early- and late-LTP, as described above (Fig. 5A).

Bonin and De Koninck recently showed that a mechanism akin to reconsolidation is engaged in the spinal dorsal horn. Using capsaicin as a noxious stimulus to induce pain plasticity they showed that inhibition of spinal protein synthesis paired with either a second capsaicin injection or optogenetic stimulation of C-fibers led to a reversal of pain plasticity.[108] In parallel experiments they examined spinal LTP and its dependence on protein synthesis following the consolidation of late-LTP (Fig. 5B). They

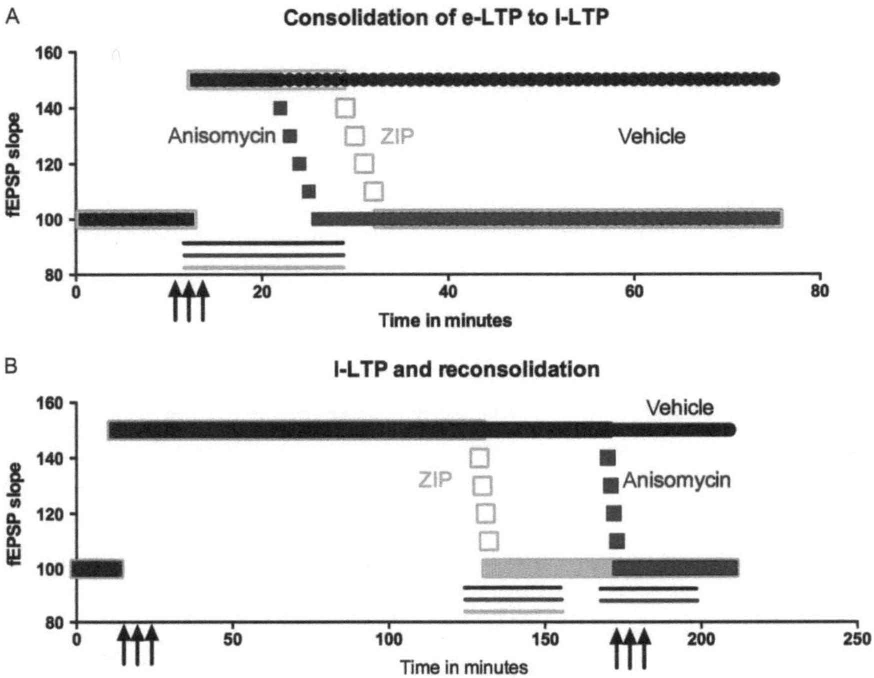

Figure 5 *Consolidation of late phase LTP (late-LTP) and reconsolidation.* (A) Following high-frequency stimulation of afferent input (three upward arrows), early-LTP (e-LTP in the figure) is induced and this consolidates to late-LTP (l-LTP in the figure) over the course of 30–60 min. Application of translation control inhibitors, such as anisomycin (red line), during early-LTP cause a failure of late-LTP consolidation. Likewise, aPKC inhibition with ZIP (green line) blocks consolidation of late-LTP. Vehicle application (blue line) has no impact on consolidation of late-LTP. (B) Once late-LTP is established administering translation inhibitors (e.g., anisomycin, red line) in the absence of high-frequency stimulation of afferents fails to reverse late-LTP while ZIP application (green line) does induce late-LTP decay. Restimulation of afferents at high frequency during late-LTP (upward arrows) opens a reconsolidation window. Application of translation inhibitors such as anisomycin (red line) during this reconsolidation period leads to late-LTP decay, an effect that is presumably linked to reversal of a chronic pain state in similar behavioral pharmacology experiments. (See the color plate.)

found that after late-LTP consolidation spinal LTP was not reversed by protein synthesis inhibition unless C-fibers were tetanized at the time of protein synthesis inhibitor application. This crucial experiment parallels reconsolidation of pain memory in a behavioral paradigm providing a strong link between behavioral manifestations of pain plasticity (hyperalgesia) and spinal

LTP. Similar experiments using the hyperalgesic priming paradigm indicate that reconsolidation of pain memory can be engaged even at late stages after the initiation of a pain memory trace (Kim and Price unpublished observations) suggesting that opening of a reconsolidation window in chronic pain patients may provide an opportunity for reversal of pain plasticity and a resolution of chronic pain.

5. CLINICAL IMPLICATIONS OF PAIN MEMORY

In order to fully grasp the importance of the research findings discussed herein, it is important to reflect on the utility of using experimental models of pain memory to gain better insight into human pain plasticity. Along these lines it is must be noted that human experimental models of perceptual LTP, and even LTD, exist and involve afferent stimulation protocols that are similar if not identical to those used in preclinical studies. There is also evidence that LTP stimulation protocols in humans can lead to a transition to late-LTP in certain individuals. Another important point is that the experimental framework of the hyperalgesic priming model, which is often used to study pain memory, provides important insight into clinical chronic pain because it captures the recurrent nature of some of the most common pathological pain conditions.[12] In 1921, Wilfred Harris described his clinical experience treating patients with presumed injuries to peripheral nerves. He described pain in these patients as episodic with pain episodes provoked by acute exacerbation.[109] Hence, from some of the earliest descriptions of pain as a disease, the notion of priming followed by subthreshold provocation of long-lived pain episodes has been apparent.

Population-based studies in prevalent chronic pain conditions have directly demonstrated the episodic yet progressive nature of chronic pain. Perhaps the best-known example is headache and, in the case of migraine, frequency of attacks is the best predictor of a transition to chronic migraine.[110] In fact, the vast majority of migraineurs move from a low-frequency episodic headache stage to a high frequency stage and eventually into chronic migraine.[111] Moreover, migraines can frequently be provoked by migraine triggers. These are, by definition, subthreshold stimuli because they fail to provoke migraines in the nonmigraineur population. This situation is not unique to migraine. Acute episodes of low back pain generally resolve[112,113] but recurrence rates over 5 years are as high as 70%[113–116] and lifetime recurrence is estimated at 85%.[117,118] The probability of low back pain recurrence increases with previous episodes of low back pain.[115]

A similar clinical picture has been found for chronic neck pain.[119,120] Finally, in the case of surgery and chronic postsurgical pain, there is evidence that preexisting pain is a major risk factor for chronic post-incision pain suggesting that a pain memory trace might already be in place in some patients causing precipitation of a very long-lasting pain state induced by incision.[121,122] Hence, the "priming" event in the hyperaglesic priming model may be viewed as an induction of a pain memory with important clinical parallels that help to explain mechanisms of disease and potential pathways to resolution of plasticity that drives chronicity.

6. CONCLUSIONS

The concept of pain memory, first proposed by Ronald Melzack and colleagues nearly 40 years ago, has inspired remarkable progress into our understanding of mechanisms that cause pain to become chronic. Here we have described two major potential mechanisms for pain memory: (1) changes in gene expression in peripheral nociceptors that permanently alter the phenotype and function of these crucial neurons for pain plasticity and (2) changes in synaptic strength in the dorsal horn governed by mechanisms that play an active role in maintaining a chronic pain state. We have highlighted several intervention points that have potential to lead to disease modifying therapeutics for the permanent reversal of chronic pain states. While these mechanisms ultimately await testing in human patients, there is little question that the concept of pain memory will continue to provoke new research questions and provide novel insight into how plasticity leads to the neurological disorder that is chronic pain.

ACKNOWLEDGMENTS

This work was supported by NIH grants NS065926 (T. J. P.) and GM102575 (T. J. P. and G. D.) and The University of Texas STARS program (T. J. P.).

REFERENCES

1. Woolf CJ, Walters ET. Common patterns of plasticity contributing to nociceptive sensitization in mammals and Aplysia. *Trends Neurosci*. 1991;14:74–78.
2. Cervero F. Spinal cord mechanisms of hyperalgesia and allodynia: role of peripheral input from nociceptors. *Prog Brain Res*. 1996;113:413–422.
3. Devor M, Janig W, Michaelis M. Modulation of activity in dorsal root ganglion neurons by sympathetic activation in nerve-injured rats. *J Neurophysiol*. 1994;71:38–47.
4. Lisney SJ, Devor M. Afterdischarge and interactions among fibers in damaged peripheral nerve in the rat. *Brain Res*. 1987;415:122–136.

5. Reichling DB, Green PG, Levine JD. The fundamental unit of pain is the cell. *Pain.* 2013;154:S2–S9.
6. Ji RR, Kohno T, Moore KA, Woolf CJ. Central sensitization and LTP: do pain and memory share similar mechanisms? *Trends Neurosci.* 2003;26:696–705.
7. Latremoliere A, Woolf CJ. Central sensitization: a generator of pain hypersensitivity by central neural plasticity. *J Pain.* 2009;10:895–926.
8. Woolf CJ. Central sensitization: uncovering the relation between pain and plasticity. *Anesthesiology.* 2007;106:864–867.
9. Ruscheweyh R, Wilder-Smith O, Drdla R, Liu XG, Sandkuhler J. Long-term potentiation in spinal nociceptive pathways as a novel target for pain therapy. *Mol Pain.* 2011;7:20.
10. Sandkuhler J. Understanding LTP in pain pathways. *Mol Pain.* 2007;3:9.
11. Price TJ, Ghosh S. ZIPping to pain relief: the role (or not) of PKMzeta in chronic pain. *Mol Pain.* 2013;9:6.
12. Reichling DB, Levine JD. Critical role of nociceptor plasticity in chronic pain. *Trends Neurosci.* 2009;32:611–618.
13. Dennis SG, Melzack R. Self-mutilation after dorsal rhizotomy in rats: effects of prior pain and pattern of root lesions. *Exp Neurol.* 1979;65:412–421.
14. Katz J, Melzack R. Pain 'memories' in phantom limbs: review and clinical observations. *Pain.* 1990;43:319–336.
15. Flor H, Nikolajsen L, Staehelin Jensen T. Phantom limb pain: a case of maladaptive CNS plasticity? *Nat Rev Neurosci.* 2006;7:873–881.
16. Vaso A, Adahan HM, Gjika A, et al. Peripheral nervous system origin of phantom limb pain. *Pain.* 2014;155:1384–1391.
17. Coderre TJ, Melzack R. Cutaneous hyperalgesia: contributions of the peripheral and central nervous systems to the increase in pain sensitivity after injury. *Brain Res.* 1987;404:95–106.
18. Kandel ER. The molecular biology of memory storage: a dialog between genes and synapses. *Biosci Rep.* 2004;24:475–522.
19. De Camilli P, Carew TJ. Nobel celebrates the neurosciences. Modulatory signaling in the brain. *Cell.* 2000;103:829–833.
20. Clatworthy AL, Walters ET. Rapid amplification and facilitation of mechanosensory discharge in Aplysia by noxious stimulation. *J Neurophysiol.* 1993;70:1181–1194.
21. Illich PA, Walters ET. Mechanosensory neurons innervating Aplysia siphon encode noxious stimuli and display nociceptive sensitization. *J Neurosci.* 1997;17:459–469.
22. Weragoda RM, Ferrer E, Walters ET. Memory-like alterations in Aplysia axons after nerve injury or localized depolarization. *J Neurosci.* 2004;24:10393–10401.
23. Walters ET, Moroz LL. Molluscan memory of injury: evolutionary insights into chronic pain and neurological disorders. *Brain Behav Evol.* 2009;74:206–218.
24. Crook RJ, Dickson K, Hanlon RT, Walters ET. Nociceptive sensitization reduces predation risk. *Curr Biol.* 2014;24:1121–1125.
25. Price TJ, Dussor G. Evolution: the advantage of 'maladaptive' pain plasticity. *Curr Biol.* 2014;24:R384–R386.
26. Steward O, Levy WB. Preferential localization of polyribosomes under the base of dendritic spines in granule cells of the dentate gyrus. *J Neurosci.* 1982;2:284–291.
27. Bliss TV, Lomo T. Long-lasting potentiation of synaptic transmission in the dentate area of the anaesthetized rabbit following stimulation of the perforant path. *J Physiol.* 1973;232:331–356.
28. Costa-Mattioli M, Sossin WS, Klann E, Sonenberg N. Translational control of long-lasting synaptic plasticity and memory. *Neuron.* 2009;61:10–26.
29. Bassell GJ, Warren ST. Fragile X syndrome: loss of local mRNA regulation alters synaptic development and function. *Neuron.* 2008;60:201–214.

30. Abraham WC, Williams JM. LTP maintenance and its protein synthesis-dependence. *Neurobiol Learn Mem.* 2008;89:260–268.
31. Price TJ, Geranton SM. Translating nociceptor sensitivity: the role of axonal protein synthesis in nociceptor physiology. *Eur J Neurosci.* 2009;29:2253–2263.
32. Sacktor TC. How does PKMzeta maintain long-term memory? *Nat Rev Neurosci.* 2011;12:9–15.
33. Brown V, Jin P, Ceman S, et al. Microarray identification of FMRP-associated brain mRNAs and altered mRNA translational profiles in fragile X syndrome. *Cell.* 2001;107:477–487.
34. Darnell JC, Jensen KB, Jin P, Brown V, Warren ST, Darnell RB. Fragile X mental retardation protein targets G quartet mRNAs important for neuronal function. *Cell.* 2001;107:489–499.
35. Willis D, Li KW, Zheng JQ, et al. Differential transport and local translation of cytoskeletal, injury-response, and neurodegeneration protein mRNAs in axons. *J Neurosci.* 2005;25:778–791.
36. Willis DE, van Niekerk EA, Sasaki Y, et al. Extracellular stimuli specifically regulate localized levels of individual neuronal mRNAs. *J Cell Biol.* 2007;178:965–980.
37. Weatheritt RJ, Gibson TJ, Babu MM. Asymmetric mRNA localization contributes to fidelity and sensitivity of spatially localized systems. *Nat Struct Mol Biol.* 2014;21:833–839.
38. Tantos A, Han KH, Tompa P. Intrinsic disorder in cell signaling and gene transcription. *Mol Cell Endocrinol.* 2012;348:457–465.
39. Boussemart L, Malka-Mahieu H, Girault I, et al. eIF4F is a nexus of resistance to anti-BRAF and anti-MEK cancer therapies. *Nature.* 2014;513:105–109.
40. Wolfe AL, Singh K, Zhong Y, et al. RNA G-quadruplexes cause eIF4A-dependent oncogene translation in cancer. *Nature.* 2014;513:65–70.
41. Bogen O, Alessandri-Haber N, Chu C, Gear RW, Levine JD. Generation of a pain memory in the primary afferent nociceptor triggered by PKCepsilon activation of CPEB. *J Neurosci.* 2012;32:2018–2026.
42. Dina OA, Khasar SG, Gear RW, Levine JD. Activation of Gi induces mechanical hyperalgesia poststress or inflammation. *Neuroscience.* 2009;160:501–507.
43. Ferrari LF, Bogen O, Alessandri-Haber N, Levine E, Gear RW, Levine JD. Transient decrease in nociceptor GRK2 expression produces long-term enhancement in inflammatory pain. *Neuroscience.* 2012;222:392–403.
44. Joseph EK, Levine JD. Multiple PKCepsilon-dependent mechanisms mediating mechanical hyperalgesia. *Pain.* 2010;150:17–21.
45. Wang H, Heijnen CJ, van Velthoven CT, et al. Balancing GRK2 and EPAC1 levels prevents and relieves chronic pain. *J Clin Invest.* 2013;123:5023–5034.
46. Topisirovic I, Sonenberg N. mRNA translation and energy metabolism in cancer: the role of the MAPK and mTORC1 pathways. *Cold Spring Harb Symp Quant Biol.* 2011;76:355–367.
47. Melemedjian OK, Mejia GL, Lepow TS, Zoph OK, Price TJ. Bidirectional regulation of P body formation mediated by eIF4F complex formation in sensory neurons. *Neurosci Lett.* 2013;563:169–174.
48. Melemedjian OK, Asiedu MN, Tillu DV, et al. IL-6- and NGF-induced rapid control of protein synthesis and nociceptive plasticity via convergent signaling to the eIF4F complex. *J Neurosci.* 2010;30:15113–15123.
49. Asiedu MN, Tillu DV, Melemedjian OK, et al. Spinal protein kinase M zeta underlies the maintenance mechanism of persistent nociceptive sensitization. *J Neurosci.* 2011;31:6646–6653.
50. Carling D, Thornton C, Woods A, Sanders MJ. AMP-activated protein kinase: new regulation, new roles? *Biochem J.* 2012;445:11–27.

51. Inoki K, Zhu T, Guan KL. TSC2 mediates cellular energy response to control cell growth and survival. *Cell.* 2003;115:577–590.
52. Jakobsen SN, Hardie DG, Morrice N, Tornqvist HE. 5′-AMP-activated protein kinase phosphorylates IRS-1 on Ser-789 in mouse C2C12 myotubes in response to 5-aminoimidazole-4-carboxamide riboside. *J Biol Chem.* 2001;276:46912–46916.
53. Shen CH, Yuan P, Perez-Lorenzo R, et al. Phosphorylation of BRAF by AMPK impairs BRAF-KSR1 association and cell proliferation. *Mol Cell.* 2013;52:161–172.
54. Price TJ, Dussor G. AMPK: an emerging target for modification of injury-induced pain plasticity. *Neurosci Lett.* 2013;557:9–18.
55. Melemedjian OK, Asiedu MN, Tillu DV, et al. Targeting adenosine monophosphate-activated protein kinase (AMPK) in preclinical models reveals a potential mechanism for the treatment of neuropathic pain. *Mol Pain.* 2011;7:70.
56. Tillu DV, Melemedjian OK, Asiedu MN, et al. Resveratrol engages AMPK to attenuate ERK and mTOR signaling in sensory neurons and inhibits incision-induced acute and chronic pain. *Mol Pain.* 2012;8:5.
57. Russe OQ, Moser CV, Kynast KL, et al. Activation of the AMP-activated protein kinase reduces inflammatory nociception. *J Pain.* 2013;14:1330–1340.
58. Price TJ, Melemedjian OK. Fragile X mental retardation protein (FMRP) and the spinal sensory system. *Results Probl Cell Differ.* 2012;54:41–59.
59. Price TJ, Rashid MH, Millecamps M, Sanoja R, Entrena JM, Cervero F. Decreased nociceptive sensitization in mice lacking the fragile X mental retardation protein: role of mGluR1/5 and mTOR. *J Neurosci.* 2007;27:13958–13967.
60. Richter JD. CPEB: a life in translation. *Trends Biochem Sci.* 2007;32:279–285.
61. Udagawa T, Swanger SA, Takeuchi K, et al. Bidirectional control of mRNA translation and synaptic plasticity by the cytoplasmic polyadenylation complex. *Mol Cell.* 2012;47:253–266.
62. Ferrari LF, Bogen O, Levine JD. Role of nociceptor alphaCaMKII in transition from acute to chronic pain (hyperalgesic priming) in male and female rats. *J Neurosci.* 2013;33:11002–11011.
63. Atkins CM, Davare MA, Oh MC, Derkach V, Soderling TR. Bidirectional regulation of cytoplasmic polyadenylation element-binding protein phosphorylation by Ca^{2+}/calmodulin-dependent protein kinase II and protein phosphatase 1 during hippocampal long-term potentiation. *J Neurosci.* 2005;25:5604–5610.
64. Si K, Lindquist S, Kandel ER. A neuronal isoform of the aplysia CPEB has prion-like properties. *Cell.* 2003;115:879–891.
65. Si K, Giustetto M, Etkin A, et al. A neuronal isoform of CPEB regulates local protein synthesis and stabilizes synapse-specific long-term facilitation in aplysia. *Cell.* 2003;115:893–904.
66. Melemedjian OK, Tillu DV, Moy JK, et al. Local translation and retrograde axonal transport of CREB regulates IL-6-induced nociceptive plasticity. *Mol Pain.* 2014;10:45.
67. Dash PK, Hochner B, Kandel ER. Injection of the cAMP-responsive element into the nucleus of Aplysia sensory neurons blocks long-term facilitation. *Nature.* 1990;345:718–721.
68. Kaang BK, Kandel ER, Grant SG. Activation of cAMP-responsive genes by stimuli that produce long-term facilitation in Aplysia sensory neurons. *Neuron.* 1993;10:427–435.
69. Rahn EJ, Guzman-Karlsson MC, David Sweatt J. Cellular, molecular, and epigenetic mechanisms in non-associative conditioning: implications for pain and memory. *Neurobiol Learn Mem.* 2013;105:133–150.
70. Cox LJ, Hengst U, Gurskaya NG, Lukyanov KA, Jaffrey SR. Intra-axonal translation and retrograde trafficking of CREB promotes neuronal survival. *Nat Cell Biol.* 2008;10:149–159.

71. Ferrari LF, Bogen O, Chu C, Levine JD. Peripheral administration of translation inhibitors reverses increased hyperalgesia in a model of chronic pain in the rat. *J Pain*. 2013;14:731–738.
72. Whitlock JR, Heynen AJ, Shuler MG, Bear MF. Learning induces long-term potentiation in the hippocampus. *Science*. 2006;313:1093–1097.
73. De Roo M, Klauser P, Garcia PM, Poglia L, Muller D. Spine dynamics and synapse remodeling during LTP and memory processes. *Prog Brain Res*. 2008;169:199–207.
74. Tan AM, Choi JS, Waxman SG, Hains BC. Dendritic spine remodeling after spinal cord injury alters neuronal signal processing. *J Neurophysiol*. 2009;102:2396–2409.
75. Tan AM, Chang YW, Zhao P, Hains BC, Waxman SG. Rac1-regulated dendritic spine remodeling contributes to neuropathic pain after peripheral nerve injury. *Exp Neurol*. 2011;232:222–233.
76. Tan AM, Stamboulian S, Chang YW, et al. Neuropathic pain memory is maintained by Rac1-regulated dendritic spine remodeling after spinal cord injury. *J Neurosci*. 2008;28:13173–13183.
77. Ikeda H, Stark J, Fischer H, et al. Synaptic amplifier of inflammatory pain in the spinal dorsal horn. *Science*. 2006;312:1659–1662.
78. Latremoliere A, Woolf CJ. Synaptic plasticity and central sensitization: author reply. *J Pain*. 2010;11:801–803.
79. Sandkuhler J. Central sensitization versus synaptic long-term potentiation (LTP): a critical comment. *J Pain*. 2010;11:798–800.
80. Fenselau H, Heinke B, Sandkuhler J. Heterosynaptic long-term potentiation at GABAergic synapses of spinal lamina I neurons. *J Neurosci*. 2011;31:17383–17391.
81. Drdla R, Gassner M, Gingl E, Sandkuhler J. Induction of synaptic long-term potentiation after opioid withdrawal. *Science*. 2009;325:207–210.
82. Drdla-Schutting R, Benrath J, Wunderbaldinger G, Sandkuhler J. Erasure of a spinal memory trace of pain by a brief, high-dose opioid administration. *Science*. 2012;335:235–238.
83. Ma W, Ribeiro-da-Silva A, De Koninck Y, Radhakrishnan V, Cuello AC, Henry JL. Substance P and enkephalin immunoreactivities in axonal boutons presynaptic to physiologically identified dorsal horn neurons. An ultrastructural multiple-labelling study in the cat. *Neuroscience*. 1997;77:793–811.
84. Ribeiro-da-Silva A, De Koninck Y, Cuello AC, Henry JL. Enkephalin-immunoreactive nociceptive neurons in the cat spinal cord. *Neuroreport*. 1992;3:25–28.
85. Schroeder JE, McCleskey EW. Inhibition of Ca^{2+} currents by a mu-opioid in a defined subset of rat sensory neurons. *J Neurosci*. 1993;13:867–873.
86. Schroeder JE, Fischbach PS, Zheng D, McCleskey EW. Activation of mu opioid receptors inhibits transient high- and low-threshold Ca^{2+} currents, but spares a sustained current. *Neuron*. 1991;6:13–20.
87. Corder G, Doolen S, Donahue RR, et al. Constitutive mu-opioid receptor activity leads to long-term endogenous analgesia and dependence. *Science*. 2013;341: 1394–1399.
88. Huganir RL, Nicoll RA. AMPARs and synaptic plasticity: the last 25 years. *Neuron*. 2013;80:704–717.
89. Pastalkova E, Serrano P, Pinkhasova D, Wallace E, Fenton AA, Sacktor TC. Storage of spatial information by the maintenance mechanism of LTP. *Science*. 2006;313: 1141–1144.
90. Melemedjian OK, Tillu DV, Asiedu MN, et al. BDNF regulates atypical PKC at spinal synapses to initiate and maintain a centralized chronic pain state. *Mol Pain*. 2013;9:12.
91. Laferriere A, Pitcher MH, Haldane A, et al. PKMzeta is essential for spinal plasticity underlying the maintenance of persistent pain. *Mol Pain*. 2011;7:99.

92. Lee AM, Kanter BR, Wang D, et al. Prkcz null mice show normal learning and memory. *Nature*. 2013;493:416–419.
93. Volk LJ, Bachman JL, Johnson R, Yu Y, Huganir RL. PKM-zeta is not required for hippocampal synaptic plasticity, learning and memory. *Nature*. 2013;493:420–423.
94. Migues PV, Hardt O, Wu DC, et al. PKMzeta maintains memories by regulating GluR2-dependent AMPA receptor trafficking. *Nat Neurosci*. 2010;13:630–634.
95. Tao YX. AMPA receptor trafficking in inflammation-induced dorsal horn central sensitization. *Neurosci Bull*. 2012;28:111–120.
96. Balkowiec A, Katz DM. Activity-dependent release of endogenous brain-derived neurotrophic factor from primary sensory neurons detected by ELISA in situ. *J Neurosci*. 2000;20:7417–7423.
97. Matayoshi S, Jiang N, Katafuchi T, et al. Actions of brain-derived neurotrophic factor on spinal nociceptive transmission during inflammation in the rat. *J Physiol*. 2005;569:685–695.
98. Garraway SM, Petruska JC, Mendell LM. BDNF sensitizes the response of lamina II neurons to high threshold primary afferent inputs. *Eur J Neurosci*. 2003;18: 2467–2476.
99. Kerr BJ, Bradbury EJ, Bennett DL, et al. Brain-derived neurotrophic factor modulates nociceptive sensory inputs and NMDA-evoked responses in the rat spinal cord. *J Neurosci*. 1999;19:5138–5148.
100. Pezet S, Malcangio M, Lever IJ, et al. Noxious stimulation induces Trk receptor and downstream ERK phosphorylation in spinal dorsal horn. *Mol Cell Neurosci*. 2002;21:684–695.
101. Mannion RJ, Costigan M, Decosterd I, et al. Neurotrophins: peripherally and centrally acting modulators of tactile stimulus-induced inflammatory pain hypersensitivity. *Proc Natl Acad Sci U S A*. 1999;96:9385–9390.
102. Zhao J, Seereeram A, Nassar MA, et al. Nociceptor-derived brain-derived neurotrophic factor regulates acute and inflammatory but not neuropathic pain. *Mol Cell Neurosci*. 2006;31:539–548.
103. Trang T, Beggs S, Salter MW. Brain-derived neurotrophic factor from microglia: a molecular substrate for neuropathic pain. *Neuron Glia Biol*. 2011;7:99–108.
104. Lu Y, Christian K, Lu B. BDNF: a key regulator for protein synthesis-dependent LTP and long-term memory? *Neurobiol Learn Mem*. 2008;89:312–323.
105. Zhou LJ, Zhong Y, Ren WJ, Li YY, Zhang T, Liu XG. BDNF induces late-phase LTP of C-fiber evoked field potentials in rat spinal dorsal horn. *Exp Neurol*. 2008;212:507–514.
106. Nader K, Hardt O. A single standard for memory: the case for reconsolidation. *Nat Rev Neurosci*. 2009;10:224–234.
107. Nader K, Schafe GE, Le Doux JE. Fear memories require protein synthesis in the amygdala for reconsolidation after retrieval. *Nature*. 2000;406:722–726.
108. Bonin RP, De Koninck Y. A spinal analog of memory reconsolidation enables reversal of hyperalgesia. *Nat Neurosci*. 2014;17:1043–1045.
109. Harris W. Persistent pain in lesions of the peripheral and central nervous system. *Br Med J*. 1921;2:896–900.
110. Lipton RB. Tracing transformation: chronic migraine classification, progression, and epidemiology. *Neurology*. 2009;72:S3–S7.
111. Bigal ME, Lipton RB. Clinical course in migraine: conceptualizing migraine transformation. *Neurology*. 2008;71:848–855.
112. Bartleson JD. Low back pain. *Curr Treat Options Neurol*. 2001;3:159–168.
113. Cassidy JD, Cote P, Carroll LJ, Kristman V. Incidence and course of low back pain episodes in the general population. *Spine (Phila Pa 1976)*. 2005;30:2817–2823.

114. Carey TS, Garrett JM, Jackman A, Hadler N. Recurrence and care seeking after acute back pain: results of a long-term follow-up study. North Carolina Back Pain Project. *Med Care*. 1999;37:157–164.
115. Kolb E, Canjuga M, Bauer GF, Laubli T. Course of back pain across 5 years: a retrospective cohort study in the general population of Switzerland. *Spine (Phila Pa 1976)*. 2011;36:E268–E273.
116. Von Korff M, Saunders K. The course of back pain in primary care. *Spine (Phila Pa 1976)*. 1996;21:2833–2837 [discussion: 2838–2839].
117. Andersson GB. Epidemiological features of chronic low-back pain. *Lancet*. 1999;354:581–585.
118. Tamcan O, Mannion AF, Eisenring C, Horisberger B, Elfering A, Muller U. The course of chronic and recurrent low back pain in the general population. *Pain*. 2010;150:451–457.
119. Croft PR, Lewis M, Papageorgiou AC, et al. Risk factors for neck pain: a longitudinal study in the general population. *Pain*. 2001;93:317–325.
120. Nolet PS, Cote P, Cassidy JD, Carroll LJ. The association between a lifetime history of a neck injury in a motor vehicle collision and future neck pain: a population-based cohort study. *Eur Spine J*. 2010;19:972–981.
121. Althaus A, Hinrichs-Rocker A, Chapman R, et al. Development of a risk index for the prediction of chronic post-surgical pain. *Eur J Pain*. 2012;16:901–910.
122. Pinto PR, McIntyre T, Ferrero R, Almeida A, Araujo-Soares V. Risk factors for moderate and severe persistent pain in patients undergoing total knee and hip arthroplasty: a prospective predictive study. *PLoS One*. 2013;8:e73917.

CHAPTER FIFTEEN

Dietary Influence on Pain via the Immune System

Stacie K. Totsch, Megan E. Waite, Robert E. Sorge[1]
Department of Psychology, University of Alabama at Birmingham, Birmingham, Alabama, USA
[1]Corresponding author: e-mail address: rsorge@uab.edu

Contents

1. Cytokines and Immunity 436
2. Cytokines and Chronic Pain 439
3. Obesity and Pain 440
4. Diet and Inflammation 441
5. Omega-3 and Omega-6 PUFAs 441
6. Saturated Fats 444
7. Grape Seed Extract 444
8. Green Tea Extract 445
9. Soy Products 446
10. Broccoli 447
11. Carotenoids 448
12. Ginger 449
13. Ginseng 451
14. Caffeine 452
15. Grains and Gluten 453
16. Ketogenic Diets 453
17. Dietary Interventions 454
18. Conclusions 456
References 457

Abstract

Obesity rates are approaching epidemic proportions and are a significant factor in annual health care costs. In addition to cardiovascular comorbidities, the presence of diabetes and/or chronic pain is extremely high in this population of individuals. It is now well accepted that the cells of the innate (and adaptive) immune system mediate both acute and chronic pain through release of cytokines into the system. In this chapter, we outline the ways in which poor food choices and elevated adipose tissue (body fat) are likely to activate the immune system and increase inflammation and pain. In addition, we explore the ways in which a variety of foods (e.g., broccoli, ginger, grapes, and fish oils) may have anti-inflammatory effects via their direct action on cells in the immune system and on the subsequent release of inflammatory cytokines. Some foods

Progress in Molecular Biology and Translational Science, Volume 131
ISSN 1877-1173
http://dx.doi.org/10.1016/bs.pmbts.2014.11.013

(green tea, ginger, and broccoli) have been found to antagonize specific cell surface receptors, whereas others (grapes, soy proteins, tomatoes and ginseng) appear to reduce nuclear translocation of the major transcription factor NFκB, thereby reducing production of inflammatory cytokines. Together, we provide data in support of the use of diet interventions to reduce pain and inflammation in patients suffering from chronic pain or other inflammation-mediated disorders.

The experience of chronic pain is highly subjective and can arise from any number of potential sources. One commonality between most patients suffering from some form of chronic pain is the presence of elevated levels of proinflammatory cytokines in blood and tissue. These cytokines are the body's defense against injury and/or pathogen invasion and serve to induce hypersensitivity and immune cell mobilization to the site. In most cases of chronic pain, the immune system comes into action and experimental suppression of various immune cell types can reduce the expression of chronic pain. One aspect that has received little attention historically is the contribution of diet to pain—aside from the impact of diet on obesity and type 2 diabetes. However, various foods interact directly with the immune system and can increase or decrease immune cell function and subsequent pain at sufficient levels. This chapter will outline the mechanisms that make these alterations possible and discuss the viability of dietary interventions for treatment of chronic pain.

1. CYTOKINES AND IMMUNITY

The immune system, as it is generally considered, involves two interacting systems—the adaptive and innate immune systems. The adaptive immune system most often is discussed in terms of antibodies and the classical T cells (and B cells). In brief, this system uses a type of immunological memory of past pathogens to mount an immune response and destroy the pathogen or infected cell. This aspect of the system forms the basis for vaccinations. Within the adaptive immune system, the primary cell types thought to be responsible for pain are the Th1 and Th2 (CD4+) cells. Th1 cells are mainly proinflammatory in nature (releasing IL-2, IFN-γ, and TNFα), whereas the Th2 cells are anti-inflammatory (IL-4 and IL-10). In addition, Th2 cells can inhibit macrophages through release of IL-4 and IL-10 (see Table 1) and both Th1 and Th2 cells can inhibit one another to maintain homeostasis.

Table 1 Some of the major immune cells that comprise the innate and adaptive immune system

Immune cell	System	Cytokines released
Mast cells	Innate	Pro: IL-6, IL-18, IFN-γ, TNFα Anti: IL-4, IL-5, IL-10, IL-13
Neutrophils	Innate	Pro: IL-1, IL-6, IL-7, IL-9, IL-16, IL-17, IL-18, TNFα, PGE_2 Anti: IL-4, IL-10, TGFβ
Macrophages (microglia, astrocytes)	Innate	Pro: IL-1, IL-6, IL-8, IL-12, IL-15, IL-18, IL-23, IL-33, TNFα, PGE_2 Anti: IL-10, TGFβ
Lymphocytes—Th1	Adaptive	Pro: IL-2, IFN-γ, TNFα
Lymphocytes—Th2	Adaptive	Anti: IL-4, IL-5, IL-10, IL-13
Lymphocytes—Th17	Adaptive	Pro: IL-17, IL-21, IL-23

For each cell type, the main pro- and anti-inflammatory cytokines are listed.

The innate immune system is comprised of a number of cell types (see Table 1) and forms the first-line defense against infection. All cells within this system function in much the same way—to recognize evolutionarily conserved aspects of bacteria and viruses to elicit an immune response to invasion. One of the most-studied cell types in the innate immune system are the microglia cells that function as macrophages within the brain and central nervous system. Microglia are found throughout the central nervous system with extended processes covered with receptors in a surveillance state, referred to as ramified. These arms of the microglia express cell surface (or cytosolic in the case of the NOD-like receptors) receptors that function to recognize specific patterns or danger signals. Upon detection of a signal, the processes retract and the microglia become activated and begin releasing molecules, including cytokines. The most common of the pattern recognition receptors (referred to as pathogen-associated molecular pattern receptors) found on microglia are the toll-like receptors that recognize bacterial components (TLR1, 2, 4, 5, 9, 11, and 13) and viral genetic material (TLR3, 7, 9, and 13). For instance, the toll-like receptor 4 (TLR4) recognizes lipopolysaccharide (LPS), a cell wall component in Gram-negative bacteria. Together, the danger- (or damage) associated molecular patterns (DAMPs) are in place to recognize conserved structures of common pathogens and mount an immediate response. On the other hand, the DAMP receptors recognize molecules that are released following tissue or cell damage. The purinergic receptors are the most-studied members of this class of receptors. From this class, the P2X (ion channel) and P2Y (G protein-coupled) subclasses respond to excessive ATP as a result of tissue damage/cell death. Regardless of the receptor type involved, activation results in immediate intracellular signals culminating with the transcription and release of cytokines (see below). As such, the innate immune system can be rapidly activated and mobilized to deal with common infectious agents without the need for antibody development (a key component of the adaptive immune system).

Each cell surface receptor (TLRs, P2X/Y, TNF, and IL-1) activates IκB kinase (IKK), which results in nuclear translocation of NFκB. NFκB is a transcription factor that is responsible for the synthesis of a variety of inflammatory mediators including IL-1β, IL-2, IL-6, IL-8, IL-12, IL-18, COX2, and TNFα. In the case of IL-1β and IL-18, the translation products are immature forms of the cytokines referred to as pro-IL-1β and pro-IL-18. Successful cleavage of the immature products into their mature and active forms is the job of caspase-1 as a result of inflammasome activation. The inflammasome is a cytosolic multiprotein complex that is assembled following recognition of danger signals. As a result of specific intracellular signals, the inflammasomes (NLRP1, NLRP3, NLRC4, AIM2, and IPAF) are formed and bind to caspase-1 to allow cleavage of pro-IL-1β and pro-IL-18 to their mature forms (IL-1β and IL-18). Therefore, it is the coordinated effort of cell surface receptors, transcriptional activation, and intracellular inflammasome assemblage that results in proinflammatory cytokine release. Importantly, aside from the rapid cytokine response, the innate immune cells actively recruit cells of the adaptive immune system to facilitate antibody development and production. Thus, both systems work in concert to recognize and remove pathogens (or damaged cells) and to maintain a balance via pro- and anti-inflammatory cytokines.

Activation of the immune system results in the release of cytokines (pro- and anti-inflammatory; see Table 1) that work to maintain homeostasis and attract cells involved in repair (if necessary). Administration of LPS, recognized by TLR4, elicits the release of prostaglandins and cytokines from macrophages and results in hypersensitivity that can be reduced by a macrophage inhibitor.[1] Local release of cytokines by immune cells attracts other cells and promotes hypersensitivity in an evolutionarily adaptive manner to avoid further damage and encourage protection of the injured area. As expected, administration of IL-1β, IL-6, or TNFα causes persistent hyperalgesia in rats[2–4] whether injected locally or into the brain ventricles. The fact that cytokines cause pain-like behaviors suggests that there may be high levels of circulating cytokines in a variety of chronic pain conditions and that seems to be the case.

2. CYTOKINES AND CHRONIC PAIN

There are a number of sources of chronic pain with a wide range of clinical symptoms. One commonality to a majority of these chronic pain conditions is the presence of elevated levels of C-reactive protein (CRP) thought to be a blood-borne marker of inflammation. It is relatively easy to study and single "snapshot" measures of CRP (as well as IL-8, TNFα) are reliable indicators of monthly serum variability.[5] Though not an exhaustive list, increased levels of CRP have been found in patients suffering from interstitial cystitis,[6] low back pain,[7] and fibromyalgia,[8] but not CRPS1.[9] In terms of specific cytokines, TNFα is elevated in patients with widespread pain[10] and neuropathic pain.[11] Interestingly, acute noxious stimulation elicits a rise in tissue levels of inflammatory cytokines in humans that is reduced by anti-inflammatory medication.[12] Thus, it may be that chronic pain patients suffer from long-term immune system activation resulting in continuous release of proinflammatory cytokines. In contrast, there is evidence that levels of anti-inflammatory cytokines are decreased in patients with widespread pain,[10] neuropathic pain,[11] and temporomandibular disorder.[13] Together, it is reasonable to assume that chronic pain may arise through an increase in proinflammatory cytokines and the resulting inhibition of anti-inflammatory cytokines.

There is one other chronic condition that has also been linked to elevated inflammatory cytokine levels—obesity. Amount of adipose tissue is directly related to CRP levels[14] and those levels are positively related to metabolic syndrome symptoms, such as presence of diabetes.[15] In fact, over years of testing, CRP levels were more often associated with obesity than any acute injury/infection.[16] Adipose tissue releases the adipokine leptin, which has a known role in hunger and metabolism,[17] with leptin levels being positively correlated with adipose tissue accumulation. Leptin also activates the immune system directly and causes the release of proinflammatory cytokines (see Ref. 18), through the classical JAK/STAT or p38/MAPK/ERK pathways.[19] Therefore, obese individuals, with high-adipose tissue and subsequent leptin release, may experience persistent immune system activation and accompanying pain that may be unrelated to increased pressure on weight-bearing joints.

3. OBESITY AND PAIN

The body mass index (BMI) measurement is the standard used to determine whether a person meets the criteria for obesity. Within this scale is a grading system for obesity from moderate (BMI 30–35) to severe (BMI 35–40) and very severe (BMI > 40). Obesity is a risk factor for many cardiovascular and metabolic disorders and is often a risk factor for chronic pain (see Ref. 20). In older adults, obese subjects were twice as likely to report chronic pain and severely obese subjects were four times more likely.[21] Abdominal obesity is a positive predictor of chronic pain, even when presence of osteoarthritis and neuropathies are accounted for.[22] As expected, inflammatory cytokine levels are elevated in obese subjects[22,23] and weight loss has been shown to reduce these levels.[24,25] Chronic inflammation is thought to contribute to the development of osteoarthritis and there is a strong correlation between BMI and osteoarthritis of the knee,[26] hands,[26a] and multiple other locations including the spine and neck (see Ref. 27). Interestingly, the hands and neck are not weight-bearing joints, which suggest that inflammation may be the root cause of the pain and not the increased load.

A chronic low-grade inflammatory condition may be expected to predispose obese subjects to hypersensitivity, but that is often not the case. Whereas obese subjects tend to be at a higher risk for chronic pain conditions like osteoarthritis, obese subjects often show higher pain thresholds in various pain tests.[28–30] This is also true of rats fed a high-fat diet—higher tail-flick latencies have been reported with obese rats.[31] Thus, increased adipose tissue accumulation results in high levels of proinflammatory cytokines, known to cause hypersensitivity when administered acutely. However, obese subjects (human or rodent) have reduced pain sensitivity in spite of presumed chronic inflammation. While somewhat counterintuitive, there is evidence that high levels of TNFα can contribute to demyelination[32] and reduce rapid pain transmission signals via A-δ fibers (first pain), which may explain the decreased sensitivity to acute pain. Furthermore, acute injection of TNFα has been shown to sensitize nociceptive C-fibers (second pain) in the rat.[33] Therefore, chronic low-grade inflammation (and increased TNFα) may decrease acute sensitivity and increase susceptibility to chronic pain states in obese subjects through action of cytokines on nociceptors directly.

4. DIET AND INFLAMMATION

There are cases of genetically based obesity; however, more often than not, obesity is a result of diet. The typical "Western" dietary pattern is one that is rich in refined grains, high-fat foods, processed meats, and sweetened beverages/desserts and is linked to obesity, metabolic disorders,[34,35] and depression.[36] Rats fed high-fat diets display high anxiety, aggression, and elevated corticosterone[37]; and rats fed a "cafeteria-like" diet showed more weight gain, insulin resistance, and inflammation than high-fat diet-fed controls.[38] With respect to inflammatory state, weight loss through diet or diet and exercise has been shown to reduce CRP and TNFα levels,[39–41] and this effect has been explained by a decrease in total fat mass.[39] Therefore, it seems that the inflammatory state caused by obesity can be partially alleviated through weight loss alone. To this end, obese mice given olive oil as a supplement in their diet showed lower levels of inflammation and reduced weight,[42] suggesting that specific foods have direct effects on inflammation.

In contrast to the Western diet, the Mediterranean diet (comprised of unrefined carbohydrates, nuts, fish, and olive oil) has been associated with increased health. Specifically, with respect to inflammation, 3 months of a "Mediterranean diet" was shown to reduce serum levels of IL-6 and CRP.[43] This suggests that this diet has either: (a) an anti-inflammatory effect, (b) lowered proinflammatory potential, or (c) a mixture of both. It is known that a diet high in whole grains can reduce TNFα and IL-6 levels,[44] but the positive effects of the Mediterranean diet are often attributed to the high levels of polyunsaturated fatty acids (PUFAs) that have a beneficial effect on inflammation.

In the following sections, we outline specific foods and food categories and their effects on the immune system. In each case, the implied or evidenced effect on pain is noted.

5. OMEGA-3 AND OMEGA-6 PUFAs

Omega-6 and omega-3 fatty acids are essential PUFAs. Both are required for proper growth and development and must be obtained through diet. In general, it is beneficial that the ratio of omega-3 to omega-6 fatty acids in the diet is close to 1:4, but the typical American diet consists of 15–20 times the amount of omega-6 fatty acids compared to omega-3.[45]

This is concerning when one considers how each of the fatty acids is used by the body. Prostaglandins (and other eicosanoids) are produced by COX2-mediated conversion of arachidonic acid during inflammation and are thought to amplify cytokine activity. Arachidonic acid is an omega-6 PUFA found in the cell membrane (phospholipid) and used during inflammation. As such, high levels of omega-6 PUFAs provide excessive amounts of prostaglandin precursor material leading to an increase in proinflammatory cytokine activity. As expected, increased prostaglandin activity is seen in patients suffering from cardiovascular disorder, diabetes, rheumatoid arthritis, inflammatory bowel disease, and obesity (see Ref. 45).

On the other hand, omega-3 PUFAs can substitute for omega-6 PUFAs in the membrane and, thus, reduce the synthesis of prostaglandins and subsequent inflammation (see Ref. 46 for a review). Omega-3 PUFAs are also involved in the production of resolvins[47] that act to reduce inflammatory cell activity. The omega-3 fatty acids are a group of fatty acids comprised of α-linoleic acid (ALA), eicosapentaenoic acid (EPA), and docosahexaenoic acid (DHA), with DHA being considered the most critical. In experiments with bone-marrow-derived macrophages, it was found that DHA blocked caspase-1 activity following LPS priming. This effect was due to inhibition of the NLRP1b and NLRP3 inflammasomes through action of DHA (and presumably EPA and ALA) on G protein-coupled receptors 40 and 120 (GPR40 and GRP120). In each case, IL-1β and IL-18 secretion was reduced following administration of DHA (or EPA and ALA), but cell death was not,[48] suggesting a specific role for omega-3 PUFAs in the reduction of proinflammatory cytokine release. Interestingly, omega-3 PUFAs have been shown not only to directly activate TRPV1 receptors in culture (at diet-relevant doses) but also to antagonize the behavioral response to capsaicin, a TRPV1 agonist[49] suggesting a complex action at TRPV1 receptors.

If omega-3 PUFAs are effective in blocking the effects of omega-6 PUFAs and attenuating the release of proinflammatory cytokines through inflammasome inhibition, it is reasonable to assume that dietary omega-3s would be beneficial for chronic pain patients. Figueroa and colleagues[50] fed rats an omega-3-rich diet for 8 weeks prior to spinal cord contusion injury and followed the rats for an additional 8 weeks with the diet present. Using radiant heat measures, they showed that the omega-3-rich diet completely abolished the hyperalgesic response following injury (Fig. 1), reduced CGRP-containing neuron sprouting and p38 MAPK gene expression in the dorsal horn.[50] To compare the effects of omega-3 and omega-6 fatty acids, mice were fed either concentrated omega-3 fish oil or safflower

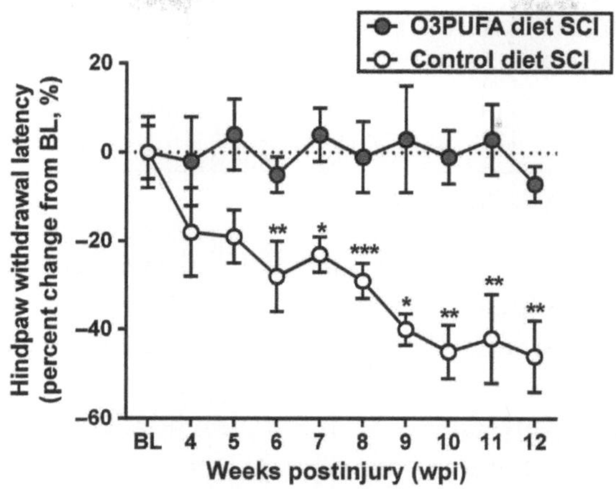

Figure 1 Thermal sensitivity on the radiant heat paw withdrawal test over the course of 12-week postinjury (spinal contusion) in rats. Rats were fed an omega-3-rich diet (O3PUFA:DHA = 12.82/100 g chow, EPA = 6.91/100 g chow) for 8 weeks prior to injury and for the remainder of the experimental time period. Control diet was matched for cholesterol content. Omega-3-rich diets were protective following spinal cord injury. *Adapted from Ref. 50 with permission (http://creativecommons.org/licenses/by-nc-sa/3.0/legalcode).*

oil (high in omega-6), and it was found that omega-3 supplementation reduced sensitivity to thermal stimuli and reduced the expression of ASICs and TRPV1,[51] suggesting a protective role of omega-3 PUFAs. In another study, mice were fed diets supplemented with DHA and/or EPA for 4 weeks prior to, and 45 days following, collagen-induced arthritis induction. DHA, but not EPA, reduced arthritis severity, IL-1β and IFN-γ levels and increased IL-10 in splenocytes.[52] While both of these studies used prophylactic treatments, there is a clear benefit of omega-3 PUFAs when it comes to prevention and/or reduction of chronic neuropathic pain.

In humans with rheumatoid arthritis, 24-week supplementation with EPA and DHA reduced the number of tender joints and swollen joints as early as 12 weeks into treatment. In addition, IL-1 secretion by macrophages was reduced by as much as 54% by the end of treatment.[53] This is in line with a more recent study that examined the effect of omega-3 PUFA supplementation in sedentary overweight adults. In a study by Kiecolt-Glaser and colleagues,[54] omega-3 supplements decreased IL-6 and TNFα levels over 16 weeks. These authors also found that plasma omega-3 fatty acids increased in the treatments groups.[54] While the previous study did not look at pain, one can infer that reduction in inflammatory cytokines would have an analgesic effect given that high inflammatory cytokine levels are comorbid with many chronic pain conditions. Importantly, serum levels of omega-3 are inversely related to CRP, IL-6, and TNFα levels, whereas serum-saturated fatty acid (SFA) levels are positively correlated.[55]

6. SATURATED FATS

SFAs are long-chain acids (saturated with hydrogen atoms) and are found in a wide variety of foods with animal products including dairy and meats. Saturated fats are known to have poor health outcomes associated with their intake, such as an increase in inflammatory cytokine gene expression.[56] There are many kinds of SFAs and the research to date has dealt with a number of them. In murine macrophage cultures, exposure to SFAs (but not PUFAs) was shown to induce expression of COX2 (involved in prostaglandin synthesis) through NFκB activation. This activation of COX2 was dependent on TLR4 and could be blocked by administration of PUFAs.[57] This suggests that, whereas SFAs may activate the TLR4 receptor to presumably cause NFκB induction, PUFAs block TLR4 or may inactivate the receptor. In further studies, it was shown that exposure of monocytes to SFAs increased inflammatory gene expression (IL-1β and TNFα) and enhanced the inflammatory response (measured by IL-6 and IL-8 secretion) to future LPS or TLR2 agonists, indicating a sensitized immune response. The amplification was hypothesized to be due to conversion of SFAs to ceramide.[58] Ceramide activity has been linked to insulin resistance[59] and is seen following administration of TNFα.[60] In human monocytes, SFAs have also been shown to activate TLR2 (which can be inhibited by DHA), leading to expression of pro-IL-1β and the assembly of the NLRP3 inflammasome.[61] Taken together, dietary intake of SFAs has a direct stimulatory effect on the immune system via TLR4 that leads to induction of proinflammatory cytokines. This activation can prime the immune system in the event of future challenge and lead to hyperactive responses to damage or invasion potentially leading to chronic pain or hypersensitivity. This is a serious consideration for those consuming large quantities of SFAs in their diet.

7. GRAPE SEED EXTRACT

Grape seeds and their extracts have high amounts of proanthocyanidins, which are believed to be powerful antioxidants that have anti-inflammatory effects through reactive oxygen species (ROS) scavenging. Recent research has indicated that grape seed proanthocyanidin extracts (GSPE) have secondary effects that may have a more direct anti-inflammatory effects through modulation of immune cells. In rodents, GSPE reduced edema and local levels of IL-1β, TNFα, and PGE_2 following carrageenan,[62] pain in the late phase of the formalin test,[63] hyperalgesia and allodynia in an osteoarthritis model,[64] hypersensitivity and TNFα levels in an arthritis model,[65] and allodynia in a streptozotocin-induced diabetic neuropathy model.[65a] Additionally, in a model of temporomandibular joint pain via complete Freund's adjuvant, GSPE reduced microglia and astrocyte activation in the brain and nerves in addition to a local reduction of MAPK,[66] suggesting an immune-cell-specific mechanism of action. In human epithelial cells, GSPE blocked increased expression of IL-6 as a result of treatment with IL-17, IL-1β, or TNFα. This action is thought to be due to inhibition of the NFκB pathway[67] and was seen in human adipocytes and macrophages as well.[68] Kim and colleagues suggest that the effect of GSPE is on inhibition of IkBα phosphorylation that is necessary for NFκB translocation to the nucleus.[67]. Therefore, the anti-inflammatory effects of GSPE are likely to be mediated through a reduction in cytokine transcription as a result of inhibition of NFκB activity.

Interestingly, other fruits with high levels of anthocyanidins have recently been examined for their potential to reduce pain. The bilberry is regularly consumed in Finnish populations and contains high amounts of anthocyanidins. In a study comprised of healthy controls with some symptoms of metabolic syndrome and no history of diabetes, a diet rich in bilberries was found to reduce CRP, IL-6, and IL-12 levels.[69] Similarly, in obese rats, a blueberry-rich diet also reduced plasma levels of TNFα, IL-6, and CRP and tissue levels of NFκB.[70] Together, the data support the notion that fruits high in anthocyanidins have a clear anti-inflammatory effect likely through inhibition of NFκB.

8. GREEN TEA EXTRACT

For many centuries, green tea has been used as a traditional medicine for a variety of illnesses. However, only in the recent past the active component (epigallocatechin gallate, EGCG) has been isolated, and the mechanism of action investigated. In terms of reduction in experimental pain models, EGCG has been shown to reduce arthritic pain,[71] fatigue-induced pain,[72] and allodynia/hyperalgesia following chronic constriction injury.[73] Mechanistically, the most recent evidence indicates that EGCG binds to the dominant laminin-binding protein in neutrophils, macrophages, and monocytes, 67LR.[74] However, at the level of inflammation, EGCG has been shown to block NLRP1 inflammasome activity,[75] P2X4-induced IFN-γ genes expression,[76] and EGFR activation by EGF.[77] In mice, 6 weeks of treatment with EGCG reduced body weight and increased the number of anti-inflammatory T cells in the spleen.[78] There have yet to be human studies examining the potential of EGCG to reduce pain in any chronic pain populations, but there is evidence that daily EGCG can reduce lymphocyte (generally, T cells and B cells) counts in patients with chronic lymphocytic leukemia.[79] Importantly, high amounts of EGCG (equivalent to 8–16 cups of green tea per day) are well tolerated by humans,[75] suggesting that consumption of green tea itself may work as an intervention in place of synthesized extract. Given the current evidence of potent anti-inflammatory effects of EGCG on immune cell receptors and pathways as well as tolerability, there is good reason to believe that chronic pain patients may benefit from green tea in the diet.

9. SOY PRODUCTS

Soy beans contain a number of proteins called isoflavones that function as phytoestrogens—plant-derived estrogen-like compounds. One of the primary soy isoflavones is genistein. In culture, genistein has been shown to bind to estrogen receptors and to activate PPAR-γ,[80] which has a role in suppression of proinflammatory genes (see Ref. 81). In human, chondrocytes primed with LPS, genistein reduced the expression of IL-1β, COX2, and NO,[82] indicating a clear anti-inflammatory effect for genistein and soy in general. In rat models of partial spinal nerve ligation, a soy-rich diet produced more pronounced reductions in allodynia and hyperalgesia when given for 2–4 weeks[83,84] as opposed to 1 week before injury.[85] Soy-based diets also had a significant effect on edema, allodynia, and hyperalgesia in response to inflammatory injury with CFA (Fig. 2) or carrageenan.[86] In mice, soy diets reduced NFκB activity by decreasing the degradation of IkBα following LPS and reduced blood levels of IL-1β, Il-6, and TNFα.[87] In humans, there has been very little done to examine the potential of soy to alleviate pain. In one study of patients suffering from osteoarthritis, consumption of soy protein significantly reduced pain over the course of 3 months and increased range of motion in the affected joint.[88] Therefore, soy-based diets have anti-inflammatory effects through activation of PPAR-γ and subsequent suppression of proinflammatory cytokine genes in immune cells.

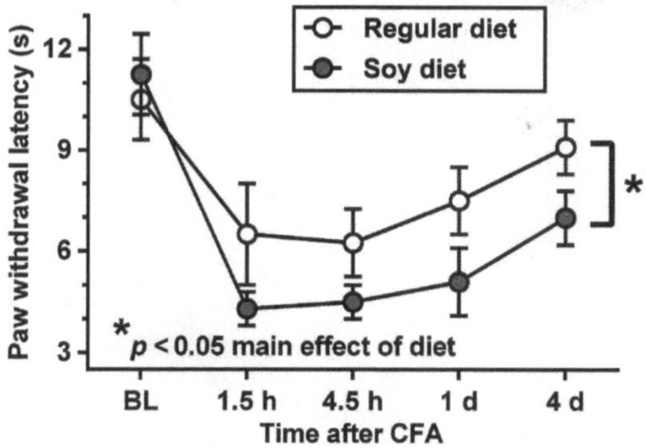

Figure 2 The effect of soy diet on paw withdrawal latencies in the radiant heat paw withdrawal test following 2 weeks of soy (338.46 g/kg diet) or casein-based diet and intraplantar injection of complete Freund's adjuvant (CFA). Soy diet significantly reduced the peak of the hyperalgesic response to CFA. *Adapted from Ref. 86 with permission.*

10. BROCCOLI

Broccoli is one of a few cruciferous vegetables that include cauliflower, bok choy, and cabbage that contain organosulfur compounds of varying amounts. In terms of immune system interactions and possible pain suppression, the most important of these compounds is the isothiocyanate sulforaphane (SULF). In human T-lymphocytes and mouse splenocytes, SULF has been shown to reduce immune cell proliferation and production of proinflammatory cytokines.[89] It has been shown to reduce inflammatory T cell activation in mouse airway albumin models[90] and reduce TLR4 expression in macrophages.[91] Specifically, SULF is thought to suppress NFκB through reduced IkB phosphorylation[92] as well as reducing expression of COX2 in cell cultures[65] and blocking activity of MD2, the coreceptor for TLR4.[93] Therefore, it seems as if SULF may have anti-inflammatory effects via three distinct mechanisms—NFκB, TLR4, and COX2.

In various pain models, SULF has been shown to be efficacious at reducing pain and inflammatory cytokine activity. In a mouse model of collagen-induced arthritis, SULF-suppressed arthritis scores (Fig. 3) as well as levels of TNFα, IL-6, IL-17, and IFN-γ while increasing levels of IL-10.[94] SULF was also given as a dietary supplement to rats with streptozotocin-induced diabetic neuropathy following induction, as a model of postsymptomatic

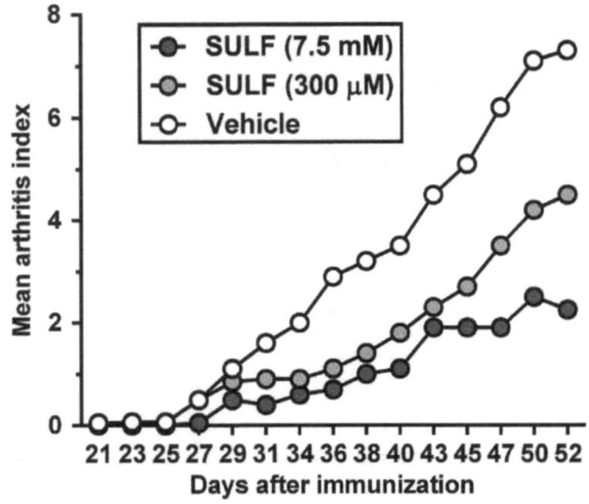

Figure 3 Mean arthritis index in mice with collagen-induced arthritis treated with various concentrations of sulforaphane (SULF) or vehicle. Three weeks after the primary immunization with collagen, mice were injected with SULF every second day for 5 weeks and assessed for arthritis score. *Adapted from Ref. 94 with permission.*

intervention. SULF reduced pain scores as well as levels of TNFα, IL-6, and COX2.[95] Finally, in a spinal nerve transection model of neuropathic pain, SULF decreased levels of ROS, IL-1β, TNFα, and Iba-1—a marker of microglia activation. Spinal treatment with SULF importantly reduced allodynia following the SNT injury, suspected to be due to SULF activation of the anti-inflammatory HO-1 pathway.[96] To date, there have been no systematic examinations of the effect of SULF on pain levels in human patients. However, it is likely that cruciferous vegetables with high levels of sulforaphane, broccoli in particular, could reduce inflammation and subsequent pain if incorporated into the diet in sufficient amounts.

11. CAROTENOIDS

Carotenoids are organic pigments found in many orange, yellow, and red vegetables and fruits. The two of most interest are β-carotene (found in carrots) and lycopene (found in tomatoes). Much of the early work that was done on β-carotene in humans found that supplementation increased cell surface expression of antigen-presentation molecules in blood monocytes.[97,98] When given mixtures of both β-carotene and lycopene for 2 weeks, male subjects showed an increased IL-2 and IL-4 response in T-lymphocytes that was subsequently reduced by carotenoid depletion.[99]

This suggested that carotenoids were important for a proper immune response, supporting a clear and vital role in the immune system.

With respect to inflammation, the role of carotenoids, and lycopene in particular, is even more clear. In various cell lines, lycopene administration has been shown to reduce chemical-induced (LPS, cigarette smoke extract, oxysterols) expression of IL-1β,[100] IL-12,[101] IL-6,[100,102] TNFα,[100,103,104] iNOS,[105] p38/JNK and/or ERK1/2 MAPKs[101,102,104,106], NFκB translocation,[100,102–104,106,107] and increase PPARγ.[100,106] In addition, a recent report showed that lycopene reduced COX2 expression through heme oxygenase-1 activation,[107] suggesting a direct anti-inflammatory mechanism. Together, it is reasonable to assume that lycopene treatment should have an analgesic effect in various pain conditions.

To that end, there is limited research investigating the role of lycopene in the treatment of chronic pain. In rat models of streptozotocin-induced diabetic neuropathy, chronic treatment with lycopene dose-dependently reduced pain scores in various tests.[108,109] To date, lycopene has not been examined in other pain models in animals. In humans, lycopene (or tomatoes) are often included in dietary interventions, and so it is difficult to assess the primary effects of lycopene on pain measures. However, it should be noted that tomato drink effectively reduced TNFα levels in human blood.[110] The limited clinical trials of lycopene in human conditions has been focused on prostate cancer and, when reported, the associated pain. Only one study has assessed pain and found that lycopene treatment reduced bone pain in these patients.[111] It is worth noting that both clinical trials using lycopene for prostate cancer showed a significant reduction in analgesic use in the patients receiving lycopene,[111,112] suggesting a general reduction in pain. Given the identified mechanisms whereby carotenoids reduce a number of proinflammatory pathways, it is not surprising that these foods are generally a large part of any dietary intervention (i.e., Mediterranean diet) and, at realistic consumable levels, should have a direct impact on pain transmission.

12. GINGER

The spice ginger often refers to the root of the ginger plant and has been used for medicines and cooking for centuries. In healthy controls, ginger has been shown to reduce muscle soreness[113] and muscle pain following exercise.[114] Ginger also reduces pain in patients with moderate to severe dysmenorrhea[115] and osteoarthritis[116] (see Ref. 117 for a review). In animal models of pain, ginger reduces pain in the acetic acid test,[118,119] thermal tests,[119,120] and following capsaicin injection (Fig. 4; 118). The results of human and animal studies point to a general effect of ginger on many forms of pain.

Mechanistically, ginger was first found to activate TRPA1 channels in cell culture in a similar manner to cinnamaldehyde and mustard oil.[121] Ginger was also determined to be a voltage-dependent calcium channel blocker in anesthetized animals[122] and, recently, ginger has been found to bind directly to the TLR4 coreceptor MD2 and reduce TNFα, IL-1β, IFN-γ, and NFκB expression following LPS treatment in animals.[123] This latter effect on inflammatory cytokine activity may also be due to the discovery that ginger directly activates IKKβ, leading to reduced IkB phosphorylation in human macrophages treated with LPS.[124]. In whole animals, this effect was confirmed when mice given LPS showed reduced levels of IFN-γ and IL-6 in serum and reduced NFκB and COX2 expression with ginger treatment.[125] Ginger reduced high-fat diet-induced increases in TNFα and IL-6 levels in rats[126] and suppressed TNFα, IL-6, and CRP levels in patients with type 2 diabetes.[127] Finally, recent research supports the long-held belief that ginger is a potent analgesic with clear anti-inflammatory effects mediated through the immune cells.

Turmeric is a member of the ginger family and the principal ingredient, curcumin, has recognized analgesic effects as well. Curcumin has been shown to reduce allodynia in streptozotocin- [128] and CCI-induced neuropathic pain,[92,129] as well as acetic acid writhing in rats.[130] Very recently,

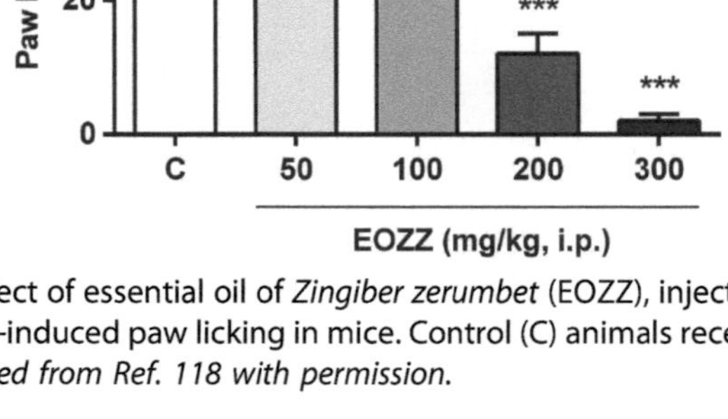

Figure 4 The effect of essential oil of *Zingiber zerumbet* (EOZZ), injected intraperitoneally, on capsaicin-induced paw licking in mice. Control (C) animals received only vehicle injections. *Adapted from Ref. 118 with permission.*

curcumin was found to reduce hypersensitivity in a postoperative pain model without affecting locomotor behavior and without the need for presurgery treatment.[131] Curcumin's effects are thought to be due to suppression of NFκB and reduced neutrophil activity,[132] though there is some indication that curcumin may function as a blocker of the noxious heat-gated and capsaicin receptor, TRPV1.[133] In a model of postoperative pain (using hind paw incision), treatment with curcumin effectively reduced hypersensitivity and swelling in mice. This effect was thought to be due to enhanced TGF-β in the curcumin-treated mice.[134] In human blood monocytes, curcumin reduces the production of IL-1β, TNFα, and IL-8 following LPS, supporting the anti-inflammatory nature of this compound.[135] The limited work in chronic pain patients demonstrates that curcumin reduces pain in osteoarthritis patients.[136] Similar to ginger, turmeric seems to work as an anti-inflammatory agent through multiple pathways and should be considered as an addition to the diet of patients suffering from chronic pain.

13. GINSENG

Ginseng has been used medicinally in Asia for centuries and is finally gaining some acceptance in Western culture. It has a number of preparations that may or may not have differential effects on the efficacy of the compound; thus, for the sake of simplicity, all studies of ginseng will be categorized together. The active chemicals in ginseng are thought to be a class of steroid glycosides known as the ginsenosides. There are a number of ginsenosides that can be classified into specific groups, but will be considered together unless otherwise specified. Very little human clinical data exist that examine the potential of ginseng to reduce pain in patients; however, one small study found that it performed as well as other common analgesics in fibromyalgia patients.[137]

However, there have been a number of animal studies that have examined the potential of ginseng to reduce pain in rodents and all show positive results to date. Importantly, one study tested 500 mg/kg but found that doses up to 5000 mg were well tolerated in animals.[138] This suggests that ginseng is safe even at very high doses. Ginseng, in one form or another, has been shown to reduce pain caused by capsaicin,[139] substance P,[140] acetic acid,[141] formalin,[141,142] carrageenan,[138] and to reduce cold-swim stress-induced analgesia.[143] In addition, administration of ginseng reduced IL-6 expression in a collagen-induced arthritis model,[138] suggesting a direct effect on inflammatory cytokine production. This suggestion is confirmed as ginseng is known to inhibit voltage-gated calcium channels in rodent sensory neurons[144] and dorsal root ganglia.[145] There is some evidence that ginseng may block TRPV1 channels[146] as well. Mechanistically, these effects may all derive from the findings that red ginseng can block LPS-induced expression of NO, PGE2, COX2, TNFα, and IL-6 in cell lines, presumably through inhibition of NFκB nuclear translocation.[147] There is also very recent evidence that red ginseng inhibits the NLRP3 (and possibly AIM2) inflammasome,[148] resulting in reduced IL-1β production. Taken together, these results suggest a clear effect of ginseng on the pain pathway and a need for future clinical studies examining the efficacy of ginseng as a natural anti-inflammatory.

14. CAFFEINE

It is estimated that up to 75% of young adults consume caffeine daily.[149] Caffeine itself is a xanthine alkaloid with very specific properties in the body where it acts as an antagonist to adenosine. Adenosine is important for the formation of ATP, for sleep, and has some actions in immune system mediation. Specifically, adenosine blocks the expression of the cellular adhesion molecule ICAM-1 which is critical for the production of TNFα and IFN-γ following stimulation with IL-18[150] or LPS.[151] This would suggest that blockade of adenosine receptors by caffeine would have a proinflammatory effect and this is the case in adult peripheral blood cells where TNFα levels are increased following LPS priming.[152] Thus, it appears that caffeine would have a negative effect on pain, but this is often not the case.

In animals, caffeine was shown to reduce nociceptive behaviors in the late phase of the formalin test,[153] but there are very few animal studies to support or challenge this finding. The bulk of studies investigating the effects of caffeine on pain have shown moderate effects on nonmigraine[154] and tension headaches[155,156] in human sufferers, but have found that caffeine is no more effective than traditional over-the-counter medications. The enhanced effects come from using caffeine as an adjuvant and using it in combination with other drugs like ibuprofen[155] or morphine.[157] While the mechanism of action in the immune system argues against the efficacy of caffeine as an analgesic, the human clinical data supports it use. It may be, however, that the positive effects in tension headaches come from the established ability of caffeine to constrict cerebral blood flow.[158] This ability may be directly linked to the headaches experienced in caffeine withdrawal.

15. GRAINS AND GLUTEN

Gluten is a protein product of grains and, as such, grains and gluten have related effects on physiology. While the actual mechanisms of action are somewhat unclear, diets that include each affect pain and cytokine levels in rodents and humans. In a correlational study, humans with high levels of whole grain or bran in their diet tended to have lower levels of TNFα and CRP than those with lower intake.[159] Other aspects of the subjects' diets may have contributed to their reduced cytokine levels, but were not examined. Gluten may cause an immune response and subsequent inflammation in the gut in patients with celiac disease. In patients with comorbid fibromyalgia and celiac disease, a full year of a gluten-free diet decreased fibromyalgia symptoms including pain and fatigue.[160] In addition, in celiac patients, the inflammation (characterized by heightened IFN-γ levels) caused by gluten was driven by increased production of IL-18. Increased IL-18 was also present in the absence of a challenge,[161] suggesting chronic immune system activity. There may be a benefit to increased whole grains and bran in the diet, but this must take into account the possibility of a gluten-induced inflammatory response in certain populations.

16. KETOGENIC DIETS

Ketogenic diets have been used for treatment of epilepsy in children[162] and are suggested as treatments for other disorders including pain (see Ref. 163). This is a high-fat, low carbohydrate diet that encourages ketosis as a form of energy consumption and is similar to the Atkins diet. Whereas the diet has been efficacious for treating some forms of epilepsy, the results for pain reduction are mixed. Early work showed that rats fed a ketogenic diet for 10 weeks had lower tail-flick latencies,[164] suggesting greater sensitivity to painful thermal stimuli. Subsequent work by Ruskin and colleagues has found that various ketogenic diet formulations resulted in less sensitivity in hot plate tests between 49 and 51 °C.[165,166] These authors also noted a reduction in CFA-induced edema in the paw, though allodynia was never measured.[165] However, ketogenic diets were shown to have no effect on neuropathic pain induced by sciatic nerve constriction or paclitaxel injection.[167] It appears that the analgesia effect of the ketogenic diet may be very limited to a small thermal range, but there is good evidence that the diet has effects on the immune system. Rodents on ketogenic diets showed reduced IL-18 in blood[168]; reduced IL-1β, Il-6, and TNFα in brain[169]; and reduced TNFα, COX2, PGE_2, and NFκB in hippocampus.[170] The reduction in NFκB was thought to be due to PPARγ activation (see earlier, Section 9) in neurons in the hippocampus.[170] Thus, there is good evidence that ketogenic diets have immune-cell-specific effects, but their use as analgesic treatments may be limited.

The above sections have outlined the currently available research on dietary items and their effect on immune cells and pain specifically. The following section will deal with dietary interventions for pain in animals and humans with various painful conditions. It is useful to keep in mind that many of the interventions outlined below include any number of foods from the above sections to round out a treatment strategy and that one aim of this chapter is to provide evidence that dietary interventions have the potential to reduce pain, primarily through immune system interactions.

17. DIETARY INTERVENTIONS

The preceding sections have outlined a number of specific foods or food groups that have anti-inflammatory effects that are mediated through interactions with the immune system cells. Given that obesity and resulting high levels of leptin are known to activate the immune system, it seems logical that weight loss would have clear beneficial effects on pain. In addition, the fact that a common complaint in obese individuals happens to be knee pain, reduction in weight would also alleviate the added stress on this joint. Thus, reduction in weight should be beneficial for the aforementioned reasons, but the diet itself may have anti-inflammatory effects on its own over and above the obvious.

Many of the foods listed above with anti-inflammatory properties consist of fruits (grapes and berries), vegetables (broccoli), and legumes (soy beans). These are staples of the classic vegetarian, or vegan, diet, and it is not surprising that there are some studies that have examined the analgesic effects of this dietary pattern. In patients with rheumatoid arthritis, a fasting period followed by 1 year of a vegetarian diet was found to reduce pain and markers of inflammation (i.e., CRP levels). In fact, there was a noticeable reduction in pain, swollen joints, tender points, and CRP by the fourth week of the diet,[171] suggesting that this diet may have almost immediate pain-relieving properties. A slightly more restrictive vegan diet was also shown to reduce pain in fibromyalgia syndrome patients after only 3 months. The analgesic effect of the diet was accompanied by a significant drop in cholesterol levels,[172] evidencing another benefit of this diet. Therefore, the limited evidence seems to support a vegetarian diet as an effective means to reduce pain in a few chronic pain conditions, but the high drop-out rates[173] suggest that patients have a difficult time with this manner of lifestyle change, limiting its usefulness.

For a number of years, the Mediterranean diet has been espoused as a beneficial diet for those struggling with cardiovascular issues or obesity. This diet is high in unrefined carbohydrates, fish oils, and nuts that coincide closely with the reported anti-inflammatory effects of the PUFAs explored previously. A Mediterranean diet supplemented with higher than normal levels of olive oil or nuts was found to reduce CRP, IL-6, CD11b (a marker for microglia), and CD40 (required for T cell activation by macrophages) levels in serum of subjects at high risk for cardiovascular disease.[174] The reduction in inflammatory biomarkers supports the finding that this diet reduced pain in females with rheumatoid arthritis after 3–6 months of the diet.[175] Finally, one additional study examined whether weight loss associated with the Mediterranean diet was necessary for the anti-inflammatory effects. It was found that the diet reduced CRP, IL-6, IL-18, and TNFα levels *prior to* actual weight loss occurring. Weight loss enhanced this reduction but was not necessary to see moderate, but significant, reductions in inflammatory cytokine levels.[176] Thus, it is clear that the Mediterranean diet has anti-inflammatory effects on its own (see Ref. 177 for a number of genes that are inhibited/induced by this diet in animals) and should be considered a viable option for reducing pain in human patients.

As noted above, the Mediterranean diet had potent anti-inflammatory effects in human patients that were separate from the effects of weight loss.[176] Because of the noted beneficial effects of weight loss on pain, a number of studies have been completed that utilize caloric restriction and/or exercise as interventions for patients with chronic pain. The first group of studies deals with primarily obese patients and examines the resulting change in cytokine levels during the course of treatment. Following bariatric[178] or specifically gastric bypass surgery,[179] levels of CRP, IL-6, TNFα, and leptin are reduced. Interestingly, one group noted that obese subjects also had low levels of the NFκB inhibitor IRAK3 that returned to control levels following surgery.[178] These results suggest that obese subjects have less inhibition of NFκB that may contribute to increased inflammation, but that these levels are normalized with weight loss following surgery. In a variety of other studies, weight loss through diet in obese subjects resulted in reductions of IL-6, IL-18, CRP, TNFα,[180,181] and increased anti-inflammatory glycoproteins.[182] These studies suggest that weight loss through diet has a clear potential as an intervention for chronic pain.

To address the possibility of using weight loss and diet to reduce pain in obese individuals, two studies have examined osteoarthritis pain in obese individuals following weight loss. After 12 weeks of weight loss intervention, there was a noted decrease in joint pain that was positively correlated with weight loss.[183] One other randomized control study used a low-energy diet over the course of 1 year in obese patients with knee osteoarthritis and found that their diet reduced weight and pain scores as early as 8 weeks into the study and persisted.[184] Other studies have looked specifically at pain levels following dietary interventions and found similar results. One group examined the effects of diet and/or exercise on pain ratings in patients with osteoarthritis after 6 or 18 months and found reductions in IL-6 and pain and an increase in knee function with diet.[185,186] Interestingly, they found that caloric restriction had more of an effect on pain than exercise, irrespective of weight loss,[185] suggesting that the diet itself was the primary driver for reducing pain. Finally, Skoldstam and colleagues[187] combined a number of studies to determine whether weight reduction contributed to the alleviation of pain in rheumatoid arthritis patients and found that diet had more of an effect than weight reduction. Together, these findings suggest that, while weight loss is clearly beneficial, the diet itself seems to have more of an effect on pain, potentially due to the components in the diets.

18. CONCLUSIONS

Obesity and chronic pain are two conditions with a high degree of comorbidity that have recently been linked at the molecular level. Whereas it was once assumed that chronic pain in obese individuals was the result of increased load on weight-bearing joints, it is now known that fat tissue itself activates the immune system, resulting in chronic low-level inflammation. Heightened immune cell activity predisposes one to the development of chronic pain and, thus, weight loss was thought to be a highly beneficial intervention for obese chronic pain patients. The previous sections have outlined ways in which specific foods can activate (SFAs) or inhibit multiple components of the immune system through direct actions on immune cells (see Fig. 5). From blockade of toll-like receptors on macrophages to inhibition of proinflammatory cytokine transcription, the foods described above

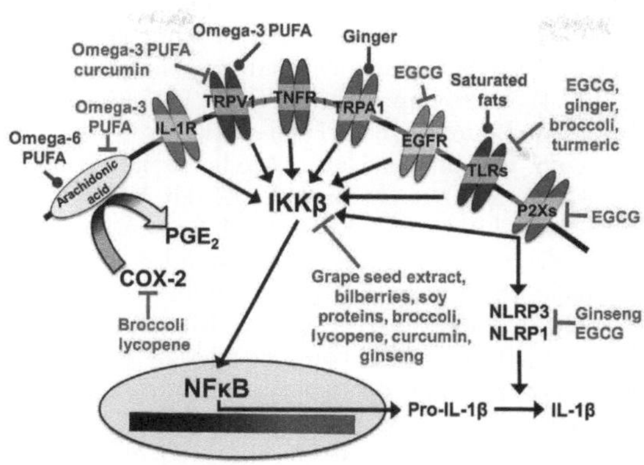

Figure 5 Location of action of the various dietary components within the cells of the immune system. Antagonism or inhibition is indicated in red, activation or agonism is indicated by green. Abbreviations are as follows: COX2 (cyclooxygenase-2), EGCG (epigallocatechin gallate), EGFR (epidermal growth factor receptor), IL-1β (interleukin-1 beta), IL-1R (interleukin-1 receptor), IKKβ (inhibitor of nuclear factor kappa-beta kinase subunit beta), NFκB (nuclear factor kappa-light-chain-enhancer of activated B cells), NLRP1 (NACHT, LRR, and PYD domains-containing protein 1), NLRP3 (NACHT, LRR, and PYD domains-containing protein 3), P2Xs (purinergic receptor type 2X), PGE_2 (prostaglandin E2), pro-IL-1β (pro-interleukin-1 beta), PUFA (polyunsaturated fatty acid), TLRs (Toll-like receptors), TNFR (tumor necrosis factor receptor), TRPA1 (transient receptor potential cation channel subfamily A, member 1), and TRPV1(transient receptor potential cation channel subfamily V member 1). (See the color plate.)

have clear anti-inflammatory effects. Therefore, though weight loss is generally beneficial to health and pain reduction, a diet with high levels of PUFAs, grape seed, soy, broccoli, ginger, and green tea may have exceptional anti-inflammatory activity and provide a nonpharmaceutical means to reduce pain via the immune system.

REFERENCES

1. Watkins LR, Wiertelak EP, Goehler LE, Smith KP, Martin D, Maier SF. Characterization of cytokine-induced hyperalgesia. *Brain Res.* 1994;654(1):15–26.
2. Cunha FQ, Poole S, Lorenzetti BB, Ferreira SH. The pivotal role of tumour necrosis factor alpha in the development of inflammatory hyperalgesia. *Br J Pharmacol.* 1992;107(3):660–664 [Research Support, Non-U.S. Gov't].
3. Ferreira SH, Lorenzetti BB, Bristow AF, Poole S. Interleukin-1 beta as a potent hyperalgesic agent antagonized by a tripeptide analogue. *Nature.* 1988;334(6184):698–700. http://dx.doi.org/10.1038/334698a0.
4. Perkins MN, Kelly D. Interleukin-1 beta induced-desArg9bradykinin-mediated thermal hyperalgesia in the rat. *Neuropharmacology.* 1994;33(5):657–660.
5. Navarro SL, Brasky TM, Schwarz Y, et al. Reliability of serum biomarkers of inflammation from repeated measures in healthy individuals. *Cancer Epidemiol Biomarkers Prev.* 2012;21(7):1167–1170. http://dx.doi.org/10.1158/1055-9965.EPI-12-0110.

6. Chung SD, Liu HT, Lin H, Kuo HC. Elevation of serum c-reactive protein in patients with OAB and IC/BPS implies chronic inflammation in the urinary bladder. *Neurourol Urodyn*. 2011;30(3):417–420. http://dx.doi.org/10.1002/nau.20938.
7. Rannou F, Ouanes W, Boutron I, et al. High-sensitivity C-reactive protein in chronic low back pain with vertebral end-plate Modic signal changes. *Arthritis Rheum*. 2007;57(7):1311–1315. http://dx.doi.org/10.1002/art.22985.
8. Lund Haheim L, Nafstad P, Olsen I, Schwarze P, Ronningen KS. C-reactive protein variations for different chronic somatic disorders. *Scand J Public Health*. 2009;37(6):640–646. http://dx.doi.org/10.1177/1403494809104358.
9. Schinkel C, Scherens A, Koller M, Roellecke G, Muhr G, Maier C. Systemic inflammatory mediators in post-traumatic complex regional pain syndrome (CRPS I)—longitudinal investigations and differences to control groups. *Eur J Med Res*. 2009;14(3):130–135.
10. Uceyler N, Valenza R, Stock M, Schedel R, Sprotte G, Sommer C. Reduced levels of antiinflammatory cytokines in patients with chronic widespread pain. *Arthritis Rheum*. 2006;54(8):2656–2664. http://dx.doi.org/10.1002/art.22026.
11. Uceyler N, Rogausch JP, Toyka KV, Sommer C. Differential expression of cytokines in painful and painless neuropathies. *Neurology*. 2007;69(1):42–49. http://dx.doi.org/10.1212/01.wnl.0000265062.92340.a5.
12. Angst MS, Clark JD, Carvalho B, Tingle M, Schmelz M, Yeomans DC. Cytokine profile in human skin in response to experimental inflammation, noxious stimulation, and administration of a COX-inhibitor: a microdialysis study. *Pain*. 2008;139(1):15–27. http://dx.doi.org/10.1016/j.pain.2008.02.028.
13. Slade GD, Conrad MS, Diatchenko L, et al. Cytokine biomarkers and chronic pain: association of genes, transcription, and circulating proteins with temporomandibular disorders and widespread palpation tenderness. *Pain*. 2011;152(12):2802–2812. http://dx.doi.org/10.1016/j.pain.2011.09.005.
14. Yudkin JS, Stehouwer CD, Emeis JJ, Coppack SW. C-reactive protein in healthy subjects: associations with obesity, insulin resistance, and endothelial dysfunction: a potential role for cytokines originating from adipose tissue? *Arterioscler Thromb Vasc Biol*. 1999;19(4):972–978.
15. Kahn SE, Zinman B, Haffner SM, O'Neill MC, Kravitz BG, Yu D. Obesity is a major determinant of the association of C-reactive protein levels and the metabolic syndrome in type 2 diabetes. *Diabetes*. 2006;55(8):2357–2364. http://dx.doi.org/10.2337/db06-0116.
16. Ishii S, Karlamangla AS, Bote M, et al. Gender, obesity and repeated elevation of C-reactive protein: data from the CARDIA cohort. *PLoS One*. 2012;7(4):e36062. http://dx.doi.org/10.1371/journal.pone.0036062.
17. Ahima RS, Antwi DA. Brain regulation of appetite and satiety. *Endocrinol Metab Clin North Am*. 2008;37(4):811–823. http://dx.doi.org/10.1016/j.ecl.2008.08.005 [Research Support, N.I.H., Extramural Review].
18. Fernandez-Riejos P, Najib S, Santos-Alvarez J, et al. Role of leptin in the activation of immune cells. *Mediators Inflamm*. 2010;2010:568343. http://dx.doi.org/10.1155/2010/568343.
19. Agrawal S, Gollapudi S, Su H, Gupta S. Leptin activates human B cells to secrete TNF-alpha, IL-6, and IL-10 via JAK2/STAT3 and p38MAPK/ERK1/2 signaling pathway. *J Clin Immunol*. 2011;31(3):472–478. http://dx.doi.org/10.1007/s10875-010-9507-1.
20. Janke EA, Collins A, Kozak AT. Overview of the relationship between pain and obesity: what do we know? Where do we go next? *J Rehabil Res Dev*. 2007;44(2):245–262.
21. McCarthy LH, Bigal ME, Katz M, Derby C, Lipton RB. Chronic pain and obesity in elderly people: results from the Einstein aging study. *J Am Geriatr Soc*. 2009;57(1):115–119. http://dx.doi.org/10.1111/j.1532-5415.2008.02089.x.

22. Ray L, Lipton RB, Zimmerman ME, Katz MJ, Derby CA. Mechanisms of association between obesity and chronic pain in the elderly. *Pain*. 2011;152(1):53–59. http://dx.doi.org/10.1016/j.pain.2010.08.043.
23. Verdam FJ, Fuentes S, de Jonge C, et al. Human intestinal microbiota composition is associated with local and systemic inflammation in obesity. *Obesity (Silver Spring)*. 2013;21(12):E607–E615. http://dx.doi.org/10.1002/oby.20466.
24. Senna MK, Sallam RA, Ashour HS, Elarman M. Effect of weight reduction on the quality of life in obese patients with fibromyalgia syndrome: a randomized controlled trial. *Clin Rheumatol*. 2012;31(11):1591–1597. http://dx.doi.org/10.1007/s10067-012-2053-x.
25. You T, Nicklas BJ. Chronic inflammation: role of adipose tissue and modulation by weight loss. *Curr Diabetes Rev*. 2006;2(1):29–37.
26. Manek NJ, Hart D, Spector TD, MacGregor AJ. The association of body mass index and osteoarthritis of the knee joint: an examination of genetic and environmental influences. *Arthritis Rheum*. 2003;48(4):1024–1029. http://dx.doi.org/10.1002/art.10884.
26a. Carman WJ, Sowers M, Hawthorne VM, Weissfeld LA. Obesity as a risk factor for osteoarthritis of the hand and wrist: a prospective study. *Am J Epidemiol*. 1994;139(2):119–129. Epub 1994/01/15. PubMed PMID:8296779.
27. Vincent HK, Heywood K, Connelly J, Hurley RW. Obesity and weight loss in the treatment and prevention of osteoarthritis. *PM R*. 2012;4(suppl 5):S59–S67. http://dx.doi.org/10.1016/j.pmrj.2012.01.005.
28. Khimich S. Level of sensitivity of pain in patients with obesity. *Acta Chir Hung*. 1997;36(1–4):166–167.
29. Zahorska-Markiewicz B, Kucio C, Pyszkowska J. Obesity and pain. *Hum Nutr Clin Nutr*. 1983;37(4):307–310.
30. Zahorska-Markiewicz B, Zych P, Kucio C. Pain sensitivity in obesity. *Acta Physiol Pol*. 1988;39(3):183–187.
31. Ramzan I, Wong BK, Corcoran GB. Pain sensitivity in dietary-induced obese rats. *Physiol Behav*. 1993;54(3):433–435.
32. Stoll G, Jung S, Jander S, van der Meide P, Hartung HP. Tumor necrosis factor-alpha in immune-mediated demyelination and Wallerian degeneration of the rat peripheral nervous system. *J Neuroimmunol*. 1993;45(1–2):175–182.
33. Richter F, Natura G, Loser S, Schmidt K, Viisanen H, Schaible HG. Tumor necrosis factor causes persistent sensitization of joint nociceptors to mechanical stimuli in rats. *Arthritis Rheum*. 2010;62(12):3806–3814. http://dx.doi.org/10.1002/art.27715.
34. Deshmukh-Taskar PR, O'Neil CE, Nicklas TA, et al. Dietary patterns associated with metabolic syndrome, sociodemographic and lifestyle factors in young adults: the Bogalusa Heart Study. *Public Health Nutr*. 2009;12(12):2493–2503. http://dx.doi.org/10.1017/S1368980009991261.
35. Liese AD, Schulz M, Moore CG, Mayer-Davis EJ. Dietary patterns, insulin sensitivity and adiposity in the multi-ethnic Insulin Resistance Atherosclerosis Study population. *Br J Nutr*. 2004;92(6):973–984.
36. Lucas M, Chocano-Bedoya P, Shulze MB, et al. Inflammatory dietary pattern and risk of depression among women. *Brain Behav Immun*. 2014;36:46–53. http://dx.doi.org/10.1016/j.bbi.2013.09.014.
37. Buchenauer T, Behrendt P, Bode FJ, et al. Diet-induced obesity alters behavior as well as serum levels of corticosterone in F344 rats. *Physiol Behav*. 2009;98(5):563–569. http://dx.doi.org/10.1016/j.physbeh.2009.09.003.
38. Sampey BP, Vanhoose AM, Winfield HM, et al. Cafeteria diet is a robust model of human metabolic syndrome with liver and adipose inflammation: comparison to high-fat diet. *Obesity (Silver Spring)*. 2011;19(6):1109–1117. http://dx.doi.org/10.1038/oby.2011.18.
39. Fisher G, Hyatt TC, Hunter GR, Oster RA, Desmond RA, Gower BA. Effect of diet with and without exercise training on markers of inflammation and fat distribution in

overweight women. *Obesity (Silver Spring)*. 2011;19(6):1131–1136. http://dx.doi.org/10.1038/oby.2010.310.

40. Imayama I, Ulrich CM, Alfano CM, et al. Effects of a caloric restriction weight loss diet and exercise on inflammatory biomarkers in overweight/obese postmenopausal women: a randomized controlled trial. *Cancer Res*. 2012;72(9):2314–2326. http://dx.doi.org/10.1158/0008-5472.CAN-11-3092.
41. Neuhouser ML, Schwarz Y, Wang C, et al. A low-glycemic load diet reduces serum C-reactive protein and modestly increases adiponectin in overweight and obese adults. *J Nutr*. 2012;142(2):369–374. http://dx.doi.org/10.3945/jn.111.149807.
42. Cintra DE, Ropelle ER, Moraes JC, et al. Unsaturated fatty acids revert diet-induced hypothalamic inflammation in obesity. *PLoS One*. 2012;7(1):e30571. http://dx.doi.org/10.1371/journal.pone.0030571.
43. Mena MP, Sacanella E, Vazquez-Agell M, et al. Inhibition of circulating immune cell activation: a molecular antiinflammatory effect of the Mediterranean diet. *Am J Clin Nutr*. 2009;89(1):248–256. http://dx.doi.org/10.3945/ajcn.2008.26094.
44. Langkamp-Henken B, Nieves Jr C, Culpepper T, et al. Fecal lactic acid bacteria increased in adolescents randomized to whole-grain but not refined-grain foods, whereas inflammatory cytokine production decreased equally with both interventions. *J Nutr*. 2012;142(11):2025–2032. http://dx.doi.org/10.3945/jn.112.164996.
45. Patterson E, Wall R, Fitzgerald GF, Ross RP, Stanton C. Health implications of high dietary omega-6 polyunsaturated Fatty acids. *J Nutr Metab*. 2012;2012:539426. http://dx.doi.org/10.1155/2012/539426.
46. Calder PC. Polyunsaturated fatty acids and inflammation. *Biochem Soc Trans*. 2005;33(pt 2):423–427. http://dx.doi.org/10.1042/BST0330423.
47. Calder PC. Polyunsaturated fatty acids and inflammatory processes: new twists in an old tale. *Biochimie*. 2009;91(6):791–795. http://dx.doi.org/10.1016/j.biochi.2009.01.008 [Review].
48. Yan Y, Jiang W, Spinetti T, et al. Omega-3 fatty acids prevent inflammation and metabolic disorder through inhibition of NLRP3 inflammasome activation. *Immunity*. 2013;38(6):1154–1163. http://dx.doi.org/10.1016/j.immuni.2013.05.015.
49. Matta JA, Miyares RL, Ahern GP. TRPV1 is a novel target for omega-3 polyunsaturated fatty acids. *J Physiol*. 2007;578(Pt 2):397–411. http://dx.doi.org/10.1113/jphysiol.2006.121988.
50. Figueroa JD, Cordero K, Serrano-Illan M, et al. Metabolomics uncovers dietary omega-3 fatty acid-derived metabolites implicated in anti-nociceptive responses after experimental spinal cord injury. *Neuroscience*. 2013;255:1–18. http://dx.doi.org/10.1016/j.neuroscience.2013.09.012.
51. Veigas JM, Williams PJ, Halade G, Rahman MM, Yoneda T, Fernandes G. Fish oil concentrate delays sensitivity to thermal nociception in mice. *Pharmacol Res*. 2011;63(5):377–382. http://dx.doi.org/10.1016/j.phrs.2011.02.004.
52. Olson MV, Liu YC, Dangi B, Paul Zimmer J, Salem Jr N, Nauroth JM. Docosahexaenoic acid reduces inflammation and joint destruction in mice with collagen-induced arthritis. *Inflamm Res*. 2013;62(12):1003–1013. http://dx.doi.org/10.1007/s00011-013-0658-4.
53. Kremer JM, Lawrence DA, Jubiz W, et al. Dietary fish oil and olive oil supplementation in patients with rheumatoid arthritis. Clinical and immunologic effects. *Arthritis Rheum*. 1990;33(6):810–820.
54. Kiecolt-Glaser JK, Belury MA, Andridge R, Malarkey WB, Hwang BS, Glaser R. Omega-3 supplementation lowers inflammation in healthy middle-aged and older adults: a randomized controlled trial. *Brain Behav Immun*. 2012;26(6):988–995. http://dx.doi.org/10.1016/j.bbi.2012.05.011.

55. Kalogeropoulos N, Panagiotakos DB, Pitsavos C, et al. Unsaturated fatty acids are inversely associated and n-6/n-3 ratios are positively related to inflammation and coagulation markers in plasma of apparently healthy adults. *Clin Chim Acta.* 2010;411(7–8):584–591. http://dx.doi.org/10.1016/j.cca.2010.01.023.
56. Joffe YT, Collins M, Goedecke JH. The relationship between dietary fatty acids and inflammatory genes on the obese phenotype and serum lipids. *Nutrients.* 2013;5(5):1672–1705. http://dx.doi.org/10.3390/nu5051672.
57. Lee JY, Sohn KH, Rhee SH, Hwang D. Saturated fatty acids, but not unsaturated fatty acids, induce the expression of cyclooxygenase-2 mediated through Toll-like receptor 4. *J Biol Chem.* 2001;276(20):16683–16689. http://dx.doi.org/10.1074/jbc.M011695200.
58. Schwartz EA, Zhang WY, Karnik SK, et al. Nutrient modification of the innate immune response: a novel mechanism by which saturated fatty acids greatly amplify monocyte inflammation. *Arterioscler Thromb Vasc Biol.* 2010;30(4):802–808. http://dx.doi.org/10.1161/ATVBAHA.109.201681.
59. Long SD, Pekala PH. Lipid mediators of insulin resistance: ceramide signalling down-regulates GLUT4 gene transcription in 3 T3-L1 adipocytes. *Biochem J.* 1996;319(pt 1):179–184.
60. Teruel T, Hernandez R, Lorenzo M. Ceramide mediates insulin resistance by tumor necrosis factor-alpha in brown adipocytes by maintaining Akt in an inactive dephosphorylated state. *Diabetes.* 2001;50(11):2563–2571.
61. Snodgrass RG, Huang S, Choi IW, Rutledge JC, Hwang DH. Inflammasome-mediated secretion of IL-1beta in human monocytes through TLR2 activation; modulation by dietary fatty acids. *J Immunol.* 2013;191(8):4337–4347. http://dx.doi.org/10.4049/jimmunol.1300298.
62. Li WG, Zhang XY, Wu YJ, Tian X. Anti-inflammatory effect and mechanism of proanthocyanidins from grape seeds. *Acta Pharmacol Sin.* 2001;22(12):1117–1120.
63. Uchida S, Hirai K, Hatanaka J, Hanato J, Umegaki K, Yamada S. Antinociceptive effects of St. John's wort, *Harpagophytum procumbens* extract and Grape seed proanthocyanidins extract in mice. *Biol Pharm Bull.* 2008;31(2):240–245.
64. Woo YJ, Joo YB, Jung YO, et al. Grape seed proanthocyanidin extract ameliorates monosodium iodoacetate-induced osteoarthritis. *Exp Mol Med.* 2011;43(10):561–570. http://dx.doi.org/10.3858/emm.2011.43.10.062.
65. Cho NP, Han HS, Leem DH, et al. Sulforaphane enhances caspase-dependent apoptosis through inhibition of cyclooxygenase-2 expression in human oral squamous carcinoma cells and nude mouse xenograft model. *Oral Oncol.* 2009;45(8):654–660. http://dx.doi.org/10.1016/j.oraloncology.2008.07.003.

65a. Cui XP, Li BY, Gao HQ, Wei N, Wang WL, Lu M. Effects of grape seed proanthocyanidin extracts on peripheral nerves in streptozocin-induced diabetic rats. *J Nutr Sci Vitaminol (Tokyo).* 2008;54(4):321–328. Epub 2008/09/18. PubMed PMID: 18797155.

66. Cady RJ, Hirst JJ, Durham PL. Dietary grape seed polyphenols repress neuron and glia activation in trigeminal ganglion and trigeminal nucleus caudalis. *Mol Pain.* 2010;6:91. http://dx.doi.org/10.1186/1744-8069-6-91.
67. Kim H, Kim JY, Song HS, Park KU, Mun KC, Ha E. Grape seed proanthocyanidin extract inhibits interleukin-17-induced interleukin-6 production via MAPK pathway in human pulmonary epithelial cells. *Naunyn Schmiedebergs Arch Pharmacol.* 2011;383(6):555–562. http://dx.doi.org/10.1007/s00210-011-0633-y.
68. Chacon MR, Ceperuelo-Mallafre V, Maymo-Masip E, et al. Grape-seed procyanidins modulate inflammation on human differentiated adipocytes *in vitro*. *Cytokine.* 2009;47(2):137–142. http://dx.doi.org/10.1016/j.cyto.2009.06.001.

69. Kolehmainen M, Mykkanen O, Kirjavainen PV, et al. Bilberries reduce low-grade inflammation in individuals with features of metabolic syndrome. *Mol Nutr Food Res*. 2012;56(10):1501–1510. http://dx.doi.org/10.1002/mnfr.201200195.
70. Vendrame S, Daugherty A, Kristo AS, Riso P, Klimis-Zacas D. Wild blueberry (Vaccinium angustifolium) consumption improves inflammatory status in the obese Zucker rat model of the metabolic syndrome. *J Nutr Biochem*. 2013;24(8):1508–1512. http://dx.doi.org/10.1016/j.jnutbio.2012.12.010.
71. Meki AR, Hamed EA, Ezam KA. Effect of green tea extract and vitamin C on oxidant or antioxidant status of rheumatoid arthritis rat model. *Indian J Clin Biochem*. 2009;24(3):280–287. http://dx.doi.org/10.1007/s12291-009-0053-7.
72. Sachdeva AK, Kuhad A, Chopra K. Epigallocatechin gallate ameliorates behavioral and biochemical deficits in rat model of load-induced chronic fatigue syndrome. *Brain Res Bull*. 2011;86(3–4):165–172. http://dx.doi.org/10.1016/j.brainresbull.2011.06.007.
73. Kuang X, Huang Y, Gu HF, et al. Effects of intrathecal epigallocatechin gallate, an inhibitor of Toll-like receptor 4, on chronic neuropathic pain in rats. *Eur J Pharmacol*. 2012;676(1–3):51–56. http://dx.doi.org/10.1016/j.ejphar.2011.11.037.
74. Byun EB, Choi HG, Sung NY, Byun EH. Green tea polyphenol epigallocatechin-3-gallate inhibits TLR4 signaling through the 67-kDa laminin receptor on lipopolysaccharide-stimulated dendritic cells. *Biochem Biophys Res Commun*. 2012;426(4):480–485. http://dx.doi.org/10.1016/j.bbrc.2012.08.096.
75. Chow HH, Cai Y, Hakim IA, et al. Pharmacokinetics and safety of green tea polyphenols after multiple-dose administration of epigallocatechin gallate and polyphenon E in healthy individuals. *Clin Cancer Res*. 2003;9(9):3312–3319.
76. Tang Y, Matsuoka I, Ono T, Inoue K, Kimura J. Selective up-regulation of P2X4-receptor gene expression by interferon-gamma in vascular endothelial cells. *J Pharmacol Sci*. 2008;107(4):419–427.
77. Sen T, Chatterjee A. Epigallocatechin-3-gallate (EGCG) downregulates EGF-induced MMP-9 in breast cancer cells: involvement of integrin receptor alpha5beta1 in the process. *Eur J Nutr*. 2011;50(6):465–478. http://dx.doi.org/10.1007/s00394-010-0158-z.
78. Pae M, Ren Z, Meydani M, et al. Dietary supplementation with high dose of epigallocatechin-3-gallate promotes inflammatory response in mice. *J Nutr Biochem*. 2012;23(6):526–531. http://dx.doi.org/10.1016/j.jnutbio.2011.02.006.
79. Shanafelt TD, Call TG, Zent CS, et al. Phase 2 trial of daily, oral Polyphenon E in patients with asymptomatic, Rai stage 0 to II chronic lymphocytic leukemia. *Cancer*. 2013;119(2):363–370. http://dx.doi.org/10.1002/cncr.27719.
80. Dang ZC, Audinot V, Papapoulos SE, Boutin JA, Lowik CW. Peroxisome proliferator-activated receptor gamma (PPARgamma) as a molecular target for the soy phytoestrogen genistein. *J Biol Chem*. 2003;278(2):962–967. http://dx.doi.org/10.1074/jbc.M209483200.
81. Martin H. Role of PPAR-gamma in inflammation. Prospects for therapeutic intervention by food components. *Mutat Res*. 2010;690(1–2):57–63.
82. Hooshmand S, Soung do Y, Lucas EA, Madihally SV, Levenson CW, Arjmandi BH. Genistein reduces the production of proinflammatory molecules in human chondrocytes. *J Nutr Biochem*. 2007;18(9):609–614. http://dx.doi.org/10.1016/j.jnutbio.2006.11.006.
83. Shir Y, Campbell JN, Raja SN, Seltzer Z. The correlation between dietary soy phytoestrogens and neuropathic pain behavior in rats after partial denervation. *Anesth Analg*. 2002;94(2):421–426, table of contents.
84. Shir Y, Sheth R, Campbell JN, Raja SN, Seltzer Z. Soy-containing diet suppresses chronic neuropathic sensory disorders in rats. *Anesth Analg*. 2001;92(4):1029–1034.
85. Perez J, Ware MA, Chevalier S, Gougeon R, Shir Y. Dietary omega-3 fatty acids may be associated with increased neuropathic pain in nerve-injured rats. *Anesth Analg*.

2005;101(2):444–448. http://dx.doi.org/10.1213/01.ANE.0000158469.11775.52 table of contents.
86. Borzan J, Tall JM, Zhao C, Meyer RA, Raja SN. Effects of soy diet on inflammation-induced primary and secondary hyperalgesia in rat. *Eur J Pain*. 2010;14(8):792–798. http://dx.doi.org/10.1016/j.ejpain.2009.12.002.
87. Burris RL, Ng HP, Nagarajan S. Soy protein inhibits inflammation-induced VCAM-1 and inflammatory cytokine induction by inhibiting the NF-kappaB and AKT signaling pathway in apolipoprotein E-deficient mice. *Eur J Nutr*. 2014;53(1):135–148. http://dx.doi.org/10.1007/s00394-013-0509-7.
88. Arjmandi BH, Khalil DA, Lucas EA, et al. Soy protein may alleviate osteoarthritis symptoms. *Phytomedicine*. 2004;11(7–8):567–575. http://dx.doi.org/10.1016/j.phymed.2003.11.001.
89. Hushmendy S, Jayakumar L, Hahn AB, Bhoiwala D, Bhoiwala DL, Crawford DR. Select phytochemicals suppress human T-lymphocytes and mouse splenocytes suggesting their use in autoimmunity and transplantation. *Nutr Res*. 2009;29(8):568–578. http://dx.doi.org/10.1016/j.nutres.2009.08.003.
90. Park JH, Kim JW, Lee CM, et al. Sulforaphane inhibits the Th2 immune response in ovalbumin-induced asthma. *BMB Rep*. 2012;45(5):311–316.
91. Kim SY, Jeong E, Joung SM, Lee JY. PI3K/Akt contributes to increased expression of Toll-like receptor 4 in macrophages exposed to hypoxic stress. *Biochem Biophys Res Commun*. 2012;419(3):466–471. http://dx.doi.org/10.1016/j.bbrc.2012.02.015.
92. Lee YR, Noh EM, Han JH, et al. Sulforaphane controls TPA-induced MMP-9 expression through the NF-kappaB signaling pathway, but not AP-1, in MCF-7 breast cancer cells. *BMB Rep*. 2013;46(4):201–206.
93. Koo JE, Park ZY, Kim ND, Lee JY. Sulforaphane inhibits the engagement of LPS with TLR4/MD2 complex by preferential binding to Cys133 in MD2. *Biochem Biophys Res Commun*. 2013;434(3):600–605. http://dx.doi.org/10.1016/j.bbrc.2013.03.123.
94. Kong JS, Yoo SA, Kim HS, et al. Inhibition of synovial hyperplasia, rheumatoid T cell activation, and experimental arthritis in mice by sulforaphane, a naturally occurring isothiocyanate. *Arthritis Rheum*. 2010;62(1):159–170. http://dx.doi.org/10.1002/art.25017.
95. Negi G, Kumar A, Sharma SS. Nrf2 and NF-kappaB modulation by sulforaphane counteracts multiple manifestations of diabetic neuropathy in rats and high glucose-induced changes. *Curr Neurovasc Res*. 2011;8(4):294–304.
96. Kim D, You B, Jo EK, Han SK, Simon MI, Lee SJ. NADPH oxidase 2-derived reactive oxygen species in spinal cord microglia contribute to peripheral nerve injury-induced neuropathic pain. *Proc Natl Acad Sci U S A*. 2010;107(33):14851–14856. http://dx.doi.org/10.1073/pnas.1009926107.
97. Hughes DA, Wright AJ, Finglas PM, et al. Beta-carotene supplementation enhances the expression of functionally associated molecules on human monocytes. *Biochem Soc Trans*. 1996;24(3):388S.
98. Hughes DA, Wright AJ, Finglas PM, et al. Comparison of effects of beta-carotene and lycopene supplementation on the expression of functionally associated molecules on human monocytes. *Biochem Soc Trans*. 1997;25(2):206S.
99. Watzl B, Bub A, Brandstetter BR, Rechkemmer G. Modulation of human T-lymphocyte functions by the consumption of carotenoid-rich vegetables. *Br J Nutr*. 1999;82(5):383–389 [Clinical Trial Comparative Study].
100. Palozza P, Simone R, Catalano A, et al. Lycopene prevention of oxysterol-induced proinflammatory cytokine cascade in human macrophages: inhibition of NF-kappaB nuclear binding and increase in PPARgamma expression. *J Nutr Biochem*. 2011;22(3):259–268. http://dx.doi.org/10.1016/j.jnutbio.2010.02.003.
101. Kim GY, Kim JH, Ahn SC, et al. Lycopene suppresses the lipopolysaccharide-induced phenotypic and functional maturation of murine dendritic cells through inhibition of

mitogen-activated protein kinases and nuclear factor-kappaB. *Immunology*. 2004;113(2):203–211. http://dx.doi.org/10.1111/j.1365-2567.2004.01945.x.
102. Feng D, Ling WH, Duan RD. Lycopene suppresses LPS-induced NO and IL-6 production by inhibiting the activation of ERK, p38MAPK, and NF-kappaB in macrophages. *Inflamm Res*. 2010;59(2):115–121. http://dx.doi.org/10.1007/s00011-009-0077-8.
103. Di Tomo P, Canali R, Ciavardelli D, et al. beta-Carotene and lycopene affect endothelial response to TNF-alpha reducing nitro-oxidative stress and interaction with monocytes. *Mol Nutr Food Res*. 2012;56(2):217–227. http://dx.doi.org/10.1002/mnfr.201100500.
104. Marcotorchino J, Romier B, Gouranton E, et al. Lycopene attenuates LPS-induced TNF-alpha secretion in macrophages and inflammatory markers in adipocytes exposed to macrophage-conditioned media. *Mol Nutr Food Res*. 2012;56(5):725–732. http://dx.doi.org/10.1002/mnfr.201100623.
105. Rafi MM, Yadav PN, Reyes M. Lycopene inhibits LPS-induced proinflammatory mediator inducible nitric oxide synthase in mouse macrophage cells. *J Food Sci*. 2007;72(1):S069–S074. http://dx.doi.org/10.1111/j.1750-3841.2006.00219.x.
106. Simone RE, Russo M, Catalano A, et al. Lycopene inhibits NF-kB-mediated IL-8 expression and changes redox and PPARgamma signalling in cigarette smoke-stimulated macrophages. *PLoS One*. 2011;6(5):e19652. http://dx.doi.org/10.1371/journal.pone.0019652.
107. Lin HY, Huang BR, Yeh WL, et al. Antineuroinflammatory effects of lycopene via activation of adenosine monophosphate-activated protein kinase-alpha1/heme oxygenase-1 pathways. *Neurobiol Aging*. 2014;35(1):191–202. http://dx.doi.org/10.1016/j.neurobiolaging.2013.06.020.
108. Kuhad A, Chopra K. Lycopene ameliorates thermal hyperalgesia and cold allodynia in STZ-induced diabetic rat. *Indian J Exp Biol*. 2008;46(2):108–111.
109. Kuhad A, Sharma S, Chopra K. Lycopene attenuates thermal hyperalgesia in a diabetic mouse model of neuropathic pain. *Eur J Pain*. 2008;12(5):624–632. http://dx.doi.org/10.1016/j.ejpain.2007.10.008.
110. Riso P, Visioli F, Grande S, et al. Effect of a tomato-based drink on markers of inflammation, immunomodulation, and oxidative stress. *J Agric Food Chem*. 2006;54(7):2563–2566. http://dx.doi.org/10.1021/jf053033c.
111. Ansari MS, Gupta NP. A comparison of lycopene and orchidectomy vs orchidectomy alone in the management of advanced prostate cancer. *BJU Int*. 2004;94(4):678. http://dx.doi.org/10.1111/j.1464-410x.2003.05067_4.x [Comment Comparative Study Letter].
112. Ansari MS, Gupta NP. A comparison of lycopene and orchidectomy vs orchidectomy alone in the management of advanced prostate cancer. *BJU Int*. 2003;92(4):375–378 [Clinical Trial Comparative Study Randomized Controlled Trial]; discussion 378.
113. Mashhadi NS, Ghiasvand R, Askari G, et al. Influence of ginger and cinnamon intake on inflammation and muscle soreness endued by exercise in Iranian female athletes. *Int J Prev Med*. 2013;4(suppl 1):S11–S15.
114. Black CD, O'Connor PJ. Acute effects of dietary ginger on muscle pain induced by eccentric exercise. *Phytother Res*. 2010;24(11):1620–1626. http://dx.doi.org/10.1002/ptr.3148.
115. Rahnama P, Montazeri A, Huseini HF, Kianbakht S, Naseri M. Effect of Zingiber officinale R. rhizomes (ginger) on pain relief in primary dysmenorrhea: a placebo randomized trial. *BMC Complement Altern Med*. 2012;12:92. http://dx.doi.org/10.1186/1472-6882-12-92.
116. Drozdov VN, Kim VA, Tkachenko EV, Varvanina GG. Influence of a specific ginger combination on gastropathy conditions in patients with osteoarthritis of the knee or hip. *J Altern Complement Med*. 2012;18(6):583–588. http://dx.doi.org/10.1089/acm.2011.0202.

117. Terry R, Posadzki P, Watson LK, Ernst E. The use of ginger (Zingiber officinale) for the treatment of pain: a systematic review of clinical trials. *Pain Med.* 2011;12(12):1808–1818. http://dx.doi.org/10.1111/j.1526-4637.2011.01261.x [Research Support, Non-U.S. Gov't Review].
118. Khalid MH, Akhtar MN, Mohamad AS, et al. Antinociceptive effect of the essential oil of Zingiber zerumbet in mice: possible mechanisms. *J Ethnopharmacol.* 2011;137(1):345–351. http://dx.doi.org/10.1016/j.jep.2011.05.043.
119. Ojewole JA. Analgesic, antiinflammatory and hypoglycaemic effects of ethanol extract of Zingiber officinale (Roscoe) rhizomes (Zingiberaceae) in mice and rats. *Phytother Res.* 2006;20(9):764–772. http://dx.doi.org/10.1002/ptr.1952.
120. Sepahvand R, Esmaeili-Mahani S, Arzi A, Rasoulian B, Abbasnejad M. Ginger (Zingiber officinale Roscoe) elicits antinociceptive properties and potentiates morphine-induced analgesia in the rat radiant heat tail-flick test. *J Med Food.* 2010;13(6):1397–1401. http://dx.doi.org/10.1089/jmf.2010.1043.
121. Bandell M, Story GM, Hwang SW, et al. Noxious cold ion channel TRPA1 is activated by pungent compounds and bradykinin. *Neuron.* 2004;41(6):849–857 [Research Support, Non-U.S. Gov't Research Support, U.S. Gov't, P.H.S.].
122. Ghayur MN, Gilani AH. Ginger lowers blood pressure through blockade of voltage-dependent calcium channels. *J Cardiovasc Pharmacol.* 2005;45(1):74–80.
123. Park SH, Kyeong MS, Hwang Y, Ryu SY, Han SB, Kim Y. Inhibition of LPS binding to MD-2 co-receptor for suppressing TLR4-mediated expression of inflammatory cytokine by 1-dehydro-10-gingerdione from dietary ginger. *Biochem Biophys Res Commun.* 2012;419(4):735–740. http://dx.doi.org/10.1016/j.bbrc.2012.02.091.
124. Lee HY, Park SH, Lee M, et al. 1-Dehydro-[10]-gingerdione from ginger inhibits IKKbeta activity for NF-kappaB activation and suppresses NF-kappaB-regulated expression of inflammatory genes. *Br J Pharmacol.* 2012;167(1):128–140. http://dx.doi.org/10.1111/j.1476-5381.2012.01980.x.
125. Choi YY, Kim MH, Hong J, Kim SH, Yang WM. Dried Ginger (Zingiber officinalis) inhibits inflammation in a lipopolysaccharide-induced mouse model. *Evid Based Complement Alternat Med.* 2013;2013:914563. http://dx.doi.org/10.1155/2013/914563.
126. Li XH, McGrath KC, Nammi S, Heather AK, Roufogalis BD. Attenuation of liver pro-inflammatory responses by Zingiber officinale via inhibition of NF-kappa B activation in high-fat diet-fed rats. *Basic Clin Pharmacol Toxicol.* 2012;110(3):238–244. http://dx.doi.org/10.1111/j.1742-7843.2011.00791.x.
127. Mahluji S, Ostadrahimi A, Mobasseri M, Ebrahimzade Attari V, Payahoo L. Anti-inflammatory effects of Zingiber officinale in type 2 diabetic patients. *Adv Pharm Bull.* 2013;3(2):273–276. http://dx.doi.org/10.5681/apb.2013.044.
128. Banafshe HR, Hamidi GA, Noureddini M, Mirhashemi SM, Mokhtari R, Shoferpour M. Effect of curcumin on diabetic peripheral neuropathic pain: possible involvement of opioid system. *Eur J Pharmacol.* 2014;723:202–206. http://dx.doi.org/10.1016/j.ejphar.2013.11.033.
129. Zhao X, Xu Y, Zhao Q, Chen CR, Liu AM, Huang ZL. Curcumin exerts antinociceptive effects in a mouse model of neuropathic pain: descending monoamine system and opioid receptors are differentially involved. *Neuropharmacology.* 2012;62(2):843–854. http://dx.doi.org/10.1016/j.neuropharm.2011.08.050.
130. Tajik H, Tamaddonfard E, Hamzeh-Gooshchi N. The effect of curcumin (active substance of turmeric) on the acetic acid-induced visceral nociception in rats. *Pak J Biol Sci.* 2008;11(2):312–314.
131. Zhu Q, Sun Y, Yun X, Ou Y, Zhang W, Li JX. Antinociceptive effects of curcumin in a rat model of postoperative pain. *Sci Rep.* 2014;4:4932. http://dx.doi.org/10.1038/srep04932.
132. Antoine F, Simard JC, Girard D. Curcumin inhibits agent-induced human neutrophil functions *in vitro* and lipopolysaccharide-induced neutrophilic infiltration

in vivo. *Int Immunopharmacol*. 2013;17(4):1101–1107. http://dx.doi.org/10.1016/j.intimp.2013.09.024.
133. Zhi L, Dong L, Kong D, et al. Curcumin acts via transient receptor potential vanilloid-1 receptors to inhibit gut nociception and reverses visceral hyperalgesia. *Neurogastroenterol Motil*. 2013;25(6):e429–e440. http://dx.doi.org/10.1111/nmo.12145.
134. Sahbaie P, Sun Y, Liang DY, Shi XY, Clark JD. Curcumin treatment attenuates pain and enhances functional recovery after incision. *Anesth Analg*. 2014;118(6):1336–1344. http://dx.doi.org/10.1213/ANE.0000000000000189 [Research Support, Non-U.S. Gov't Research Support, U.S. Gov't, Non-P.H.S.].
135. Abe Y, Hashimoto S, Horie T. Curcumin inhibition of inflammatory cytokine production by human peripheral blood monocytes and alveolar macrophages. *Pharmacol Res*. 1999;39(1):41–47. http://dx.doi.org/10.1006/phrs.1998.0404.
136. Belcaro G, Cesarone MR, Dugall M, et al. Efficacy and safety of Meriva(R), a curcumin-phosphatidylcholine complex, during extended administration in osteoarthritis patients. *Altern Med Rev*. 2010;15(4):337–344.
137. Braz AS, Morais LC, Paula AP, Diniz MF, Almeida RN. Effects of Panax ginseng extract in patients with fibromyalgia: a 12-week, randomized, double-blind, placebo-controlled trial. *Rev Bras Psiquiatr*. 2013;35(1):21–28.
138. Lee JH, Lee JH, Lee YM, Kim PN, Jeong CS. Potential analgesic and anti-inflammatory activities of Panax ginseng head butanolic fraction in animals. *Food Chem Toxicol*. 2008;46(12):3749–3752. http://dx.doi.org/10.1016/j.fct.2008.09.055.
139. Nah JJ, Hahn JH, Chung S, Choi S, Kim YI, Nah SY. Effect of ginsenosides, active components of ginseng, on capsaicin-induced pain-related behavior. *Neuropharmacology*. 2000;39(11):2180–2184.
140. Yoon SR, Nah JJ, Shin YH, et al. Ginsenosides induce differential antinociception and inhibit substance P induced-nociceptive response in mice. *Life Sci*. 1998;62(21):PL 319–PL 325.
141. Mogil JS, Shin YH, McCleskey EW, Kim SC, Nah SY. Ginsenoside Rf, a trace component of ginseng root, produces antinociception in mice. *Brain Res*. 1998;792(2):218–228.
142. Yang JC, Pang CS, Tsang SF, Ng KF. Effect of American ginseng extract (Panax quinquefolius) on formalin-induced nociception in mice. *Am J Chin Med*. 2001;29(1):149–154. http://dx.doi.org/10.1142/S0192415X01000162.
143. Choi SS, Lee JK, Suh HW. Effect of ginsenosides administered intrathecally on the antinociception induced by cold water swimming stress in the mouse. *Biol Pharm Bull*. 2003;26(6):858–861.
144. Nah SY, McCleskey EW. Ginseng root extract inhibits calcium channels in rat sensory neurons through a similar path, but different receptor, as mu-type opioids. *J Ethnopharmacol*. 1994;42(1):45–51.
145. Rhim H, Kim H, Lee DY, Oh TH, Nah SY. Ginseng and ginsenoside Rg3, a newly identified active ingredient of ginseng, modulate Ca2+ channel currents in rat sensory neurons. *Eur J Pharmacol*. 2002;436(3):151–158.
146. Hahn J, Nah SY, Nah JJ, Uhm DY, Chung S. Ginsenosides inhibit capsaicin-activated channel in rat sensory neurons. *Neurosci Lett*. 2000;287(1):45–48.
147. Bak MJ, Hong SG, Lee JW, Jeong WS. Red ginseng marc oil inhibits iNOS and COX-2 via NFkappaB and p38 pathways in LPS-stimulated RAW 264.7 macrophages. *Molecules*. 2012;17(12):13769–13786. http://dx.doi.org/10.3390/molecules171213769.
148. Kim J, Ahn H, Han BC, et al. Korean red ginseng extracts inhibit NLRP3 and AIM2 inflammasome activation. *Immunol Lett*. 2014;158(1–2):143–150. http://dx.doi.org/10.1016/j.imlet.2013.12.017.

149. Branum AM, Rossen LM, Schoendorf KC. Trends in caffeine intake among U.S. children and adolescents. *Pediatrics*. 2014;133(3):386–393. http://dx.doi.org/10.1542/peds.2013-2877.
150. Takahashi HK, Iwagaki H, Hamano R, et al. Effects of adenosine on adhesion molecule expression and cytokine production in human PBMC depend on the receptor subtype activated. *Br J Pharmacol*. 2007;150(6):816–822. http://dx.doi.org/10.1038/sj.bjp.0707126.
151. Hamano R, Takahashi HK, Iwagaki H, et al. Stimulation of adenosine A2A receptor inhibits LPS-induced expression of intercellular adhesion molecule 1 and production of TNF-alpha in human peripheral blood mononuclear cells. *Shock*. 2008;29(2):154–159. http://dx.doi.org/10.1097/shk.0b013e31812385da.
152. Chavez-Valdez R, Wills-Karp M, Ahlawat R, Cristofalo EA, Nathan A, Gauda EB. Caffeine modulates TNF-alpha production by cord blood monocytes: the role of adenosine receptors. *Pediatr Res*. 2009;65(2):203–208. http://dx.doi.org/10.1203/PDR.0b013e31818d66b1.
153. Diaz-Reval MI, Carrillo-Munguia N, Martinez-Casas M, Gonzalez-Trujano ME. Tramadol and caffeine produce synergistic interactions on antinociception measured in a formalin model. *Pharmacol Biochem Behav*. 2010;97(2):357–362. http://dx.doi.org/10.1016/j.pbb.2010.09.004 [Research Support, Non-U.S. Gov't].
154. Ward N, Whitney C, Avery D, Dunner D. The analgesic effects of caffeine in headache. *Pain*. 1991;44(2):151–155.
155. Diamond S, Balm TK, Freitag FG. Ibuprofen plus caffeine in the treatment of tension-type headache. *Clin Pharmacol Ther*. 2000;68(3):312–319. http://dx.doi.org/10.1067/mcp.2000.109353 [Clinical Trial Multicenter Study Randomized Controlled Trial Research Support, Non-U.S. Gov't].
156. Diamond S, Freitag FG. The use of ibuprofen plus caffeine to treat tension-type headache. *Curr Pain Headache Rep*. 2001;5(5):472–478.
157. Mercadante S, Serretta R, Casuccio A. Effects of caffeine as an adjuvant to morphine in advanced cancer patients. A randomized, double-blind, placebo-controlled, crossover study. *J Pain Symptom Manage*. 2001;21(5):369–372 [Clinical Trial Randomized Controlled Trial].
158. Mathew RJ, Wilson WH. Caffeine induced changes in cerebral circulation. *Stroke*. 1985;16(5):814–817.
159. Qi L, van Dam RM, Liu S, Franz M, Mantzoros C, Hu FB. Whole-grain, bran, and cereal fiber intakes and markers of systemic inflammation in diabetic women. *Diabetes Care*. 2006;29(2):207–211.
160. Rodrigo L, Blanco I, Bobes J, de Serres FJ. Clinical impact of a gluten-free diet on health-related quality of life in seven fibromyalgia syndrome patients with associated celiac disease. *BMC Gastroenterol*. 2013;13(1):157. http://dx.doi.org/10.1186/1471-230X-13-157.
161. Leon AJ, Garrote JA, Blanco-Quiros A, et al. Interleukin 18 maintains a long-standing inflammation in coeliac disease patients. *Clin Exp Immunol*. 2006;146(3):479–485. http://dx.doi.org/10.1111/j.1365-2249.2006.03239.x.
162. Neal EG, Chaffe H, Schwartz RH, et al. The ketogenic diet for the treatment of childhood epilepsy: a randomised controlled trial. *Lancet Neurol*. 2008;7(6):500–506. http://dx.doi.org/10.1016/S1474-4422(08)70092-9.
163. Stafstrom CE, Rho JM. The ketogenic diet as a treatment paradigm for diverse neurological disorders. *Front Pharmacol*. 2012;3:59. http://dx.doi.org/10.3389/fphar.2012.00059.
164. Ziegler DR, Gamaro GD, Araujo E, et al. Nociception and locomotor activity are increased in ketogenic diet fed rats. *Physiol Behav*. 2005;84(3):421–427. http://dx.doi.org/10.1016/j.physbeh.2005.01.003.

165. Ruskin DN, Kawamura M, Masino SA. Reduced pain and inflammation in juvenile and adult rats fed a ketogenic diet. *PLoS One*. 2009;4(12):e8349. http://dx.doi.org/10.1371/journal.pone.0008349.
166. Ruskin DN, Suter TA, Ross JL, Masino SA. Ketogenic diets and thermal pain: dissociation of hypoalgesia, elevated ketones, and lowered glucose in rats. *J Pain*. 2013;14(5):467–474. http://dx.doi.org/10.1016/j.jpain.2012.12.015.
167. Masino SA, Ruskin DN. Ketogenic diets and pain. *J Child Neurol*. 2013;28(8):993–1001. http://dx.doi.org/10.1177/0883073813487595.
168. Poplawski MM, Mastaitis JW, Isoda F, Grosjean F, Zheng F, Mobbs CV. Reversal of diabetic nephropathy by a ketogenic diet. *PLoS One*. 2011;6(4):e18604. http://dx.doi.org/10.1371/journal.pone.0018604.
169. Yang X, Cheng B. Neuroprotective and anti-inflammatory activities of ketogenic diet on MPTP-induced neurotoxicity. *J Mol Neurosci*. 2010;42(2):145–153. http://dx.doi.org/10.1007/s12031-010-9336-y.
170. Jeong EA, Jeon BT, Shin HJ, et al. Ketogenic diet-induced peroxisome proliferator-activated receptor-gamma activation decreases neuroinflammation in the mouse hippocampus after kainic acid-induced seizures. *Exp Neurol*. 2011;232(2):195–202. http://dx.doi.org/10.1016/j.expneurol.2011.09.001.
171. Kjeldsen-Kragh J, Haugen M, Borchgrevink CF, et al. Controlled trial of fasting and one-year vegetarian diet in rheumatoid arthritis. *Lancet*. 1991;338(8772):899–902.
172. Kaartinen K, Lammi K, Hypen M, Nenonen M, Hanninen O, Rauma AL. Vegan diet alleviates fibromyalgia symptoms. *Scand J Rheumatol*. 2000;29(5):308–313.
173. Hagen KB, Byfuglien MG, Falzon L, Olsen SU, Smedslund G. Dietary interventions for rheumatoid arthritis. *Cochrane Database Syst Rev*. 2009;(1). Art. No.: CD006400. http://dx.doi.org/10.1002/14651858.CD006400.pub2.
174. Estruch R. Anti-inflammatory effects of the Mediterranean diet: the experience of the PREDIMED study. *Proc Nutr Soc*. 2010;69(3):333–340. http://dx.doi.org/10.1017/S0029665110001539.
175. McKellar G, Morrison E, McEntegart A, et al. A pilot study of a Mediterranean-type diet intervention in female patients with rheumatoid arthritis living in areas of social deprivation in Glasgow. *Ann Rheum Dis*. 2007;66(9):1239–1243. http://dx.doi.org/10.1136/ard.2006.065151.
176. Richard C, Couture P, Desroches S, Lamarche B. Effect of the Mediterranean diet with and without weight loss on markers of inflammation in men with metabolic syndrome. *Obesity (Silver Spring)*. 2013;21(1):51–57. http://dx.doi.org/10.1002/oby.20239.
177. Eletto D, Leone A, Bifulco M, Tecce MF. Effect of unsaturated fat intake from Mediterranean diet on rat liver mRNA expression profile: selective modulation of genes involved in lipid metabolism. *Nutr Metab Cardiovasc Dis*. 2005;15(1):13–23. http://dx.doi.org/10.1016/j.numecd.2004.07.001.
178. Hulsmans M, Geeraert B, De Keyzer D, et al. Interleukin-1 receptor-associated kinase-3 is a key inhibitor of inflammation in obesity and metabolic syndrome. *PLoS One*. 2012;7(1):e30414. http://dx.doi.org/10.1371/journal.pone.0030414.
179. Dalmas E, Rouault C, Abdennour M, et al. Variations in circulating inflammatory factors are related to changes in calorie and carbohydrate intakes early in the course of surgery-induced weight reduction. *Am J Clin Nutr*. 2011;94(2):450–458. http://dx.doi.org/10.3945/ajcn.111.013771.
180. Fisher G, Hyatt TC, Hunter GR, Oster RA, Desmond RA, Gower BA. Markers of inflammation and fat distribution following weight loss in African-American and white women. *Obesity (Silver Spring)*. 2012;20(4):715–720. http://dx.doi.org/10.1038/oby.2011.85.
181. Tajik N, Keshavarz SA, Masoudkabir F, et al. Effect of diet-induced weight loss on inflammatory cytokines in obese women. *J Endocrinol Invest*. 2013;36(4):211–215. http://dx.doi.org/10.3275/8465.

182. Schulte DM, Muller N, Neumann K, et al. Pro-inflammatory wnt5a and anti-inflammatory sFRP5 are differentially regulated by nutritional factors in obese human subjects. *PLoS One*. 2012;7(2):e32437. http://dx.doi.org/10.1371/journal.pone.0032437.
183. de Luis DA, Izaola O, Garcia Alonso M, Aller R, Cabezas G, de la Fuente B. Effect of a hypocaloric diet with a commercial formula in weight loss and quality of life in obese patients with chronic osteoarthritis. *Nutr Hosp*. 2012;27(5):1648–1654. http://dx.doi.org/10.3305/nh.2012.27.5.6053.
184. Bliddal H, Leeds AR, Stigsgaard L, Astrup A, Christensen R. Weight loss as treatment for knee osteoarthritis symptoms in obese patients: 1-year results from a randomised controlled trial. *Ann Rheum Dis*. 2011;70(10):1798–1803. http://dx.doi.org/10.1136/ard.2010.142018.
185. Messier SP, Loeser RF, Miller GD, et al. Exercise and dietary weight loss in overweight and obese older adults with knee osteoarthritis: the Arthritis, Diet, and Activity Promotion Trial. *Arthritis Rheum*. 2004;50(5):1501–1510. http://dx.doi.org/10.1002/art.20256.
186. Messier SP, Mihalko SL, Legault C, et al. Effects of intensive diet and exercise on knee joint loads, inflammation, and clinical outcomes among overweight and obese adults with knee osteoarthritis: the IDEA randomized clinical trial. *JAMA*. 2013;310(12):1263–1273. http://dx.doi.org/10.1001/jama.2013.277669.
187. Skoldstam L, Brudin L, Hagfors L, Johansson G. Weight reduction is not a major reason for improvement in rheumatoid arthritis from lacto-vegetarian, vegan or Mediterranean diets. *Nutr J*. 2005;4:15. http://dx.doi.org/10.1186/1475-2891-4-15.

CHAPTER SIXTEEN

Chemotherapy-Induced Peripheral Neuropathy

Jill C. Fehrenbacher[1]
Department of Pharmacology and Toxicology, Indiana University School of Medicine, Indianapolis, Indiana, USA
Stark Neuroscience Research Institute, Indiana University School of Medicine, Indianapolis, Indiana, USA
Department of Anesthesiology, Indiana University School of Medicine, Indianapolis, Indiana, USA
[1]Corresponding author: e-mail address: jfehrenb@iu.edu

Contents

1. Chemotherapy-Induced Neuropathy: An Introduction 472
2. Classes of Chemotherapeutics 473
   2.1 Microtubule-targeting agents 473
   2.2 Platinum-containing agents 474
   2.3 Proteasome inhibitors 475
   2.4 Angiogenesis inhibitors 475
3. Clinical Assessment of CIPN 476
4. Experimental Studies: Animal Models of CIPN 478
5. Experimental Studies: *In Vitro* Models of CIPN 481
6. Proposed Mechanisms Underlying CIPN 483
   6.1 Mitochondrial dysfunction 484
   6.2 Nitroxidative stress 486
   6.3 DNA damage 487
   6.4 Ion channel modulation 487
   6.5 Inflammation 489
   6.6 Neurotrophic factors 490
   6.7 Microtubule alterations 492
7. Challenges to CIPN Research 492
References 493

Abstract

Chemotherapy-induced peripheral neuropathy (CIPN) is common in patients receiving anticancer treatment and can affect survivability and long-term quality of life of the patient following treatment. The symptoms of CIPN primarily include abnormal sensory discrimination of touch, vibration, thermal information, and pain. There is currently a paucity of pharmacological agents to prevent or treat CIPN. The lack of efficacious therapeutics is due, at least in part, to an incomplete understanding of the mechanisms by which chemotherapies alter the sensitivity of sensory neurons. Although the clinical presentation of CIPN can be similar with the various classes of chemotherapeutic agents,

Progress in Molecular Biology and Translational Science, Volume 131
ISSN 1877-1173
http://dx.doi.org/10.1016/bs.pmbts.2014.12.002

there are subtle differences, suggesting that each class of drugs might induce neuropathy via different mechanisms. Multiple mechanisms have been proposed to underlie the development and maintenance of neuropathy; however, most pharmacological agents generated from preclinical experiments have failed to alleviate the symptoms of CIPN in the clinic. Further research is necessary to identify the specific mechanisms by which each class of chemotherapeutics induces neuropathy.

1. CHEMOTHERAPY-INDUCED NEUROPATHY: AN INTRODUCTION

With considerable advances in the development of anticancer agents over the past 20–30 years, patients diagnosed with cancer are surviving and living longer following treatment.[1] With this increase in survival, researchers and patients are now shifting their focus toward a primary side effect of anticancer treatment, chemotherapy-induced peripheral neuropathy (CIPN). Peripheral neuropathy is an adverse effect following treatment with multiple classes of chemotherapeutics, including platinum drugs, microtubule-targeting agents (MTAs), proteasome inhibitors, and angiogenesis inhibitors. The reported incidence of CIPN varies widely among patients and chemotherapeutic drugs, but often ranges between 30% and 40%.[2] CIPN can be painful and persist as a significant disability following anticancer treatment, resulting in a decreased quality of life for survivors.

The symptoms of peripheral neuropathy are primarily sensory and can be divided into gain in function manifestations including burning pain, tingling, and hypersensitivity to cold or touch and loss-of-function characteristics including loss of proprioception, decreased perception of vibration and pinprick, and numbness.[3–8] Numbness, the loss of vibratory sense, and deep tendon ankle reflexes are some of the first signs of neuropathy in patients receiving chemotherapy treatment, followed by the development of paresthesia and loss of positional sense.[7,9,10] The neuropathy induced by chemotherapeutics is thought to be induced by varying degrees of axonopathy.[11–13] As such, symptoms develop in a stocking and glove distribution,[14] suggesting that sensory neurons with long fibers such as the axons innervating the hands and feet, are especially susceptible to neurotoxic insult. At high doses, the effects of chemotherapeutics on the autonomic and motor nervous systems can be observed,[7,8,15,16] but these are not observed frequently and are not thought to mediate the symptoms of CIPN most commonly reported by patients.

There are a number of factors which can contribute to the development of neuropathy in patients receiving chemotherapy, including dose intensity,[17] cumulative dose,[9,18] concurrent treatment with other neurotoxic agents,[19] and preexisting conditions, such as diabetic- and/or alcohol-induced peripheral neuropathy.[14,20–22] The onset and reversibility of CIPN symptoms are variable between different chemotherapeutic classes and are also dependent on the cumulative dose of drug administered.

In some cases, neuropathy symptoms may worsen following completion of treatment, and this phenomenon is called coasting.[23,24] Resolution of peripheral neuropathy following the discontinuation of chemotherapeutic treatment is quite variable among patients and between the drugs used for treatment, but can also be dependent on the cumulative dose of the chemotherapeutics administered[25,26] and on the intensity of the neuropathic symptoms at their peak.[26,27] Some patients experience persistent disabling neuropathy for 9–13 years following the cessation of therapy.[28,29] There are also some patients who report mild neuropathy during treatment and an absence of neuropathy symptoms following the cessation of drug treatment. Even in these asymptomatic patients, however, objective analyses of neuropathy (standardized examinations or nerve conduction studies) indicate the persistence of neuropathy,[30] suggesting long-term effects of chemotherapy. Unfortunately, no established biomarkers have been identified to predict the development, intensity, or reversibility of CIPN.

2. CLASSES OF CHEMOTHERAPEUTICS

2.1 Microtubule-targeting agents

Alterations in microtubule dynamics are a common mechanism of blocking mitosis within dividing cells, which lead to subsequent apoptosis in cancer cells,[31] and MTAs have been developed and used extensively as anticancer drugs.[32] There are two main classes of MTAs, which are compounds that bind directly to the tubulin subunits to alter the dynamic rearrangement of microtubules. Compounds that bind free β-tubulin subunits and prevent polymerization into microtubules are called depolymerizing or destabilizing agents and include vincristine, vinblastine, and vinorelbine.[33–35] The other class of MTAs binds polymerized β-tubulin comprising the microtubule polymer and prevents the depolymerization of the microtubule. These agents are called polymerizing or stabilizing agents and include paclitaxel, docetaxel, and the epothilones.[36,37] Although they have opposite effects on microtubules, both MTA classes interfere with the dynamic rearrangement of microtubules.

The taxanes are used as treatments for several types of cancer including breast, ovarian, gastric, and lung cancers.[22] The clinical onset of symptoms of neuropathy occurs 3–6 weeks following the first dose of paclitaxel.[7] Prior to the development of CIPN, many patients also experience paclitaxel-associated acute pain syndrome which presents as pain in the large axial muscles in the shoulders, pelvis, and thighs.[38,39] Importantly, recent patients data indicate that patients with intense pain in the days following paclitaxel infusion have a higher risk for the development of CIPN 3–18 weeks following the first paclitaxel infusion.[39,40] This association suggests that acute increases in the excitability and/or sensitivity of nociceptive sensory neurons could underlie the chronic changes in neuronal sensitivity induced by paclitaxel. Many questions remain regarding the mechanisms by which sensitization of one population of somatic sensory neurons (those innervating the muscles) can predict enhanced sensitivity of cutaneous sensory neurons.

The vinca alkaloids are used to treat acute lymphocytic leukemia, lymphomas, and neuroblastoma. The onset and incidence of vinca-induced neuropathy are similar to the taxanes; however, the vinca alkaloids are also associated with increases in autonomic neuropathy, which is rarely observed with the taxanes.[41–44] The only epothilone currently used to treat cancer is ixabepilone, which is administered to patients with metastatic or locally advanced breast cancer. The incidence, dose dependence, and reversibility of neuropathy with ixabepilone treatment are similar to that of the taxanes and vinca alkaloids.[45]

2.2 Platinum-containing agents

The platinum-derived chemotherapeutics, cisplatin, carboplatin, and oxaliplatin induce intrastrand and interstrand DNA cross-linkage, which causes the denaturation of nuclear and mitochondrial DNA.[46] This interaction with DNA results in damage to the mitochondria and subsequent inhibition of ATPase activity, cellular arrest, and ultimate death of cancer cells via apoptosis and necrosis.[47] In addition to cross-linkage of DNA, both cisplatin and oxaliplatin increase the intracellular production of reactive oxygen species (ROS),[48,49] which may or may not contribute to the anticancer activity of the drugs. Cisplatin is administered to treat bladder, ovarian, and testicular cancer. Patients receiving cisplatin generally develop peripheral neuropathy when the cumulative dose of cisplatin reaches 250–300 mg/m^2.[6,18] Once the dosage of cisplatin exceeds 500–600 mg/m^2, almost all patients demonstrate symptoms of neuropathy.[50] Carboplatin, utilized for ovarian cancer, does not typically produce a significant neuropathy in patients. Doses of carboplatin up to 300 mg/m^2, which are effective for anticancer treatment, elicit only a mild transient neuropathy.[51,52]

Oxaliplatin is used for the treatment of colon and advanced colorectal cancer. Patients treated with oxaliplatin may experience both acute and chronic peripheral neuropathies. For example, ~86% of patients receiving oxaliplatin describe immediate neurological symptoms, including, but not limited to, cold-induced perioral or pharyngolaryngeal dysesthesias, jaw pain, and muscle stiffness/cramping.[53,54] Approximately, 50% of patients receiving a cumulative dose of oxaliplatin of 1000 mg/m^2 or higher, which is the standard treatment protocol, develop chronic peripheral neuropathy.[5,55] As observed with paclitaxel, the incidence and severity of acute neurological symptoms induced by oxaliplatin can predict the incidence and severity of chronic neuropathy induced by oxaliplatin in patients.[56]

2.3 Proteasome inhibitors

Inhibition of the ubiquitin-proteasome pathway is another mechanism that is being utilized to kill cancer cells. The proteasome pathway is the primary mechanism for the elimination of damaged or unneeded endogenous proteins in eukaryotic cells.[57] This regulatory pathway is critical for numerous cellular functions, including regulation of the cell cycle and gene transcription. Inhibition of the proteasome, especially in cancer cells, increases the rate of apoptosis.[58] Bortezomib is the first proteasome inhibitor to be examined in human clinical studies as an anticancer agent and is currently used to treat multiple myeloma.[59] Patients receiving bortezomib generally develop chronic sharp, burning pain of moderate-to-severe intensity, self-reporting pain values of 7.8 ± 0.7 out of 10 on a 10-point VAS scale.[60] It is common that dose reductions of bortezomib must be made to accommodate for the development of chronic neuropathy symptoms.[61] Bortezomib also induces an autonomic neuropathy in a subset of patients.[62]

2.4 Angiogenesis inhibitors

In the past years, angiogenesis has been highlighted as a potential anticancer target because of the reliance of tumor growth and metastatic spread on the development of a vascular supply (see Ref. 63). Vascular endothelial growth factor (VEGF) and fibroblast growth factor 2 (FGF2) are two proangiogenic factors which have been shown to promote tumor angiogenesis.[64–66] Bevacizumab and thalidomide are two chemotherapeutics which target the signaling of VEGF and FGF2. Bevacizumab is approved for the treatment of colorectal cancer, glioblastoma, renal cell cancer, and nonsmall cell lung carcinoma, while thalidomide is used for multiple myeloma[67] and investigations into efficacy for combination treatment with thalidomide for the treatment of prostate cancer are ongoing.[68,69] Bevacizumab is a recombinant, humanized monoclonal antibody, which binds to and neutralizes VEGF, preventing its association with endothelial receptors. The exact anticancer mechanism of thalidomide is still debated, but it has been shown to reduce tumor levels of both VEGF and FGF2 to hinder angiogenesis.[70,71] Bevacizumab has not been documented to cause extensive neuropathy as a monotherapy, but reports indicate that coadministration of bevacizumab with oxaliplatin can intensify peripheral neuropathy induced by the platin.[72] Thalidomide elicits the development of peripheral neuropathy symptoms in 50% of patients within 4 months of the initiation of thalidomide treatment.[73,74]

3. CLINICAL ASSESSMENT OF CIPN

A limitation to the interpretation of some clinical studies is the subjective nature by which the presence and intensity of neuropathy is determined by the clinicians and patients. There is much debate regarding the proper endpoints to use to determine the severity of neuropathy in patients and the intensity of neuropathy is often difficult to gage. Several grading scales have been developed by the World Health Organization, Eastern Cooperative Oncology Group, and the National Cancer Institute (NCI) to categorize the intensity of symptoms experienced by patients.[75–78] Using the newly revised NCI-Common Terminology Criteria for Adverse Events (version 4.0), a grade 1 neuropathy indicates mild or asymptomatic neuropathy with loss of deep tendon reflexes or the presence of paresthesia that do not interfere with daily life, grade 2 indicates a moderate neuropathy with some interference with complex activities of daily living, grade 3 describes severe symptoms of neuropathy that limit everyday activities of daily living, and grade 4 indicates life threatening neuropathy.[77–79] A limitation of these neuropathy scales, however, is that they do not provide descriptive data on the quality of neuropathic symptoms, which is especially problematic in a clinical trial setting. To overcome the subjective nature of symptom reporting, a composite measure of peripheral nerve function, the total neuropathy score

(TNS), was established.[19,80,81] The TNS is a tool that clinicians can use to objectively assess neurological endpoints of sensory and motor function in the upper and lower limbs, autonomic symptoms, vibration sensory threshold, sensitivity to pinprick, and changes in deep tendon reflexes or grip strength. In addition to clinical examination findings, the TNS also has a neurophysiological component, in which the conduction velocities in the sural and peroneal nerves are determined and factored into the total score.[82] Chemotherapy sometimes induces a decrease in the action potential amplitude (SNAP) and/or conduction velocity (sCV) of sensory nerve fibers, with occasional effects on the compound muscle action potential and conduction velocity of motor nerves (CMAP and mCV, respectively).[83–90] A major limitation to sensory neurophysiological studies is that the action potential amplitude and conduction velocity of sensory nerves mostly reflect the function of large-diameter sensory neurons. Dysfunction of small- and medium-diameter nerve fibers, which may contribute heavily to burning pain sensations in CIPN patients, is not detected by the recorded evoked potentials.[91] Although neurophysiological outcomes do not always correlate with painful neuropathy symptoms in patients,[91,92] thus questioning their relevance for the diagnosis and/or grading of CIPN, recent studies have demonstrated a strong correlation between early changes in oxaliplatin-induced changes in neurophysiological outcomes and the severity of neuropathy following discontinuation of drug treatment.[83]

A consistent pathological finding in patients receiving chemotherapy and exhibiting numbness and loss of vibratory senses is intraepidermal nerve fiber (IENF) loss in the hands and feet. The IENF is comprised of unmyelinated axons from small-diameter sensory neurons.[93] Thus, observing the degree of change in intraepidermal innervation has been measured using immunohistochemistry and novel functional techniques to specifically assess small-fiber neuropathy in CIPN patients. These novel techniques include nerve excitability-threshold tracking[94] and Doppler flowmetry to determine axon reflex flare areas.[95] Evaluating these endpoints is preferable for testing the ability of drugs to prevent the development of functional neuropathy, because they focus the evaluation toward the site of primary damage, the peripheral nervous system. A limitation of these techniques is that structural loss of innervation does not always correlate with the presence of painful neuropathy symptoms in patients.[96] In addition, chemotherapeutics are not the only cause of IENF loss; it is also observed with diabetes mellitus,[97] HIV infection[98], and other nerve damage. Paradoxically, a complete loss of IENF is observed both in patients with persistent pain[93] and genetic insensitivity to pain,[99] suggesting that epidermal innervation does not necessarily correlate with pain values. The question remains whether the degeneration of IENF results from changes in the sensitivity of sensory neurons or whether the IENF damage causes an increase in the sensitivity of nerve fibers, contributing to the primary symptom of neuropathy that burdens cancer patients: pain.

4. EXPERIMENTAL STUDIES: ANIMAL MODELS OF CIPN

Experimental animal models have been developed to examine the mechanisms of CIPN following exposure to paclitaxel, vincristine, epothilones, cisplatin, oxaliplatin, bortezomib, or thalidomide. Although these animal models have evolved over time to closely mimic the behavioral and neurophysical symptoms observed in patients receiving chemotherapy, the initial animal studies were performed using high doses of the chemotherapeutic drugs. These high doses of drugs caused sensory neuron death, overt axonal damage and involvement of the motor and autonomic nervous systems,[100–103] and produced impairments in pain-like behaviors, decreases in movement and coordination and decreases in nerve conduction velocity in the tail nerve, suggesting a functional decrease in the sensitivity of sensory neurons.[104–110] Since a common complaint of patients with CIPN is augmented pain, rather than analgesia, models with high doses of drug do not optimally reflect the patient condition. The second generation of animal models utilizes lower dosage treatment protocols. This low-dose treatment paradigm elicits mechanical and cold hypersensitivity without inducing sensory neuron loss and gross axonal degeneration.[106,111,112] However, the newer treatment paradigms still produce similar distal morphological changes as observed in patients receiving drug treatment, such as degeneration of the IENFs.[4,113–116] Dose–symptom relationships in patients and animal models will always be complicated by the fact that the sensory neural tissues, including dorsal root ganglia (DRG), nerve roots and nerves, accumulate the chemotherapeutics,[6,18,117] thus discontinuation of the drug treatments does not necessarily mean that chemotherapeutic exposure ceases. Accumulation of the drugs in the DRG has been proposed to occur via uptake in the terminals and soma and is enabled by the dense vascularization and high permeability of capillaries which surround the ganglia.[118–121] The amount of drug accumulation has functional consequences, as the DRG concentration of drug is related to the degree of clinical neurotoxicity in patients.[6]

The onset of behavioral changes in preclinical models varies with the type of chemotherapeutic administered (Table 1). With oxaliplatin and

Table 1 Chemotherapeutics associated with the development of peripheral neuropathy

			Low-dose animal model			
Anticancer agent class	**Agent**	**Clinical onset of sensory symptoms**	**Mechanical sensitivity**	**Cold sensitivity**	**Heat sensitivity**	**IENF density**
Microtubule-targeting agents	Taxanes	Acute and chronic	↑[111]	↑[111]	↑[111]	↓[115]
	Vinca alkaloids	Chronic	↑[112,140]	↑[140]	↓[112]	↓[115]
	Epothilones	Chronic	ND	ND	ND	↓[130]
DNA-cross-linking agents	Cisplatin[139,144,145]	Chronic	↑[106,128]	No change[145]	↑,[128] No change[106]	↓[93]
	Oxaliplatin	Acute and chronic	↑[12]	↑[12]	No change[12]	↓[12]
Proteasome inhibitor	Bortezomib	Chronic	↑[116]	↑[116]	No change[116]	↓[116]
Angiogenesis inhibitors	Thalidomide	Chronic	Animal model not yet developed			
	Bevacizumab	Chronic	Animal model not yet developed			

ND: not determined.

paclitaxel, some studies show an acute behavioral effect within 60 min, whereas others observe effects at 24–48 h after initial dosing.[122–126] Long-term effects generally present around 6–10 days following the first drug injection and persist for at least 3–7 weeks following the last injection.[105,127,128] Consistent modalities affected by various chemotherapeutics include mechanical allodynia and cold allodynia.[105,106,111,112] Sensitivity to warm/heat is quite variable, depending on the type and dose of drug administration and on the investigating laboratory.[101,105,111,129]

The neurophysiological effects of chemotherapeutic treatment in animal models mimics that observed in patients. Lower doses of chemotherapeutics generate functional effects which are limited to sensory fibers, whereas higher doses also involve motor and autonomic nerves. Endpoints utilized to measure the effects of chemotherapeutics on nerve activity include: nerve conduction velocity,[130–133] amplitude of action potentials,[134] and changes in blood flow induced by neurostimulation.[135] Investigators examining the neurophysiological effects of chemotherapy have reported a slowing of the sensory nerve conduction velocity (SNCV)[130,131,136] and an increase in the action potential amplitude when examining Aδ fibers[137]; however, the effects on C fibers are more variable, with high-dose vincristine causing a decrease in the SNCV[136] and high-dose paclitaxel causing no change in the SNCV.[122] The data from skin-nerve experiments support an effect of high-dose paclitaxel to decrease the SNCV in all types of sensory neurons (Aβ, Aδ, and C fibers).[134]

As observed in patients receiving lower doses of chemotherapeutics, the morphological changes in the axonal diameter or myelination of the nerve bundle following lower dose treatment of animals with the various chemotherapeutics are unimpressive.[12,111,116,138–140] However, lower doses of chemotherapeutics, which cause an enhanced sensitivity to nociceptive stimuli, cause a degeneration or loss of IENF, similar to that observed in patients.[115,116,141–143] One advantage of using animal models is that studies can be designed to determine whether this IENF loss is a cause or an effect of changes in neuronal sensitivity following chemotherapy treatment, although, to date, these studies have not been performed. The data from the animal models suggest that both increases and decreases in neuronal sensitivity occur following treatment with chemotherapeutics, depending on the endpoint measured and the dosing paradigm administered. There is still a lack of understanding, however, as to what population of neurons is affected and the mechanisms by which these changes in sensitivity develop (Figure 1).

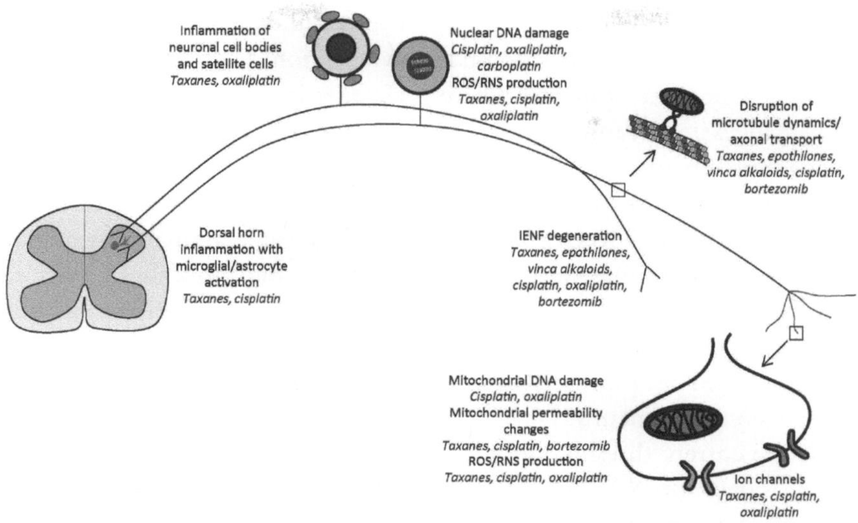

Figure 1 Putative sites of chemotherapeutic toxicity in the peripheral nervous system. (See the color plate.)

5. EXPERIMENTAL STUDIES: *IN VITRO* MODELS OF CIPN

To accurately delineate the intracellular signaling mechanisms by which chemotherapeutics alter the sensitivity of sensory neurons, investigators have broadened their research to include *in vitro* models of CIPN. Cultures of sensory neurons provide a tightly controlled system which allows relatively easy manipulation of protein expression and function via genetic and pharmacological tools. The limitation of neuronal cultures is that the neurons have been removed from their native environment and thus, communication with other cell types (direct or indirect), is largely altered. Furthermore, the addition of growth factors to the growth media could alter the effect of chemotherapeutics on neuronal sensitivity and outgrowth (this will be discussed below). *In vitro* experiments with chemotherapeutics have largely been limited to morphological studies to determine the effects of the drugs on neurite lengths in DRG cultures.[130,146–151] Chemotherapy-induced toxicity, as measured by decreasing neurite length, is dependent on distal axonal exposure to the drug.[148,150] For example, local application of paclitaxel or vincristine to the soma of cultured neurons does not affect the neurite length, whereas application to the distal terminals causes a shortening of the neurites. It is known that alterations of microtubule dynamics can lead to both neurite retraction and inhibition of neurite outgrowth by altering growth cones.[152,153] The specific mechanisms underlying chemotherapy-induced decreases in IENFs are under investigation. Regardless of the specific mechanism, it is unclear whether changes in neurite length correlate with changes in the sensitivity of sensory neurons in culture and mechanistic *in vitro* studies to examine the effects of chemotherapeutics on the sensitivity of subpopulations of sensory neurons are limited. Several investigators have isolated the lumbar DRG from chemotherapy-treated animals to examine the effects of the drugs on electrical excitability and intracellular calcium signaling within the soma of the lumbar neurons, which predominantly innervate the hindpaws.[154–156] Increases in the excitability of medium- and large-sized neurons isolated from paclitaxel-treated rats were observed: the percentage of neurons which exhibited spontaneous firing increased, while the current threshold for activation of the remaining quiescent neurons was decreased.[156] The contribution of voltage-gated calcium currents to these increases in excitability were investigated by Kawakami and colleagues, who demonstrated that paclitaxel treatment, enhanced voltage-dependent calcium current in both small- and medium-sized neurons.[155] When neurons were isolated from paclitaxel-treated rats and examined for ATP-stimulated increases in intracellular calcium accumulation; however, a decrease in stimulated calcium levels was observed.[154] It is unclear whether stimulation of intracellular calcium accumulation with a general depolarizing stimulus, such as high extracellular potassium, would elicit the same decrease in intracellular calcium. Overall, these data suggest a chemotherapy-induced increase in the excitability of sensory neurons.

An effect of chemotherapeutics to modulate integrated neuronal responses, such as the stimulated release of neurotransmitters from the neurites and soma of sensory neurons, has been investigated for paclitaxel.[157] Paclitaxel alters the sensitivity of isolated sensory neurons in concentration- and time-dependent manner without altering the viability of the sensory neurons. Paclitaxel can enhance or reduce TRPV1- and TRPA1-mediated release of calcitonin gene-related peptide (CGRP), depending on the magnitude and duration of exposure to the chemotherapeutic. However, paclitaxel augments CGRP release evoked by high extracellular potassium, regardless of the concentration or duration of paclitaxel exposure.[149] Exposing sensory neurons in culture to cisplatin results in a concentration-dependent decrease in the TRPV1-mediated release of CGRP, even after accounting for the loss of content via cell death.[158] These experiments investigate long-term effects of chemotherapeutic exposure on neuropeptide release, but some studies demonstrate a direct excitatory effect of chemotherapeutics on neuropeptide release at higher drug concentrations, suggesting the possibility for an acute effect of the drugs on neuronal sensitivity.[159,160] To date, the different effects of the various classes of chemotherapeutics on neuropeptide release suggest that each class of drugs likely has a different mechanism of action to elicit changes in neuronal sensitivity. Using neurophysiological endpoints, such as the release of neurotransmitters from sensory neurons, which integrate the contributions of possible changes in function of ion channels, mitochondria, DNA damage, oxidative stress, and axonal transport, will be critical to determining whether all classes of chemotherapeutics alter neuronal sensitivity via the same mechanisms and identifying the specific signaling mechanisms by which neuronal sensitivity is altered by chemotherapeutics.

6. PROPOSED MECHANISMS UNDERLYING CIPN

There have been multiple theories on the mechanisms by which each of the chemotherapeutics alters neuronal function. In this section, the use of pharmacological interventions that have been used for other types of neuropathic pain will be discussed. In addition, various putative mechanisms will be discussed in the context of the clinical and experimental evidence supporting or refuting that mechanism. It is appreciated that many of these proposed mechanisms may overlap, and thus, it is probable that neuronal function is altered via a combination of the proposed cellular mechanisms.

Despite differences in disease pathophysiology and symptoms experienced by patients with CIPN,[161] the initial approach to investigate CIPN focused on administering drugs utilized for other pain syndromes to treat the painful symptoms of peripheral neuropathy. As such, there have been many clinical studies to examine the effects of opioids,[60,162] antidepressants,[163–165] and antiepileptics[166,167] on the symptoms of neuropathy induced by chemotherapeutics. Some of these drugs have been investigated and shown to be efficacious in reversing mechanical and/or cold allodynia in animal models of CIPN, including opioids,[168] antidepressants,[169–171] and antiepileptics.[172,173] Of all the aforementioned drugs, only duloxetine, a serotonin/norepinephrine reuptake inhibitor, has been demonstrated to have clinical efficacy in the treatment of chronic neuropathy symptoms induced by oxaliplatin administration.[165] In fact, a recent set of guidelines published by the American Society for Clinical

Oncologists moderately recommends duloxetine for the treatment of established CIPN, but does not endorse any other treatments for the prevention or treatment of CIPN.[174]

6.1 Mitochondrial dysfunction

Arguably, the most investigated putative mechanism by which most chemotherapeutics are thought to alter neuronal function is via the perturbation of neuronal mitochondrial function. Changes in mitochondrial localization, altered fission, and fusion rates and changes in mitochondrial membrane permeability have all been investigated following neuronal exposure to chemotherapeutics. These alterations in mitochondria are associated with changes in calcium handling, release of cytochrome *c*, mitochondrial DNA damage, and enhanced production of ROS, which can all contribute to neurotoxicity. The localization of mitochondria is dependent on axonal transport via cytoskeletal components, including microtubules and actin filaments,[175] and paclitaxel and cisplatin have been shown to impair the axonal movement of organelles, including mitochondria, in neurons or neuronal cells lines.[176–178] Multiple different chemotherapeutics have been shown to induce mitochondrial swelling within sensory neuronal axons,[116,138,179] sensory neuronal soma,[180] and Schwann cells.[127] Mitochondrial swelling occurs in normal neurons, but an increased incidence of swelling is associated with changes in mitochondrial function, and can be a result of mitochondrial permeability transition.[181,182] Multiple chemotherapeutics have been shown to promote mitochondrial permeability transition and depolarize the mitochondria,[146,183–185] theoretically resulting in a decrease in the ability of the mitochondria to respire and generate ATP for the neuron.[116,117,182] Depolarization of the mitochondria also enhances the production of ROS,[186] which is another putative mechanism by which chemotherapeutics alter neuronal sensitivity (see below). An alternate mechanism by which chemotherapeutics may alter mitochondrial localization and energy production is through the disruption of the balance between mitochondrial fission and fusion. Fission, the defragmentation of mitochondria, leads to a decrease in energy production and enhances the production of ROS.[187] Although a change in mitochondrial fission was not observed in DRG from cisplatin-treated rats,[188] paclitaxel altered mitochondrial dynamics in differentiating neuroblastoma cells.[189]

Because of the demonstrated effects of anticancer drugs, of multiple mechanisms, on the localization and function of mitochondria, several different mitoprotective drug strategies have been examined to determine whether preventing mitochondrial dysfunction might reverse the signs and symptoms of neurotoxicity. The premiere mitoprotective candidate was acetyl-L-carnitine (ALCAR). ALCAR plays a role in intermediary metabolism and is thought to be neuroprotective via an increase in fatty acid β-oxidation to subsequently increase ATP production.[190] ALCAR has also been shown to reverse the depletion of plasma nerve growth factor (NGF) levels induced by cisplatin treatment.[131] In preclinical experiments, ALCAR administration reversed mechanical and cold allodynia induced by paclitaxel,[131,191] cisplatin,[131] and bortezomib.[116] ALCAR also reversed the chemotherapeutic-induced swelling and dysfunction of mitochondria.[116,131,179] These preclinical data and evidence that ALCAR have therapeutic effects on neuropathy in patients with diabetes[192] and HIV,[193,194] prompted clinical trials to examine the effects of ALCAR on paclitaxel-induced neuropathy in patients. In a phase II study with 25 patients, the effects of ALCAR were promising, providing improvement of sensory neuropathy in 60% of patients.[195] In order to confirm these hopeful findings, a prospective randomized, double-blind, placebo-controlled trial of 409 breast cancer patients receiving taxane-based adjuvant therapy were administered ALCAR. In contrast to the findings of the small-clinical trial, ALCAR treatment in the large trial increased CIPN at 24 weeks following treatment, as determined by patient self-reporting and clinician grading of sensory and motor neuropathy adverse effects.[196]

Another putative mitoprotective drug, olesoxime, has also been investigated to determine whether it reverses hypersensitivity in animal models of CIPN. Olesoxime binds directly to two different components of the mitochondrial membrane permeability transition pore[197] to regulate depolarization of the mitochondria. In addition, treatment with olesoxime restores mitochondrial motility, compromised by treatment with paclitaxel, suggesting that olesoxime can interact with microtubule function.[198] Systemic administration of olesoxime reversed mechanical allodynia in animal models of vincristine, paclitaxel, and oxaliplatin-induced CIPN,[12,143,197] but did not attenuate the spontaneous activity of the peripheral C- and Aδ-fibers induced by the drugs, suggesting that the site of action for olesoxime is in the CNS.[143] The translation of these protective preclinical effects of olesoxime into a clinically relevant therapeutic for CIPN has yet to be determined. A phase II clinical study (NCT00876538) to examine the effects of olesoxime on paclitaxel-induced pain and/or dysesthesia has been completed, but the study outcomes have not been released.

Despite observations in experimental animals after anticancer drug treatment, the question of mitochondrial involvement in CIPN remains controversial. It is unclear whether mitochondrial dysfunction precedes changes in neuronal sensitivity and/or changes in neuronal morphology. Furthermore, mitochondrial changes such as those listed above have also been associated with general nerve injury[199] or insult, such as that resulting from streptozotocin treatment as a preclinical model for diabetes mellitus,[200] suggesting that damage to mitochondria might be a common downstream effect of nonspecific damage to sensory neurons rather than a causative mechanism by which chemotherapeutics alter nerve function.

6.2 Nitroxidative stress

Chemotherapeutics have also been shown to enhance the production of both ROS and reactive nitrogen species (RNS) in sensory neurons and in the spinal cord.[158,201,202] In fact, the generation of nitroxidative stress, induced by ROS and RNS, has been shown to enhance the sensitivity of nociceptive sensory neurons.[203,204] Chemotherapy-induced ROS and RNS are generated by neurons, supporting satellite cells and inflammatory cells.[205,206] Intracellular generators of ROS and RNS include the mitochondria, via the electron transport chain and an increased incidence of fission, and an increased activity of plasma membrane NADPH oxidase.[207] Unfortunately, nitroxidative stress can compromise the survival and function of sensory neurons and supporting cells by causing DNA damage, nerve fiber demyelination, mitochondrial damage and dysfunction, activation of signal transduction pathways, and neuronal death via apoptosis.[208–210] Multiple different strategies have been employed preclinically to try reverse chemotherapy-induced nociceptive behaviors with scavengers of ROS (phenyl-*N*-*tert*-butylnitrone and TEMPOL) or peroxynitrite decomposition catalysts ($FeTMPyP^{5+}$ and $MnTE\text{-}2\text{-}PyP^{5+}$),[201,210–212] and these experiments demonstrate moderate attenuation of chemotherapy-induced hindpaw hypersensitivity to mechanical or cold stimulation. The effects of these compounds on nerve conduction velocity or IENF loss, as other indicators of peripheral neuropathy, have not yet been explored. Glutathione is another antioxidant which facilitates the reduction of the enzyme, glutathione peroxidase, to reduce lipid and hydrogen peroxides.[213] *N*-acetylcysteine (NAC) functions to regenerate the reduced form of glutathione for continued reduction of ROS.[214] In preclinical models of CIPN, exogenous administration of glutathione and NAC reduced the severity of neuropathy using nociceptive behaviors,[202] sensory nerve conduction velocities,[215,216] and neuronal morphology[217] as endpoints. Following these preclinical studies, several small trials with glutathione and NAC were conducted in cancer patients receiving either cisplatin or oxaliplatin. The trials all reported a reduction in neurotoxicity, albeit the degree of protection varied greatly.[218–224] In a large randomized, double-blind, placebo-controlled trial to investigate the ability of glutathione to reverse the neuropathy associated with paclitaxel/carboplatin combination treatment; however, glutathione did not prevent CIPN.[225]

Several small-clinical trials also demonstrated that vitamin E might have efficacy to decrease the incidence or severity of CIPN following paclitaxel or cisplatin treatment[226–229]; however, a phase III clinical trial failed to reproduce the beneficial effects of the putative antioxidant.[230]

There are some concerns about further development of antioxidant therapies to treat CIPN, since the production of ROS and RNS has been shown to be essential for the anticancer effects of chemotherapeutics.[49,231]

6.3 DNA damage

Multiple strategies have been utilized to demonstrate the importance of DNA damage in mediating neurotoxicity induced by the platinum-derived compounds. Enhancing the activity of DNA repair enzymes, involved in either base excision or nucleotide excision repair pathways, mitigates the neurotoxicity associated with exposure to cisplatin.[158,180,232] Whether enhanced repair of the nuclear or mitochondrial DNA adducts is critical for the protection against neurotoxicity is still under investigation.[180] Mitochondrial DNA damage by cisplatin has been shown to induce the generation of ROS, which can be attenuated by overexpression of APE1, a DNA repair enzyme of the base excision repair pathway.[158] Interestingly, carboplatin, which causes the formation of nuclear DNA adducts, but does not generate ROS or mitochondrial DNA damage,[49] induces a very mild neuropathy,[48,233] leading to speculation that increases in ROS generation secondary to mitochondrial DNA damage is a critical component of neuropathy induced by cisplatin and oxaliplatin.

6.4 Ion channel modulation

Changes in ion channel conductance are also a putative target for chemotherapeutics to alter neuronal sensitivity. Both voltage- and ligand-gated ion channels have been shown to play a role in chemotherapy-induced changes in nociceptive behaviors. Included in these are sodium channels,[137,234,235] calcium channels,[155,173,236–238] ATP-gated channels,[239] acid-sensing channels,[239] and TRP channels,[240–242] although much controversy persists regarding which channels mediate different modalities of chemotherapy-induced behaviors. Gabapentin, which binds to the $\alpha_2\delta$ subunit of the calcium channel,[243] has been investigated in preclinical studies and found to have variable effects on nociceptive behaviors, depending on the class of chemotherapy administered. Sensitivity to gabapentin was correlated with chemotherapy-induced upregulation of the $\alpha_2\delta$ subunit of the calcium channel in the DRG and spinal cord.[244,245] Both taxanes and platinum-containing agents elicit an upregulation of the $\alpha_2\delta$ subunit, and nociceptive behaviors induced by treatment with these chemotherapeutics are reversed by gabapentin administration.[12,105,237,244] In contrast, vinca alkaloids do not upregulate the expression of the $\alpha_2\delta$ subunit, and nociceptive behaviors associated with vinca treatments are not reversed by gabapentin.[244,245] There has been only one clinical trial to examine whether gabapentin administration reverses established CIPN in patients receiving vinca alkaloids, taxanes, or platinum-containing agents.[166] The results of the trial failed to show an effect of gabapentin to reverse pain as a CIPN symptom.[166] Given the dependence of gabapentin efficacy on the class of chemotherapy administered in preclinical studies, it would be interesting to determine whether gabapentin could reverse CIPN in a study with sufficient power to be able to analyze the outcomes based upon the class of chemotherapy which elicited the CIPN. Alternative strategies are being developed to target the pathophysiological function of specific ion channels without complete blockade of ion channel conductance, via the modulation of ion channel-binding proteins.[246] This strategy, which is promising in the treatment of pain derived from various etiologies,[247] might be a useful approach for the treatment of CIPN once the chemotherapy-induced changes in ion channels are understood.

Another cause of ion channel dysfunction, specifically following treatment with oxaliplatin,[235,248] is the chelation of extracellular calcium and magnesium by oxalate, a metabolic product of oxaliplatin.[249] It was hypothesized that replacement of these ions, via Ca^{2+}/Mg^{2+} infusions, could help to reverse acute and chronic oxaliplatin-induced neuropathy symptoms.[250] Preclinical experiments demonstrated prevention of the hypersensitivity to cold,[105,126] but ion replacement did not alter mechanical hypersensitivity induced by either oxaliplatin or oxalate.[126] An initial retrospective review of patients receiving oxaliplatin in the absence or presence of Ca^{2+}/Mg^{2+} infusions supported the theory that Ca^{2+}/Mg^{2+} therapy prevented neuropathy, as patients receiving Ca^{2+}/Mg^{2+} were administered larger cumulative doses of oxaliplatin (650 vs. 910 mg/m^2 for placebo vs. Ca^{2+}/Mg^{2+}, respectively), and reported less acute neuropathic symptoms (laryngopharangeal dysesthesia in 9% vs. 0% of patients), and less development of grade 3 chronic neuropathy (20% vs. 8% of patients).[250] Based on this retrospective data, two clinical trials were initiated to examine the efficacy of Ca^{2+}/Mg^{2+} infusions to prevent oxaliplatin-induced neuropathy (CONcePT trial and N04C7). Trial progress was halted due to concerns that Ca^{2+}/Mg^{2+} infusions reduced the oncolytic activity of oxaliplatin,[251] however, those concerns have since been refuted.[252] The data generated from the two discontinued trials were conflicting; in the CONcePT trial, Ca^{2+}/Mg^{2+} infusions did prevent acute symptoms or chronic neuropathy[253]; whereas in the early results of the N04C7 trial,[254] Ca^{2+}/Mg^{2+} infusions decreased oxaliplatin-induced acute muscle spasms and the development of grade 2 or higher chronic neuropathy.[254] A third trial was initiated (N08CB) to resolve whether Ca^{2+}/Mg^{2+} infusions were efficacious to prevent oxaliplatin-induced neuropathy.[255] An effect of infusions to prevent the development of acute or chronic oxaliplatin-induced neuropathy was not observed; however, there have been several critiques of the data interpretation from the N08CB study. Under questions are the selected choice of sensory scales used as endpoints for neuropathy and the power of the study to determine the reported negative efficacy outcomes.[256,257] In light of the preclinical data, which demonstrate an effect of Ca^{2+}/Mg^{2+} to reverse cold, but not mechanical, hypersensitivity, it would be useful to analyze the clinical data based on individual sensory symptoms. In this manner, information regarding whether Ca^{2+}/Mg^{2+} has efficacy to prevent specific modalities of the neuropathy could be obtained.

6.5 Inflammation

Another plausible mechanism by which chemotherapeutic treatments alter the sensitivity of sensory neurons is through the activation of the immune system and subsequent induction of inflammation. Several investigators have observed the activation of resident immune cells, such as the satellite cells in the DRG[258] and the microglia in the dorsal horn of the spinal cord[188,259] following chemotherapeutic administrations. Infiltration of macrophages in the DRG and antigen-presenting Langerhans cells in the skin has also been observed following the administration of paclitaxel.[115,260] Furthermore, modulation of known inflammatory signaling pathways, such as

NFκB[141] and Toll-like receptor (TLR) signaling[261,262] attenuate the behavioral effects of chemotherapy administration in preclinical models. Indeed, paclitaxel administration to rats enhances the expression of the TLR4 in the DRG and spinal cord.[262] Intrathecal administration of a TLR4 antagonist, lipopolysaccharide derived from *Rhodobacter sphaeroides,* partially reversed the mechanical hypersensitivity associated with paclitaxel treatment.[262] Several investigators have also demonstrated an upregulation of chemokine (C–C motif) ligand 2 (CCL2 or MCP-1) following paclitaxel treatment.[259,263] This increase in chemokine levels was shown to have functional importance, since intrathecal inhibition of the receptor for CCL2, the C–C chemokine receptor type 2 (CCR2), reverses mechanical allodynia, and the degeneration of IENF induced by paclitaxel administration.[263] The signaling pathways by which paclitaxel upregulates CCL2 levels have not been elucidated, but CCL2 is enhanced primarily in small-diameter sensory neurons within the DRG and in astrocytes of the dorsal horn.[263] The physiological function of CCL2 is to recruit and activate monocytic cell types, including macrophages and microglia. This function is apparent in the dorsal horn of the spinal cord, where a paclitaxel-induced upregulation of CCL2 activates microglia, reversible by intrathecal application of an anti-CCL2 antibody.[259,263] What remains unclear, however, is how paclitaxel-induced activation of the CCL2/CCR2 pathway, which occurs in the entire dorsal column regardless of axonal length,[263] can initiate neuronal sensitivity in a "stocking and glove" distribution. Further investigations into possible interactions between inflammation and axonal damage, both induced by chemotherapeutic treatment, to elicit symptoms of CIPN might uncover additional therapeutic targets to prevent or treat the neuropathy.

6.6 Neurotrophic factors

An approach to reverse the symptoms of CIPN has been to ameliorate the neuropathy through the addition of neurotrophic factors in an attempt to overcome the nerve damaging effects of chemotherapeutics with growth promoting factors. The first neurotrophin to be investigated was NGF. NGF was discovered in the late 1950s as a nerve growth promoting factor and is essential for the survival of spinal sensory ganglion cells and sympathetic neurons (see review by Zhang *et al.* [264]). NGF levels have been shown to correlate with the intensity of neuropathy symptoms induced by paclitaxel in the clinic and in the laboratory. In patients receiving paclitaxel and cisplatin therapy, treatment-induced decreases in circulating NGF levels were associated with the development of CIPN.[265] Although this chemotherapy-induced decrease in circulating NGF levels could not be reproduced in preclinical models,[110] exogenous administration of NGF with high-dose paclitaxel reversed paclitaxel-induced thermal hypoalgesia.[266] In studies investigating the effects of paclitaxel and NGF on cell survival and neurite outgrowth, NGF treatment reversed neuronal degeneration, and death in embryonic DRG explants exposed to paclitaxel (1 μM) for 1–4 days.[147,267] Together, these data suggest a protective effect of NGF against neurotoxicity induced by exposure to chemotherapeutics; however, NGF treatment to prevent or treat CIPN has not been tried in the clinic. One limitation to the use of NGF as a treatment for CIPN is the effect that the neurotrophin has on the sensitivity of small-diameter sensory neurons. Inflammation-induced increases in the levels of NGF have been shown to mediate the thermal and mechanical hypersensitivity associated with inflammation,[268,269] thus NGF treatment may cause inflammatory-like pain in cancer patients who are already suffering from neuropathy. Indeed, anti-NGF pharmacological agents have been investigated for the treatment of inflammatory pain.[270] Another limiting factor in the translation of preclinical NGF effects is centered on the interpretation of the experimental data. Embryonic neurons, which were used in many of the preclinical studies, are dependent on NGF for survival.[271] NGF is trafficked from the nerve terminals to the nucleus and this retrograde transport is blocked by the MTA, colchicine.[272,273] Thus, the decrease in survival observed with paclitaxel exposure could be a result of possible NGF deprivation due to decreased axonal transport, rather than a positive effect of NGF to protect the neurons.[274,275] Furthermore, since embryonic neurons are dependent on NGF for survival, it is difficult to interpret the survival findings, since low concentrations of paclitaxel which alter neuronal sensitivity do not commonly cause neuronal death in cultures of DRG neurons derived from adult animals.[157] Other neurotrophic agents which have been investigated and have had positive results in preclinical animals include: prosaptide,[129] xaliproden,[276] retinoic acid,[277] and recombinant human glial growth factor 2[278], but so far none have demonstrated appreciable efficacy in clinical trials.[279] Neurotrophic factors play an integral role not only in the survival of sensory neurons but also in the maintenance of the neurons, thus helping to establish the setpoints for neuronal sensitivity and growth.[280,281] Targeting this class of compounds, without a comprehensive understanding of how chemotherapy alters intracellular signaling pathways and the axonal transport mechanisms used to traffic the products of neurotrophin signaling, is proving to be very challenging.

6.7 Microtubule alterations

An obvious mechanism by which MTAs may alter neuronal function is through the perturbation of microtubule function. Microtubules are dynamic cellular structures that have critical roles in many functions including cell signaling, mitosis and cell division, intracellular organelle and vesicular transport, cell shape and motility, and intracellular organization.[32,282] The dynamic nature of microtubules, with constant growth and shortening, is necessary for microtubule physiology.[283] In nondividing neurons, the MTAs have the potential to induce neurotoxicity in a very specific way by altering axonal transport along the microtubules,[11,284,285] by changing the posttranslational status of microtubules,[286–288] and by changing protein–protein interactions between the microtubules and cellular proteins.[289–291] Bortezomib has also been shown to change microtubule dynamics and thus could mediate changes in neuronal sensitivity via a mechanism similar to the MTAs.[292] Unfortunately, the anticancer efficacy of the MTAs is directly proportional to their ability to alter microtubule dynamics,[293] thus modulating the microtubule interactions to prevent CIPN would most likely compromise the anticancer properties of the drugs.

7. CHALLENGES TO CIPN RESEARCH

There has been very little success in translating preclinical findings to efficacious therapies for patients suffering with CIPN. One caveat that has been largely unaddressed in CIPN research is the possibility that the presence of cancer may predispose patients to neuropathy or induce novel interactions with anticancer drugs to "prime" patients to develop a more robust neuropathy than that which is observed in preclinical models. The presence of cancer alone has been shown to induce central nervous system toxicity, using deficits in cognitive dysfunction as an endpoint for patients.[294] To address the possibility that the presence of cancer may also contribute to peripheral neurotoxicity, researchers are shifting their preclinical models from rats to mice.[294] The mouse is well developed as a disease model for cancer research and could facilitate the necessary experiments to examine the development of neuropathy in the absence and presence of anticancer treatment in tumor-burdened animals. While a possible interaction between cancer and anticancer drugs in preclinical models could underlie the paucity of translational successes from the lab to the clinic, there are other putative pitfalls that may also contribute. CIPN research was originally focused on the development of animal models; efforts then shifted to try to reverse evoked nociceptive behaviors induced by anticancer treatment with a variety of "pain" drugs to see if they would alleviate nociceptive behaviors in the animals. Currently, the nociceptive tests that are used to discriminate the symptoms of neuropathy and the success and/or failure of putative chemotherapy drugs in preclinical models, do not assess this modality of spontaneous pain. Indeed, this small-fiber neuropathy is proving to be very difficult to study. Research by Kalliomaki and colleagues has demonstrated that the endpoints commonly used as surrogates for neuronal sensitivity in preclinical models: thermal behaviors, axon flare reflexes, and IENF loss, do not necessarily correlate with pain levels in patients who experience pain as a symptom of the neuropathy,[96] suggesting that using appropriate endpoints to monitor neuropathy, even in *in vitro* studies, is critical for the translational ability of the research. The neuropathy induced by chemotherapeutics is multifaceted and deriving the mechanisms for the various symptoms will take much effort. One of the distinct differences between CIPN and neuropathies induced by physical nerve damage is that the timing of the insult is defined. This predictability allows for the introduction of pharmacological agents to inhibit the neuronal effects of the chemotherapeutics. Only after, we understand the neuronal signaling pathways altered by chemotherapeutics; however, we can subsequently develop therapeutics to specifically prevent the development or maintenance of neurotoxicity without compromising the oncolytic activity of the drug.

REFERENCES

1. Burton AW, Fanciullo GJ, Beasley RD, Fisch MJ. Chronic pain in the cancer survivor: a new frontier. *Pain Med.* 2007;8(2):189–198.
2. Wolf S, Barton D, Kottschade L, Grothey A, Loprinzi C. Chemotherapy-induced peripheral neuropathy: prevention and treatment strategies. *Eur J Cancer.* 2008;44(11):1507–1515.
3. Iniguez C, Larrode P, Mayordomo JI, et al. Reversible peripheral neuropathy induced by a single administration of high-dose paclitaxel. *Neurology.* 1998;51(3):868–870.
4. Boyette-Davis JA, Cata JP, Driver LC, et al. Persistent chemoneuropathy in patients receiving the plant alkaloids paclitaxel and vincristine. *Cancer Chemother Pharmacol.* 2013;71(3):619–626.
5. Krishnan AV, Goldstein D, Friedlander M, Kiernan MC. Oxaliplatin-induced neurotoxicity and the development of neuropathy. *Muscle Nerve.* 2005;32(1):51–60.
6. Gregg RW, Molepo JM, Monpetit VJ, et al. Cisplatin neurotoxicity: the relationship between dosage, time, and platinum concentration in neurologic tissues, and morphologic evidence of toxicity. *J Clin Oncol.* 1992;10(5):795–803.
7. Forsyth PA, Balmaceda C, Peterson K, Seidman AD, Brasher P, DeAngelis LM. Prospective study of paclitaxel-induced peripheral neuropathy with quantitative sensory testing. *J Neurooncol.* 1997;35(1):47–53.

8. Lipton RB, Apfel SC, Dutcher JP, et al. Taxol produces a predominantly sensory neuropathy. *Neurology*. 1989;39(3):368–373.
9. Postma TJ, Vermorken JB, Liefting AJ, Pinedo HM, Heimans JJ. Paclitaxel-induced neuropathy. *Ann Oncol*. 1995;6(5):489–494.
10. Park SB, Lin CS, Krishnan AV, Friedlander ML, Lewis CR, Kiernan MC. Early, progressive, and sustained dysfunction of sensory axons underlies paclitaxel-induced neuropathy. *Muscle Nerve*. 2011;43(3):367–374.
11. Theiss C, Meller K. Taxol impairs anterograde axonal transport of microinjected horseradish peroxidase in dorsal root ganglia neurons in vitro. *Cell Tissue Res*. 2000;299(2):213–224.
12. Xiao WH, Zheng H, Bennett GJ. Characterization of oxaliplatin-induced chronic painful peripheral neuropathy in the rat and comparison with the neuropathy induced by paclitaxel. *Neuroscience*. 2012;203:194–206.
13. Cavaletti G, Tredici G, Marmiroli P, Petruccioli MG, Barajon I, Fabbrica D. Morphometric study of the sensory neuron and peripheral nerve changes induced by chronic cisplatin (DDP) administration in rats. *Acta Neuropathol*. 1992;84(4):364–371.
14. Rowinsky EK, Chaudhry V, Cornblath DR, Donehower RC. Neurotoxicity of Taxol. *J Natl Cancer Inst Monogr*. 1993;15:107–115.
15. Wiernik PH, Schwartz EL, Strauman JJ, Dutcher JP, Lipton RB, Paietta E. Phase I clinical and pharmacokinetic study of taxol. *Cancer Res*. 1987;47(9):2486–2493.
16. Dougherty PM, Cata JP, Cordella JV, Burton A, Weng HR. Taxol-induced sensory disturbance is characterized by preferential impairment of myelinated fiber function in cancer patients. *Pain*. 2004;109(1–2):132–142.
17. Nabholtz JM, Gelmon K, Bontenbal M, et al. Multicenter, randomized comparative study of two doses of paclitaxel in patients with metastatic breast cancer. *J Clin Oncol*. 1996;14(6):1858–1867.
18. Thompson SW, Davis LE, Kornfeld M, Hilgers RD, Standefer JC. Cisplatin neuropathy. Clinical, electrophysiologic, morphologic, and toxicologic studies. *Cancer*. 1984;54(7):1269–1275.
19. Chaudhry V, Rowinsky EK, Sartorius SE, Donehower RC, Cornblath DR. Peripheral neuropathy from taxol and cisplatin combination chemotherapy: clinical and electrophysiological studies. *Ann Neurol*. 1994;35(3):304–311.
20. Lee JJ, Swain SM. Peripheral neuropathy induced by microtubule-stabilizing agents. *J Clin Oncol*. 2006;24(10):1633–1642.
21. Windebank AJ, Grisold W. Chemotherapy-induced neuropathy. *J Peripher Nerv Syst*. 2008;13(1):27–46.
22. Rowinsky EK, Gilbert MR, McGuire WP, et al. Sequences of taxol and cisplatin: a phase I and pharmacologic study. *J Clin Oncol*. 1991;9(9):1692–1703.
23. Siegal T, Haim N. Cisplatin-induced peripheral neuropathy. Frequent off-therapy deterioration, demyelinating syndromes, and muscle cramps. *Cancer*. 1990;66(6):1117–1123.
24. von Schlippe M, Fowler CJ, Harland SJ. Cisplatin neurotoxicity in the treatment of metastatic germ cell tumour: time course and prognosis. *Br J Cancer*. 2001; 85(6):823–826.
25. Brydoy M, Oldenburg J, Klepp O, et al. Observational study of prevalence of long-term Raynaud-like phenomena and neurological side effects in testicular cancer survivors. *J Natl Cancer Inst*. 2009;101(24):1682–1695.
26. Park SB, Lin CS, Krishnan AV, Goldstein D, Friedlander ML, Kiernan MC. Long-term neuropathy after oxaliplatin treatment: challenging the dictum of reversibility. *Oncologist*. 2011;16(5):708–716.
27. Haim N, Epelbaum R, Ben-Shahar M, Yarnitsky D, Simri W, Robinson E. Full dose vincristine (without 2-mg dose limit) in the treatment of lymphomas. *Cancer*. 1994;73(10):2515–2519.

28. Moser EC, Noordijk EM, Carde P, et al. Late non-neoplastic events in patients with aggressive non-Hodgkin's lymphoma in four randomized European Organisation for Research and Treatment of Cancer trials. *Clin Lymphoma Myeloma.* 2005;6(2):122–130.
29. Strumberg D, Brugge S, Korn MW, et al. Evaluation of long-term toxicity in patients after cisplatin-based chemotherapy for non-seminomatous testicular cancer. *Ann Oncol.* 2002;13(2):229–236.
30. Ramchandren S, Leonard M, Mody RJ, et al. Peripheral neuropathy in survivors of childhood acute lymphoblastic leukemia. *J Peripher Nerv Syst.* 2009;14(3):184–189.
31. Jordan MA, Wendell K, Gardiner S, Derry WB, Copp H, Wilson L. Mitotic block induced in HeLa cells by low concentrations of paclitaxel (Taxol) results in abnormal mitotic exit and apoptotic cell death. *Cancer Res.* 1996;56(4):816–825.
32. Wilson L, Jordan MA. New microtubule/tubulin-targeted anticancer drugs and novel chemotherapeutic strategies. *J Chemother.* 2004;16(suppl 4):83–85.
33. Johnson IS, Armstrong JG, Gorman M, Burnett Jr JP. The vinca alkaloids: a new class of oncolytic agents. *Cancer Res.* 1963;23:1390–1427.
34. Sartorelli AC, Creasey WA. Cancer chemotherapy. *Annu Rev Pharmacol.* 1969;9:51–72.
35. Zhou XJ, Rahmani R. Preclinical and clinical pharmacology of vinca alkaloids. *Drugs.* 1992;44(suppl 4):1–16 [discussion 66–69].
36. Altaha R, Fojo T, Reed E, Abraham J. Epothilones: a novel class of non-taxane microtubule-stabilizing agents. *Curr Pharm Des.* 2002;8(19):1707–1712.
37. Sackett DL, Fojo T. Taxanes and other microtubule stabilizing agents. *Cancer Chemother Biol Response Modif.* 1999;18:59–80.
38. Loprinzi CL, Maddocks-Christianson K, Wolf SL, Rao RD, Dyck PJ, Mantyh P. The paclitaxel acute pain syndrome: sensitization of nociceptors as the putative mechanism. *Cancer J.* 2007;13(6):399–403.
39. Loprinzi CL, Reeves BN, Dakhil SR, et al. Natural history of paclitaxel-associated acute pain syndrome: prospective cohort study NCCTG N08C1. *J Clin Oncol.* 2011;29(11):1472–1478.
40. Reeves BN, Dakhil SR, Sloan JA, et al. Further data supporting that paclitaxel-associated acute pain syndrome is associated with development of peripheral neuropathy: North Central Cancer Treatment Group trial N08C1. *Cancer.* 2012;118(20):5171–5178.
41. Legha SS. Vincristine neurotoxicity. Pathophysiology and management. *Med Toxicol.* 1986;1(6):421–427.
42. Rowinsky EK, Eisenhauer EA, Chaudhry V, Arbuck SG, Donehower RC. Clinical toxicities encountered with paclitaxel (Taxol). *Semin Oncol.* 1993;20(4 suppl 3): 1–15.
43. Jerian SM, Sarosy GA, Link Jr CJ, Fingert HJ, Reed E, Kohn EC. Incapacitating autonomic neuropathy precipitated by taxol. *Gynecol Oncol.* 1993;51(2):277–280.
44. Miltenburg NC, Boogerd W. Chemotherapy-induced neuropathy: a comprehensive survey. *Cancer Treat Rev.* 2014;40(7):872–882.
45. Vahdat LT, Thomas ES, Roche HH, et al. Ixabepilone-associated peripheral neuropathy: data from across the phase II and III clinical trials. *Support Care Cancer.* 2012;20(11):2661–2668.
46. Roberts JJ, Knox RJ, Friedlos F, Lydall DA. DNA as the target for the cytotoxic and anti-tumour action of platinum coordination complexes: comparative in vitro and in vivo studies of cisplatin and carboplatin. In: McBrien DCH, Slater TF, eds. *Biochemical Mechanisms of Platinum Antitumour Drugs.* Oxford: IRL Press; 1986:29–64.
47. Eastman A. Activation of programmed cell death by anticancer agents: cisplatin as a model system. *Cancer Cells.* 1990;2(8–9):275–280.

48. Kelley MR, Jiang Y, Guo C, Reed A, Meng H, Vasko MR. Role of the DNA base excision repair protein, APE1 in cisplatin, oxaliplatin, or carboplatin induced sensory neuropathy. *PLoS One*. 2014;9(9):e106485.
49. Marullo R, Werner E, Degtyareva N, et al. Cisplatin induces a mitochondrial-ROS response that contributes to cytotoxicity depending on mitochondrial redox status and bioenergetic functions. *PLoS One*. 2013;8(11):e81162.
50. van der Hoop RG, van der Burg ME, ten Bokkel Huinink WW, van Houwelingen C, Neijt JP. Incidence of neuropathy in 395 patients with ovarian cancer treated with or without cisplatin. *Cancer*. 1990;66(8):1697–1702.
51. Anderson H, Wagstaff J, Crowther D, et al. Comparative toxicity of cisplatin, carboplatin (CBDCA) and iproplatin (CHIP) in combination with cyclophosphamide in patients with advanced epithelial ovarian cancer. *Eur J Cancer Clin Oncol*. 1988;24(9):1471–1479.
52. Gurney H, Crowther D, Anderson H, et al. Five year follow-up and dose delivery analysis of cisplatin, iproplatin or carboplatin in combination with cyclophosphamide in advanced ovarian carcinoma. *Ann Oncol*. 1990;1(6):427–433.
53. Argyriou AA, Cavaletti G, Briani C, et al. Clinical pattern and associations of oxaliplatin acute neurotoxicity: a prospective study in 170 patients with colorectal cancer. *Cancer*. 2013;119(2):438–444.
54. Wilson RH, Lehky T, Thomas RR, Quinn MG, Floeter MK, Grem JL. Acute oxaliplatin-induced peripheral nerve hyperexcitability. *J Clin Oncol*. 2002;20(7): 1767–1774.
55. Andre T, Boni C, Mounedji-Boudiaf L, et al. Oxaliplatin, fluorouracil, and leucovorin as adjuvant treatment for colon cancer. *N Engl J Med*. 2004;350(23):2343–2351.
56. Park SB, Goldstein D, Lin CS, Krishnan AV, Friedlander ML, Kiernan MC. Acute abnormalities of sensory nerve function associated with oxaliplatin-induced neurotoxicity. *J Clin Oncol*. 2009;27(8):1243–1249.
57. Sorokin AV, Kim ER, Ovchinnikov LP. Proteasome system of protein degradation and processing. *Biochemistry (Mosc)*. 2009;74(13):1411–1442.
58. Testa U. Proteasome inhibitors in cancer therapy. *Curr Drug Targets*. 2009; 10(10):968–981.
59. Doss S, Hay N, Sutcliffe F. NICE guidance on bortezomib and thalidomide for first-line treatment of multiple myeloma. *Lancet Oncol*. 2011;12(9):837–838.
60. Cata JP, Weng HR, Burton AW, Villareal H, Giralt S, Dougherty PM. Quantitative sensory findings in patients with bortezomib-induced pain. *J Pain*. 2007;8(4):296–306.
61. Cavaletti G, Jakubowiak AJ. Peripheral neuropathy during bortezomib treatment of multiple myeloma: a review of recent studies. *Leuk Lymphoma*. 2010;51(7):1178–1187.
62. Stratogianni A, Tosch M, Schlemmer H, et al. Bortezomib-induced severe autonomic neuropathy. *Clin Auton Res*. 2012;22(4):199–202.
63. Folkman J. Tumor angiogenesis: therapeutic implications. *N Engl J Med*. 1971; 285(21):1182–1186.
64. Folkman J, Merler E, Abernathy C, Williams G. Isolation of a tumor factor responsible for angiogenesis. *J Exp Med*. 1971;133(2):275–288.
65. Leung DW, Cachianes G, Kuang WJ, Goeddel DV, Ferrara N. Vascular endothelial growth factor is a secreted angiogenic mitogen. *Science*. 1989;246(4935):1306–1309.
66. Grose R, Dickson C. Fibroblast growth factor signaling in tumorigenesis. *Cytokine Growth Factor Rev*. 2005;16(2):179–186.
67. Singhal S, Mehta J, Desikan R, et al. Antitumor activity of thalidomide in refractory multiple myeloma. *N Engl J Med*. 1999;341(21):1565–1571.
68. Drake MJ, Robson W, Mehta P, Schofield I, Neal DE, Leung HY. An open-label phase II study of low-dose thalidomide in androgen-independent prostate cancer. *Br J Cancer*. 2003;88(6):822–827.

69. Chen L, Qiu X, Wang R, Xie X. The efficacy and safety of docetaxel plus thalidomide vs. docetaxel alone in patients with androgen-independent prostate cancer: a systematic review. *Sci Rep*. 2014;4:4818.
70. D'Amato RJ, Loughnan MS, Flynn E, Folkman J. Thalidomide is an inhibitor of angiogenesis. *Proc Natl Acad Sci USA*. 1994;91(9):4082–4085.
71. Verheul HM, Panigrahy D, Yuan J, D'Amato RJ. Combination oral antiangiogenic therapy with thalidomide and sulindac inhibits tumour growth in rabbits. *Br J Cancer*. 1999;79(1):114–118.
72. Alejandro LM, Behrendt CE, Chen K, Openshaw H, Shibata S. Predicting acute and persistent neuropathy associated with oxaliplatin. *Am J Clin Oncol*. 2013;36(4):331–337.
73. Plasmati R, Pastorelli F, Cavo M, et al. Neuropathy in multiple myeloma treated with thalidomide: a prospective study. *Neurology*. 2007;69(6):573–581.
74. Glasmacher A, Hahn C, Hoffmann F, et al. A systematic review of phase-II trials of thalidomide monotherapy in patients with relapsed or refractory multiple myeloma. *Br J Haematol*. 2006;132(5):584–593.
75. Miller AB, Hoogstraten B, Staquet M, Winkler A. Reporting results of cancer treatment. *Cancer*. 1981;47(1):207–214.
76. Oken MM, Creech RH, Tormey DC, et al. Toxicity and response criteria of the Eastern Cooperative Oncology Group. *Am J Clin Oncol*. 1982;5(6):649–655.
77. Postma TJ, Heimans JJ. Grading of chemotherapy-induced peripheral neuropathy. *Ann Oncol*. 2000;11(5):509–513.
78. National Cancer Institute Common Terminology Criteria for Adverse Events Criteria (NCI CTCAE) for Peripheral Neuropathy Version 4.0. <http://www.evs.nci.nih.gov/ftp1/CTCAE/About.html>; Accessed 01.08.14.
79. Carlson K, Ocean AJ. Peripheral neuropathy with microtubule-targeting agents: occurrence and management approach. *Clin Breast Cancer*. 2011;11(2):73–81.
80. Cavaletti G, Bogliun G, Marzorati L, et al. Grading of chemotherapy-induced peripheral neurotoxicity using the Total Neuropathy Scale. *Neurology*. 2003;61(9):1297–1300.
81. Cornblath DR, Chaudhry V, Carter K, et al. Total neuropathy score: validation and reliability study. *Neurology*. 1999;53(8):1660–1664.
82. Cavaletti G, Jann S, Pace A, et al. Multi-center assessment of the total neuropathy score for chemotherapy-induced peripheral neurotoxicity. *J Peripher Nerv Syst*. 2006;11(2):135–141.
83. Velasco R, Bruna J, Briani C, et al. Early predictors of oxaliplatin-induced cumulative neuropathy in colorectal cancer patients. *J Neurol Neurosurg Psychiatry*. 2014;85(4):392–398.
84. Briani C, Campagnolo M, Lucchetta M, et al. Ultrasound assessment of oxaliplatin-induced neuropathy and correlations with neurophysiologic findings. *Eur J Neurol*. 2013;20(1):188–192.
85. McHugh JC, Tryfonopoulos D, Fennelly D, Crown J, Connolly S. Electroclinical biomarkers of early peripheral neurotoxicity from oxaliplatin. *Eur J Cancer Care*. 2012;21(6):782–789.
86. Stubblefield MD, Vahdat LT, Balmaceda CM, Troxel AB, Hesdorffer CS, Gooch CL. Glutamine as a neuroprotective agent in high-dose paclitaxel-induced peripheral neuropathy: a clinical and electrophysiologic study. *Clin Oncol (R Coll Radiol)*. 2005;17(4):271–276.
87. Riggs JE, Ashraf M, Snyder RD, Gutmann L. Prospective nerve conduction studies in cisplatin therapy. *Ann Neurol*. 1988;23(1):92–94.
88. Boogerd W, ten Bokkel Huinink WW, Dalesio O, Hoppenbrouwers WJ, van der Sande JJ. Cisplatin induced neuropathy: central, peripheral and autonomic nerve involvement. *J Neurooncol*. 1990;9(3):255–263.

89. Sarosy G, Kohn E, Stone DA, et al. Phase I study of taxol and granulocyte colony-stimulating factor in patients with refractory ovarian cancer. *J Clin Oncol.* 1992;10(7):1165–1170.
90. New PZ, Jackson CE, Rinaldi D, Burris H, Barohn RJ. Peripheral neuropathy secondary to docetaxel (Taxotere). *Neurology.* 1996;46(1):108–111.
91. Devigili G, Tugnoli V, Penza P, et al. The diagnostic criteria for small fibre neuropathy: from symptoms to neuropathology. *Brain.* 2008;131(Pt 7):1912–1925.
92. Cavaletti G, Beronio A, Reni L, et al. Thalidomide sensory neurotoxicity: a clinical and neurophysiologic study. *Neurology.* 2004;62(12):2291–2293.
93. Lauria G, Morbin M, Lombardi R, et al. Expression of capsaicin receptor immunoreactivity in human peripheral nervous system and in painful neuropathies. *J Peripher Nerv Syst.* 2006;11(3):262–271.
94. Park SB, Lin CS, Kiernan MC. Nerve excitability assessment in chemotherapy-induced neurotoxicity. *J Vis Exp.* 2012;62:e3439.
95. Namer B, Pfeffer S, Handwerker HO, Schmelz M, Bickel A. Axon reflex flare and quantitative sudomotor axon reflex contribute in the diagnosis of small fiber neuropathy. *Muscle Nerve.* 2013;47(3):357–363.
96. Kalliomaki M, Kieseritzky JV, Schmidt R, et al. Structural and functional differences between neuropathy with and without pain? *Exp Neurol.* 2011;231(2):199–206.
97. Lauria G, McArthur JC, Hauer PE, Griffin JW, Cornblath DR. Neuropathological alterations in diabetic truncal neuropathy: evaluation by skin biopsy. *J Neurol Neurosurg Psychiatry.* 1998;65(5):762–766.
98. Polydefkis M, Yiannoutsos CT, Cohen BA, et al. Reduced intraepidermal nerve fiber density in HIV-associated sensory neuropathy. *Neurology.* 2002;58(1):115–119.
99. Nolano M, Crisci C, Santoro L, et al. Absent innervation of skin and sweat glands in congenital insensitivity to pain with anhidrosis. *Clin Neurophysiol.* 2000;111(9):1596–1601.
100. Cavaletti G, Tredici G, Braga M, Tazzari S. Experimental peripheral neuropathy induced in adult rats by repeated intraperitoneal administration of taxol. *Exp Neurol.* 1995;133(1):64–72.
101. Cliffer KD, Siuciak JA, Carson SR, et al. Physiological characterization of Taxol-induced large-fiber sensory neuropathy in the rat. *Ann Neurol.* 1998;43(1):46–55.
102. Roytta M, Raine CS. Taxol-induced neuropathy: further ultrastructural studies of nerve fibre changes in situ. *J Neurocytol.* 1985;14(1):157–175.
103. Roytta M, Raine CS. Taxol-induced neuropathy: chronic effects of local injection. *J Neurocytol.* 1986;15(4):483–496.
104. Cavaletti G, Cavalletti E, Montaguti P, Oggioni N, De Negri O, Tredici G. Effect on the peripheral nervous system of the short-term intravenous administration of paclitaxel in the rat. *Neurotoxicology.* 1997;18(1):137–145.
105. Ling B, Authier N, Balayssac D, Eschalier A, Coudore F. Behavioral and pharmacological description of oxaliplatin-induced painful neuropathy in rat. *Pain.* 2007;128(3):225–234.
106. Cata JP, Weng HR, Dougherty PM. Behavioral and electrophysiological studies in rats with cisplatin-induced chemoneuropathy. *Brain Res.* 2008;1230:91–98.
107. Authier N, Gillet JP, Fialip J, Eschalier A, Coudore F. Description of a short-term Taxol-induced nociceptive neuropathy in rats. *Brain Res.* 2000;887(2):239–249.
108. Contreras PC, Vaught JL, Gruner JA, et al. Insulin-like growth factor-I prevents development of a vincristine neuropathy in mice. *Brain Res.* 1997;774(1–2):20–26.
109. Verdu E, Vilches JJ, Rodriguez FJ, Ceballos D, Valero A, Navarro X. Physiological and immunohistochemical characterization of cisplatin-induced neuropathy in mice. *Muscle Nerve.* 1999;22(3):329–340.

110. Persohn E, Canta A, Schoepfer S, et al. Morphological and morphometric analysis of paclitaxel and docetaxel-induced peripheral neuropathy in rats. *Eur J Cancer*. 2005;41(10):1460–1466.
111. Polomano RC, Mannes AJ, Clark US, Bennett GJ. A painful peripheral neuropathy in the rat produced by the chemotherapeutic drug, paclitaxel. *Pain*. 2001;94(3): 293–304.
112. Aley KO, Reichling DB, Levine JD. Vincristine hyperalgesia in the rat: a model of painful vincristine neuropathy in humans. *Neuroscience*. 1996;73(1):259–265.
113. Boyette-Davis J, Xin W, Zhang H, Dougherty PM. Intraepidermal nerve fiber loss corresponds to the development of Taxol-induced hyperalgesia and can be prevented by treatment with minocycline. *Pain*. 2011;152(2):308–313.
114. Boyette-Davis JA, Cata JP, Zhang H, et al. Follow-up psychophysical studies in bortezomib-related chemoneuropathy patients. *J Pain*. 2011;12(9):1017–1024.
115. Siau C, Xiao W, Bennett GJ. Paclitaxel- and vincristine-evoked painful peripheral neuropathies: loss of epidermal innervation and activation of Langerhans cells. *Exp Neurol*. 2006;201(2):507–514.
116. Zheng H, Xiao WH, Bennett GJ. Mitotoxicity and bortezomib-induced chronic painful peripheral neuropathy. *Exp Neurol*. 2012;238(2):225–234.
117. Xiao WH, Zheng H, Zheng FY, Nuydens R, Meert TF, Bennett GJ. Mitochondrial abnormality in sensory, but not motor, axons in paclitaxel-evoked painful peripheral neuropathy in the rat. *Neuroscience*. 2011;199:461–469.
118. Olsson Y. Topographical differences in the vascular permeability of the peripheral nervous system. *Acta Neuropathol*. 1968;10(1):26–33.
119. Somjen GG, Herman SP, Klein R, et al. The uptake of methyl mercury (203Hg) in different tissues related to its neurotoxic effects. *J Pharmacol Exp Ther*. 1973;187(3):602–611.
120. Arvidson B, Tjalve H. Distribution of 109Cd in the nervous system of rats after intravenous injection. *Acta Neuropathol*. 1986;69(1–2):111–116.
121. Jimenez-Andrade JM, Herrera MB, Ghilardi JR, Vardanyan M, Melemedjian OK, Mantyh PW. Vascularization of the dorsal root ganglia and peripheral nerve of the mouse: implications for chemical-induced peripheral sensory neuropathies. *Mol Pain*. 2008;4:10.
122. Dina OA, Chen X, Reichling D, Levine JD. Role of protein kinase Cepsilon and protein kinase A in a model of paclitaxel-induced painful peripheral neuropathy in the rat. *Neuroscience*. 2001;108(3):507–515.
123. Joseph EK, Chen X, Bogen O, Levine JD. Oxaliplatin acts on IB4-positive nociceptors to induce an oxidative stress-dependent acute painful peripheral neuropathy. *J Pain*. 2008;9(5):463–472.
124. Rigo FK, Dalmolin GD, Trevisan G, et al. Effect of omega-conotoxin MVIIA and Phalpha1beta on paclitaxel-induced acute and chronic pain. *Pharmacol Biochem Behav*. 2013;114–115:16–22.
125. Ferrier J, Bayet-Robert M, Pereira B, et al. A polyamine-deficient diet prevents oxaliplatin-induced acute cold and mechanical hypersensitivity in rats. *PLoS One*. 2013;8(10):e77828.
126. Sakurai M, Egashira N, Kawashiri T, Yano T, Ikesue H, Oishi R. Oxaliplatin-induced neuropathy in the rat: involvement of oxalate in cold hyperalgesia but not mechanical allodynia. *Pain*. 2009;147(1–3):165–174.
127. Cavaletti G, Gilardini A, Canta A, et al. Bortezomib-induced peripheral neurotoxicity: a neurophysiological and pathological study in the rat. *Exp Neurol*. 2007;204(1):317–325.
128. Joseph EK, Levine JD. Comparison of oxaliplatin- and cisplatin-induced painful peripheral neuropathy in the rat. *J Pain*. 2009;10(5):534–541.

129. Campana WM, Eskeland N, Calcutt NA, Misasi R, Myers RR, O'Brien JS. Prosaptide prevents paclitaxel neurotoxicity. *Neurotoxicology*. 1998;19(2):237–244.
130. Chiorazzi A, Nicolini G, Canta A, et al. Experimental epothilone B neurotoxicity: results of in vitro and in vivo studies. *Neurobiol Dis*. 2009;35(2):270–277.
131. Pisano C, Pratesi G, Laccabue D, et al. Paclitaxel and cisplatin-induced neurotoxicity: a protective role of acetyl-L-carnitine. *Clin Cancer Res*. 2003;9(15):5756–5767.
132. Wang MS, Davis AA, Culver DG, Wang Q, Powers JC, Glass JD. Calpain inhibition protects against Taxol-induced sensory neuropathy. *Brain*. 2004;127(Pt 3):671–679.
133. Xiao WH, Bennett GJ. Effects of mitochondrial poisons on the neuropathic pain produced by the chemotherapeutic agents, paclitaxel and oxaliplatin. *Pain*. 2012;153(3):704–709.
134. Zhang J, Tuckett RP. Comparison of paclitaxel and cisplatin effects on the slowly adapting type I mechanoreceptor. *Brain Res*. 2008;1214:50–57.
135. Gracias NG, Cummins TR, Kelley MR, Basile DP, Iqbal T, Vasko MR. Vasodilatation in the rat dorsal hindpaw induced by activation of sensory neurons is reduced by paclitaxel. *Neurotoxicology*. 2011;32(1):140–149.
136. Tanner KD, Reichling DB, Levine JD. Nociceptor hyper-responsiveness during vincristine-induced painful peripheral neuropathy in the rat. *J Neurosci*. 1998;18(16):6480–6491.
137. Adelsberger H, Quasthoff S, Grosskreutz J, Lepier A, Eckel F, Lersch C. The chemotherapeutic oxaliplatin alters voltage-gated Na(+) channel kinetics on rat sensory neurons. *Eur J Pharmacol*. 2000;406(1):25–32.
138. Flatters SJ, Bennett GJ. Studies of peripheral sensory nerves in paclitaxel-induced painful peripheral neuropathy: evidence for mitochondrial dysfunction. *Pain*. 2006; 122(3):245–257.
139. Authier N, Gillet JP, Fialip J, Eschalier A, Coudore F. An animal model of nociceptive peripheral neuropathy following repeated cisplatin injections. *Exp Neurol*. 2003;182(1):12–20.
140. Authier N, Gillet JP, Fialip J, Eschalier A, Coudore F. A new animal model of vincristine-induced nociceptive peripheral neuropathy. *Neurotoxicology*. 2003;24(6): 797–805.
141. Boyette-Davis J, Dougherty PM. Protection against oxaliplatin-induced mechanical hyperalgesia and intraepidermal nerve fiber loss by minocycline. *Exp Neurol*. 2011;229(2):353–357.
142. Melli G, Jack C, Lambrinos GL, Ringkamp M, Hoke A. Erythropoietin protects sensory axons against paclitaxel-induced distal degeneration. *Neurobiol Dis*. 2006;24(3):525–530.
143. Xiao WH, Zheng FY, Bennett GJ, Bordet T, Pruss RM. Olesoxime (cholest-4-en-3-one, oxime): analgesic and neuroprotective effects in a rat model of painful peripheral neuropathy produced by the chemotherapeutic agent, paclitaxel. *Pain*. 2009;147(1–3):202–209.
144. Boehmerle W, Huehnchen P, Peruzzaro S, Balkaya M, Endres M. Electrophysiological, behavioral and histological characterization of paclitaxel, cisplatin, vincristine and bortezomib-induced neuropathy in C57Bl/6 mice. *Sci Rep*. 2014;4:6370.
145. Deuis JR, Lim YL, Rodrigues de Sousa S, et al. Analgesic effects of clinically used compounds in novel mouse models of polyneuropathy induced by oxaliplatin and cisplatin. *Neuro Oncol*. 2014;16(10):1324–1332.
146. Melli G, Taiana M, Camozzi F, et al. Alpha-lipoic acid prevents mitochondrial damage and neurotoxicity in experimental chemotherapy neuropathy. *Exp Neurol*. 2008;214(2):276–284.
147. Scuteri A, Nicolini G, Miloso M, et al. Paclitaxel toxicity in post-mitotic dorsal root ganglion (DRG) cells. *Anticancer Res*. 2006;26(2A):1065–1070.

148. Yang IH, Siddique R, Hosmane S, Thakor N, Hoke A. Compartmentalized microfluidic culture platform to study mechanism of paclitaxel-induced axonal degeneration. *Exp Neurol.* 2009;218(1):124–128.
149. Anand U, Otto WR, Anand P. Sensitization of capsaicin and icilin responses in oxaliplatin treated adult rat DRG neurons. *Mol Pain.* 2010;6:82.
150. Silva A, Wang Q, Wang M, Ravula SK, Glass JD. Evidence for direct axonal toxicity in vincristine neuropathy. *J Peripher Nerv Syst.* 2006;11(3):211–216.
151. Ustinova EE, Shurin GV, Gutkin DW, Shurin MR. The role of TLR4 in the paclitaxel effects on neuronal growth in vitro. *PLoS One.* 2013;8(2):e56886.
152. Erturk A, Hellal F, Enes J, Bradke F. Disorganized microtubules underlie the formation of retraction bulbs and the failure of axonal regeneration. *J Neurosci.* 2007;27(34):9169–9180.
153. Gracias NG. *Paclitaxel Alters the Function of the Small Diameter Sensory Neurons* [dissertation]. Indianapolis, IN: University Graduate School, Indiana University; 2011.
154. Mo M, Erdelyi I, Szigeti-Buck K, Benbow JH, Ehrlich BE. Prevention of paclitaxel-induced peripheral neuropathy by lithium pretreatment. *FASEB J.* 2012; 26(11):4696–4709.
155. Kawakami K, Chiba T, Katagiri N, et al. Paclitaxel increases high voltage-dependent calcium channel current in dorsal root ganglion neurons of the rat. *J Pharmacol Sci.* 2012;120(3):187–195.
156. Zhang H, Dougherty PM. Enhanced excitability of primary sensory neurons and altered gene expression of neuronal ion channels in dorsal root ganglion in paclitaxel-induced peripheral neuropathy. *Anesthesiology.* 2014;120(6):1463–1475.
157. Pittman SK, Gracias NG, Vasko MR, Fehrenbacher JC. Paclitaxel alters the evoked release of calcitonin gene-related peptide from rat sensory neurons in culture. *Exp Neurol.* 2014;253:146–153.
158. Jiang Y, Guo C, Vasko MR, Kelley MR. Implications of apurinic/apyrimidinic endonuclease in reactive oxygen signaling response after cisplatin treatment of dorsal root ganglion neurons. *Cancer Res.* 2008;68(15):6425–6434.
159. Materazzi S, Fusi C, Benemei S, et al. TRPA1 and TRPV4 mediate paclitaxel-induced peripheral neuropathy in mice via a glutathione-sensitive mechanism. *Pflugers Arch.* 2012;463(4):561–569.
160. Miyano K, Tang HB, Nakamura Y, Morioka N, Inoue A, Nakata Y. Paclitaxel and vinorelbine, evoked the release of substance P from cultured rat dorsal root ganglion cells through different PKC isoform-sensitive ion channels. *Neuropharmacology.* 2009;57(1):25–32.
161. Wolf SL, Barton DL, Qin R, et al. The relationship between numbness, tingling, and shooting/burning pain in patients with chemotherapy-induced peripheral neuropathy (CIPN) as measured by the EORTC QLQ-CIPN20 instrument, N06CA. *Support Care Cancer.* 2012;20(3):625–632.
162. Nagashima M, Ooshiro M, Moriyama A, et al. Efficacy and tolerability of controlled-release oxycodone for oxaliplatin-induced peripheral neuropathy and the extension of FOLFOX therapy in advanced colorectal cancer patients. *Support Care Cancer.* 2014;22(6):1579–1584.
163. Kautio AL, Haanpaa M, Saarto T, Kalso E. Amitriptyline in the treatment of chemotherapy-induced neuropathic symptoms. *J Pain Symptom Manag.* 2008;35(1):31–39.
164. Hammack JE, Michalak JC, Loprinzi CL, et al. Phase III evaluation of nortriptyline for alleviation of symptoms of cis-platinum-induced peripheral neuropathy. *Pain.* 2002;98(1–2):195–203.
165. Smith EM, Pang H, Cirrincione C, et al. Effect of duloxetine on pain, function, and quality of life among patients with chemotherapy-induced painful peripheral neuropathy: a randomized clinical trial. *J Am Med Assoc.* 2013;309(13):1359–1367.

166. Rao RD, Michalak JC, Sloan JA, et al. Efficacy of gabapentin in the management of chemotherapy-induced peripheral neuropathy: a phase 3 randomized, double-blind, placebo-controlled, crossover trial (N00C3). *Cancer.* 2007;110(9):2110–2118.
167. Rao RD, Flynn PJ, Sloan JA, et al. Efficacy of lamotrigine in the management of chemotherapy-induced peripheral neuropathy: a phase 3 randomized, double-blind, placebo-controlled trial, N01C3. *Cancer.* 2008;112(12):2802–2808.
168. Pascual D, Goicoechea C, Burgos E, Martin MI. Antinociceptive effect of three common analgesic drugs on peripheral neuropathy induced by paclitaxel in rats. *Pharmacol Biochem Behav.* 2010;95(3):331–337.
169. Sada H, Egashira N, Ushio S, Kawashiri T, Shirahama M, Oishi R. Repeated administration of amitriptyline reduces oxaliplatin-induced mechanical allodynia in rats. *J Pharmacol Sci.* 2012;118(4):547–551.
170. Xiao W, Naso L, Bennett GJ. Experimental studies of potential analgesics for the treatment of chemotherapy-evoked painful peripheral neuropathies. *Pain Med.* 2008;9(5):505–517.
171. Balayssac D, Ling B, Ferrier J, Pereira B, Eschalier A, Authier N. Assessment of thermal sensitivity in rats using the thermal place preference test: description and application in the study of oxaliplatin-induced acute thermal hypersensitivity and inflammatory pain models. *Behav Pharmacol.* 2014;25(2):99–111.
172. Chogtu B, Bairy KL, Smitha D, Dhar S, Himabindu P. Comparison of the efficacy of carbamazepine, gabapentin and lamotrigine for neuropathic pain in rats. *Indian J Pharm.* 2011;43(5):596–598.
173. Xiao W, Boroujerdi A, Bennett GJ, Luo ZD. Chemotherapy-evoked painful peripheral neuropathy: analgesic effects of gabapentin and effects on expression of the alpha-2-delta type-1 calcium channel subunit. *Neuroscience.* 2007;144(2):714–720.
174. Hershman DL, Lacchetti C, Dworkin RH, et al. Prevention and management of chemotherapy-induced peripheral neuropathy in survivors of adult cancers: American Society of Clinical Oncology clinical practice guideline. *J Clin Oncol.* 2014;32(18):1941–1967.
175. Morris RL, Hollenbeck PJ. Axonal transport of mitochondria along microtubules and F-actin in living vertebrate neurons. *J Cell Biol.* 1995;131(5):1315–1326.
176. Shemesh OA, Spira ME. Paclitaxel induces axonal microtubules polar reconfiguration and impaired organelle transport: implications for the pathogenesis of paclitaxel-induced polyneuropathy. *Acta Neuropathol.* 2010;119(2):235–248.
177. Shprung T, Gozes I. A novel method for analyzing mitochondrial movement: inhibition by paclitaxel in a pheochromocytoma cell model. *J Mol Neurosci.* 2009;37(3):254–262.
178. Russell JW, Windebank AJ, McNiven MA, Brat DJ, Brimijoin WS. Effect of cisplatin and ACTH4-9 on neural transport in cisplatin induced neurotoxicity. *Brain Res.* 1995;676(2):258–267.
179. Jin HW, Flatters SJ, Xiao WH, Mulhern HL, Bennett GJ. Prevention of paclitaxel-evoked painful peripheral neuropathy by acetyl-L-carnitine: effects on axonal mitochondria, sensory nerve fiber terminal arbors, and cutaneous Langerhans cells. *Exp Neurol.* 2008;210(1):229–237.
180. Podratz JL, Knight AM, Ta LE, et al. Cisplatin induced mitochondrial DNA damage in dorsal root ganglion neurons. *Neurobiol Dis.* 2011;41(3):661–668.
181. Nowikovsky K, Schweyen RJ, Bernardi P. Pathophysiology of mitochondrial volume homeostasis: potassium transport and permeability transition. *Biochim Biophys Acta.* 2009;1787(5):345–350.
182. Zheng H, Xiao WH, Bennett GJ. Functional deficits in peripheral nerve mitochondria in rats with paclitaxel- and oxaliplatin-evoked painful peripheral neuropathy. *Exp Neurol.* 2011;232(2):154–161.

183. Kidd JF, Pilkington MF, Schell MJ, et al. Paclitaxel affects cytosolic calcium signals by opening the mitochondrial permeability transition pore. *J Biol Chem.* 2002;277(8):6504–6510.
184. Andre N, Braguer D, Brasseur G, et al. Paclitaxel induces release of cytochrome c from mitochondria isolated from human neuroblastoma cells'. *Cancer Res.* 2000; 60(19):5349–5353.
185. Evtodienko YV, Teplova VV, Sidash SS, Ichas F, Mazat JP. Microtubule-active drugs suppress the closure of the permeability transition pore in tumour mitochondria. *FEBS Lett.* 1996;393(1):86–88.
186. Batandier C, Leverve X, Fontaine E. Opening of the mitochondrial permeability transition pore induces reactive oxygen species production at the level of the respiratory chain complex I. *J Biol Chem.* 2004;279(17):17197–17204.
187. Knott AB, Bossy-Wetzel E. Impairing the mitochondrial fission and fusion balance: a new mechanism of neurodegeneration. *Ann N Y Acad Sci.* 2008;1147:283–292.
188. Corsetti G, Rodella L, Rezzani R, Stacchiotti A, Bianchi R. Cytoplasmic changes in satellite cells of spinal ganglia induced by cisplatin treatment in rats. *Ultrastruct Pathol.* 2000;24(4):259–265.
189. Voccoli V, Colombaioni L. Mitochondrial remodeling in differentiating neuroblasts. *Brain Res.* 2009;1252:15–29.
190. Fritz IB, Yue KT. Long-chain carnitine acyltransferase and the role of acylcarnitine derivatives in the catalytic increase of fatty acid oxidation induced by carnitine. *J Lipid Res.* 1963;4:279–288.
191. Flatters SJ, Xiao WH, Bennett GJ. Acetyl-L-carnitine prevents and reduces paclitaxel-induced painful peripheral neuropathy. *Neurosci Lett.* 2006;397(3):219–223.
192. De Grandis D, Minardi C. Acetyl-L-carnitine (levacecarnine) in the treatment of diabetic neuropathy. A long-term, randomised, double-blind, placebo-controlled study. *Drugs R&D.* 2002;3(4):223–231.
193. Youle M, Osio M. A double-blind, parallel-group, placebo-controlled, multicentre study of acetyl L-carnitine in the symptomatic treatment of antiretroviral toxic neuropathy in patients with HIV-1 infection. *HIV Med.* 2007;8(4):241–250.
194. Osio M, Muscia F, Zampini L, et al. Acetyl-L-carnitine in the treatment of painful antiretroviral toxic neuropathy in human immunodeficiency virus patients: an open label study. *J Peripher Nerv Syst.* 2006;11(1):72–76.
195. Bianchi G, Vitali G, Caraceni A, et al. Symptomatic and neurophysiological responses of paclitaxel- or cisplatin-induced neuropathy to oral acetyl-L-carnitine. *Eur J Cancer.* 2005;41(12):1746–1750.
196. Hershman DL, Unger JM, Crew KD, et al. Randomized double-blind placebo-controlled trial of acetyl-L-carnitine for the prevention of taxane-induced neuropathy in women undergoing adjuvant breast cancer therapy. *J Clin Oncol.* 2013;31(20):2627–2633.
197. Bordet T, Buisson B, Michaud M, et al. Identification and characterization of cholest-4-en-3-one, oxime (TRO19622), a novel drug candidate for amyotrophic lateral sclerosis. *J Pharmacol Exp Ther.* 2007;322(2):709–720.
198. Rovini A, Carre M, Bordet T, Pruss RM, Braguer D. Olesoxime prevents microtubule-targeting drug neurotoxicity: selective preservation of EB comets in differentiated neuronal cells. *Biochem Pharmacol.* 2010;80(6):884–894.
199. Renehan WE, Munger BL. Degeneration and regeneration of peripheral nerve in the rat trigeminal system. II. Response to nerve lesions. *J Comp Neurol.* 1986;249(4): 429–459.
200. Huang TJ, Price SA, Chilton L, et al. Insulin prevents depolarization of the mitochondrial inner membrane in sensory neurons of type 1 diabetic rats in the presence of sustained hyperglycemia. *Diabetes.* 2003;52(8):2129–2136.

201. Doyle T, Chen Z, Muscoli C, et al. Targeting the overproduction of peroxynitrite for the prevention and reversal of paclitaxel-induced neuropathic pain. *J Neurosci.* 2012;32(18):6149–6160.
202. Barriere DA, Rieusset J, Chanteranne D, et al. Paclitaxel therapy potentiates cold hyperalgesia in streptozotocin-induced diabetic rats through enhanced mitochondrial reactive oxygen species production and TRPA1 sensitization. *Pain.* 2012;153(3): 553–561.
203. Ferrari LF, Chum A, Bogen O, Reichling DB, Levine JD. Role of Drp1, a key mitochondrial fission protein, in neuropathic pain. *J Neurosci.* 2011;31(31):11404–11410.
204. Ridger VC, Greenacre SA, Handy RL, et al. Effect of peroxynitrite on plasma extravasation, microvascular blood flow and nociception in the rat. *Br J Pharmacol.* 1997;122(6):1083–1088.
205. Hamanaka RB, Chandel NS. Mitochondrial reactive oxygen species regulate cellular signaling and dictate biological outcomes. *Trends Biochem Sci.* 2010;35(9):505–513.
206. Kim D, You B, Jo EK, Han SK, Simon MI, Lee SJ. NADPH oxidase 2-derived reactive oxygen species in spinal cord microglia contribute to peripheral nerve injury-induced neuropathic pain. *Proc Natl Acad Sci USA.* 2010;107(33):14851–14856.
207. Jang HJ, Hwang S, Cho KY, Kim do K, Chay KO, Kim JK. Taxol induces oxidative neuronal cell death by enhancing the activity of NADPH oxidase in mouse cortical cultures. *Neurosci Lett.* 2008;443:17–22.
208. Russell JW, Golovoy D, Vincent AM, et al. High glucose-induced oxidative stress and mitochondrial dysfunction in neurons. *FASEB J.* 2002;16(13):1738–1748.
209. Hamilton RT, Bhattacharya A, Walsh ME, et al. Elevated protein carbonylation, and misfolding in sciatic nerve from db/db and Sod1(−/−) mice: plausible link between oxidative stress and demyelination. *PLoS One.* 2013;8(6):e65725.
210. Janes K, Doyle T, Bryant L, et al. Bioenergetic deficits in peripheral nerve sensory axons during chemotherapy-induced neuropathic pain resulting from peroxynitrite-mediated post-translational nitration of mitochondrial superoxide dismutase. *Pain.* 2013;154(11):2432–2440.
211. Fidanboylu M, Griffiths LA, Flatters SJ. Global inhibition of reactive oxygen species (ROS) inhibits paclitaxel-induced painful peripheral neuropathy. *PLoS One.* 2011;6(9): e25212.
212. Kim HK, Zhang YP, Gwak YS, Abdi S. Phenyl N-tert-butylnitrone, a free radical scavenger, reduces mechanical allodynia in chemotherapy-induced neuropathic pain in rats. *Anesthesiology.* 2010;112(2):432–439.
213. Arthur JR. The glutathione peroxidases. *Cell Mol Life Sci.* 2000;57(13–14):1825–1835.
214. Moldeus P, Cotgreave IA, Berggren M. Lung protection by a thiol-containing antioxidant: N-acetylcysteine. *Respiration.* 1986;50(suppl 1):31–42.
215. Tredici G, Cavaletti G, Petruccioli MG, Fabbrica D, Tedeschi M, Venturino P. Low-dose glutathione administration in the prevention of cisplatin-induced peripheral neuropathy in rats. *Neurotoxicology.* 1994;15(3):701–704.
216. Hamers FP, Brakkee JH, Cavalletti E, et al. Reduced glutathione protects against cisplatin-induced neurotoxicity in rats. *Cancer Res.* 1993;53(3):544–549.
217. Cavaletti G, Minoia C, Schieppati M, Tredici G. Protective effects of glutathione on cisplatin neurotoxicity in rats. *Int J Radiat Oncol Biol Phys.* 1994;29(4):771–776.
218. Bogliun G, Marzorati L, Cavaletti G, Frattola L. Evaluation by somatosensory evoked potentials of the neurotoxicity of cisplatin alone or in combination with glutathione. *Ital J Neurol Sci.* 1992;13(8):643–647.
219. Cascinu S, Cordella L, Del Ferro E, Fronzoni M, Catalano G. Neuroprotective effect of reduced glutathione on cisplatin-based chemotherapy in advanced gastric cancer: a randomized double-blind placebo-controlled trial. *J Clin Oncol.* 1995;13(1):26–32.

220. Colombo N, Bini S, Miceli D, et al. Weekly cisplatin +/− glutathione in relapsed ovarian carcinoma. *Int J Gynecol Cancer*. 1995;5(2):81–86.
221. Smyth JF, Bowman A, Perren T, et al. Glutathione reduces the toxicity and improves quality of life of women diagnosed with ovarian cancer treated with cisplatin: results of a double-blind, randomised trial. *Ann Oncol*. 1997;8(6):569–573.
222. Cascinu S, Catalano V, Cordella L, et al. Neuroprotective effect of reduced glutathione on oxaliplatin-based chemotherapy in advanced colorectal cancer: a randomized, double-blind, placebo-controlled trial. *J Clin Oncol*. 2002;20(16):3478–3483.
223. Milla P, Airoldi M, Weber G, Drescher A, Jaehde U, Cattel L. Administration of reduced glutathione in FOLFOX4 adjuvant treatment for colorectal cancer: effect on oxaliplatin pharmacokinetics, Pt-DNA adduct formation, and neurotoxicity. *Anticancer Drugs*. 2009;20(5):396–402.
224. Lin PC, Lee MY, Wang WS, et al. N-acetylcysteine has neuroprotective effects against oxaliplatin-based adjuvant chemotherapy in colon cancer patients: preliminary data. *Support Care Cancer*. 2006;14(5):484–487.
225. Leal AD, Qin R, Atherton PJ, et al. North Central Cancer Treatment Group/Alliance trial N08CA-the use of glutathione for prevention of paclitaxel/carboplatin-induced peripheral neuropathy: a phase 3 randomized, double-blind, placebo-controlled study. *Cancer*. 2014;120(12):1890–1897.
226. Argyriou AA, Chroni E, Koutras A, et al. Preventing paclitaxel-induced peripheral neuropathy: a phase II trial of vitamin E supplementation. *J Pain Symptom Manag*. 2006;32(3):237–244.
227. Argyriou AA, Chroni E, Koutras A, et al. A randomized controlled trial evaluating the efficacy and safety of vitamin E supplementation for protection against cisplatin-induced peripheral neuropathy: final results. *Support Care Cancer*. 2006;14(11):1134–1140.
228. Pace A, Savarese A, Picardo M, et al. Neuroprotective effect of vitamin E supplementation in patients treated with cisplatin chemotherapy. *J Clin Oncol*. 2003;21(5):927–931.
229. Pace A, Giannarelli D, Galie E, et al. Vitamin E neuroprotection for cisplatin neuropathy: a randomized, placebo-controlled trial. *Neurology*. 2010;74(9):762–766.
230. Kottschade LA, Sloan JA, Mazurczak MA, et al. The use of vitamin E for the prevention of chemotherapy-induced peripheral neuropathy: results of a randomized phase III clinical trial. *Support Care Cancer*. 2011;19(11):1769–1777.
231. Alexandre J, Batteux F, Nicco C, et al. Accumulation of hydrogen peroxide is an early and crucial step for paclitaxel-induced cancer cell death both in vitro and in vivo. *Int J Cancer*. 2006;119(1):41–48.
232. Dzagnidze A, Katsarava Z, Makhalova J, et al. Repair capacity for platinum-DNA adducts determines the severity of cisplatin-induced peripheral neuropathy. *J Neurosci*. 2007;27(35):9451–9457.
233. Screnci D, McKeage MJ. Platinum neurotoxicity: clinical profiles, experimental models and neuroprotective approaches. *J Inorg Biochem*. 1999;77(1–2):105–110.
234. Nieto FR, Entrena JM, Cendan CM, Pozo ED, Vela JM, Baeyens JM. Tetrodotoxin inhibits the development and expression of neuropathic pain induced by paclitaxel in mice. *Pain*. 2008;137(3):520–531.
235. Grolleau F, Gamelin L, Boisdron-Celle M, Lapied B, Pelhate M, Gamelin E. A possible explanation for a neurotoxic effect of the anticancer agent oxaliplatin on neuronal voltage-gated sodium channels. *J Neurophysiol*. 2001;85(5):2293–2297.
236. Flatters SJ, Bennett GJ. Ethosuximide reverses paclitaxel- and vincristine-induced painful peripheral neuropathy. *Pain*. 2004;109(1–2):150–161.
237. Matsumoto M, Inoue M, Hald A, Xie W, Ueda H. Inhibition of paclitaxel-induced A-fiber hypersensitization by gabapentin. *J Pharmacol Exp Ther*. 2006;318(2):735–740.

238. Okubo K, Takahashi T, Sekiguchi F, et al. Inhibition of T-type calcium channels and hydrogen sulfide-forming enzyme reverses paclitaxel-evoked neuropathic hyperalgesia in rats. *Neuroscience*. 2011;188:148–156.
239. Hori K, Ozaki N, Suzuki S, Sugiura Y. Upregulations of P2X(3) and ASIC3 involve in hyperalgesia induced by cisplatin administration in rats. *Pain*. 2010;149(2):393–405.
240. Hara T, Chiba T, Abe K, et al. Effect of paclitaxel on transient receptor potential vanilloid 1 in rat dorsal root ganglion. *Pain*. 2013;154(6):882–889.
241. Ta LE, Bieber AJ, Carlton SM, Loprinzi CL, Low PA, Windebank AJ. Transient Receptor Potential Vanilloid 1 is essential for cisplatin-induced heat hyperalgesia in mice. *Mol Pain*. 2010;6:15.
242. Chen Y, Yang C, Wang ZJ. Proteinase-activated receptor 2 sensitizes transient receptor potential vanilloid 1, transient receptor potential vanilloid 4, and transient receptor potential ankyrin 1 in paclitaxel-induced neuropathic pain. *Neuroscience*. 2011;193:440–451.
243. Gee NS, Brown JP, Dissanayake VU, Offord J, Thurlow R, Woodruff GN. The novel anticonvulsant drug, gabapentin (Neurontin), binds to the alpha2delta subunit of a calcium channel. *J Biol Chem*. 1996;271(10):5768–5776.
244. Luo ZD, Calcutt NA, Higuera ES, et al. Injury type-specific calcium channel alpha 2 delta-1 subunit up-regulation in rat neuropathic pain models correlates with anti-allodynic effects of gabapentin. *J Pharmacol Exp Ther*. 2002;303(3):1199–1205.
245. Gauchan P, Andoh T, Ikeda K, et al. Mechanical allodynia induced by paclitaxel, oxaliplatin and vincristine: different effectiveness of gabapentin and different expression of voltage-dependent calcium channel alpha(2)delta-1 subunit. *Biol Pharm Bull*. 2009;32(4):732–734.
246. Brittain JM, Duarte DB, Wilson SM, et al. Suppression of inflammatory and neuropathic pain by uncoupling CRMP-2 from the presynaptic Ca(2)(+) channel complex. *Nat Med*. 2011;17(7):822–829.
247. Wilson SM, Brittain JM, Piekarz AD, et al. Further insights into the antinociceptive potential of a peptide disrupting the N-type calcium channel-CRMP-2 signaling complex. *Channels (Austin)*. 2011;5(5):449–456.
248. Armstrong CM, Cota G. Calcium block of Na^+ channels and its effect on closing rate. *Proc Natl Acad Sci USA*. 1999;96(7):4154–4157.
249. Jacobsen D, McMartin KE. Methanol and ethylene glycol poisonings. Mechanism of toxicity, clinical course, diagnosis and treatment. *Med Toxicol*. 1986;1(5):309–334.
250. Gamelin L, Boisdron-Celle M, Delva R, et al. Prevention of oxaliplatin-related neurotoxicity by calcium and magnesium infusions: a retrospective study of 161 patients receiving oxaliplatin combined with 5-fluorouracil and leucovorin for advanced colorectal cancer. *Clin Cancer Res*. 2004;10(12 Pt 1):4055–4061.
251. Hochster HS, Grothey A, Childs BH. Use of calcium and magnesium salts to reduce oxaliplatin-related neurotoxicity. *J Clin Oncol*. 2007;25(25):4028–4029.
252. Gamelin L, Boisdron-Celle M, Morel A, et al. Oxaliplatin-related neurotoxicity: interest of calcium–magnesium infusion and no impact on its efficacy. *J Clin Oncol*. 2008;26(7):1188–1189 [author reply 1189–1190].
253. Grothey A, Hart L, Rowland KM. Intermittent oxaliplatin administration and time-to-treatment-failure in metastatic colorectal cancer: final results of the phase III CONcePT trial. *J Clin Oncol*. 2008;26(Suppl):4010.
254. Grothey A, Nikcevich DA, Sloan JA, et al. Intravenous calcium and magnesium for oxaliplatin-induced sensory neurotoxicity in adjuvant colon cancer: NCCTG N04C7. *J Clin Oncol*. 2011;29(4):421–427.
255. Loprinzi CL, Qin R, Dakhil SR, et al. Phase III randomized, placebo-controlled, double-blind study of intravenous calcium and magnesium to prevent oxaliplatin-induced sensory neurotoxicity (N08CB/Alliance). *J Clin Oncol*. 2014;32(10):997–1005.

256. Atkins CD. Estimating the value of intravenous calcium and magnesium in ameliorating oxaliplatin-induced neuropathy. *J Clin Oncol.* 2014;32:3341.
257. Avan A, Giovannetti E, Peters GJ. Calcium/magnesium infusion for oxaliplatin-induced neuropathy: protective or not? *J Clin Oncol.* 2014;32(29):3341.
258. Warwick RA, Hanani M. The contribution of satellite glial cells to chemotherapy-induced neuropathic pain. *Eur J Pain.* 2013;17(4):571–580.
259. Pevida M, Lastra A, Hidalgo A, Baamonde A, Menendez L. Spinal CCL2 and microglial activation are involved in paclitaxel-evoked cold hyperalgesia. *Brain Res Bull* 2013;95:21–27.
260. Peters CM, Jimenez-Andrade JM, Jonas BM, et al. Intravenous paclitaxel administration in the rat induces a peripheral sensory neuropathy characterized by macrophage infiltration and injury to sensory neurons and their supporting cells. *Exp Neurol.* 2007;203(1):42–54.
261. Park HJ, Stokes JA, Corr M, Yaksh TL. Toll-like receptor signaling regulates cisplatin-induced mechanical allodynia in mice. *Cancer Chemother Pharmacol.* 2014;73(1):25–34.
262. Li Y, Zhang H, Kosturakis AK, Jawad AB, Dougherty PM. Toll-like receptor 4 signaling contributes to paclitaxel-induced peripheral neuropathy. *J Pain.* 2014;15(7):712–725.
263. Zhang H, Boyette-Davis JA, Kosturakis AK, et al. Induction of monocyte chemoattractant protein-1 (MCP-1) and its receptor CCR2 in primary sensory neurons contributes to paclitaxel-induced peripheral neuropathy. *J Pain.* 2013;14(10): 1031–1044.
264. Levi-Montalcini R, Angeletti PU. Nerve growth factor. *Physiol Rev.* 1968;48(3):534–569.
265. Cavaletti G, Bogliun G, Marzorati L, et al. Early predictors of peripheral neurotoxicity in cisplatin and paclitaxel combination chemotherapy. *Ann Oncol.* 2004;15(9): 1439–1442.
266. Apfel SC, Lipton RB, Arezzo JC, Kessler JA. Nerve growth factor prevents toxic neuropathy in mice. *Ann Neurol.* 1991;29(1):87–90.
267. Peterson ER, Crain SM. Nerve growth factor attenuates neurotoxic effects of taxol on spinal cord-ganglion explants from fetal mice. *Science.* 1982;217(4557):377–379.
268. Woolf CJ, Ma QP, Allchorne A, Poole S. Peripheral cell types contributing to the hyperalgesic action of nerve growth factor in inflammation. *J Neurosci.* 1996;16(8):2716–2723.
269. Fukuoka T, Kondo E, Dai Y, Hashimoto N, Noguchi K. Brain-derived neurotrophic factor increases in the uninjured dorsal root ganglion neurons in selective spinal nerve ligation model. *J Neurosci.* 2001;21(13):4891–4900.
270. Holmes D. Anti-NGF painkillers back on track? *Nat Rev Drug Discov.* 2012;11(5): 337–338.
271. Zhu W, Galoyan SM, Petruska JC, Oxford GS, Mendell LM. A developmental switch in acute sensitization of small dorsal root ganglion (DRG) neurons to capsaicin or noxious heating by NGF. *J Neurophysiol.* 2004;92(5):3148–3152.
272. Claude P, Hawrot E, Dunis DA, Campenot RB. Binding, internalization, and retrograde transport of 125I-nerve growth factor in cultured rat sympathetic neurons. *J Neurosci.* 1982;2(4):431–442.
273. Hendry IA, Stockel K, Thoenen H, Iversen LL. The retrograde axonal transport of nerve growth factor. *Brain Res.* 1974;68(1):103–121.
274. Acosta CG, Fabrega AR, Masco DH, Lopez HS. A sensory neuron subpopulation with unique sequential survival dependence on nerve growth factor and basic fibroblast growth factor during development. *J Neurosci.* 2001;21(22):8873–8885.
275. Buchman VL, Davies AM. Different neurotrophins are expressed and act in a developmental sequence to promote the survival of embryonic sensory neurons. *Development.* 1993;118(3):989–1001.

276. Andoh T, Sakamoto A, Kuraishi Y. Effects of xaliproden, a 5-HT(1)A agonist, on mechanical allodynia caused by chemotherapeutic agents in mice. *Eur J Pharmacol.* 2013;721(1–3):231–236.
277. Arrieta O, Hernandez-Pedro N, Fernandez-Gonzalez-Aragon MC, et al. Retinoic acid reduces chemotherapy-induced neuropathy in an animal model and patients with lung cancer. *Neurology.* 2011;77(10):987–995.
278. ter Laak MP, Hamers FP, Kirk CJ, Gispen WH. rhGGF2 protects against cisplatin-induced neuropathy in the rat. *J Neurosci Res.* 2000;60(2):237–244.
279. Porzner M, Muller T, Seufferlein T. SR 57746A/xaliproden, a non-peptide neurotrophic compound: prospects and constraints for the treatment of nervous system diseases. *Expert Opin Invest Drugs.* 2009;18(11):1765–1772.
280. Park KA, Fehrenbacher JC, Thompson EL, Duarte DB, Hingtgen CM, Vasko MR. Signaling pathways that mediate nerve growth factor-induced increase in expression and release of calcitonin gene-related peptide from sensory neurons. *Neuroscience.* 2010;171(3):910–923.
281. Schmutzler BS, Roy S, Hingtgen CM. Glial cell line-derived neurotrophic factor family ligands enhance capsaicin-stimulated release of calcitonin gene-related peptide from sensory neurons. *Neuroscience.* 2009;161(1):148–156.
282. Burnside B. The form and arrangement of microtubules: an historical, primarily morphological, review. *Ann N Y Acad Sci.* 1975;253:14–26.
283. Gelfand VI, Bershadsky AD. Microtubule dynamics: mechanism, regulation, and function. *Annu Rev Cell Biol.* 1991;7:93–116.
284. Nennesmo I, Reinholt FP. Effects of intraneural injection of taxol on retrograde axonal transport and morphology of corresponding nerve cell bodies. *Virchows Archiv B Cell Pathol Incl Mol Pathol.* 1988;55(4):241–246.
285. Nakata T, Yorifuji H. Morphological evidence of the inhibitory effect of taxol on the fast axonal transport. *Neurosci Res.* 1999;35(2):113–122.
286. Zhou J, Liu M, Aneja R, Chandra R, Joshi HC. Enhancement of paclitaxel-induced microtubule stabilization, mitotic arrest, and apoptosis by the microtubule-targeting agent EM012. *Biochem Pharmacol.* 2004;68(12):2435–2441.
287. Zhuang SH, Hung YE, Hung L, et al. Evidence for microtubule target engagement in tumors of patients receiving ixabepilone. *Clin Cancer Res.* 2007;13(24):7480–7486.
288. Hammond JW, Huang CF, Kaech S, Jacobson C, Banker G, Verhey KJ. Posttranslational modifications of tubulin and the polarized transport of kinesin-1 in neurons. *Mol Biol Cell.* 2010;21(4):572–583.
289. Nishio K, Arioka H, Ishida T, et al. Enhanced interaction between tubulin and microtubule-associated protein 2 via inhibition of MAP kinase and CDC2 kinase by paclitaxel. *Int J Cancer.* 1995;63(5):688–693.
290. Goswami C, Dreger M, Jahnel R, Bogen O, Gillen C, Hucho F. Identification and characterization of a Ca^{2+}-sensitive interaction of the vanilloid receptor TRPV1 with tubulin. *J Neurochem.* 2004;91(5):1092–1103.
291. Maldonado EN, Patnaik J, Mullins MR, Lemasters JJ. Free tubulin modulates mitochondrial membrane potential in cancer cells. *Cancer Res.* 2010;70(24):10192–10201.
292. Staff NP, Podratz JL, Grassner L, et al. Bortezomib alters microtubule polymerization and axonal transport in rat dorsal root ganglion neurons. *Neurotoxicology.* 2013;39:124–131.
293. Orr GA, Verdier-Pinard P, McDaid H, Horwitz SB. Mechanisms of Taxol resistance related to microtubules. *Oncogene.* 2003;22(47):7280–7295.
294. Carozzi VA, Canta A, Oggioni N, et al. Neurophysiological and neuropathological characterization of new murine models of chemotherapy-induced chronic peripheral neuropathies. *Exp Neurol.* 2010;226(2):301–309.

CHAPTER SEVENTEEN

Stress and Chronic Pelvic Pain

Angela N. Pierce, Julie A. Christianson[1]

Department of Anatomy and Cell Biology, School of Medicine, University of Kansas Medical Center, Kansas City, Kansas, USA

[1]Corresponding author: e-mail address: jchristianson@kumc.edu

Contents

1. Introduction 509
2. Central and Peripheral Regulation of the Stress Pathway 510
3. Consequences of Early Life Stress 513
4. Irritable Bowel Syndrome 514
5. Interstitial Cystitis/Painful Bladder Syndrome 517
6. Vulvodynia 520
7. Chronic Prostatitis/Chronic Pelvic Pain Syndrome 522
8. Conclusions 524

References 524

Abstract

Chronic pelvic pain is the number one reason that patients suffering from irritable bowel syndrome, interstitial cystitis/painful bladder syndrome, vulvodynia, or chronic prostatitis/chronic pelvic pain syndrome seek medical attention. These syndromes generally have no associated pathology or identified underlying etiology, although dysfunction within the immune system, central nervous system, and peripheral nervous system has been identified. Due to the lack of pathology, chronic pelvic pain syndromes are often diagnosed by exclusion, and the high degree of comorbid symptomology among these and other functional pain disorders complicate identifying appropriate treatment strategies. Chronic stress exposure early in life has been shown to increase the likelihood of pelvic pain later in life, and acute stress exposure can induce or increase symptom severity. In this chapter, we describe the individual chronic pelvic pain syndromes and how stress influences the likelihood of diagnosis and the severity of symptoms experienced by patients.

1. INTRODUCTION

Chronic pelvic pain is not in itself a disease, but rather a term associated with the ongoing spontaneous and/or evoked pain experienced by patients diagnosed with irritable bowel syndrome (IBS), interstitial cystitis/painful bladder syndrome (IC/PBS), vulvodynia, or chronic prostatitis/chronic

Progress in Molecular Biology and Translational Science, Volume 131
ISSN 1877-1173
http://dx.doi.org/10.1016/bs.pmbts.2014.11.009

pelvic pain syndrome (CP/CPPS). These syndromes are often comorbid and share many characteristics in that they have no associated pathology or identified underlying etiology, although dysfunction within the immune system, central nervous system, and peripheral nervous system has been shown to contribute toward the maintenance and progression of these disorders.[1–3] Diagnosis of these chronic pelvic pain syndromes is largely done by exclusion, as patients present with positive symptoms despite a lack of underlying pathology. Not only are these syndromes comorbid with one another, patients with chronic pelvic pain are also more likely to present with symptoms of additional, nonpelvic-related functional pain disorders, such as migraine, fibromyalgia, and temporomandibular joint disorder, as well as certain mood disorders, including anxiety, depression, and panic disorder. Lack of pathology, disparate etiologies, and comorbid symptomology complicates identifying appropriate treatment strategies, which are few and less than optimal.

In this chapter, we describe the individual disorders that generate chronic pelvic pain and how stress influences the likelihood of diagnosis and the severity of symptoms experienced by patients. We also discuss how animal models are being used to understand the mechanisms underlying the influence of stress on these common syndromes.

2. CENTRAL AND PERIPHERAL REGULATION OF THE STRESS PATHWAY

Stress can have dichotomous effects on pain signaling. In instances of acute stress, the effect can be used to diminish the perception of pain, termed stress-induced hypoalgesia.[4] However, when stress becomes chronic, increased circulation of glucocorticoids and dysregulation of regulatory systems controlling the stress response pathway can initiate or increase the perception of pain. As schematized in Fig. 1, corticotropin-releasing factor (CRF) is the primary initiator of the stress response and, in the brain, is primarily expressed in the paraventricular nucleus (PVN) of the hypothalamus, central nucleus of the amygdala (ceA),[5] and Barrington's nucleus, the pontine micturition center.[6,7] Under stressful conditions, CRF and arginine vasopressin are secreted from the PVN and travel through the hypophysial portal veins to reach the anterior pituitary corticotrophs and induce the release of adrenocorticotropic hormone (ACTH). Systemic circulation of ACTH induces glucocorticoid (cortisol in humans; corticosterone in

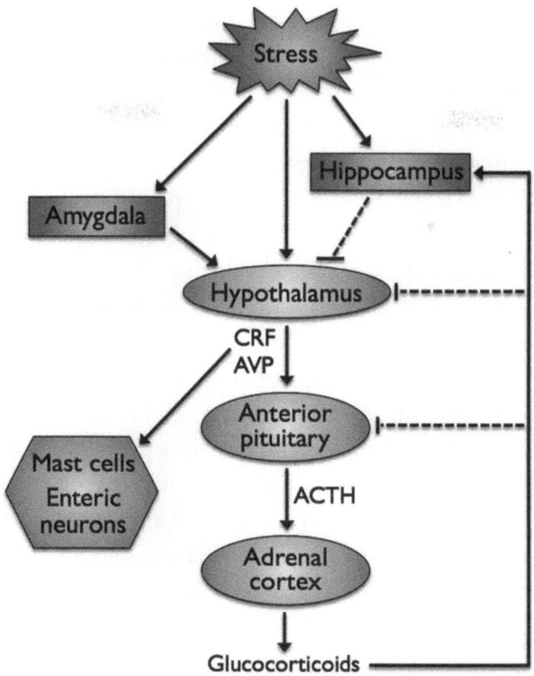

Figure 1 Acute stress induces activation of the parvocellular neurons in the PVN of the hypothalamus and synthesis and release of corticotropin-releasing factor (CRF) and arginine vasopressin (AVP) into the hypophysial portal system. Both CRF and AVP will reach the anterior pituitary and synergistically promote the release of adrenocorticotropic hormone (ACTH), which will act at the adrenal cortex to produce glucocorticoids (cortisol in humans; corticosterone in rodents). Glucocorticoids act in a negative feedback loop at the hypothalamus and anterior pituitary to cease production of CRF and ACTH, respectively, as well as at higher limbic structures, including the amygdala and hippocampus, which respectively activate and inhibit the hypothalamus. Peripheral release of CRF also acts on mast cells and enteric neurons.

rodents) synthesis and secretion from the adrenal cortex.[8,9] Glucocorticoid release initiates an overall immune suppression through the inhibition of inflammatory cells and mediators and also prompts a negative feedback loop within the hypothalamic–pituitary–adrenal (HPA) axis by suppressing the production of both CRF and ACTH.[10,11]

Glucocorticoid-driven feedback occurs through two receptors, which are relatively slow-acting and effect long-term changes in gene transcription.[12] Mineralocorticoid receptor (MR) has a high affinity for glucocorticoids, is extensively bound (even at basal levels of glucocorticoid release), and has been proposed to mediate tonic/proactive feedback to the HPA axis.

Glucocorticoid receptor (GR) has a lower affinity for glucocorticoids than MR, is largely unbound at basal glucocorticoid release levels, and is proposed to mediate feedback onto the HPA axis during phases of acute and chronic stress. In addition to the relatively slow, genomic regulation of glucocorticoid binding, evidence of fast, nongenomic glucocorticoid actions at the membrane have been reported. The receptor involved (yet to be identified, although thought to be GR) binds glucocorticoid and initiates an intracellular signaling cascade that results in endocannabinoid synthesis. The subsequent release of endocannabinoid reduces presynaptic glutamate release, thereby diminishing the neural activity of parvocellular neurons.[13]

Modulation of the HPA axis also occurs through the G-protein-coupled CRF receptors, CRF_1 and CRF_2 (Fig. 2). CRF and its family members Urocortin (Ucn) 1–3 bind the two CRF receptors with varying affinity. CRF binds CRF_1 with a 10-fold higher affinity than CRF_2, Ucn 1 binds CRF_1 and CRF_2 with equal affinity, and Ucn2 and Ucn3 both preferentially bind CRF_2.[5] Through gene deletion and pharmacological studies, opposing roles of CRF_1 and CRF_2 in stress-related behaviors have emerged. Deletion or pharmacological blockade of CRF_1 is largely anxiolytic, resulting in a

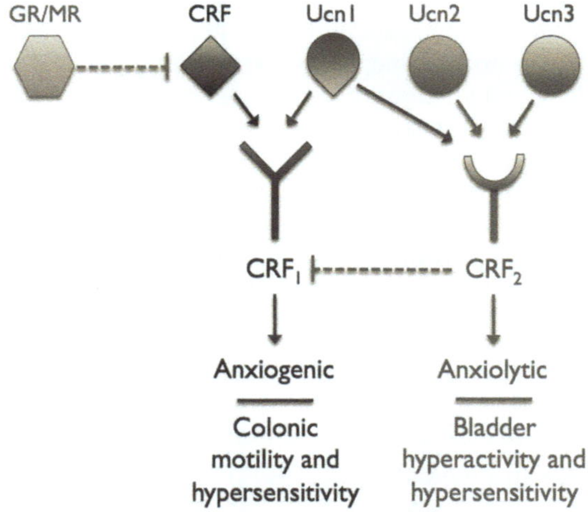

Figure 2 Corticotropin-releasing factor (CRF) and the related Urocortins (Ucn) 1–3 act via a two-receptor system to regulate the hypothalamic–pituitary–adrenal axis. Activation of CRF receptor 1 (CRF_1) occurs via CRF or Ucn1 and activation of CRF_2 occurs via all three Ucns. Activation of CRF_1 has been shown to increase anxiety-like behaviors in rodents as well as increase colonic motility and hypersensitivity. In contrast, activation of CRF_2 has largely been shown to be anxiolytic and can increase urinary bladder hyperactivity and hypersensitivity. Activation of CRF_2 has also been shown to inhibit CRF_1 and activation of either glucocorticoid receptor (GR) or mineralocorticoid receptor (MR) can inhibit CRF production and signaling. (See the color plate.)

significant decrease in anxiety-like behaviors.[14] In contrast, deletion or blockade of CRF_2 significantly enhances behavioral indicators of anxiety as well as prevents the HPA axis from returning to a homeostatic state following a stressful event.[15,16]

Limbic structures, including the amygdala and hippocampus, provide descending regulatory input to the HPA axis through interneurons that project to the PVN (Fig. 1). Under normal conditions, the amygdala and hippocampus stimulate and inhibit CRF production/secretion from the PVN, respectively, working in opposition to one another to control the activation of the HPA axis.[8,9] Upon exposure to chronic stress, glucocorticoids activate GR receptors on the amygdala to increase CRF production in the ceA, which in turn increases CRF release in the PVN and subsequently enhances glucocorticoid production.[17] The glucocorticoid-induced increase in CRF production in the ceA can be blocked by treatment with antagonists to GR or CRF_1.[18] In the hippocampus, exposure to chronic stress can induce both a decrease in GR expression and GR resistance to ligand binding,[19] which reduces the extent of descending inhibition onto the PVN and in turn increases CRF release and glucocorticoid production.[8,9]

Downstream activation of the HPA axis also affects peripheral changes in neuroimmune function (Fig. 1). One of the primary players in this cascade is the mast cell, which is a highly granulated, stem-cell-derived immune cell that contains numerous cytokines, proteases, histamine, and other potent algesic agents.[20] Mast cells are exquisitely responsive to HPA axis activation, as they express five isoforms of the CRF_1 receptor, a single isoform of CRF_2, and contain one of the largest peripheral stores of CRF found in the human body.[21] Unlike allergic and/or anaphylactic mast cell responses, which generally occur via complete release of intracellular granules, stress can induce mast cell "activation," meaning release of cytokines or growth factors in the absence of partial or complete degranulation.[21–23] The close apposition of mast cells with sensory nerve endings and associated vasculature allows for activation-induced sensitization and endothelial leakage, respectively.[24–27] Specific actions of mast cells will be outlined in more detail as they pertain to each chronic pelvic pain syndrome discussed.

3. CONSEQUENCES OF EARLY LIFE STRESS

Patients with chronic pelvic pain commonly report stress-related symptom onset or increased severity; have difficulty coping with stressful situations; and many suffer from depression, anxiety, and panic disorders.[1,28–39] Comorbidity among chronic pelvic pain syndromes and mood disorders has been associated with altered functioning of the HPA axis,[40–45] and exposure to early life stress or trauma is a significant risk factor for developing HPA abnormalities and associated chronic pain syndromes.[40,42,46–52] This is a growing concern in the United States, as the national rate of child maltreatment has consistently risen over the past decade,[53] and although medical advancements have allowed for prematurely born babies to survive at increasingly earlier gestational stages, prolonged stays in the neonatal intensive care unit (NICU) provide chronic exposure to numerous stressors, including repeated invasive procedures and prolonged periods of maternal separation.[54] The likelihood of NICU admission is higher for full-term babies born to women aged 40–44, an age group whose birth rate has steadily risen at a 2% annual rate since 2000.[55–58]

Rodent models of neonatal maternal separation (NMS), which involves removing the pups from their dam for a set period of time during the preweaning period, have been used for the past two decades to study the outcomes of early life stress. Anxiety-like behaviors and the duration of ACTH and corticosterone release following a stressful event were significantly increased in NMS rodents.[59–63] Production of CRF in both the PVN of the hypothalamus and/or the ceA of the amygdala was increased in NMS rodents at baseline, as well as following acute stress exposure.[59,61,62,64] Accordingly, increased baseline CRF_1 expression in the PVN and amygdala has been reported in NMS rats[62,65]; however, the expression of CRF_1 in the PVN has been shown to either remain increased[62] or significantly decrease[65] following acute stress exposure. Limbic feedback onto the HPA axis through CRF and glucocorticoid receptors was also disrupted in NMS rodents. Specifically, CRF_2 expression in the amygdala was significantly decreased at baseline and increased following acute stress in NMS rats, well above naïve expression levels.[65] In the hippocampus, CRF_1 and CRF_2 were both found to be increased in NMS rats, both prior to and following acute stress exposure,[65] whereas GR expression was significantly decreased in NMS rats[59,60] and both GR and MR were significantly decreased in NMS mice.[61] Taken together, NMS largely increases HPA axis activation, and resulting anxiety-like behaviors, by directly affecting gene expression within the hypothalamus as well as disrupting downstream regulation from limbic structures.

4. IRRITABLE BOWEL SYNDROME

IBS is the most commonly diagnosed and well-recognized chronic pelvic pain disorder. Diagnosis using Rome III criteria requires recurrent abdominal pain or discomfort at least 3 days/month during the previous 3 months that is associated with at least two of the following: (1) improvement with defecation, (2) onset associated with a change in stool frequency, and (3) onset associated with change in stool consistency.[66] It is estimated that IBS affects approximately 20% of the US population and generates up to $30 billion in indirect and direct medical costs on an annual basis.[67,68] As with most functional pain disorders, IBS is diagnosed in twice as many women as men and symptoms often fluctuate with the menstrual cycle.[69,70] In a study of monozygotic and dizygotic twins, low birth weight significantly impacted the likelihood of IBS diagnosis and a genetic component was observed only among female patients.[71] Comorbidity with other chronic pelvic pain conditions is remarkably high among IBS patients with up to 80% of diagnosed patients presenting with symptoms of IC/PBS, vulvodynia, and/or CP/CPPS.[1,28,33,72–74] Comorbidity with other nonpelvic functional pain disorders, including fibromyalgia, migraine, chronic fatigue syndrome, and temporomandibular joint disorder,[1,74,75] is also common and, most relevant to this chapter, up to 60% of patients with IBS suffer from anxiety and/or depression.[76] A Norwegian study revealed that IBS patients with comorbid symptomology have health care costs ten times that of patients with IBS alone.[77]

Patients with IBS commonly report that their symptoms worsen during or are brought on by periods of heightened stress.[78–81] Having a history of adverse events during infancy or childhood, such as premature birth, abuse, neglect, and parental discord, divorce, or death, increases the likelihood of developing IBS later in life.[45,82,83] This is thought to occur due to dysregulation of the HPA axis, which has been shown to be overactive in subpopulations of IBS patients.[45,84] Indeed, treatment with a CRF_1 antagonist produced significant inhibitory effects within the locus coeruleus and hypothalamus of high-anxiety female IBS patients who were anticipating a painful stimulus.[85] In a generalized diarrhea-predominant IBS patient cohort, treatment with CRF_1 antagonist did not significantly reduce colonic transit or bowel symptoms; however, the authors noted that the study size was not of sufficient power to detect a significant effect in patients with clinically high anxiety scores.[86] The results of these studies highlight that, despite exhibiting similar bowel symptoms, the underlying etiology of IBS is not the same for every patient and that identifying comorbidities might aid in designing appropriate, personalized therapeutic strategies that will have a greater chance for success.

Recognizing the impact that early life stress has on the likelihood of developing IBS has led researchers to manipulate the early developmental environment of rodents in an effort to create an appropriate model of IBS. As mentioned above, NMS in rodents in a well-established model of early life stress that recapitulates many of the abnormalities observed in human patients with a history of early adverse events. As such, NMS has been used to study the long-term effects of early life stress on colorectal sensitivity and motility, primarily in rats. Colorectal sensitivity has most commonly been assessed by measuring abdominal muscle contraction (termed the visceromotor response, VMR), either visually or electromyographically, during balloon distension of the colorectum (colorectal distension, CRD), described in detail in Christianson and Gebhart.[87] Adult rats that have undergone NMS generally have exhibited increased colorectal sensitivity at baseline and following an additional stressor they displayed a further increase in VMR during CRD when compared to non-NMS rats that underwent the same adult stress paradigm.[88–90] The most commonly used adult stress paradigm in these studies has been water avoidance stress (WAS), which selectively increases colorectal sensitivity primarily through mast cell degranulation.[91] Rats that underwent NMS not only displayed a larger VMR immediately following WAS, they also displayed a significantly longer period of WAS-induced hypersensitivity than non-NMS rats.[88,92] Treatment with CRF_1 and/or CRF_2 antagonists, prior to WAS exposure, was shown to reduce colorectal sensitivity in NMS rats.[4,92] This effect was observed both centrally and peripherally, as administration of α-helical CRF [9–14], which predominantly blocks CRF_2 and does not cross the blood–brain barrier, was capable of preventing increased VMR when administered prior to WAS exposure.[92] The authors attribute this result to preventing mast cell destabilization following stress exposure, as α-helical CRF [9–14] was unable to reverse WAS-induced colorectal hypersensitivity when given after WAS exposure. The efficacy of CRF_1 antagonist to reverse stress-induced colorectal sensitivity in NMS rats has not been appropriately tested, as Schwetz *et al.*[4] treated with CRF_1 antagonist both prior to WAS and 30 min before CRD.

In addition to the central effects of NMS on gene expression (outlined previously), peripheral effects of NMS on gene expression and protein levels have been reported in both rats and mice. Peripheral expression of CRF_1 and CRF_2 has been reported in the colon, specifically within the mucosal layer, enteric nervous system, and innate inflammatory cells.[93] Expression of both CRF_1 and CRF_2 mRNA was increased in distal colon from NMS rats,[93] and we reported a significant increase in CRF_2, but not CRF_1, mRNA in colon from NMS mice compared to naïve.[61] Rodents exposed to NMS have also demonstrated increased growth factor and cytokine expression in the distal colon, including nerve growth factor (NGF), interleukin (IL)-6, IL-1β, IL-2, IL-4, IL-10, and interferon (IFN)-γ.[94–97] A follow-up study revealed that CRF and IL-6 interact to potentiate calcium responses in submucosal neurons and thereby alter colonic secretory activity.[98] Colons from NMS rats have also demonstrated increased paracellular permeability, myeloperoxidase activity, and mucosal mast cell infiltration.[95] Our study of NMS in mice revealed no change in colonic cytokine mRNA levels; however, transient receptor potential ankyrin 1 (TRPA1) protein was significantly increased compared to naïve mice.[61] Taken together, these rodent models of NMS have demonstrated changes in the local colonic environment that are capable of sensitizing peripheral nociceptors, which may be contributing toward the observed colorectal hypersensitivity in these models.

Direct irritation of the colon during neonatal development (neonatal colon irritation, NCI) has also been used as a model of IBS in rodents. Both intracolonic administration of an irritant[99–102] and repeated CRD[99,103] during neonatal development resulted in increased VMR during CRD in adult rodents. Interestingly, the pattern of colon sensitivity exhibited by NCI rodents differed from that of NMS rodents. In general, NCI rodents exhibit an increase in VMR only at the highest intraballoon pressures applied, which are considered to be noxious. In contrast, NMS rodents display increased VMR at both noxious and non-noxious intraballoon pressures.[88,90] Studies on NCI rodents also reported an increase in TRPA1[100] or its family member TRP vanilloid 1 (TRPV1)[101] expression in colon-specific dorsal root ganglion neurons and an increase in the number of CRD-responsive dorsal horn neurons.[103] Neonatal irritation of other organs, including the bladder[104] and stomach,[105] has also been shown to increase colorectal sensitivity in adult mice, suggesting that the colon may be exquisitely sensitive to neonatal perturbations, which may contribute to the greater prevalence of IBS among the general population, in comparison to other related chronic pelvic pain disorders.

5. INTERSTITIAL CYSTITIS/PAINFUL BLADDER SYNDROME

An estimated 11% of women and 5% of men meet the high sensitivity definition of IC/PBS.[106–108] Direct medical costs for IC/PBS treatment are estimated to total $4000/year,[109] indicating that, at minimum, $20–40 billion dollars are spent each year in the United States to treat IC/PBS alone. Diagnosis of IC/PBS requires chronic pelvic pain or discomfort for at least

6 weeks duration perceived to be related to the urinary bladder with at least one additional urological symptom that cannot be attributed to known causes of bladder pain such as infection or organic disease.[110–112] Comorbidity with other chronic pelvic pain disorders occurs in up to 40% of IC/PBS patients,[1,2,67,73,113] and a significant proportion of patients also suffer from fibromyalgia, allergies, and mood disorders.[34,72,114–116] Like many idiopathic pain disorders, IC/PBS is polysyndromic and patients may exhibit one of multiple subtypes. Classically, IC referred to observable findings on cystoscopic examinations: namely, glomerulations, overt histopathology, or Hunner's ulcers[110,111]; however, the patient population was far more heterogeneous than the strict criteria used by research qualifications. Therefore, diagnoses today are primarily attributed symptomatically where IC/PBS subtypes may be differentiated by additional features.[117] The only FDA-approved treatments for IC/PBS are oral pentosan polysulfate sodium, whose mechanism of action is unknown, and intravesicular installation of dimethyl sulfoxide; off-label treatment options include oral tricyclic antidepressants or antihistamines and intravesicular instillation of heparin or lidocaine.[118–120]

The pathophysiology underlying IC/PBS is likely multifactorial as a reflection upon the broad spectrum of clinical subtypes. A wide array of etiologies has been explored in IC/PBS that when taken together proposes a cyclical cascade of events involving neurogenic inflammation, hyper-responsive immune system, urothelial lining dysfunction, and chronic pain. The initial insult that triggers IC/PBS pathophysiology is less clear. As IC/PBS is primarily an adult-onset disorder, identification of the prodrome will need to be a priority in order to halt the circular nature of the early pathophysiology. To that end, a number of urine biomarkers are associated with IC/PBS, such as IL-6,[121–124] histamine,[123] and NGF.[125,126]

A substantial percentage (~25%) of IC/PBS patients report a history of childhood sexual or physical abuse, which can significantly affect presentation of symptoms, including voiding and urgency.[42,127–129] Animal models that incorporate neonatal bladder irritation or stress exposure have been developed to study the effects on bladder sensitivity and function. Rats that received intravesicular zymosan as neonates displayed an increase in VMR during urinary bladder distension (UBD), had an increased micturition frequency with reduced micturition volume thresholds, and displayed greater plasma extravasation and neuropeptide release following intravesicular mustard oil application.[130,131] Our recent study revealed that NMS significantly increased both IL-10 and NGF mRNA levels in the bladder of female mice.[61] Additional work is underway to determine the effect of NMS on bladder sensitivity and function.

Symptom onset and/or increased severity during periods of heightened stress is also a hallmark of IC/PBS.[3,32,33] Unpredictable footshock in adult rats generated bladder hypersensitivity that was attenuated by intrathecal administration of CRF_2 antagonist, but not CRF_1 antagonist, prior to UBD.[132] A later study reported that genetic deletion or pharmacological blockade of CRF_2 prevented stress-induced bladder vascular permeability, whereas the loss of CRF_1 had no effect.[133] Intrathecal administration of CRF or Ucn2 has been shown to increase bladder and micturition volumes, whereas treatment with CRF_1 antagonist decreased both bladder and micturition volumes.[134] However, a study by Klausner *et al.*[135] reported the opposite effect, showing that CRF decreased micturition volume in normal Wistar rats and intrathecal administration of astressin, a nonselective CRF_1/CRF_2 antagonist, increased micturition volume of high-anxiety Wistar Kyoto rats. A recent study using adeno-associated vector-mediated transfer of CRF into Barrington's nucleus, which controls the micturition reflex, reported a decrease in micturition volume in rats transvected with a forward reading copy of CRF.[136] Together, these studies suggest that CRF, Ucn2, CRF_1, and CRF_2 are involved in bladder sensitivity, but their mechanisms of action within the urinary system remain unclear.

Stress can also impact luminal barrier integrity in the viscera; for instance, downregulation of mRNA encoding the tight junctional protein occludin in the gastrointestinal tract is stress-mediated.[137] Integrity of bladder urothelium in IC/PBS has also come into question. Microarray analysis of low-volume bladder capacity from IC/PBS tissue revealed downregulation of genes for urothelial tight junction proteins and upregulation of genes involved in the inflammatory response.[138] Transcript levels of proinflammatory cytokines are also increased in IC/PBS bladder.[139] A heightened inflammatory response can be related to bladder urothelial dysfunction as many components of urine metabolites, such as potassium, which is highly concentrated in urine, are toxic to underlying tissue. As described above, stress has a profound effect on visceral sensitivity, but its impact on urothelium specifically remains to be elucidated. Permeability changes themselves may also be attributed to disruption of the protective glycosaminoglycan (GAG) layer, which lines the luminal surface of the urothelium.[140,141] Intravesicular replenishment of GAG as a treatment for IC/PBS has yielded mixed results.[142–146] Additionally, the clinical relevance of urothelial permeability in IC/PBS remains controversial.[147–148]

Finally, the role of mast cells in IC/PBS has been investigated for several decades. Multiple independent studies have confirmed that mast cell infiltration is increased in biopsies from IC/PBS patients.[27,149–153] These observations have been correlated with increased release of granular contents,[154] increased stem cell factor, IL-6, histamine, and NGF expression within the bladder wall,[150,152,155] elevated NGF, histamine, and proinflammatory cytokine protein levels in patient serum[156] and urine[124,157–159] samples, and increased density of substance P (SP)-immunopositive nerve fibers and juxtaposition to mast cells in patient biopsies.[160] Secreted mediators can lead to, among other signal transduction pathways, activation of TRP channels, the chronic activation of which can drive sensitization.[161] Mast cell paracrine functions involving degranulated tryptase and histamine binding to protease-activated receptors have been shown to sensitize TRPA1 and TRPV1 *in vivo*[162,163] and likely contributes to chronic functional pain.[24,164] Furthermore, early life stress increased TRPA1 protein expression in bladder,[61] while experimental cystitis induced a TRPA1-, but not TRPV1-, dependent hyperalgesia.[165] Expression of metabotropic glutamate receptors were also upregulated primarily in bladder, rather than lumbosacral spinal cord, of mice with CYP-cystitis.[166] Together, these studies suggest that while there are etiological similarities between IBS and IC/PBS, the specific molecular mechanisms contributing towards bladder pain are likely distinct from those contributing towards the more predominantly studied gastrointestinal pain syndromes.

6. VULVODYNIA

The International Society for the Study of Vulvovaginal Disease (ISSVD) defines vulvodynia as "vulvar discomfort, most often described as a burning pain without relevant visible findings or a specific, clinically identifiable, neurologic disorder".[167] Patients with vulvodynia may present with a variety of symptom characteristics, and the ISSVD distinguishes the associated pain of localized and generalized subtypes of vulvodynia between provoked, unprovoked, or of mixed exacerbation and symptom onset can be either sudden or gradual.[168] Despite the variability in clinical presentation, allodynia of the vulva, vestibule, or vaginal canal is the hallmark symptom and qualitatively described as burning, stinging, or itching.[169,170] Diagnosis is primarily made by cotton-swab examination of the vulva as a means to reproduce severe pain or discomfort.[168] Observed erythema is minor and biopsy of the vulva is pathologically unremarkable; therefore, the diagnosis of vulvodynia more accurately reflects symptomatology as an idiopathic pain disorder rather than a disease process.

At any given time, vulvodynia affects 4% of women in the United States and cumulative evidence predicts 16% of women among the general population will have vulvodynia at some point in their lives.[171,172] Conservative estimates suggest the national costs associated with vulvodynia range $31–72 billion annually.[173] Similar to idiopathic pain disorders in general, vulvodynia is proposed to be of a multifactorial nature. The magnitude of treatment options span nonspecific lifestyle interventions, topical or oral medications, injections, physical therapy, and surgical procedures.[174] Recently, epidemiological studies have found predispositions between the onset of vulvodynia as triggered by a traumatic life event or a history of early life stress.[175] Antidepressant[176–178] and anticonvulsant[179,180] medication has been shown to alleviate symptoms associated with vulvodynia suggesting a psychosomatic pathophysiologic component to the disease process.

Clinically, vulvodynia has been associated with other chronic pelvic pain syndromes, most often IC/PBS, and increased rates of mood disorders such as depression or anxiety.[28,127,169,181–185] Strong evidence has linked early life adverse events with an increased likelihood of vulvodynia in adulthood, which has been attributed to dysfunctional regulation of the HPA axis.[28,30,175,186] Vulvodynia patients demonstrate blunted serum cortisol cycles[186] and acute stress exposure has been shown to increase symptom severity.[187] We recently investigated the long-term effects of NMS in mice as it pertains to vaginal sensitivity and gene expression within and downstream of the HPA axis.[61] Adult female mice that underwent NMS displayed significantly increased VMR during vaginal balloon distension at both low (presumably non-noxious) and high intraballoon pressures, similar to the pattern of CRD-evoked VMR in NMS rats.[88,90] Adult female NMS mice also displayed evidence of diminished negative feedback onto the HPA axis from higher brain structures and an altered cytokine response to acute stress.[61] Recently, other animal models have investigated various characteristics of vulvodynia such as peripheral nerve sprouting under low estrogenic conditions,[188] vulvar allodynia following repeated vulvovaginal fungal infection,[189] and oxazolone-induced delayed-type contact hypersensitivity of the vulva.[190]

Women with vulvodynia have also shown defective regulation of the inflammatory response related to downstream activation of the HPA axis. Vulvodynia patients have increased mast cell degranulation and infiltration within vestibular biopsies when compared to controls.[191] Specifically, increased mast cell-derived heparanase has been observed in the vestibule of vulvodynia patients,[191] where it serves to degrade the heparin sulfate component of basement membranes and the extracellular matrix as needed for leukocyte infiltration.[192] Mast cell activation increases in post- compared to premenopausal vulvodynia patients, although expression of estrogen receptor α and evidence of neural hyperplasia were similar between groups,[193] suggesting that additional regulatory processes may contribute to the age-related discrepancy.

Activation of mast cells not only leads to release of preformed mediators, but also to *de novo* synthesis and subsequent release of other bioactive molecules of the inflammatory response, including mast cell-derived cytokines, eicosinoids, and chemokines. The proinflammatory cytokines IL-1 and tumor necrosis factor-α have both been reported to be increased in vulvodynia biopsy samples[194,195] and are produced by mast cells following simulation.[196] Increased activity of IL-1β is associated with diminished activity of the IL-1 receptor antagonist (IL-1RA), and polymorphisms in allele 2 of the *IL-1RA* gene are more prevalent in patients with chronic vulvar pain[195,197,198] than controls. Circulating serum levels of IL-1RA are decreased in vulvodynia patients and inflammatory challenge results in a blunted IL-1RA response in vulvodynia patients *in vivo*[199] and increases of IL-1β[200] among tissues derived from vulvodynia patients.

7. CHRONIC PROSTATITIS/CHRONIC PELVIC PAIN SYNDROME

Despite having a lifetime prevalence of approximately 14%[201] and annual patient costs estimated at $4400 (twice that of low back pain or rheumatoid arthritis[202]), CP/CPPS is perhaps the least well-recognized and characterized chronic pelvic pain syndrome. The National Institutes of Health chronic prostatitis symptom index (NIH-CPSI) provides four main categories for diagnosis: (I) acute bacterial, (II) chronic bacterial, (III) the chronic pelvic pain syndrome (CP/CPPS; formerly known as chronic abacterial prostatitis), and (IV) asymptomatic inflammatory prostatitis.[203] Category III prostatitis is predominantly characterized by pain in the perineum, rectum, prostate, penis, testicles, and/or abdomen and has been further divided into inflammatory (IIIa) and noninflammatory (IIIb) CP/CPPS, with the former being associated with the presence of white cells in prostatic secretions.[204] From 1990 to 1994, CP/CPPS accounted for nearly 2 million outpatient visits and was the most common urological diagnosis in men under the age of 50, representing 8% of urology office visits.[205] Later, population-based screening methods revealed that CP/CPPS symptoms are likely more widespread and that this condition is underdiagnosed among men in the United States.[108,206]

Similar to other chronic pelvic pain syndromes, the etiology of CP/CPPS is largely unknown and patients commonly present with symptoms of or are diagnosed with comorbid chronic pelvic pain or mood disorders.[1,3,29,207] In fact, the significant overlap of presenting signs and symptoms between IC/PBS and CP/CPPS has led several investigators to propose that they are different manifestations of the same underlying syndrome.[206,208–210] Recurrent infection, leaky epithelium, neurogenic inflammation, and autoimmunity have all been surmised as potential underlying causes of IC/PBS and CP/CPPS, as well as mast cell activation and degranulation.[208] Expressed prostatic secretions from men with CP/CPPS had increased mast cell tryptase and NGF levels,[211] and a later study confirmed that tryptase and carboxypeptidase A (CPA3), a marker of mast cell activation, were also increased in the urine of CP/CPPS patients.[212] The extent of mast cell degranulation or activation has varied between studies as both altered granular structure[213] and a decrease in the number of intact mast cells[214] have been observed in biopsies from CP/CPPS patients, implying that mast cell activation without complete degranulation, as well as a higher rate of complete mast cell degranulation, respectively, could occur in CP/CPPS.

The potential role for mast cells in the onset and maintenance of CP/CPPS has been a major focus of animal research on this syndrome thus far. The most commonly employed rodent model used to study CP/CPPS is an experimental autoimmune prostatitis (EAP) model generated by subcutaneous injection of prostate antigen in Complete Freund's adjuvant, which results in varied degrees of prostatic inflammation depending on species and strain used.[211,215–218] Mast cell infiltration and activation/degranulation has been shown to increase following induction of EAP[211,212,219] and the appearance of intact mast cells decreased over time,[211] suggestive of an increase in complete degranulation as has been observed in biopsies from CP/CPPS patients.[214] Mast cell deficient $Kit^{W\text{-}sh}/Kit^{W\text{-}sh}$ mice did not develop pelvic mechanical allodynia 5 days following EAP unlike wild-type mice and treatment with cromolyn sodium, a mast cell stabilizer, or histamine receptor antagonists also significantly reduced pain behaviors in EAP mice.[211] The murine ortholog of tryptase, mMCP-6, and its cognate receptor PAR2, were both increased in the prostate of EAP mice.[212] Additionally, PAR2−/− mice did not develop prostatic tactile sensitivity following EAP induction despite mounting similar levels of prostatic inflammation compared to wild-type EAP mice.[212]

8. CONCLUSIONS

The negative effect of stress exposure on chronic pelvic pain syndromes has long been documented in clinical settings. The prevalence of early adverse event exposure among chronic pelvic pain patients, as well as comorbidity between functional pain syndromes and mood disorders, suggests an underlying role of permanently altered stress response and regulation. In particular, the mast cell, which is exquisitely responsive to CRF, appears to play a critical role in every clinically recognized chronic pelvic pain syndrome. Considering the disparate etiologies that have been proposed to underlie these similar disorders, future therapeutic investigation might best be directed toward identifying shared characteristics among patients with similar early life histories and stress-related symptomologies.

REFERENCES

1. Bullones Rodriguez MA, Afari N, Buchwald DS, National Institute of Diabetes and Digestive and Kidney Diseases Working Group on Urological Chronic Pelvic Pain. Evidence for overlap between urological and nonurological unexplained clinical conditions. *J Urol.* 2013;189:S66–S74.
2. Mathias SD, Kuppermann M, Liberman RF, Lipschutz RC, Steege JF. Chronic pelvic pain: prevalence, health-related quality of life, and economic correlates. *Obstet Gynecol.* 1996;87:321–327.
3. Rodriguez MA, Afari N, Buchwald DS, National Institute of Diabetes and Digestive and Kidney Diseases Working Group on Urological Chronic Pelvic Pain. Evidence for overlap between urological and nonurological unexplained clinical conditions. *J Urol.* 2009;182:2123–2131.
4. Schwetz I, McRoberts JA, Coutinho SV, et al. Corticotropin-releasing factor receptor 1 mediates acute and delayed stress-induced visceral hyperalgesia in maternally separated Long-Evans rats. *Am J Physiol Gastrointest Liver Physiol.* 2005;289:G704–G712.
5. Bale TL, Vale WW. CRF and CRF receptors: role in stress responsivity and other behaviors. *Annu Rev Pharmacol Toxicol.* 2004;44:525–557.
6. Imaki T, Nahan JL, Rivier C, Sawchenko PE, Vale W. Differential regulation of corticotropin-releasing factor mRNA in rat brain regions by glucocorticoids and stress. *J Neurosci.* 1991;11:585–599.
7. Pavcovich LA, Valentino RJ. Central regulation of micturition in the rat the corticotropin-releasing hormone from Barrington's nucleus. *Neurosci Lett.* 1995;196: 185–188.
8. Herman JP, Ostrander MM, Mueller NK, Figueiredo H. Limbic system mechanisms of stress regulation: hypothalamo-pituitary-adrenocortical axis. *Prog Neuropsychopharmacol Biol Psychiatry.* 2005;29:1201–1213.
9. Ulrich-Lai YM, Herman JP. Neural regulation of endocrine and autonomic stress responses. *Nat Rev Neurosci.* 2009;10:397–409.

10. Kageyama K, Suda T. Regulatory mechanisms underlying corticotropin-releasing factor gene expression in the hypothalamus. *Endocr J.* 2009;56:335–344.
11. Tasker JG, Herman JP. Mechanisms of rapid glucocorticoid feedback inhibition of the hypothalamic-pituitary-adrenal axis. *Stress.* 2011;14:398–406.
12. De Kloet ER, Vreugdenhil E, Oitzl MS, Joels M. Brain corticosteroid receptor balance in health and disease. *Endocr Rev.* 1998;19:269–301.
13. Evanson NK, Tasker JG, Hill MN, Hillard CJ, Herman JP. Fast feedback inhibition of the HPA axis by glucocorticoids is mediated by endocannabinoid signaling. *Endocrinology.* 2010;151:4811–4819.
14. Smith GW, Aubry JM, Dellu F, et al. Corticotropin releasing factor receptor 1-deficient mice display decreased anxiety, impaired stress response, and aberrant neuroendocrine development. *Neuron.* 1998;20:1093–1102.
15. Bale TL, Contarino A, Smith GW, et al. Mice deficient for corticotropin-releasing hormone receptor-2 display anxiety-like behaviour and are hypersensitive to stress. *Nat Genet.* 2000;24:410–414.
16. Bale TL, Picetti R, Contarino A, Koob GF, Vale WW, Lee KF. Mice deficient for both corticotropin-releasing factor receptor 1 (CRFR1) and CRFR2 have an impaired stress response and display sexually dichotomous anxiety-like behavior. *J Neurosci.* 2002;22:193–199.
17. Cook CJ. Glucocorticoid feedback increases the sensitivity of the limbic system to stress. *Physiol Behav.* 2002;75:455–464.
18. Myers DA, Gibson M, Schulkin J, Greenwood Van-Meerveld B. Corticosterone implants to the amygdala and type 1 CRH receptor regulation: effects on behavior and colonic sensitivity. *Behav Brain Res.* 2005;161:39–44.
19. Silverman MN, Sternberg EM. Glucocorticoid regulation of inflammation and its functional correlates: from HPA axis to glucocorticoid receptor dysfunction. *Ann N Y Acad Sci.* 2012;1261:55–63.
20. Black PH. Stress and the inflammatory response: a review of neurogenic inflammation. *Brain Behav Immun.* 2002;16:622–653.
21. Theoharides TC, Donelan JM, Papadopoulou N, Cao J, Kempuraj D, Conti P. Mast cells as targets of corticotropin-releasing factor and related peptides. *Trends Pharmacol Sci.* 2004;25:563–568.
22. Anand P, Singh B, Jaggi AS, Singh N. Mast cells: an expanding pathophysiological role from allergy to other disorders. *Naunyn Schmiedebergs Arch Pharmacol.* 2012;385:657–670.
23. Galli SJ, Tsai M. Mast cells in allergy and infection: versatile effector and regulatory cells in innate and adaptive immunity. *Eur J Immunol.* 2010;40:1843–1851.
24. Barbara G, Stanghellini V, De Giorgio R, et al. Activated mast cells in proximity to colonic nerves correlate with abdominal pain in irritable bowel syndrome. *Gastroenterology.* 2004;126:693–702.
25. Barreau F, Salvador-Cartier C, Houdeau E, Bueno L, Fioramonti J. Long-term alterations of colonic nerve-mast cell interactions induced by neonatal maternal deprivation in rats. *Gut.* 2008;57:582–590.
26. Singh LK, Boucher W, Pang X, et al. Potent mast cell degranulation and vascular permeability triggered by urocortin through activation of corticotropin-releasing hormone receptors. *J Pharmacol Exp Ther.* 1999;288:1349–1356.
27. Tomaszewski JE, Landis JR, Russack V, et al. Biopsy features are associated with primary symptoms in interstitial cystitis: results from the interstitial cystitis database study. *Urology.* 2001;57:67–81.
28. Arnold LD, Bachmann GA, Rosen R, Kelly S, Rhoads GG. Vulvodynia: characteristics and associations with comorbidities and quality of life. *Obstet Gynecol.* 2006;107:617–624.

29. Clemens JQ, Brown SO, Calhoun EA. Mental health diagnoses in patients with interstitial cystitis/painful bladder syndrome and chronic prostatitis/chronic pelvic pain syndrome: a case/control study. *J Urol.* 2008;180:1378–1382.
30. Groysman V. Vulvodynia: new concepts and review of the literature. *Dermatol Clin.* 2010;28:681–696.
31. Lutgendorf SK, Kreder KJ, Rothrock NE, Ratliff TL, Zimmerman B. Stress and symptomatology in patients with interstitial cystitis: a laboratory stress model. *J Urol.* 2000;164:1265–1269.
32. Nickel JC, Shoskes D, Irvine-Bird K. Clinical phenotyping of women with interstitial cystitis/painful bladder syndrome: a key to classification and potentially improved management. *J Urol.* 2009;182:155–160.
33. Nickel JC, Tripp DA, Pontari M, et al. Interstitial cystitis/painful bladder syndrome and associated medical conditions with an emphasis on irritable bowel syndrome, fibromyalgia and chronic fatigue syndrome. *J Urol.* 2010;184:1358–1363.
34. Nickel JC, Tripp DA, Pontari M, et al. Psychosocial phenotyping in women with interstitial cystitis/painful bladder syndrome: a case control study. *J Urol.* 2010;183:167–172.
35. Novi JM, Jeronis S, Srinivas S, Srinivasan R, Morgan MA, Arya LA. Risk of irritable bowel syndrome and depression in women with interstitial cystitis: a case–control study. *J Urol.* 2005;174:937–940.
36. Rothrock NE, Lutgendorf SK, Kreder KJ, Ratliff T, Zimmerman B. Stress and symptoms in patients with interstitial cystitis: a life stress model. *Urology.* 2001;57:422–427.
37. Theoharides TC. Panic disorder, interstitial cystitis, and mast cells. *J Clin Psychopharmacol.* 2004;24:361–364.
38. Watkins KE, Eberhart N, Hilton L, et al. Depressive disorders and panic attacks in women with bladder pain syndrome/interstitial cystitis: a population-based sample. *Gen Hosp Psychiatry.* 2011;33:143–149.
39. Weissman MM, Gross R, Fyer A, et al. Interstitial cystitis and panic disorder: a potential genetic syndrome. *Arch Gen Psychiatry.* 2004;61:273–279.
40. Heim C, Ehlert U, Hanker JP, Hellhammer DH. Abuse-related posttraumatic stress disorder and alterations of the hypothalamic-pituitary-adrenal axis in women with chronic pelvic pain. *Psychosom Med.* 1998;60:309–318.
41. Heim C, Newport DJ, Bonsall R, Miller AH, Nemeroff CB. Altered pituitary-adrenal axis responses to provocative challenge tests in adult survivors of childhood abuse. *Am J Psychiatry.* 2001;158:575–581.
42. Mayson BE, Teichman JM. The relationship between sexual abuse and interstitial cystitis/painful bladder syndrome. *Curr Urol Rep.* 2009;10:441–447.
43. Rao U, Hammen C, Ortiz LR, Chen LA, Poland RE. Effects of early and recent adverse experiences on adrenal response to psychosocial stress in depressed adolescents. *Biol Psychiatry.* 2008;64:521–526.
44. Tyrka AR, Wier L, Price LH, et al. Childhood parental loss and adult hypothalamic-pituitary-adrenal function. *Biol Psychiatry.* 2008;63:1147–1154.
45. Videlock EJ, Adeyemo M, Licudine A, et al. Childhood trauma is associated with hypothalamic-pituitary-adrenal axis responsiveness in irritable bowel syndrome. *Gastroenterology.* 2009;137:1954–1962.
46. Anand KJ. Clinical importance of pain and stress in preterm neonates. *Biol Neonate.* 1998;73:1–9.
47. Anda RF, Felitti VJ, Bremner JD, et al. The enduring effects of abuse and related adverse experiences in childhood. *Eur Arch Psychiatry Clin Neurosci.* 2006;256:174–186.
48. Bennett EJ, Tennant CC, Piesse C, Badcock CA, Kellow JE. Level of chronic life stress predicts clinical outcome in irritable bowel syndrome. *Gut.* 1998;43:256–261.

49. Grunau RV, Whitfield MF, Petrie JH, Fryer EL. Early pain experience, child and family factors, as precursors of somatization: a prospective study of extremely premature and fullterm children. *Pain*. 1994;56:353–359.
50. Heim C, Nemeroff CB. The impact of early adverse experiences on brain systems involved in the pathophysiology of anxiety and affective disorders. *Biol Psychiatry*. 1999;46:1509–1522.
51. Kamm MA. The role of psychosocial factors in functional gut disease. *Eur J Surg Suppl*. 1998;37–40.
52. Whitfield MF, Grunau RE. Behavior, pain perception, and the extremely low-birth weight survivor. *Clin Perinatol*. 2000;27:363–379.
53. U.S. Department of Health and Human Services, Administration for Children and Families, Administration on Children, Youth and Families, Children's Bureau. *Child Maltreatment, 2011*. Washington, D.C.: Administration for Children and Families; 2012.
54. Simons SH, van Dijk M, Anand KS, Roofthooft D, van Lingen RA, Tibboel D. Do we still hurt newborn babies? A prospective study of procedural pain and analgesia in neonates. *Arch Pediatr Adolesc Med*. 2003;157:1058–1064.
55. Hamilton BE, Martin JA, Ventura SJ. Births: preliminary data for 2012. *Natl Vital Stat Rep*. 2013;62:1–20.
56. Martin JA, Hamilton BE, Ventura SJ, et al. Births: final data for 2009. *Natl Vital Stat Rep*. 2011;60:72.
57. Martin JA, Hamilton BE, Ventura SJ, Osterman MJK, Wilson EC, Mathews TJ. Births: final data for 2010. *Natl Vital Stat Rep*. 2012;61:100.
58. Osterman MJK, Martin JA, Mathews MS, Hamilton BE. Expanded data from the new birth certificate, 2008. *Natl Vital Stat Rep*. 2011;59:1–29.
59. Aisa B, Tordera R, Lasheras B, Del Rio J, Ramirez MJ. Effects of maternal separation on hypothalamic-pituitary-adrenal responses, cognition and vulnerability to stress in adult female rats. *Neuroscience*. 2008;154:1218–1226.
60. Ladd CO, Huot RL, Thrivikraman KV, Nemeroff CB, Plotsky PM. Long-term adaptations in glucocorticoid receptor and mineralocorticoid receptor mRNA and negative feedback on the hypothalamo-pituitary-adrenal axis following neonatal maternal separation. *Biol Psychiatry*. 2004;55:367–375.
61. Pierce AN, Ryals JM, Wang R, Christianson JA. Vaginal hypersensitivity and hypothalamic-pituitary-adrenal axis dysfunction as a result of neonatal maternal separation in female mice. *Neuroscience*. 2014;263:216–230.
62. Plotsky PM, Thrivikraman KV, Nemeroff CB, Caldji C, Sharma S, Meaney MJ. Long-term consequences of neonatal rearing on central corticotropin-releasing factor systems in adult male rat offspring. *Neuropsychopharmacology*. 2005;30:2192–2204.
63. Romeo RD, Mueller A, Sisti HM, Ogawa S, McEwen BS, Brake WG. Anxiety and fear behaviors in adult male and female C57BL/6 mice are modulated by maternal separation. *Horm Behav*. 2003;43:561–567.
64. Desbonnet L, Garrett L, Daly E, McDermott KW, Dinan TG. Sexually dimorphic effects of maternal separation stress on corticotrophin-releasing factor and vasopressin systems in the adult rat brain. *Int J Dev Neurosci*. 2008;26:259–268.
65. O'Malley D, Dinan TG, Cryan JF. Neonatal maternal separation in the rat impacts on the stress responsivity of central corticotropin-releasing factor receptors in adulthood. *Psychopharmacology (Berl)*. 2011;214:221–229.
66. Ford AC, Bercik P, Morgan DG, Bolino C, Pintos-Sanchez MI, Moayyedi P. Characteristics of functional bowel disorder patients: a cross-sectional survey using the Rome III criteria. *Aliment Pharmacol Ther*. 2014;39:312–321.
67. Hulisz D. The burden of illness of irritable bowel syndrome: current challenges and hope for the future. *J Manag Care Pharm*. 2004;10:299–309.

68. Lembo AJ. The clinical and economic burden of irritable bowel syndrome. *Pract Gastroenterol*. 2007;31:3–9.
69. Heitkemper M, Jarrett M, Bond EF, Chang L. Impact of sex and gender on irritable bowel syndrome. *Biol Res Nurs*. 2003;5:56–65.
70. Mulak A, Tache Y, Larauche M. Sex hormones in the modulation of irritable bowel syndrome. *World J Gastroenterol*. 2014;20:2433–2448.
71. Bengtson MB, Ronning T, Vatn MH, Harris JR. Irritable bowel syndrome in twins: genes and environment. *Gut*. 2006;55:1754–1759.
72. Alagiri M, Chottiner S, Ratner V, Slade D, Hanno PM. Interstitial cystitis: unexplained associations with other chronic disease and pain syndromes. *Urology*. 1997;49:52–57.
73. Fenton BW, Brobeck L, Witten E, Von Gruenigen V. Chronic pelvic pain syndrome-related diagnoses in an outpatient office setting. *Gynecol Obstet Investig*. 2012;74:64–67.
74. Kim SE, Chang L. Overlap between functional GI disorders and other functional syndromes: what are the underlying mechanisms? *Neurogastroenterol Motil*. 2012;24:895–913.
75. Henningsen P, Zipfel S, Herzog W. Management of functional somatic syndromes. *Lancet*. 2007;369:946–955.
76. Fond G, Loundou A, Hamdani N, et al. Anxiety and depression comorbidities in irritable bowel syndrome (IBS): a systematic review and meta-analysis. *Eur Arch Psychiatry Clin Neurosci*. 2014;264:651–660.
77. Johansson PA, Farup PG, Bracco A, Vandvik PO. How does comorbidity affect cost of health care in patients with irritable bowel syndrome? A cohort study in general practice. *BMC Gastroenterol*. 2010;10:31.
78. Blanchard EB, Lackner JM, Jaccard J, et al. The role of stress in symptom exacerbation among IBS patients. *J Psychosom Res*. 2008;64:119–128.
79. Dufton LM, Konik B, Colletti R, et al. Effects of stress on pain threshold and tolerance in children with recurrent abdominal pain. *Pain*. 2008;136:38–43.
80. Mayer EA, Naliboff BD, Chang L, Coutinho SV. V. Stress and irritable bowel syndrome. *Am J Physiol Gastrointest Liver Physiol*. 2001;280:G519–G524.
81. Stasi C, Rosselli M, Bellini M, Laffi G, Milani S. Altered neuro-endocrine-immune pathways in the irritable bowel syndrome: the top-down and the bottom-up model. *J Gastroenterol*. 2012;47:1177–1185.
82. Barreau F, Ferrier L, Fioramonti J, Bueno L. New insights in the etiology and pathophysiology of irritable bowel syndrome: contribution of neonatal stress models. *Pediatr Res*. 2007;62:240–245.
83. Chitkara DK, van Tilburg MA, Blois-Martin N, Whitehead WE. Early life risk factors that contribute to irritable bowel syndrome in adults: a systematic review. *Am J Gastroenterol*. 2008;103:765–774 [quiz 775].
84. Chang L, Sundaresh S, Elliott J, et al. Dysregulation of the hypothalamic-pituitary-adrenal (HPA) axis in irritable bowel syndrome. *Neurogastroenterol Motil*. 2009;21:149–159.
85. Hubbard CS, Labus JS, Bueller J, et al. Corticotropin-releasing factor receptor 1 antagonist alters regional activation and effective connectivity in an emotional-arousal circuit during expectation of abdominal pain. *J Neurosci*. 2011;31:12491–12500.
86. Sweetser S, Camilleri M, Linker Nord SJ, et al. Do corticotropin releasing factor-1 receptors influence colonic transit and bowel function in women with irritable bowel syndrome? *Am J Physiol Gastrointest Liver Physiol*. 2009;296:G1299–G1306.
87. Christianson JA, Gebhart GF. Assessment of colon sensitivity by luminal distension in mice. *Nat Protoc*. 2007;2:2624–2631.
88. Coutinho SV, Plotsky PM, Sablad M, et al. Neonatal maternal separation alters stress-induced responses to viscerosomatic nociceptive stimuli in rat. *Am J Physiol Gastrointest Liver Physiol*. 2002;282:G307–G316.

89. Moloney RD, O'Leary OF, Felice D, Bettler B, Dinan TG, Cryan JF. Early-life stress induces visceral hypersensitivity in mice. *Neurosci Lett.* 2012;512:99–102.
90. van den Wijngaard RM, Klooker TK, Welting O, et al. Essential role for TRPV1 in stress-induced (mast cell-dependent) colonic hypersensitivity in maternally separated rats. *Neurogastroenterol Motil.* 2009;21:1107-e94.
91. Bradesi S, Schwetz I, Ennes HS, et al. Repeated exposure to water avoidance stress in rats: a new model for sustained visceral hyperalgesia. *Am J Physiol Gastrointest Liver Physiol.* 2005;289:G42–G53.
92. van den Wijngaard RM, Stanisor OI, van Diest SA, et al. Peripheral alpha-helical CRF (9–41) does not reverse stress-induced mast cell dependent visceral hypersensitivity in maternally separated rats. *Neurogastroenterol Motil.* 2012;24:274–282, e111.
93. O'Malley D, Dinan TG, Cryan JF. Alterations in colonic corticotropin-releasing factor receptors in the maternally separated rat model of irritable bowel syndrome: differential effects of acute psychological and physical stressors. *Peptides.* 2010;31:662–670.
94. Barreau F, Cartier C, Ferrier L, Fioramonti J, Bueno L. Nerve growth factor mediates alterations of colonic sensitivity and mucosal barrier induced by neonatal stress in rats. *Gastroenterology.* 2004;127:524–534.
95. Barreau F, Ferrier L, Fioramonti J, Bueno L. Neonatal maternal deprivation triggers long term alterations in colonic epithelial barrier and mucosal immunity in rats. *Gut.* 2004;53:501–506.
96. O'Malley D, Dinan TG, Cryan JF. Altered expression and secretion of colonic interleukin-6 in a stress-sensitive animal model of brain-gut axis dysfunction. *J Neuroimmunol.* 2011;235:48–55.
97. O'Malley D, Liston M, Hyland NP, Dinan TG, Cryan JF. Colonic soluble mediators from the maternal separation model of irritable bowel syndrome activate submucosal neurons via an interleukin-6-dependent mechanism. *Am J Physiol Gastrointest Liver Physiol.* 2011;300:G241–G252.
98. O'Malley D, Cryan JF, Dinan TG. Crosstalk between interleukin-6 and corticotropin-releasing factor modulate submucosal plexus activity and colonic secretion. *Brain Behav Immun.* 2013;30:115–124.
99. Al-Chaer ED, Kawasaki M, Pasricha PJ. A new model of chronic visceral hypersensitivity in adult rats induced by colon irritation during postnatal development. *Gastroenterology.* 2000;119:1276–1285.
100. Christianson JA, Bielefeldt K, Malin SA, Davis BM. Neonatal colon insult alters growth factor expression and TRPA1 responses in adult mice. *Pain.* 2010;151:540–549.
101. Winston J, Shenoy M, Medley D, Naniwadekar A, Pasricha PJ. The vanilloid receptor initiates and maintains colonic hypersensitivity induced by neonatal colon irritation in rats. *Gastroenterology.* 2007;132:615–627.
102. Xu GY, Shenoy M, Winston JH, Mittal S, Pasricha PJ. P2X receptor-mediated visceral hyperalgesia in a rat model of chronic visceral hypersensitivity. *Gut.* 2008;57: 1230–1237.
103. Lin C, Al-Chaer ED. Long-term sensitization of primary afferents in adult rats exposed to neonatal colon pain. *Brain Res.* 2003;971:73–82.
104. Miranda A, Mickle A, Schmidt J, et al. Neonatal cystitis-induced colonic hypersensitivity in adult rats: a model of viscero-visceral convergence. *Neurogastroenterol Motil.* 2011;23:683-e281.
105. Smith C, Nordstrom E, Sengupta JN, Miranda A. Neonatal gastric suctioning results in chronic visceral and somatic hyperalgesia: role of corticotropin releasing factor. *Neurogastroenterol Motil.* 2007;19:692–699.
106. Berry SH, Elliott MN, Suttorp M, et al. Prevalence of symptoms of bladder pain syndrome/interstitial cystitis among adult females in the United States. *J Urol.* 2011;186:540–544.

107. Clemens JQ, Meenan RT, O'Keeffe Rosetti MC, Brown SO, Gao SY, Calhoun EA. Prevalence of interstitial cystitis symptoms in a managed care population. *J Urol.* 2005;174:576–580.
108. Suskind AM, Berry SH, Ewing BA, Elliott MN, Suttorp MJ, Clemens JQ. The prevalence and overlap of interstitial cystitis/bladder pain syndrome and chronic prostatitis/chronic pelvic pain syndrome in men: results of the RAND Interstitial Cystitis Epidemiology male study. *J Urol.* 2013;189:141–145.
109. Clemens JQ, Meenan RT, Rosetti MC, Kimes T, Calhoun EA. Costs of interstitial cystitis in a managed care population. *Urology.* 2008;71:776–780, discussion 780–781.
110. Baranowski AP, Abrams P, Berger RE, et al. Urogenital pain–time to accept a new approach to phenotyping and, as a consequence, management. *Eur Urol.* 2008;53:33–36.
111. Bogart LM, Berry SH, Clemens JQ. Symptoms of interstitial cystitis, painful bladder syndrome and similar diseases in women: a systematic review. *J Urol.* 2007;177:450–456.
112. Hanno PM, Burks DA, Clemens JQ, et al. AUA guideline for the diagnosis and treatment of interstitial cystitis/bladder pain syndrome. *J Urol.* 2011;185:2162–2170.
113. Nickel JC. Interstitial cystitis: a chronic pelvic pain syndrome. *Med Clin N Am.* 2004;88:467–481.
114. Chrysanthopoulou EL, Doumouchtsis SK. Challenges and current evidence on the management of bladder pain syndrome. *Neurourol Urodyn.* 2013;33:1193–1201.
115. Clemens JQ, Meenan RT, O'Keeffe Rosetti MC, Kimes TA, Calhoun EA. Case–control study of medical comorbidities in women with interstitial cystitis. *J Urol.* 2008;179:2222–2225.
116. Warren JW, van de Merwe JP, Nickel JC. Interstitial cystitis/bladder pain syndrome and nonbladder syndromes: facts and hypotheses. *Urology.* 2011;78:727–732.
117. Offiah I, McMahon SB, O'Reilly BA. Interstitial cystitis/bladder pain syndrome: diagnosis and management. *Int Urogynecol J.* 2013;24:1243–1256.
118. Diniz S, Dinis P, Cruz F, Pinto R. Bladder pain syndrome/interstitial cystitis: present and future treatment perspectives. *Minerva Urol Nefrol.* 2013;65:263–276.
119. French LM, Bhambore N. Interstitial cystitis/painful bladder syndrome. *Am Fam Physician.* 2011;83:1175–1181.
120. Nordling J, Fall M, Hanno P. Global concepts of bladder pain syndrome (interstitial cystitis). *World J Urol.* 2012;30:457–464.
121. Erickson DR, Belchis DA, Dabbs DJ. Inflammatory cell types and clinical features of interstitial cystitis. *J Urol.* 1997;158:790–793.
122. Erickson DR, Xie SX, Bhavanandan VP, et al. A comparison of multiple urine markers for interstitial cystitis. *J Urol.* 2002;167:2461–2469.
123. Lamale LM, Lutgendorf SK, Zimmerman MB, Kreder KJ. Interleukin-6, histamine, and methylhistamine as diagnostic markers for interstitial cystitis. *Urology.* 2006;68:702–706.
124. Lotz M, Villiger P, Hugli T, Koziol J, Zuraw BL. Interleukin-6 and interstitial cystitis. *J Urol.* 1994;152:869–873.
125. Kuo H-C. Potential urine and serum biomarkers for patients with bladder pain syndrome/interstitial cystitis. *Int J Urol.* 2014;21(suppl 1):34–41.
126. Qu HC, Yan S, Zhang XL, Zhu XW, Liu YL, Wang P. Urinary nerve growth factor levels could be a biomarker for overactive bladder symptom: a meta-analysis. *Genet Mol Res.* 2014;13:8609–8619.
127. Carrico DJ, Sherer KL, Peters KM. The relationship of interstitial cystitis/painful bladder syndrome to vulvodynia. *Urol Nurs.* 2009;29:233–238.
128. Seth A, Teichman JM. Differences in the clinical presentation of interstitial cystitis/painful bladder syndrome in patients with or without sexual abuse history. *J Urol.* 2008;180:2029–2033.

129. Tietjen GE, Brandes JL, Peterlin BL, et al. Childhood maltreatment and migraine (part III). Association with comorbid pain conditions. *Headache*. 2010;50:42–51.
130. DeBerry J, Randich A, Shaffer AD, Robbins MT, Ness TJ. Neonatal bladder inflammation produces functional changes and alters neuropeptide content in bladders of adult female rats. *J Pain*. 2010;11:247–255.
131. Randich A, Uzzell T, DeBerry JJ, Ness TJ. Neonatal urinary bladder inflammation produces adult bladder hypersensitivity. *J Pain*. 2006;7:469–479.
132. Robbins MT, Ness TJ. Footshock-induced urinary bladder hypersensitivity: role of spinal corticotropin-releasing factor receptors. *J Pain*. 2008;9:991–998.
133. Boucher W, Kempuraj D, Michaelian M, Theoharides TC. Corticotropin-releasing hormone-receptor 2 is required for acute stress-induced bladder vascular permeability and release of vascular endothelial growth factor. *BJU Int*. 2010;106:1394–1399.
134. Kiddoo DA, Valentino RJ, Zderic S, et al. Impact of state of arousal and stress neuropeptides on urodynamic function in freely moving rats. *Am J Physiol Regul Integr Comp Physiol*. 2006;290:R1697–R1706.
135. Klausner AP, Streng T, Na YG, et al. The role of corticotropin releasing factor and its antagonist, astressin, on micturition in the rat. *Auton Neurosci*. 2005;123:26–35.
136. McFadden K, Griffin TA, Levy V, Wolfe JH, Valentino RJ. Overexpression of corticotropin-releasing factor in Barrington's nucleus neurons by adeno-associated viral transduction: effects on bladder function and behavior. *Eur J Neurosci*. 2012;36:3356–3364.
137. Demaude J, Salvador-Cartier C, Fioramonti J, Ferrier L, Bueno L. Phenotypic changes in colonocytes following acute stress or activation of mast cells in mice: implications for delayed epithelial barrier dysfunction. *Gut*. 2006;55:655–661.
138. Colaco M, Koslov DS, Keys T, et al. Correlation of gene expression with bladder capacity in interstitial cystitis/bladder pain syndrome. *J Urol*. 2014;192(4):1123–1129.
139. Logadottir Y. Cytokine expression in patients with bladder pain syndrome/interstitial cystitis ESSIC Type 3C. *J Urol*. 2014;192(5):1564–1568.
140. Costantini E. GAGs and GAGs diseases: when pathophysiology supports the clinic. *Urologia*. 2013;80:173–178.
141. Janssen DA, van Wijk XM, Jansen KC, van Kuppevelt TH, Heesakkers JP, Schalken JA. The distribution and function of chondroitin sulfate and other sulfated glycosaminoglycans in the human bladder and their contribution to the protective bladder barrier. *J Urol*. 2013;189:336–342.
142. Costantini E, Lazzeri M, Pistolesi D, et al. Morphological changes of bladder mucosa in patients who underwent instillation with combined sodium hyaluronic acid-chondroitin sulphate (Ialuril(R)). *Urol Int*. 2013;91:81–88.
143. Fiander N. Painful bladder syndrome and interstitial cystitis: treatment options. *Br J Nurs*. 2013;22, S26, S28–33.
144. Iacovelli V. Intravesical glycosaminoglycans in the management of chronic cystitis. *Ital J Urol Nephrol*. 2013;65:249–262.
145. Jiang YH, Liu HT, Kuo HC. Decrease of urinary nerve growth factor but not brain-derived neurotrophic factor in patients with interstitial cystitis/bladder pain syndrome treated with hyaluronic acid. *PLoS One*. 2014;9:e91609.
146. Lai MC, Kuo YC, Kuo HC. Intravesical hyaluronic acid for interstitial cystitis/painful bladder syndrome: a comparative randomized assessment of different regimens. *Int J Urol*. 2013;20:203–207.
147. Chelsky MJ, Rosen SI, Knight LC, Maurer AH, Hanno PM, Ruggieri MR. Bladder permeability in interstitial cystitis is similar to that of normal volunteers: direct measurement by transvesical absorption of 99mtechnetium-diethylenetriaminepentaacetic acid. *J Urol*. 1994;151:346–349.
148. Parsons CL, Greene RA, Chung M, Stanford EJ, Singh G. Abnormal urinary potassium metabolism in patients with interstitial cystitis. *J Urol*. 2005;173:1182–1185.

149. Christmas TJ, Rode J. Characteristics of mast cells in normal bladder, bacterial cystitis and interstitial cystitis. *Br J Urol.* 1991;68:473–478.
150. Kastrup J, Hald T, Larsen S, Nielsen VG. Histamine content and mast cell count of detrusor muscle in patients with interstitial cystitis and other types of chronic cystitis. *Br J Urol.* 1983;55:495–500.
151. Larsen MS, Mortensen S, Nordling J, Horn T. Quantifying mast cells in bladder pain syndrome by immunohistochemical analysis. *BJU Int.* 2008;102:204–207 [discussion 207].
152. Peeker R, Enerback L, Fall M, Aldenborg F. Recruitment, distribution and phenotypes of mast cells in interstitial cystitis. *J Urol.* 2000;163:1009–1015.
153. Spanos C, Pang X, Ligris K, et al. Stress-induced bladder mast cell activation: implications for interstitial cystitis. *J Urol.* 1997;157:669–672.
154. Theoharides TC, Sant GR, el-Mansoury M, Letourneau R, Ucci Jr AA, Meares Jr EM. Activation of bladder mast cells in interstitial cystitis: a light and electron microscopic study. *J Urol.* 1995;153:629–636.
155. Lowe EM, Anand P, Terenghi G, Williams-Chestnut RE, Sinicropi DV, Osborne JL. Increased nerve growth factor levels in the urinary bladder of women with idiopathic sensory urgency and interstitial cystitis. *Br J Urol.* 1997;79:572–577.
156. Jiang YH, Peng CH, Liu HT, Kuo HC. Increased pro-inflammatory cytokines, C-reactive protein and nerve growth factor expressions in serum of patients with interstitial cystitis/bladder pain syndrome. *PLoS One.* 2013;8:e76779.
157. Corcoran AT, Yoshimura N, Tyagi V, Jacobs B, Leng W, Tyagi P. Mapping the cytokine profile of painful bladder syndrome/interstitial cystitis in human bladder and urine specimens. *World J Urol.* 2013;31:241–246.
158. Jacobs BL, Smaldone MC, Tyagi V, et al. Increased nerve growth factor in neurogenic overactive bladder and interstitial cystitis patients. *Can J Urol.* 2010;17:4989–4994.
159. Yun SK, Laub DJ, Weese DL, Lad PM, Leach GE, Zimmern PE. Stimulated release of urine histamine in interstitial cystitis. *J Urol.* 1992;148:1145–1148.
160. Pang X, Marchand J, Sant GR, Kream RM, Theoharides TC. Increased number of substance P positive nerve fibres in interstitial cystitis. *Br J Urol.* 1995;75:744–750.
161. Cortright DN, Krause JE, Broom DC. TRP channels and pain. *Biochim Biophys Acta (BBA)—Mol Basis Dis.* 2007;1772:978–988.
162. Chen Y, Yang C, Wang ZJ. Proteinase-activated receptor 2 sensitizes transient receptor potential vanilloid 1, transient receptor potential vanilloid 4, and transient receptor potential ankyrin 1 in paclitaxel-induced neuropathic pain. *Neuroscience.* 2011;193:440–451.
163. Sipe WE, Brierley SM, Martin CM, et al. Transient receptor potential vanilloid 4 mediates protease activated receptor 2-induced sensitization of colonic afferent nerves and visceral hyperalgesia. *Am J Physiol Gastrointest Liver Physiol.* 2008;294:G1288–G1298.
164. Steinhoff M, Vergnolle N, Young SH, et al. Agonists of proteinase-activated receptor 2 induce inflammation by a neurogenic mechanism. *Nat Med.* 2000;6:151–158.
165. DeBerry JJ. TRPA1 mediates bladder hyperalgesia in a mouse model of cystitis. *Pain.* 2014;155(7):1280–1287.
166. Golubeva AV. The mouse cyclophosphamide model of bladder pain syndrome: tissue characterization, immune profiling, and relationship to metabotropic glutamate receptors. *Physiol Rep.* 2014;2:e00260.
167. Moyal-Barracco M, Lynch PJ. 2003 ISSVD terminology and classification of vulvodynia: a historical perspective. *J Reprod Med.* 2004;49:772–777.
168. Andrews JC. Vulvodynia interventions–systematic review and evidence grading. *Obstet Gynecol Surv.* 2011;66:299–315.
169. Edwards L. Subsets of vulvodynia: overlapping characteristics. *J Reprod Med.* 2004;49:883–887.

170. McKay M. Subsets of vulvodynia. *J Reprod Med.* 1988;33:695–698.
171. Harlow BL, Stewart EG. A population-based assessment of chronic unexplained vulvar pain: have we underestimated the prevalence of vulvodynia? *J Am Med Womens Assoc.* 2003;58:82–88.
172. Stockdale CK, Lawson HW. 2013 vulvodynia guideline update. *J Low Genit Tract Dis.* 2014;18:93–100.
173. Xie Y, Shi L, Xiong X, Wu E, Veasley C, Dade C. Economic burden and quality of life of vulvodynia in the United States. *Curr Med Res Opin.* 2012;28:601–608.
174. Ventolini G. Vulvar pain: anatomic and recent pathophysiologic considerations. *Clin Anat.* 2013;26:130–133.
175. Plante AF, Kamm MA. Life events in patients with vulvodynia. *BJOG.* 2008;115:509–514.
176. McKay M. Dysesthetic ("essential") vulvodynia. Treatment with amitriptyline. *J Reprod Med.* 1993;38:9–13.
177. Reed BD, Caron AM, Gorenflo DW, Haefner HK. Treatment of vulvodynia with tricyclic antidepressants: efficacy and associated factors. *J Low Genit Tract Dis.* 2006;10:245–251.
178. Updike GM, Wiesenfeld HC. Insight into the treatment of vulvar pain: a survey of clinicians. *Am J Obstet Gynecol.* 2005;193:1404–1409.
179. Boardman LA, Cooper AS, Blais LR, Raker CA. Topical gabapentin in the treatment of localized and generalized vulvodynia. *Obstet Gynecol.* 2008;112:579–585.
180. Spoelstra SK, Borg C, Weijmar Schultz WC. Anticonvulsant pharmacotherapy for generalized and localized vulvodynia: a critical review of the literature. *J Psychosom Obstet Gynaecol.* 2013;34:133–138.
181. Edwards L. New concepts in vulvodynia. *Am J Obstet Gynecol.* 2003;189:S24–S30.
182. Gardella B, Porru D, Nappi RE, Dacco MD, Chiesa A, Spinillo A. Interstitial cystitis is associated with vulvodynia and sexual dysfunction–a case–control study. *J Sex Med.* 2011;8:1726–1734.
183. Nylanderlundqvist E, Bergdahl J. Vulvar vestibulitis: evidence of depression and state anxiety in patients and partners. *Acta Derm Venereol.* 2003;83:369–373.
184. Parsons CL, Dell J, Stanford EJ, Bullen M, Kahn BS, Willems JJ. The prevalence of interstitial cystitis in gynecologic patients with pelvic pain, as detected by intravesical potassium sensitivity. *Am J Obstet Gynecol.* 2002;187:1395–1400.
185. Reed BD, Harlow SD, Sen A, Edwards RM, Chen D, Haefner HK. Relationship between vulvodynia and chronic comorbid pain conditions. *Obstet Gynecol.* 2012;120:145–151.
186. Ehrstrom S, Kornfeld D, Rylander E, Bohm-Starke N. Chronic stress in women with localised provoked vulvodynia. *J Psychosom Obstet Gynaecol.* 2009;30:73–79.
187. Gordon AS, Panahian-Jand M, McComb F, Melegari C, Sharp S. Characteristics of women with vulvar pain disorders: responses to a Web-based survey. *J Sex Marital Ther.* 2003;29(suppl 1):45–58.
188. Bhattacherjee A, Rumi MA, Staecker H, Smith PG. Bone morphogenetic protein 4 mediates estrogen-regulated sensory axon plasticity in the adult female reproductive tract. *J Neurosci.* 2013;33:1050-61a.
189. Farmer MA, Taylor AM, Bailey AL, et al. Repeated vulvovaginal fungal infections cause persistent pain in a mouse model of vulvodynia. *Sci Transl Med.* 2011;3:101ra91.
190. Martinov T, Glenn-Finer R, Burley S, et al. Contact hypersensitivity to oxazolone provokes vulvar mechanical hyperalgesia in mice. *PLoS One.* 2013;8:e78673.
191. Bornstein J, Cohen Y, Zarfati D, Sela S, Ophir E. Involvement of heparanase in the pathogenesis of localized vulvodynia. *Int J Gynecol Pathol.* 2008;27:136–141.
192. Goldberg R, Meirovitz A, Hirshoren N, et al. Versatile role of heparanase in inflammation. *Matrix Biol.* 2013;32:234–240.

193. Leclair CM. Histopathologic characteristics of menopausal vestibulodynia. *Obstet Gynecol.* 2013;122:787–793.
194. Foster DC, Hasday JD. Elevated tissue levels of interleukin-1 beta and tumor necrosis factor-alpha in vulvar vestibulitis. *Obstet Gynecol.* 1997;89:291–296.
195. Jeremias J, Ledger WJ, Witkin SS. Interleukin 1 receptor antagonist gene polymorphism in women with vulvar vestibulitis. *Am J Obstet Gynecol.* 2000;182:283–285.
196. Galli SJ, Nakae S, Tsai M. Mast cells in the development of adaptive immune responses. *Nat Immunol.* 2005;6:135–142.
197. Hurme M, Lahdenpohja N, Santtila S. Gene polymorphisms of interleukins 1 and 10 in infectious and autoimmune diseases. *Ann Med.* 1998;30:469–473.
198. Santtila S, Savinainen K, Hurme M. Presence of the IL-1RA allele 2 (IL1RN*2) is associated with enhanced IL-1beta production in vitro. *Scand J Immunol.* 1998;47:195–198.
199. Gerber S. Defective regulation of the proinflammatory immune response in women with vulvar vestibulitis syndrome. *Am J Obstet Gynecol.* 2002;186:696–700.
200. Foster DC. Enhanced synthesis of proinflammatory cytokines by vulvar vestibular fibroblasts: implications for vulvar vestibulitis. *Am J Obstet Gynecol.* 2007;196:346. e1–346.e8.
201. Pontari MA, Joyce GF, Wise M, McNaughton-Collins M, Urologic Diseases in America Project. Prostatitis. *J Urol.* 2007;177:2050–2057.
202. Calhoun EA, McNaughton Collins M, Pontari MA, et al. The economic impact of chronic prostatitis. *Arch Intern Med.* 2004;164:1231–1236.
203. Litwin MS, McNaughton-Collins M, Fowler Jr FJ, et al. The National Institutes of Health chronic prostatitis symptom index: development and validation of a new outcome measure. Chronic prostatitis collaborative research network. *J Urol.* 1999;162:369–375.
204. Nickel JC, Alexander RB, Anderson R, et al. Category III chronic prostatitis/chronic pelvic pain syndrome: insights from the National Institutes of Health Chronic Prostatitis Collaborative Research Network studies. *Curr Urol Rep.* 2008;9:320–327.
205. Collins MM, Meigs JB, Barry MJ, Walker Corkery E, Giovannucci E, Kawachi I. Prevalence and correlates of prostatitis in the health professionals follow-up study cohort. *J Urol.* 2002;167:1363–1366.
206. Forrest JB, Schmidt S. Interstitial cystitis, chronic nonbacterial prostatitis and chronic pelvic pain syndrome in men: a common and frequently identical clinical entity. *J Urol.* 2004;172:2561–2562.
207. Mehik A, Hellstrom P, Sarpola A, Lukkarinen O, Jarvelin MR. Fears, sexual disturbances and personality features in men with prostatitis: a population-based cross-sectional study in Finland. *BJU Int.* 2001;88:35–38.
208. Eisenberg ER, Moldwin RM. Etiology: where does prostatitis stop and interstitial cystitis begin? *World J Urol.* 2003;21:64–69.
209. Hassan AA, Elgamal SA, Sabaa MA, Salem K. Evaluation of intravesical potassium sensitivity test and bladder biopsy in patients with chronic prostatitis/chronic pelvic pain syndrome. *Int J Urol.* 2007;14:738–742.
210. Moldwin RM. Similarities between interstitial cystitis and male chronic pelvic pain syndrome. *Curr Urol Rep.* 2002;3:313–318.
211. Done JD, Rudick CN, Quick ML, Schaeffer AJ, Thumbikat P. Role of mast cells in male chronic pelvic pain. *J Urol.* 2012;187:1473–1482.
212. Roman K, Done JD, Schaeffer AJ, Murphy SF, Thumbikat P. Tryptase-PAR2 axis in experimental autoimmune prostatitis, a model for chronic pelvic pain syndrome. *Pain.* 2014;155:1328–1338.
213. Theoharides TC, Flaris N, Cronin CT, Ucci A, Meares E. Mast cell activation in sterile bladder and prostate inflammation. *Int Arch Allergy Appl Immunol.* 1990;92:281–286.

214. Amir T, Pai RR, Raghuveer CV. Mast cell profile in prostatic lesions. *Indian J Med Sci.* 1998;52:507–513.
215. Donadio AC, Depiante-Depaoli M. Inflammatory cells and MHC class II antigens expression in prostate during time-course experimental autoimmune prostatitis development. *Clin Immunol Immunopathol.* 1997;85:158–165.
216. Keetch DW, Humphrey P, Ratliff TL. Development of a mouse model for nonbacterial prostatitis. *J Urol.* 1994;152:247–250.
217. Rivero VE, Cailleau C, Depiante-Depaoli M, Riera CM, Carnaud C. Non-obese diabetic (NOD) mice are genetically susceptible to experimental autoimmune prostatitis (EAP). *J Autoimmun.* 1998;11:603–610.
218. Rudick CN, Schaeffer AJ, Thumbikat P. Experimental autoimmune prostatitis induces chronic pelvic pain. *Am J Physiol Regul Integr Comp Physiol.* 2008;294:R1268–R1275.
219. Rivero VE, Iribarren P, Riera CM. Mast cells in accessory glands of experimentally induced prostatitis in male Wistar rats. *Clin Immunol Immunopathol.* 1995;74:236–242.

CHAPTER EIGHTEEN

Meningeal Afferent Signaling and the Pathophysiology of Migraine

Carolina Burgos-Vega, Jamie Moy, Gregory Dussor[1]
Behavioral and Brain Sciences, The University of Texas at Dallas, Richardson, Texas, USA
[1]Corresponding author: e-mail address: gregory.dussor1@utdallas.edu

Contents

1. Introduction 538
2. Features of Migraine 538
3. Current Migraine Treatments 541
4. Anatomy of the Meningeal Afferent System 542
5. Migraine Pathophysiology 544
6. Potential Mechanisms of Dural Afferent Activation 545
7. Ion Channels and Dural Afferent Activation 546
8. Dural Afferent Input May Lead to Neuroplasticity 549
9. Neuroplasticity and Migraine 550
10. CGRP and Migraine Plasticity 553
11. BDNF and Migraine 554
12. Conclusion 555
Acknowledgments 556
References 557

Abstract

Migraine is the most common neurological disorder. Attacks are complex and consist of multiple phases but are most commonly characterized by intense, unilateral, throbbing headache. The pathophysiology contributing to migraine is poorly understood and the disorder is not well managed with currently available therapeutics, often rendering patients disabled during attacks. The mechanisms most likely to contribute to the pain phase of migraine require activation of trigeminal afferent signaling from the cranial meninges and subsequent relay of nociceptive information into the central nervous system in a region of the dorsal brainstem known as the trigeminal nucleus caudalis. Events leading to activation of meningeal afferents are unclear, but nerve endings within this tissue are mechanosensitive and also express a variety of ion channels including acid-sensing ion channels and transient receptor-potential channels. These properties may provide clues into the pathophysiology of migraine by suggesting that decreased extracellular pH and environmental irritant exposure in the meninges contributes to headache. Neuroplasticity is also likely to play a role in migraine given that attacks are triggered by routine events that are typically nonnoxious in healthy patients and clear

Progress in Molecular Biology and Translational Science, Volume 131
ISSN 1877-1173
http://dx.doi.org/10.1016/bs.pmbts.2015.01.001

evidence of sensitization occurs during an attack. Where and how plasticity develops is also not clear but may include events directly on the afferents and/or within the TNC. Among the mediators potentially contributing to plasticity, calcitonin gene-related peptide has received the most attention within the migraine field but other mechanisms may also contribute. Ultimately, greater understanding of the molecules and mechanisms contributing to migraine will undoubtedly lead to better therapeutics and relief for the large number of patients across the globe who suffer from this highly disabling neurological disorder.

1. INTRODUCTION

Migraine is a subtype of headache characterized by repetitive episodes of intense unilateral throbbing head pain lasting many hours to days. However, it is a multiphasic disorder that comprises complex symptomology that can vary greatly among patients. These complex features of migraine differentiate it from other types of chronic pain and include the presence of premonitory symptoms, an aura, and during the headache phase, severe nausea, vomiting, and hypersensitivity to light and sound. Although migraine is episodic in nature, together these collections of symptoms cause an entire attack to last close to a week, and given the severity, migraine has a significant negative impact on a patient's quality of life. The World Health Organization's Global Burden of Disease Study conducted in 2010 found migraine headache to be the third most prevalent disease across the globe, preceded only by dental caries and tension type headache.[1] Given the current understanding of migraine as a pathological condition of the nervous system, migraine is thus the most common neurological disorder with an estimated 33% of women and 10% of men affected by the condition.[1–3] Despite the prevalence, the basic physiology and underlying contributing factors to the development of migraine headache is still poorly understood.

2. FEATURES OF MIGRAINE

Migraines comprise four phases: premonitory phase (also commonly referred to as the prodrome), aura, headache, and postdrome[2,3] (Fig. 1). Studies have estimated that the number of patients with migraine who experience premonitory symptoms may range from as low as 37% to as high as 80%.[4,5] The premonitory phase occurs hours to days before the actual headache and can be a reliable predictor of an upcoming migraine for some

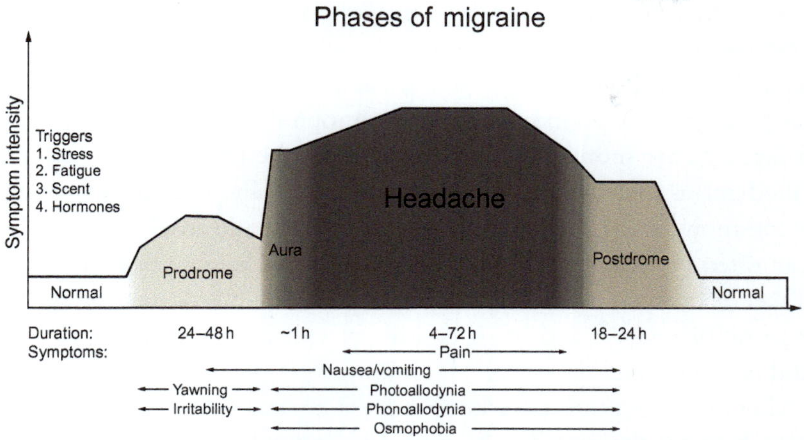

Figure 1 Timeline of migraine attacks. Migraine typically consists of four phases although not all patients experience all phases. Attacks often begin with triggers, which can be routine events such as stress or fatigue. Whether or not triggers provoke an attack, the first phase, prodrome, lasts between 24 and 48 h and consists of premonitory symptoms such as yawning or irritability. The next phase to develop is aura, which affects less than 30% of migraine patients and consists of sensory disturbances, mostly visual. The most debilitating phase of a migraine attack is the headache phase. This unilateral and throbbing head pain can last between 4 and 72 h and is often accompanied by photo- and phonoallodynia as well as nausea and vomiting. The fourth phase of a migraine attack is the postdrome. Symptoms related to the postdrome phase often vary; however, patients commonly report being exhausted and irritable. Together, all four phases of a migraine attack can last between 3 and 5 days.

patients.[4] This phase consists of symptoms that range in severity such as excessive yawning, food cravings, mood changes, fatigue, sore neck, and confusion among others.[4–6] Following the premonitory phase is the aura phase that takes place up to an hour prior to the headache phase, but is only experienced by 15–30% of migraineurs. Aura involves sensory disturbances, often visual, with moving and intensifying regions of flashing lights or scintillations accompanied by partial vision loss or scotomas (areas of decreased acuity in the visual field). Scotomas are reported as typically originating in the center of a patient's visual field and progressing toward the edges.[7] Aura may present other somatosensory symptoms that can affect normal motor function, perception of language, or even production of discernible language. The next phase is headache which must, by definition, be a minimum of 4 h but can last up to 72 h without medical intervention.[7–9] It is described as moderate to severe throbbing pain, initially localized to one side of the head.[7] The headache can be aggravated by repositioning of the head, by normal physical activity, or by changes in intracranial pressure, e.g., coughing or sneezing. Common symptoms that accompany the throbbing pain include sensitivity to light, sounds, and smells, also commonly referred to as photophobia, phonophobia, and osmophobia, respectively, although the first two are more correctly described by the terms photo- and phonoallodynia as these stimuli cause pain. Additionally, nausea and vomiting are common during the headache phase. After the headache has resolved, it can often be followed by a postdrome phase where the patient still does not feel recovered or entirely back to normal. The postdrome can typically last from 18 to 24 h and often consists of exhaustion, irritability, and an inability to concentrate.[7]

Migraine currently lacks a known cause or source that can be identified as the driver of the disorder. Nonetheless, individual patients may be susceptible to distinct stimuli that are capable of evoking a migraine. These stimuli, referred to as triggers (Fig. 1), can vary between individuals and yet they share the commonality of eliciting a migraine attack. However, it should be noted that these triggers do not exhibit a noxious effect in nonmigraine sufferers and therefore are not the true source of the condition. Sensitivity to triggers in migraine patients suggests the presence of maladaptive plasticity in the nervous system (discussed below) that contributes to the development of attacks in response to stimuli that are innocuous in normal individuals.

One extremely common trigger in those susceptible to migraine is stress. In fact, studies demonstrate that anywhere from 59% to over 80% of patients indicate that stress is their primary trigger for migraine,[10–12] thus making stress the most commonly reported trigger. The effect of stress exposure may be cumulative. Stress appears to have a greater influence with two consecutive days of moderate or greater stress in a row. On the other hand, low stress followed by a day with moderate or greater stress on the subsequent day is associated with less risk.[9] Conversely, this study also found that if stress is moderate or greater on the first day, but low on the following day the migraine risk is still increased, thus suggesting a complex relationship between stress and migraine. Recent reports also show that the peak susceptibility to migraine is in the 18–24 h after a stressful event,[13] further supporting the complex nature of the contribution of stress to migraine. It is hypothesized that stress itself may not be the trigger; however, it may evoke changes in normal behaviors such as sleep or food intake that can aggravate the condition.[14] Patients able to identify their triggers can benefit and potentially reduce migraine attacks by minimizing their encounters with that particular stimulus (or stimuli).[15] Other common triggers include foods (and alcohol), environmental irritants (cigarette smoke, odors), exercise, changes in the environment (barometric pressure), too much/lack of sleep, hormone fluctuations, sensory stimuli (such as intense light), and skipping meals. Notably, drugs such as nitric oxide donors (e.g., nitroglycerin) are capable of eliciting headaches reliably among migraineurs,[16] an important observation since NO donors are often used to experimentally trigger attacks in clinical studies.

3. CURRENT MIGRAINE TREATMENTS

Migraine is an extremely disabling disorder,[1] a fact made more troubling by the lack of efficacy of currently available therapeutics. Treatment typically falls in one of the two categories, abortive and prophylactic. Abortive agents are taken at the onset of headache or during aura with the goal being termination or at least a decrease in intensity of the building attack. Prophylactic therapies are given daily to patients who have frequent migraines (e.g., >5 per month) with the goal of decreasing the frequency of attacks. The most commonly prescribed abortive therapies are the triptans and nonsteroidal anti-inflammatory drugs (NSAIDs).[17,18] Triptans are a family of $5HT_{1b}$ and $5HT_{1d}$ agonists that include members such as sumatriptan, zolmitriptan, rizatriptan, eletriptan, and naritriptan. Triptans alone account for up to 80% of medications prescribed for migraine.[2,19,20] Despite the high propensity of their use, triptans are not always effective. For example, the percentage of patients that report being completely pain-free 2 h after dosing ranges from 12% (frovatriptan 2.5 mg) to 40% (rizatriptan 10 mg).[21] Further, triptans possess vasoconstrictive properties and are therefore contraindicated in patients at risk for cardiovascular complications or abnormal blood pressure.[22] Another issue with triptan use (this can also occur with many migraine medications) is a clinical concern that triptan overuse can lead to phenomenon known as medication-overuse headache (MOH).[2,23,24] MOH is characterized by worsened or more frequent headache caused by regular triptan intake that resolves after removal of the offending medication. Thus, acute treatments such as triptans are not recommended for daily use and cannot be used as prophylactic agents.[18] Opioids coupled with NSAIDs or acetaminophen, as well as barbiturates are also used by migraine patients, even though they too can induce MOH.[25,26]

As mentioned above, prophylactic migraine treatment is given on a daily basis in order to attempt to prevent the onset of migraine attacks or decrease frequency. Commonly used prophylactic agents are antiepileptics, such as topiramate, gabapentin, or valproate, beta blockers such as propranolol, or antidepressants such as amitriptyline. Use of propranolol, valproate, and topiramate, for example, can result in a 50% reduction in migraine frequency in 40–50% of patients.[21] However, prophylactic treatment is generally not well tolerated by patients (with the exception of propranolol) due to numerous adverse side effects. These side effects can include significant decreases in cognitive function, nausea, vomiting, weight gain, and even withdrawal symptoms upon discontinued use of the medication.[18,19,27] Taking into account the statistics noted above for both types of agents (<40% of patients pain-free with triptans and 50% of patients achieving 50% frequency reduction with prophylactics), it is clear that many patients are unable to achieve relief.

4. ANATOMY OF THE MENINGEAL AFFERENT SYSTEM

The meninges are innervated predominantly by afferents whose cell bodies reside in the trigeminal ganglion (Fig. 2). The trigeminal nerve is the fifth cranial nerve that exits the brainstem at the level of the pons as a single nerve root, passes through the trigeminal ganglion, and continues distally from the ganglion as separate nerve branches.[28,29] There are three major branches that emerge from the ganglion into V1, V2, and V3 subdivisions. Each branch innervates a distinct dermatome. The first division, V1, is also referred to as the ophthalmic nerve, and it innervates the nose, areas above the eye, and the scalp. The second division is V2, or the maxillary nerve, and it innervates areas below the eye and above the mouth as well as the mandible. Lastly, V3 is the mandibular nerve that innervates the mouth and lower portions of the jaw and face. Most relevant to migraine is the V1 branch as this is the branch that provides the majority of the innervation of the meninges (dura and blood vessels). These trigeminal afferents send afferent input from the head into the brainstem (at the pons) and they descend into the medulla where they synapse on second-order neurons and interneurons within a region of the dorsal medulla known as the spinal trigeminal nucleus or trigeminal nucleus caudalis (TNC) (Fig. 2). The nucleus caudalis lies below the obex of the brainstem and is the most caudal part of the trigeminal nucleus.[30] Multiple

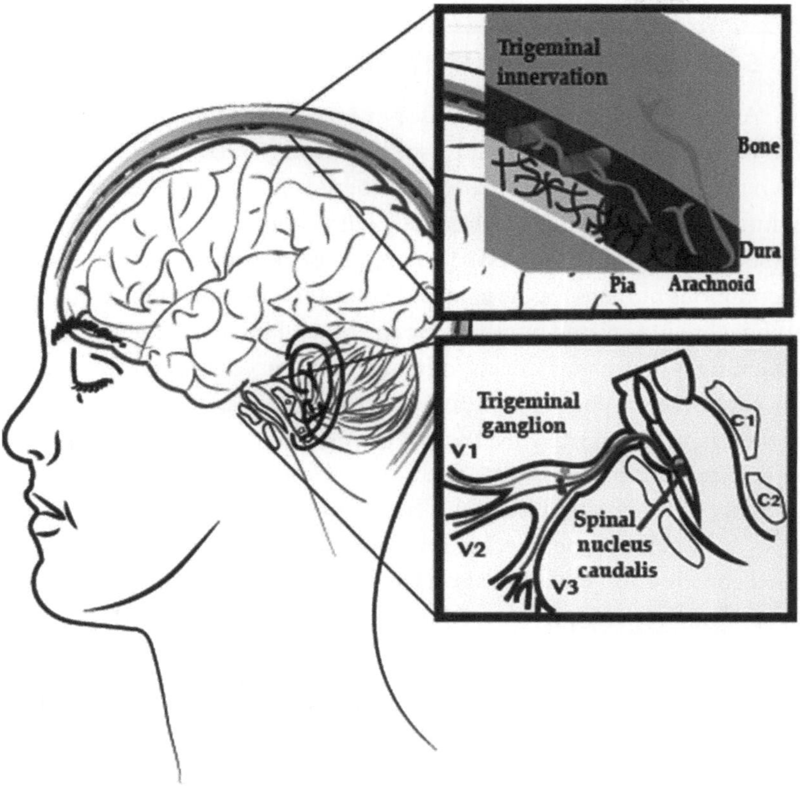

Figure 2 Anatomy of the meningeal afferent system. Illustration of the meningeal layer and its relationship to the overlying bone and brain below. The dura is highly innervated by sensory afferents projecting from the trigeminal ganglion. Three major branches emerge from the ganglion into V1, V2, and V3 subdivisions. Each branch innervates a distinct dermatome of the head with V1 providing innervation to the meninges. Trigeminal afferents send input into the brainstem (at the pons) and they descend into the medulla where they synapse on second-order neurons and interneurons within a region of the dorsal medulla known as the spinal trigeminal nucleus or trigeminal nucleus caudalis.

projections into the caudalis may share the same interneuron, allowing crosstalk between afferents and providing the basis for referred pain seen in the trigeminal system.[31] The axons of second-order neurons form the trigeminothalamic pathway then decussate and ascend to the ventroposteromedial nucleus of the thalamus. From the thalamus, connections are made to cortical sites.

5. MIGRAINE PATHOPHYSIOLOGY

Although the brain lacks sensory innervation, which renders it incapable of sensing noxious stimuli, studies dating as far back as 1940 have shown that sensory afferents that innervate the cranial meninges, particularly the dura mater, are in fact mechanically (and chemically) sensitive.[32–38] The earliest of these studies showed that in response to stimulation of the meninges, conscious humans report pain and no other sensation, indicating that the meninges are innervated purely by nociceptors. However, the events within the meninges that occur during migraine are not yet known. Migraine has been extensively hypothesized to arise from vascular swelling in the brain and meninges (supported by the triggering ability of vasodilators such as NO donors) that produced the distinct throbbing characteristic of the pain, thus many antimigraine treatments have possessed vasoconstrictive properties.[39–41] Together, the mechanical sensitivity of dural afferents and the alleged contribution of vasodilation has been a leading hypothesis in the field for decades. However, it is becoming increasingly evident that vasodilation of dural vessels during migraine is a secondary effect of the disorder rather than the clinical cause.[41,42] Several recent imaging studies in humans, the most recent examining natural (i.e., nonexperimentally triggered) migraines, found no extracranial dilation of vessels and only slight intracranial dilation.[10,43,44] The recent study also showed efficacy for headache by the vasoconstrictor sumatriptan, even in the absence of any vasodilation present. These studies cast serious doubt on vasodilation being the primary mediator of migraine. Further evidence supporting this concept is that functional imaging studies have concluded that migraine headache appears to arise within the brain and does not require vasodilation in order to occur.[43,45–49] In fact, the hypothalamus is now regularly cited as a potential central origin of migraine attacks[50] and activation of the hypothalamus is known to occur in the premonitory phase of migraine.[51] In relation to pain, the hypothalamus is also known to modulate afferent trafficking in the TNC and can enhance afferent input from the meninges.[52,53] Although attacks are potentially driven centrally, the pain phase of migraine still requires activation of peripheral afferents and these afferents can be modulated by central outputs. Importantly, activation of afferents need not be in response to vasodilation. Thus, migraine is now becoming increasingly accepted as neurological disorder where maladaptive responses (plasticity) of dural afferents innervating the vasculature or the dura itself, possibly modulated by descending pathways from the brain, can provide the basis of the headache.[7,9,54] Attention is now focused on identifying mechanisms of activation and plasticity in dural afferents that may contribute to the pain of migraine.

6. POTENTIAL MECHANISMS OF DURAL AFFERENT ACTIVATION

Dural afferents are capable of sensing noxious stimuli generated within the dura and the pain that arises during migraine is likely to involve the activation of these primary afferent nociceptors.[55,56] However, the events that lead to the activation and/or sensitization of these neurons are not fully known.[8,57,58] The dura is often overlooked as just one of the three meningeal layers; however, the dura mater regulates events in the overlying skull during development. Ultimately, the dura will control the size and shape of the cranial bones, along with suture patency, and the proper maturation of these bones and sutures is dependent on interactions with the dura mater.[43,59] Whether there is any interaction between the dura and the skull that contributes to migraine is not known but, interestingly, some afferents that innervate the dura send branches that penetrate into the skull, particularly at sutures.[60–62] There are thus sensory afferents that innervate both tissues suggesting importance of both in sensory signaling from the dura.

Within the dura mater are fibroblasts, which generate the fibrous tissue and accompanying elastic fibers. These cells have recently been shown to play a potential role in headache as they release factors that both sensitize dural afferents and cause headache behaviors in rodents when applied to the dura.[63] Among the factors released by these cells is the proinflammatory cytokine interleukin-6 (IL-6). The dura also contains other nonneuronal cell types such as mast cells and macrophages. There are numerous arteries within the dura mater, the most prominent being the middle meningeal artery (MMA) and the branches originating from this artery. Mast cells are known to aggregate around blood vessels within the dura and may play potential role in migraine pathophysiology.[64,65] Mast cells have also been found in direct contact with afferents within the dura, which could allow for potential modulation.[64–66] The degranulation of meningeal mast cells has been shown to activate and sensitize dural afferents and this is due, at least in part, to activation of the protease-activated receptor 2 (PAR_2) by mast-cell-derived tryptase.[66–69] Mast cell degranulation can occur as a result of factors known to be associated with migraine in humans including: the release of the stress hormone corticotropin releasing factor (CRF), calcitonin gene-related peptide (CGRP) release, nitroglycerin infusion, and also increases in estrogen.[67,69–71] Degranulation of mast cells could result in the release of factors such as histamine, serotonin, and protons capable of contributing to sensitization of dural afferents. Among this list of chemical mediators, many can sensitize dural afferents resulting in increased excitability. Application of a cocktail of inflammatory mediators that included histamine, serotonin, bradykinin, and PGE_2 all at pH 5.0 cause widespread cutaneous allodynia in rodents.[72,73] This has been shown recently following application of bradykinin, PGE_2, and histamine, leading to sensitization of ionic currents that contribute to the increased excitability of dural afferents.[63,74]

7. ION CHANNELS AND DURAL AFFERENT ACTIVATION

Acid-sensing ion channels or ASICs are a family of four ion channels consisting of, ASIC1–ASIC4, with several splice variants.[75–77] ASIC1–3 are sensitive to various ranges of pH from approximately 4.5 to just above 7.0. ASIC4 is not a pH-sensitive channel. Although ASICs are abundant throughout the nervous system, including expression primary sensory neurons, ASIC3 is largely restricted to the periphery.[77,78] ASICs on dural afferents may act as an indicator of changes in extracellular pH. However, a change in extracellular pH within the dura remains to be demonstrated. Cortical spreading depression, which may contribute to migraine headache through activation of the trigeminovascular system,[79,80] was found to be associated with a decrease in dural blood flow[81,82] as well as cortical hypoxia,[80,82,83] events that may decrease the pH of the cortex and/or meninges.

ASICs may play a role in dural afferent signaling and as a result in the pain of migraine. Decreased pH has been shown to activate dural afferents in several studies using *in vivo* electrophysiological recording techniques. These studies found the following: pH 4.7 activated 75% of rat dural afferents 1, pH 5.0 activated 37% in guinea pigs,[84] and pH 6.1 activated 64% in rats.[85] Although these studies show pH sensitivity of dural afferents, they did not identify mechanisms. Electrical stimulation of the dura activates afferent signaling that is blocked by the ASIC blocker amiloride.[86] And pH 5.4–5.9 promotes CGRP release from dura and trigeminal ganglia that is sensitive to ASIC blockers.[87,88]

However, one of the best examples supporting a role for ASICs in migraine is the finding of ASIC currents in 80% of dural afferents at pH

6.0 and 50% at pH 7.0 that were blocked by amiloride or activated by an ASIC3-specific agonist.[89] When low pH (from 5.0 to 6.4) was applied to the rat dura, animals displayed headache-related behaviors that were blocked by ASIC antagonists. These findings provide some of the most direct evidence that ASICs contribute to processes related to migraine.

Recently, it was also demonstrated that CSDs are sensitive to ASIC blockers,[86] implicating these channels in migraine-related pathophysiology. Also included in this study was an exciting clinical trial on seven difficult to treat migraine patients. When given the ASIC blocker amiloride, four out of seven patients achieved significant relief of both headache and aura symptoms with minimal side effects. These studies suggest that targeting ASICs in humans may be an effective strategy for novel migraine therapeutics.

Mast cell degranulation (discussed above) may decrease meningeal pH as the contents of mast cell granules are approximately pH 5.5.[90] This may lead to pH-dependent activation of dural afferents, and these nerve endings may also be sensitized by additional factors released from mast cells. Supporting this idea is the observation that the number of dural afferents that fire action potentials is doubled at pH 7.0 in the presence of mast cell mediators (40% vs. 20% in controls).[89] This study also showed that higher pH values applied to the dura can produce headache behaviors in the presence of mast cell mediators than under control conditions. Together, mast cell degranulation may provide both a drop in pH and a sensitizing stimulus enhancing the afferent input mediated by dural afferents. These studies argue for further investigation of ASICs for a role in migraine and ultimately as novel therapeutic targets.

TRP channels are a family of membrane ion channels known to be responsive to stimuli such as temperature, changes in extracellular osmolarity, pH, and an extensive list of natural products.[91–95] There are six subfamilies of TRP channels, denoted with a letter corresponding to the primary amino acid sequences: TRPA, TRPC, TRPM, TRPP, TRPL, and TRPMV.[96–98] TRP channels are nonselective ion channels that aid in depolarization as well as activation of second messenger signaling pathways via influx of Na^+ and Ca^{2+}.[99] They are proposed to contribute to many types of sensory input in varying tissues from cutaneous to visceral and have been implicated in many pain conditions.[92,100,101,102]

Relevant to migraine, the TRP channel most implicated in headache pathophysiology is TRPA1. TRPA1 is activated by a diverse array of substances including the environmental irritants acrolein,[103,104] chlorine,[105,106] formaldehyde,[107,108] and cigarette smoke,[109,110] natural products like cinnamaldehyde,[74,111] isothiocyanates,[112] and allicin,[113] and endogenous substances like 4-hydroxynonenal,[39,114] nitro-oleic acid,[115] and reactive prostaglandins.[116] Many of these are linked to migraine, commonly as triggers.[12,117–119]

Preclinical studies continue to build a case that TRPA1 contributes to the pathophysiology of headache. TRPA1 is expressed [120] and generates currents[121] in dural afferents in rodents. When TRPA1 agonists, including acrolein, are introduced into the nasal cavity, there is an increase in meningeal blood flow that is blocked by either a CGRP or a TRPA1 antagonist. This study implicates TRPA1 in headaches due to inhaled environmental irritants.[122] More direct evidence comes from studies showing that dural application of mustard oil produces headache behaviors (both cutaneous facial hypersensitivity and a decrease in exploratory activity) that are blocked by an oral TRPA1 antagonist.

Interestingly, a tree native to northern California and southern Oregon known as the "headache tree" (*Umbellularia californica*) can trigger cluster headache-like attacks in individual susceptible to headaches.[119] A primary component of the extracts of this tree is umbellulone, a TRPA1 activator.[119] Dural application of umbellulone leads to headache behaviors identical to those described above with mustard oil.[121] How umbellulone gains access to the meninges to produce headache is not clear. It may be via the nasal route since preclinical studies found that intranasal delivery can produce meningeal vasodilation in a TRPA1-dependent mechanism.[104] Together, the studies described here provide an intriguing case for a contribution of TRPA1 to headache and like ASICs, argue for the development of TRPA1 therapeutics to test in migraine patients.

As described above, dural afferents are mechanically sensitive but they are also sensitive to changes in extracellular osmolarity,[38] suggesting that they express the mechano- and osmosensitive channel TRPV4.[123–126] Prior studies have shown that TRPV4 mRNA in trigeminal ganglion neurons[127] and the channel appears to be functional, at least *in vitro*.[128,129] Deletion of TRPV4 in mice results in a loss of the ability to detect both osmotic and pressure stimuli. Additionally,[130,131] TRPV4 can be sensitized by proinflammatory mediators and its mechanosensitivity may be enhanced under inflammatory conditions as administration of IM to the hindpaw leads to osmotic and mechanical sensitivity that is lost in TRPV4 null mice.[132]

Recently, TRPV4 currents were found on dural afferents *in vitro* in response to hypotonic stimulation and the TRPV4 activator 4αPDD.[133] Using the behavioral model described above, application of TRPV4 activators to the dura of rats caused headache behaviors,[133] indicating that these channels in the dura can contribute functionally to headache. Changes in osmolarity, at least to the degree required to activate TRPV4, do not likely contribute to migraine. But as mentioned previously, changes in intracranial pressure (head movements, coughing, etc.) can worsen headache in migraine patients[56] and this may be mediated by the mechanically sensitive TRPV4. More work is still necessary to determine whether TRPV4 contributes to headache in humans and whether targeting this channel may have efficacy for migraine.

The TRPM subfamily contains eight members: TRPM1–M8. The TRPM8 channel is sensitive to both cold temperature as well as chemical cooling agents such as menthol.[134–136] TRPM8 is activated by both noxious and nonnoxious temperatures from ~28 to 8 °C.[137] TRPM8 transcripts are expressed in a small percentage (<15%) of small diameter sensory neurons.[134,136] TRPM8 expression within the nervous system is thought to confer innocuous cold sensitivity to the somatosensory system,[138,139] especially in sensory neurons innervating cutaneous tissues. TRPM8 is not isolated to the cutaneous afferents and has been found in neurons innervating deep tissues such as the bladder and colon.[140,141] As these neurons are not directly exposed to external cold stimuli, cold may not be an exclusive activator in these tissues supporting the hypothesis that the channel may respond to endogenous activators.[142,143] Possible endogenous activators/sensitizers in visceral tissues include lysophospholipids, cyclopentenone prostaglandins, and phosphatidylinositol biphosphate among others.[94] Ultimately, TRPM8 may be able to sense various internal and external stimuli. Additionally, recent genome wide association studies have consistently revealed single nucleotide polymorphisms in and around the TRPM8 locus in several populations of migaineurs,[144–146] further strengthening the hypothesis of possible TRP channel involvement in migraine. It is not currently known how these mutations impact channel function, but these data nonetheless provide exciting support for a potential role of this channel in the pathophysiology of migraine.

8. DURAL AFFERENT INPUT MAY LEAD TO NEUROPLASTICITY

Although the events that actually initiate a migraine attack remain unknown, the ultimate activation of the trigeminovascular system is considered to be essential[147,148]. Trigeminovascular activation may provoke release of multiple excitatory neurotransmitters including substance P, neurokinin A, and CGRP from dural afferent terminals resulting in neurogenic vasodilation of dural blood vessels, release of proinflammatory mediators, degranulation of mast cells, and plasma protein extravasation.[9,149,150] Activation of these primary nociceptors innervating the meninges may also play a role in maladaptive changes in central circuits that contribute to migraine. However, how these nociceptors might contribute to central plasticity is not well understood. Dural afferents project into the TNC and activation, or repetitive activation of these neurons may contribute to plasticity in the central circuitry in the TNC that contributes to the pathophysiology of migraine. Preclinical studies have demonstrated widespread cutaneous allodynia following dural afferent stimulation with a variety of compounds,[89,121,151] an effect that implicates plasticity at central sites following stimulation of the dura. Migraineurs also experience hypersensitivity that can diverge from the initial or primary site of the spontaneous head pain over the course of an attack.[152] For example, patients commonly have facial allodynia initially that over time expands to extracephalic regions as well,[56] again strongly implicating central plasticity during migraine attacks. The mechanisms that contribute to the plasticity likely required for these observed changes have yet to be fully elucidated. Elevated cytokines, neuropeptides, and growth factors such as IL-6, CGRP, and brain-derived neurotrophic factor (BDNF) have been shown in patients during migraine attacks compared to migraine free periods,[109,153,154] and these mediators, among others, may contribute to the development of neuronal plasticity.

9. NEUROPLASTICITY AND MIGRAINE

As discussed above, there are a number of proposed mechanisms that may contribute to the pathophysiology of migraine, one of which involves "neuroplasticity" in the peripheral and central nervous systems. There are two different kinds of neuroplasticity, one that is beneficial, contributing processes within the brain that lead to learning and memory, and one that is harmful, contributing to pathological pain or migraine. The underlying mechanisms mediating these forms of plasticity are thought to be similar. Plasticity in the nervous system manifests as a change in the response of a neuron or circuit to a stimulus such that after plasticity develops, a given stimulus produces a different response than before the onset of plasticity. In the case of pain, this can present as an increase in sensitivity of primary sensory neurons or a painful response in an individual to a normally nonnoxious stimulus. For example, cutaneous allodynia or pain in response to touching of the skin is thought to be mediated by plasticity within the nociceptive system. Migraine is another example of maladaptive plasticity in the nervous system, especially since it seems to play no beneficial role in warning affected patients of ongoing tissue damage. Migraine is a chronic, episodic, pain disorder in which there is no identifiable injury or pathology. Consequently, the changes within the body that lead to migraine are likely subtle. This strongly suggests maladaptive plasticity to relatively normal events. Additionally, the rate at which migraine episodes occur often increases. Approximately, 14% of patients who develop episodic migraines will transition to chronic migraines (15 or more headache days per month), suggesting a pathophysiology based on plasticity.[2,155,156] These data suggest that the presence of migraine may contribute to further plasticity that perpetuates the disorder.

In general, neuroplasticity can occur throughout the central and peripheral nervous systems at many sites within the neuron including at synapses, near the cell soma (e.g., at the axon initial segment), or in axons. Relevant to the pain of migraine, the peripheral neurons, where plasticity is likely to occur are meningeal afferents (innervating the dura and pia), and the central endings of these neurons at synapses in the TNC. Importantly, prior studies implicate both of these anatomical locations as potential sites of plasticity contributing to migraine.

The dura mater is almost entirely innervated by pain-sensing fibers, Aδ and C,[88] generally considered to be responsible for migraine pain.[7,8,148] As mentioned above, the pain of migraine is commonly described as unilateral, spontaneous, and throbbing in nature, and it is also well documented that changes in intracranial pressure such as head movements or coughing can exacerbate headache pain. This clinical picture fits with the preclinical observation that meningeal afferents are responsive to mechanical stimulation.[35,37,148] However, it is not clear whether baseline mechanical sensitivity contributes to the pain of migraine or whether plasticity in these fibers plays a greater role. Plasticity in meningeal afferents is supported by the increase in their mechanical sensitivity after application of noxious stimulants.[157] After sensitization, these neurons may respond to stimuli that normally do not initiate activity. They may then generate afferent input in response to normal events such as coughing, which under normal circumstances does not cause headaches. These neurons may also begin to respond to normal arterial pulsations following sensitization that contributes to the throbbing nature of migraine pain.

Headache-producing events within the meninges can lead not only to activation/sensitization of primary afferents but also to sensitization of the central terminals of these neurons in the TNC.[158] The second-order neurons in the TNC, where meningeal afferents synapse, receive input from all primary trigeminal neurons innervating tissues throughout the head (Fig. 2). The second-order neurons that receive input from the dura can also receive input from the facial skin, particularly after sensitization when the receptive fields of these neruons expand. It has been shown that after applying an inflammatory soup to the dura, there is a development of cutaneous sensitivity in the periorbital region.[158] Chemical stimulation of dural afferents is shown to expand cutaneous trigeminal and cervical receptive fields in the TNC,[159] lowering the threshold for pain sensitivity in the skin of the face and cervical regions. When recording excitability from second-order neurons in the TNC that receive input from the dural meninges, there is a development of facial input from these neurons (in response to tactile stimulation) after applying chemical mediators such as capsaicin, inflammatory soup, or mustard oil to the dura.[38,84,121,133] This demonstrates that peripheral activation of afferents in the dura leads to plasticity at second-order neurons in the TNC such that input from the facial skin is now unmasked. These mechanisms described above may explain the cutaneous cephalic allodynia observed in the majority of migraine patients during the headache phase.[148]

While the exact mechanism of how these plastic changes happen is unknown, it is clear that there is a change in the mechanical threshold of peripheral afferents and also in the receptive fields of second-order neurons in the TNC receiving input from the dura. One potential explanation for these events is that activation of dural afferents could cause the release of neurotransmitters that can lead to a maladaptive state. This may be due to the release of substance P and CGRP from afferent endings in the dura, promoting the degranulation of mast cells, and sensitizing afferents to mechanical and other stimuli. Central release of transmitters in the TNC may also lead to loss of inhibitory networks such that afferent input from distant sites (such as facial skin), which would normally not gain access to second-order neurons with dural input, begin to activate these pathways leading to facial allodynia along with headache. The candidate transmitters, leading to plasticity in migraine are not fully known, but among them CGRP has received the most attention.

10. CGRP AND MIGRAINE PLASTICITY

Numerous studies show that CGRP plays a critical role in migraine headache. CGRP is a neuropeptide produced by peripheral neurons commonly affiliated with nociceptive signals entering the spinal cord due to its release from C-fibers at this site.[32,34,36] The role of CGRP in migraine is still evolving, but many advances have been made recently. As described above, it was previously hypothesized that the cause of migraine pain was due to vasodilation of blood vessels innervating the cerebrum[39,99] and CGRP released from trigeminal sensory neurons caused this dilation.[43] While it is no longer thought that vasodilation is the cause of migraine pain, CGRP is proved to play a key role in the pathophysiology of migraine.[43] In migraine patients, levels of blood CGRP are elevated during migraine attacks,[101] and in animals, dural activation precipitates the release of CGRP.[55] It has also been shown that intravenous delivery of CGRP triggers migraine headaches in migraineurs, however, no headaches were observed in healthy individuals.[57,58] These studies have lead to a number of studies looking at the different therapies for migraine patients by blocking the effects of CGRP.[43] While it is now widely accepted that CGRP plays a role in migraine, its mechanism is still unknown. The question remains as to how and where CGRP is acting to contribute to migraine pain.

There are two main sites where CGRP can act to contribute to the mechanisms of migraine: at the level of meninges where it could be released from the peripheral terminals of nociceptors, or at the spinal trigeminal nucleus where it could be released from the central terminals of nociceptors. Blocking CGRP receptors via intravenous delivery of an antagonist, olcegepant, led to a decrease in markers of afferent activity in the trigeminal ganglion.[64] CGRP receptor antagonists at the central site in the TNC were also able to block markers of afferent activation in second-order neurons.[64] However, dural application of the CGRP antagonist failed to be effective in preclinical models.[67] These studies add to the uncertainty of where CGRP antagonists act in migraine. CGRP can trigger the release of inflammatory agents,[67] and these agents may participate in the activation/sensitization of dural afferents. The lack of efficacy of dural CGRP antagonists is thus not easy to explain. However, recently published clinical data demonstrate efficacy of a systemically delivered antibody to CGRP in migraine.[160] Given the unlikely access of antibodies to central sites, these data support a peripheral role for CGRP in migraine.

CGRP has been shown to contribute to synaptic plasticity at both peripheral and central sites. In the periphery, CGRP released presynaptically (at the neuromuscular junction) produces an augmented response to acetylcholine in the postsynaptic terminal through protein kinase A mechanisms,[78] possibly altering the development of new synapses. CGRP release has been shown to be increased in the spinal cord eight days after applying IM at peripheral sites,[79] suggesting that upregulation of CGRP release from the central terminals of primary afferents in the spinal cord contributes to pain signaling induced by peripheral inflammation. Moreover, a CGRP antagonist was able to reverse synaptic plasticity in dorsal horn synapses in arthritic rats but had no effect on synaptic responses in normal rats.[81] This same study showed a greater enhancement of postsynaptic response by CGRP in neurons from arthritic rats compared to control rats. These effects were found to be post- rather than presynaptic. Similarly, CGRP antagonists acting at postsynaptic sites have shown to be effective in inhibiting pain plasticity in the amygdala.[83] These studies establish a role for CGRP in plasticity in peripheral and central synapses, but a role for CGRP-induced plasticity in migraine has not yet been established with these types of studies.

11. BDNF AND MIGRAINE

Long-term potentiation (LTP) is an example of neuroplasticity, where the persistent activation of a synapse leads to long-lasting changes in transmission between two neurons. While LTP has long been implicated in learning and the formation of memories at central synapses in the cortex and hippocampus, LTP can play an important role in the dorsal horn of the spinal cord leading to pain plasticity.[91] LTP is separated into two phases: an early and late or maintenance phase. While early-phase LTP is known to require the contribution of protein kinases to regulate AMPA receptors,[96] the mechanisms involved in the maintenance of LTP have different requirements. The maintenance phase of LTP is thought to involve a member of the neurotrophin family of growth factors: BDNF.[115]

BDNF is widely accepted to play a role in the development of LTP in the hippocampus[100] and in LTP-like processes in the spinal dorsal horn.[101] This links BDNF to pain plasticity through memory-like mechanisms. BDNF is released from the central terminals of nociceptive afferent neurons to act on its postsynaptic receptor, tyrosine receptor kinase B (TrkB) in the spinal cord.[103,105] Thus, the activity of BDNF on the spinal cord may be crucial for the development of pain plasticity. This hypothesis is supported by several findings. Application of BDNF to spinal cord slices potentiated synaptic transmission in an NMDA-dependent manner.[161] BDNF is also able to increase functional connectivity of sensory afferents with motor neurons, indicating a role for BDNF in plasticity in the circuitry mediating pain reflexes.[162] When BDNF is genetically deleted from nociceptors, a variety of acute and prolonged pain behaviors are significantly attenuated,[163] implicating a role for the release of BDNF in pain signaling. Further, pain plasticity fails to develop and is also reversed in the presence of an injection of a BDNF scavenger, TrkB/fc, into the spinal cord.[107]

While the majority of BDNF-related pain plasticity is focused on the dorsal root ganglion and the spinal cord, this concept may be similar in the trigeminal system and the TNC for the pain of migraine. In fact, similar to CGRP, BDNF levels are increased during migraine attacks in episodic migraineurs.[109] Additionally, CGRP facilitates the release of BDNF from trigeminal neurons, [111] suggesting an interaction between the two peptides where release of CGRP during migraine can enhance the release of BDNF. These studies together suggest that BDNF is a contributor to the plastic changes involved in the transition to chronic migraine.

12. CONCLUSION

Migraine is the most common neurological disorder and is one of the most disabling disorders worldwide. Surprisingly, the pathophysiological mechanisms contributing to this prevalent disorder remain poorly understood. One primary result of the poor understanding of the condition is slow development of novel therapeutics. The gold standard in the treatment of migraine, triptans, was developed on an outdated hypothesis for the mechanism contributing to migraine, vasodilation. Although they have efficacy in some patients, the percentage of patients that achieve complete relief is quite low. In those that do achieve relief with triptans, concerns over MOH with repetitive use can limit the utility of these drugs. Prophylactic agents can have efficacy in the reduction of migraine frequency in some patients, but side effects are often intolerable. As a result, patients must often choose between frequent migraines and extreme side effects of medications. The lack of adequate treatment of migraine causes a significant negative impact on the quality of life for patients.

Better understanding of the pathophysiology of migraine will undoubtedly lead to more efficacious therapeutics. What was once considered a vascular disorder is now considered a pathology of the nervous system involving the brain and peripheral nociceptive system. While there is much that is still unknown, progress is being made toward uncovering mechanisms that may contribute to migraine, particularly in relation to the meningeal afferent system. It has been known for decades that stimulation of the dura causes pain in humans, that meningeal afferents mediate this pain, and that these neurons are mechanically sensitive. More recent work has expanded on the understanding of these neurons now implicating a number of processes in their activation. These processes include mast cell degranulation and activation of several families of ion channels such as ASICs and TRPs. With the discovery of these mechanisms comes potential for development of novel therapeutics based on these targets.

There is also a likely role for neuroplasticity in migraine that needs further exploration. Plasticity has been demonstrated, for example, in meningeal afferents with enhanced mechanosensitivity and in the TNC with the expansion of receptive fields in second-order neurons. But the molecules and mechanisms that mediate plasticity are largely unexplored in this context. CGRP has long been implicated in migraine and it can potentially contribute to plasticity in the meninges, via proinflammatory mechanisms, and in the TNC where it acts as a transmitter across the first synapse. It is now clear from late-phase clinical trials that blocking CGRP receptors has efficacy for migraine but it is still not clear how and where CGRP contributes to migraine. Understanding even the location of action of these drugs could help provide insights into critical regions of the nervous system necessary for migraine but this awaits the results of future studies. Additional mechanisms of plasticity in the TNC could involve LTP-like mechanisms, as these have been shown to contribute to pain plasticity in the spinal cord. Mechanisms have yet to be determined but may involve BDNF, a plasticity-related neuropeptide that is elevated in migraine patients. As with afferents, discovery of mechanisms contributing to plasticity in the TNC may lead to novel therapeutics.

Ultimately, more research is clearly necessary to further uncover the pathophysiology of migraine. Given that migraine is the most common neurological disorder, one that has a significant negative impact on the lives of patients, and one that is poorly treated by current therapeutics, it is difficult to provide a more compelling rationale for the need for research to better understand this complex disorder.

ACKNOWLEDGMENTS

This work was supported by funding from the National Institutes of Health (NS072204) and the Migraine Research Foundation.

REFERENCES

1. Vos T, Flaxman AD, Naghavi M, et al. Years lived with disability (YLDs) for 1160 sequelae of 289 diseases and injuries 1990–2010: a systematic analysis for the Global Burden of Disease Study 2010. *Lancet*. 2012;380(9859):2163–2196.
2. Diener H-C, Dodick DW, Goadsby PJ, Lipton RB, Olesen J, Silberstein SD. Chronic migraine—classification, characteristics and treatment. *Nat Rev Neurol*. 2011;8(3): 162–171.
3. Lipton RB. Chronic migraine, classification, differential diagnosis, and epidemiology. *Headache*. 2011;51(suppl 2):77–83.
4. Giffin NJ, Ruggiero L, Lipton RB, et al. Premonitory symptoms in migraine: an electronic diary study. *Neurology*. 2003;60(6):935–940.
5. Becker WJ. The premonitory phase of migraine and migraine management. *Cephalalgia*. 2013;33(13):1117–1121.
6. Blau JN, MacGregor EA. Migraine and the neck. *Headache*. 1994;34(2):88–90.
7. Goadsby PJ. Pathophysiology of migraine. *Ann Indian Acad Neurol*. 2012;15(suppl 1): S15–S22.
8. Levy D. Migraine pain and nociceptor activation—where do we stand? *Headache*. 2010;50(5):909–916.
9. Goadsby PJ, Charbit AR, Andreou AP, Akerman S. Neurobiology of migraine. *Neuroscience*. 2009;161(2):327–341.
10. Schoonman GG, Evers DJ, Ballieux BE, et al. Is stress a trigger factor for migraine? *Psychoneuroendocrinology*. 2007;32(5):532–538.
11. Andress-Rothrock D, King W, Rothrock J. An analysis of migraine triggers in a clinic-based population. *Headache*. 2009;50(8):1366–1370.
12. Kelman L. The triggers or precipitants of the acute migraine attack. *Cephalalgia*. 2007;27(5):394–402.
13. Lipton RB, Buse DC, Hall CB, et al. Reduction in perceived stress as a migraine trigger: testing the "let-down headache" hypothesis. *Neurology*. 2014;82(16):1395–1401.
14. Goadsby PJ. Stress and migraine: something expected, something unexpected. *Neurology*. 2014;82(16):1388–1389.
15. Martin PR, Reece J, Callan M, et al. Behavioral management of the triggers of recurrent headache: a randomized controlled trial. *Behav Res Ther*. 2014;61:1–11.
16. Olesen J. The role of nitric oxide (NO) in migraine, tension-type headache and cluster headache. *Pharmacol Ther*. 2008;120(2):157–171.
17. Silberstein SD, Stirpe JC. COX inhibitors for the treatment of migraine. *Expert Opin Pharmacother*. 2014;15(13):1863–1874.
18. Edvinsson L, Linde M. New drugs in migraine treatment and prophylaxis: telcagepant and topiramate. *Lancet*. 2010;376(9741):645–655.
19. Stovner LJ, Tronvik E, Hagen K. New drugs for migraine. *J Headache Pain*. 2009;10(6):395–406.
20. Monteith TS, Goadsby PJ. Acute migraine therapy: new drugs and new approaches. *Curr Treat Options Neurol*. 2010;13(1):1–14.
21. Tfelt-Hansen P, Olesen J. Taking the negative view of current migraine treatments. *CNS Drugs*. 2012;26(5):375–382.
22. Dodick D, Lipton RB, Martin V, et al. Consensus statement: cardiovascular safety profile of triptans (5-HT agonists) in the acute treatment of migraine. *Headache*. 2004;44(5):414–425.
23. Evers S, Marziniak M. Clinical features, pathophysiology, and treatment of medication-overuse headache. *Lancet Neurol*. 2010;9(4):391–401.
24. Hazard E, Munakata J, Bigal ME, Rupnow MFT, Lipton RB. The burden of migraine in the United States: current and emerging perspectives on disease management and economic analysis. *Value Health*. 2009;12(1):55–64.

25. Kristoffersen ES, Lundqvist C. Medication-overuse headache: epidemiology, diagnosis and treatment. *Ther Adv Drug Saf.* 2014;5(2):87–99.
26. Kristoffersen ES, Lundqvist C. Medication-overuse headache: a review. *J Pain Res.* 2014;7:367–378.
27. Chambers JB, Seed PT, Ridsdale L. Clopidogrel as prophylactic treatment for migraine: a pilot randomised, controlled study. *Cephalalgia.* 2014;34(14):1163–1168.
28. Walker AE. Anatomy, physiology and surgical con-siderations of the spinal tract of the trigeminal nerve. *J Neurophysiol.* 1939;2:234–248.
29. Gudmundsson K, Rhoton Jr AL, Rushton JG. Detailed anatomy of the intracranial portion of the trigeminal nerve. *J Neurosurg.* 1971;35(5):592–600. http://dxdoiorg/103171/jns19713550592.
30. Woolfall P, Coulthard A. Trigeminal nerve: anatomy and pathology. *Br J Radiol.* 2001;74(881):458–467.
31. Markman S. Referred pain. *J N J Dent Assoc.* 2014;85(2):26–29.
32. Powell KJ, Ma W, Sutak M, Doods H, Quirion R, Jhamandas K. Blockade and reversal of spinal morphine tolerance by peptide and non-peptide calcitonin gene-related peptide receptor antagonists. *Br J Pharmacol.* 2000;131(5):875–884.
33. Ray BS, Wolff HG. Experimental studies on headache: pain-sensitive structures of the head and their significance in headache. *Arch Surg.* 1940;41(4):813–856.
34. van Rossum D, Hanisch UK, Quirion R. Neuroanatomical localization, pharmacological characterization and functions of CGRP, related peptides and their receptors. *Neurosci Biobehav Rev.* 1997;21(5):649–678.
35. Levy D, Strassman AM. Mechanical response properties of A and C primary afferent neurons innervating the rat intracranial dura. *J Neurophysiol.* 2002;88(6):3021–3031.
36. Chen L-J, Zhang F-G, Li J, et al. Expression of calcitonin gene-related peptide in anterior and posterior horns of the spinal cord after brachial plexus injury. *J Clin Neurosci.* 2010;17(1):87–91.
37. Kaube H, Hoskin KL, Goadsby PJ. Activation of the trigeminovascular system by mechanical distension of the superior sagittal sinus in the cat. *Cephalalgia.* 1992;12(3): 133–136.
38. Strassman AM, Raymond SA, Burstein R. Sensitization of meningeal sensory neurons and the origin of headaches. *Nature.* 1996;384(6609):560–564.
39. Jansen-Olesen I, Mortensen A, Edvinsson L. Calcitonin gene-related peptide is released from capsaicin-sensitive nerve fibres and induces vasodilatation of human cerebral arteries concomitant with activation of adenylyl cyclase. *Cephalalgia.* 1996;16(5): 310–316.
40. Goadsby PJ. The vascular theory of migraine—a great story wrecked by the facts. *Brain.* 2009;132(pt 1):6–7.
41. Panconesi A, Bartolozzi ML, Guidi L. Migraine pain: reflections against vasodilatation. *J Headache Pain.* 2009;10(5):317–325.
42. Shevel E. The extracranial vascular theory of migraine—a great story confirmed by the facts. *Headache.* 2011;51(3):409–417.
43. Villalón CM, Olesen J. The role of CGRP in the pathophysiology of migraine and efficacy of CGRP receptor antagonists as acute antimigraine drugs. *Pharmacol Ther.* 2009;124(3):309–323.
44. Amin FM, Asghar MS, Hougaard A, et al. ArticlesMagnetic resonance angiography of intracranial andextracranial arteries in patients with spontaneous migrainewithout aura: a cross-sectional study. *Lancet Neurol.* 2013;7:1–8.
45. Cohen AS, Goadsby PJ. Functional neuroimaging of primary headache disorders. *Curr Neurol Neurosci Rep.* 2004;4(2):105–110.
46. Cohen AS, Goadsby PJ. Functional neuroimaging of primary headache disorders. *Expert Rev Neurother.* 2006;6(8):1159–1171.

47. Vanmolkot FH, de Hoon JN. Endothelial function in migraine: a cross-sectional study. *BMC Neurol.* 2010;10:119.
48. Sprenger T, Goadsby PJ. What has functional neuroimaging done for primary headache . . . and for the clinical neurologist? *J Clin Neurosci.* 2010;17(5):547–553.
49. Marziniak M, Herzog AL, Herzog A, Mössner R, Sommer C. Investigation of the functional brain-derived neurotrophic factor gene variant Val66MET in migraine. *J Neural Transm.* 2008;115(9):1321–1325.
50. Alstadhaug KB. Migraine and the hypothalamus. *Cephalalgia.* 2009;29(8):809–817.
51. Maniyar FH, Sprenger T, Monteith T, Schankin C, Goadsby PJ. Brain activations in the premonitory phase of nitroglycerin-triggered migraine attacks. *Brain.* 2014;137(1): 232–241.
52. Holland P, Goadsby PJ. The hypothalamic orexinergic system: pain and primary headaches. *Headache.* 2007;47(6):951–962.
53. Noseda R, Burstein R. Migraine pathophysiology: anatomy of the trigeminovascular pathway and associated neurological symptoms, CSD, sensitization and modulation of pain. *Pain.* 2013;154(suppl 1):1–10.
54. Bolay H, Moskowitz MA. The neurobiology of migraine and transformation of headache therapy. In: Waxman S, ed. *From neuroscience to neurology: neuroscience, molecular medicine, and the therapeutic transformation of neurology.* San Diego: Elsevier; 2004:107–123.
55. Zagami AS, Goadsby PJ, Edvinsson L. Stimulation of the superior sagittal sinus in the cat causes release of vasoactive peptides. *Neuropeptides.* 1990;16(2):69–75.
56. Burstein R, Cutrer MF, Yarnitsky D. The development of cutaneous allodynia during a migraine attack clinical evidence for the sequential recruitment of spinal and supraspinal nociceptive neurons in migraine. *Brain.* 2000;123(pt 8):1703–1709.
57. Lassen LH, Haderslev PA, Jacobsen VB, Iversen HK, Sperling B, Olesen J. CGRP may play a causative role in migraine. *Cephalalgia.* 2002;22(1):54–61.
58. Olesen J, Iversen HK, Thomsen LL. Nitric oxide supersensitivity: a possible molecular mechanism of migraine pain. *Neuroreport.* 1993;4(8):1027–1030.
59. Gagan JR, Tholpady SS, Ogle RC. Cellular dynamics and tissue interactions of the dura mater during head development. *Birth Defects Res C Embryo Today.* 2007; 81(4):297–304.
60. Zhao J, Levy D. The sensory innervation of the calvarial periosteum is nociceptive and contributes to headache-like behavior. *Pain.* 2014;155(7):1392–1400.
61. Kosaras B, Jakubowski M, Kainz V, Burstein R. Sensory innervation of the calvarial bones of the mouse. *J Comp Neurol.* 2009;515(3):331–348.
62. Schueler M, Neuhuber WL, De Col R. Innervation of rat and human Dura mater and pericranial tissues in the parieto-temporal region by meningeal afferents. *Headache.* 2014;54(6):996–1009.
63. Wei X, Melemedjian OK, Ahn DD-U, Weinstein N, Dussor G. Dural fibroblasts play a potential role in headache pathophysiology. *Pain.* 2014;155(7):1238–1244. Available from: http://eutils.ncbi.nlm.nih.gov/entrez/eutils/elink.fcgi?dbfrom=pubmed&id=24657451&retmode=ref&cmd=prlinks.
64. Sixt M-L, Messlinger K, Fischer MJM. Calcitonin gene-related peptide receptor antagonist olcegepant acts in the spinal trigeminal nucleus. *Brain.* 2009;132(pt 11): 3134–3141.
65. Rozniecki JJ, Dimitriadou V, Lambracht-Hall M, Pang X, Theoharides TC. Morphological and functional demonstration of rat dura mater mast cell-neuron interactions in vitro and in vivo. *Brain Res.* 1999;849(1–2):1–15.
66. Zhang X-C, Levy D. Modulation of meningeal nociceptors mechanosensitivity by peripheral proteinase-activated receptor-2: the role of mast cells. *Cephalalgia.* 2008; 28(3):276–284.

67. Messlinger K, Lennerz JK, Eberhardt M, Fischer MJM. CGRP and NO in the trigeminal system: mechanisms and role in headache generation. *Headache*. 2012; 52(9):1411–1427.
68. Zhang X-C, Strassman AM, Burstein R, Levy D. Sensitization and activation of intracranial meningeal nociceptors by mast cell mediators. *J Pharmacol Exp Ther*. 2007;322(2):806–812.
69. Baun M, Pedersen MHF, Olesen J, Jansen-Olesen I. Dural mast cell degranulation is a putative mechanism for headache induced by PACAP-38. *Cephalalgia*. 2012; 32(4):337–345.
70. Smith JH, Butterfield JH, Cutrer FM. Primary headache syndromes in systemic mastocytosis. *Cephalalgia*. 2011;31(15):1522–1531.
71. Levy D. Endogenous mechanisms underlying the activation and sensitization of meningeal nociceptors: the role of immuno-vascular interactions and cortical spreading depression. *Curr Pain Headache Rep*. 2012;16(3):270–277.
72. Levy D, Burstein R, Strassman AM. Calcitonin gene-related peptide does not excite or sensitize meningeal nociceptors: implications for the pathophysiology of migraine. *Ann Neurol*. 2005;58(5):698–705.
73. Edelmayer RM, Ossipov MH, Porreca F. An experimental model of headache-related pain. *Methods Mol Biol*. 2012;851:109–120.
74. Bandell M, Story GM, Hwang SW, et al. Noxious cold ion channel TRPA1 is activated by pungent compounds and bradykinin. *Neuron*. 2004;41(6):849–857.
75. Geppetti P, Capone JG, Trevisani M, Nicoletti P, Zagli G, Tola MR. CGRP and migraine: neurogenic inflammation revisited. *J Headache Pain*. 2005;6(2):61–70.
76. Deval E, Gasull X, Noël J, Salinas M, Baron A. Acid-sensing ion channels (ASICs): pharmacology and implication in pain. *Pharmacol Ther*. 2010;128(3):549–558.
77. Wemmie JA, Taugher RJ, Kreple CJ. Acid-sensing ion channels in pain and disease. *Nat Rev Neurosci*. 2013;14(7):461–471.
78. Lu B, Fu WM, Greengard P, Poo MM. Calcitonin gene-related peptide potentiates synaptic responses at developing neuromuscular junction. *Nature*. 1993;363(6424): 76–79.
79. Seybold VS, Galeazza MT, Garry MG, Hargreaves KM. Plasticity of calcitonin gene related peptide neurotransmission in the spinal cord during peripheral inflammation. *Can J Physiol Pharmacol*. 1995;73(7):1007–1014.
80. Eikermann-Haerter K, Ayata C. Cortical spreading depression and migraine. *Curr Neurol Neurosci Rep*. 2010;10(3):167–173.
81. Bird GC, Han JS, Fu Y, Adwanikar H, Willis WD, Neugebauer V. Pain-related synaptic plasticity in spinal dorsal horn neurons: role of CGRP. *Mol Pain*. 2006;2(1):31.
82. Otori T, Greenberg JH, Welsh FA. Cortical spreading depression causes a long-lasting decrease in cerebral blood flow and induces tolerance to permanent focal ischemia in rat brain. *J Cereb Blood Flow Metab*. 2003;23(1):43–50.
83. Han JS, Li W, Neugebauer V. Critical role of calcitonin gene-related peptide 1 receptors in the amygdala in synaptic plasticity and pain behavior. *J Neurosci*. 2005; 25(46):10717–10728.
84. Bove GM, Moskowitz MA. Primary afferent neurons innervating guinea pig dura. *J Neurophysiol*. 1997;77(1):299–308.
85. Schepelmann K, Ebersberger A, Pawlak M, Oppmann M, Messlinger K. Response properties of trigeminal brain stem neurons with input from dura mater encephali in the rat. *Neuroscience*. 1999;90(2):543–554.
86. Holland PR, Akerman S, Andreou AP, Karsan N, Wemmie JA, Goadsby PJ. Acid-sensing ion channel 1: a novel therapeutic target for migraine with aura. *Ann Neurol*. 2012;72(4):559–563.
87. Durham PL, Masterson CG. Two mechanisms involved in trigeminal CGRP release: implications for migraine treatment. *Headache*. 2013;53(1):67–80.

88. Zimmermann K, Reeh PW, Averbeck B. ATP can enhance the proton-induced CGRP release through P2Y receptors and secondary PGE(2) release in isolated rat dura mater. *Pain*. 2002;97(3):259–265.
89. Yan J, Edelmayer RM, Wei X, De Felice M, Porreca F, Dussor G. Dural afferents express acid-sensing ion channels: a role for decreased meningeal pH in migraine headache. *Pain*. 2011;152(1):106–113. Available from: http://eutils.ncbi.nlm.nih.gov/entrez/eutils/elink.fcgi?dbfrom=pubmed&id=20971560&retmode=ref&cmd=prlinks.
90. De Young MB, Nemeth EF, Scarpa A. Measurement of the internal pH of mast cell granules using microvolumetric fluorescence and isotopic techniques. *Arch Biochem Biophys*. 1987;254(1):222–233.
91. Sandkühler J. Understanding LTP, in pain pathways. *Mol Pain*. 2007;3(1):9.
92. Voets T. TRP channels and thermosensation. In: Nilius B, Flockerzi V, eds. *Handbook of Experimental Pharmacology*. Cham: Springer International Publishing; 2014:729–741. Available from: http://eutils.ncbi.nlm.nih.gov/entrez/eutils/elink.fcgi?dbfrom=pubmed&id=24961967&retmode= ref&cmd=prlinks.
93. Premkumar LS, Abooj M. TRP channels and analgesia. *Life Sci*. 2013;92(8–9):415–424. Available from: http://eutils.ncbi.nlm.nih.gov/entrez/eutils/elink.fcgi?dbfrom=pubmed&id=22910182&retmode=ref&cmd=prlinks.
94. Sousa-Valente J, Andreou AP, Urban L, Nagy I. Transient receptor potential ion channels in primary sensory neurons as targets for novel analgesics. *Br J Pharmacol*. 2014;171(10):2508–2527. Available from: http://eutils.ncbi.nlm.nih.gov/entrez/eutils/elink.fcgi?dbfrom=pubmed&id=24283624&retmode=ref&cmd=prlinks.
95. Nilius B, Szallasi A. Transient receptor potential channels as drug targets: from the science of basic research to the art of medicine. *Pharmacol Rev*. 2014;66(3):676–814.
96. Huganir RL, Nicoll RA. AMPARs and synaptic plasticity: the last 25 years. *Neuron*. 2013;80(3):704–717.
97. Vriens J, Appendino G, Nilius B. Pharmacology of vanilloid transient receptor potential cation channels. *Mol Pharmacol*. 2009;75(6):1262–1279.
98. Tominaga M. Nociception and TRP channels. *Transient Receptor Potential (TRP) Channels*. 2007;179:489–505.
99. Ramsey IS, Delling M, Clapham DE. An introduction to TRP channels. *Annu Rev Physiol*. 2006;68:619–647.
100. Lu Y, Christian K, Lu B. BDNF: a key regulator for protein synthesis-dependent LTP and long-term memory? *Neurobiol Learn Mem*. 2008;89(3):312–323.
101. Zhou L-J, Zhong Y, Ren W-J, Li Y-Y, Zhang T, Liu X-G. BDNF induces late-phase LTP of C-fiber evoked field potentials in rat spinal dorsal horn. *Exp Neurol*. 2008; 212(2):507–514.
102. Cortright DN, Szallasi A. TRP channels and pain. *Curr Pharm Des*. 2009; 15(15):1736–1749.
103. Balkowiec A, Katz DM. Activity-dependent release of endogenous brain-derived neurotrophic factor from primary sensory neurons detected by ELISA in situ. *J Neurosci*. 2000;20(19):7417–7423.
104. Bautista DM, Jordt S-E, Nikai T, et al. TRPA1 mediates the inflammatory actions of environmental irritants and proalgesic agents. *Cell*. 2006;124(6):1269–1282.
105. Matayoshi S, Jiang N, Katafuchi T, et al. Actions of brain-derived neurotrophic factor on spinal nociceptive transmission during inflammation in the rat. *J Physiol Lond*. 2005;569(pt 2):685–695.
106. Bessac BF, Jordt S-E. Breathtaking TRP channels: TRPA1 and TRPV1 in airway chemosensation and reflex control. *Physiology (Bethesda)*. 2008;23:360–370.
107. Melemedjian OK, Tillu DV, Asiedu MN, et al. BDNF regulates atypical PKC at spinal synapses to initiate and maintain a centralized chronic pain state. *Mol Pain*. 2013;9:12.
108. McNamara CR, Mandel-Brehm J. TRPA1 mediates formalin-induced pain. *Proc Natl Acad Sci U S A*. 2007;104(33):13525–13530.

109. Tanure MTA, Gomez RS, Hurtado RCL, Teixeira AL, Domingues RB. Increased serum levels of brain-derived neurotropic factor during migraine attacks: a pilot study. *J Headache Pain*. 2010;11(5):427–430.
110. Andrè E, Campi B, Materazzi S, et al. Cigarette smoke-induced neurogenic inflammation is mediated by alpha, beta-unsaturated aldehydes and the TRPA1 receptor in rodents. *J Clin Invest*. 2008;118(7):2574–2582.
111. Buldyrev I, Tanner NM, Hsieh H-Y, Dodd EG, Nguyen LT, Balkowiec A. Calcitonin gene-related peptide enhances release of native brain-derived neurotrophic factor from trigeminal ganglion neurons. *J Neurochem*. 2006;99(5):1338–1350.
112. Wang L, Cvetkov TL, Chance MR, Moiseenkova-Bell VY. Identification of in vivo disulfide conformation of TRPA1 ion channel. *J Biol Chem*. 2012;287(9):6169–6176.
113. Bautista DM, Movahed P, Hinman A, et al. Pungent products from garlic activate the sensory ion channel TRPA1. *Proc Natl Acad Sci U S A*. 2005;102(34):12248–12252.
114. Trevisani M, Siemens J, Materazzi S, et al. 4-Hydroxynonenal, an endogenous aldehyde, causes pain and neurogenic inflammation through activation of the irritant receptor TRPA1. *Proc Natl Acad Sci U S A*. 2007;104(33):13519–13524.
115. Taylor-Clark TE, Ghatta S, Bettner W, Undem BJ. Nitrooleic acid, an endogenous product of nitrative stress, activates nociceptive sensory nerves via the direct activation of TRPA1. *Mol Pharmacol*. 2009;75(4):820–829.
116. Materazzi S, Nassini R, Andrè E, et al. Cox-dependent fatty acid metabolites cause pain through activation of the irritant receptor TRPA1. *Proc Natl Acad Sci U S A*. 2008;105(33):12045–12050.
117. Wantke F, Focke M, Hemmer W, et al. Exposure to formaldehyde and phenol during an anatomy dissecting course: sensitizing potency of formaldehyde in medical students. *Allergy*. 2000;55(1):84–87.
118. Farber RE. Nasally triggered headache. *Neurology*. 2002;59(1):151 (author reply 151).
119. Nassini R, Materazzi S, Vriens J, et al. The "headache tree" via umbellulone and TRPA1 activates the trigeminovascular system. *Brain*. 2012;135(pt 2):376–390.
120. Huang D, Li SY, Dhaka A, Story GM, Cao Y-QQ. Expression of the transient receptor potential channels TRPV1, TRPA1 and TRPM8 in mouse trigeminal primary afferent neurons innervating the dura. *Mol Pain*. 2012;8(1):66. Available from: http://eutils.ncbi.nlm.nih.gov/entrez/eutils/elink.fcgi?dbfrom=pubmed&id=22971321&retmode=ref&cmd=prlinks.
121. Edelmayer RM, Le LN, Yan J, et al. Activation of TRPA1 on dural afferents: a potential mechanism of headache pain. *Pain*. 2012;153(9):1949–1958. Available from: http://eutils.ncbi.nlm.nih.gov/entrez/eutils/elink.fcgi?dbfrom=pubmed&id=22809691&retmode=ref&cmd=prlinks.
122. Kunkler PE, Ballard CJ, Oxford GS, Hurley JH. TRPA1 receptors mediate environmental irritant-induced meningeal vasodilatation. *Pain*. 2011;152(1):38–44.
123. Liedtke W, Tobin DM, Bargmann CI, Friedman JM. Mammalian TRPV4 (VR-OAC) directs behavioral responses to osmotic and mechanical stimuli in Caenorhabditis elegans. *Proc Natl Acad Sci U S A*. 2003;100(suppl 2):14531–14536.
124. Vriens J, Watanabe H, Janssens A, Droogmans G, Voets T, Nilius B. Cell swelling, heat, and chemical agonists use distinct pathways for the activation of the cation channel TRPV4. *Proc Natl Acad Sci U S A*. 2004;101(1):396–401.
125. Watanabe H, Davis JB, Smart D, et al. Activation of TRPV4 channels (hVRL-2/mTRP12) by phorbol derivatives. *J Biol Chem*. 2002;277(16):13569–13577.
126. Liedtke W, Choe Y, Martí-Renom MA, et al. Vanilloid receptor-related osmotically activated channel (VR-OAC), a candidate vertebrate osmoreceptor. *Cell*. 2000;103(3):525–535.
127. Kitahara T, Li H-S, Balaban CD. Changes in transient receptor potential cation channel superfamily V (TRPV) mRNA expression in the mouse inner ear ganglia after kanamycin challenge. *Hear Res*. 2005;201(1–2):132–144.

128. Chen L, Liu C, Liu L. The modulation of voltage-gated potassium channels by anisotonicity in trigeminal ganglion neurons. *Neuroscience*. 2008;154(2):482–495.
129. Chen L, Liu C, Liu L. Changes in osmolality modulate voltage-gated calcium channels in trigeminal ganglion neurons. *Brain Res*. 2008;1208:56–66.
130. Mizuno A, Matsumoto N, Imai M, Suzuki M. Impaired osmotic sensation in mice lacking TRPV4. *Am J Physiol Cell Physiol*. 2003;285(1):C96–C101.
131. Suzuki M, Mizuno A, Kodaira K, Imai M. Impaired pressure sensation in mice lacking TRPV4. *J Biol Chem*. 2003;278(25):22664–22668.
132. Alessandri-Haber N, Joseph E, Dina OA, Liedtke W, Levine JD. TRPV4 mediates pain-related behavior induced by mild hypertonic stimuli in the presence of inflammatory mediator. *Pain*. 2005;118(1–2):70–79.
133. Wei X, Edelmayer RM, Yan J, Dussor G. Activation of TRPV4 on dural afferents produces headache-related behavior in a preclinical rat model. *Cephalalgia*. 2011;31(16):1595–1600.
134. McKemy DD, Neuhausser WM, Julius D. Identification of a cold receptor reveals a general role for TRP channels in thermosensation. *Nature*. 2002;416(6876): 52–58. Available from: http://eutils.ncbi.nlm.nih.gov/entrez/eutils/elink.fcgi?dbfrom=pubmed&id=11882888&retmode=ref&cmd=prlinks.
135. Knowlton WM, Palkar R, Lippoldt EK, et al. A sensory-labeled line for cold: TRPM8-expressing sensory neurons define the cellular basis for cold, cold pain, and cooling-mediated analgesia. *J Neurosci*. 2013;33(7):2837–2848. Available from: http://www.jneurosci.org/cgi/doi/10.1523/JNEUROSCI.1943-12.2013.
136. Peier AM, Moqrich A, Hergarden AC, et al. A TRP channel that senses cold stimuli and menthol. *Cell*. 2002;108(5):705–715. Available from: http://eutils.ncbi.nlm.nih.gov/entrez/eutils/elink.fcgi?dbfrom=pubmed&id=11893340&retmode=ref&cmd=prlinks.
137. McKemy DD. How cold is it? TRPM8 and TRPA1 in the molecular logic of cold sensation. *Mol Pain*. 2005;1(1):16. Available from: http://www.molecularpain.com/content/1/1/16.
138. Dhaka A, Earley TJ, Watson J, Patapoutian A. Visualizing cold spots: TRPM8-expressing sensory neurons and their projections. *J Neurosci*. 2008;28(3):566–575. Available from: http://eutils.ncbi.nlm.nih.gov/entrez/eutils/elink.fcgi?dbfrom=pubmed&id=18199758&retmode=ref&cmd=prlinks.
139. Chung M-K, Caterina MJ. TRP channel knockout mice lose their cool. *Neuron*. 2007;54(3):345–347. Available from: http://eutils.ncbi.nlm.nih.gov/entrez/eutils/elink.fcgi?dbfrom=pubmed&id=17481385&retmode=ref&cmd=prlinks.
140. Mukerji G, Yiangou Y, Corcoran S, et al. Cool and menthol receptor TRPM8 in human urinary bladder disorders and clinical correlations. *BMC Urol*. 2006;6(1):6. Available from: http://eutils.ncbi.nlm.nih.gov/entrez/eutils/elink.fcgi?dbfrom=pubmed&id=16519806&retmode=ref&cmd=prlinks.
141. Harrington AM, Hughes PA, Martin CM, et al. A novel role for TRPM8 in visceral afferent function. *Pain*. 2011;152(7):1459–1468. Available from: http://eutils.ncbi.nlm.nih.gov/entrez/eutils/elink.fcgi?dbfrom=pubmed&id=21489690&retmode=ref&cmd=prlinks.
142. Nealen ML. TRPM8 mRNA is expressed in a subset of cold-responsive trigeminal neurons from rat. *J Neurophysiol*. 2003;90(1):515–520. Available from: http://eutils.ncbi.nlm.nih.gov/entrez/eutils/elink.fcgi?dbfrom=pubmed&id=12634279&retmode=ref&cmd=prlinks.
143. Lippoldt EK, Elmes RR, McCoy DD, Knowlton WM, McKemy DD. Artemin, a glial cell line-derived neurotrophic factor family member, induces TRPM8-dependent cold pain. *J Neurosci*. 2013;33(30):12543–12552. Available from: http://eutils.ncbi.nlm.nih.gov/entrez/eutils/elink.fcgi?dbfrom=pubmed&id=23884957&retmode=ref&cmd prlinks.

144. Fan X, Wang J, Fan W, et al. Replication of migraine GWAS susceptibility loci in Chinese Han population. *Headache*. 2014;54(4):709–715.
145. Ghosh J, Pradhan S, Mittal B. Genome-wide-associated variants in migraine susceptibility: a replication study from North India. *Headache*. 2013;53(10):1583–1594.
146. An X-K, Ma Q-L, Lin Q, Zhang X-R, Lu C-X, Qu H-L. PRDM16 rs2651899 variant is a risk factor for Chinese common migraine patients. *Headache*. 2013; 53(10):1595–1601. Available from: http://eutils.ncbi.nlm.nih.gov/entrez/eutils/elink.fcgi?dbfrom=pubmed&id=24021092&retmode=ref&cmd=prlinks.
147. Strassman AM, Levy D. Response properties of dural nociceptors in relation to headache. *J Neurophysiol*. 2006;95(3):1298–1306.
148. Bernstein C, Burstein R. Sensitization of the trigeminovascular pathway: perspective and implications to migraine pathophysiology. *J Clin Neurol*. 2012;8(2):89–99.
149. Levy D. Migraine pain, meningeal inflammation, and mast cells. *Curr Pain Headache Rep*. 2009;13(3):237–240.
150. Goadsby PJ, Edvinsson L. The trigeminovascular system and migraine: studies characterizing cerebrovascular and neuropeptide changes seen in humans and cats. *Ann Neurol* 1993;33(1):48–56.
151. Oshinsky ML, Gomonchareonsiri S. Episodic dural stimulation in awake rats: a model for recurrent headache. *Headache*. 2007;47(7):1026–1036. Available from: http://eutils.ncbi.nlm.nih.gov/entrez/eutils/elink.fcgi?dbfrom=pubmed&id=17635594&retmode=ref&cmd=prlinks.
152. Dodick D, Silberstein S. Central sensitization theory of migraine: clinical implications. *Headache*. 2006;46(suppl 4):S182–S191.
153. Uzar E, Evliyaoglu O, Yucel Y, et al. Serum cytokine and pro-brain natriuretic peptide (BNP) levels in patients with migraine. *Eur Rev Med Pharmacol Sci*. 2011; 15(10):1111–1116.
154. Wang F, He Q, Ren Z, et al. Association of serum levels of intercellular adhesion molecule-1 and interleukin-6 with migraine. *Neurol Sci*. 2014;23:1–6.
155. Katsarava Z, Manack A, Yoon M-S, et al. Chronic migraine: classification and comparisons. *Cephalalgia*. 2011;31(5):520–529.
156. IHS HCCOTIHS. The International Classification of Headache Disorders, 3rd edition (beta version). *Cephalalgia*. 2013;33(9):629–808.
157. Bolton S, O'Shaughnessy CT, Goadsby PJ. Properties of neurons in the trigeminal nucleus caudalis responding to noxious dural and facial stimulation. *Brain Res*. 2005;1046(1–2):122–129.
158. Burstein R, Yamamura H, Malick A, Strassman AM. Chemical stimulation of the intracranial dura induces enhanced responses to facial stimulation in brain stem trigeminal neurons. *J Neurophysiol*. 1998;79(2):964–982.
159. Bartsch T, Goadsby PJ. Increased responses in trigeminocervical nociceptive neurons to cervical input after stimulation of the dura mater. *Brain*. 2003;126(pt 8):1801–1813.
160. Dodick DW, Goadsby PJ, Spierings ELH, Scherer JC, Sweeney SP, Grayzel DS. Safety and efficacy of LY2951742, a monoclonal antibody to calcitonin gene-related peptide, for the prevention of migraine: a phase 2, randomised, double-blind, placebo-controlled study. *Lancet Neurol*. 2014;13(9):885–892.
161. Garraway SM, Petruska JC, Mendell LM. BDNF sensitizes the response of lamina II neurons to high threshold primary afferent inputs. *Eur J Neurosci*. 2003; 18(9):2467–2476.
162. Arvanian VL, Mendell LM. Acute modulation of synaptic transmission to motoneurons by BDNF in the neonatal rat spinal cord. *Eur J Neurosci*. 2001;14(11):1800–1808.
163. Zhao J, Seereeram A, Nassar MA, et al. Nociceptor-derived brain-derived neurotrophic factor regulates acute and inflammatory but not neuropathic pain. *Mol Cell Neurosci*. 2006;31(3):539–548.

CHAPTER NINETEEN

Chronic Pain Syndromes, Mechanisms, and Current Treatments

Justin Sirianni, Mohab Ibrahim, Amol Patwardhan[1]
Department of Anesthesiology, University of Arizona, Tucson, Arizona, USA
[1]Corresponding author: e-mail address: apatwardhan@anesth.arizona.edu

Contents

1. Cancer Pain 566
2. Central Poststroke Pain 572
3. Chronic Pelvic Pain 573
4. Complex Regional Pain Syndrome 575
5. Failed Back Syndrome 576
6. Fibromyalgia 578
7. Irritable Bowel Syndrome 579
8. Migraine 581
9. Multiple Sclerosis 582
10. Peripheral Neuropathic Pain 583
11. Persistent Postsurgical Pain 586
12. Phantom Limb Pain 587
13. Postherpetic Neuralgia 588
14. Rheumatoid Arthritis 589
15. Trigeminal Neuralgia 589
16. Future Directions 590
References 591

Abstract

Although acute pain is a physiological response warning the human body of possible harm, chronic pain can be a pathological state associated with various diseases or a disease in itself. In the United States alone, around one-third of the population has experienced a chronic pain condition and annual cost to the society is in the range of 500–600 billion dollars.[1] It should be noted that if at all this is a very modest estimate, it surpasses the costs associated with cancer, heart disease, and diabetes combined.[1] Unfortunately, despite these humongous costs, the treatment of chronic pain is inadequate.[1] Chronic pain affects individuals in a variety of forms, and below we highlight some of the most common chronic pain conditions seen in a pain clinic. Most of these disorders are difficult to treat and typically require multimodal therapy including pharmacotherapy, behavioral modification, and targeted interventions. We have summarized the scope of each disorder, clinical features, proposed mechanisms, and current therapies for them (Table 1).

Progress in Molecular Biology and Translational Science, Volume 131
ISSN 1877-1173
http://dx.doi.org/10.1016/bs.pmbts.2015.01.004

1. CANCER PAIN

It is estimated that every year, cancer is diagnosed in more than 12 million people worldwide and as many as 75% of these patients will suffer from cancer-related pain.[2] Cancer pain can be categorized as acute or chronic pain. Acute pain is largely related to possible diagnostic methods, hemorrhage into a tumor, pathologic fractures, or oral mucositis. The type of chronic cancer pain is mostly related to the site of the tumor. There are several sites where tumors may take up residence and result in severe pain.

Bone metastases are the most common cause of chronic pain in cancer patients.[3] The etiology of cancer bone pain may be the result of direct tumor burden, pathological fracture, or damage to a nearby structure. The role of growth factors has been implicated in this process.[4,5] Most of the metastatic disease involves red marrow areas.[6] The most common bone site is the vertebral column.[7] A patient with a history of cancer who complains of skeletal pain should raise suspicion for bone metastasis. Another type of chronic cancer pain is visceral pain syndrome. It can be caused by obstruction of any hollow viscus or injury to any other visceral structure innervated by nociceptors, such as the hepatic capsule or peritoneum.[8] Visceral pain is usually seen in patients with gastrointestinal and gynecological malignancies. The anatomical location of the tumor and inflammation determines the distribution of pain; e.g., hepatic distention causes pain on the right side of the abdomen and at times may radiate to the shoulder. In retroperitoneal tumors, patients may complain of dull or boring pain that is exasperated when they lay down and improves with sitting. In such a clinical scenario, tumor invasion of the pancreas or other midline structure below the diaphragm should be suspected.[9–11] Diffuse continuous or colicky pain may represent intestinal obstruction.[12] Peritoneal seeding of certain cancers can cause diffuse pain that is related to eating.[13] Peritoneal cancers can cause pain that is positional and may result in bladder spasms.[14] Adrenal metastases may result in pain that radiates to ipsilateral upper and lower quadrants of the abdomen.[15] Obstruction of a ureter presents as pain in the flank region with radiation to the groin. It is colicky (associated with spasms along the gastrointestinal track) in nature and may be associated with nausea and vomiting.[16] Patients having their spinal cords, nerve roots, plexuses, or peripheral nerves invaded by tumor cells experience neuropathic pain. The mechanism of cancer pain is multifactorial and involves factors released by tumor cells, inflammatory response, and physical compression by the tumor, among others.[17]

Table 1 Summary of chronic pain syndromes

Chronic pain syndrome	Clinical features	Possible mechanism of pain	Treatment
Cancer pain	– Acute pain caused by diagnostic methods, tumor hemorrhage, pathologic fractures, or oral mucositis – Chronic pain related to tumor site – Most commonly caused by bone metastases, especially to vertebral column	– Inflammatory response, physical compression by tumor, factors released by tumor cells	– Opioids – Radiation therapy – Bisphosphonates – Cannabinoids – Adjuvants based upon symptoms – Surgery to reduce tumor burden
Central poststroke pain	– Severe, burning pain – Thermosensory disturbances	– Most commonly caused by lesions of thalamus and posterior limb of the internal capsule – Altered thalamocortical processing – Hyperexcitable neurons – Loss of inhibition	– Tricyclics – Anticonvulsants – Opioids – Motor cortex or deep brain stimulation as last resort
CRPS	– Characterized by pain, swelling, movement abnormalities, and vasomotor instability including temperature change – Hyperalgesia, allodynia	– Exaggerated local inflammatory reaction – Sympathetic nervous dysfunction – Cortical reorganization and neurologic damage	– Physical therapy is the cornerstone, other treatments aimed at allowing maximal participation – Mirror therapy – Anticonvulsants – Ketamine infusion – Sympathetic blockade

Continued

Table 1 Summary of chronic pain syndromes—cont'd

Chronic pain syndrome	Clinical features	Possible mechanism of pain	Treatment
Failed back syndrome	– Mainly affects lumbosacral spine – High rates of failure for back surgery – Inappropriate patient selection a major problem	– Spinal stenosis – Arachnoiditis – Painful discs – Fibrosis	– Physical therapy – NSAIDs – Opioids despite weak efficacy – Interventional procedures including epidural injections and facet procedures – Spinal cord stimulation
Fibromyalgia	– Multiple trigger points – Associated with psychiatric comorbidities such as depression/ anxiety	– Alteration in central pain processing – Peripheral nervous system dysregulation	– Aerobic exercise – SNRI, tricyclics – Gabapentinoids – Trigger point injections
Irritable bowel syndrome	– Abdominal discomfort and altered bowel habits for at least 3 months without structural or biochemical abnormalities – Can be diarrhea or constipation predominant	– Motility abnormalities – Visceral hypersensitivity and hyperalgesia – Intestinal inflammation	– Dietary and lifestyle modifications – Symptomatic treatment – Antispasmodics – Tricyclics
Migraine	– Severe, debilitating headache associated with nausea, vomiting, photophobia, and phonophobia with or without aura	– Vascular theory questionable – Cortical spreading depression – Trigeminovascular system activation – Serotonin and CGRP	– Abortive – NSAIDs, acetaminophen, triptans, ergots – Preventive – Anticonvulsants, botulinum toxin injections, coenzyme Q10, antihypertensives – Neuromodulation

Multiple sclerosis	– Central demyelinating disorder – Pain from optic neuritis, muscle spasms, or spasticity – Central neuropathic pain	– Neuroinflammation – Ectopic sodium channel activity – Altered synaptic transmission – Neuroelectric blocking factors	– Baclofen – Diazepam – Tricyclics – Opioids – Intrathecal baclofen or morphine – Botulinum toxin – Cannabinoids
Chronic pelvic pain	– Pain below the umbilicus for at least 6 months needing treatment or causing disability	– Endometriosis – Chronic pelvic inflammatory disease – IBS – Interstitial cystitis/painful bladder syndrome – Myofascial pain – Neuralgia – Pelvic floor myalgia	– Hysterectomy – Focused treatment on underlying pain generator – Anticonvulsants and tricyclics
Peripheral neuropathic pain	– Most commonly caused by chemotherapy, HIV, and diabetes – Sensory loss and paresthesias	– Altered DNA synthesis or axonal transport – Mitochondrial abnormalities – Proinflammatory markers – Increased oxidative stress	– SNRIs – NSAIDs – Anticonvulsants – Opioids – Topical creams with local anesthetics, ketamine – Spinal cord stimulators – TENS

Continued

Table 1 Summary of chronic pain syndromes—cont'd

Chronic pain syndrome	Clinical features	Possible mechanism of pain	Treatment
Persistent postsurgical pain	– Pain persisting after surgical wound has healed, usually 3–6 months beyond the surgery – Neuropathic pain – Hyperalgesia and allodynia	– Neuronal injury during surgery – Persistent inflammation – Neuroimmune interactions	– Antineuropathic agents – Foreign body removal
Phantom limb pain	– Perception of pain in body parts that have been lost	– Ectopic neuroma activity – Dysregulation of descending inhibitory pathways – Neuroplasticity dysfunction	– Sodium channel blockers – Anticonvulsants – NMDA antagonists – Opioids – Peripheral nerve blockade – Mirror therapy – Spinal cord stimulator
Postherpetic neuralgia	– Sequela of herpes zoster (shingles) infection – Dermatomal neuropathic pain that is burning or electric shock-like and associated with hyperesthesia, dysesthesia, paresthesia, allodynia, and windup	– Sensitized nociceptor – Neuronal sprouting – Central sensitization – Abnormal A-beta fiber wiring – Synaptic plasticity	– Topical local anesthetics – Anticonvulsants – Neuromodulation – Capsaicin patch – Opioids

Rheumatoid arthritis	– Synovial joint inflammation, autoantibody production, and cartilage and bone destruction – Multiple small, peripheral joints affected – Extra-articular manifestations	– Sensitization of sensory terminals by proinflammatory cytokines and mediators	– NSAIDs – Immune suppressors—methotrexate and steroids – Biologic agents
Trigeminal neuralgia	– Paroxysmal, high intensity, electric shock-like facial pain in the distribution of the fifth cranial nerve	– Compression by microvasculature – Demyelination – Pons lesions – Idiopathic	– Sodium channel blockers – Anticonvulsants – Gasserian ganglia neurolysis – Percutaneous rhizotomy – Gamma knife – Surgical decompression

Because opioids, such as morphine, have predictable efficacy in cancer pain and they can be administered via multiple routes, they are the first line of therapy in moderate to severe pain.[18] However, chronic opioid therapy is associated with significant side effects, including mental clouding, constipation, and respiratory depression, that have detrimental effects on the quality of life of a cancer patient.[19] Hence, the overall goal is to minimize pain with the least amount of opioids to lower possible side effects. To accomplish such a goal, the use of adjuvant therapy is encouraged. Radiation therapy can have a significant effect on pain.[20,21] More recently, bisphosphonates have become an essential component for the management of bone cancer. They were once thought to arrest bone loss by preventing the dissolution of hydroxyapatite. Now, it is believed that they exhibit their action by decreasing bone resorption and increase mineralization by inhibiting osteoclast activity.[22–25] Aside from a pharmacological approach, a surgical approach may be utilized to lessen the tumor burden. Additionally, interventional pain management can offer other modalities to lessen the pain associated with cancer. Trigger point injections with local anesthetics may offer a temporary solution for soft tissue with localized and focal pain, though the evidence is mostly anecdotal.[26,27] For vertebral pathological fracture secondary to bone metastasis, a procedure in which cement is injected inside the vertebral body in a procedure known as vertebroplasty has shown to be very effective in controlling pain.[28] Celiac plexus neurolysis is one of the effective methods to control pancreatic cancer pain, though it lasts for around 3–4 months.[29] In intractable cases, intrathecal pumps delivering a variety of medications in the intrathecal space can have significant pain control with relatively low side effects. Common intrathecal medications used to treat bone cancer pain include opioids, ziconotide, local anesthetics, and clonidine.[30] Spinal cord stimulation can be used in selected cases, although the mechanism of action is not completely understood.[31–33]

The treatment of cancer itself can result in chronic pain conditions including persistent postsurgical pain and chemotherapy-induced pain. These conditions are discussed under separate headings below.

2. CENTRAL POSTSTROKE PAIN

Central poststroke pain (CPSP) is described as "among the most spectacular, distressing, and intractable of pain syndromes."[34] In their famous paper "Le syndrome thalamique," Dejerine and Roussy described CPSP as "…severe, persistent, paroxysmal, often intolerable, pains on the hemiplegic side, not yielding to any analgesic treatment."[35] They coined the term "thalamic syndrome" because the most common cause of CPSP is occlusion of the thalamogeniculate branches of the posterior cerebral artery and pathology most commonly reveals lesions of the thalamus and posterior limb of the internal capsule.[35–38] The resultant severe, burning pain may be caused by the thalamus losing cortical inhibition.[38] Other associated symptoms include thermosensory disturbances, dysesthesias, hemiataxia, hemiastereognosia, and choreoathetoid movements.[35] CPSP originating supratentorially (area of the brain above the cerebellar tentorium, e.g., cerebrum) has been reported to have larger deficits in sharp and cold sensation.[39] Infratentorial (area of the brain below the cerebellar tentorium, e.g., cerebellum) origins produce greater deficits in hot pain.[39] The incidence of CPSP is between 1% and 12% with no consistent differences related to age, sex, or laterality of the lesion.[40–58] Onset of pain is usually gradual and within the first few months after stroke but can be delayed.[34,51]

The etiology of CPSP is attributed to lesions along the spinothalamocortical tract.[59] The resultant spinothalamic dysfunction corresponds to the abnormal thermal sensations of CPSP.[40] However, the precise mechanism of CPSP is unknown, hence the need for targeted drug development.[60] Suggested mechanisms include alterations in thalamocortical processing such as loss of inhibition or sensitization, subthreshold activation of thalamic nociceptive neurons, hyperexcitability of thalamic neurons after interruption of ascending spinothalamic fibers, loss of inhibition from lesions in the lemniscal or spinothalamic tract, or degeneration of corticothalamic neurons projecting to the reticular nucleus and activating GABAergic inhibitory neurons.[60–65]

Current pharmacologic treatment response is typically moderate and dosing is limited by side effects, especially in the elderly, making CPSP challenging to manage.[60,66] Amitriptyline, a tricyclic antidepressant, and antiepileptics, such as lamotrigine, gabapentin, pregabalin, and carbamazepine, are usually first-line treatments.[50,67,68] Opioids have been used for refractory pain.[69,70] Chronic motor cortex stimulation and deep brain stimulation are last resort therapies for patients who have failed or are intolerant of other modalities.[71–76] Sleep disturbances, anxiety, and depression are common with CPSP; thus physical therapy, behavior therapy, and acupuncture may be helpful.[77]

3. CHRONIC PELVIC PAIN

Chronic pelvic pain (CPP) is defined as pain below the umbilicus that lasts at least 6 months and requires treatment or causes functional disability.[78]

CPP can be caused by a variety factors, but about 65% of women are never given a specific diagnosis.[79] Women with more than one contributing condition usually have greater pain.[80] The prevalence of CPP is about 15% in the United States.[79] The relative frequency of the etiologies of CPP varies based on the population or practice setting. For instance, practices that specialize in endometriosis may give this diagnosis to over 70% of women with CPP, but, in a different population, a gynecologic cause may be diagnosed in only 20% of patients.[81,82] The other most common causes of CPP include chronic pelvic inflammatory disease, interstitial cystitis/painful bladder syndrome, irritable bowel syndrome (IBS), pelvic floor myalgia, myofascial pain or trigger points, and neuropathy of the iliohypogastric, ilioinguinal, genitofemoral, or pudendal nerves.[80,83–87]

Hysterectomy can significantly improve or eliminate CPP in up to 96% of women; however, it makes a poor choice for those women who still wish to bear children.[88–92] History and physical exam usually can elucidate at least one potential etiology of CPP. Therapy can then be focused on treating that potential pain generator. If there is improvement in the patient's pain, then the suspected pain generator has an increased likelihood of being the cause of the patient's CPP. However, treatment effects are nonspecific and may improve other pain generators that were not recognized initially; thus, a definitive diagnosis of the causative factor cannot be made based on the treatment response alone. Using intensive testing to try to identify a specific etiology can prove costly as it requires laboratory tests, imaging studies, and, usually, exploratory surgery. First-line treatment of endometriosis involves NSAIDs or oral contraceptive pills, either cyclic or continuous.[93–95] Patients unresponsive to first-line therapy can try continuous progestin or gonadotropin-releasing hormone agonists.[96,97] Patients with chronic pelvic inflammatory disease should be treated with antibiotics if there is a suspected ongoing or subclinical infection. There are no curative treatments for interstitial cystitis/bladder pain syndrome. Symptom control is attempted with self-care and behavior modification, physical therapy, and oral medications, including amitriptyline and pentosan polysulfate sodium.[98–100] Acute flares can be treated with urinary analgesics, such as phenazopyridine or methenamine, intravesical lidocaine, or other oral or transdermal nonnarcotic and narcotic pain medications used in other chronic pain conditions.[99,101] Treatment for IBS is discussed elsewhere in the chapter. Myofascial pain is most effectively treated using physical therapy, trigger point injections, or local anesthetic patches.[85,102,103] Patients with symptoms of neuropathic pain have responded well to gabapentin and amitriptyline.[104] More data are needed regarding the use of nerve transection procedures and neuromodulation in CPP.[105,106]

4. COMPLEX REGIONAL PAIN SYNDROME

Complex regional pain syndrome (CRPS) results when the normally well-organized and predictable healing process following injury, usually to an extremity, goes awry. CRPS is subdivided into CRPS type I, formerly reflex sympathetic dystrophy, where there is no identifiable nerve injury, and CRPS type II, formerly causalgia, where there is an obvious nerve injury.[107] CRPS can manifest in as little as 2 weeks following injury but is usually seen within a month.[108,109] The spectrum of severity is quite broad with some patients having mild symptoms and recovering quickly, while others with more severe symptoms may not recover at all.[109]

There are four cardinal features of CRPS: pain, swelling, movement abnormalities, and vasomotor instability. The pain is nagging, throbbing, burning or searing, and nondermatomal.[108,109] One of the defining characteristics of the disease is that the pain is out of proportion (hyperalgesia) to the severity of the initial injury.[108,109] The pain can be so severe that movement of or contact with the affected limb is avoided.[110,111] Even sudden, loud noises can cause allodynia.[110,111] Touching the affected limb can sometimes cause pain in the other limb (synchiria). Due to the distribution of the sensory homunculus, lower extremity CRPS can cause pain in the genitals and bladder and upper extremity CRPS can affect the mouth and face.[112] The increased heat, redness, and edema seen early in the disease course are sometimes referred to as "high flow" or warm CRPS.[108,109] As with the pain, the swelling is out of proportion to the injury and generally involves the whole foot or hand.[113] Even after the swelling subsides, a "body perception change" can occur where the patient believes that the extremity is still swollen and bigger than the other side.[113] Later in the disease course, "low flow" or cold CRPS describes joint stiffness and contracture.[108,109] A sensation of decreased range of motion can be felt at all stages of the disease and may be related to limitations from pain or swelling.[114] Other motor abnormalities include muscle weakness, tremor, and, more rarely, dystonic movements.[115,116] The thought of movement alone (motor imagery) is enough to cause pain in some CRPS patients.[117,118] The final cardinal feature of CRPS is vasomotor instability. Patients can show changes in color, temperature, hair/nail growth, and sudomotor activity of the limbs. Affected extremities will sweat more and hair/nail growth is often accelerated.[119]

The incidence of CRPS varies from 5.5 to 26 per 10^5 person-years depending on the case definition.[120,121] Women are two to four times more likely to suffer from CRPS, the median age of onset is between 40 and 53 years, and the upper extremities are most likely to be affected.[120,122–124] Trauma is the most common inciting event.[124]

The pathogenesis of CRPS remains unclear and controversial. There are numerous theories, but the most commonly accepted are that the inciting injury causes an exaggerated local inflammatory reaction or neurologic damage.[114] In the early stages of the disease, clinical features including swelling, pain, heat, redness, and loss of function suggest an inflammatory response.[114] Systemic inflammatory markers including TNF-alpha, CGRP, IL-2, and substance P have also been isolated.[125] On the other hand, cortical reorganization may explain the phenomenon of pain being experienced in other body parts whose cortical areas are adjacent to the area representing the injured body part.[114] The neuralgic pain of CRPS can be explained by damage to A-delta, C-fibers, and sympathetic axons.[114] These small fibers release CGRP and substance P and are very sensitive to injury.[126] Immunolabeling and histological examination have confirmed small fiber damage in CRPS.[127,128]

Due to the uncertainty of the disease's etiology, a vast array of treatment modalities has been applied. Physical therapy is the cornerstone of the treatment plan with the goal of other modalities being to allow maximal participation in therapy.[114] Physical therapy is used to prevent joint contractures, weakness, and pathologic cortical remodeling.[129] Mirror therapy, whereby a patient moves their normal hand and sees the mirror image move as if it were the affected hand leading to more movement of the actual affected hand, has also been shown to be efficacious.[130] Medications used include free radical scavengers like DMSO cream and *N*-acetyl cysteine, bisphosphonates, calcitonin, capsaicin cream, steroids, ketamine infusion, and gabapentin.[109,131–133] Local anesthetic injections of sympathetic ganglia have provided longer durations of symptom improvement compared to saline and, while not directly treating the disease, allow the patient to participate more fully in physical therapy.[133] Spinal cord stimulation has been used for refractory cases but long-term efficacy has not been proven.[134]

5. FAILED BACK SYNDROME

Failed back syndrome (FBS) can affect any level of the spine, but the majority of available literature focuses on the lumbosacral spine. Annually, hundreds of thousands of lumbar surgeries occur in the United States, and that number has been on the rise over the past few decades.[135] Unfortunately, these surgeries carry a significant failure rate.[136,137] After lumbar laminectomy for spinal stenosis, patients have failure rates of approximately 20–35%.[136] Long-term success is even more dismal. Almost 75% of patients had residual low back pain at 10- to 22-year follow-up after lumbar discectomy.[137]

Advancements in radiologic imaging have enabled the etiology for FBS to be determined in up to 95% of patients.[138,139] The most common etiologies of FBS are stenosis, painful discs (defined by reproduction of a patient's pain after increasing mechanical pressure in a suspected disc) or internal disc disruption (IDD), arachnoiditis, and fibrosis.[138–140] A number of factors contribute to the development of FBS and can be grouped into preoperative, intraoperative, and postoperative factors.[141] A significant portion of FBS may be related to inappropriate preoperative patient selection.[142–145] Herniated disc pain can be mimicked by gluteal, iliopsoas, or quadratus lumborum muscle syndromes, fibromuscular dysplasia, or myofascial pain.[142] Up to 95% of FBS may be caused by inappropriate surgery on patients presenting with pain mimicking a herniated disc pain distribution.[142] Other patient selection factors that are predictors of poor surgical success include psychological risk factors, workers' compensation and other litigation, and poor communication regarding expectations.[143–145] Intraoperative factors include operating on the wrong spinal level and overlooking other pain generators.[141,146] Incomplete surgery or poor technique, including misplaced screws, bony fragments, loose spinous processes, conjoint nerve roots, residual disc material, inadequate decompression of stenosis, and malunion or pseudarthrosis, is also contributory.[139,147] Postoperative factors of FBS can be directly related to the surgery or indirectly from unavoidable structural changes.[141] Nerve irritation from central spinal or neural foraminal stenosis can result from degenerative changes.[140] Disc pain from residual disc material, increasing degeneration, and new disc herniation can occur as well.[138,139] Direct tissue trauma from surgery can trigger epidural scarring, adhesions, and fibrosis, which can irritate surrounding nerves and serve as another pain generator.[138,140] Given the increasing rates of surgery paired with a low success rate and multiple etiologies of FBS, the need for adequate treatment is substantial.

Current treatment of FBS begins with conservative physical therapy and medical management, not unlike any other chronic low back pain.[141,148] NSAIDs have strong evidence for analgesic efficacy.[149] Despite weak evidence for efficacy, opioids continue to be used as well.[149] Interventional pain procedures including epidural injections, facet procedures, adhesiolysis, and disc interventions can be used depending on the underlying etiology of the patient's pain.[141] Spinal cord stimulation has shown benefits for patients with refractory pain and appears to be more cost-effective compared to repeat surgery.[150–152] Peripheral nerve stimulation and peripheral nerve field stimulation also show promising results for patients with hard to treat pain.[153,154]

6. FIBROMYALGIA

Fibromyalgia is a condition in which patients feel widespread pain, loss of energy, possible cognitive, and psychiatric deviation from baseline.[155] People with fibromyalgia present with widespread musculoskeletal pain,[156] fatigue, cognitive disturbance, psychiatric symptoms, and headaches[157,158] The diagnosis of fibromyalgia can be somewhat challenging because there are no laboratory tests or imaging studies that can definitively point to it. However, most patients will present with specific point tenderness on physical examinations to palpation. Recently, some investigators are advocating the use of the symptoms only instead of the point tenderness palpation as a diagnostic measure.[159] In the United States, the prevalence of fibromyalgia is 2–3% and is considered one of the common causes for chronic pain in women between the ages of 20–55.[160,161] Women are more predisposed to be inflicted with fibromyalgia six times more than men.[162]

The pathophysiology of fibromyalgia is not very well understood. In the periphery, abnormal muscle metabolism and altered ratio of sensory and sympathetic fibers have been implicated in the pathogenesis.[163,164] More recently, elegant imaging studies in humans have demonstrated altered CNS processing in fibromyalgia patients.[165] Other mechanisms involved in fibromyalgia are thought to be central sensitization, dysregulation of descending inhibitory pathways, and glial activation.[166]

Patients usually respond better when pharmacological, interventional, and behavioral modalities are employed in a multidisciplinary fashion.[167] Before any treatment starts, patients should be educated about the nature of the disease and the uncertainty associated with both the diagnosis and the management.[167] In terms of nonpharmacotherapy approach, the first step is patient education that has been shown to reduce medication consumption by patients.[168,169] Furthermore, intense educational courses for patients provided by multidisciplinary team resulted significantly less pain reported by patients as well as improved function, less fatigue, less stiffness, less anxiety, and less depressed mood.[170] Aside from patient education, another very successful technique is structured aerobic exercise. Multiple studies have shown that low impact aerobic exercise such as swimming, biking, or fast walking resulted in significantly improved pain and other symptoms associated with fibromyalgia.[171]

The pharmacotherapy of fibromyalgia includes tricyclic antidepressants, serotonin and norepinephrine reuptake inhibitors (SNRI), and anticonvulsants. Randomized clinical trials have demonstrated that 25–45% of patients obtained significant relief from TCA.[172–174] Those patient who do not respond well to TCA may find better results with SNRI. The meta-analysis of six randomized clinical trials involving more than 2200 patients showed that 60 mg/day of duloxetine reduced pain significantly in 50% of the patient population.[175] Anticonvulsants such as gabapentin and pregabalin have shown some efficacy in fibromyalgia.[176,177] Interventional pain procedures such as trigger point injections and dry needling have shown some efficacy.[178]

7. IRRITABLE BOWEL SYNDROME

Irritable bowel syndrome (IBS) is characterized by abdominal discomfort and altered bowel habits of at least 3 months duration, without structural or biochemical abnormalities.[179] The abdominal pain is usually crampy in nature, of variable severity, subject to periodic exacerbations, and relieved with defecation.[180,181] Altered bowel habits can be either diarrhea or constipation.[179] Other gastrointestinal symptoms including dysphagia, gastroesophageal reflux, nausea, and chest pain are common with IBS.[180]

IBS is a frequently diagnosed gastrointestinal condition. The prevalence in North America and Europe is about 10–15%.[179,182–187] While it affects all demographics, women (2:1 predominance) and younger patients are most at risk.[179,188] Reported prevalence may underestimate the true impact of the disease because only about 15% of those affected seek medical attention.[182,186,189–191] Health-care costs associated with IBS approach $30 billion annually.[192] It is the second most common reason to miss work behind the common cold.[193]

The pathogenesis of IBS is multifactorial and of much interest secondary to the possibility of targeted therapy development.[179,194] Abnormalities in gastrointestinal motility including an exaggerated motor response to cholecystokinin have been examined.[195] Visceral hypersensitivity and hyperalgesia are common findings in IBS and may be related to increased spinal cord excitability secondary to NMDA receptor activation.[196] Intestinal inflammation may be another mechanism. Increased numbers of mast cells, lymphocytes, and proinflammatory cytokines, such as tumor necrosis factor, have been found in intestinal biopsies of IBS patients.[197–199] The risk of developing IBS is increased sixfold after a bout of infectious enteritis.[200] Alterations in fecal microflora have led to the study of probiotics in IBS but more data are needed.[201,202] Other studies into the pathophysiology of IBS have looked at food sensitivity, gene polymorphisms, and psychosocial factors.[203–205]

In patients with mild or intermittent symptoms, dietary and lifestyle modifications alone may be sufficient. In patients with moderate to severe symptoms or those who have failed conservative management, adjunctive pharmacologic therapy targets the predominant IBS subtype, i.e., diarrhea versus constipation. IBS with constipation is treated with osmotic laxatives like polyethylene glycol; however, no effect on abdominal pain has been demonstrated.[206,207] For constipation unresponsive to polyethylene glycol, lubiprostone, a chloride channel activator that stimulates intestinal fluid secretion, or linaclotide, a guanylate cyclase agonist that stimulates intestinal transit and fluid secretion, has been used and were more effective at providing pain relief.[208–210] Serotonin 5-HT4 agonist tegaserod, which increased colonic motility, was removed from the market due to cardiac side effects.[211]

Loperamide, an antidiarrheal, has been used for diarrhea-predominant IBS; however, like polyethylene glycol, no improvement over placebo was seen for abdominal pain.[179,212–214] Bile acid sequestrants, such as cholestyramine, colestipol, and colesevelam, may improve diarrhea but come with the side effects of increased bloating and abdominal pain.[215,216] The 5-HT3 antagonists including alosetron, ondansetron, and cilansetron have been shown to improve abdominal discomfort and diarrhea by decreasing colonic motility and secretion; however, caution must be taken as alosetron withdrawn from the market for a period of time, due to adverse effects of severe constipation and ischemic colitis, before being made available again under restricted conditions.[217–220]

Abdominal pain and bloating can be treated on an as needed basis with anticholinergics (e.g., dicyclomine or hyoscyamine, muscarinic antagonist) that relieve smooth muscle spasms.[179] Tricyclic antidepressants may be beneficial in IBS with diarrhea due to ability to slow intestinal transit time, secondary to their anticholinergic effects.[221,222] Selective serotonin reuptake inhibitors have also been shown to be effective in improving patients' overall well-being.[223]

8. MIGRAINE

Migraine is a neurological disorder resulting in severe debilitating headache, sometimes associated with nausea, vomiting, photophobia, and phonophobia. Migraine is a clinical diagnosis that is based on following symptoms according to the International Headache Society guidelines: (1) Five or more attacks for migraines without aura, or two attacks with aura. (2) Four hours to 3 days in duration. (3) Two or more of the following: Unilateral; pulsating; moderate or severe intensity; and resulting in avoidance of routine daily activity. (4) One or more of the following: Nausea and/or vomiting; photophobia; and phonophobia.[224] Patients suffering from migraine typically experience four stages: Prodrome (affective symptoms such as euphoria), aura, migraine headache, and, finally, postdrome (e.g., feeling exhausted). However, not all patients experience all phases.[224] There are several types of migraines including migraine with aura, migraine without aura, childhood, and chronic migraine.[225] The earliest recorded description of migraine is documented in the Ebers Papyrus written around 1500 BC in Egypt.[226] Migraine is a common disorder that affects about 12% of the general population. Women are more likely to experience migraine than men with rates of 17% and 6% each year, respectively. Migraine tends to run in families and takes place most commonly during the third decade of life.[227–229]

The exact cause for migraine is not fully understood. The once popular vascular theory suggesting vasodilatation to be the cause of migraine is no longer viable.[227] Today, there are several hypotheses to explain the cause of migraine. The following is a summary of the current theories:

1. Cortical spreading depression: Self-propagating wave of neuronal and glial depolarization that spreads across the cerebral cortex.[230]
2. Trigeminovascular system: Activation of this system containing peripheral neurons that innervates large cerebral vessels, pial vessels, dura mater, and venous sinuses sends a pain signal to the brain.[231–233]
3. Sensitization: Peripheral sensitization in the primary afferent neurons and central sensitization within second-order neurons in the trigeminal nucleus caudalis and higher order neurons on the central nervous system play a significant part in the development of migraine attacks and possibly chronic migraine.[234–236]
4. Serotonin and calcitonin gene-related peptide: These two agents have some correlation with migraines, but their role is not very clear at this time.[237–240]

There are two therapeutic approaches to manage migraines: abortive and preventive. Abortive therapy contains several families of medications. Several NSAIDs have shown promising results in some patients.[241–243] Acetaminophen as a single agent was also effective as an abortive therapy in some patients.[244] Triptans are a class of medications that activate the serotonin 1b/1d receptors. They inhibit the release of vasoactive peptides, promote vasoconstriction, and block the pain pathway in the brainstem. They also block afferent input to second-order neurons.[245,246] Antiemetics are a class of medication given for nausea and vomiting control. The intravenous administration of antiemetics has been shown to control migraines.[247] Ergots are a class of medication with some similarities to triptans. They bind the serotonin 1b/d receptors resulting in migraine control.[248] Opioids and barbiturates should be considered as a last resort for headache management.[249]

There are several classes of medications that are used for preventive migraine therapy. Antihypertensive medications, such as beta-blockers, calcium channel blockers, and ACE inhibitors/ARBs, have been shown to reduce the frequency of migraines attacks.[250–252] Anticonvulsants, especially valproate and topiramate, were shown to reduce the frequency of migraines compared to placebo.[250,253] Botulinum toxin injections have mixed results.[254–256] Coenzyme Q10 was found to be effective in reducing the frequency of migraine attacks.[257]

Neuromodulation has been investigated for the management of migraines. Some of the early clinical trials using neuromodulation were less than significant.[258] Subsequent trials showed promising results.[259]

9. MULTIPLE SCLEROSIS

Multiple sclerosis (MS) is a central demyelinating disorder in which as many as 86% of patients suffer from chronic pain.[260,261] Multiple types of pain often coexist in MS patients. Pain from optic neuritis is observed in up to 8% of MS patients. Moreover, painful muscle spasms and spasticity routinely affect these patients.[261] Central neuropathic pain, defined as pain distribution consistent with a CNS lesion and absence of peripheral neuropathy, is widespread in MS patients.[262,263] The clinical presentation of pain in MS patients varies substantially and comprises of headaches, trigeminal neuralgia, painful extremity spasms, back pain, extremity dysesthesia, and eye pain.[261] Disease-modifying agents, such as interferon beta, may exacerbate the patient's headaches.[264]

Multiple mechanisms have been proposed to describe pain in MS. Demyelination resulting in ectopic activity of sodium channels, neuroinflammation secondary to disrupted blood–brain barrier, disturbance of synaptic transmission, and presence of neuroelectric blocking factors (serum factor that has neuromodulatory properties) are some of the changes thought to explain MS pain.[265]

The pharmacological treatment for pain and spasticity of MS includes multiple agents such as baclofen (GABA-B agonist), diazepam (GABA-A agonist), tricyclic antidepressants, and mu opioid agonists, such as morphine, among others.[266] Moreover, advanced therapies such as intrathecal baclofen and morphine have been successfully used in MS patients.[266] Botulinum toxin is also used to treat spasticity in MS patients.[267] Interestingly, multiple recent studies have demonstrated efficacy of cannabinoids in the treatment of both MS pain and spasticity.[268–270]

10. PERIPHERAL NEUROPATHIC PAIN

Painful peripheral neuropathy can be caused by a number of systemic factors including chemotherapy, HIV, and diabetes. While cancer in and of itself can cause pain, the chemotherapy agents can cause additional pain. This phenomenon is known as chemotherapy-induced neuropathic pain (CIPN).[271] There are several chemotherapy agents with different mechanisms of actions and different pain manifestations. Typically, chemotherapy-induced pain exhibits itself as peripheral neuropathy mainly affecting distal extremities.[271] The onset of the chemotherapy-induced pain usually takes several months, and it continues despite the cessation of the treatment.[271] Chemotherapy-induced peripheral neuropathy is seen in approximately 30–40% of patients receiving chemotherapy.[272]

The mechanism of chemotherapy-induced pain is not well understood. However, most of these drugs affect either DNA synthesis or axonal transport and can change ion conductance and axoplasmic transport in neurons, evoke toxicity in Schwann cells leading to axonal degeneration, and inhibit growth factors necessary for proper functioning of the nerve terminals.[273]

Multiple studies have evaluated the role of preventive strategy for CIPN. Chemoprotectant agents, such as amifostine, nimodipine, rhuLIF, neurotropin, and diethyldithiocarbamate, and antioxidants, such as vitamin E, have had mixed results at best with no strong recommendations for their routine use.[272,274–289] The pharmacotherapy of already established CIPN usually includes dual (serotonin and norepinephrine) reuptake inhibitors, such as duloxetine, anticonvulsants, such as gabapentin, and opioids in most resistant cases. Topical compounded creams containing amitriptyline, baclofen, and ketamine have had a significant effect on controlling tingling, cramping, and burning pain.[290,291] Recent studies have demonstrated usefulness of TRPM8 agonist menthol in controlling some of the symptoms of CIPN.[292] More advanced interventions, such as spinal cord stimulators, have shown significant promise in the treatment of CIPN.[293]

In HIV, peripheral neuropathy can develop as a complication of antiretroviral therapy, the infection itself, or other host factors.[294] Furthermore, a variety of painful peripheral neuropathies can affect HIV patients, including mononeuritis multiplex, HIV-associated neuromuscular weakness syndrome, diffuse infiltrative lymphocytosis syndrome, and, most commonly, distal symmetrical polyneuropathy (DSPN).[294] Significant overlap in clinical symptoms presents a challenge for treating physicians.[295] Patients with DSPN typically present with hyperalgesia in both feet.[294] Sensory symptoms spread proximally over weeks to months but rarely involve the upper extremities.[296] Pain and temperature sensations are impaired in the majority of patients.[296] Nerve conduction studies reveal length-dependent axonal polyneuropathy.[296] Partial denervation of distal lower extremity muscle groups is evident on electromyography.[296] Myelinated and unmyelinated sensory nerve fibers are lost through axonal degeneration and macrophage activation.[297]

Like CIPN, the pathogenesis of DSPN is unclear but believed to be multifactorial.[294] Nucleoside reverse transcription inhibitors are neurotoxic and, like chemotherapy drugs, alter DNA synthesis and also cause mitochondrial abnormalities.[298] The incidence of DSPN is dose and duration dependent and varies between 13% and 30% depending on the degree of disease progression.[299,300] The virus itself can also cause axonal injury triggered by envelope protein gp120, indirectly by neuronal apoptosis or directly via mitochondrial caspase activation.[301] Viral antigens create a toxic environment for peripheral nerves via immune activation. Activated cytokines, macrophages, and T cells have been found in peripheral nerves and dorsal root ganglia.[302,303] HIV gp120's chemokine receptor, CXCR4, causes neuronal apoptosis and production and release of neurotoxic tumor necrosis factor and other proinflammatory cytokines via Schwann cell activation.[301,304,305]

As with CIPN, treatment of DSPN is limited to symptom management and includes NSAIDs, topical analgesics, tricyclics, anticonvulsants, and narcotics.[294] Recombinant human nerve growth factor has significantly improved pain scores despite no improvement in neuropathy severity in experimental trials.[306–308]

Diabetic polyneuropathy is the most common neuropathy in the world.[309] Like HIV patients, diabetics can suffer from myriad neuropathies but the most common is distal symmetric polyneuropathy, which is essentially synonymous with diabetic neuropathy.[309] In one study, 50% of patients had distal symmetric polyneuropathy and 14% were symptomatic.[309] Progressive sensory loss and paresthesias are in the classic "stocking-glove" distribution.[310] The loss of motor axons can follow in severe cases.[310] Like HIV DPSN, there is "dying back," where the distal portions of the longest myelinated and unmyelinated sensory axons degenerate.[311–313] Hyperglycemia causes cellular damage in a number of ways. Excess glycolysis overloads the electron transport chain leading to reactive oxygen species generation.[314] Increased oxidative stress also occurs from increased flux through the polyol pathway.[315] Inflammatory injury takes place as a result of increased flux through the hexosamine pathway.[316] Advanced glycation end products trigger inflammatory signaling cascades generating more oxidative stress.[317] Increased free fatty acids from dyslipidemia initiate inflammatory cytokine release from macrophages and adipocytes and also damage Schwann cells directly.[318,319] Insulin promotes neuronal growth and survival when not involved in glucose uptake in neurons.[320,321] The reduction in these neurotrophic effects with insulin deficiency (type 1 diabetes) or resistance (type 2 diabetes) may contribute to diabetic neuropathy.[322]

Symptomatic treatment of diabetic neuropathy follows the same course as previously discussed with CIPN and DSPN. Additionally, alpha-lipoic acid, an antioxidant, has been used to reduce oxidative stress in patients intolerant or unresponsive to anticonvulsants or antidepressants.[323–326] Transcutaneous electrical nerve stimulation is another effective modality, especially for patients intolerant or unresponsive to pharmacologic treatment.[327–329] However, the first step is achieving optimal glycemic control.[330] Glucose control delays onset, slows progression, and can even make some improvements in diabetic neuropathy.[331–333] Components of metabolic syndrome have recently been investigated as potentially modifiable risk factors.[334–336] Future targets for disease-modifying treatments include obesity, HDL, and triglyceride levels, which have all been found to be independently associated with neuropathy.[337]

11. PERSISTENT POSTSURGICAL PAIN

Although medical science has made significant advances in the treatment of acute postoperative pain with the use of multimodal analgesia, persistent postsurgical pain remains a common debilitating chronic pain condition.[338,339] Persistent postsurgical pain is defined as pain that persists even after the surgical wound has healed, typically lasting 3–6 months beyond the original surgery.[338] While immediate postoperative pain tends to be both nociceptive and inflammatory in nature, chronic postoperative pain tends to be neuropathic in nature. The signs and symptoms include ongoing pain, hyperalgesia to deep palpation, and tactile allodynia.[338] Because of the prevalence of neuropathic symptoms in many persistent postsurgical pain patients, nerve injury during the surgery is thought to be one of the major etiologies of chronic postsurgical pain. This is further supported by the fact that the incidence of chronic postoperative pain varies significantly by the type of surgery; e.g., 30–50% in thoracotomy surgeries where intercostal nerve injury is almost impossible to avoid versus much lower incidence (2%) in laparoscopic hernia repair where ilioinguinal nerve damage can be reduced.[338,340] Common surgeries that are associated with a high incidence of persistent postsurgical pain include thoracotomy, mastectomy, coronary artery bypass surgery, caesarian section, and open inguinal hernia repair.[338] The risk factors for development of chronic postoperative pain include preexisting pain, age, sex, type of surgery, psychosocial factors, and genetic susceptibility.[338]

The mechanism of persistent postsurgical pain is thought to be secondary to neuronal injury during surgery, persistent inflammation caused by a foreign body (e.g., hernia mesh), and neuroimmune interactions. Chronic nociceptor activation due to nerve injury and/or inflammation results in changes in synaptic plasticity in the CNS that may result in persistence of the pain state.[338,341]

The treatment of persistent postsurgical pain depends upon the patient's symptoms; where neuropathic pain can be treated with antineuropathic agents[342] and if foreign body reaction is thought to be behind the chronic pain, removal of the offending agent (e.g., vaginal mesh) may be advisable.[343] Because of the difficulty in treating chronic postoperative pain when it is established, numerous attempts have been made to formulate a perioperative strategy that would potentially decrease the incidence and severity of persistent postsurgical pain. These strategies include nerve sparing surgical approaches, use of regional anesthesia with local anesthetics, and preemptive analgesia with drugs acting at receptors involved in central sensitization such as ketamine, gabapentin, and NSAIDs.[344–348]

12. PHANTOM LIMB PAIN

Phantom limb pain (PLP) is a phenomenon that refers to the perception of pain in certain body parts by individuals who have lost these parts of their bodies. This is different from phantom limb sensation in which individuals feel that an amputated body part is still present. PLP is not exclusively related to a trauma or surgical procedure resulting in amputation of a body part. It is a phenomenon that is also experienced by some individuals who are born with congenital amputations.[349,350] Surgical procedures that result in the removal of body parts have been associated with PLP. For example, tongue, eyes, breasts, or teeth removal has been associated with PLP.[351–354] In the United States alone, there are currently over 1.6 million individuals with amputated limbs, and this number is expected to double by the year 2050.[355] On average, 80% of amputees will experience either PLP or PLS after their amputations.[356]

Historically, PLP was thought to be psychogenic in nature. It is still unclear how PLP develops, but there are several theories that attempt to explain this interesting phenomenon. Ectopic activity from neuromas that develop after peripheral nerves are cut has been recorded in amputees. These signals may be interpreted as pain when they reach the sensory cortex.[357] Another hypothesis is dysregulation of descending inhibitory pathways. It is theorized that the initial input from the amputation or trauma coupled with the ectopic discharge from the formed neuromas renders the ascending pathway less responsive to the inhibitory drive of the descending pathway. Thus, nociceptive signals reach the brain unopposed.[358–360] Additionally, observed changes in sensitivity and distribution of the NMDA and AMPA receptors may contribute to the long-term neuroplasticity leading to chronic ongoing discharge of the ascending pain pathways.[357] Other mechanisms have been hypothesized to play a part in PLP such as sympathetic dysregulation.[361,362]

Regardless of the mechanism with which PLP develops, the effect on patients can be devastating. Pharmacological approaches include sodium channel blockers, antiepileptics (gabapentin and topiramate), NMDA antagonists, and opioids.[363] Prolonged peripheral nerve blockade via a regional anesthetic infusion has been shown to substantially decrease PLP.[364]

Epidural application of sodium channel blockers and opioids had mixed effects.[365–367] Nonpharmacological approaches have been tried to manage PLP. Mirror box therapy may be helpful for some patients.[368] Mirror box therapy is proposed to work via reorganizing the sensory cortex.[363] Some promising results have been observed with the use of spinal cord stimulator for PLP.[369]

13. POSTHERPETIC NEURALGIA

Postherpetic neuralgia (PHN) is a common sequela of herpes zoster (shingles) that causes debilitating neuropathic pain in a dermatomal distribution.[370] Dormant varicella zoster virus reactivation causes degeneration of primary afferent sensory neurons, where it was seeded during the earlier primary infection.[371–374] Pain is the most common symptom of zoster reactivation and can precede the dermatomal eruption by days to weeks.[375–378] The associated pain is often described as burning, electric shock-like, or a deep ache and is associated with exaggerated responses to stimuli (hyperesthesia) and altered sensation with light touch (paresthesia) that may be unpleasant (dysesthesia).[379–382] Pain increasing over time with repeated stimulus (windup pain) and pain caused by normally nonnoxious stimuli (allodynia) are also seen.[383] Acute zoster pain and PHN are most commonly differentiated by the persistence of pain for greater than a month after the onset of zoster eruption.[377,384] Up to 70% of patients with acute zoster will develop PHN, which may occur after a pain-free interval.[385–388] The incidence of PHN increases with age.[377,389,390] Untreated adults over 55, 60, and 70 have PHN incidences of 27%, 47%, and 73%, respectively.[377,386,391] With increasing age and concomitant immunocompromise, cellular immunity wanes and is no longer able to prevent clinical reexpression of the disease.[392,393]

Multiple mechanisms have been proposed for PHN. They include "irritable nociceptor" or primary sensitization, neuronal sprouting or neuroma formation, central sensitization, abnormal wiring of A-beta fibers, and possibly synaptic plasticity in the somatosensory pain circuits in the brain.[394]

Current modalities of treatment for PHN include topical medications, such as local anesthetics, systemic antineuropathic agents, such as gabapentin and pregabalin, and antiepileptic agents such as carbamazepine.[370,395,396] In resistant cases, chemical neurolysis using a high-dose capsaicin patch and neuromodulation therapies can be employed.[370] Chronic opioid therapy can be used as a last resort.[397] Certain studies have demonstrated that preventive strategies, such as aggressive treatment with an antiviral agent and/or a corticosteroid during active shingles infection, may decrease incidence of PHN.[370]

14. RHEUMATOID ARTHRITIS

Rheumatoid arthritis (RA) is an autoimmune disease that is characterized by synovial joint inflammation, autoantibody production, and cartilage and bone destruction with accompanying systemic manifestations.[398] With ongoing joint inflammation and destruction, pain becomes one of the cardinal features of RA.[399] The disease typically involves multiple small, peripheral joints, such as interphalangeal joints, presence of antiantibodies, and has prolonged duration of symptoms.[399] Joints are swollen, stiff, and painful and severely limit the range of motion. Due to cartilaginous damage and bone destruction, joints may become deformed over the years. Joint deformities can result in peripheral neuropathies such as carpel tunnel syndrome.[400] Extra-articular manifestations of the disease include rheumatoid nodules, pleural effusions, and pericardial effusions.[398]

The mechanism of pain in RA involves activation and sensitization of sensory terminals innervating various joints by a variety of cytokines and inflammatory mediators as well as increased pressure in the joint.[401,402] Autoantibodies including rheumatoid factor, antikeratin, anticitrulline, and antifilaggrin among others are elevated in patients suffering from RA.[403] Similar autoantibody-induced arthritis is also observed in other diseases such as systemic lupus erythematous and Sjogren's syndrome.[403]

Given the significant role that cytokines and the immune system in general play in the development of RA, the treatment of RA is directed toward reducing the activity of these mediators. Traditionally, anti-inflammatories, such as NSAIDs, and immune suppressors, like methotrexate and corticosteroids, were used in the management of pain in RA.[404] More recently, biologic agents specifically targeting certain cytokines, e.g., TNF-alpha, interleukin 1 and 6, are more frequently used in the management of RA as they have a lesser side-effect profile.[405–409]

15. TRIGEMINAL NEURALGIA

Trigeminal neuralgia is a facial pain disorder in the distribution of fifth cranial nerve that is characterized by paroxysmal, high-intensity pain that may or may not have an underlying cause.[410] Trigeminal neuralgia has been described in ancient literature as old as the first century AD.[410] The pain of trigeminal neuralgia is typically described as "stabbing" or "electric shock-like" in nature by patients.[411] The pain may or may not have inciting factors such as light touch or cold temperature.[412] The incidence of trigeminal neuralgia is thought to be around 3–4 in 100,000 people in the United States.[410–412]

The pathogenesis of trigeminal neuralgia may involve morphological compression of the trigeminal entry root by microvasculature, demyelination, infiltration of the trigeminal ganglia or nerve, or lesions of pons; however, a demonstrable cause is certainly not necessary as evident in the many idiopathic trigeminal neuralgia cases.[412,413]

The pharmacotherapy of trigeminal neuralgia comprises carbamazepine (sodium channel blocker) or its analogues as first-line therapy followed by multiple anticonvulsants such as gabapentin, lamotrigine, and topiramate.[414] Interventional modalities to treat trigeminal neuralgia involve neurolysis of the gasserian ganglia, percutaneous rhizotomy, gamma knife lesion of the trigeminal root, and microvascular decompression.[414]

16. FUTURE DIRECTIONS

The chronic pain syndromes described above are some of the most challenging diseases to treat. Even though the presenting symptom for all these disorders is the same, i.e., pain, the pathophysiology is substantially different. Overall paucity of definitive knowledge about the mechanisms behind these chronic pain states is one of the major deterrents in formulating effective treatment strategy. Although the inciting events may vary, some of these disorders probably share a convergence of mechanism in regard to changes in the pain pathway, e.g., peripheral and central sensitization, alteration of descending pathways, and affective processing. This is further supported by the fact that drugs that modulate these processes have some efficacy in a variety of pain syndromes (e.g., local anesthetics, gabapentin, opioids etc.). Better understanding of the events leading to long-term changes in the pain pathway and how to reverse them once they are established will be the key to successful management of chronic pain conditions.

Sophisticated molecular techniques and assays have discovered a variety of novel therapeutic targets. *In vivo* verification of these targets should be done in a variety of behavioral assays that evaluate not just nociceptive reflexes but more affective dimension of pain as pain in humans is an affective state. Apart from pharmacological therapies, better interventional techniques are needed as they can provide long-lasting relief without the side effects associated with daily medications. Unfortunately, very little is known about the mechanism behind even one of the most commonly used intervention such as neuromodulation. Further research in delineating the mechanisms for therapies that already work may lead us to better interventions.

Finally, the psychosocial aspect of chronic pain cannot be ignored. Chronic pain routinely coexists with various psychiatric disorders that are in turn influenced by the social support structure of a patient. Further research is needed how best to organize behavioral and social support, lifestyle changes in combination with pharmacotherapy and interventions for the betterment of chronic pain patients. We may not be able to cure chronic pain in near future; but if we take "all of the above" approach present it to the patient in the form of an interactive multidisciplinary team, we may help the patients manage their pain better.

REFERENCES

1. Pizzo PA, Clark NM. Alleviating suffering 101—pain relief in the United States. *N Engl J Med*. 2012;366(3):197–199.
2. Jemal A, Bray F, Center MM, Ferlay J, Ward E, Forman D. Global cancer statistics. *CA Cancer J Clin*. 2011;61(2):69–90.
3. Coleman RE. Clinical features of metastatic bone disease and risk of skeletal morbidity. *Clin Cancer Res*. 2006;12(20 Pt 2):6243s–6249s.
4. Mercadante S. Malignant bone pain: pathophysiology and treatment. *Pain*. 1997;69(1–2):1–18.
5. Schweizerhof M, Stosser S, Kurejova M, et al. Hematopoietic colony-stimulating factors mediate tumor-nerve interactions and bone cancer pain. *Nat Med*. 2009;15(7):802–807.
6. Miric A, Banks M, Allen D, et al. Cortical metastatic lesions of the appendicular skeleton from tumors of known primary origin. *J Surg Oncol*. 1998;67(4):255–260.
7. Choi J, Raghavan M. Diagnostic imaging and image-guided therapy of skeletal metastases. *Cancer Control*. 2012;19(2):102–112.
8. Doyle D. *Oxford Textbook of Palliative Medicine*. Oxford; New York: Oxford University Press; 2004.
9. Grahm AL, Andren-Sandberg A. Prospective evaluation of pain in exocrine pancreatic cancer. *Digestion*. 1997;58(6):542–549.
10. Kelsen DP, Portenoy R, Thaler H, Tao Y, Brennan M. Pain as a predictor of outcome in patients with operable pancreatic carcinoma. *Surgery*. 1997;122(1):53–59.
11. Schonenberg P, Bastid C, Guedes J, Sahel J. Percutaneous echography-guided alcohol block of the celiac plexus as treatment of painful syndromes of the upper abdomen: study of 21 cases. *Schweiz Med Wochenschr*. 1991;121(15):528–531.
12. Ripamonti C. Management of bowel obstruction in advanced cancer. *Curr Opin Oncol*. 1994;6(4):351–357.
13. Archer AG, Sugarbaker PH, Jelinek JS. Radiology of peritoneal carcinomatosis. *Cancer Treat Res*. 1996;82:263–288.

14. Foley KM, Bonica JJ, Ventafridda V. *Proceedings of the Second International Congress on Cancer Pain*. New York: Raven Press; 1990.
15. Berger MS, Cooley ME, Abrahm JL. A pain syndrome associated with large adrenal metastases in patients with lung cancer. *J Pain Symptom Manag*. 1995;10(2):161–166.
16. Karanikiotis C, Tentes AA, Markakidis S, Vafiadis K. Large bilateral adrenal metastases in non-small cell lung cancer. *World J Surg Oncol*. 2004;2:37.
17. Mantyh PW. Cancer pain and its impact on diagnosis, survival and quality of life. *Nat Rev Neurosci*. 2006;7(10):797–809.
18. Quigley C. Opioids in people with cancer-related pain. *Clin Evid*. 2008;2008.
19. Caraceni A, Hanks G, Kaasa S, et al. Use of opioid analgesics in the treatment of cancer pain: evidence-based recommendations from the EAPC. *Lancet Oncol*. 2012;13(2): e58–e68.
20. Hartsell WF, Scott CB, Bruner DW, et al. Randomized trial of short- versus long-course radiotherapy for palliation of painful bone metastases. *J Natl Cancer Inst*. 2005;97(11):798–804.
21. Lutz S, Berk L, Chang E, et al. Palliative radiotherapy for bone metastases: an ASTRO evidence-based guideline. *Int J Radiat Oncol Biol Phys*. 2011;79(4):965–976.
22. Coleman RE, McCloskey EV. Bisphosphonates in oncology. *Bone*. 2011;49(1):71–76.
23. Domchek SM, Younger J, Finkelstein DM, Seiden MV. Predictors of skeletal complications in patients with metastatic breast carcinoma. *Cancer*. 2000;89(2):363–368.
24. Fleisch H. Bisphosphonates: mechanisms of action. *Endocr Rev*. 1998;19(1):80–100.
25. Gralow JR, Biermann JS, Farooki A, et al. NCCN task force report: bone health in cancer care. *Natl Compr Canc Netw*. 2009;7(Suppl 3):S1–S32, quiz S33–35.
26. Sist T, Miner M, Lema M. Characteristics of postradical neck pain syndrome: a report of 25 cases. *J Pain Symptom Manag*. 1999;18(2):95–102.
27. Sist T, Wong C. Difficult problems and their solutions in patients with cancer pain of the head and neck areas. *Curr Rev Pain*. 2000;4(3):206–214.
28. Chew C, Craig L, Edwards R, Moss J, O'Dwyer PJ. Safety and efficacy of percutaneous vertebroplasty in malignancy: a systematic review. *Clin Radiol*. 2011;66(1):63–72.
29. Eisenberg E, Carr DB, Chalmers TC. Neurolytic celiac plexus block for treatment of cancer pain: a meta-analysis. *Anesth Analg*. 1995;80(2):290–295.
30. Deer TR, Smith HS, Burton AW, et al. Comprehensive consensus based guidelines on intrathecal drug delivery systems in the treatment of pain caused by cancer pain. *Pain Physician*. 2011;14(3):E283–E312.
31. Ferrante FM. Neuraxial infusion in the management of cancer pain. *Oncology*. 1999;13(5 Suppl 2):30–36.
32. Landau B, Levy RM. Neuromodulation techniques for medically refractory chronic pain. *Annu Rev Med*. 1993;44:279–287.
33. Smith TJ, Staats PS, Deer T, et al. Randomized clinical trial of an implantable drug delivery system compared with comprehensive medical management for refractory cancer pain: impact on pain, drug-related toxicity, and survival. *J Clin Oncol*. 2002;20(19):4040–4049.
34. Tasker RR. Central pain states. In: Loeser JD, ed. *Bonica's Management of Pain*. Philadelphia: Lippincott Williams & Wilkins; 2001:433–457.
35. Dejerine J, Roussy G. Thalamic syndrome (1). *Rev Neurol*. 1906;14:521–532.
36. Foix C, Hillemand P. Les syndromes de la region thalamique. *Presse Med*. 1925;33:113–117.
37. Foix C, Masson A. Le syndrome de l'artere cerebrale posterieure. *Presse Med*. 1923;31:361–365.
38. Langworthy OR, Fox HM. Thalamic syndrome—syndrome of the posterior cerebral artery, a review. *Arch Intern Med*. 1937;60(2):203–224.

39. Bowsher D, Leijon G, Thoumas KA. Central poststroke pain—correlation of MRI with clinical pain characteristics and sensory abnormalities. *Neurology*. 1998;51(5):1352–1358.
40. Andersen G, Vestergaard K, Ingemannielsen M, Jensen TS. Incidence of central post-stroke pain. *Pain*. 1995;61(2):187–193.
41. Appelros P. Prevalence and predictors of pain and fatigue after stroke: a population-based study. *Int J Rehabil Res*. 2006;29(4):329–333.
42. Bogousslavsky J, Regli F, Uske A. Thalamic infarcts—clinical syndromes, etiology, and prognosis. *Neurology*. 1988;38(6):837–848.
43. Bowsher D. Central pain: clinical and physiological characteristics. *J Neurol Neurosurg Psychiatry*. 1996;61(1):62–69.
44. Bowsher D. Stroke and central poststroke pain in an elderly population. *J Pain*. 2001;2(5):258–261.
45. Indredavik B, Rohweder G, Naalsund E, Lydersen S. Medical complications in a comprehensive stroke unit and an early supported discharge service. *Stroke*. 2008;39(2):414–420.
46. Jonsson AC, Lindgren I, Hallstrom B, Norrving B, Lindgren A. Prevalence and intensity of pain after stroke: a population based study focusing on patients' perspectives. *J Neurol Neurosurg Psychiatry*. 2006;77(5):590–595.
47. Kim JS, Choi-Kwon S. Sensory sequelae of medullary infarction—differences between lateral and medial medullary syndrome. *Stroke*. 1999;30(12):2697–2703.
48. Kong KH, Woon VC, Yang SY. Prevalence of chronic pain and its impact on health-related quality of life in stroke survivors. *Arch Phys Med Rehabil*. 2004;85(1):35–40.
49. Kuptniratsaikul V, Kovindha A, Suethanapornkul S, Manimmanakorn N, Archongka Y. Complications during the rehabilitation period in Thai patients with stroke a multicenter prospective study. *Am J Phys Med Rehabil*. 2009;88(2):92–99.
50. Lampl C, Yazdi K, Roper C. Amitriptyline in the prophylaxis of central poststroke pain—preliminary results of 39 patients in a placebo-controlled, long-term study. *Stroke*. 2002;33(12):3030–3032.
51. Leijon G, Boivie J, Johansson I. Central post-stroke pain—neurological symptoms and pain characteristics. *Pain*. 1989;36(1):13–25.
52. Lundstrom E, Smits A, Terent A, Borg J. Risk factors for stroke-related pain 1 year after first-ever stroke. *Eur J Neurol*. 2009;16(2):188–193.
53. MacGowan DJL, Janal MN, Clark MC, et al. Central poststroke pain and Wallenberg's lateral medullary infarction: frequency, character, and determinants in 63 patients. *Neurology*. 1997;49(1):120–125.
54. Misra UK, Kalita J, Kumar B. A study of clinical, magnetic resonance imaging, and somatosensory-evoked potential in central post-stroke pain. *J Pain*. 2008;9(12):1116–1122.
55. Nasreddine ZS, Saver JL. Pain after thalamic stroke: right diencephalic predominance and clinical features in 180 patients. *Neurology*. 1997;48(5):1196–1199.
56. Sackley C, Brittle N, Patel S, et al. The prevalence of joint contractures, pressure sores, painful shoulder, other pain, falls, and depression in the year after a severely disabling stroke. *Stroke*. 2008;39(12):3329–3334.
57. Weimar C, Kloke M, Schlott M, Katsarava Z, Diener HC. Central poststroke pain in a consecutive cohort of stroke patients. *Cerebrovasc Dis*. 2002;14(3–4):261–263.
58. Widar M, Samuelsson L, Karlsson-Tivenius S, Ahlstrom G. Long-term pain conditions after a stroke. *J Rehabil Med*. 2002;34(4):165–170.
59. Cassinari V, Pagni CA. *Central Pain: A Neurosurgical Survey*. Cambridge, MA: Harvard University Press; 1969.
60. Henry JL, Lalloo C, Yashpal K. Central poststroke pain: an abstruse outcome. *Pain Res Manag*. 2008;13(1):41–49.

61. Boivie J, Leijon G, Johansson I. Central post-stroke pain—a study of the mechanisms through analyses of the sensory abnormalities. *Pain*. 1989;37(2):173–185.
62. Hansson P. Post-stroke pain case study: clinical characteristics, therapeutic options and long-term follow-up. *Eur J Neurol*. 2004;11(Suppl 1):22–30.
63. Koyama S, Katayama Y, Maejima S, Hirayama T, Fujii M, Tsubokawa T. Thalamic neuronal hyperactivity following transection of the spinothalamic tract in the cat: involvement of N-methyl-D-aspartate receptor. *Brain Res*. 1993;612(1–2):345–350.
64. Maruno N, Kaminaga T, Mikami M, Furui S. Activation of supplementary motor area during imaginary movement of phantom toes. *Neurorehabil Neural Repair*. 2000;14(4):345–349.
65. Roux FE, Ibarrola D, Lazorthes Y, Berry I. Chronic motor cortex stimulation for phantom limb pain: a functional magnetic resonance imaging study: technical case report. *Neurosurgery*. 2001;48(3):681–687, discussion 687–688.
66. Hanada T, Kurihara T, Tokudome M, Tokimura H, Arita K, Miyata A. Development and pharmacological verification of a new mouse model of central post-stroke pain. *Neurosci Res*. 2014;78:72–80.
67. Bowsher D. Central post-stroke ('thalamic syndrome') and other central pains. *Am J Hosp Palliat Care*. 1999;16(4):593–597.
68. Leijon G, Boivie J. Central post-stroke pain—a controlled trial of amitriptyline and carbamazepine. *Pain*. 1989;36(1):27–36.
69. Attal N, Guirimand F, Brasseur L, Gaude V, Chauvin M, Bouhassira D. Effects of IV morphine in central pain: a randomized placebo-controlled study. *Neurology*. 2002;58(4):554–563.
70. Rowbotham MC, Twilling L, Davies PS, Reisner L, Taylor K, Mohr D. Oral opioid therapy for chronic peripheral and central neuropathic pain. *N Engl J Med*. 2003;348(13):1223–1232.
71. Hosobuchi Y. Motor cortical stimulation for control of central deafferentation pain. *Adv Neurol*. 1993;63:215–217.
72. Katayama Y, Fukaya C, Yamamoto T. Poststroke pain control by chronic motor cortex stimulation: neurological characteristics predicting a favorable response. *J Neurosurg*. 1998;89(4):585–591.
73. Nandi D, Aziz TZ. Deep brain stimulation in the management of neuropathic pain and multiple sclerosis tremor. *J Clin Neurophysiol*. 2004;21(1):31–39.
74. Nguyen JP, Lefaucheur JP, Decq P, et al. Chronic motor cortex stimulation in the treatment of central and neuropathic pain. Correlations between clinical, electrophysiological and anatomical data. *Pain*. 1999;82(3):245–251.
75. Pirotte B, Voordecker P, Neugroschl C, et al. Combination of functional magnetic resonance imaging-guided neuronavigation and intraoperative cortical brain mapping improves targeting of motor cortex stimulation in neuropathic pain. *Neurosurgery*. 2005;56(2 Suppl):344–359, discussion 344–359.
76. Tsubokawa T, Katayama Y, Yamamoto T, Hirayama T, Koyama S. Chronic motor cortex stimulation for the treatment of central pain. *Acta Neurochir Suppl*. 1991;52:137–139.
77. Backonja MM, Serra J. Pharmacologic management part 2: lesser-studied neuropathic pain diseases. *Pain Med*. 2004;5(Suppl 1):S48–S59.
78. Reiter RC. A profile of women with chronic pelvic pain. *Clin Obstet Gynecol*. 1990;33(1):130–136.
79. Mathias SD, Kuppermann M, Liberman RF, Lipschutz RC, Steege JF. Chronic pelvic pain: prevalence, health-related quality of life, and economic correlates. *Obstet Gynecol*. 1996;87(3):321–327.
80. Zondervan KT, Yudkin PL, Vessey MP, et al. Chronic pelvic pain in the community—symptoms, investigations, and diagnoses. *Am J Obstet Gynecol*. 2001;184(6):1149–1155.

81. Howard FM. The role of laparoscopy in the evaluation of chronic pelvic pain: pitfalls with a negative laparoscopy. *J Am Assoc Gynecol Laparosc.* 1996;4(1):85–94.
82. Zondervan KT, Yudkin PL, Vessey MP, Dawes MG, Barlow DH, Kennedy SH. Patterns of diagnosis and referral in women consulting for chronic pelvic pain in UK primary care. *Br J Obstet Gynaecol.* 1999;106(11):1156–1161.
83. Loos MJ, Scheltinga MR, Mulders LG, Roumen RM. The Pfannenstiel incision as a source of chronic pain. *Obstet Gynecol.* 2008;111(4):839–846.
84. Ness RB, Soper DE, Holley RL, et al. Effectiveness of inpatient and outpatient treatment strategies for women with pelvic inflammatory disease: results from the Pelvic Inflammatory Disease Evaluation and Clinical Health (PEACH) Randomized Trial. *Am J Obstet Gynecol.* 2002;186(5):929–937.
85. Sharp HT. Myofascial pain syndrome of the abdominal wall for the busy clinician. *Clin Obstet Gynecol.* 2003;46(4):783–788.
86. Stanford EJ, Dell JR, Parsons CL. The emerging presence of interstitial cystitis in gynecologic patients with chronic pelvic pain. *Urology.* 2007;69(4 Suppl):53–59.
87. Tu FF, Fitzgerald CM, Kuiken T, Farrell T, Harden RN. Comparative measurement of pelvic floor pain sensitivity in chronic pelvic pain. *Obstet Gynecol.* 2007;110(6):1244–1248.
88. Brandsborg B, Nikolajsen L, Hansen CT, Kehlet H, Jensen TS. Risk factors for chronic pain after hysterectomy: a nationwide questionnaire and database study. *Anesthesiology.* 2007;106(5):1003–1012.
89. Carlson KJ, Miller BA, Fowler Jr FJ. The Maine Women's health study: I. Outcomes of hysterectomy. *Obstet Gynecol.* 1994;83(4):556–565.
90. Hartmann KE, Ma C, Lamvu GM, Langenberg PW, Steege JF, Kjerulff KH. Quality of life and sexual function after hysterectomy in women with preoperative pain and depression. *Obstet Gynecol.* 2004;104(4):701–709.
91. Stovall TG, Ling FW, Crawford DA. Hysterectomy for chronic pelvic pain of presumed uterine etiology. *Obstet Gynecol.* 1990;75(4):676–679.
92. Tay SK, Bromwich N. Outcome of hysterectomy for pelvic pain in premenopausal women. *Aust N Z J Obstet Gynaecol.* 1998;38(1):72–76.
93. Kennedy S, Bergqvist A, Chapron C, et al. ESHRE guideline for the diagnosis and treatment of endometriosis. *Hum Reprod.* 2005;20(10):2698–2704.
94. Vercellini P, Frontino G, De Giorgi O, Pietropaolo G, Pasin R, Crosignani PG. Continuous use of an oral contraceptive for endometriosis-associated recurrent dysmenorrhea that does not respond to a cyclic pill regimen. *Fertil Steril.* 2003;80(3):560–563.
95. Vercellini P, Trespidi L, Colombo A, Vendola N, Marchini M, Crosignani PG. A gonadotropin-releasing hormone agonist versus a low-dose oral contraceptive for pelvic pain associated with endometriosis. *Fertil Steril.* 1993;60(1):75–79.
96. Ling FW. Randomized controlled trial of depot leuprolide in patients with chronic pelvic pain and clinically suspected endometriosis. Pelvic pain study group. *Obstet Gynecol.* 1999;93(1):51–58.
97. Vercellini P, Cortesi I, Crosignani PG. Progestins for symptomatic endometriosis: a critical analysis of the evidence. *Fertil Steril.* 1997;68(3):393–401.
98. FitzGerald MP, Payne CK, Lukacz ES, et al. Randomized multicenter clinical trial of myofascial physical therapy in women with interstitial cystitis/painful bladder syndrome and pelvic floor tenderness. *J Urol.* 2012;187(6):2113–2118.
99. Hanno PM, Burks DA, Clemens JQ, et al. AUA guideline for the diagnosis and treatment of interstitial cystitis/bladder pain syndrome. *J Urol.* 2011;185(6):2162–2170.
100. van Ophoven A, Pokupic S, Heinecke A, Hertle L. A prospective, randomized, placebo controlled, double-blind study of amitriptyline for the treatment of interstitial cystitis. *J Urol.* 2004;172(2):533–536.

101. Parsons CL. Successful downregulation of bladder sensory nerves with combination of heparin and alkalinized lidocaine in patients with interstitial cystitis. *Urology*. 2005;65(1):45–48.
102. Affaitati G, Fabrizio A, Savini A, et al. A randomized, controlled study comparing a lidocaine patch, a placebo patch, and anesthetic injection for treatment of trigger points in patients with myofascial pain syndrome: evaluation of pain and somatic pain thresholds. *Clin Ther*. 2009;31(4):705–720.
103. FitzGerald MP, Anderson RU, Potts J, et al. Randomized multicenter feasibility trial of myofascial physical therapy for the treatment of urological chronic pelvic pain syndromes. *J Urol*. 2009;182(2):570–580.
104. Sator-Katzenschlager SM, Scharbert G, Kress HG, et al. Chronic pelvic pain treated with gabapentin and amitriptyline: a randomized controlled pilot study. *Wien Klin Wochenschr*. 2005;117(21–22):761–768.
105. Chen FP, Chang SD, Chu KK, Soong YK. Comparison of laparoscopic presacral neurectomy and laparoscopic uterine nerve ablation for primary dysmenorrhea. *J Reprod Med*. 1996;41(7):463–466.
106. Mayer RD, Howard FM. Sacral nerve stimulation: neuromodulation for voiding dysfunction and pain. *Neurotherapeutics*. 2008;5(1):107–113.
107. Stanton-Hicks M, Janig W, Hassenbusch S, Haddox JD, Boas R, Wilson P. Reflex sympathetic dystrophy: changing concepts and taxonomy. *Pain*. 1995;63(1):127–133.
108. Field J, Atkins RM. Algodystrophy is an early complication of Colles' fracture. What are the implications? *J Hand Surg (Br)*. 1997;22(2):178–182.
109. Zyluk A, Puchalski P. Complex regional pain syndrome: observations on diagnosis, treatment and definition of a new subgroup. *J Hand Surg Eur Vol*. 2013;38(6):599–606.
110. Galer BS, Jensen M. Neglect-like symptoms in complex regional pain syndrome: results of a self-administered survey. *J Pain Symptom Manage*. 1999;18(3):213–217.
111. Lewis JS, Kersten P, McCabe CS, McPherson KM, Blake DR. Body perception disturbance: a contribution to pain in complex regional pain syndrome (CRPS). *Pain*. 2007;133(1–3):111–119.
112. Penfield W, Rasmussen T. *The cerebral Cortex of Man: A Clinical Study of Localization of Function*. New York: Macmillan; 1950.
113. Moseley GL. Distorted body image in complex regional pain syndrome. *Neurology*. 2005;65(5):773.
114. Field J. Complex regional pain syndrome: a review. *J Hand Surg Eur Vol*. 2013;38(6):616–626.
115. Janig W, Baron R. Complex regional pain syndrome: mystery explained? *Lancet Neurol*. 2003;2(11):687–697.
116. Wasner G, Schattschneider J, Binder A, Baron R. Complex regional pain syndrome—diagnostic, mechanisms, CNS involvement and therapy. *Spinal Cord*. 2003;41(2):61–75.
117. Moseley GL. Imagined movements cause pain and swelling in a patient with complex regional pain syndrome. *Neurology*. 2004;62(9):1644.
118. Moseley GL, Zalucki N, Birklein F, Marinus J, van Hilten JJ, Luomajoki H. Thinking about movement hurts: the effect of motor imagery on pain and swelling in people with chronic arm pain. *Arthritis Rheum*. 2008;59(5):623–631.
119. McCabe CS, Blake DR. An embarrassment of pain perceptions? Towards an understanding of and explanation for the clinical presentation of CRPS type 1. *Rheumatology (Oxford)*. 2008;47(11):1612–1616.
120. de Mos M, de Bruijn AG, Huygen FJ, Dieleman JP, Stricker BH, Sturkenboom MC. The incidence of complex regional pain syndrome: a population-based study. *Pain*. 2007;129(1–2):12–20.

121. Sandroni P, Benrud-Larson LM, McClelland RL, Low PA. Complex regional pain syndrome type I: incidence and prevalence in Olmsted county, a population-based study. *Pain.* 2003;103(1–2):199–207.
122. Allen G, Galer BS, Schwartz L. Epidemiology of complex regional pain syndrome: a retrospective chart review of 134 patients. *Pain.* 1999;80(3):539–544.
123. Duman I, Dincer U, Taskaynatan MA, Cakar E, Tugcu I, Dincer K. Reflex sympathetic dystrophy: a retrospective epidemiological study of 168 patients. *Clin Rheumatol.* 2007;26(9):1433–1437.
124. Veldman PH, Reynen HM, Arntz IE, Goris RJ. Signs and symptoms of reflex sympathetic dystrophy: prospective study of 829 patients. *Lancet.* 1993;342(8878):1012–1016.
125. Schinkel C, Scherens A, Koller M, Roellecke G, Muhr G, Maier C. Systemic inflammatory mediators in post-traumatic complex regional pain syndrome (CRPS I)—longitudinal investigations and differences to control groups. *Eur J Med Res.* 2009;14(3):130–135.
126. Llewelyn JG, Gilbey SG, Thomas PK, King RH, Muddle JR, Watkins PJ. Sural nerve morphometry in diabetic autonomic and painful sensory neuropathy. A clinicopathological study. *Brain.* 1991;114(Pt 2):867–892.
127. Oaklander AL, Rissmiller JG, Gelman LB, Zheng L, Chang Y, Gott R. Evidence of focal small-fiber axonal degeneration in complex regional pain syndrome-I (reflex sympathetic dystrophy). *Pain.* 2006;120(3):235–243.
128. van der Laan L, ter Laak HJ, Gabreels-Festen A, Gabreels F, Goris RJ. Complex regional pain syndrome type I (RSD): pathology of skeletal muscle and peripheral nerve. *Neurology.* 1998;51(1):20–25.
129. Maihofner C, Handwerker HO, Neundorfer B, Birklein F. Cortical reorganization during recovery from complex regional pain syndrome. *Neurology.* 2004;63(4):693–701.
130. Ezendam D, Bongers RM, Jannink MJ. Systematic review of the effectiveness of mirror therapy in upper extremity function. *Disabil Rehabil.* 2009;31(26):2135–2149.
131. Geertzen JH, de Bruijn H, de Bruijn-Kofman AT, Arendzen JH. Reflex sympathetic dystrophy: early treatment and psychological aspects. *Arch Phys Med Rehabil.* 1994;75(4):442–446.
132. Goris RJ, Dongen LM, Winters HA. Are toxic oxygen radicals involved in the pathogenesis of reflex sympathetic dystrophy? *Free Radic Res Commun.* 1987;3(1–5):13–18.
133. Price DD, Long S, Wilsey B, Rafii A. Analysis of peak magnitude and duration of analgesia produced by local anesthetics injected into sympathetic ganglia of complex regional pain syndrome patients. *Clin J Pain.* 1998;14(3):216–226.
134. Kemler MA, de Vet HC, Barendse GA, van den Wildenberg FA, van Kleef M. Effect of spinal cord stimulation for chronic complex regional pain syndrome Type I: five-year final follow-up of patients in a randomized controlled trial. *J Neurosurg.* 2008;108(2):292–298.
135. Deyo RA, Gray DT, Kreuter W, Mirza S, Martin BI. United States trends in lumbar fusion surgery for degenerative conditions. *Spine (Phila Pa 1976).* 2005;30(12):1441–1445, discussion 1446–1447.
136. Javid MJ, Hadar EJ. Long-term follow-up review of patients who underwent laminectomy for lumbar stenosis: a prospective study. *J Neurosurg.* 1998;89(1):1–7.
137. Yorimitsu E, Chiba K, Toyama Y, Hirabayashi K. Long-term outcomes of standard discectomy for lumbar disc herniation: a follow-up study of more than 10 years. *Spine (Phila Pa 1976).* 2001;26(6):652–657.
138. Slipman CW, Shin CH, Patel RK, et al. Etiologies of failed back surgery syndrome. *Pain Med.* 2002;3(3):200–214, discussion 214–207.
139. Waguespack A, Schofferman J, Slosar P, Reynolds J. Etiology of long-term failures of lumbar spine surgery. *Pain Med.* 2002;3(1):18–22.

140. Burton CV, Kirkaldy-Willis WH, Yong-Hing K, Heithoff KB. Causes of failure of surgery on the lumbar spine. *Clin Orthop Relat Res.* 1981;157:191–199.
141. Hussain A, Erdek M. Interventional pain management for failed back surgery syndrome. *Pain Pract.* 2014;14(1):64–78.
142. Burton CV. Causes of failure of surgery on the lumbar spine: ten-year follow-up. *Mt Sinai J Med.* 1991;58(2):183–187.
143. Klekamp J, McCarty E, Spengler DM. Results of elective lumbar discectomy for patients involved in the workers' compensation system. *J Spinal Disord.* 1998;11(4):277–282.
144. Schofferman J, Reynolds J, Herzog R, Covington E, Dreyfuss P, O'Neill C. Failed back surgery: etiology and diagnostic evaluation. *Spine J.* 2003;3(5):400–403.
145. Trief PM, Grant W, Fredrickson B. A prospective study of psychological predictors of lumbar surgery outcome. *Spine (Phila Pa 1976).* 2000;25(20):2616–2621.
146. Williams RW. Microlumbar discectomy: a conservative surgical approach to the virgin herniated lumbar disc. *Spine (Phila Pa 1976).* 1978;3(2):175–182.
147. Krishna M, Pollock RD, Bhatia C. Incidence, etiology, classification, and management of neuralgia after posterior lumbar interbody fusion surgery in 226 patients. *Spine J.* 2008;8(2):374–379.
148. Philadelphia P. Philadelphia Panel evidence-based clinical practice guidelines on selected rehabilitation interventions for low back pain. *Phys Ther.* 2001;81(10):1641–1674.
149. White AP, Arnold PM, Norvell DC, Ecker E, Fehlings MG. Pharmacologic management of chronic low back pain: synthesis of the evidence. *Spine.* 2011;36(21 Suppl): S131–S143.
150. Kumar K, Taylor RS, Jacques L, et al. Spinal cord stimulation versus conventional medical management for neuropathic pain: a multicentre randomised controlled trial in patients with failed back surgery syndrome. *Pain.* 2007;132(1–2):179–188.
151. Kumar K, Taylor RS, Jacques L, et al. The effects of spinal cord stimulation in neuropathic pain are sustained: a 24-month follow-up of the prospective randomized controlled multicenter trial of the effectiveness of spinal cord stimulation. *Neurosurgery.* 2008;63(4):762–770, discussion 770.
152. North RB, Kidd D, Shipley J, Taylor RS. Spinal cord stimulation versus reoperation for failed back surgery syndrome: a cost effectiveness and cost utility analysis based on a randomized, controlled trial. *Neurosurgery.* 2007;61(2):361–368, discussion 368–369.
153. Verrills P, Mitchell B, Vivian D, Sinclair C. Peripheral nerve stimulation: a treatment for chronic low back pain and failed back surgery syndrome? *Neuromodulation.* 2009;12(1):68–75.
154. Yakovlev AE, Resch BE, Yakovleva VE. Peripheral nerve field stimulation in the treatment of postlaminectomy syndrome after multilevel spinal surgeries. *Neuromodulation.* 2011;14(6):534–538, discussion 538.
155. Bennett RM. Clinical manifestations and diagnosis of fibromyalgia. *Rheum Dis Clin N Am.* 2009;35(2):215–232.
156. Goldenberg DL. Diagnosis and differential diagnosis of fibromyalgia. *Am J Med.* 2009; 122(12 Suppl):S14–S21.
157. Glass JM. Cognitive dysfunction in fibromyalgia and chronic fatigue syndrome: new trends and future directions. *Curr Rheumatol Rep.* 2006;8(6):425–429.
158. Fuller-Thomson E, Nimigon-Young J, Brennenstuhl S. Individuals with fibromyalgia and depression: findings from a nationally representative Canadian survey. *Rheumatol Int.* 2012;32(4):853–862.
159. Wolfe F, Clauw DJ, Fitzcharles MA, et al. The American College of Rheumatology preliminary diagnostic criteria for fibromyalgia and measurement of symptom severity. *Arthritis Care Res.* 2010;62(5):600–610.

160. Goldenberg DL. Fibromyalgia syndrome a decade later: what have we learned? *Arch Intern Med.* 1999;159(8):777–785.
161. Goldenberg DL. Fibromyalgia syndrome: an emerging but controversial condition. *JAMA.* 2014;311(15):1570.
162. Vincent A, Lahr BD, Wolfe F, et al. Prevalence of fibromyalgia: a population-based study in Olmsted County, Minnesota, utilizing the Rochester Epidemiology Project. *Arthritis Care Res.* 2013;65(5):786–792.
163. Albrecht PJ, Hou Q, Argoff CE, Storey JR, Wymer JP, Rice FL. Excessive peptidergic sensory innervation of cutaneous arteriole-venule shunts (AVS) in the palmar glabrous skin of fibromyalgia patients: implications for widespread deep tissue pain and fatigue. *Pain Med.* 2013;14(6):895–915.
164. Bengtsson A, Henriksson KG, Larsson J. Reduced high-energy phosphate levels in the painful muscles of patients with primary fibromyalgia. *Arthritis Rheum.* 1986;29(7):817–821.
165. Gracely RH, Petzke F, Wolf JM, Clauw DJ. Functional magnetic resonance imaging evidence of augmented pain processing in fibromyalgia. *Arthritis Rheum.* 2002;46(5):1333–1343.
166. Abeles AM, Pillinger MH, Solitar BM, Abeles M. Narrative review: the pathophysiology of fibromyalgia. *Ann Intern Med.* 2007;146(10):726–734.
167. Hassett AL, Gevirtz RN. Nonpharmacologic treatment for fibromyalgia: patient education, cognitive-behavioral therapy, relaxation techniques, and complementary and alternative medicine. *Rheum Dis Clin N Am.* 2009;35(2):393–407.
168. Annemans L, Wessely S, Spaepen E, et al. Health economic consequences related to the diagnosis of fibromyalgia syndrome. *Arthritis Rheum.* 2008;58(3):895–902.
169. Hughes G, Martinez C, Myon E, Taieb C, Wessely S. The impact of a diagnosis of fibromyalgia on health care resource use by primary care patients in the UK: an observational study based on clinical practice. *Arthritis Rheum.* 2006;54(1):177–183.
170. Pfeiffer A, Thompson JM, Nelson A, et al. Effects of a 1.5-day multidisciplinary outpatient treatment program for fibromyalgia: a pilot study. *Am J Phys Med Rehabil.* 2003;82(3):186–191.
171. Busch AJ, Schachter CL, Overend TJ, Peloso PM, Barber KA. Exercise for fibromyalgia: a systematic review. *J Rheumatol.* 2008;35(6):1130–1144.
172. Carette S, Bell MJ, Reynolds WJ, et al. Comparison of amitriptyline, cyclobenzaprine, and placebo in the treatment of fibromyalgia. A randomized, double-blind clinical trial. *Arthritis Rheum.* 1994;37(1):32–40.
173. Carette S, McCain GA, Bell DA, Fam AG. Evaluation of amitriptyline in primary fibrositis. A double-blind, placebo-controlled study. *Arthritis Rheum.* 1986;29(5):655–659.
174. Goldenberg DL, Felson DT, Dinerman H. A randomized, controlled trial of amitriptyline and naproxen in the treatment of patients with fibromyalgia. *Arthritis Rheum.* 1986;29(11):1371–1377.
175. Lunn MP, Hughes RA, Wiffen PJ. Duloxetine for treating painful neuropathy, chronic pain or fibromyalgia. *Cochrane Database Syst Rev.* 2014;1, CD007115.
176. Moore A, Wiffen P, Kalso E. Antiepileptic drugs for neuropathic pain and fibromyalgia. *JAMA.* 2014;312(2):182–183.
177. Wiffen PJ, Derry S, Moore RA, et al. Antiepileptic drugs for neuropathic pain and fibromyalgia—an overview of Cochrane reviews. *Cochrane Database Syst Rev.* 2013;11, CD010567.
178. Hong CZ, Hsueh TC. Difference in pain relief after trigger point injections in myofascial pain patients with and without fibromyalgia. *Arch Phys Med Rehabil.* 1996;77(11):1161–1166.

179. American College of Gastroenterology Task Force on Irritable Bowel Syndrome, Brandt LJ, Chey WD, et al. An evidence-based position statement on the management of irritable bowel syndrome. *Am J Gastroenterol.* 2009;104(Suppl 1):S1–S35.
180. Longstreth GF, Thompson WG, Chey WD, Houghton LA, Mearin F, Spiller RC. Functional bowel disorders. *Gastroenterology.* 2006;130(5):1480–1491.
181. Swarbrick ET, Hegarty JE, Bat L, Williams CB, Dawson AM. Site of pain from the irritable bowel. *Lancet.* 1980;2(8192):443–446.
182. Drossman DA, Li Z, Andruzzi E, et al. U.S. householder survey of functional gastrointestinal disorders. Prevalence, sociodemography, and health impact. *Dig Dis Sci.* 1993;38(9):1569–1580.
183. Hahn BA, Saunders WB, Maier WC. Differences between individuals with self-reported irritable bowel syndrome (IBS) and IBS-like symptoms. *Dig Dis Sci.* 1997;42(12):2585–2590.
184. Hungin AP, Whorwell PJ, Tack J, Mearin F. The prevalence, patterns and impact of irritable bowel syndrome: an international survey of 40,000 subjects. *Aliment Pharmacol Ther.* 2003;17(5):643–650.
185. Saito YA, Locke GR, Talley NJ, Zinsmeister AR, Fett SL, Melton 3rd LJ. A comparison of the Rome and Manning criteria for case identification in epidemiological investigations of irritable bowel syndrome. *Am J Gastroenterol.* 2000;95(10):2816–2824.
186. Talley NJ, Zinsmeister AR, Van Dyke C, Melton 3rd LJ. Epidemiology of colonic symptoms and the irritable bowel syndrome. *Gastroenterology.* 1991;101(4):927–934.
187. Thompson WG, Irvine EJ, Pare P, Ferrazzi S, Rance L. Functional gastrointestinal disorders in Canada: first population-based survey using Rome II criteria with suggestions for improving the questionnaire. *Dig Dis Sci.* 2002;47(1):225–235.
188. Lovell RM, Ford AC. Global prevalence of and risk factors for irritable bowel syndrome: a meta-analysis. *Clin Gastroenterol Hepatol.* 2012;10(7):712–721, e714.
189. Ford AC, Forman D, Bailey AG, Axon AT, Moayyedi P. Irritable bowel syndrome: a 10-yr natural history of symptoms and factors that influence consultation behavior. *Am J Gastroenterol.* 2008;103(5):1229–1239, quiz 1240.
190. Heaton KW, O'Donnell LJ, Braddon FE, Mountford RA, Hughes AO, Cripps PJ. Symptoms of irritable bowel syndrome in a British urban community: consulters and nonconsulters. *Gastroenterology.* 1992;102(6):1962–1967.
191. Jones R, Lydeard S. Irritable bowel syndrome in the general population. *BMJ.* 1992;304(6819):87–90.
192. Sandler RS, Everhart JE, Donowitz M, et al. The burden of selected digestive diseases in the United States. *Gastroenterology.* 2002;122(5):1500–1511.
193. Schuster MM. Diagnostic evaluation of the irritable bowel syndrome. *Gastroenterol Clin N Am.* 1991;20(2):269–278.
194. Camilleri M. Peripheral mechanisms in irritable bowel syndrome. *N Engl J Med.* 2012;367(17):1626–1635.
195. Chey WY, Jin HO, Lee MH, Sun SW, Lee KY. Colonic motility abnormality in patients with irritable bowel syndrome exhibiting abdominal pain and diarrhea. *Am J Gastroenterol.* 2001;96(5):1499–1506.
196. Willert RP, Woolf CJ, Hobson AR, Delaney C, Thompson DG, Aziz Q. The development and maintenance of human visceral pain hypersensitivity is dependent on the N-methyl-D-aspartate receptor. *Gastroenterology.* 2004;126(3):683–692.
197. Barbara G, Stanghellini V, De Giorgio R, et al. Activated mast cells in proximity to colonic nerves correlate with abdominal pain in irritable bowel syndrome. *Gastroenterology.* 2004;126(3):693–702.
198. Chadwick VS, Chen W, Shu D, et al. Activation of the mucosal immune system in irritable bowel syndrome. *Gastroenterology.* 2002;122(7):1778–1783.

199. Liebregts T, Adam B, Bredack C, et al. Immune activation in patients with irritable bowel syndrome. *Gastroenterology*. 2007;132(3):913–920.
200. Thabane M, Kottachchi DT, Marshall JK. Systematic review and meta-analysis: the incidence and prognosis of post-infectious irritable bowel syndrome. *Aliment Pharmacol Ther*. 2007;26(4):535–544.
201. Maccaferri S, Candela M, Turroni S, et al. IBS-associated phylogenetic unbalances of the intestinal microbiota are not reverted by probiotic supplementation. *Gut Microbes*. 2012;3(5):406–413.
202. Nobaek S, Johansson ML, Molin G, Ahrne S, Jeppsson B. Alteration of intestinal microflora is associated with reduction in abdominal bloating and pain in patients with irritable bowel syndrome. *Am J Gastroenterol*. 2000;95(5):1231–1238.
203. Chang L. The role of stress on physiologic responses and clinical symptoms in irritable bowel syndrome. *Gastroenterology*. 2011;140(3):761–765.
204. Saito YA, Petersen GM, Locke 3rd GR, Talley NJ. The genetics of irritable bowel syndrome. *Clin Gastroenterol Hepatol*. 2005;3(11):1057–1065.
205. Sanders DS, Carter MJ, Hurlstone DP, et al. Association of adult coeliac disease with irritable bowel syndrome: a case-control study in patients fulfilling ROME II criteria referred to secondary care. *Lancet*. 2001;358(9292):1504–1508.
206. Chapman RW, Stanghellini V, Geraint M, Halphen M. Randomized clinical trial: macrogol/PEG 3350 plus electrolytes for treatment of patients with constipation associated with irritable bowel syndrome. *Am J Gastroenterol*. 2013;108(9):1508–1515.
207. Khoshoo V, Armstead C, Landry L. Effect of a laxative with and without tegaserod in adolescents with constipation predominant irritable bowel syndrome. *Aliment Pharmacol Ther*. 2006;23(1):191–196.
208. Chey WD, Lembo AJ, Lavins BJ, et al. Linaclotide for irritable bowel syndrome with constipation: a 26-week, randomized, double-blind, placebo-controlled trial to evaluate efficacy and safety. *Am J Gastroenterol*. 2012;107(11):1702–1712.
209. Drossman DA, Chey WD, Johanson JF, et al. Clinical trial: lubiprostone in patients with constipation-associated irritable bowel syndrome—results of two randomized, placebo-controlled studies. *Aliment Pharmacol Ther*. 2009;29(3):329–341.
210. Rao S, Lembo AJ, Shiff SJ, et al. A 12-week, randomized, controlled trial with a 4-week randomized withdrawal period to evaluate the efficacy and safety of linaclotide in irritable bowel syndrome with constipation. *Am J Gastroenterol*. 2012;107(11):1714–1724, quiz p 1725.
211. Scott LJ, Perry CM. Tegaserod. *Drugs*. 1999;58(3):491–496, discussion 497–498.
212. Cann PA, Read NW, Holdsworth CD, Barends D. Role of loperamide and placebo in management of irritable bowel syndrome (IBS). *Dig Dis Sci*. 1984;29(3):239–247.
213. Efskind PS, Bernklev T, Vatn MH. A double-blind placebo-controlled trial with loperamide in irritable bowel syndrome. *Scand J Gastroenterol*. 1996;31(5):463–468.
214. Hovdenak N. Loperamide treatment of the irritable bowel syndrome. *Scand J Gastroenterol Suppl*. 1987;130:81–84.
215. Fernandez-Banares F, Esteve M, Salas A, et al. Systematic evaluation of the causes of chronic watery diarrhea with functional characteristics. *Am J Gastroenterol*. 2007;102(11):2520–2528.
216. Wedlake L, A'Hern R, Russell D, Thomas K, Walters JR, Andreyev HJ. Systematic review: the prevalence of idiopathic bile acid malabsorption as diagnosed by SeHCAT scanning in patients with diarrhoea-predominant irritable bowel syndrome. *Aliment Pharmacol Ther*. 2009;30(7):707–717.
217. Andresen V, Montori VM, Keller J, West CP, Layer P, Camilleri M. Effects of 5-hydroxytryptamine (serotonin) type 3 antagonists on symptom relief and constipation in nonconstipated irritable bowel syndrome: a systematic review and meta-analysis of randomized controlled trials. *Clin Gastroenterol Hepatol*. 2008;6(5):545–555.

218. Gershon MD. Serotonin and its implication for the management of irritable bowel syndrome. *Rev Gastroenterol Disord.* 2003;3(Suppl 2):S25–S34.
219. Prior A, Read NW. Reduction of rectal sensitivity and post-prandial motility by granisetron, a 5 HT3-receptor antagonist, in patients with irritable bowel syndrome. *Aliment Pharmacol Ther.* 1993;7(2):175–180.
220. Zighelboim J, Talley NJ, Phillips SF, Harmsen WS, Zinsmeister AR. Visceral perception in irritable bowel syndrome. Rectal and gastric responses to distension and serotonin type 3 antagonism. *Dig Dis Sci.* 1995;40(4):819–827.
221. Clouse RE, Lustman PJ, Geisman RA, Alpers DH. Antidepressant therapy in 138 patients with irritable bowel syndrome: a five-year clinical experience. *Aliment Pharmacol Ther.* 1994;8(4):409–416.
222. Gorard DA, Libby GW, Farthing MJ. Effect of a tricyclic antidepressant on small intestinal motility in health and diarrhea-predominant irritable bowel syndrome. *Dig Dis Sci.* 1995;40(1):86–95.
223. Tabas G, Beaves M, Wang J, Friday P, Mardini H, Arnold G. Paroxetine to treat irritable bowel syndrome not responding to high-fiber diet: a double-blind, placebo-controlled trial. *Am J Gastroenterol.* 2004;99(5):914–920.
224. Olesen J, Lipton RB. Migraine classification and diagnosis. International Headache Society criteria. *Neurology.* 1994;44(6 Suppl 4):S6–S10.
225. Nappi G. Introduction to the new international classification of headache disorders. *J Headache Pain.* 2005;6(4):203–204.
226. Miller NR, Walsh FB, Hoyt WF. *Walsh and Hoyt's Clinical Neuro-Ophthalmology.* Philadelphia: Lippincott Williams & Wilkins; 2005.
227. Charles A. Vasodilation out of the picture as a cause of migraine headache. *Lancet Neurol.* 2013;12(5):419–420.
228. Lipton RB, Bigal ME, Diamond M, et al. Migraine prevalence, disease burden, and the need for preventive therapy. *Neurology.* 2007;68(5):343–349.
229. Lipton RB, Stewart WF, Diamond S, Diamond ML, Reed M. Prevalence and burden of migraine in the United States: data from the American Migraine Study II. *Headache.* 2001;41(7):646–657.
230. Leao AA. Further observations on the spreading depression of activity in the cerebral cortex. *J Neurophysiol.* 1947;10(6):409–414.
231. Arbab MA, Wiklund L, Svendgaard NA. Origin and distribution of cerebral vascular innervation from superior cervical, trigeminal and spinal ganglia investigated with retrograde and anterograde WGA-HRP tracing in the rat. *Neuroscience.* 1986;19(3):695–708.
232. Bartsch T, Goadsby PJ. Stimulation of the greater occipital nerve induces increased central excitability of dural afferent input. *Brain.* 2002;125(Pt 7):1496–1509.
233. Bartsch T, Goadsby PJ. Increased responses in trigeminocervical nociceptive neurons to cervical input after stimulation of the dura mater. *Brain.* 2003;126(Pt 8):1801–1813.
234. Burstein R. Deconstructing migraine headache into peripheral and central sensitization. *Pain.* 2001;89(2–3):107–110.
235. Kaube H, Katsarava Z, Przywara S, Drepper J, Ellrich J, Diener HC. Acute migraine headache: possible sensitization of neurons in the spinal trigeminal nucleus? *Neurology.* 2002;58(8):1234–1238.
236. Strassman AM, Raymond SA, Burstein R. Sensitization of meningeal sensory neurons and the origin of headaches. *Nature.* 1996;384(6609):560–564.
237. Benarroch EE. CGRP: sensory neuropeptide with multiple neurologic implications. *Neurology.* 2011;77(3):281–287.
238. Hamel E. Serotonin and migraine: biology and clinical implications. *Cephalalgia.* 2007;27(11):1293–1300.

239. Lassen LH, Haderslev PA, Jacobsen VB, Iversen HK, Sperling B, Olesen J. CGRP may play a causative role in migraine. *Cephalalgia*. 2002;22(1):54–61.
240. Panconesi A. Serotonin and migraine: a reconsideration of the central theory. *J Headache Pain*. 2008;9(5):267–276.
241. Kirthi V, Derry S, Moore RA, McQuay HJ. Aspirin with or without an antiemetic for acute migraine headaches in adults. *Cochrane Database Syst Rev*. 2010;4, CD008041.
242. Law S, Derry S, Moore RA. Naproxen with or without an antiemetic for acute migraine headaches in adults. *Cochrane Database Syst Rev*. 2013;10, CD009455.
243. Rabbie R, Derry S, Moore RA, McQuay HJ. Ibuprofen with or without an antiemetic for acute migraine headaches in adults. *Cochrane Database Syst Rev*. 2010;10, CD008039.
244. Derry S, Moore RA, McQuay HJ. Paracetamol (acetaminophen) with or without an antiemetic for acute migraine headaches in adults. *Cochrane Database Syst Rev*. 2010;11, CD008040.
245. Bartsch T, Knight YE, Goadsby PJ. Activation of 5-HT(1B/1D) receptor in the periaqueductal gray inhibits nociception. *Ann Neurol*. 2004;56(3):371–381.
246. Tfelt-Hansen P, De Vries P, Saxena PR. Triptans in migraine: a comparative review of pharmacology, pharmacokinetics and efficacy. *Drugs*. 2000;60(6):1259–1287.
247. Kelley NE, Tepper DE. Rescue therapy for acute migraine, part 2: neuroleptics, antihistamines, and others. *Headache*. 2012;52(2):292–306.
248. Capobianco DJ, Cheshire WP, Campbell JK. An overview of the diagnosis and pharmacologic treatment of migraine. *Mayo Clin Proc*. 1996;71(11):1055–1066.
249. Langer-Gould AM, Anderson WE, Armstrong MJ, et al. The American Academy of Neurology's top five choosing wisely recommendations. *Neurology*. 2013;81(11):1004–1011.
250. Pringsheim T, Davenport WJ, Becker WJ. Prophylaxis of migraine headache. *Can Med Assoc J*. 2010;182(7):E269–E276.
251. Schrader H, Stovner LJ, Helde G, Sand T, Bovim G. Prophylactic treatment of migraine with angiotensin converting enzyme inhibitor (lisinopril): randomised, placebo controlled, crossover study. *BMJ*. 2001;322(7277):19–22.
252. Silberstein SD, Holland S, Freitag F, et al. Evidence-based guideline update: pharmacologic treatment for episodic migraine prevention in adults: report of the Quality Standards Subcommittee of the American Academy of Neurology and the American Headache Society. *Neurology*. 2012;78(17):1337–1345.
253. Chronicle E, Mulleners W. Anticonvulsant drugs for migraine prophylaxis. *Cochrane Database Syst Rev*. 2004;3, CD003226.
254. Elkind AH, O'Carroll P, Blumenfeld A, DeGryse R, Dimitrova R, Bo NTASG. A series of three sequential, randomized, controlled studies of repeated treatments with botulinum toxin type A for migraine prophylaxis. *J Pain*. 2006;7(10):688–696.
255. Evers S, Vollmer-Haase J, Schwaag S, Rahmann A, Husstedt IW, Frese A. Botulinum toxin A in the prophylactic treatment of migraine—a randomized, double-blind, placebo-controlled study. *Cephalalgia*. 2004;24(10):838–843.
256. Silberstein S, Mathew N, Saper J, Jenkins S. Botulinum toxin type A as a migraine preventive treatment. For the BOTOX Migraine Clinical Research Group. *Headache*. 2000;40(6):445–450.
257. Sandor PS, Di Clemente L, Coppola G, et al. Efficacy of coenzyme Q10 in migraine prophylaxis: a randomized controlled trial. *Neurology*. 2005;64(4):713–715.
258. Saper JR, Dodick DW, Silberstein SD, et al. Occipital nerve stimulation for the treatment of intractable chronic migraine headache: ONSTIM feasibility study. *Cephalalgia*. 2011;31(3):271–285.
259. Silberstein SD, Dodick DW, Saper J, et al. Safety and efficacy of peripheral nerve stimulation of the occipital nerves for the management of chronic migraine: results from a

randomized, multicenter, double-blinded, controlled study. *Cephalalgia.* 2012;32(16):1165–1179.
260. Moulin DE, Foley KM, Ebers GC. Pain syndromes in multiple sclerosis. *Neurology.* 1988;38(12):1830–1834.
261. O'Connor AB, Schwid SR, Herrmann DN, Markman JD, Dworkin RH. Pain associated with multiple sclerosis: systematic review and proposed classification. *Pain.* 2008;137(1):96–111.
262. Osterberg A, Boivie J, Thuomas KA. Central pain in multiple sclerosis—prevalence and clinical characteristics. *Eur J Pain.* 2005;9(5):531–542.
263. Svendsen KB, Jensen TS, Hansen HJ, Bach FW. Sensory function and quality of life in patients with multiple sclerosis and pain. *Pain.* 2005;114(3):473–481.
264. Pollmann W, Erasmus LP, Feneberg W, Then Bergh F, Straube A. Interferon beta but not glatiramer acetate therapy aggravates headaches in MS. *Neurology.* 2002;59(4):636–639.
265. Smith KJ, McDonald WI. The pathophysiology of multiple sclerosis: the mechanisms underlying the production of symptoms and the natural history of the disease. *Philos Trans R Soc Lond B Biol Sci.* 1999;354(1390):1649–1673.
266. Beard S, Hunn A, Wight J. Treatments for spasticity and pain in multiple sclerosis: a systematic review. *Health Technol Assess.* 2003;7(40):iii, ix–x, 1–111.
267. Snow BJ, Tsui JK, Bhatt MH, Varelas M, Hashimoto SA, Calne DB. Treatment of spasticity with botulinum toxin: a double-blind study. *Ann Neurol.* 1990;28(4):512–515.
268. Rog DJ, Nurmikko TJ, Friede T, Young CA. Randomized, controlled trial of cannabis-based medicine in central pain in multiple sclerosis. *Neurology.* 2005;65(6):812–819.
269. Svendsen KB, Jensen TS, Bach FW. Does the cannabinoid dronabinol reduce central pain in multiple sclerosis? Randomised double blind placebo controlled crossover trial. *BMJ.* 2004;329(7460):253.
270. Zajicek J, Fox P, Sanders H, et al. Cannabinoids for treatment of spasticity and other symptoms related to multiple sclerosis (CAMS study): multicentre randomised placebo-controlled trial. *Lancet.* 2003;362(9395):1517–1526.
271. Mols F, Beijers T, Vreugdenhil G, van de Poll-Franse L. Chemotherapy-induced peripheral neuropathy and its association with quality of life: a systematic review. *Support Care Cancer.* 2014;22(8):2261–2269.
272. Wolf S, Barton D, Kottschade L, Grothey A, Loprinzi C. Chemotherapy-induced peripheral neuropathy: prevention and treatment strategies. *Eur J Cancer.* 2008;44(11):1507–1515.
273. Quasthoff S, Hartung HP. Chemotherapy-induced peripheral neuropathy. *J Neurol.* 2002;249(1):9–17.
274. Cassidy J, Paul J, Soukop M, et al. Clinical trials of nimodipine as a potential neuroprotector in ovarian cancer patients treated with cisplatin. *Cancer Chemother Pharmacol.* 1998;41(2):161–166.
275. Castiglione F, Dalla Mola A, Porcile G. Protection of normal tissues from radiation and cytotoxic therapy: the development of amifostine. *Tumori.* 1999;85(2):85–91.
276. Davis ID, Kiers L, MacGregor L, et al. A randomized, double-blinded, placebo-controlled phase II trial of recombinant human leukemia inhibitory factor (rhuLIF, emfilermin, AM424) to prevent chemotherapy-induced peripheral neuropathy. *Clin Cancer Res.* 2005;11(5):1890–1898.
277. De Vos FY, Bos AM, Schaapveld M, et al. A randomized phase II study of paclitaxel with carboplatin +/- amifostine as first line treatment in advanced ovarian carcinoma. *Gynecol Oncol.* 2005;97(1):60–67.
278. Gallardo D, Mohar A, Calderillo G, et al. Cisplatin, radiation, and amifostine in carcinoma of the uterine cervix. *Int J Gynecol Cancer.* 1999;9(3):225–230.

279. Gandara DR, Nahhas WA, Adelson MD, et al. Randomized placebo-controlled multicenter evaluation of diethyldithiocarbamate for chemoprotection against cisplatin-induced toxicities. *J Clin Oncol.* 1995;13(2):490–496.
280. Gelmon K, Eisenhauer E, Bryce C, et al. Randomized phase II study of high-dose paclitaxel with or without amifostine in patients with metastatic breast cancer. *J Clin Oncol.* 1999;17(10):3038–3047.
281. Glover D, Ibrahim J, Kirkwood J, et al. Phase II randomized trial of cisplatin and WR-2721 versus cisplatin alone for metastatic melanoma: an Eastern Cooperative Oncology Group Study (E1686). *Melanoma Res.* 2003;13(6):619–626.
282. Hilpert F, Stahle A, Tome O, et al. Neuroprotection with amifostine in the first-line treatment of advanced ovarian cancer with carboplatin/paclitaxel-based chemotherapy—a double-blind, placebo-controlled, randomized phase II study from the Arbeitsgemeinschaft Gynakologische Onkologoie (AGO) Ovarian Cancer Study Grou. *Support Care Cancer.* 2005;13(10):797–805.
283. Kanat O, Evrensel T, Baran I, et al. Protective effect of amifostine against toxicity of paclitaxel and carboplatin in non-small cell lung cancer: a single center randomized study. *Med Oncol.* 2003;20(3):237–245.
284. Kemp G, Rose P, Lurain J, et al. Amifostine pretreatment for protection against cyclophosphamide-induced and cisplatin-induced toxicities: results of a randomized control trial in patients with advanced ovarian cancer. *J Clin Oncol.* 1996;14(7):2101–2112.
285. Leong SS, Tan EH, Fong KW, et al. Randomized double-blind trial of combined modality treatment with or without amifostine in unresectable stage III non-small-cell lung cancer. *J Clin Oncol.* 2003;21(9):1767–1774.
286. Lorusso D, Ferrandina G, Greggi S, et al. Phase III multicenter randomized trial of amifostine as cytoprotectant in first-line chemotherapy in ovarian cancer patients. *Ann Oncol.* 2003;14(7):1086–1093.
287. Moore DH, Donnelly J, McGuire WP, et al. Limited access trial using amifostine for protection against cisplatin- and three-hour paclitaxel-induced neurotoxicity: a phase II study of the Gynecologic Oncology Group. *J Clin Oncol.* 2003;21(22):4207–4213.
288. Planting AS, Catimel G, de Mulder PH, et al. Randomized study of a short course of weekly cisplatin with or without amifostine in advanced head and neck cancer. EORTC Head and Neck Cooperative Group. *Ann Oncol.* 1999;10(6):693–700.
289. Zhang RX, Lu ZH, Wan DS, et al. Neuroprotective effect of neurotropin on chronic oxaliplatin-induced neurotoxicity in stage II and stage III colorectal cancer patients: results from a prospective, randomised, single-centre, pilot clinical trial. *Int J Color Dis.* 2012;27(12):1645–1650.
290. Barton DL, Wos EJ, Qin R, et al. A double-blind, placebo-controlled trial of a topical treatment for chemotherapy-induced peripheral neuropathy: NCCTG trial N06CA. *Support Care Cancer.* 2011;19(6):833–841.
291. Gewandter JS, Mohile SG, Heckler CE, et al. A phase III randomized, placebo-controlled study of topical amitriptyline and ketamine for chemotherapy-induced peripheral neuropathy (CIPN): a University of Rochester CCOP study of 462 cancer survivors. *Support Care Cancer.* 2014;22(7):1807–1814.
292. Richardson PG, Briemberg H, Jagannath S, et al. Frequency, characteristics, and reversibility of peripheral neuropathy during treatment of advanced multiple myeloma with bortezomib. *J Clin Oncol.* 2006;24(19):3113–3120.
293. Cata JP, Cordella JV, Burton AW, Hassenbusch SJ, Weng HR, Dougherty PM. Spinal cord stimulation relieves chemotherapy-induced pain: a clinical case report. *J Pain Symptom Manag.* 2004;27(1):72–78.
294. Ferrari S, Vento S, Monaco S, et al. Human immunodeficiency virus-associated peripheral neuropathies. *Mayo Clin Proc.* 2006;81(2):213–219.

295. Manji H, Miller R. The neurology of HIV infection. *J Neurol Neurosurg Psychiatry*. 2004;75(Suppl 1):i29–i35.
296. Cornblath DR, McArthur JC. Predominantly sensory neuropathy in patients with AIDS and AIDS-related complex. *Neurology*. 1988;38(5):794–796.
297. de la Monte SM, Gabuzda DH, Ho DD, et al. Peripheral neuropathy in the acquired immunodeficiency syndrome. *Ann Neurol*. 1988;23(5):485–492.
298. Lewis W, Dalakas MC. Mitochondrial toxicity of antiviral drugs. *Nat Med*. 1995;1(5):417–422.
299. Lichtenstein KA, Armon C, Baron A, et al. Modification of the incidence of drug-associated symmetrical peripheral neuropathy by host and disease factors in the HIV outpatient study cohort. *Clin Infect Dis*. 2005;40(1):148–157.
300. So YT, Holtzman DM, Abrams DI, Olney RK. Peripheral neuropathy associated with acquired immunodeficiency syndrome. Prevalence and clinical features from a population-based survey. *Arch Neurol*. 1988;45(9):945–948.
301. Melli G, Keswani SC, Fischer A, Chen W, Hoke A. Spatially distinct and functionally independent mechanisms of axonal degeneration in a model of HIV-associated sensory neuropathy. *Brain*. 2006;129(Pt 5):1330–1338.
302. Bradley WG, Shapshak P, Delgado S, Nagano I, Stewart R, Rocha B. Morphometric analysis of the peripheral neuropathy of AIDS. *Muscle Nerve*. 1998;21(9):1188–1195.
303. Rizzuto N, Cavallaro T, Monaco S, et al. Role of HIV in the pathogenesis of distal symmetrical peripheral neuropathy. *Acta Neuropathol*. 1995;90(3):244–250.
304. Keswani SC, Polley M, Pardo CA, Griffin JW, McArthur JC, Hoke A. Schwann cell chemokine receptors mediate HIV-1 gp120 toxicity to sensory neurons. *Ann Neurol*. 2003;54(3):287–296.
305. Yoshioka M, Bradley WG, Shapshak P, et al. Role of immune activation and cytokine expression in HIV-1-associated neurologic diseases. *Adv Neuroimmunol*. 1995;5(3):335–358.
306. McArthur JC, Yiannoutsos C, Simpson DM, et al. A phase II trial of nerve growth factor for sensory neuropathy associated with HIV infection. AIDS Clinical Trials Group Team 291. *Neurology*. 2000;54(5):1080–1088.
307. Schifitto G, Yiannoutsos C, Simpson DM, et al. Long-term treatment with recombinant nerve growth factor for HIV-associated sensory neuropathy. *Neurology*. 2001;57(7):1313–1316.
308. Webber CA, Salame J, Luu GL, et al. Nerve growth factor acts through the TrkA receptor to protect sensory neurons from the damaging effects of the HIV-1 viral protein, Vpr. *Neuroscience*. 2013;252:512–525.
309. Dyck PJ, Kratz KM, Karnes JL, et al. The prevalence by staged severity of various types of diabetic neuropathy, retinopathy, and nephropathy in a population-based cohort: the Rochester Diabetic Neuropathy Study. *Neurology*. 1993;43(4):817–824.
310. Callaghan BC, Cheng HT, Stables CL, Smith AL, Feldman EL. Diabetic neuropathy: clinical manifestations and current treatments. *Lancet Neurol*. 2012;11(6):521–534.
311. Kles KA, Vinik AI. Pathophysiology and treatment of diabetic peripheral neuropathy: the case for diabetic neurovascular function as an essential component. *Curr Diabetes Rev*. 2006;2(2):131–145.
312. Malik RA. The pathology of human diabetic neuropathy. *Diabetes*. 1997;46(Suppl 2): S50–S53.
313. Malik RA, Veves A, Walker D, et al. Sural nerve fibre pathology in diabetic patients with mild neuropathy: relationship to pain, quantitative sensory testing and peripheral nerve electrophysiology. *Acta Neuropathol*. 2001;101(4):367–374.
314. Vincent AM, Russell JW, Low P, Feldman EL. Oxidative stress in the pathogenesis of diabetic neuropathy. *Endocr Rev*. 2004;25(4):612–628.

315. Obrosova IG. Increased sorbitol pathway activity generates oxidative stress in tissue sites for diabetic complications. *Antioxid Redox Signal.* 2005;7(11–12):1543–1552.
316. Vincent AM, Callaghan BC, Smith AL, Feldman EL. Diabetic neuropathy: cellular mechanisms as therapeutic targets. *Nat Rev Neurol.* 2011;7(10):573–583.
317. Vincent AM, Perrone L, Sullivan KA, et al. Receptor for advanced glycation end products activation injures primary sensory neurons via oxidative stress. *Endocrinology.* 2007;148(2):548–558.
318. Padilla A, Descorbeth M, Almeyda AL, Payne K, De Leon M. Hyperglycemia magnifies Schwann cell dysfunction and cell death triggered by PA-induced lipotoxicity. *Brain Res.* 2011;1370:64–79.
319. Wiggin TD, Sullivan KA, Pop-Busui R, Amato A, Sima AA, Feldman EL. Elevated triglycerides correlate with progression of diabetic neuropathy. *Diabetes.* 2009;58(7):1634–1640.
320. Toth C, Brussee V, Martinez JA, McDonald D, Cunningham FA, Zochodne DW. Rescue and regeneration of injured peripheral nerve axons by intrathecal insulin. *Neuroscience.* 2006;139(2):429–449.
321. Xu QG, Li XQ, Kotecha SA, Cheng C, Sun HS, Zochodne DW. Insulin as an *in vivo* growth factor. *Exp Neurol.* 2004;188(1):43–51.
322. Kim B, Feldman EL. Insulin resistance in the nervous system. *Trends Endocrinol Metab.* 2012;23(3):133–141.
323. Ametov AS, Barinov A, Dyck PJ, et al. The sensory symptoms of diabetic polyneuropathy are improved with alpha-lipoic acid: the SYDNEY trial. *Diabetes Care.* 2003;26(3):770–776.
324. Ruhnau KJ, Meissner HP, Finn JR, et al. Effects of 3-week oral treatment with the antioxidant thioctic acid (alpha-lipoic acid) in symptomatic diabetic polyneuropathy. *Diabet Med.* 1999;16(12):1040–1043.
325. Ziegler D, Ametov A, Barinov A, et al. Oral treatment with alpha-lipoic acid improves symptomatic diabetic polyneuropathy: the SYDNEY 2 trial. *Diabetes Care.* 2006;29(11):2365–2370.
326. Ziegler D, Nowak H, Kempler P, Vargha P, Low PA. Treatment of symptomatic diabetic polyneuropathy with the antioxidant alpha-lipoic acid: a meta-analysis. *Diabet Med.* 2004;21(2):114–121.
327. Dubinsky RM, Miyasaki J. Assessment: efficacy of transcutaneous electric nerve stimulation in the treatment of pain in neurologic disorders (an evidence-based review): report of the Therapeutics and Technology Assessment Subcommittee of the American Academy of Neurology. *Neurology.* 2010;74(2):173–176.
328. Forst T, Nguyen M, Forst S, Disselhoff B, Pohlmann T, Pfützner A. Impact of low frequency transcutaneous electrical nerve stimulation on symptomatic diabetic neuropathy using the new Salutaris device. *Diabetes Nutr Metab.* 2004;17(3):163–168.
329. Kumar D, Marshall HJ. Diabetic peripheral neuropathy: amelioration of pain with transcutaneous electrostimulation. *Diabetes Care.* 1997;20(11):1702–1705.
330. Boulton AJ, Vinik AI, Arezzo JC, et al. Diabetic neuropathies: a statement by the American Diabetes Association. *Diabetes Care.* 2005;28(4):956–962.
331. Effect of intensive diabetes treatment on nerve conduction in the Diabetes Control and Complications Trial. *Ann Neurol.* 1995;38(6):869–880.
332. Callaghan BC, Little AA, Feldman EL, Hughes RA. Enhanced glucose control for preventing and treating diabetic neuropathy. *Cochrane Database Syst Rev.* 2012;6, CD007543.
333. Martin CL, Albers JW, Pop-Busui R, Group DER. Neuropathy and related findings in the diabetes control and complications trial/epidemiology of diabetes interventions and complications study. *Diabetes Care.* 2014;37(1):31–38.

334. Costa LA, Canani LH, Lisboa HR, Tres GS, Gross JL. Aggregation of features of the metabolic syndrome is associated with increased prevalence of chronic complications in Type 2 diabetes. *Diabet Med.* 2004;21(3):252–255.
335. Isomaa B, Henricsson M, Almgren P, Tuomi T, Taskinen MR, Groop L. The metabolic syndrome influences the risk of chronic complications in patients with type II diabetes. *Diabetologia.* 2001;44(9):1148–1154.
336. Metascreen Writing C, Bonadonna R, Cucinotta D, Fedele D, Riccardi G, Tiengo A. The metabolic syndrome is a risk indicator of microvascular and macrovascular complications in diabetes: results from Metascreen, a multicenter diabetes clinic-based survey. *Diabetes Care.* 2006;29(12):2701–2707.
337. Van Acker K, Bouhassira D, De Bacquer D, et al. Prevalence and impact on quality of life of peripheral neuropathy with or without neuropathic pain in type 1 and type 2 diabetic patients attending hospital outpatients clinics. *Diabetes Metab.* 2009;35(3):206–213.
338. Kehlet H, Jensen TS, Woolf CJ. Persistent postsurgical pain: risk factors and prevention. *Lancet.* 2006;367(9522):1618–1625.
339. Werner MU, Kongsgaard UE. I. Defining persistent post-surgical pain: is an update required? *Br J Anaesth.* 2014;113(1):1–4.
340. Grant AM, Scott NW, O'Dwyer PJ, Group MRCLGHT. Five-year follow-up of a randomized trial to assess pain and numbness after laparoscopic or open repair of groin hernia. *Br J Surg.* 2004;91(12):1570–1574.
341. Scholz J, Yaksh TL. Preclinical research on persistent postsurgical pain: what we don't know, but should start studying. *Anesthesiology.* 2010;112(3):511–513.
342. Solak O, Metin M, Esme H, et al. Effectiveness of gabapentin in the treatment of chronic post-thoracotomy pain. *Eur J Cardiothorac Surg.* 2007;32(1):9–12.
343. Margulies RU, Lewicky-Gaupp C, Fenner DE, McGuire EJ, Clemens JQ, Delancey JO. Complications requiring reoperation following vaginal mesh kit procedures for prolapse. *Am J Obstet Gynecol.* 2008;199(6):678, e671–e674.
344. Brennan TJ, Kehlet H. Preventive analgesia to reduce wound hyperalgesia and persistent postsurgical pain: not an easy path. *Anesthesiology.* 2005;103(4):681–683.
345. Clarke H, Bonin RP, Orser BA, Englesakis M, Wijeysundera DN, Katz J. The prevention of chronic postsurgical pain using gabapentin and pregabalin: a combined systematic review and meta-analysis. *Anesth Analg.* 2012;115(2):428–442.
346. Grosen K, Laue Petersen G, Pfeiffer-Jensen M, Hoejsgaard A, Pilegaard HK. Persistent post-surgical pain following anterior thoracotomy for lung cancer: a cross-sectional study of prevalence, characteristics and interference with functioning. *Eur J Cardiothorac Surg.* 2013;43(1):95–103.
347. Haroutiunian S, Nikolajsen L, Finnerup NB, Jensen TS. The neuropathic component in persistent postsurgical pain: a systematic literature review. *Pain.* 2013;154(1):95–102.
348. Kehlet H, Rathmell JP. Persistent postsurgical pain: the path forward through better design of clinical studies. *Anesthesiology.* 2010;112(3):514–515.
349. Melzack R, Israel R, Lacroix R, Schultz G. Phantom limbs in people with congenital limb deficiency or amputation in early childhood. *Brain.* 1997;120(Pt 9):1603–1620.
350. Price EH. A critical review of congenital phantom limb cases and a developmental theory for the basis of body image. *Conscious Cogn.* 2006;15(2):310–322.
351. Hansen DM, Kehlet H, Gartner R. Phantom breast sensations are frequent after mastectomy. *Dan Med Bull.* 2011;58(4):A4259.
352. Markopoulos CJ, Spyropoulou AC, Zervas IM, Christodoulou GN, Papageorgiou C. Phantom breast syndrome: the effect of *in situ* breast carcinoma. *Psychiatry Res.* 2010;179(3):333–337.
353. Rasmussen ML, Prause JU, Toft PB. Phantom pain after eye amputation. *Acta Ophthalmol (Copenh).* 2011;89(1):10–16.

354. Wade NJ, Finger S. Phantom penis: historical dimensions. *J Hist Neurosci.* 2010;19(4):299–312.
355. Sherman RA, Sherman CJ, Parker L. Chronic phantom and stump pain among American veterans: results of a survey. *Pain.* 1984;18(1):83–95.
356. Wall PD, McMahon SB. *Wall and Melzack's Textbook of Pain.* Philadelphia: Elsevier/Churchill Livingstone; 2006.
357. Wall PD, Gutnick M. Ongoing activity in peripheral nerves: the physiology and pharmacology of impulses originating from a neuroma. *Exp Neurol.* 1974;43(3):580–593.
358. Flor H, Nikolajsen L, Staehelin Jensen T. Phantom limb pain: a case of maladaptive CNS plasticity? *Nat Rev Neurosci.* 2006;7(11):873–881.
359. Latremoliere A, Woolf CJ. Central sensitization: a generator of pain hypersensitivity by central neural plasticity. *J Pain.* 2009;10(9):895–926.
360. Woodhouse A. Phantom limb sensation. *Clin Exp Pharmacol Physiol.* 2005;32(1–2):132–134.
361. Cohen SP, Gambel JM, Raja SN, Galvagno S. The contribution of sympathetic mechanisms to postamputation phantom and residual limb pain: a pilot study. *J Pain.* 2011;12(8):859–867.
362. Harden RN, Gagnon CM, Gallizzi M, Khan AS, Newman D. Residual limbs of amputees are significantly cooler than contralateral intact limbs. *Pain Pract.* 2008;8(5):342–347.
363. Flor H. Phantom-limb pain: characteristics, causes, and treatment. *Lancet Neurol.* 2002;1(3):182–189.
364. Ilfeld BM, Moeller-Bertram T, Hanling SR, et al. Treating intractable phantom limb pain with ambulatory continuous peripheral nerve blocks: a pilot study. *Pain Med.* 2013;14(6):935–942.
365. Elbert T, Flor H, Birbaumer N, et al. Extensive reorganization of the somatosensory cortex in adult humans after nervous system injury. *Neuroreport.* 1994;5(18):2593–2597.
366. Jahangiri M, Jayatunga AP, Bradley JW, Dark CH. Prevention of phantom pain after major lower limb amputation by epidural infusion of diamorphine, clonidine and bupivacaine. *Ann R Coll Surg Engl.* 1994;76(5):324–326.
367. Woolf CJ, Chong MS. Preemptive analgesia—treating postoperative pain by preventing the establishment of central sensitization. *Anesth Analg.* 1993;77(2):362–379.
368. Ramachandran VS, Rogers-Ramachandran D. Synaesthesia in phantom limbs induced with mirrors. *Proc Biol Sci.* 1996;263(1369):377–386.
369. Krainick J-U, Thoden U, Riechert T. Pain reduction in amputees by long-term spinal cord stimulation. *J Neurosurg.* 1980;52(3):346–350.
370. Kost RG, Straus SE. Postherpetic neuralgia—pathogenesis, treatment, and prevention. *N Engl J Med.* 1996;335(1):32–42.
371. Ebert MH. Histologic changes in sensory nerves of the skin in herpes zoster. *Arch Derm Syphilol.* 1949;60(5, Pt. 1):641–648.
372. Head H, Campbell AW, Kennedy PG. The pathology of Herpes Zoster and its bearing on sensory localisation. *Rev Med Virol.* 1997;7(3):131–143.
373. Watson CP, Deck JH, Morshead C, Van der Kooy D, Evans RJ. Post-herpetic neuralgia: further post-mortem studies of cases with and without pain. *Pain.* 1991;44(2):105–117.
374. Watson CP, Morshead C, Van der Kooy D, Deck J, Evans RJ. Post-herpetic neuralgia: post-mortem analysis of a case. *Pain.* 1988;34(2):129–138.
375. Gilden DH, Dueland AN, Cohrs R, Martin JR, Kleinschmidt-DeMasters BK, Mahalingam R. Preherpetic neuralgia. *Neurology.* 1991;41(8):1215–1218.
376. Huff JC, Bean B, Balfour Jr HH, et al. Therapy of herpes zoster with oral acyclovir. *Am J Med.* 1988;85(2A):84–89.

377. Rogers 3rd RS, Tindall JP. Geriatric herpes zoster. *J Am Geriatr Soc.* 1971;19(6):495–504.
378. Wood MJ, Ogan PH, McKendrick MW, Care CD, McGill JI, Webb EM. Efficacy of oral acyclovir treatment of acute herpes zoster. *Am J Med.* 1988;85(2A):79–83.
379. Nurmikko T, Bowsher D. Somatosensory findings in postherpetic neuralgia. *J Neurol Neurosurg Psychiatry.* 1990;53(2):135–141.
380. Rowbotham MC, Fields HL. Post-herpetic neuralgia: the relation of pain complaint, sensory disturbance, and skin temperature. *Pain.* 1989;39(2):129–144.
381. Rowbotham MC, Fields HL. Topical lidocaine reduces pain in post-herpetic neuralgia. *Pain.* 1989;38(3):297–301.
382. Watson CP. Postherpetic neuralgia. *Neurol Clin.* 1989;7(2):231–248.
383. Eide PK, Jorum E, Stubhaug A, Bremnes J, Breivik H. Relief of post-herpetic neuralgia with the N-methyl-D-aspartic acid receptor antagonist ketamine: a double-blind, cross-over comparison with morphine and placebo. *Pain.* 1994;58(3):347–354.
384. Hope-Simpson RE. The nature of herpes zoster: a long-term study and a new hypothesis. *Proc R Soc Med.* 1965;58:9–20.
385. Balfour Jr HH. Varicella zoster virus infections in immunocompromised hosts. A review of the natural history and management. *Am J Med.* 1988;85(2A):68–73.
386. De Moragas JM, Kierland RR. The outcome of patients with herpes zoster. *AMA Arch Derm.* 1957;75(2):193–196.
387. Epstein E. Treatment of zoster and postzoster neuralgia by the intralesional injection of triamcinolone: a computer analysis of 199 cases. *Int J Dermatol.* 1976;15:762–769.
388. Ragozzino MW, Melton 3rd LJ, Kurland LT, Chu CP, Perry HO. Population-based study of herpes zoster and its sequelae. *Medicine (Baltimore).* 1982;61(5):310–316.
389. Glynn C, Crockford G, Gavaghan D, Cardno P, Price D, Miller J. Epidemiology of shingles. *J R Soc Med.* 1990;83(10):617–619.
390. Hope-Simpson RE. Postherpetic neuralgia. *J R Coll Gen Pract.* 1975;25(157):571–575.
391. Clemmensen OJ, Andersen KE. ACTH versus prednisone and placebo in herpes zoster treatment. *Clin Exp Dermatol.* 1984;9(6):557–563.
392. Burke BL, Steele RW, Beard OW, Wood JS, Cain TD, Marmer DJ. Immune responses to varicella-zoster in the aged. *Arch Intern Med.* 1982;142(2):291–293.
393. Gershon AA, Steinberg SP. Cellular and humoral immune responses to varicella-zoster virus in immunocompromised patients during and after varicella-zoster infections. *Infect Immun.* 1979;25(1):170–174.
394. Fields HL, Rowbotham M, Baron R. Postherpetic neuralgia: irritable nociceptors and deafferentation. *Neurobiol Dis.* 1998;5(4):209–227.
395. Dworkin RH, Corbin AE, Young Jr JP, et al. Pregabalin for the treatment of postherpetic neuralgia: a randomized, placebo-controlled trial. *Neurology.* 2003;60(8):1274–1283.
396. Rowbotham M, Harden N, Stacey B, Bernstein P, Magnus-Miller L. Gabapentin for the treatment of postherpetic neuralgia: a randomized controlled trial. *JAMA.* 1998;280(21):1837–1842.
397. Raja SN, Haythornthwaite JA, Pappagallo M, et al. Opioids versus antidepressants in postherpetic neuralgia: a randomized, placebo-controlled trial. *Neurology.* 2002;59(7):1015–1021.
398. McInnes IB, Schett G. The pathogenesis of rheumatoid arthritis. *N Engl J Med.* 2011;365(23):2205–2219.
399. Aletaha D, Neogi T, Silman AJ, et al. 2010 Rheumatoid arthritis classification criteria: an American College of Rheumatology/European League Against Rheumatism collaborative initiative. *Arthritis Rheum.* 2010;62(9):2569–2581.
400. Chamberlain MA, Corbett M. Carpal tunnel syndrome in early rheumatoid arthritis. *Ann Rheum Dis.* 1970;29(2):149–152.

401. Kidd BL, Morris VH, Urban L. Pathophysiology of joint pain. *Ann Rheum Dis.* 1996;55(5):276–283.
402. Schaible HG, Grubb BD. Afferent and spinal mechanisms of joint pain. *Pain.* 1993;55(1):5–54.
403. Steiner G, Smolen J. Autoantibodies in rheumatoid arthritis and their clinical significance. *Arthritis Res.* 2002;4(Suppl 2):S1–S5.
404. Smolen JS, Landewe R, Breedveld FC, et al. EULAR recommendations for the management of rheumatoid arthritis with synthetic and biological disease-modifying antirheumatic drugs: 2013 update. *Ann Rheum Dis.* 2014;73(3):492–509.
405. Bresnihan B, Alvaro-Gracia JM, Cobby M, et al. Treatment of rheumatoid arthritis with recombinant human interleukin-1 receptor antagonist. *Arthritis Rheum.* 1998;41(12):2196–2204.
406. Emery P, Fleischmann R, Filipowicz-Sosnowska A, et al. The efficacy and safety of rituximab in patients with active rheumatoid arthritis despite methotrexate treatment: results of a phase IIB randomized, double-blind, placebo-controlled, dose-ranging trial. *Arthritis Rheum.* 2006;54(5):1390–1400.
407. Klareskog L, van der Heijde D, de Jager JP, et al. Therapeutic effect of the combination of etanercept and methotrexate compared with each treatment alone in patients with rheumatoid arthritis: double-blind randomised controlled trial. *Lancet.* 2004;363(9410):675–681.
408. Lipsky PE, van der Heijde DM, St Clair EW, et al. Infliximab and methotrexate in the treatment of rheumatoid arthritis anti-tumor necrosis factor trial in rheumatoid arthritis with concomitant therapy study group. *N Engl J Med.* 2000;343(22):1594–1602.
409. Smolen JS, Beaulieu A, Rubbert-Roth A, et al. Effect of interleukin-6 receptor inhibition with tocilizumab in patients with rheumatoid arthritis (OPTION study): a double-blind, placebo-controlled, randomised trial. *Lancet.* 2008;371(9617):987–997.
410. Katusic S, Beard CM, Bergstralh E, Kurland LT. Incidence and clinical features of trigeminal neuralgia, Rochester, Minnesota, 1945-1984. *Ann Neurol.* 1990;27(1):89–95.
411. Sindrup SH, Jensen TS. Pharmacotherapy of trigeminal neuralgia. *Clin J Pain.* 2002;18(1):22–27.
412. Love S, Coakham HB. Trigeminal neuralgia: pathology and pathogenesis. *Brain.* 2001;124(Pt 12):2347–2360.
413. List CF, Williams JR. Pathogenesis of trigeminal neuralgia: a review. *AMA Arch Neurol Psychiat.* 1957;77(1):36–43.
414. Gronseth G, Cruccu G, Alksne J, et al. Practice parameter: the diagnostic evaluation and treatment of trigeminal neuralgia (an evidence-based review): report of the Quality Standards Subcommittee of the American Academy of Neurology and the European Federation of Neurological Societies. *Neurology.* 2008;71(15):1183–1190.

INDEX

Note: Page numbers followed by "*f*" indicate figures, "*t*" indicate tables and "*b*" indicate boxes.

A

Acetyl-L-carnitine (ALCAR), 484–485
Acid-sensing ion channel 3 (ASIC3) expression, 234
Adenosine monophosphate activated protein kinase (AMPK), 416–417, 418*f*
Adenosine triphosphate, 32–33
A-kinase anchoring protein 79/150 (AKAP 79/150), 42–45
ALCAR. *See* Acetyl-L-carnitine (ALCAR)
α-Amino-3-hydroxy-5-methyl-4-isoxazolepropionic acid (AMPA) receptor, 15, 156–157
AMP-activated protein kinase (AMPK)
  chemotherapy-induced neuropathy, 198
  injury-induced dysregulation, 197
  IRS S636/639 residues, 197
  mTOR and ERK signaling pathways, 197
  PNI-induced neuropathic pain, 197
  Ser789 phosphorylation, 197
  and translational control, 193–194
  treatment, 200–201
  Tyr895 phosphorylation, 197
Anandamide, 87–88
Angiogenesis inhibitors, 475–476
Ankyrin repeat domains (ARD), 75–77
  K_v channels and, 79–80
  role, 79–80
  X-ray crystallographic structure, 79–80
Anterior cingulate cortex (ACC), 61–62
Anti-miRNA oligonucleotide (AMO), 223, 224*b*, 231*f*
aPKC. *See* Atypical PKCs (aPKC)
Arachidonic acid metabolites, TRP channel modulators, 103
ARD. *See* Ankyrin repeat domains (ARD)
β-Arrestin, 45–46
Arthritis-like symptoms, TRPV1, 87
Astrocytes
  CCL2/MCP-1, 334–335
  endothelial cells, 327–328
  glial markers in pain conditions, 347*t*
  HIV-positive patients, 340–341
  homeostatic conditions, 330
  human and mice, 342
  microglia, 326–327
  p-JNK, 332
  rat spinal cord (*see* Rat spinal cord)
  reactivity, 337
  TLR-4 agonist, 335–336
  YKL-40, 338
Atypical PKCs (aPKC)
  hyperalgesic priming, 423, 424*f*
  phosphorylation, AMPA receptors, 422–423
  PKMζ, 422–423
Axonal translation
  local proteome homeostasis, 188
  sensory neurons, 188
  single mRNA molecule, 188
  temporal and spatial precision, 188

B

BDNF. *See* Brain-derived neurotrophic factor (BDNF)
Bdnf gene
  DNA methylation, 162
  neural plasticity, 154, 162–163
  primary afferent nociceptors, 166–170
  Zif268 and *Homer1*, 158
Bortezomib
  in human clinical studies, 475
  MTAs, 492
Bradykinin receptor, 32–33, 36
Brain-derived neurotrophic factor (BDNF), 149
  CGRP, 555
  hippocampus, 423–425
  hyperalgesic priming, 423–425, 424*f*

Brain-derived neurotrophic factor (BDNF) (*Continued*)
LTP, 554
pain plasticity, 423–425, 555
spinal cord, 554–555
spinal dorsal horn, 423–425

C

Caffeine, 452–453
Calcitonin gene-related peptide (CGRP), 39–40, 83–84, 335–336
BDNF and IL-6, 549–550
dural afferent terminals, 549–550
migraine plasticity, 553–554
Calcitonin receptor, 39–40
Calcium channels, 12
cAMP response element binding protein (CREB), 156–157, 159
Cancer pain, 566–572, 567*t*
Cap-dependent translation
algogens, 194–195
AMPK, 197–198
CREB regulates nociception, 195
eIF4E complex, 189
macromolecular complexes, 189
mRNA structures, 189
mTORC1 inhibition and nociception, 196–197
peripheral nerve injury, 195–196
Capsaicin, 81–82
Carotenoids, 448–449
CCI. *See* Chronic constriction-induced (CCI)
Central nervous system (CNS)
astrocytes and microglia, 326–327
BBB, 327
endothelial cells, 327–328
human chronic pain conditions, 328–330
leukocytes infiltration, 327–328
lymphocyte infiltration, 327
"macrophages", 326–327
neuroinflammatory conditions, 327–328
neuronal synapses, 326–327
nociceptive behaviors, 327
pathophysiology, 328–330
peripheral nerve injury, 327–328
periphery, 327–328
persistent neuropathic pain, 327–328
presynaptic and postsynaptic neurons, 327, 328–330
rat spinal cord (*see* Rat spinal cord)
rodent models, 327
Central poststroke pain (CPSP), 567*t*, 572–573
Cerebrospinal fluid (CSF)
fibromyalgia patients, 340
GFAP, 341
glial markers in pain conditions, 347*t*
humans, 345–346
IL-8 level, 340
in vitro incubation, 341–342
nociceptive cytokine IL-1, 341
proinflammatory factors, 340
sclerosis patients, 338
and serum, 338
CGRP. *See* Calcitonin gene-related peptide (CGRP)
Channel desensitization, 86
Chemoprotectant agents, 583–584
Chemotherapeutics
angiogenesis inhibitors, 475–476
autonomic and motor nervous systems, 472
MTAs, 473–474
neuropathy induced, 472
platinum-containing agents, 474–475
proteasome inhibitors, 475
Chemotherapy-induced peripheral neuropathy (CIPN), 583
animal models, 478–480
antidepressants, 483–484
antiepileptics, 483–484
chemotherapeutics (*see* Chemotherapeutics)
clinical assessment, 476–478
coasting, 473
DNA damage, 487
inflammation, 489–490
in vitro models, 481–483
ion channel modulation, 487–489
mechanisms, 483
microtubule alterations, 492
mitochondrial dysfunction, 484–486
neurotrophic factors, 490–491
nitroxidative stress, 486–487
opioids, 483–484
in patients, 473

primary side effect, 472
symptoms, 472
Chromosome conformation capture (3C), 156
Chronic constriction-induced (CCI), 262–263
Chronic pain, 18
activity-dependent plasticity, 206–207
AMPK, 193–194
axonal translation (*see* Axonal translation)
cap-dependent translation (*see* Cap-dependent translation)
cap-independent mechanism, 198
de novo synthesis, 200–201
eIF2 and the 43S preinitiation complex, 192–193
eIF4E, 189–192
elucidation, 186
injured tissue, repair (*see* Injured tissue)
maladaptive changes, 186
mRNAs (*see* Messenger RNA (mRNAs))
neuropathic animals in conjunction, 200–201
S6K1/IRS-1/ERK signaling, 206–207
synaptic plasticity, 206–207
translational control (*see* Translational control)
3′ UTR and poly-A tail, mRNAs, 198
Chronic pelvic pain (CPP), 567*t*, 573–575
allergic and anaphylactic mast cell responses, 513
CRF, 510–511, 511*f*, 512–513
description, 509–510
dichotomous effects, 510–511
glucocorticoid-driven feedback, 511–512
HPA axis, neuroimmune function, 513
limbic structures, 513
MR, 511–512
Chronic prostatitis/chronic pelvic pain syndrome (CP/CPPS)
EAP, 523–524
mast cell degranulation, 523
NIH-CPSI, 522–523
population-based screening methods, 522–523
signs and symptoms, 523
CIPN. *See* Chemotherapy-induced peripheral neuropathy (CIPN)
Cisplatin
ALCAR, 484–485
DNA damage, 487
oxaliplatin, 474–475
patients, 474–475
c-Jun N-terminal kinases (JNK), 332–333, 335–336
Clinical trials, glial modulator
acute postsurgical pain, 332
AZD2423, 345
chronic neuropathic pain, 332
drugs, 343, 344*t*
CMAP. *See* Conduction velocity of motor nerves (CMAP)
CNS. *See* Central nervous system (CNS)
Complete Freund's adjuvant (CFA), 87, 166, 170–171
Complex regional pain syndrome (CRPS), 341, 347*t*, 567*t*, 575–576
Conduction velocity of motor nerves (CMAP), 476–477
Cortical spreading depression, 581
Corticotropin-releasing factor (CRF), 545–546
amygdala and hippocampus, 513
and arginine vasopressin, 510–511
intrathecal administration, 519
locus coeruleus and hypothalamus, 515
CPEB. *See* Cytoplasmic polyadenylation element binding protein (CPEB)
CPP. *See* Chronic pelvic pain (CPP)
CPSP. *See* Central poststroke pain (CPSP)
C-reactive protein (CRP), 439
Cross-sensitization, TRPV1, 84
CRPS. *See* Complex regional pain syndrome (CRPS)
CSF. *See* Cerebrospinal fluid (CSF)
Cystathionine-β-synthase (CBS), 170–171
Cytokines
cell surface receptor, 438
and chronic pain, 439
DAMP receptors, 437–438
hyperalgesia, 438–439
innate and adaptive immune system, 436, 437*t*
macrophage inhibitor, 438–439
microglia cells, 437–438
purinergic receptors, 437–438
toll-like receptor 4 (TLR4), 437–438

Cytoplasmic polyadenylation element binding protein (CPEB), 417–420, 417*f*

D

Damage-associated molecular pattern molecules (DAMPs)
  acute and protective pain, 268
  acute inflammatory disorder, 268–269
  chronic pain, 252
  cytokine blockade, 268–269
  dysregulation, inflammation and immune reactions, 268–269
  extracellular matrix, 254
  HMGB1 (*see* High-mobility group box 1 (HMGB1))
  HSPs (*see* Heat-shock proteins (HSPs))
  lipopolysaccharide, 252
  necrotic cells, 252–254
  nociceptive system, 268–269
  nociceptors, 268
  PAMPs, 252
  pathogenic infection, 252–254
  physiological conditions, 252–254
  plasticity mechanisms, 268
  PRRs, 252, 253*t*
  purine metabolites (*see* Purine metabolites)
  RAGE, 252–254
  secretory reticulum-Golgi pathway, 252–254
  S100 proteins (*see* S100 proteins)
  TNF blockade, 268–269
Dendritic spine dysgenesis
  mental retardation, 390–391
  NSC23766 treatment, 397–398, 399–400
  Rac1, 396–398
  spinal memory mechanism, 400–401
Dendritic spines
  dendritic branch, 387, 388*f*
  description, 386, 386*f*
  in neuropathic pain, 391
  in pathology, 390–391
  plasticity after peripheral nerve injury, 394
  real anomalies on neurons, 387
  remodeling in SCI, 392–393, 393*f*
  spiny neurons, 387
  stubby and thin/mushroom shapes, 387
  STZ-induced diabetes, 394–395
  in synaptic function, 389–390 (*see also* Long-term potentiation (LTP))
  thermal burn injury, 395–396
Diabetic polyneuropathy, 585
Dietary interventions
  analgesic effect, 454–455
  anti-inflammatory properties, 454–455
  knee pain, 454
  Mediterranean diet, 455–456
  NFkB inhibitor, 455–456
  randomized control study, 456
  rheumatoid arthritis, 454–455
DiGeorge syndrome chromosome region (DGCR8), 217–218, 219*f*
Distal symmetrical polyneuropathy (DSPN), 584
DLPFC. *See* Dorsolateral prefrontal cortex (DLPFC)
DNA damage, 487
DNA methylation
  DNA methyltransferases, 161–162
  learning and memory, 162
  memory formation, 161*f*, 162–163
DNA methyltransferases (DNMTs)
  antagonizing, 171
  cortical infusion, 162
  DNMT3a and 3b, 154
  DNMT3a2 levels, 161
  DNMTis, 155
  methylation, 154
  zinc fingers, 172*f*
Dorsal root ganglia (DRG) neurons, 258, 260–261, 265–266
  bladder pain, intraperitoneal (IP) inhibition, 291–292
  coexpression analysis, 289–290
  colorectal visceral pain, 291–292
  group I activity, 290–292
  group II activity, 292–293
  group III activity, 293–294
  intraplantar inflammation, 290–291
  mGluR activation, behavioral effects, 290*t*
  mGluR expression, 291*f*
  mGluR1-positive neurons, 289–290
  skin-applied capsaicin, 294
  transient receptor potential vanilloid 1 (TRPV1), 291–292

Dorsolateral prefrontal cortex (DLPFC), 61–62
Double-knot toxin (DkTx), 81
DRG neurons. *See* Dorsal root ganglia (DRG) neurons
Drug development
- mGluRs, 313–315
- TRPA1, 104–105
- TRPM8, 98
- TRPV1, 90–91

DSPN. *See* Distal symmetrical polyneuropathy (DSPN)
Dural afferent activation
- application, 545–546
- fibroblasts, 545–546
- ion channels
  - ASIC1–ASIC4, 546
  - ASIC blocker amiloride, 547
  - endogenous activators/sensitizers, 549
  - extracellular pH, 546
  - *in vivo* electrophysiological recording techniques, 546
  - mast cell degranulation, 547
  - preclinical studies, 548
  - TRP channels, 547–548
  - TRPM subfamily, 549
  - TRPM8 transcripts, 549
  - TRPV4 activator, 548–549
  - *Umbellularia californica*, 548
- mast cell degranulation, 545–546
- neuroplasticity, 549–550
- primary afferent nociceptors, 545

E

Endogenous HMGB1
- anti-HMGB1 neutralizing antibodies, 258–260
- bone cancer model, 260
- cyclophosphamide-induced cystitis, 258–260
- in d*b*/d*b* mice, 260
- endothelial anticoagulant, 258–260
- experimental models, 259*t*
- glycyrrhizin, 258
- hypernociception, 258–260
- immunohistochemistry, 258
- lumbar spinal extranuclear fraction, 260
- naïve mice and rats, 258
- nerve injury-induced pain, 258
- neurodegenerative diseases, 260
- neuroinflammation, 260
- neurons and astrocytes, 260
- neuropathic pain, 258
- protein expression levels, 258
- satellite cells, 258
- Schwann cells, 258

Endothelin receptors, 40
Epigallocatechin gallate (EGCG), 445–446
Epigenetic mechanisms
- behavioral epigenetics, 150–151
- description, 149–150
- DNA methylation, 150–151, 153–155
- epigenome, 155–156
- gene expression in chromatin compaction, 151*f*
- genome-environment interactions, 149–150
- histone modifications, 152–153
- MeCP2 gene, 151
- methyl-CpG DNA-binding protein family, 151
- miRNAs, 152
- molecular level, 150–151
- neuroepigenetics, 150–151
- and pain (*see* Pain, epigenetic mechanisms)
- postmitotic neural function, 150–151
- posttranslational modification, 150–151
- Rett syndrome, 151

ERK. *See* Extracellular signal-regulated kinase (ERK)
ET type A receptor (ET_AR), 40
ET type B receptor (ET_BR), 40
Eukaryotic initiation factor 4F (eIF4F)
- complex and initiation, 189–190, 190*f*
- CPEB-Maskin-eIF4E, 198
- enhancement, 205–206
- mRNA cap, 189
- NGF-mediated effects, 194
- PNI-induced neuropathic pain, 197

Experimental autoimmune prostatitis (EAP) model, 523–524
Extracellular signal-regulated kinase (ERK), 149, 153, 156–157, 158, 416–417, 417*f*, 418*f*
- ERKMnk1-eIF4E pathway, 194

Extracellular signal-regulated kinase (ERK) (*Continued*)
mTOR pathway (*see* Mechanistic target of rapamycin (mTOR) pathway)
phosphorylation, 196–197
Ras-Raf-Mek-ERK signaling, 196–197
S6K1/IRS-1/ERK signaling, 206–207

F

Failed back syndrome (FBS), 567*t*, 576–578
Familial episodic pain syndrome (FEPS), 103–104
Fibroblast growth factor 2 (FGF2), 475–476
Fibromyalgia, 567*t*, 578–579
Formalin, irritants, 101
Fragile X mental retardation protein (FMRP), 413–416, 417–419

G

GABA receptors
astrocytes, 362
benzodiazepines, 362–363
bicuculline/gabazine, 363
gephyrin, 362–363
metabotropic receptors, 364
spinal diazepam, 363–364
GAD. *See* Glutamic acid decarboxylase (GAD)
GAD65. *See* Glutamic acid decarboxylase 65 (GAD65)
GAG. *See* Glycosaminoglycan (GAG)
γ-aminobutyric acid (GABA)
chloride channels, 367–368
chloride regulation, 374
depolarization, 365
downregulation, KCC2, 373
and GAD, 360–361
glycine, 360
succinic semialdehyde, 362
and VGAT, 360–361
GFAP. *See* Glial fibrillary acidic protein (GFAP)
Ginger
capsaicin injection, 449–450, 450*f*
dysmenorrhea and osteoarthritis, 449–450
inflammatory cytokine activity, 450
TRPA1 channels, 450
Turmeric and curcumin, 450–451
Glia
activation/reactivity (*see* Glial activation)
chemokines, 334–335
chronic pain research, 326
CNS (*see* Central nervous system (CNS))
CX3CR1 expression, 334–335
fractalkine, 334–335
GFAP and IL-8, 345–346
glial markers studies, 345–346, 347*t*
glial modulators (*see* Glial modulators)
human *vs.* rodents, 338–342
microglia and astrocyte cells, 346–349
NK1 knockout mice, 333–334
nonneuronal central mechanisms, 345–346
pain patients, 349
peripheral tissue, 333–334
preclinical models, 326
P2X4, P2X7/P2Y12 receptors, 334
rodent models, 333
spinal ED2/CD163, 346–349
spinal NMDA receptor, 333–334
spinal phosphatases, 346–349
TLR-2 activation, 346–349
TLR-4 agonists, 335–336
Glial activation
astrocytes, 332
brain-derived neurotrophic factor, 333
cytokines and chemokines production, 332–333
description, 330
GFAP, 332
homeostatic environment, 330
immunostaining, 330–331
MAPKs, 332
microglial cells, 330–331
morphology and expression, 330
NF-kB signaling, 332–333
nociceptive signaling, 330–331
nonhypertrophic, 330
nonphysiological conditions, 330–331
p-ERK in neurons, 332
physiologic/pathologic, 330
p-JNK in astrocytes, 332
p-p38 in microglia, 332
spinal cord astrocytic S-100b, 332
Glial cells in humans
astrocyte phenotype, 338–339
chronic conditions, 341–342

[11C]-(R)-PK11195, 339, 340
CSF and serum, 338, 340
GFAP, 338–339, 341
glial marker expression, 339
HIV-positive patients, 340–341
IL-8 levels, 340
in vitro incubation, 341–342
in vivo imaging approach, 339, 340
neuroimmune pathological conditions, 339
patients with brain trauma, 338–339
proinflammatory/nociceptive cytokine IL-1, 341
rheumatoid arthritis, 341–342
YKL-40, 338
Glial fibrillary acidic protein (GFAP)
CNS, 338–339
CSF, 341
glial markers in pain conditions, 347*t*
immunolabeling, 328*f*, 329*f*, 331*f*
knockdown, 338–339
nonhuman primates, 343
serum in patients, 338–339
Glial modulators
AZD2423, 345
chronic pain conditions, 343
clinical trials, 343, 344*t*
[11C](R)-PK11195-PET, 343
human subject studies, 343
minocycline, 336–337, 343–345
neurodegeneration, 336–337
neuroimmune mechanisms, 345
nonhuman primates, 343
preclinical pain models, 336
propentofylline, 337, 345
rodent/animal models, 345
treatment, chronic pain, 338
Glutamate receptors, 14–15, 33–36, 38–39
Glutamic acid decarboxylase (GAD), 360–362, 361*f*, 370–371, 372–373
Glutamic acid decarboxylase 65 (GAD65), 166
Glycine receptors
bicarbonate flows, 368
bicuculline and strychnine, 372–373
$GABA_A$ receptors, 363
gephyrin, 363–364
nicotinic acetylcholine and type-3 serotonin receptors, 363
Glycosaminoglycan (GAG), 519
G-protein-coupled glutamate receptors (mGluRs)
cytoplasmic C-terminal domains, 283
dimers, schematic illustration, 284*f*
group I, 284–287
group II, 288
group III, 288
NMDA and AMPA ionotropic glutamate receptors, 283
N-terminal extracellular domains, 283
in pain neuraxis, expression, 284–287, 287*f*
pharmacological agents, targets and names, 283, 285*t*
Grape seed proanthocyanidin extracts (GSPE), 444–445

H

HATs. *See* Histone acetyltransferases (HATs)
HDAC inhibitors (HDACis)
CBP deficient mice, 160
chronic intraperitoneal, 159–160
class II HDACi SAHA, 166
DNMTis, 171
epigenetic drugs, 166
gene array study, 160
hyperacetylate histones, 153
hypersensitivity, 170
inflammatory and neuropathic pain states, 166
Kv4.3 gene, 166–170
memory formation, 159–160
Nav1.8 in DRGs, 166–170
neuropathic injury, 166–170
neuropathic pain models, 166–170
presynaptic GABAergic synaptic function, 166
sodium butyrate, 159
suberoyl anilide hydroxamic acid, 159–160
trichostatin A, 159
valproate, 159

Heat-shock proteins (HSPs)
DAMPs, 265
extracellular, 264–265
HSP27, HSP70 and HSP90, 265–266
intracellular chaperones, 264–265
lethal-7 miRNA, 266–267
microRNA, 266–267
proinflammatory roles, 265
High-mobility group box 1 (HMGB1)
endogenous (*see* Endogenous HMGB1)
inflammatory mediator, 255–256
nociceptive action, 260–261
nociceptive properties, 256–257
structure, localization and mechanisms, 254–255
Histone acetyltransferases (HATs), 153, 160, 164, 166–170
Histone deacetylases (HDACs). *See* HDAC inhibitors (HDACis)
Histone modifications
epigenetic mechanisms, 152–153
synaptic plasticity, 157–158
H3K4me3, 155, 172*f*
H3K27me3, 155
HMGB1. *See* High-mobility group box 1 (HMGB1)
Homo sapiens, 219–220
HSPs. *See* Heat-shock proteins (HSPs)
Human miRNA Disease Database (HMDD), 220
Hydrogen sulfide, gasotransmitter, 100–101

I

IBS. *See* Irritable bowel syndrome (IBS)
IC/PBS. *See* Interstitial cystitis/painful bladder syndrome (IC/PBS)
IENF. *See* Intraepidermal nerve fiber (IENF)
Immune system
antagonism/inhibition, 456–457, 457*f*
broccoli, 447–448, 448*f*
caffeine, 452–453
carotenoids, 448–449
cytokines, 436–439
description, 436
diet and inflammation, 441
dietary interventions, 454–456
ginger, 449–451
ginseng, 451–452
grains and gluten, 453
grape seed extract, 444–445
green tea extract, 445–446
ketogenic diets, 453–454
obesity and pain, 440
omega-3 fatty acids, 442–443, 443*f*
omega-6 fatty acids, 441–442
rheumatoid arthritis, 443–444
saturated fats, 444
soy products, 446, 447*f*
Inflammation
hyperalgesia and allodynia, 63*f*
nociceptors, 62–65
sensitization, MSCs, 64*f*
Inflammatory hyperalgesia
abnormal responses, 46–47
AKAP 79/150, 42–45
β-arrestin, 45–46
homeostatic, 46–47
human physiology, 32
mediators (*see* Inflammatory mediators)
posttranslational modifications, 32
primary afferent fibers, 32
proinflammatory receptor systems, 32
protein/protein interactions, 42
signaling mechanisms, 40–42
tissue injury, 42
Inflammatory mediators
adenosine triphosphate, 32–33
bradykinin receptor, 32–33, 36
calcitonin receptor, 39–40
damaged epithelial cells, 32–33
endothelin receptors, 32–33, 40
glutamate receptors, 33–36, 38–39
neurokinin receptor, 38
nociceptor-originated mediators, 33
peripheral tissue, 33
primary afferent nociceptors, 33–36
prostaglandin receptor, 32–33, 36–37
protease-activated receptors, 39
purinergic receptor, 33–36, 37
RAMP1, 39–40
receptors expression, 33, 34*t*
sensitization, 33–36
serotonin receptor, 37
somatosensory and chemical activation, 32–33
tyrosine kinases, 38

Inflammatory pain, 7–9
Inflammatory thermal hyperalgesia, 93
Inhibition. *See* Synaptic inhibition
Injured tissue
ATP and collagen, 186–187
axons, 187
collagenous, 187
extracellular matrix, 187
lacking nociceptive sensitization, 187–188
mast cells, 186–187
NGF and ECM proteins, 187
peripheral nervous system, 187
repair and regeneration mechanisms, 186–187
Schwann cells, 187
with/without nociceptive sensitization, 187–188
Interstitial cystitis/painful bladder syndrome (IC/PBS)
adult-onset disorder, 518
bladder urothelial dysfunction, urine metabolites, 519
childhood sexual/physical abuse, 518–519
dimethyl sulfoxide, 517–518
GAG, 519
infection/organic disease, 517–518
intrathecal administration, 519
intravesicular mustard oil application, 518–519
luminal barrier integrity, 519
mast cell infiltration, 520
metabotropic glutamate receptors, 520
signal transduction pathways, 520
urine biomarkers, 518
Intraepidermal nerve fiber (IENF), 477–478, 480
Ion channels, 9–10, 487–489
Irritable bowel syndrome (IBS), 567*t*, 579–580
adverse events, infancy/childhood, 515
bowel symptoms, 515
colorectal sensitivity, 515–516
functional pain disorders, 514–515
neonatal development, 517
NGF, 516–517
nonpelvic functional pain disorders, 514–515
Rome III criteria, 514–515
WAS, 515–516
Irritants, 101
Isolectin B4 (IB4), 292–293

K

KCC2. *See* Potassium-chloride cotransporter 2 (KCC2)
K^+ channels, 12–13

L

Lipopolysaccharide (LPS), 254–255, 258–261, 265–266
Long-term potentiation (LTP)
definition, 389
downstream kinases, activation of, 389
induction protocols, 389–390, 512–513

M

Mammalian noncoding RNA-disease database (MNDR), 220
MAPKs. *See* Mitogen activator protein kinases (MAPKs)
MCP-1. *See* Monocyte chemoattractant protein-1 (MCP-1)
Mechanical allodynia
action potential threshold, 62–65
hyperalgesia, 62–65
innocuous stimuli, 62–65
MSCs, 62–65, 63*f*
voltage-gated ion channels, 62–65
Mechanistic target of rapamycin complex 1 (mTORC1)
and ERK, 416–417
rapamycin/polyadenylation inhibitor cordycepin, 419–420
sensory neurons, 416
Mechanistic target of rapamycin (mTOR) pathway
de novo protein synthesis, 202–203
4E-BP2 and eIF2α pathways, 206
inhibition and nociception, 196–197, 206–207
mTORC1 and mTORC2, 189–190
peripheral inflammation- and injury-induced activation, 203–204

Mechanistic target of rapamycin (mTOR) pathway (*Continued*)
PI3K-AKT-mTORC1 pathway, 192, 194
spinal neuronal sensitization, 205–206
translation initiation, 191*f*
Mechanosensation
endothelial cells, 54–55
planet, 54–55
primary function, 54–55
supraoptic nucleus, 54–55
thigmotropic and thigmonastic behaviors, 54–55
Mechanosensitive ion channels (MSCs)
across species, 54*f*
bona fide MSC, 57–58
Caenorhabditis elegans, 57–58
candidate proteins, identification, 57–58
environment, 53–54
Escherichia coli, 57–58
gating mechanisms, 59–60
mechanical allodynia, 62–65
nociceptors (*see* Nociceptors)
noxious mechanical inputs, 60–62
physiological functions, 53–54
Piezo2, 58–59
potassium, 65
Mechanotransduction, 53–54, 57–58, 62–65
MeDIP-Seq. *See* Methylated-DNA immunoprecipitation (MeDIP-Seq)
Memory
DNA methylation (*see* DNA methylation)
DNA methyltransferases, 161–162
environmental adaptation, nervous system, 160
G9a/GLP histone methyltransferase complex, 160
genetic modification, HDACs expression, 159
HDAC inhibitors, 159–160
histone acetylation, 158–159
synaptic plasticity, 160
Ted Abel's group, 160
Messenger RNA (mRNAs)
decapping/repression, 200
eIF4F complex, 189, 190*f*, 200
eukaryotic mRNA 5′ cap structure, 190*f*
FMRP, 199
mGluR1-mediated activation, 199
RBP, 199
RNP granules, 200
3′ UTR and poly-A tail, 198, 199*f*
Metabotropic glutamate receptors (mGluRs), 15–16, 38–39
activity in brain, 305–312
in brainstem and brain, 300–305
cell-signaling cascades, 282–283
drug development, 313–315
G-protein-coupled glutamate receptors, 282–288
presynaptic and postsynaptic neurons, 288–289
role, 282
spinal cord (secondary sensory neurons and interneurons), 294–300
Methylated-DNA immunoprecipitation (MeDIP-Seq), 156
5-Methylcytosine (5mC), 153–154, 156
mGluR activity in brain
in amygdala, 305, 306*t*
in cortex, 305, 307*t*
group I activity, 307–310
group II activity, 310–311
group III activity, 311–312
in thalamus, 305, 305*t*
Microglia
astrocytes, 327 (*see also* Astrocytes)
AZD2423, 345
carbon-11 ([11C]-(R)-PK11195), 339
CCR3/CD11b, 330–331
CSF, 341–342
CX3CR1 receptor, 334–335
homeostatic environment, 330
ibudilast, 337
immune surveillance and orchestrate, 326–327
and neuronal interactions, 345
NK1 receptors, 333–334
p-p38, 332, 336–337
propentofylline, 337
purinergic receptor, 334
P2X4, P2X7/P2Y12 receptors, 334
rat spinal cord (*see* Rat spinal cord)
TLR-2 agonist, 346–349
TLR-4 agonist, 335–336
YKL-40, 338

MicroRNAs (miRNAs), 152, 163–164
  biogenesis, 216–218
  challenges in research, 240–242
  chronic pain conditions, 242
  circulating miRNAs, 224–227
  exosome biogenesis, 224–227
  extracellular, 242
  genomic location, 218
  nomenclature, 219–220
  and pain (*see* Pain)
  quantification and functional studies, 221–223
  seed sequence, 216
  targets, 220–221
  3′ UTR, 216
Microtubule
  anticancer efficacy, 492
  bortezomib, 492
  depolymerizing/destabilizing agents, 473–474
  MTAs (*see* Microtubuletargeting agents (MTAs))
  posttranslational status, 492
  β-tubulin, 473–474
Microtubuletargeting agents (MTAs)
  classes, 473–474
  depolymerizing/destabilizing agents, 473–474
  microtubule, 492
  neuronal sensitivity, 474
  polymerizing/stabilizing agents, 473–474
  taxanes, 474
  β-tubulin subunits and, 473–474
  vinca alkaloids, 474
Middle meningeal artery (MMA), 545–546
Migraine, 567*t*, 581–582
  ASICs and TRPs, 555–556
  BDNF, 554–555
  CGRP and synaptic plasticity, 553–554
  dural afferent activation (*see* Dural afferent activation)
  head pain, 538
  individual patients, 540
  intracranial pressure, 538–540
  ion channels, 546–549
  mechanosensitivity, 556
  meningeal afferent system, 542–543
  nausea and vomiting, 538–540
  neuroplasticity (*see* Neuroplasticity)
  nociceptive system, 555–556
  pathological condition, 538
  pathophysiology, 544–545
  phases, 538–540
  prophylactic agents, 555
  scotomas, 538–540
  treatments, 541–542
  triggers, 540–541
Mineralocorticoid receptor (MR), 511–512
Mitochondrial dysfunction, 484–486
Mitogen activator protein kinases (MAPKs), 149, 153, 156–157, 158, 332–333, 340–341
MMA. *See* Middle meningeal artery (MMA)
Monocyte chemoattractant protein-1 (MCP-1), 332–333, 334–335, 338
Monosodium urate crystals (MSU), 268
μ opioid receptors (MORs), 421–422
MR. *See* Mineralocorticoid receptor (MR)
MSU. *See* Monosodium urate crystals (MSU)
MTAs. *See* Microtubuletargeting agents (MTAs)
mTORC1. *See* Mechanistic target of rapamycin complex 1 (mTORC1)
Multiple noxious stimuli, 74
Multiple sclerosis (MS), 567*t*, 582–583

N

N-acetylcysteine (NAC), 486–487
NAD. *See* Nicotinamide adenine dinucleotide (NAD)
National Cancer Institute (NCI), 476–477
National Institutes of Health chronic prostatitis symptom index (NIH-CPSI), 522–523
Neonatal intensive care unit (NICU), 513–514
Nerve growth factor (NGF), 38, 149, 490–491, 516–517, 518, 520, 523
Neurokinin receptor, 38
Neuropathic pain. *See also* Persistent inflammatory pain
  dendritic spines, 391
  diabetic, 394–395
  IASP definition, 5
  inflammatory and neuroimmune mechanisms, 5

Neuropathic pain (*Continued*)
pathobiology, 5
peripheral nerve injury, 6–7
persistent ongoing pain, 6–7
Rac1 inhibition, mode of action, 399–400
rodent models, 5–6
spinal memory mechanism, 400–401
thermal injury, 395–396
zoster virus-induced, 6
Neuroplasticity
chemical stimulation, dural afferents, 552
cutaneous allodynia, 550–551
cytokines, neuropeptides and growth factors, 549–550
episodic migraines, 550–551
facial allodynia, 549–550
headache, 552
mechanism, 552
neurotransmitters, 552
pain-sensing fibers, 551–552
peripheral and central nervous systems, 550–551
TNC and activation, 549–550
trigeminovascular activation, 549–550
Neurotransmitters, 360–362
Neurotrophic factors
embryonic neurons, 490–491
neurotrophin, 490–491
NGF, 490–491
paclitaxel and cisplatin therapy, 490–491
preclinical animals, 490–491
sensory neurons, 490–491
NGF. *See* Nerve growth factor (NGF)
Nicotinamide adenine dinucleotide (NAD), 153
NICU. *See* Neonatal intensive care unit (NICU)
NIH-CPSI. *See* National Institutes of Health chronic prostatitis symptom index (NIH-CPSI)
Nitroxidative stress
chemotherapeutics, 486–487
N-acetylcysteine (NAC) functions, 486–487
ROS and RNS, 486–487
TEMPOL, 486–487
NKCC1. *See* Sodium-potassium-chloride-cotransporter 1 (NKCC1)
N-methyl-D-aspartate receptor (NMDA) receptor, 15, 156–157, 158
Nociception
capsaicin's excitatory effects, 78
noxious stimuli, encoding and processing, 74
and pain, 74, 77–78, 91–92
TRPV1, activation/potentiation, 87–88
Nociceptors
AMPK activation pathway, 416–417, 418*f*
anisomycin/rapamycin, 419–420
brain regions, 414
CPEB, 417–419
Drosophila larvae, 57
electrophysiological studies, 55–57
FMRP, 414–416
gene expression, 414
genetic studies, 57
hyperalgesic priming models, 416
microfluidic devices, 414–416, 415*f*
MSCs, 55–56
mTORC1 and ERK, 416–417
neurons, 55
nociceptor peripheral terminals, 55–56
originated mediators, 33
pharmacological characterization, 56–57
postsynaptic density, 414–416, 415*f*
potassium, 65
protein–protein interaction, 414–416
rheumatoid arthritis and osteoarthritis, 55
ribosomes, dendritic spines, 413–414
RNA binding proteins, 417–419
single-channel recordings, 56–57
translational control pathways, 416, 417*f*
TRPV1, 55–56
whole-cell voltage-clamp, 56–57
Nociceptor sensitization, 87
Nonsteroidal anti-inflammatory drugs (NSAIDs), 541
Nucleus Raphe Magnus (NRM), 164–166

O

Opioid receptors, 16
Oral squamal cell carcinomas (OSCCs), 40
Osteoarthritis, 55, 61–62, 65

Oxaliplatin
  acute behavioral effect, 478–480
  bevacizumab, 475–476
  Ca^{2+}/Mg^{2+} infusions, 488–489
  CIPN, 485
  cisplatin, 474–475
  ion channel dysfunction, 488–489
  patients, 475
  treatment, 475

P

Paclitaxel
  acute pain syndrome, 474
  ALCAR, 484–485
  application, 481–482
  CCL2/CCR2 pathway, 489–490
  CIPN, 478, 487
  NGF levels, 490–491
  oxaliplatin, 478–480
  pain and dysesthesia, 485
  SNCV, 480
  treatment, 481–482
  TRPV1-mediated release, 482–483
Pain
  acute syndrome, 474
  in bone cancers, 87
  CFA-induced inflammatory, 93
  CIPN patients, 476–477
  epigenetic mechanisms
    acute to chronic pain, 173–174
    DNA methylation, 170–171
    early life injury, 174
    epigenetic machinery and epigenome, 164–166
    HDACs and HATs, 166–170
    in infancy, 150*f*
    miRNA expression, 163–164
    molecular and behavioral response, 163–164
    treatment, persistent pain, 171–173
  IENF loss, 477–478
  inflammatory, 490–491
  miRNAs, 216–218, 219–223, 224–242
  olesoxime on paclitaxel-induced, 485
  oxaliplatin, 475
  sensory TRP channels (*see* Transient receptor potential (TRP) channel)
  TRPA1 role, 103–104
  TRPM8 role, 97–98
  TRPV1, 86–88
Pain and pain pathways
  activity-dependent synaptic plasticity, 2–4
  C-fibers, 4–5
  definition, 2
  inflammatory pain, 7–9
  International Association for the Study of Pain (IASP), 2
  neuropathic pain, 5–7
  nociceptive, 2–4
  pronociceptive neurotransmitters, 4
  schematic diagram, 3*f*
  tissue injury and inflammation, 4
  TRPV1-expressing C-fibers, 4–7
  vasoactive mediators, 4
Painful peripheral neuropathy, 583
Pain in brainstem and brain
  anatomic loci, 300–301
  group I activity, 303
  group II activity, 304
  group III activity, 304–305
  mGluR activation, 301–302, 302*t*
  mGluR expression, 300–301, 301*f*
Pain memory
  acute and chronic inflammatory pain, 410–411
  chronic neuropathic pain, 410
  clinical experience, 427
  hippocampus and cerebral cortex, 410–411
  hyperalgesic priming, 411–412
  mechanisms, 411–412
  neuronal plasticity, 410
  pain amplification, 410
  peripheral nervous system, 411
  peripheral nociceptor (*see* Nociceptors)
  phantom limb pain, 411
  spinal dorsal horn (*see* Spinal dorsal horn)
Pain neuraxis, mGluRs analysis
  in brainstem and brain, 300–305
  conventional knockout and systemic effects, 312–313
  mGluR activity in brain, 305–312
  mGluR drug development, 313–315
  periphery, 289–294
  presynaptic and postsynaptic neurons, 288–289

Pain neuraxis, mGluRs analysis (*Continued*)
  spinal cord (secondary sensory neurons and interneurons), 294–300
Pain plasticity, 412–413
Pathogen-associated molecular pattern molecules (PAMPs), 252
Pattern recognition receptors (PRRs), 252, 253*t*, 268–269
Peripheral nerve injury
  dendritic spine remodeling, 394
  NSC23766 treatment, 399–400
  and SCI, 395
  STZ-induced diabetes, 395
Peripheral neuropathic pain
  chemoprotectant agents, 583–584
  CIPN, 583
  diabetic polyneuropathy, 585
  distal symmetrical polyneuropathy (DSPN), 584
  DRG neurons (*see* Dorsal root ganglia (DRG) neurons)
  factors, 583
  in HIV, 584
  hyperglycemia, 585
  mechanisms and treatments, 567*t*
  nucleoside reverse transcription inhibitors, 584
  viral antigens, 584
Persistent inflammatory pain
  AMPA receptors, 15
  calcium channels, 12
  glutamate receptors, 14–15
  ion channels, 9–10
  K^+ channels, 12–13
  metabotropic glutamate receptors, 15–16
  NMDA receptors, 15
  opioid receptors, 16
  PAR receptors, 14
  pronociceptive neurotransmitters, 17–18
  prostaglandin (prostanoid) E_2, 17
  purinergic receptors, 13
  receptors, 13
  sodium channels, 10–11
  toll-like receptors, 13–14
  TRPV receptors, 16–17
Persistent postsurgical pain, 567*t*, 586–587
PET. *See* Positron emission tomography (PET)
Phantom limb pain (PLP), 567*t*, 587–588
PHN. *See* Postherpetic neuralgia (PHN)
Phosphatidylinositol-bis-phosphate (PIP2), 82–83
Phosphodiesterase 4 D3 (PDE4D3), 43
Phospholipase C-β (PLC-β), 84–85
Platinum-containing agents, 474–475
PLP. *See* Phantom limb pain (PLP)
Pore loop, 74–75
Positron emission tomography (PET), 339, 343
Postherpetic neuralgia (PHN), 567*t*, 588–589
Potassium-chloride cotransporter 2 (KCC2)
  and NKCC1, 367–368
  spinal neurons, 373
Presynaptic TRPV1, 79
Pronociceptive neurotransmitters, 17–18
  nerve growth factor, 18
  nitric oxide, 17
Prostaglandin (prostanoid) E_2, 17
Prostaglandin receptor, 32–33, 36–37
Protease-activated receptors (PAR), 14, 39
Proteasome inhibitors, 475
Protein kinase Ce (PKCe), 173–174
Protein kinase Mz (PKMz), 173–174
PRRs. *See* Pattern recognition receptors (PRRs)
Purine metabolites
  ATP, 267–268
  description, 267
  uric acid, 268
Purinergic receptors, 13, 33–36, 37

R

RA. *See* Rheumatoid arthritis (RA)
Rat spinal cord
  cellular interactions, glial cells, 329*f*
  glial cells and neurons, physical interactions, 328*f*
  microglia and astrocytes, physical interactions, 329*f*
  naïve and peripheral nerve injury conditions, 331*f*
Receptor activity-modifying protein 1 (RAMP1), 39–40

Receptor for advanced glycation end products (RAGE), 252–254, 255, 256–257, 257*f*, 261, 264
Resiniferatoxin (RTX), 81
Rheumatoid arthritis (RA), 567*t*, 589
RNA-induced silencing complex (RISC), 217–218, 228

S

SAHA. *See* Suberoyl anilide hydroxamic acid (SAHA)
Sensitization, 581
Sensory nerve conduction velocity (SNCV), 480
Serotonin and calcitonin gene-related peptide, 581
Serotonin receptor, 37
Sodium channels, 10–11
Sodium-potassium-chloride-cotransporter 1 (NKCC1)
  and KCC2, 373–374
  neuronal chloride homeostasis, 367–368
Spinal cord
  astrocytic S-100b, 332
  glial fibrillary acidic protein, 332
  HIV-positive patients, 340–341
  MAPKs, 332
  mGluRs analysis
    acute nociception, 294–295
    dynamic modulation, 294–295
    group I, 295–298
    group II, 298–299
    group III, 299–300
    mGluR activation, behavioral effects, 295*t*
    mGluR1 and mGluR5 activation, 298
    postsynaptic modulators, 297–298
  microglial CCR3/CD11b, 330–331
  microglial p-p38, 336–337
  neuronal excitatory milieu, 333
  NMDA receptor hyperactivity and neuronal sensitization, 332–333
  peripheral lymphocytes and monocytes, 327
  rat spinal cord (*see* Rat spinal cord)
Spinal cord injury (SCI)
  dendritic spine remodeling, 392–393
  density and altered spine distribution, 392–393, 393*f*
  NSC23766 treatment, 397–398, 399–400
  and peripheral nerve injury, 395
Spinal dorsal horn
  acetazolamide and benzodiazepines, 375
  allodynia, 420–421
  atypical PKCs and BDNF, 422–425
  bicarbonate efflux, 368
  capsaicin and formalin, 421, 425–427
  central sensitization, 420–421
  chloride dysregulation, 374–375
  chloride flows, 365
  chloride regulation, 366–367, 366*f*
  ClC-2 channels, 366–367
  excitatory interneurons, 371–372
  $GABA_A$ and glycine receptors, 364
  Gate Control Theory, 360
  GHK equation, 364–365
  hippocampal/cortical memory mechanisms, 425
  hyperalgesic priming paradigm, 425–427
  impaired synaptic transmission *vs.* impaired inhibitory effect, 374–375
  inhibitory neurons, 370–371
  intracellular chloride concentration, 368
  ion concentration, 365–366
  late phase LTP (late-LTP) and reconsolidation, 425, 426*f*
  long-lasting memory, 425
  microglia and inflammation, 374
  MORs, 421
  Nernst equation, 364–365
  nerve injury/inflammation, 375
  neuropathic pain, 371–372
  opioid-dependent mechanisms, 421–422
  perforated patch technique, 368
  potassium-chloride cotransporter 2 (KCC2), 367
  sodium-potassium-chloride-cotransporter 1 (NKCC1), 367–368
  superficial dorsal horn, 370, 371*f*
Spinal nerve ligation (SNL) model, 229–232, 233
Spiny neurons, 387
S100 proteins
  calcium-binding EF-hand motifs, 262
  characterization, 262

S100 proteins (*Continued*)
DAMP molecules, 262
intracellular functions, 262
S100A8 and S100A9, 262–263
S100B, 263–264
Streptozotocin (STZ), 394–395, 399
Stress
child maltreatment, 513–514
comorbidity, 513–514
NICU, 513–514
NMS, 514
Suberoyl anilide hydroxamic acid (SAHA), 166, 167*t*
Substance P (SP)-expressing C-fibers, 8–9
Synaptic inhibition
disinhibition, 374
$GABA_A$ receptors, 369
glycinergic transmission, 373
inhibitory neurotransmitters (*see* Neurotransmitters)
KCC2 and NKCC1, 373–374
nerve injury, 372–373
pain processing, 372–373
Synaptic plasticity
acute and prolonged pain states, 149
α-amino-3-hydroxy-5-methyl-4-isoxazolepropionic acid, 156–157
brain-derived neurotrophic factor, 149
cAMP response elementbinding protein, 156–157
epigenetic mechanisms (*see* Epigenetic mechanisms)
epigenome, 163
extracellular signal-regulated kinases, 149
histone modifications, 157–158
learning and memory, 149
memory formation, 156–157 (*see also* Memory)
mitogen-activated protein, 149
neuroepigenetics, 149
neurotrophins, 149
N-methyl-D-aspartate receptor, 156–157
pain, 149
posttranslational regulation, 156–157

T

Taqman low-density array (TLDA), 221–222
Thermal burn injury, 395–396
TLR signals. *See* Toll-like receptor (TLR) signals
TNC. *See* Trigeminal nucleus caudalis (TNC)
TNF. *See* Tumor necrosis factor alpha (TNF-α)
Toll-like receptor (TLR) signals, 13–14, 489–490
Total internal reflective fluorescence-Forster resonance energy transfer (TIRF-FRET), 43–45
Total neuropathy score (TNS), 476–477
Transcriptional activator-like effectors (TALENS), 172–173
Transient receptor potential (TRP) channel
cytoplasmic regions, structural–functional features, 77*f*
functional, 76*f*, 77–78
in phototransduction, 74
tetramers of 6-transmembrane (6-TM), 74–75
thermosensing nociceptive, 75*f*
TRP box, 75–77
Transient receptor potential subfamily A member 1 (TRPA1)
in autonomic reflexes and disorders, 99–100
channel expression and function modulation, 102–103
channel, functional properties, 100–102
cold-sensing ion channel, 102
drug development targeting, 104–105
human, 101
human lung fibroblasts, 98–99
in pain conditions, 103–104
sensory afferents expressing, 99
structure, 100
Transient receptor potential subfamily A member 2 (TRPA2)
phagocytosis, 91
in rodent sensory neurons, 91
Transient receptor potential subfamily A member 3 (TRPA3), 93
antagonists, 91–92
camphor, 91–92
expression, 91–92
rodent PNS and CNS neurons, 91–92
sequence homology cloning, 91–92

Transient receptor potential subfamily A member 4 (TRPA4)
activation, 92–93
in DRG and TG neurons, 92–93
energy homeostasis maintenance, 92–93
irritable bowel syndrome (IBS), rodent model, 92–93
in neurons and muscles, 92–93
pancreatitis, rodent model, 92–93
polyunsaturated fatty acids (PUFAs), 92–93
Transient receptor potential subfamily A member 8 (TRPA8)
channel expression and function modulation, 96
in cold hypersensitivity, 98
drug development targeting, 98
functional properties, 95–96
knockout mice, 97–98
in nervous system, expression and distribution, 94
in pain conditions, 97–98
in prostate epithelial cells, 94
role, 95–96
structure, 94–95
Transient receptor potential vanilloid 1 (TRPV1), 16–17, 55–56, 58
activation, 78
capsaicin's excitatory effects, 78
channel desensitization, 86
channel expression and function, modulation, 83–86
channel, functional properties, 80–83
cryo-EM structure, 81–82
drug development targeting, 90–91
in nervous system, expression and distribution, 78–79
in pain conditions, 86–88
phosphorylation, 84–85
in physiological and pathological conditions, 88–90
in sensory neurons, 78
S4 segment, voltage-dependent movement, 81–82
structure, 79–80
toxins VaTx and DkTx, 81–82
Translational control
capsaicin, 202–203
definition, 189
formalin, 202–203
long-term depression (LTD), 201–202
long-term potentiation (LTP), 201–202
mechanisms, 201
mTOR pathway, 201–202
nociceptive pathways, 201
opioid-induced hyperalgesia, 204
peripheral C-fiber activation, 202–203
reconsolidation, 205
sensory neurons, 201
spinal LTP, 201–202
spinal neuronal sensitization, 205–206
spinal neurons, 203–204
spinal nociceptive circuit, 201
spinal nonneuronal cells, 204–205
synaptic plasticity in CNS, 201–202
Trigeminal neuralgia, 567*t*, 589–590
Trigeminal nucleus caudalis (TNC), 549–550, 552, 556
Trigeminovascular system, 581
TRPA1. *See* Transient receptor potential subfamily A member 1 (TRPA1)
TRPV1. *See* Transient receptor potential vanilloid 1 (TRPV1)
Tumor necrosis factor alpha (TNF-α), 332–333, 337, 340–341
Tyrosine kinases, 38

U

3′ untranslated region (3′ UTR)
lin-14 3′UTR, 216–217
mRNAs, 216
target mRNAs, 220–221

V

Vanilloid receptor subtype-1 (VR1), 78
Vanillotoxins (VaTx), 81
Vascular endothelial growth factor (VEGF), 475–476
Vertebroplasty, 572
Vesicular GABA transporter (VGAT), 360–362, 361*f*
Vinca alkaloids, 474, 487–488
Visceral pain, 566
Voltage-gated calcium channels (VGCC), 9

Vulvodynia
adult female mice, 521
allodynia, 520–521
antidepressant and anticonvulsant medication, 521
erythema, 520–521
idiopathic pain disorders, 521
IL-1RA, 522
leukocyte infiltration, 521–522
mood disorders, 521
neurologic disorder, 520–521
proinflammatory cytokines IL-1, 522
tumor necrosis factor-α, 522

W

Water avoidance stress (WAS), 515–516
Whole-genome bisulphite sequencing (WGBS), 156

Z

Zif268 gene, 156–157, 158, 163

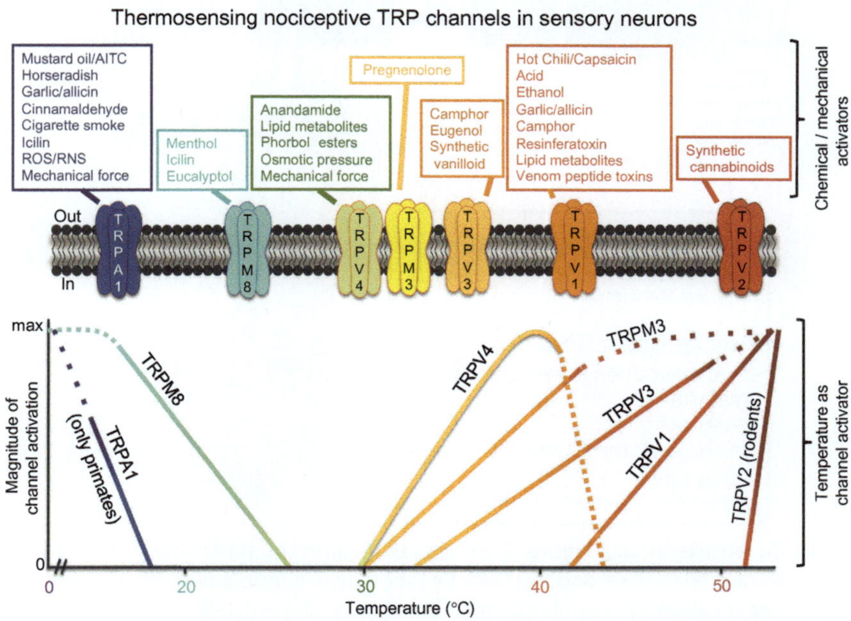

Aaron D. Mickle *et al.*, Figure 1 Classification of thermosensing nociceptive TRP channels in mammalian sensory neurons. The upper row of individual boxes denotes chemical/mechanical activators of marked TRP channels. The lower panel depicts the magnitude of channel activity upon activation by temperature of the independent TRP channels shown.

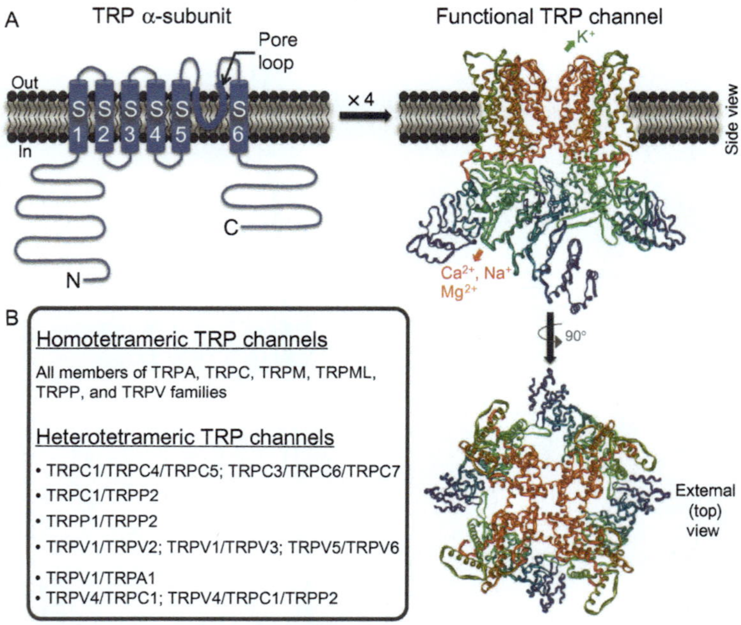

Aaron D. Mickle *et al.*, Figure 2 Structural features of functional TRP channels. (A) Model of predicted secondary structure topology of a monomeric TRP channel subunit (called α-subunits) on a plasma membrane lipid bilayer. S1–S6 denote transmembrane segments/domains 1 through 6, with a hydrophilic pore loop between S5 and S6 transmembrane segments, and N and C denote cytoplasmic amino- and carboxy-terminals, respectively. Four monomeric α-subunits assemble in a three-dimensional arrangement to make a functional channel, which is shown on the right. The structure was constructed from the PDB structure file obtained from the protein data bank, originally submitted by Refs. 8,9 from their cryoelectron microscopic structure of TRPV1. (B) List of homo- and heterotetrameric TRP channels formed within individual TRP subfamily member channels, as well as in combination of inter-subfamily TRP channel members.

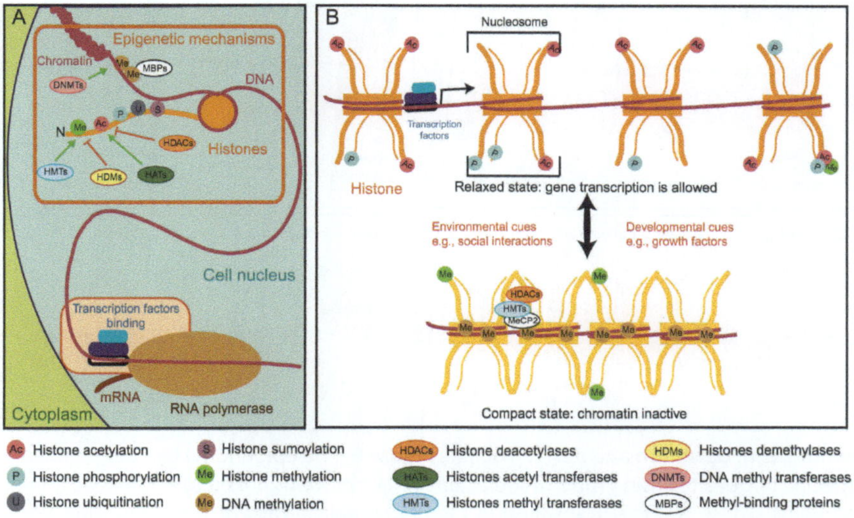

Sandrine M. Géranton and Keri K. Tochiki, Figure 2 Regulation of gene expression via changes in chromatin compaction. (A) Epigenetic mechanisms control the compaction of chromatin by regulating DNA methylation and modifying histone tails by addition or removal of chemical groups. As a direct result, the binding of transcription factors is permitted or prevented. (B) While histone acetylation and phosphorylation have been shown to lead to an open state of the chromatin and therefore facilitate gene expression, histone methylation can have positive and negative impact on gene expression. DNA methylation can inhibit transcription by recruiting methylated DNA-binding proteins such as MeCP2, a member of the methyl-CpG DNA-binding protein family, and other proteins from the chromatin.

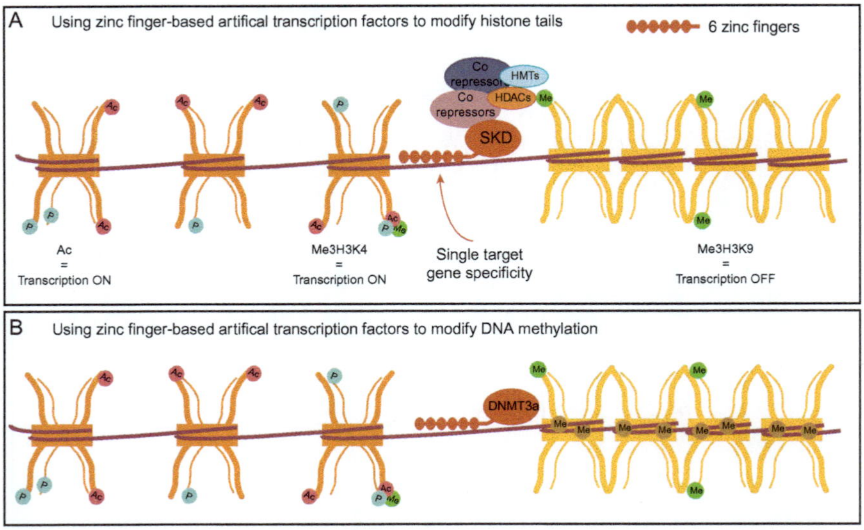

Sandrine M. Géranton and Keri K. Tochiki, Figure 6 The epigenome can be modified at specific target sites using multimodular zinc finger proteins linked to epigenetic editing domains. These artificial transcription factors contain arrays of Cys2-His2 zinc finger domains, which specifically interact with 3 bp (triplet of recognition) of DNA. By combining a number of zing finger domains, multimodular proteins have been engineered to recognize specific sequences of promoters and regulate gene expression. 6 zinc finger proteins (18 bps targeted) represent optimized, state-of-the-art designs that regulate single target genes with high selectivity. (A) Zinc fingers linked to the Kruppel-associated box (SKD) can be used to change histone modifications at a specific target site. Upon recruitment of corepressors such as KRAB-associated protein 1 and nucleosome remodeling and deacetylase (NuRD), a repressive complex is assembled including HDACs, histone methyltransferases such as SETDB1 and the heterochromatin protein 1. This repressive complex could catalyze the formation of condensed chromatin by deacetylation of histones, demethylation of H3K4me3, and incorporation of H3K9me3.[129] (B) Zinc fingers linked to DNMT3a have been used previously to increase DNA methylation and silenced the oncogene SOX2.[130]

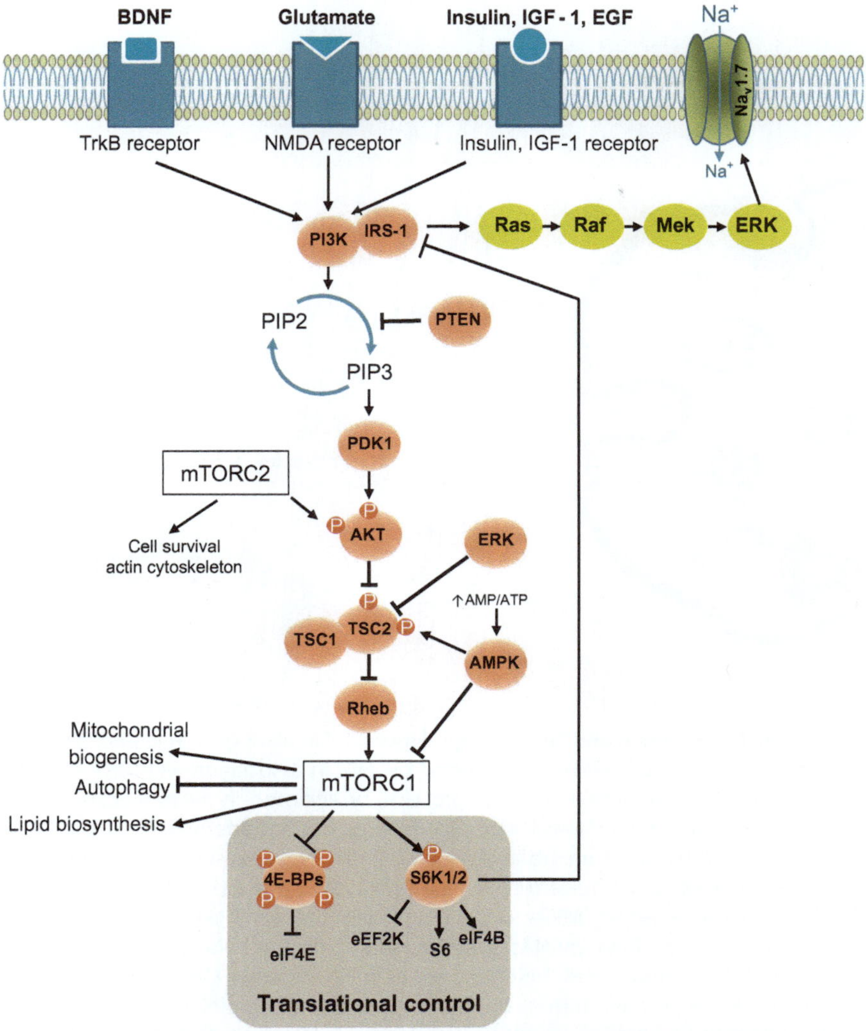

Ohannes K. Melemedjian and Arkady Khoutorsky, Figure 2 Regulation of translation initiation by mTOR signaling pathway. mTOR is a critical downstream target of the phosphatidylinositol-3 kinase (PI3K) signaling pathway. Insulin response element (IRS1) "transmits" signals from the TrkB, NMDA, insulin, and insulin-like growth factor-1 (IGF-1) receptors through PI3K and PI3K-dependent kinase 1 (PDK1), which phosphorylates Akt. mTOR forms two distinct protein complexes: mTORC1, which regulates translation, is sensitive to rapamycin and is defined by the scaffolding protein Raptor, and mTORC2, which is defined by Rictor and is generally rapamycin insensitive. mTORC1 phosphorylates 4E-BPs, S6K1, and S6K2. Akt can activate mTORC1 through phosphorylation and inhibition of the tuberous sclerosis complex (TSC).

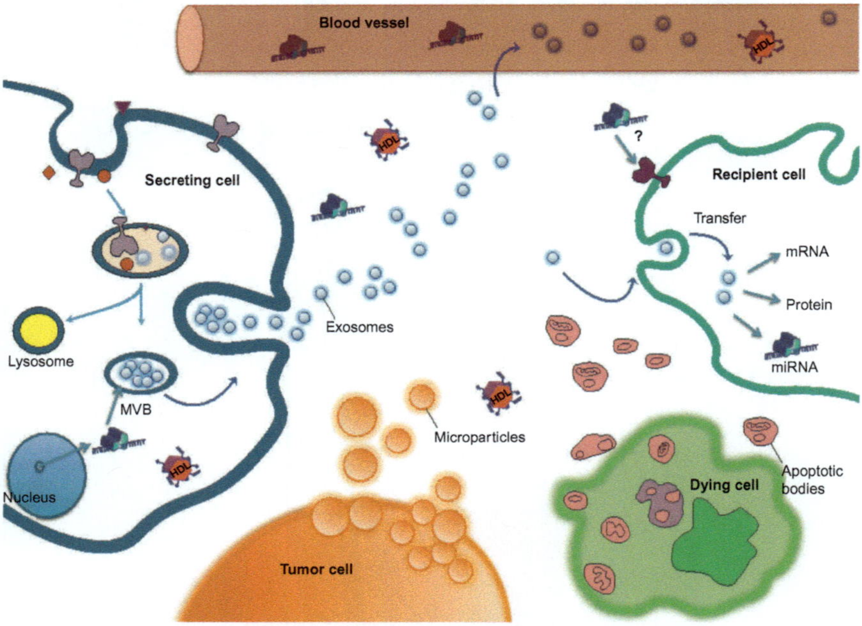

Marguerite K. McDonald and Seena K. Ajit, Figure 3 Circulating miRNAs and exosome biogenesis. Circulating miRNAs are secreted by cells in association with RNA-binding proteins such as Ago, high-density lipoproteins, or enclosed in microvesicles. Microvesicles include lipid-bound vesicles that are released during apoptosis (pink), shedding vesicles released by direct membrane budding (orange), or exosomes (blue) released from the multivesicular bodies (MVBs). Exosomes, the smallest of the microvesicles (30–100 nm), form within MVBs originating from the early endosomal pathway. Exosomes contain miRNAs, mRNAs, proteins, and lipids, the composition of which is reflective of the secreting cell. However, not all miRNAs present within the secreting cells are packaged into exosomes, and the mechanisms determining what is included in the exosomes for secretion is not well understood. After secretion, exosomes enter the blood stream and are transferred into recipient cells, releasing functional contents. Mechanisms similar to receptor-ligand mediated recognition have been postulated to guide exosomal uptake by recipient cells. It has been shown that extracellular Ago-bound miRNAs influence cellular function through receptor binding and hypothesized that HDL-associated miRNAs are secreted and taken up in a receptor-dependent mechanism.

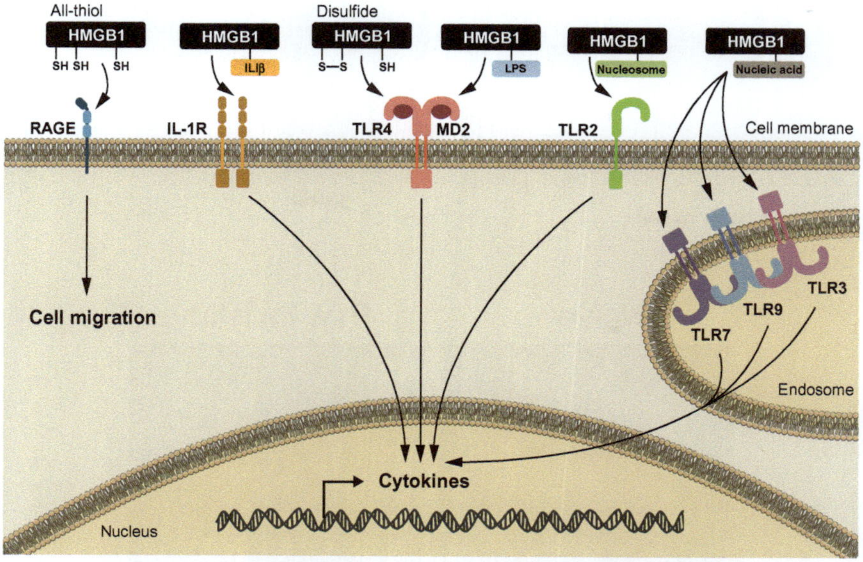

Jungo Kato and Camilla I. Svensson, Figure 2 HMGB1 activates multiple signaling pathways. HMGB1 interaction with TLRs and RAGE induces cytokine production, migration, proliferation, and differentiation. When HMGB1 is in the all-thiol form it activates RAGE and when in the disulfide form it activates TLR4. The HMGB1 redox dependence in interactions with the other receptors is presently unknown. In addition to direct receptor interactions, HMGB1 can form complexes with different ligands and thereby enhance the responses induced by the bound ligand.

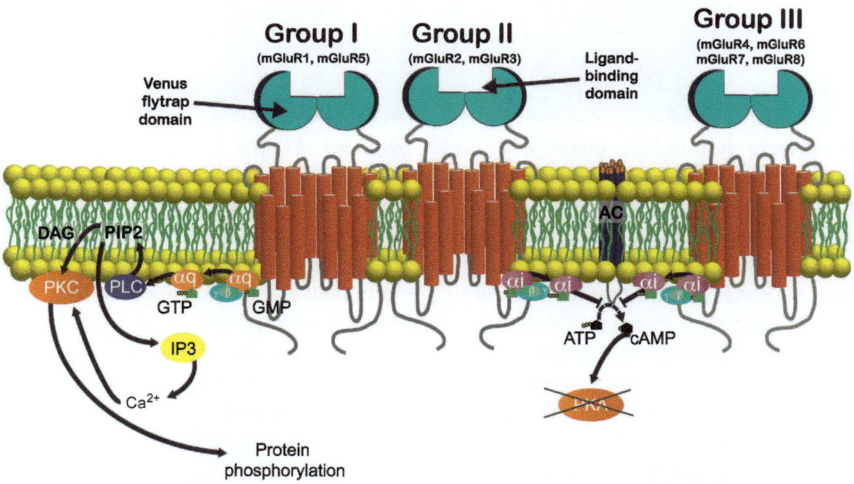

Benedict J. Kolber, Figure 1 Schematic illustration of mGluR dimers from Group I, II, and III receptors. All mGluRs contain large extracellular domains known as Venus flytrap domains. These domains contain the binding site for glutamate and other agents. Group I mGluRs (mGluR1 and mGluR5) couple to the G-protein $G_{\alpha q}$. $G_{\alpha q}$ activates phospholipase C (PLC) which hydrolyzes phospholipid phosphatidylinositol 4,5-bisphosphate ($PI(4,5)P_2$) to generate DAG and IP3 to increase intracellular calcium and activate PKC. Both Group II (mGluR2 and mGluR3) and Group III (mGluR4, mGluR6, mGluR7, mGluR8) mGluRs couple to the G-protein $G_{\alpha i}$. Activated $G_{\alpha i}$ inhibits the conversion of ATP to cAMP by adenylyl cyclase (AC), which ultimately lowers protein kinase A (PKA) activation.

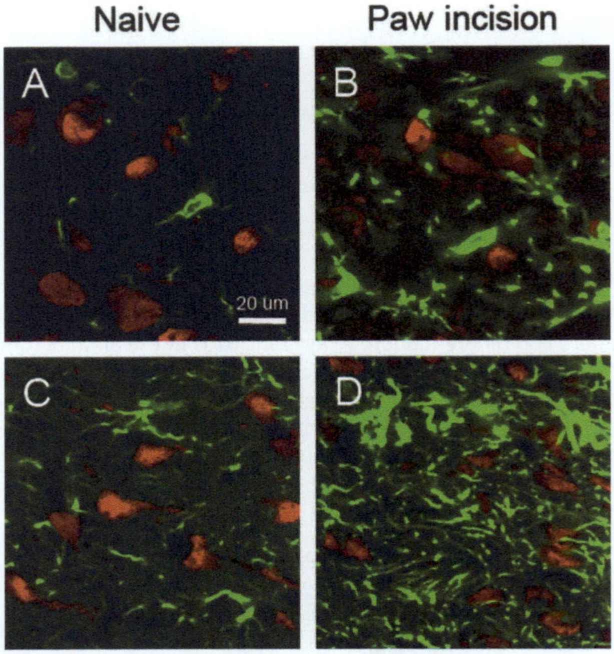

E. Alfonso Romero-Sandoval and Sarah Sweitzer, Figure 1 Physical interaction between glial cells and neurons in the rat spinal cord. This figure displays confocal microscopy images of dorsal horn (superficial laminae) spinal cord of rats. (A) Immunolabeling for Iba1 (green, microglial marker) and NeuN (red, neuronal nuclear marker) in a naïve rat. (B) Immunolabeling for Iba1 (green, microglial marker) and NeuN (red, neuronal nuclear marker) in a rat undergone paw incision (postoperative day 3); note the apparent increase in the physical interaction with neurons compared to naïve conditions. (C) Immunolabeling for GFAP (green, astrocytic marker) and NeuN (red, neuronal nuclear marker) in a naïve rat. (D) Immunolabeling for GFAP (green, astrocytic marker) and NeuN (red, neuronal nuclear marker) in a rat undergone paw incision (postoperative day 3); note the apparent loss of territoriality of astrocytes, which appears to increase the physical interaction with neurons.

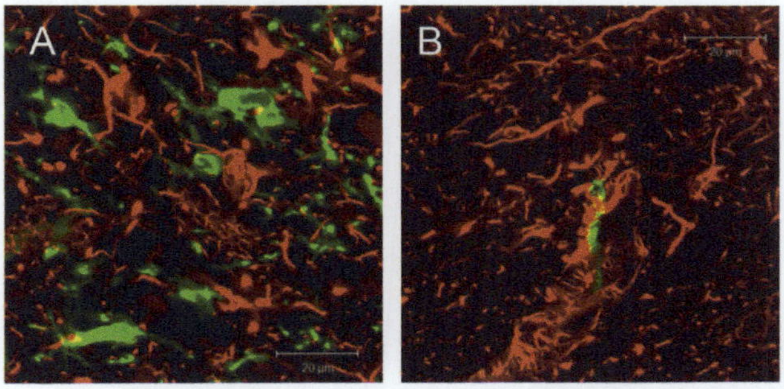

E. Alfonso Romero-Sandoval and Sarah Sweitzer, Figure 2 Physical interactions between microglia and astrocytes in the rat spinal cord. This figure displays confocal microscopy images of dorsal horn (superficial laminae) spinal cord of rats. (A) Immunolabeling of parenchymal microglia (Iba1, green) and astrocytes (GFAP, red) in a rat undergone paw incision (postoperative day 3). A "reactive" glial morphology is observed. (B) Immunolabeling of a perivascular microglial cell (ED2, green) and astrocytes (GFAP, red). Note the intimate interaction of astrocytes and a perivascular cell in what appears to be a capillary.

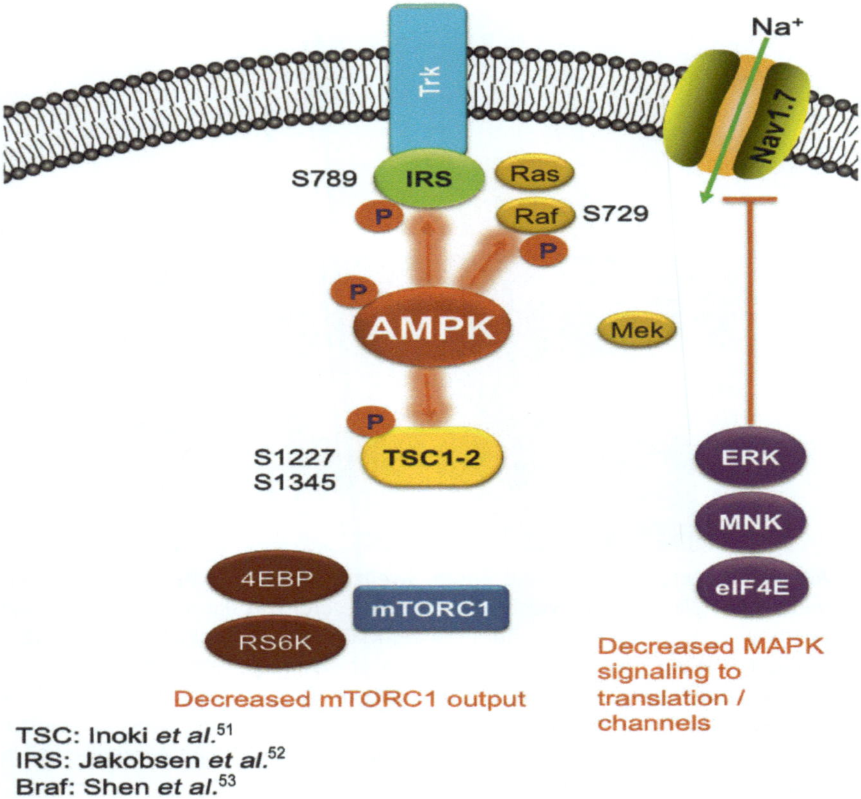

Theodore J. Price and Kufreobong E. Inyang, Figure 3 *AMPK activation pathway*. AMPK activation phosphorylates TSCs at Ser 1227 and 1345 leading to the inhibition of mTORC1. This is shown in the figure as an uncoupling of mTORC1 from Trk signaling via phosphorylation of TSC1/2 by AMPK. AMPK activation also phosphorylates Braf (Raf) at Ser 729 leading to inhibition of ERK signaling. Again, this is shown in the figure by an uncoupling of Raf/Mek signaling to ERK, MNK, and eIF4E via phosphorylation of Raf at Ser 729. Finally, AMPK phosphorylates IRS1 at Ser 789 leading to further inhibition of tyrosine kinase receptor signaling.

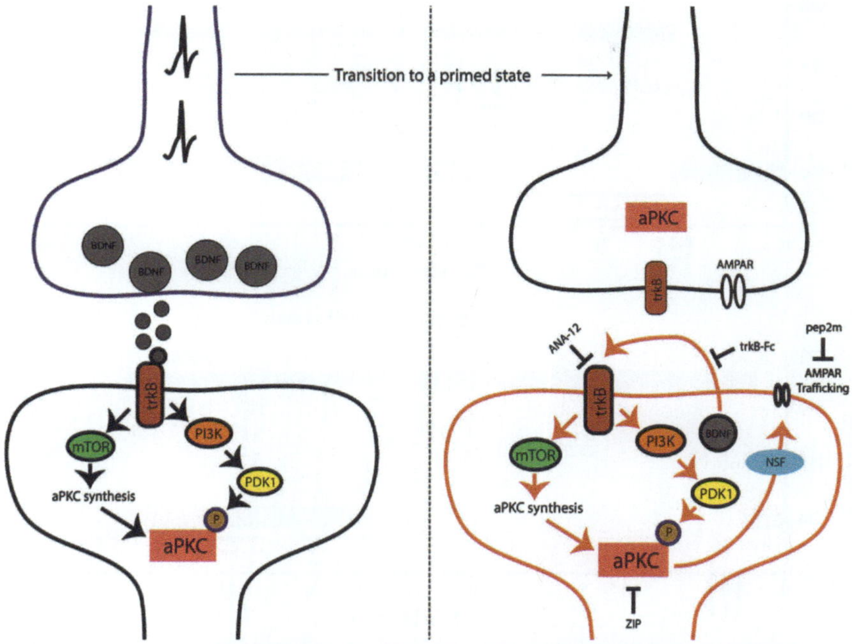

Theodore J. Price and Kufreobong E. Inyang, Figure 4 *The role of aPKCs and BDNF in hyperalgesic priming initiation and maintenance.* Nociceptor activation leads to spinal BDNF release and a postsynaptic mTORC1-dependent translation of aPKC protein. These newly synthesized aPKCs are then phosphorylated by PDK1. Increased levels and phosphorylation of aPKCs are thought to be involved in initiating priming. Once priming is established (right panel), increased aPKC protein and phosphorylation leads to a constitutive increase in AMPAR trafficking to the postsynaptic membrane. This appears to be regulated by BDNF signaling via trkB with BDNF potentially being released from postsynaptic dendrites in the maintenance stage of priming. Presynaptic trkB may also be activated by increased BDNF action in primed animals. Once established, hyperalgesic priming can be permanently reversed by inhibition of aPKCs with ZIP, disruption of AMPAR trafficking with pep2M or via inhibition of trkB/BDNF signaling with ANA-12 or trkB-Fc, respectively.

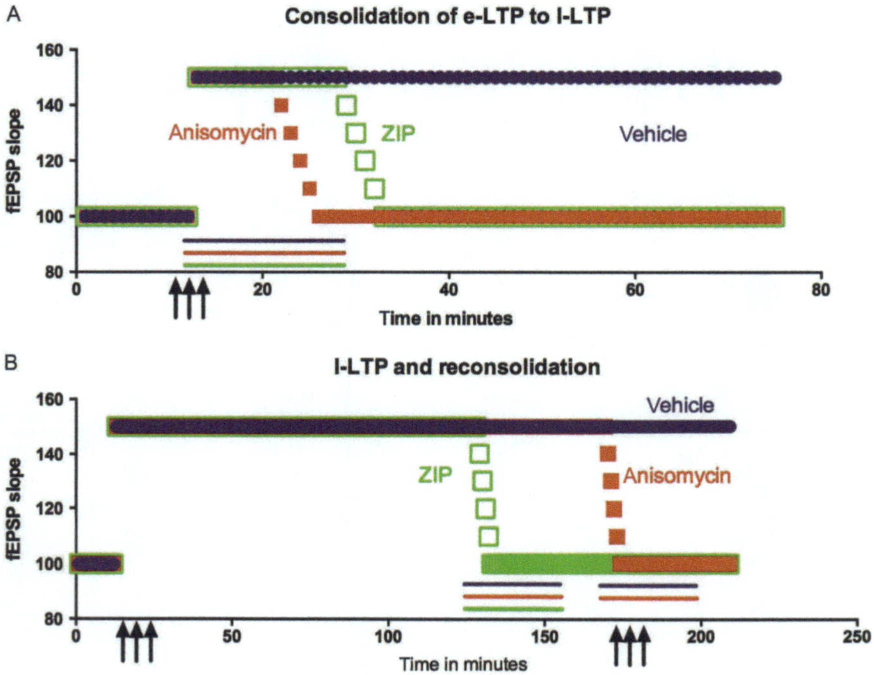

Theodore J. Price and Kufreobong E. Inyang, Figure 5 *Consolidation of late phase LTP (late-LTP) and reconsolidation*. (A) Following high-frequency stimulation of afferent input (three upward arrows), early-LTP (e-LTP in the figure) is induced and this consolidates to late-LTP (l-LTP in the figure) over the course of 30–60 min. Application of translation control inhibitors, such as anisomycin (red line), during early-LTP cause a failure of late-LTP consolidation. Likewise, aPKC inhibition with ZIP (green line) blocks consolidation of late-LTP. Vehicle application (blue line) has no impact on consolidation of late-LTP. (B) Once late-LTP is established administering translation inhibitors (e.g., anisomycin, red line) in the absence of high-frequency stimulation of afferents fails to reverse late-LTP while ZIP application (green line) does induce late-LTP decay. Restimulation of afferents at high frequency during late-LTP (upward arrows) opens a reconsolidation window. Application of translation inhibitors such as anisomycin (red line) during this reconsolidation period leads to late-LTP decay, an effect that is presumably linked to reversal of a chronic pain state in similar behavioral pharmacology experiments.

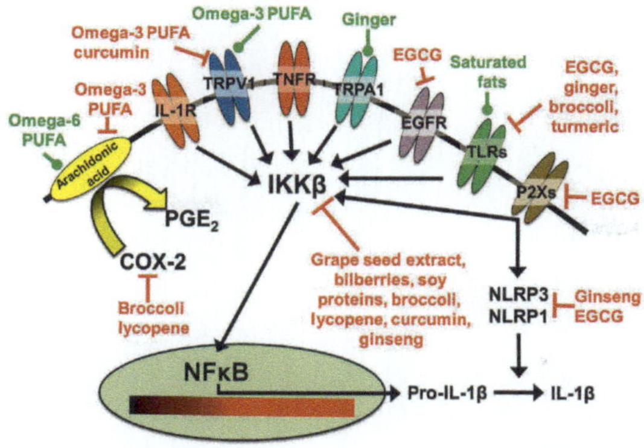

Stacie K. Totsch *et al.*, Figure 5 Location of action of the various dietary components within the cells of the immune system. Antagonism or inhibition is indicated in red, activation or agonism is indicated by green. Abbreviations are as follows: COX2 (cyclooxygenase-2), EGCG (epigallocatechin gallate), EGFR (epidermal growth factor receptor), IL-1β (interleukin-1 beta), IL-1R (interleukin-1 receptor), IKKβ (inhibitor of nuclear factor kappa-beta kinase subunit beta), NFκB (nuclear factor kappa-light-chain-enhancer of activated B cells), NLRP1 (NACHT, LRR, and PYD domains-containing protein 1), NLRP3 (NACHT, LRR, and PYD domains-containing protein 3), P2Xs (purinergic receptor type 2X), PGE_2 (prostaglandin E2), pro-IL-1β (pro-interleukin-1 beta), PUFA (polyunsaturated fatty acid), TLRs (Toll-like receptors), TNFR (tumor necrosis factor receptor), TRPA1 (transient receptor potential cation channel subfamily A, member 1), and TRPV1(transient receptor potential cation channel subfamily V member 1).

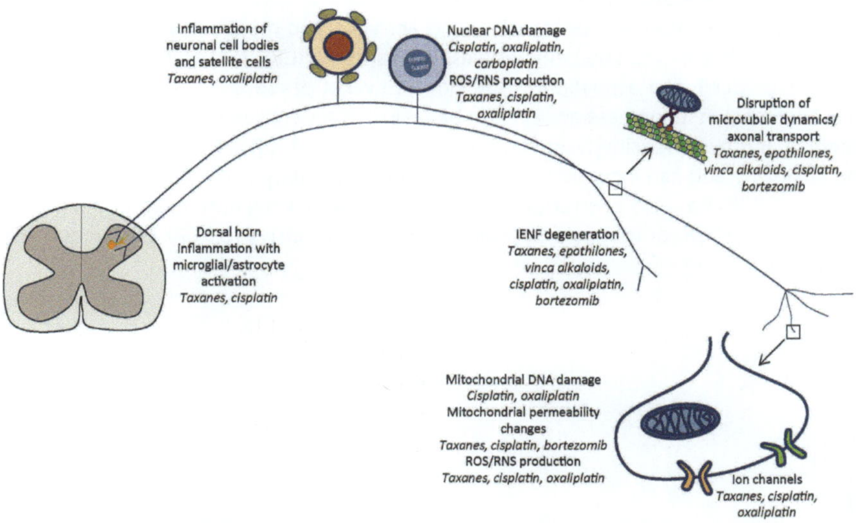

Jill C. Fehrenbacher, Figure 1 Putative sites of chemotherapeutic toxicity in the peripheral nervous system.

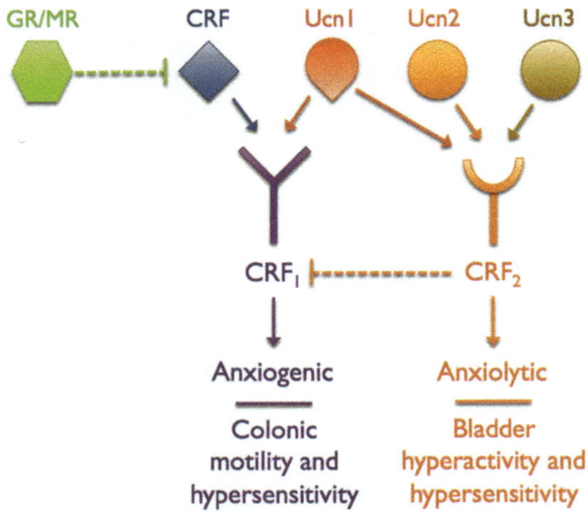

Angela N. Pierce and Julie A. Christianson, Figure 2 Corticotropin-releasing factor (CRF) and the related Urocortins (Ucn) 1–3 act via a two-receptor system to regulate the hypothalamic–pituitary–adrenal axis. Activation of CRF receptor 1 (CRF_1) occurs via CRF or Ucn1 and activation of CRF_2 occurs via all three Ucns. Activation of CRF_1 has been shown to increase anxiety-like behaviors in rodents as well as increase colonic motility and hypersensitivity. In contrast, activation of CRF_2 has largely been shown to be anxiolytic and can increase urinary bladder hyperactivity and hypersensitivity. Activation of CRF_2 has also been shown to inhibit CRF_1 and activation of either glucocorticoid receptor (GR) or mineralocorticoid receptor (MR) can inhibit CRF production and signaling.

Kolibri 360 GmbH
Ferdinand-Jühlke-Straße 7
99095 Erfurt, Deutschland
produktsicherheit@kolibri360.de